Classics in Mathematics

Joram Lindenstrauss   Lior Tzafriri   Classical Banach Spaces I and II

Springer
Berlin
Heidelberg
New York
Barcelona
Budapest
Hong Kong
London
Milan
Paris
Santa Clara
Singapore
Tokyo

Joram Lindenstrauss was born in Tel Aviv on October 28, 1936. He studied at the Hebrew University in Jerusalem, where he received his doctoral degree in 1962. He has been on the faculty of this institution ever since 1965.

Lindenstrauss works in various areas of functional analysis and geometry. Besides structure theory of Banach spaces, which is the subject of this book, he has contributed prominently to the theory of convex sets in finite and infinite dimensions and to geometric nonlinear functional analysis.

Lior Tzafriri was born in Bucharest on May 9, 1936. He began his studies in Romania but completed them only in Jerusalem after emigrating to Israel in 1961. He received his doctoral degree from the Hebrew University in Jerusalem in 1966 and has been on the faculty of this institution ever since 1970.

Tzafriri's work reaches into various areas of functional analysis. Besides his influential work on the geometry of Banach spaces, he has made important contributions to operator theory and to problems in harmonic analysis.

Joram Lindenstrauss   Lior Tzafriri

Classical Banach Spaces I

Sequence Spaces

Reprint of the 1977 Edition

Classical Banach Spaces II

Function Spaces

Reprint of the 1979 Edition

Springer

Joram Lindenstrauss
Lior Tzafriri
Department of Mathematics
The Hebrew University of Jerusalem
Jerusalem 91904
Israel

Originally published as Vol. 92 and Vol. 97 of the
Ergebnisse der Mathematik und ihre Grenzgebiete

Cataloging-in-Publication Data applied for

Die Deutsche Bibliothek - CIP-Einheitsaufnahme

Lindenstrauss, Joram:
Classical Banach spaces / Joram Lindenstrauss ; Lior Tzafriri. -
Reprint of the 1977, 1979 ed. - Berlin ; Heidelberg ; New York ;
Barcelona ; Budapest ; Hong Kong ; London ; Milan ; Paris ;
Santa Clara ; Singapore ; Tokyo : Springer, 1996
(Ergebnisse der Mathematik und ihrer Grenzgebiete ; Vol. 92 und 97)
(Classics in mathematics)
Enth.: 1. Sequence spaces. 2. Function spaces

NE: Tzafriri, Lior:; 1. GT

Mathematics Subject Classification (1991):
46-02, 46A40, 46A45, 46BXX, 46JXX

ISBN 978-3-540-60628-4 ISBN 978-3-662-53294-2 (eBook)
DOI 10.1007/978-3-662-53294-2

SPIN 10485236 41/3144- 5 4 3 2 1 0 – Printed on acid-free paper

Joram Lindenstrauss   Lior Tzafriri

Classical Banach Spaces I

Sequence Spaces

Springer-Verlag
Berlin Heidelberg New York 1977

Joram Lindenstrauss
Lior Tzafriri
Department of Mathematics, The Hebrew University of Jerusalem
Jerusalem, Israel

AMS Subject Classification (1970): 46-02, 46 A45, 46Bxx, 46Jxx

ISBN 3-540-08072-4 Springer-Verlag Berlin Heidelberg New York
ISBN 0-387-08072-4 Springer-Verlag New York Heidelberg Berlin

Library of Congress Cataloging in Publication Data. Lindenstrauss, Joram, 1936-. Classical Banach spaces. (Ergebnisse der Mathematik und ihrer Grenzgebiete; 92). Bibliography: v. H1, p. Includes index. CONTENTS: 1. Sequence spaces. 1. Banach spaces. 2. Sequence spaces. 3. Function spaces. I. Tzafriri, Lior, 1936- joint author. II. Title. III. Series QA322.2.L56. 1977- 515'.73. 77-23131.

Typesetting: William Clowes & Sons Ltd., London, Beccles and Colchester.
Printing and bookbinding: K. Triltsch, Würzburg.
2141/3140-543210

To Naomi and Marianne

Preface

The appearance of Banach's book [8] in 1932 signified the beginning of a systematic study of normed linear spaces, which have been the subject of continuous research ever since.

In the sixties, and especially in the last decade, the research activity in this area grew considerably. As a result, Banach space theory gained very much in depth as well as in scope. Most of its well known classical problems were solved, many interesting new directions were developed, and deep connections between Banach space theory and other areas of mathematics were established.

The purpose of this book is to present the main results and current research directions in the geometry of Banach spaces, with an emphasis on the study of the structure of the classical Banach spaces, that is $C(K)$ and $L_p(\mu)$ and related spaces. We did not attempt to write a comprehensive survey of Banach space theory, or even only of the theory of classical Banach spaces, since the amount of interesting results on the subject makes such a survey practically impossible.

A part of the subject matter of this book appeared in outline in our lecture notes [96]. In contrast to those notes, most of the results presented here are given with complete proofs. We therefore hope that it will be possible to use the present book both as a text book on Banach space theory and as a reference book for research workers in the area. It contains much material which was not discussed in [96], a large part of which being the result of very recent research work. An indication to the rapid recent progress in Banach space theory is the fact that most of the many problems stated in [96] have been solved by now.

In the present volume we also state some open problems. It is reasonable to expect that many of these will be solved in the not too far future. We feel, however, that most of the topics discussed here have reached a relatively final form, and that their presentation will not be radically affected by the solution of the open problems. Among the topics discussed in detail in this volume, the one which seems to us to be the least well understood and which might change the most in the future, is that of the approximation property.

We divided our book into four volumes. The present volume deals with sequence spaces. The notion of a Schauder basis plays a central role here. The classical spaces which are in the most natural way sequence spaces are c_0 and l_p, $1 \leqslant p \leqslant \infty$. Volumes II and III will deal with function spaces. In Volume II we shall present the general theory of Banach lattices with an emphasis on those notions concerning lattices which are related to $L_p(\mu)$ spaces. Volume III will be devoted to a study of the structure of the spaces $L_p(0, 1)$, $C(K)$ and general preduals of

$L_1(\mu)$ spaces. The division of the common Banach spaces into sequence and function spaces is made according to the usual practice. It should be remembered, however, that several spaces have natural representations both as sequence and function spaces. The best known example is the separable Hilbert space, which can be represented both as the sequence space l_2 and as the function space $L_2(0, 1)$. A less trivial example is the space l_p, $1 \leq p \leq \infty$, which is isomorphic to the function space $H_p(D)$ of the analytic functions on the disc $D = \{z;\ |z| < 1\}$ with $\|f\| = (\iint |f(z)|^p \ dx \ dy)^{1/p} < \infty$ (cf. [88]). Also, the spaces $C(0, 1)$ and $L_p(0, 1)$, $1 \leq p < \infty$, have Schauder bases, and thus it is convenient sometimes to use their representations as sequence spaces.

In Volume IV we intend to present the local theory of Banach spaces. This theory deals with the structure of finite-dimensional Banach spaces and the relation between an infinite-dimensional Banach space and its finite-dimensional subspaces. A central part in this approach to Banach space theory is played by the evaluation of various parameters of finite-dimensional Banach spaces. The role of the classical finite-dimensional spaces, that is of the spaces l_p^n, $1 \leq p \leq \infty$, $n = 1, 2, \ldots$ in the local theory of Banach spaces is even more central than the role of the classical spaces in the general theory of Banach sequence spaces and function spaces.

We sketch now briefly the contents of this volume. Chapter 1 contains a quite complete account of the main results on Schauder bases in general Banach spaces. Several notions related to Schauder bases—the various approximation properties, general biorthogonal systems and Schauder decompositions—as well as some examples are discussed in detail.

Chapter 2 is devoted to a study of the spaces c_0 and l_p, $1 \leq p < \infty$, and to some extent also of l_∞. Section *a* is devoted to an examination of the basic properties of these spaces, some of which are shown to characterize these spaces among general Banach spaces. The other sections of Chapter 2 are basically independent of each other and can thus be read in any order. In Sections *b* and *c* we discuss certain ideals of operators on general Banach spaces and show how they can be used in the study of the structure of the classical sequence spaces. Section *d* contains a structure theorem for "nice" subspaces of c_0 and l_p as well as examples of subspaces which are not "nice" (i.e. subspaces which fail to have the approximation property). This section contains also a discussion of general results related to the approximation property which complement the treatment of this property in Section *e* of Chapter 1. Section *f* contains an example of an infinite-dimensional Banach space which fails to have any of the classical sequence spaces as a subspace and also criteria for general Banach spaces to have subspaces isomorphic to c_0 and especially to l_1. The final section of Chapter 2 deals with the extension properties of c_0 and l_∞, the lifting property of l_1, and the closely related topic of the automorphisms of these spaces.

In Chapter 3 we discuss the special properties of symmetric bases and the relation between symmetric bases and general unconditional bases. A large part of this chapter is devoted to results and examples related to the possible characterizations of c_0 and l_p, $1 \leq p < \infty$, in the class of all spaces with a symmetric basis. The final chapter of this volume is devoted to a detailed study of the structure of some particular classes of spaces with symmetric bases, mainly Orlicz sequence spaces. The main emphasis is again on the relation between these spaces and the spaces c_0 and l_p. Several examples given there demonstrate how much more complicated the structure of general Orlicz sequence spaces is, as compared to that of l_p spaces. In section *d* it is shown that Orlicz sequence spaces enter naturally into the study of spaces like $l_p \oplus l_r$ with $p \neq r$. In Vol. III it will be shown that Orlicz sequence spaces arise naturally in the study of the structure of subspaces of $L_1(0, 1)$.

We assume that the reader is familiar with the basic results of real analysis and functional analysis which are usually covered in first year graduate courses in these subjects. An acquaintance with the main results in chapters I–VI of [33] will certainly suffice (much less is actually needed for being able to read this book).

The bibliography contains only those papers which are actually quoted in the text. We tried to indicate in the text the source of the main results which we present. The reference list is, however, far from being complete. Reference to papers where the basic results in Banach space theory were first proved can be found, for example, in [28] and [33]. Further references on bases may be found in [135]. References to further literature on Orlicz spaces may be found in [75].

The overlap between this book and existing books on related topics is very small.

We would like to acknowledge the contribution of G. Schechtman, who read the entire manuscript and made several very valuable suggestions, and of J. Arazy, who was very helpful in proofreading this volume and eliminated several mistakes. We also wish to thank Z. Altshuler and Y. Sternfeld for their help. We are grateful to Danit Sharon who expertly carried out the task of typing the manuscript of this book. We are also indebted to the U.S.-Israel Binational Science Foundation for partial support.

Finally, we would like to thank Springer-Verlag and especially Roberto Minio for their cooperation and help in all the stages of the preparation of this book.

Jerusalem — Joram Lindenstrauss
January 1977 — Lior Tzafriri

Table of Contents

1. Schauder Bases 1

a. Existence of Bases and Examples 1
b. Schauder Bases and Duality 7
c. Unconditional Bases 15
d. Examples of Spaces Without an Unconditional Basis 24
e. The Approximation Property 29
f. Biorthogonal Systems 42
g. Schauder Decompositions 47

2. The Spaces c_0 and l_p 53

a. Projections in c_0 and l_p and Characterizations of these Spaces . . . 53
b. Absolutely Summing Operators and Uniqueness of Unconditional Bases 63
c. Fredholm Operators, Strictly Singular Operators and Complemented Subspaces of $l_p \oplus l_r$ 75
d. Subspaces of c_0 and l_p and the Approximation Property, Complementably Universal Spaces 84
e. Banach Spaces Containing l_p or c_0 95
f. Extension and Lifting Properties, Automorphisms of l_∞, c_0 and l_1 . 104

3. Symmetric Bases 113

a. Properties of Symmetric Bases, Examples and Special Block Bases . 113
b. Subspaces of Spaces with a Symmetric Basis 123

4. Orlicz Sequence Spaces 137

a. Subspaces of Orlicz Sequence Spaces which have a Symmetric Basis . 137
b. Duality and Complemented Subspaces 147
c. Examples of Orlicz Sequence Spaces 156
d. Modular Sequence Spaces and Subspaces of $l_p \oplus l_r$ 166
e. Lorentz Sequence Spaces 175

References 180

Subject Index 185

Standard Definitions, Notations and Conventions

For most of the results presented in this book it does not matter whether the field of scalars is real or complex. In the isometric theory there are some differences (usually minor) between real and complex spaces. As a rule we shall work with real scalars and, in a few places, we shall indicate the changes needed in the complex case. In a few instances e.g. where spaces of analytic functions are involved or where spectral theory is used we shall use complex scalars.

By $L_p(\mu)=L_p(\Omega, \Sigma, \mu)$, $1 \leqslant p \leqslant \infty$ we denote the Banach space of equivalence classes of measurable functions on (Ω, Σ, μ) whose p'th power is integrable (respectively, which are essentially bounded if $p=\infty$). The norm in $L_p(\mu)$ is defined by $||f||=(\int|f(\omega)|^p\, d\mu(\omega))^{1/p}$ (ess sup $|f(\omega)|$ if $p=\infty$). If (Ω, Σ, μ) is the usual Lebesgue measure space on $[0, 1]$ we denote $L_p(\mu)$ by $L_p(0, 1)$. If (Γ, Σ, μ) is the discrete measure space on a set Γ, with $\mu(\{\gamma\})=1$ for every $\gamma \in \Gamma$, we denote $L_p(\mu)$ by $l_p(\Gamma)$. If Γ is the set of positive integers we denote $l_p(\Gamma)$ also by l_p while if $\Gamma=\{1, 2, \dots, n\}$, for some $n<\infty$, we denote $l_p(\Gamma)$ by l_p^n. The subspace of $l_\infty(\Gamma)$, of those functions which vanish at ∞, is denoted by $c_0(\Gamma)$ (if Γ is the set of positive integers we denote this space by c_0). The subspace of l_∞ consisting of convergent sequences is denoted by c. For a compact Hausdorff space K we denote by $C(K)$ the Banach space of all continuous scalar-valued functions on K with the supremum norm. If K is the unit interval $[0, 1]$ in its usual topology we denote $C(K)$ by $C(0, 1)$.

In a Banach space X we denote the ball with center x and radius r, i.e. $\{y; ||y-x|| \leqslant r\}$, by $B_X(x, r)$. If the space X is clear from the context, we simply write $B(x, r)$. The unit ball $B_X(0, 1)$ of X is denoted also by B_X. For a sequence $\{x_n\}_{n=1}^\infty$ of elements of X we denote by span $\{x_n\}_{n=1}^\infty$ the algebraic linear span of $\{x_n\}_{n=1}^\infty$ i.e. the set of all finite linear combinations of $\{x_n\}_{n=1}^\infty$. The closure of span $\{x_n\}_{n=1}^\infty$ is denoted by $[x_n]_{n=1}^\infty$. A similar notation is used for the span of a set other than a sequence. For a set $A \subset X$ its norm closure is denoted by $\overline{A}$, e.g. $[x_n]_{n=1}^\infty=\overline{\text{span}\,\{x_n\}_{n=1}^\infty}$. The convex hull of a sequence $\{x_n\}_{n=1}^\infty$ is denoted by conv $\{x_n\}_{n=1}^\infty$; the closed convex hull by $\overline{\text{conv}}\,\{x_n\}_{n=1}^\infty$.

The term "operator" means a bounded linear operator unless specified otherwise. The space of all operators from X to Y with the usual operator norm is denoted by $L(X, Y)$. An operator $T \in L(X, Y)$ is called *compact* if $\overline{T B_X}$ is a norm compact subset of Y. The identity operator of a Banach space X is denoted by I_X (or simply by I if X is clear from the context). For an operator $T \in L(X, Y)$ the notation $T_{|Z}$ denotes the restriction of T to the subspace Z of X.

Two Banach spaces X and Y are called *isomorphic* (denoted by $X \approx Y$) if there exists an invertible operator from X onto Y. *The Banach-Mazur distance coefficient*

$d(X, Y)$ is defined by $\inf \|T\| \|T^{-1}\|$, the infimum being taken over all invertible operators from X onto Y (if X is not isomorphic to Y we put $d(X, Y)=\infty$). Notice that $d(X, Y) \geq 1$, for every X and Y, and that $d(X, Y)\, d(Y, Z) \geq d(X, Z)$, for every X, Y and Z. If there exists an invertible operator T from X onto Y so that $\|T\|=\|T^{-1}\|=1$ (i.e. $\|Tx\|=\|x\|$, for every $x \in X$) we say that *X is isometric to Y*. In this case $d(X, Y)=1$ (the converse is false in general; it is possible that $d(X, Y)=1$ but that the infimum in the definition of $d(X, Y)$ is not attained i.e. X is not isometric to Y). An operator $T \in L(X, Y)$ is said to be an isomorphism into Y if there is some constant $C>0$ so that $\|Tx\| \geq C\|x\|$ for every $x \in X$. In this case T^{-1} is a well defined element in $L(TX, X)$.

A closed linear subspace Y of a Banach space X is said to be a *complemented subspace* of X if there is a bounded linear projection from X onto Y, or what is the same, if there exists a closed linear subspace Z of X so that X is the direct sum of Y and Z, i.e. $X = Y \oplus Z$. We shall also use some direct sums of infinite sequences of Banach spaces. If $\{X_n\}_{n=1}^{\infty}$ is a sequence of Banach spaces we define the direct sum of these spaces in the sense of l_p, $1 \leqslant p < \infty$, namely $\left(\sum_{n=1}^{\infty} \oplus X_n\right)_p$, as the space of all sequences $x=(x_1, x_2, \ldots)$, with $x_n \in X_n$ for all n, for which $\|x\| = \left(\sum_{n=1}^{\infty} \|x_n\|^p\right)^{1/p} < \infty$. Similarly, $\left(\sum_{n=1}^{\infty} \oplus X_n\right)_0$ denotes the direct sum of $\{X_n\}_{n=1}^{\infty}$ in the sense of c_0 i.e. the space of all sequences $x=(x_1, x_2, \ldots)$, with $x_n \in X_n$ for all n, for which $\lim_n \|x_n\|=0$. The norm in this direct sum is taken as $\|x\| = \max_n \|x_n\|$. We shall occasionally use also other types of infinite direct sums. These will be defined in the proper places in the text.

Besides the norm (or strong) topology of a Banach space X we often use some other topologies. If Y is a subspace of the dual X^* of X then the Y-topology of X is the weakest topology making all the elements of Y continuous. A basis for the Y topology is obtained by taking all the sets of the form $V(x, \varepsilon, A) = \{u;\ |x^*(u) - x^*(x)| < \varepsilon,\ x^* \in A\}$, where $x \in X$, $\varepsilon > 0$ and A is a finite subset of Y. If $Y = X^*$ the Y topology is called the weak topology (w topology). If $X = Z^*$ and we take as Y the canonical image of Z in $Z^{**} = X^*$ we obtain the w^* topology induced by Z (if Z is clear from the context we simply talk of the w^* topology). Convergence of sequences in the w topology (resp. w^* topology) is denoted by $x_n \xrightarrow{w} x$ or $w \lim_n x_n = x$ (resp. $x_n \xrightarrow{w^*} x$ or $w^* \lim_n x_n = x$). An operator $T \in L(X, Y)$ is said to be *w compact* if $\overline{T B_X}$ is a compact set in Y, in its w topology (i.e. a w compact set in Y).

Whenever we consider a Banach space X as a subspace of its second dual X^{**} we assume that it is embedded canonically. For a subset $A \subset X$ we denote by $A^{\perp}$ the subspace $\{x^*;\ x^*(x)=0,\ x \in A\}$ of X^*. For a subset $A \subset X^*$ we denote by $A^{\top}$ the subspace $\{x;\ x^*(x)=0,\ x^* \in A\}$ of X. For every subset $A \subset X$ we have $A^{\perp\top} \supset A$ and equality holds if and only if A is a closed linear subspace.

Besides subspaces of Banach spaces we shall also study quotient spaces. An operator $T\colon X \to Y$ is called a *quotient map* if $\overline{T B_X} = B_Y$. A Banach space Y is isomorphic to a quotient space of a space X if and only if there exists an operator T from X onto Y. If such a T exists then $Y \approx X/\ker T$, where $\ker T = \{x;\ Tx=0\}$, and Y^* is isomorphic to the subspace $(\ker T)^\perp$ of X^*. Similarly, if Z is a subspace of X then Z^* is isometric to the quotient space $X/(Z^\perp)$.

Among the general notations used in this book we want to single out the following. For a positive number S we denote by $[S]$ the largest integer $\leqslant S$. For a set A we denote by $\bar{\bar{A}}$ the cardinality of A. If A and B are sets we put $A \sim B = \{x, x \in A, x \notin B\}$.

1. Schauder Bases

a. Existence of Bases and Examples

The aim of this volume is to describe some results concerning sequence spaces, i.e. those Banach spaces which can be presented in some natural manner as spaces of sequences. In general, such a representation is achieved by introducing in the space a sort of "coordinate system". There are, obviously, many different ways of giving a precise meaning to the terms "Banach sequence spaces" and "coordinate systems". The best known and most useful approach is by using the notion of a Schauder basis.

Definition 1.a.1. A sequence $\{x_n\}_{n=1}^{\infty}$ in a Banach space X is called a *Schauder basis* of X if for every $x \in X$ there is a unique sequence of scalars $\{a_n\}_{n=1}^{\infty}$ so that $x = \sum_{n=1}^{\infty} a_n x_n$. A sequence $\{x_n\}_{n=1}^{\infty}$ which is a Schauder basis of its closed linear span is called a *basic sequence.*

In this book we shall not consider any type of bases in infinite-dimensional Banach spaces besides Schauder bases. We shall therefore often omit the word Schauder. In addition to Schauder bases we shall only encounter algebraic bases in finite-dimensional spaces. This should not cause any confusion. As a matter of fact, quantitative notions concerning Schauder bases (like the basis constant defined below) have a meaning and will be used also in the context of algebraic bases in finite dimensional spaces.

Evidently, a space X with a Schauder basis $\{x_n\}_{n=1}^{\infty}$ can be considered as a sequence space by identifying each $x = \sum_{n=1}^{\infty} a_n x_n$ with the unique sequence of coefficients $(a_1, a_2, a_3, \ldots)$. It is important to note that for describing a Schauder basis one has to define the basis vectors not only as a set but as an ordered sequence.

Let $(X, \|\ \|)$ be a Banach space with a basis $\{x_n\}_{n=1}^{\infty}$. For every $x = \sum_{n=1}^{\infty} a_n x_n$ in X the expression $|||x||| = \sup_n \left\| \sum_{i=1}^{n} a_i x_i \right\|$ is finite. Evidently, $|||\cdot|||$ is a norm on X and $\|x\| \leq |||x|||$ for every $x \in X$. A simple argument shows that X is complete also with respect to $|||\cdot|||$ and thus, by the open mapping theorem, the norms $\|\cdot\|$ and $|||\cdot|||$ are equivalent. These remarks prove the following proposition [8].

Proposition 1.a.2. *Let X be a Banach space with a Schauder basis $\{x_n\}_{n=1}^{\infty}$. Then the* *projections* $P_n: X \to X$, *defined by* $P_n\left(\sum_{i=1}^{\infty} a_i x_i\right) = \sum_{i=1}^{n} a_i x_i$, *are bounded linear operators and* $\sup_n \|P_n\| < \infty$.

The projections $\{P_n\}_{n=1}^{\infty}$ are called the natural projections associated to $\{x_n\}_{n=1}^{\infty}$; the number $\sup_n \|P_n\|$ is called the *basis constant* of $\{x_n\}_{n=1}^{\infty}$. A basis whose basis constant is 1 is called a *monotone basis*. In other words, a basis is monotone if, for every choice of scalars $\{a_n\}_{n=1}^{\infty}$, the sequence of numbers $\left\{\left\|\sum_{i=1}^{n} a_i x_i\right\|\right\}_{n=1}^{\infty}$ is non-decreasing. Every Schauder basis $\{x_n\}_{n=1}^{\infty}$ is monotone with respect to the norm $|||x||| = \sup_n \|P_n x\|$ which was already used above. Indeed,

$$|||P_n x||| = \sup_m \|P_m P_n x\| = \sup_{1 \le m \le n} \|P_m x\| \le |||x||| .$$

Thus, given any Schauder basis $\{x_n\}_{n=1}^{\infty}$ of X, we can pass to an equivalent norm in X for which the given basis is monotone.

There is a simple and useful criterion for checking whether a given sequence is a Schauder basis.

Proposition 1.a.3. *Let* $\{x_n\}_{n=1}^{\infty}$ *be a sequence of vectors in* X. *Then* $\{x_n\}_{n=1}^{\infty}$ *is a Schauder basis of* X *if and only if the following three conditions hold.*

(i) $x_n \neq 0$ *for all* n.

(ii) *There is a constant* K *so that, for every choice of scalars* $\{a_i\}_{i=1}^{\infty}$ *and integers* $n < m$, *we have*

$$\left\|\sum_{i=1}^{n} a_i x_i\right\| \le K \left\|\sum_{i=1}^{m} a_i x_i\right\| .$$

(iii) *The closed linear span of* $\{x_n\}_{n=1}^{\infty}$ *is all of* X.

The proof is easy. The necessity of (i) and (iii) is clear from the definition, while that of (ii) follows from 1.a.2. Conversely, if (i) and (ii) hold then $\sum_{n=1}^{\infty} a_n x_n = 0$ implies that $a_n = 0$ for all n. This proves the uniqueness of the expansion in terms of $\{x_n\}_{n=1}^{\infty}$. In order to prove that every $x \in X$ has such an expansion it is enough, in view of (iii), to show that the space of all elements of the form $\sum_{n=1}^{\infty} a_n x_n$ is a closed linear space. This latter fact can be easily proved by using (ii). □

Obviously conditions (i) and (ii) of 1.a.3, by themselves, form a necessary and sufficient condition for a sequence $\{x_n\}_{n=1}^{\infty}$ to be a basic sequence. It is also worthwhile to observe that in case we can take $K=1$ it is enough to verify (ii) for $m = n+1$.

A basis $\{x_n\}_{n=1}^{\infty}$ is called *normalized* if $\|x_n\| = 1$ for all n. Clearly, whenever $\{x_n\}_{n=1}^{\infty}$ is a Schauder basis of X, the sequence $\{x_n/\|x_n\|\}_{n=1}^{\infty}$ is a normalized basis in X.

Before proceeding with the general discussion we present some examples of bases. The unit vectors $e_n=(0, 0, 0,..., \overset{n}{1}, 0,...)$ form a monotone and normalized basis in each of the spaces c_0 and l_p, $1\leq p<\infty$. An example of a basis in the space c, of convergent sequences of scalars, is given by

$$x_1=(1, 1, 1,...) \quad \text{and, for } n>1,\ x_n=e_{n-1}.$$

The expansion of $x=(a_1, a_2,...)\in c$ with respect to this basis is

$$x=(\lim_n a_n)x_1+(a_1-\lim_n a_n)x_2+(a_2-\lim_n a_n)x_3+\cdots.$$

An important example of a Schauder basis is the Haar system in $L_p(0, 1)$, $1\leq p<\infty$.

Definition 1.a.4. The sequence of functions $\{\chi_n(t)\}_{n=1}^\infty$ defined by $\chi_1(t)\equiv 1$ and, for $k=0, 1, 2,..., l=1, 2,..., 2^k$,

$$\chi_{2^k+l}(t)=\begin{cases} 1 & \text{if } t\in[(2l-2)2^{-k-1}, (2l-1)2^{-k-1}] \\ -1 & \text{if } t\in((2l-1)2^{-k-1}, 2l\cdot 2^{-k-1}] \\ 0 & \text{otherwise} \end{cases}$$

is called the *Haar system*.

The Haar system is (in its given order) a monotone (but obviously not normalized) basis of $L_p(0, 1)$ for every $1\leq p<\infty$. Indeed, since the linear span of the Haar system contains all the characteristic functions of dyadic intervals (i.e. intervals of the form $[l\cdot 2^{-k}, (l+1)\cdot 2^{-k}]$), it is clear that (iii) of 1.a.3 holds. We have only to verify that (ii) holds with $K=1$. Let $\{a_i\}_{i=1}^\infty$ be any sequence of scalars, let n be an integer and let $f(t)=\sum_{i=1}^{n} a_i\chi_i(t)$ and $g(t)=\sum_{i=1}^{n+1} a_i\chi_i(t)$. The only difference between f and g is that on some dyadic interval I where f has the constant value b, say, g has the value $b+a_{n+1}$ on the first half of I and $b-a_{n+1}$ on the second half. Since, for every $p\geq 1$, $|b+a_{n+1}|^p+|b-a_{n+1}|^p\geq 2|b|^p$ we get that $\|f\|\leq\|g\|$.

By integrating the Haar system or more precisely by putting

$$\varphi_1(t)\equiv 1; \qquad \varphi_n(t)=\int_0^t \chi_{n-1}(u)\,du, \quad n>1$$

we obtain another famous and important basis. The sequence $\{\varphi_n\}_{n=1}^\infty$ is called the *Schauder system*. The Schauder system is a monotone basis of $C(0, 1)$. Indeed, the linear span of the $\{\varphi_n\}_{n=1}^\infty$ consists exactly of the continuous piecewise linear functions on [0, 1] whose nodes are dyadic points. This shows that (iii) of 1.a.3 is satisfied. Since, for every integer n, the interval on which the function $\varphi_{n+1}(t)$ is different from 0 is such that on it all the functions $\{\varphi_i(t)\}_{i=1}^n$ are linear it follows immediately that (ii) of 1.a.3 holds with $K=1$.

Schauder bases have been constructed in many other important Banach spaces appearing in analysis. Of particular interest in this direction are the results of Z. Ciesielski and J. Domsta [18] and S. Schonefeld [132] who proved the existence of a basis in $C^k(I^n)$ (=the space of all real functions $f(t_1, t_2, \ldots, t_n)$, $t_i \in [0, 1]$ which are k times continuously differentiable, with the obvious norm) and the result of S. V. Botschkariev [13] who proved the existence of a basis in the disc algebra A (=the space consisting of all the functions $f(z)$ which are analytic on $|z|<1$ and continuous on $|z|\leqslant 1$, with the sup norm). In these papers an important role is played by the *Franklin system*. The Franklin system consists of the sequence $\{f_n(t)\}_{n=1}^\infty$ of functions on $[0, 1]$ which are obtained from the Schauder system $\{\varphi_n\}_{n=1}^\infty$ by applying the Gram–Schmidt orthogonalization procedure (with respect to the Lesbegue measure on $[0, 1]$). The Franklin system is (by definition) an orthonomal sequence which turns out to be also a Schauder basis of $C(0, 1)$. For a detailed study of the Franklin system we refer to the above mentioned papers as well as to [17].

The fact that in the common spaces there exists a Schauder basis led Banach to pose the question whether every separable Banach space has a basis. This problem (known as the basis problem) remained open for a long time and was solved in the negative by P. Enflo [37]. We shall present later on in this book (in Section 2.d) a variant of Enflo's solution.

The question whether every infinite-dimensional Banach space contains a basic sequence has, however, a positive answer. This simple fact was known already to Banach.

Theorem 1.a.5. *Every infinite dimensional Banach space contains a basic sequence.*

The proof, due to S. Mazur, is based on the following lemma.

Lemma 1.a.6. *Let X be an infinite dimensional Banach space. Let $B\subset X$ be a finite-dimensional subspace and let $\varepsilon>0$. Then there is an $x\in X$ with $\|x\|=1$ so that $\|y\|\leqslant(1+\varepsilon)\|y+\lambda x\|$ for every $y\in B$ and every scalar λ.*

Proof of 1.a.6. We may clearly assume that $\varepsilon<1$. Let $\{y_i\}_{i=1}^m$ be elements of norm 1 in B such that for every $y\in B$ with $\|y\|=1$ there is an i for which $\|y-y_i\|<\varepsilon/2$. Let $\{y_i^*\}_{i=1}^m$ be elements of norm 1 in X^* so that $y_i^*(y_i)=1$ for all i, and let $x\in X$ with $\|x\|=1$ and $y_i^*(x)=0$ for all i. This x has the desired property. Indeed, let $y\in Y$ with $\|y\|=1$, let i be such that $\|y-y_i\|\leqslant\varepsilon/2$ and let λ be a scalar. Then

$$\|y+\lambda x\|\geqslant\|y_i+\lambda x\|-\varepsilon/2\geqslant y_i^*(y_i+\lambda x)-\varepsilon/2=1-\varepsilon/2\geqslant\|y\|/(1+\varepsilon). \quad \square$$

Proof of 1.a.5. Let ε be any positive number and let $\{\varepsilon_n\}_{n=1}^\infty$ be positive numbers such that $\prod_{n=1}^{\infty}(1+\varepsilon_n)\leqslant 1+\varepsilon$. Let x_1 be any element in X with norm 1. By 1.a.6 we can construct inductively a sequence of unit vectors $\{x_n\}_{n=2}^\infty$ so that for every $n\geqslant 1$

$$\|y\|\leqslant(1+\varepsilon_n)\|y+\lambda x_{n+1}\| \quad \text{for all } y\in\operatorname{span}\{x_1,\ldots,x_n\} \text{ and every scalar } \lambda.$$

The sequence $\{x_n\}_{n=1}^{\infty}$ is a basic sequence in X whose basis constant is $\leqslant 1+\varepsilon$ (observe that $\|P_n\| \leqslant \prod_{i=n}^{\infty}(1+\varepsilon_i)$, $n=1, 2, \ldots$). □

Remark. It is useful to note that in the proof of 1.a.6 it is enough to take a vector x with $\|x\|=1$ and $|y_i^*(x)|<\varepsilon/4$ for $i=1,\ldots, m$. Indeed, if $\|y\|=1$ and $|\lambda|\geqslant 2$ then $\|y+\lambda x\|\geqslant\|y\|$ while if $|\lambda|<2$ the computation in the proof of 1.a.6 gives that $\|y+\lambda x\|>(1-\varepsilon)\|y\|$. It follows from this observation and the proof of 1.a.5 that *if $\{x_n\}_{n=1}^{\infty}$ is a sequence of vectors in X such that* $\liminf_n \|x_n\|>0$ *and* $x_n \xrightarrow{w} 0$ *then $\{x_n\}_{n=1}^{\infty}$ has a subsequence $\{x_{n_k}\}_{k=1}^{\infty}$ which is a basic sequence.*

Once it is known that a Banach space has a Schauder basis it is natural to raise the question of its uniqueness. In order to study this question properly we introduce first the notion of equivalence of bases.

Definition 1.a.7. Two bases, $\{x_n\}_{n=1}^{\infty}$ of X and $\{y_n\}_{n=1}^{\infty}$ of Y, are called *equivalent* provided a series $\sum_{n=1}^{\infty} a_n x_n$ converges if and only if $\sum_{n=1}^{\infty} a_n y_n$ converges.

Thus the bases are equivalent if the sequence space associated to X by $\{x_n\}_{n=1}^{\infty}$ is identical to the sequence space associated to Y by $\{y_n\}_{n=1}^{\infty}$. It follows immediately from the closed graph theorem that $\{x_n\}_{n=1}^{\infty}$ is equivalent to $\{y_n\}_{n=1}^{\infty}$ if and only if there is an isomorphism T from X onto Y for which $Tx_n=y_n$ for all n.

Using the notion of equivalence, the uniqueness question can be given a meaningful formulation. It turns out however that even up to equivalence bases, if they exist at all, are never unique.

Theorem 1.a.8 [120]. *Let X be an infinite dimensional Banach space with a Schauder basis. Then there are uncountably many mutually non-equivalent normalized bases in X.*

We shall discuss in detail some aspects of uniqueness of bases in Chapters 2 and 3 below. We shall show there that if we restrict the discussion to bases which have some nice properties then it is possible to have uniqueness in some interesting special cases. In this context we shall also present a proof of a weak version of 1.a.8 (namely that there are at least two non-equivalent normalized bases in every space having a basis).

Schauder bases have certain stability properties. If we perturb each element of a basis by a sufficiently small vector we still get a basis. The perturbed basis is equivalent to the original one. The simplest result in this direction is the following useful proposition [76].

Proposition 1.a.9. (i) *Let $\{x_n\}_{n=1}^{\infty}$ be a normalized basis of a Banach space X with basis constant K. Let $\{y_n\}_{n=1}^{\infty}$ be a sequence of vectors in X with $\sum_{n=1}^{\infty}\|x_n-y_n\|<1/2K$. Then $\{y_n\}_{n=1}^{\infty}$ is a basis of X which is equivalent to $\{x_n\}_{n=1}^{\infty}$ (if $\{x_n\}_{n=1}^{\infty}$ is just a basic sequence then $\{y_n\}_{n=1}^{\infty}$ will also be a basic sequence which is equivalent to $\{x_n\}_{n=1}^{\infty}$).*

(ii) *Let $\{x_n\}_{n=1}^{\infty}$ be a normalized basic sequence in a Banach space X with a basis constant K. Assume that there is a projection P from X onto $[x_n]_{n=1}^{\infty}$. Let $\{y_n\}_{n=1}^{\infty}$ be a sequence of vectors in X such that $\sum_{n=1}^{\infty} \|x_n - y_n\| \leq 1/8K\|P\|$. Then $Y = [y_n]_{n=1}^{\infty}$ is complemented in X.*

Proof. For $x = \sum_{n=1}^{\infty} a_n x_n \in X$ define $Tx = \sum_{n=1}^{\infty} a_n y_n$. The series converges and

$$(*) \qquad \|x - Tx\| \leq \sum_{n=1}^{\infty} |a_n| \, \|x_n - y_n\| \leq \max_n |a_n| \sum_{n=1}^{\infty} \|x_n - y_n\|$$
$$\leq 2K\|x\| \sum_{n=1}^{\infty} \|x_n - y_n\| .$$

To prove (i) we have just to observe that under its assumptions $\|I - T\| < 1$ and hence T is an automorphism of X.

To prove (ii) we have to observe that if we put $y = Tx$, then $\|y - x\| < \|x\|/4$ and in particular $\|x\| < 2\|y\|$ and $\|T\| < 2$. Thus

$$\|TPy - y\| = \left\|TP\left(\sum_{n=1}^{\infty} a_n(y_n - x_n)\right)\right\| < 8K\|P\| \, \|y\| \sum_{n=1}^{\infty} \|x_n - y_n\| = \delta\|y\|$$

for some $\delta < 1$. Thus $S = TP_{|Y}$ is an invertible operator on Y and $S^{-1}TP$ is a projection from X onto Y. □

A very useful method to obtain new basic sequences, starting from a given basis or basic sequence, is by considering block bases.

Definition 1.a.10. Let $\{x_n\}_{n=1}^{\infty}$ be a basic sequence in a Banach space X. A sequence of non-zero vectors $\{u_j\}_{j=1}^{\infty}$ in X of the form $u_j = \sum_{n=p_j+1}^{p_{j+1}} a_n x_n$, with $\{a_n\}_{n=1}^{\infty}$ scalars and $p_1 < p_2 < \cdots$ an increasing sequence of integers, is called a *block basic sequence* or briefly a *block basis* of the $\{x_n\}_{n=1}^{\infty}$.

It is obvious that a block basis $\{u_j\}_{j=1}^{\infty}$ of $\{x_n\}_{n=1}^{\infty}$ is a basic sequence whose basis constant does not exceed that of $\{x_n\}_{n=1}^{\infty}$. The usefulness of the notion of block basis rests very much on the following simple observation [9].

Proposition 1.a.11. *Let X be a Banach space with a Schauder basis $\{x_n\}_{n=1}^{\infty}$. Let Y be a closed infinite dimensional subspace of X. Then there is a subspace Z of Y which has a basis which is equivalent to a block basis of $\{x_n\}_{n=1}^{\infty}$.*

Proof. Observe first that since Y is infinite dimensional there is, for every integer p, an element $y \in Y$ with $\|y\| = 1$ of the form $y = \sum_{n=p+1}^{\infty} a_n x_n$. We construct the block basis of $\{x_n\}_{n=1}^{\infty}$ inductively. Pick any $y_1 = \sum_{n=1}^{\infty} a_{n,1} x_n \in Y$ with $\|y_1\| = 1$. Let p_1 be an integer so that $||y_1-u_1||<1/4K$ where $u_1=\sum_{n=1}^{p_1} a_{n,1}x_n$ and K is the basis constant of $\{x_n\}_{n=1}^\infty$. Next we pick a $y_2=\sum_{n=p_1+1}^\infty a_{n,2}x_n \in Y$ with $||y_2||=1$ and an integer p_2 so that $||y_2-u_2||<1/4^2K$ where $u_2=\sum_{n=p_1+1}^{p_2} a_{n,2}x_n$. We continue in an obvious manner. The sequence $\{u_j\}_{j=1}^\infty$ obtained in this way is a block basis of $\{x_n\}_{n=1}^\infty$. Since $\sum_{j=1}^\infty ||y_j-u_j||<1/3K$ it follows by 1.a.9 that $\{y_j\}_{j=1}^\infty$ is a basic sequence which is equivalent to $\{u_j\}_{j=1}^\infty$. The space $Z=[y_j]_{j=1}^\infty$ has the desired property. □

In the proof of 1.a.11 we used the vectors $y_j \in Y$ whose expansions with respect to the basis $\{x_n\}_{n=1}^\infty$ started arbitrarily far. In some instances it is important to be able to choose a basic sequence out of a subset Y of X which is not a subspace. For this purpose it is of interest to observe that what we actually need in the proof of 1.a.11 is the following. For every $\varepsilon>0$ and every integer p there is a $y=\sum_{n=1}^\infty a_nx_n$ in Y with $||y||\geqslant 1$ and $\left\|\sum_{n=1}^p a_nx_n\right\|\leqslant\varepsilon$. This remark proves the following.

Proposition 1.a.12. *Let $\{x_n\}_{n=1}^\infty$ be a Schauder basis of a Banach space X. Let $y_k=\sum_{n=1}^\infty a_{n,k}x_n$, $k=1, 2,\ldots$, be a sequence of vectors such that* $\limsup_k ||y_k||>0$ *and* $\lim_k a_{n,k}=0$ *for every n (this is the case in particular if $y_k \xrightarrow{w} 0$ and $||y_k|| \not\to 0$). Then there is a subsequence $\{y_{k_j}\}_{j=1}^\infty$ of $\{y_k\}_{k=1}^\infty$ which is equivalent to a block basis of $\{x_n\}_{n=1}^\infty$.*

Proposition 1.a.11 enables us to give an alternative proof of 1.a.5. It is clearly enough to prove 1.a.5 for separable Banach spaces X. Every such X is isometric to a subspace of $C(0, 1)$. Hence, by 1.a.11, X has a subspace with a basis which is equivalent to a block basis of the Schauder system in $C(0, 1)$.

b. Schauder Bases and Duality

Let X be a Banach space with a Schauder basis $\{x_n\}_{n=1}^\infty$. For every integer n the linear functional x_n^* on X defined by $x_n^*\left(\sum_{i=1}^\infty a_ix_i\right)=a_n$ is, by 1.a.2, a bounded linear functional. In fact $||x_n^*||\leqslant 2K/||x_n||$ where K is the basis constant of $\{x_n\}_{n=1}^\infty$. These functionals $\{x_n^*\}_{n=1}^\infty$, which are characterized by the relation $x_n^*(x_m)=\delta_n^m$, are called the *biorthogonal functionals* associated to the basis $\{x_n\}_{n=1}^\infty$. Let $\{P_n\}_{n=1}^\infty$ be the natural projections associated to the basis, i.e. $P_n\left(\sum_{i=1}^\infty a_ix_i\right)=\sum_{i=1}^n a_ix_i$. For every choice of scalars $\{a_i\}_{i=1}^\infty$ and for all integers $n<m$ we have $P_n^*\left(\sum_{i=1}^m a_ix_i^*\right)=\sum_{i=1}^n a_ix_i^*$.

Hence, by 1.a.3, the sequence $\{x_n^*\}_{n=1}^\infty$ is a basic sequence in X^* whose basis constant is identical to that of $\{x_n\}_{n=1}^\infty$. Since $\lim_n \|P_n x - x\| = 0$, for every $x \in X$, we get that, in the sense of convergence in the w^* topology, $x^* = \sum_{n=1}^\infty x^*(x_n)x_n^*$ for every $x^* \in X^*$. In general, this expansion does not converge in norm. We have convergence in norm for every $x^* \in X^*$ if and only if the sequence $\{x_n^*\}_{n=1}^\infty$ is a basis of X^* i.e., (by 1.a.3) if and only if the closed linear span of $\{x_n^*\}_{n=1}^\infty$ is all of X^*. For this to happen X^* must in particular be separable; thus, for $X = l_1$ or $X = C(0, 1)$ this cannot happen for any basis. On the other hand this is always the case if X is reflexive. The following proposition gives a very simple but useful criterion for checking whether $\{x_n^*\}_{n=1}^\infty$ is a basis of X^*.

Proposition 1.b.1. *Let $\{x_n\}_{n=1}^\infty$ be a basis of a Banach space X. The biorthogonal functionals $\{x_n^*\}_{n=1}^\infty$ form a basis of X^* if and only if, for every $x^* \in X^*$, the norm of $x^*_{|[x_i]_{i=n}^\infty}$ (= the restriction of x^* to the span of $\{x_i\}_{i=n}^\infty$) tends to 0 as $n \to \infty$. A basis $\{x_n\}_{n=1}^\infty$ which has this property is called shrinking.*

Proof. If $\{x_n^*\}_{n=1}^\infty$ is a basis of X^* then, for every $x^* \in X^*$, $\|P_n^* x^* - x^*\| \to 0$. Since $(P_{n-1}^* x^*)_{|[x_i]_{i=n}^\infty} = 0$ it follows that $\lim_n \|x^*_{|[x_i]_{i=n}^\infty}\| = 0$. Conversely, assume that $\|x^*_{|[x_i]_{i=n}^\infty}\| \to 0$ and let $x \in X$ be any element of norm 1. Then,

$$(x^* - P_n^* x^*)(x) = x^*((I - P_n)x) \leqslant \|x^*_{|[x_i]_{i=n+1}^\infty}\|(K+1)$$

where K is the basis constant of $\{x_n\}_{n=1}^\infty$. Hence, $\|P_n^* x^* - x^*\| \to 0$ □

If X has a shrinking basis it is possible to give a convenient representation of X^{**} by using the basis.

Proposition 1.b.2. *Let $\{x_n\}_{n=1}^\infty$ be a shrinking basis of a Banach space X. Then X^{**} can be identified with the space of all sequences of scalars $\{a_n\}_{n=1}^\infty$ such that $\sup_n \left\|\sum_{i=1}^n a_i x_i\right\| < \infty$. This correspondence is given by $x^{**} \leftrightarrow (x^{**}(x_1^*), x^{**}(x_2^*), \ldots)$. The norm of x^{**} is equivalent (and in case the basis constant is 1 even equal) to $\sup_n \left\|\sum_{i=1}^n x^{**}(x_i^*)x_i\right\|$.*

Proof. We may clearly assume that the basis constant of $\{x_n\}_{n=1}^\infty$ is 1. If $x^{**} \in X^{**}$ and $\{P_n\}_{n=1}^\infty$ are the projections on X associated to the basis then $P_n^{**} x^{**} = \sum_{i=1}^n x^{**}(x_i^*)x_i$ and $\|x^{**}\| = \lim_n \|P_n^{**} x^{**}\| = \sup_n \|P_n^{**} x^{**}\|$. Conversely, if $\{a_n\}_{n=1}^\infty$ is such that $\sup_n \left\|\sum_{i=1}^n a_i x_i\right\| < \infty$ then any w^* limit point x^{**} of the bounded set $\left\{\sum_{i=1}^n a_i x_i\right\}_{n=1}^\infty$ satisfies $x^{**}(x_i^*) = a_i$ for all i. (In particular, the uniqueness of x^{**} implies that $\left\{\sum_{i=1}^n a_i x_i\right\}_{n=1}^\infty$ must tend w^* to x^{**}.) □

Observe that the canonical image of X in X^{**} corresponds to those sequences $\{a_n\}_{n=1}^\infty$ for which $\left\{\sum_{i=1}^n a_i x_i\right\}_{n=1}^\infty$ is not only bounded but converges in norm.

Another important notion concerning bases, which is in a sense dual to "shrinking", is that of "boundedly complete".

Definition 1.b.3. A basis $\{x_n\}_{n=1}^\infty$ of a Banach space is called *boundedly complete* if, for every sequence of scalars $\{a_n\}_{n=1}^\infty$ such that $\sup_n \left\|\sum_{i=1}^n a_i x_i\right\| < \infty$, the series $\sum_{n=1}^\infty a_n x_n$ converges.

A typical example of a non-boundedly complete basis is the unit vector basis of c_0. The unit vector basis is boundedly complete in all the l_p spaces, $1 \leq p < \infty$.

If $\{x_n\}_{n=1}^\infty$ is a shrinking basis in X then $\{x_n^*\}_{n=1}^\infty$ is a boundedly complete basis in X^*. Indeed, if $\sup_n \left\|\sum_{i=1}^n a_i x_i^*\right\| < \infty$ then the expansion of a w^* limit point x^* of $\left\{\sum_{i=1}^n a_i x_i^*\right\}_{n=1}^\infty$ with respect to the basis $\{x_n^*\}_{n=1}^\infty$ must be $\sum_{n=1}^\infty a_n x_n^*$. The converse of this remark is also valid:

Proposition 1.b.4. *A Banach space X with a boundedly complete basis $\{x_n\}_{n=1}^\infty$ is isomorphic to a conjugate space. More precisely, X is isomorphic to the dual of the subspace $[x_n^*]_{n=1}^\infty$ of X^*.*

Proof. Let $Z = [x_n^*]_{n=1}^\infty$ and let J be the canonical map from X to Z^* defined by $Jx(z) = z(x)$. We claim that J is an isomorphism onto. Indeed, let $x \in \text{span}\,\{x_i\}_{i=1}^n$, for some integer n, and $x^* \in X^*$ be such that $\|x^*\| = 1$ and $x^*(x) = \|x\|$. Then, $P_n^* x^*(x) = x^*(x)$, $P_n^* x^* \in Z$ and $\|P_n^* x^*\| \leq K$, where K is the basis constant of $\{x_n\}_{n=1}^*$. Hence, $\|x\|/K \leq \|Jx\| \leq \|x\|$ and this shows that J is an isomorphism. To show that J is onto observe that $\{Jx_n\}_{n=1}^\infty$ are functionals biorthogonal to $\{x_n^*\}_{n=1}^*$ in Z^*. Let $z^* \in Z^*$. The sequence $\left\{\sum_{i=1}^n z^*(x_i^*) Jx_i\right\}_{n=1}^\infty$ is bounded in norm (by $K\|z^*\|$) and thus, since $\{x_n\}_{n=1}^\infty$ is boundedly complete, the series $\sum_{n=1}^\infty z^*(x_n^*) x_n$ converges in X to an element x. Clearly $z^* = Jx$. □

By combining the notions of shrinking and boundedly complete we get the following elegant characterization of reflexivity in terms of bases.

Theorem 1.b.5 [53]. *Let X be a Banach space with a Schauder basis $\{x_n\}_{n=1}^\infty$. Then X is reflexive if and only if $\{x_n\}_{n=1}^\infty$ is both shrinking and boundedly complete.*

Proof. Assume first that X is reflexive. We already observed above that $\{x_n\}_{n=1}^\infty$ must be shrinking. Also if $\sup_n \left\|\sum_{i=1}^n a_i x_i\right\| < \infty$ then any w limit point x of $\left\{\sum_{i=1}^n a_i x_i\right\}_{n=1}^\infty$ must be of the form $\sum_{n=1}^\infty a_n x_n$ and in particular this series converges.

The converse assertion follows immediately from 1.b.2. We shall give an additional proof of the converse assertion since it is somewhat more convenient to use it in more general situations in which an analogue of 1.b.5 is valid. Let $\{y_k\}_{k=1}^\infty$ be a sequence of vectors of norm 1 in X. By the diagonal procedure we can find a subsequence $\{y_{k_j}\}_{j=1}^\infty$ of $\{y_k\}_{k=1}^\infty$ so that $a_n = \lim_j x_n^*(y_{k_j})$ exists for every n. We have that $\sum_{i=1}^n a_i x_i = \lim_j P_n y_{k_j}$ and hence, $\sup_n \left\| \sum_{i=1}^n a_i x_i \right\| \leq K$. Since $\{x_n\}_{n=1}^\infty$ is boundedly complete $y = \sum_{n=1}^\infty a_n x_n$ exists and, by definition, $\lim_j x_n^*(y_{k_j}) = x_n^*(y)$, for every n. Since the basis $\{x_n\}_{n=1}^\infty$ is also shrinking $[x_n^*]_{n=1}^\infty = X^*$ and therefore $w \lim_j y_{k_j} = y$. This proves that X is reflexive. □

In 1.b.5 we considered a single basis in X. In [147] M. Zippin showed that if we consider all the bases in a given space it is enough to use only one of the two properties appearing in 1.b.5. More precisely: a Banach space X with a basis is reflexive if (and clearly only if) every basis in X is shrinking or, alternatively, if every basis in X is boundedly complete.

We pass now to questions relating existence of bases and duality. These questions are non-trivial (and thus of interest) only for non-reflexive Banach spaces. If a Banach space X has a basis its dual X^* need not have a basis even if X^* is separable (cf. Section e below). An interesting and rather deep result of W. B. Johnson, H. P. Rosenthal and M. Zippin [61] shows that the existence of a basis in X^* does imply that also X has a basis.

Theorem 1.b.6. *Let X be a Banach space such that X^* has a basis. Then X has a shrinking basis and therefore X^* has a boundedly complete basis.*

The proof of this theorem is closely related to the local theory of Banach spaces and therefore we shall give it only in Vol. IV.

In Section a we proved that every Banach space X contains a basic sequence. For a reflexive X this result implies that X has a quotient space with a basis. W. B. Johnson and H. P. Rosenthal proved in [60] that the same holds for general separable Banach spaces.

Theorem 1.b.7. *Every separable infinite dimensional Banach space has an infinite-dimensional quotient space with a basis.*

For the proof of 1.b.7 we introduce first the following.

Definition 1.b.8. A basic sequence $\{x_n^*\}_{n=1}^\infty$ in the dual X^* of a Banach space X is called a w^* *basic sequence* if there exists a sequence $\{x_n\}_{n=1}^\infty$ in X for which $x_n^*(x_m) = \delta_n^m$ and such that, for every x^* in the w^* closure of span $\{x_n^*\}_{n=1}^\infty$, we have

$$x^* = w^* \lim_n \sum_{i=1}^n x^*(x_i) x_i^*.$$

The next proposition clarifies the meaning of the notion of a w^* basic sequence and its relation to 1.b.7.

Proposition 1.b.9. *A sequence $\{x_n^*\}_{n=1}^\infty \in X^*$ is a w^* basic sequence if and only if there is a basis $\{y_n\}_{n=1}^\infty$ of $Y=X/([x_n^*]_{n=1}^\infty)^\top$ so that $x_n^*=T^*y_n^*$, $n=1, 2,\dots$, where $T\colon X\to Y$ is the quotient map and $\{y_n^*\}_{n=1}^\infty$ are the functionals biorthogonal to $\{y_n\}_{n=1}^\infty$.*

This proposition is proved by a straightforward verification. Note that T^* is an isometry from Y^* onto the w^* closure of $[x_n^*]_{n=1}^\infty$ since it is w^* continuous.

In the proof of 1.b.7 we shall use also the following fact which is a direct consequence of the w^* density of the unit ball of a Banach space X in the unit ball of X^{**}. Let B be a finite dimensional subspace of X^* and let $\varepsilon>0$. Then, there exists a finite set F of elements of norm 1 in X so that, for every $f\in B^*$ with $\|f\|=1$, there is an $x\in F$ which satisfies $|f(x^*)-x^*(x)|\leq\varepsilon\|x^*\|$, for all $x^*\in B$.

Proof of 1.b.7. For every infinite-dimensional Banach space X the set $\{x^*\in X^*;\ \|x^*\|=1\}$ is w^* dense in the unit ball of X^*. Since X is separable the unit ball in X^* is w^* metrizable and hence there is a sequence $\{x_k^*\}_{k=1}^\infty$ in X^* so that $\|x_k^*\|=1$ for all k and $w^*\lim x_k^*=0$. We shall construct a subsequence of $\{x_k^*\}_{k=1}^\infty$ which is w^* basic (this will conclude the proof in view of 1.b.9).

Let $\{\varepsilon_n\}_{n=1}^\infty$ be a sequence of numbers so that $0<\varepsilon_n<1$ and $\sum_{n=1}^\infty \varepsilon_n<\infty$. By the remark above and the separability of X we can choose inductively a sequence of integers $k_1<k_2<\cdots$ and a sequence of finite sets $F_1\subset F_2\subset\cdots$ of elements of norm 1 in X so that

(i) $X=\overline{\operatorname{span}\bigcup_{n=1}^\infty F_n}$

(ii) For every $f\in([x_{k_i}^*]_{i=1}^n)^*$ with $\|f\|=1$ there is an $x\in F_n$ so that $|f(x^*)-x^*(x)|<\varepsilon_n\|x^*\|/3$ for every $x^*\in[x_{k_i}^*]_{i=1}^n$.

(iii) $|x_{k_{n+1}}^*(x)|\leq\varepsilon_n/3$ for all $x\in F_n$.

We claim that $\{x_{k_n}^*\}_{n=1}^\infty$ is a w^* basic sequence. Note first that, by the proof of 1.a.5 (and the remark following it), this is a basic sequence. Moreover, if $\{P_n\}_{n=1}^\infty$ denote the natural projections (on $[x_{k_n}^*]_{n=1}^\infty$) which are associated to this basic sequence, then

$$(*)\qquad \|P_n\|\leq\prod_{j=n}^\infty(1-\varepsilon_j)^{-1}\xrightarrow[n\to\infty]{}1\,.$$

Let $\{y_n\}_{n=1}^\infty\subset([x_{k_n}^*]_{n=1}^\infty)^*$ be the functionals biorthogonal to $\{x_{k_n}^*\}_{n=1}^\infty$. In view of 1.b.9 it suffices to show that the operator $T\colon X\to([x_{k_n}^*]_{n=1}^\infty)^*$ defined by $Tx(x^*)=x^*(x)$, $x^*\in[x_{k_n}^*]_{n=1}^\infty$ maps X onto $[y_n]_{n=1}^\infty$ and thus, in particular, is a quotient map (note that the kernel of T is $([x_{k_n}^*]_{n=1}^\infty)^\top$).

We show first that $TX\subset[y_n]_{n=1}^\infty$. This follows from (i) and (iii) since if $x\in F_n$ for some n then $\sum_{i=1}^\infty|x_{k_i}^*(x)|<\infty$ and hence $Tx=\sum_{i=1}^\infty x_{k_i}^*(x)y_i\in[y_n]_{n=1}^\infty$.

To prove that TX exhausts all of $[y_n]_{n=1}^\infty$ it suffices to show that, for every

$y \in \operatorname{span}\{y_n\}_{n=1}^\infty$ of norm 1 and every $\varepsilon>0$, there exists an $x \in X$ with $||x||=1$ and $||Tx-y||<4\varepsilon$ (the desired result will then follow by successive approximation). We may assume that $y \in \operatorname{span}\{y_i\}_{i=1}^n$ and that n is so large that $\varepsilon>\sum_{i=n}^\infty \varepsilon_i$ and $||P_m||<1+\varepsilon$ for $m \geqslant n$ (use (*)). For $u \in \operatorname{span}\{y_i\}_{i=1}^n$ we denote $||u_{|[x^*_{k_i}]_{i=1}^n}||$ by $||u||_1$. Observe that for every such u

$$||u||_1 \leqslant ||u|| \leqslant ||P_n|| \, ||u||_1 \leqslant (1+\varepsilon)||u||_1 .$$

Choose an $x \in F_n$ which satisfies (ii) for $z=y/||y||_1$. We get that $\left\|\sum_{i=1}^n x^*_{k_i}(x)y_i - z\right\|_1 < \varepsilon_n/3<\varepsilon/3$ and hence $\left\|\sum_{i=1}^n x^*_{k_i}(x)y_i - z\right\| < 2\varepsilon/3$. Since $||y_i||=||P_i-P_{i-1}||\leqslant 4$ for $i \geqslant n$ we deduce from (iii) that $\left\|\sum_{i=n+1}^\infty x^*_{k_i}(x)y_i\right\|<4\varepsilon/3$. Hence

$$\begin{aligned} ||Tx-y|| &\leqslant ||Tx-z||+||y-z|| \leqslant \left\|\sum_{i=1}^\infty x^*_{k_i}(x)y_i - z\right\| + 2\varepsilon \\ &\leqslant \left\|\sum_{i=1}^n x^*_{k_i}(x)y_i - z\right\| + \left\|\sum_{i=n+1}^\infty x^*_{k_i}(x)y_i\right\| + 2\varepsilon \\ &\leqslant 2\varepsilon/3+4\varepsilon/3+2\varepsilon \leqslant 4\varepsilon . \end{aligned}$$

This concludes the proof of 1.b.7. □

It is not known whether 1.b.7 is true without the separability assumption on X. This question is clearly equivalent to the following: Does every infinite-dimensional Banach space have an infinite dimensional and separable quotient space?

Another open problem which arises naturally in view of 1.a.5 and 1.b.7 is the following

Problem 1.b.10. *Let X be an infinite-dimensional separable Banach space. Does there exist a subspace Y of X so that both Y and X/Y have a Schauder basis?*

In Section g below we shall present a partial positive answer to 1.b.10.

The construction used in the proof of 1.b.7 yields under an additional assumption a boundedly complete basis. Before stating this precisely we prove a useful renorming result due to Kadec [65] and Klee [71]. The proof we present is taken from [25].

Proposition 1.b.11. *Let X be a separable Banach space and let Y be a separable subspace of X^*. Then, there is an equivalent norm $|||\cdot|||$ on X so that, whenever $x_n^* \xrightarrow{w^*} x^*$ with $\{x_n^*\}_{n=1}^\infty \subset X^*$, $x^* \in Y$ and $|||x_n^*||| \to |||x^*|||$, we have also that $|||x_n^*-x^*||| \to 0$. ($|||\cdot|||$ denotes here the new norm in X as well as the new norm induced by it in X^*.)*

Proof. Let $B_1 \subset B_2 \subset \cdots$ be a sequence of finite dimensional subspaces of Y whose union is dense in Y. Define a new norm on X^* by putting

$$|||x^*||| = ||x^*|| + \sum_{n=1}^{\infty} 2^{-n} d(x^*, B_n)$$

where $d(x^*, B_n)$ denotes the distance with respect to $||\cdot||$ of x^* from B_n (i.e., the norm of the canonical image of x^* in the quotient space X^*/B_n). Clearly $|||\cdot|||$ is an equivalent norm on X^* and since its unit ball is w^* closed (observe that each B_n is w^* closed) it is induced by an equivalent norm in X which is also denoted by $|||\cdot|||$.

Assume now that $x_n^* \xrightarrow{w^*} x^*$ with $x^* \in Y$ and $|||x_n^*||| \to |||x^*|||$. Then, since $\liminf_n d(x_n^*, B_k) \geq d(x^*, B_k)$ for all k and $\liminf_n ||x_n^*|| \geq ||x^*||$, we deduce that $d(x^*, B_k) = \lim_n d(x_n^*, B_k)$. From this and the fact that $\lim_k d(x^*, B_k) = 0$ it follows that for every $\varepsilon > 0$ there exists an integer k and elements $u_n^* \in B_k$, $n = 1, 2, \ldots$, such that $||x_n^* - u_n^*|| < \varepsilon/4$ for n sufficiently large. Since B_k is finite-dimensional, by passing to a subsequence if needed, we can assume that $u_n^* \to u^*$ for some $u^* \in B_k$. Thus, for n large enough, $||x_n^* - u^*|| < \varepsilon/2$ which, by taking the w^*-limit, gives $||x^* - u^*|| \leq \varepsilon/2$ i.e., $||x^* - x_n^*|| \leq \varepsilon$. This proves that $||x_n^* - x^*|| \to 0$ or, equivalently, that $|||x_n^* - x^*||| \to 0$. □

Observe that if X^* is separable we can take $Y = X^*$. In this case we obtain a renorming of X so that in X^*, w^* convergence on the boundary of the new unit ball is equivalent to norm convergence.

We can now prove the result ensuring the existence of boundedly complete basic sequences [60] at which we hinted above.

Proposition 1.b.12. *Let X be a Banach space whose dual X^* is separable. Then every sequence $\{x_k^*\}_{k=1}^{\infty}$ in X^* such that $x_k^* \xrightarrow{w^*} 0$ and $\limsup_k ||x_k^*|| > 0$ has a boundedly complete basic subsequence $\{x_{k_n}^*\}_{n=1}^{\infty}$.*

Proof. By 1.b.11 there is no loss of generality to assume that in X^*, $y_n^* \xrightarrow{w^*} y^*$ and $||y_n^*|| \to ||y^*|| \Rightarrow ||y_n^* - y^*|| \to 0$. In the proof of 1.b.7 we showed that $\{x_k^*\}_{k=1}^{\infty}$ has a w^* basic subsequence $\{x_{k_n}^*\}_{n=1}^{\infty}$ for which $||P_n|| \to 1$ (see (*)). Thus, if $y_n^* = \sum_{i=1}^{n} a_i x_{k_i}^*$, $n = 1, 2, \ldots$, is a bounded sequence it follows from the fact that $\{x_{k_n}^*\}_{n=1}^{\infty}$ is w^* basic that y_n^* converges w^* to a limit y^*. Since

$$||y^*|| \leq \liminf_n ||y_n^*|| \leq \limsup_n ||y_n^*|| = \limsup_n ||P_n^* y^*|| \leq ||y^*||$$

it follows that $||y_n^* - y^*|| \to 0$ i.e., $\sum_{n=1}^{\infty} a_n x_{k_n}^*$ converges. □

The dual result to 1.b.12 is also true. This was observed in [30].

Proposition 1.b.13. *Let X be an infinite-dimensional Banach space with a separable dual. Then X contains a shrinking basic sequence.*

Proof. Let $\{y_k^*\}_{k=1}^\infty$ be a sequence which is norm dense in the unit ball of X^*. The construction of $\{x_n\}_{n=1}^\infty$ in the proof of 1.a.5 can clearly be carried out so that in addition to the properties required there we have also that $y_k^*(x_n)=0$ for $n>k$. The basic sequence $\{x_n\}_{n=1}^\infty$ obtained in this manner is obviously shrinking. □

The preceding propositions imply an interesting result whose statement has nothing to do with bases. This result, stated by V. D. Milman [106] and first proved by W. B. Johnson and H. P. Rosenthal [60], is a very good illustration of the use of bases in the investigation of linear topological properties of Banach spaces.

Theorem 1.b.14. (i) *Let X be a Banach space whose dual X^* is separable. Let Y be an infinite dimensional subspace of X^* with a separable dual Y^*. Then Y has an infinite-dimensional reflexive subspace.*

(ii) *Let X be an infinite-dimensional Banach space whose second dual X^{**} is separable. Then every infinite dimensional subspace of X or of X^* contains an infinite-dimensional reflexive subspace.*

Proof [60]. (i). By 1.b.13 Y contains a shrinking basic sequence $\{y_k\}_{k=1}^\infty$ with $\|y_k\|=1$ for all k. Clearly, $y_k \xrightarrow{w} 0$ and thus also $y_k \xrightarrow{w^*} 0$ (as an element of X^*). By 1.b.11 there is a subsequence $\{y_{k_n}\}_{n=1}^\infty$ of $\{y_k\}_{k=1}^\infty$ which is a boundedly complete basic sequence. Since $\{y_{k_n}\}_{n=1}^\infty$ is of course also shrinking we deduce from 1.b.5 that $[y_{k_n}]_{n=1}^\infty$ is reflexive.

(ii) Both assertions are immediate consequences of (i). For example, let Y be an infinite-dimensional subspace of X. Then Y^* is separable and Y is a subspace of the separable conjugate X^{**} so (i) applies to Y. □

We conclude this section by mentioning the following result [24].

Theorem 1.b.15. *Let X be a Banach space whose dual X^* is separable. Then there is a Banach space Y with a shrinking basis which has X as a quotient space. In particular, X^* is isomorphic to a subspace of a space with a boundedly complete basis, namely Y^*.*

The proof of this theorem is quite long. Since the theorem will not be used in the sequel we omit its proof and refer the interested reader to [24].

Note that every separable Banach space is a quotient space of l_1 which has a boundedly complete basis. Thus, in one sense, the dual to 1.b.15 is trivially true. In another sense the dualization of 1.b.15 leads to the following open problem.

Problem 1.b.16. *Let X be a Banach space with a separable dual. Is X isomorphic to a subspace of a space with a shrinking basis?*

c. Unconditional Bases

The existence of a Schauder basis in a Banach space does not give very much information on the structure of the space. If one wants to study in more detail the structure of a Banach space by using bases one is led to consider bases with various special properties. In Section b we encountered already two such useful types of bases, namely shrinking and boundedly complete bases. Undoubtedly, the most useful and widely studied special class of bases is that of unconditional bases.

Before studying unconditional bases we present some general facts concerning unconditional convergence.

Proposition 1.c.1. *Let $\{x_n\}_{n=1}^{\infty}$ be a sequence of vectors in a Banach space X. Then the following conditions are equivalent.*

(i) *The series $\sum_{n=1}^{\infty} x_{\pi(n)}$ converges for every permutation π of the integers.*

(ii) *The series $\sum_{i=1}^{\infty} x_{n_i}$ converges for every choice of $n_1<n_2<n_3\ldots$.*

(iii) *The series $\sum_{n=1}^{\infty} \theta_n x_n$ converges for every choice of signs θ_n (i.e. $\theta_n=\pm 1$).*

(iv) *For every $\varepsilon>0$ there exists an integer n so that $\left\|\sum_{i\in\sigma} x_i\right\|<\varepsilon$ for every finite set of integers σ which satisfies* $\min\{i\in\sigma\}>n$.

A series $\sum_{n=1}^{\infty} x_n$ which satisfies one, and thus all of the above conditions, is said to be unconditionally convergent.

Proof. The equivalence of (ii) and (iii) is obvious. If (iv) holds then the partial sums of the series appearing in (i) and in (ii) satisfy the Cauchy condition and thus (iv) $\Rightarrow$ (i) and (iv) $\Rightarrow$ (ii). Assume that (iv) is not satisfied. Then, there is an $\varepsilon>0$ and finite sets $\{\sigma_n\}_{n=1}^{\infty}$ of integers so that

$$q_n=\max\{i,\ i\in\sigma_n\}<p_{n+1}=\min\{i;\ i\in\sigma_{n+1}\}$$

and $\left\|\sum_{i\in\sigma_n} x_i\right\|\geqslant\varepsilon$, for all n. It is clear that $\sigma=\bigcup_{n=1}^{\infty}\sigma_n$ is a subsequence of the integers for which $\sum_{i\in\sigma} x_i$ does not converge (hence (ii) $\Rightarrow$ (iv)). Also if π is a permutation of the integers which, for every n, maps the set $\{i;\ p_n\leqslant i\leqslant q_n\}$ onto itself in such a manner that $\pi^{-1}(\sigma_n)=\{p_n, p_n+1,\ldots, p_n+k_n\}$, where k_n is the cardinality of σ_n, then $\sum_{i=1}^{\infty} x_{\pi(i)}$ does not converge (and hence (i) $\Rightarrow$ (iv)). $\square$

It is easily verified that if $\sum_{n=1}^{\infty} x_n$ converges unconditionally then the sum of $\sum_{n=1}^{\infty} x_{\pi(n)}$ does not depend on the permutation π. The set of vectors of the form

$\sum_{n=1}^{\infty} \theta_n x_n$, $\theta_n = \pm 1$ forms a norm compact set (by (iv) the map from $\{-1, 1\}^N$ into X which assigns to $\{\theta_n\}_{n=1}^{\infty}$ the point $\sum_{n=1}^{\infty} \theta_n x_n$ is continuous). It is also easy to verify that if $\sum_{n=1}^{\infty} x_n$ converges unconditionally then, for every bounded sequence of scalars $\{a_n\}_{n=1}^{\infty}$, the series $\sum_{n=1}^{\infty} a_n x_n$ converges and the operator $T: l_\infty \to X$, defined by $T(a_1, a_2, \ldots) = \sum_{n=1}^{\infty} a_n x_n$, is a bounded linear operator.

In finite-dimensional spaces a series $\sum_{n=1}^{\infty} x_n$ converges unconditionally if and only if it *converges absolutely*, i.e. $\sum_{n=1}^{\infty} \|x_n\| < \infty$. In every infinite-dimensional space there exists a series $\sum_{n=1}^{\infty} x_n$ which converges unconditionally but not absolutely. More precisely we have the following result of Dvoretzky and Rogers [34].

Theorem 1.c.2. *Let X be an infinite-dimensional Banach space. Let $\{\lambda_n\}_{n=1}^{\infty}$ be a sequence of positive numbers such that $\sum_{n=1}^{\infty} \lambda_n^2 < \infty$. Then, there is an unconditionally convergent series $\sum_{n=1}^{\infty} x_n$ in X such that $\|x_n\| = \lambda_n$, for every n.*

For the proof of 1.c.2 we need the following result, due to Auerbach, which is useful in many contexts.

Proposition 1.c.3. *Let B be a Banach space of dimension n. Then, there exist n vectors $\{x_i\}_{i=1}^{n}$ of norm 1 in B and n vectors $\{x_i^*\}_{i=1}^{n}$ of norm 1 in X^* so that $x_i^*(x_i) = \delta_i^j$.*

Proof. Introduce a coordinate system in B and, for $y_1, \ldots, y_n$ in the unit ball of B, let $V(y_1, y_2, \ldots, y_n)$ be the determinant of $(a_{i,j})_{i,j=1}^{n}$, where $(a_{i,1}, a_{i,2}, \ldots, a_{i,n})$ denote the coordinates of y_i, $1 \leq i \leq n$. The function V attains its maximum at an n-tuple $\{x_1, x_2, \ldots, x_n\}$ of vectors of norm 1. Put

$$x_i^*(x) = V(x_1, x_2, \ldots, x_{i-1}, x, x_{i+1}, \ldots, x_n) / V(x_1, x_2, \ldots, x_n).$$

The n-tuples $\{x_i\}_{i=1}^{n}$ and $\{x_i^*\}_{i=1}^{n}$ have the desired property. □

The n-tuples of vectors $\{x_i\}_{i=1}^{n}$, whose existence is ensured by 1.c.3, is called an *Auerbach system*.

The main step in the proof of 1.c.2 is the next lemma (the proof we present is taken from [43]).

Lemma 1.c.4. *Let B be a Banach space of dimension n^2 and norm $\|\cdot\|$. Then there is an n-dimensional subspace C of B and an inner product norm $|||\ |||$ on C so that $\|y\| \leq |||y|||$, for all $y \in C$, and an orthonormal (with respect to $|||\cdot|||$) basis $\{y_i\}_{i=1}^{n}$ of C with $\|y_i\| \geq 1/8$, for every i.*

Proof. Let $\{x_j\}_{j=1}^{n^2}$ be an Auerbach system in B and put $|||x|||_1 = n\left(\sum_{j=1}^{n^2} x_j^*(x)^2\right)^{1/2}$. Then, $|||\cdot|||_1$ is an inner product norm on B and

$$|||x|||_1/n^2 \leqslant \max_j |x_j^*(x)| \leqslant \|x\| \leqslant \sum_{j=1}^{n^2} |x_j^*(x)| \leqslant |||x|||_1 .$$

Consider the following statement

(†) Every subspace C of B with $\dim C > \dim B/2$ contains a vector y with $|||y|||_1 = 1$ and $\|y\| > 1/8$.

If (†) is true we can construct inductively at least $n^2/2 - 1$ elements y_i which are orthonormal with respect to $|||\cdot|||_1$ and satisfy $\|y_i\| \geqslant 1/8$, for every i. In this case there is nothing more to prove.

If (†) does not hold there is a subspace B_2 of B with $\dim B_2 > \dim B/2$ and an inner product $|||\cdot|||_2 = |||\cdot|||_1/8$ on B_2 so that $8|||x|||_2/n^2 \leqslant \|x\| \leqslant |||x|||_2$, for every $x \in B_2$. Consider now the statement obtained from (†) by replacing B by B_2 and $|||\cdot|||_1$ by $|||\cdot|||_2$. If the statement we get is true we have nothing more to prove. If it fails there is a subspace B_3 of B_2 with $\dim B_3 > \dim B_2/2$ and an inner product norm $|||\cdot|||_3$ on B_3 so that $8^2|||x|||_3/n^2 \leqslant \|x\| \leqslant |||x|||_3$, for every $x \in B_3$.

We continue in an obvious way. The process must end after $l-1$ steps for an integer l such that $8^l \leqslant n^2$. The space B_l will be of dimension $\geqslant n^2/2^{l-1}$. In this space we can find at least $\dim B_l/2 - 1$ vectors y_i which are orthonormal with respect to $|||\cdot|||_l$ and satisfy $\|y_i\| > 1/8$, for every i. Since $n^2 \geqslant 8^l$ we get that $n^2 \cdot 2^{-l} - 1 > n$ and this concludes the proof. □

Observe that the unit vectors $u_i = y_i/\|y_i\|$, $1 \leqslant i \leqslant n$, whose existence was proved in 1.c.4, satisfy

$$\left\|\sum_{i=1}^n a_i u_i\right\| \leqslant \left|\left|\left|\sum_{n=1}^n a_i u_i\right|\right|\right| = \left(\sum_{i=1}^n |a_i|^2 |||u_i|||^2\right)^{1/2} \leqslant 8\left(\sum_{i=1}^n |a_i|^2\right)^{1/2}$$

for every choice of scalars $\{a_i\}_{i=1}^n$.

Proof of 1.c.2. Let λ_i be positive numbers such that $\sum_{i=1}^\infty \lambda_i^2 < \infty$. Choose an increasing sequence of integers $\{n_k\}_{k=1}^\infty$ such that $\sum_{i=n_k}^\infty \lambda_i^2 \leqslant 2^{-2k}$, $k=1, 2, \ldots$. By the preceding lemma we can find in any Banach space of dimension $\geqslant (n_{k+1} - n_k)^2$ (and thus in every infinite-dimensional Banach space) unit vectors $\{u_i\}_{i=n_k}^{n_{k+1}-1}$ so that if we put $x_i = \lambda_i u_i$ then, for every choice of signs θ_i,

$$\left\|\sum_{i=n_k}^{n_{k+1}-1} \theta_i x_i\right\| \leqslant 8\left(\sum_{i=n_k}^{n_{k+1}-1} \lambda_i^2\right)^{1/2} \leqslant 8 \cdot 2^{-k} .$$

For $i < n_1$ we take as x_i any vector in X of norm λ_i. The series $\sum_{i=1}^\infty x_i$ converges unconditionally and clearly $\|x_i\| = \lambda_i$, for all i. □

Theorem 1.c.2 is the strongest possible general result in this direction. It follows readily from the parallelogram identity in Hilbert space that if $\{x_i\}_{i=1}^n$ are any n vectors in l_2 then the average of $\left\|\sum_{i=1}^n \theta_i x_i\right\|^2$ taken over all 2^n choices of signs $\{\theta_i\}_{i=1}^n$ is equal to $\sum_{i=1}^n \|x_i\|^2$. Consequently, if $\sum_{i=1}^\infty x_i$ is an unconditionally convergent series in l_2 then $\sum_{i=1}^\infty \|x_i\|^2<\infty$. We shall discuss possible converse statements to 1.c.2 in other examples of Banach spaces and their relation to uniform convexity in Vol. II (see also Remark 2 following 2.b.9 below).

We pass now to unconditional bases.

Definition 1.c.5. A basis $\{x_n\}_{n=1}^\infty$ of a Banach space X is said to be *unconditional* if for every $x \in X$, its expansion in terms of the basis $\sum_{n=1}^\infty a_n x_n$ converges unconditionally.

The following proposition is an immediate consequence of 1.c.1.

Proposition 1.c.6. *A basic sequence $\{x_n\}_{n=1}^\infty$ is unconditional if and only if any of the following conditions holds.*

(i) *For every permutation π of the integers the sequence $\{x_{\pi(n)}\}_{n=1}^\infty$ is a basic sequence.*

(ii) *For every subset σ of the integers the convergence of $\sum_{n=1}^\infty a_n x_n$ implies the convergence of $\sum_{n\in\sigma} a_n x_n$.*

(iii) *The convergence of $\sum_{n=1}^\infty a_n x_n$ implies the convergence of $\sum_{n=1}^\infty b_n x_n$ whenever $|b_n|\leqslant|a_n|$, for all n.*

It follows from (ii) and the closed graph theorem that if $\{x_n\}_{n=1}^\infty$ is an unconditional basic sequence and σ is a subset of the integers then there is a bounded linear projection P_σ defined on $[x_n]_{n=1}^\infty$ by $P_\sigma\left(\sum_{n=1}^\infty a_n x_n\right)=\sum_{n\in\sigma} a_n x_n$. These projections are called the *natural projections associated to the unconditional basic sequence.* For the sets σ of the form $\sigma=\{1, 2,\dots, n\}$ we get that the projections P_σ coincide with the projections P_n which are the natural projections associated to the basic sequence $\{x_n\}_{n=1}^\infty$. Similarly, for every choice of signs $\theta=\{\theta_n\}_{n=1}^\infty$, we have a bounded linear operator M_θ on $[x_n]_{n=1}^\infty$ defined by $M_\theta\left(\sum_{n=1}^\infty a_n x_n\right)=\sum_{n=1}^\infty a_n\theta_n x_n$. Observe that if $\sigma=\{n;\ \theta_n=1\}$ then $P_\sigma=(I+M_\sigma)/2$, and if $\eta=\{\eta_n\}_{n=1}^\infty$ is another sequence of signs then $M_\theta M_\eta=M_{\theta\eta}$ where $(\theta\eta)_n=\theta_n\eta_n$. The uniform boundedness principle implies that $\sup_\sigma \|P_\sigma\|$ and $\sup_\sigma \|M_\theta\|$ are finite. These numbers are related by the inequality

$$\sup_\sigma \|P_\sigma\|\leqslant\sup_\theta \|M_\theta\|\leqslant 2\sup_\sigma \|P_\sigma\|\,.$$

The number $\sup_\theta \|M_\theta\|$ is called *the unconditional constant* of $\{x_n\}_{n=1}^\infty$. Observe that the unconditional constant of a basis is always larger or equal to the basis constant. If $\{x_n\}_{n=1}^\infty$ is an unconditional basis of X we can always define on X an equivalent norm so that the unconditional constant becomes 1. We have simply to take as a new norm the expression $|||x|||=\sup_\theta \|M_\theta x\|$. Every block basis of an unconditional basis is again unconditional. The unconditional constant of a block basis is smaller or equal to the unconditional constant of the original basis. If $\{x_n\}_{n=1}^\infty$ is an unconditional basis of X then the biorthogonal functionals $\{x_n^*\}_{n=1}^\infty$ form an unconditional basic sequence in X^* whose unconditional constant is the same as that of $\{x_n\}_{n=1}^\infty$.

Another often used trivial observation concerning the unconditional constant is the following.

Proposition 1.c.7. *Let $\{x_n\}_{n=1}^\infty$ be an unconditional basic sequence with an unconditional constant K. Then, for every choice of scalars $\{a_n\}_{n=1}^\infty$ such that $\sum_{n=1}^\infty a_n x_n$ converges and every choice of bounded scalars $\{\lambda_n\}_{n=1}^\infty$, we have*

$$\left\|\sum_{n=1}^\infty \lambda_n a_n x_n\right\| \leq 2K \sup_n |\lambda_n| \left\|\sum_{n=1}^\infty a_n x_n\right\|$$

(in the real case we can take K instead of 2K).

Proof. Assume the scalars are real and pick an $x^* \in X^*$, with $\|x^*\|=1$, so that $\sum_{n=1}^\infty \lambda_n a_n x^*(x_n) = \left\|\sum_{n=1}^\infty \lambda_n a_n x_n\right\|$. Let $\{\theta_n\}_{n=1}^\infty$ be defined by $\theta_n=1$ if $a_n x^*(x_n) \geq 0$ and $\theta_n=-1$ if $a_n x^*(x_n)<0$. Then

$$\left\|\sum_{n=1}^\infty \lambda_n a_n x_n\right\| \leq \sum_{n=1}^\infty |\lambda_n|\,|a_n x^*(x_n)| \leq \sup_n |\lambda_n| \sum_{n=1}^\infty \theta_n a_n x^*(x_n)$$

$$\leq \sup_n |\lambda_n| x^*\left(M_\theta\left(\sum_{n=1}^\infty a_n x_n\right)\right) \leq \sup_n |\lambda_n| \cdot K \left\|\sum_{n=1}^\infty a_n x_n\right\| .$$

If the scalars are complex we get the desired result by considering separately the real and imaginary parts of $\sum_{n=1}^\infty a_n x^*(x_n)$. □

The simplest examples of unconditional bases are the unit vector bases in c_0 or in l_p, $1 \leq p < \infty$. A much more interesting example of an unconditional basis is the Haar system in $L_p(0,1)$, $1<p<\infty$. We shall prove the unconditionality of this basis in Vol. II. There is a general simple procedure of constructing an unconditional basis. Let $\{x_n\}_{n=1}^\infty$ be a sequence of non-zero vectors in a Banach space X. Let X_0 be the completion of the space of all sequences of scalars $y=(a_1, a_2, \ldots)$, which are eventually zero, with respect to the norm

$$\|y\| = \sup\left\{\left\|\sum_{n=1}^\infty \theta_n a_n x_n\right\|,\ \theta_n = \pm 1,\ n=1,2,\ldots\right\} .$$

The unit vectors form an unconditional basis of X_0 with unconditional constant 1.

Obviously, this basis is equivalent to $\{x_n\}_{n=1}^\infty$ if and only if $\{x_n\}_{n=1}^\infty$ is itself an unconditional basic sequence.

A simple and important example of a basis which is not unconditional is the *summing basis* in c. It consists of the vectors

$$x_n=(\overbrace{0, 0,\dots, 0}^{n-1}, 1, 1,\dots),\quad n=1, 2, 3,\dots.$$

The norm of $\sum_{n=1}^m a_n x_n$ is $\sup_{1\le n\le m}\left|\sum_{i=1}^n a_i\right|$. The basis $\{x_n\}_{n=1}^\infty$ is a monotone and normalized basis of c which is not unconditional since $\left\|\sum_{i=1}^n x_i\right\|=n$ while $\left\|\sum_{i=1}^n (-1)^i x_i\right\|=1$, for all n. It is of interest to note that if we apply the previously described general procedure of constructing unconditional bases to the sequence $\{x_n\}_{n=1}^\infty$ in c we get as X_0 the space l_1. As we shall see in the next section, the Schauder system in $C(0, 1)$ and the Haar basis of $L_1(0, 1)$ are not unconditional. In the next section we shall present several other examples of bases which are not unconditional.

Bounded linear operators which map one sequence space into another sequence space have a natural representation by an infinite matrix. If $\{x_i\}_{i=1}^\infty$ is a basis of X and $\{y_j\}_{j=1}^\infty$ is a basis of Y, the matrix $A=(\alpha_{i,j})$ corresponding to a bounded linear operator $T: X\to Y$ is defined by the relation $Tx_i=\sum_{j=1}^\infty \alpha_{i,j}y_j$. This representation is especially useful in the case when both bases are unconditional. We shall prove here only one simple fact (cf. [139]), concerning matrix representation, which will be applied in the sequel. A study of some related and deeper questions may be found in [78].

Proposition 1.c.8. *Let the matrix $A=(\alpha_{i,j})$ represent a bounded linear operator T from a Banach space X into a Banach space Y with unconditional bases $\{x_i\}_{i=1}^\infty$ and $\{y_j\}_{j=1}^\infty$, respectively. Then the diagonal of A (i.e. the matrix $(\delta_j^i\alpha_{i,j})$) also represents a bounded linear operator D from X into Y. If the unconditional constants of $\{x_i\}_{i=1}^\infty$ and $\{y_j\}_{j=1}^\infty$ are 1 then $\|D\|\le\|T\|$.*

Proof. It is clearly enough to prove only the second part of the statement. Assume therefore that the unconditional constants are 1. Notice that the matrices

$$A_1=\begin{pmatrix} -\alpha_{1,1} & -\alpha_{1,2} & -\alpha_{1,3} & \cdots \\ \alpha_{2,1} & \alpha_{2,2} & \alpha_{2,3} & \cdots \\ \alpha_{3,1} & \alpha_{3,2} & \alpha_{3,3} & \cdots \\ \cdot & \cdot & \cdot & \cdot \end{pmatrix}\qquad A_2=\begin{pmatrix} -\alpha_{1,1} & \alpha_{1,2} & \alpha_{1,3} & \cdots \\ -\alpha_{2,1} & \alpha_{2,2} & \alpha_{2,3} & \cdots \\ -\alpha_{3,1} & \alpha_{3,2} & \alpha_{3,3} & \cdots \\ \cdot & \cdot & \cdot & \cdot \end{pmatrix}$$

represent operators with the same norm as that of T. Hence, the matrix $(A_1+A_2)/2$ i.e.,

$$\begin{pmatrix} -\alpha_{1,1} & 0 & 0 & \cdots \\ 0 & \alpha_{2,2} & \alpha_{2,3} & \cdots \\ 0 & \alpha_{3,2} & \alpha_{3,3} & \cdots \\ \cdot & \cdot & \cdot & \cdot \end{pmatrix}$$

represents an operator of norm $\leqslant \|T\|$. Applying a similar procedure to this matrix, using the second column and row, we get that

$$\begin{pmatrix} -\alpha_{1,1} & 0 & 0 & 0 & \cdots \\ 0 & -\alpha_{2,2} & 0 & 0 & \cdots \\ 0 & 0 & \alpha_{3,3} & \alpha_{3,4} & \cdots \\ 0 & 0 & \alpha_{4,3} & \alpha_{4,4} & \cdots \\ \cdot & \cdot & \cdot & \cdot & \cdot \end{pmatrix}$$

also represents an operator of norm $\leqslant \|T\|$. By continuing inductively we obtain the desired result. □

Remarks 1. The same method of proof shows that 1.c.8 is also valid for "block diagonal" matrices. More precisely, if $\{m_k\}_{k=1}^\infty$ and $\{n_k\}_{k=1}^\infty$ are increasing sequences of integers and

$$d_{i,j}=\begin{cases}\alpha_{i,j} & \text{for } m_k\leqslant i<m_{k+1},\ n_k\leqslant j<n_{k+1},\ k=1,2,\ldots\\ 0 & \text{otherwise}\end{cases}$$

then $(d_{i,j})$ represents a bounded linear operator (whose norm does not exceed that of T if the unconditional constants are 1).

2. A variant of 1.c.8 which is valid even when only one of the spaces has an unconditional basis will also be used later on. Let Z be a Banach space with an unconditional basis $\{z_n\}_{n=1}^\infty$ and let $\{y_n\}_{n=1}^\infty$ be a sequence of non-zero vectors in Z. Let V be the completion of the space of all finite sequences of scalars $v=(a_1, a_2,\ldots)$, which are eventually zero, equipped with the norm

$$|||v|||=\sup\left\{\left\|\sum_{n=1}^\infty a_n\theta_n y_n\right\|,\ \theta_n=\pm1,\ n=1,2,\ldots\right\}.$$

Put $y_i=\sum_{j=1}^\infty \alpha_{i,j}z_j$ and observe that the operator $T\colon V\to Z$, defined by $Tv=\sum_{n=1}^\infty a_n y_n$, has norm $\leqslant 1$. Since the unit vectors form an unconditional basis of V it follows from 1.c.8 that the diagonal of $(\alpha_{i,j})_{i,j=1}^\infty$ defines a bounded operator from V into Z. More precisely, for each sequence of scalars $\{a_n\}_{n=1}^\infty$, we have

$$\left\|\sum_{n=1}^\infty a_n\alpha_{n,n}z_n\right\|\leqslant K\sup_{\theta_n=\pm1}\left\|\sum_{n=1}^\infty a_n\theta_n y_n\right\|,$$

where K denotes the unconditional constant of $\{z_n\}_{n=1}^\infty$.

For Banach spaces having an unconditional basis there are two fundamental structure theorems which are due to R. C. James [53].

Theorem 1.c.9. *Let X be a Banach space with an unconditional basis $\{x_n\}_{n=1}^\infty$. Then $\{x_n\}_{n=1}^\infty$ is shrinking if and only if X does not have a subspace isomorphic to l_1.*

Proof. If l_1 is isomorphic to a subspace of X then X^* is non-separable and therefore no basis of X can be shrinking. Conversely, if $\{x_n\}_{n=1}^\infty$ is not shrinking there is an $x^* \in X^*$ with $\|x^*\|=1$, an $\varepsilon>0$ and a normalized block basis $\{u_j\}_{j=1}^\infty$ of $\{x_n\}_{n=1}^\infty$ so that $x^*(u_j)\geqslant \varepsilon$ for every j. Hence, for every choice of positive $\{a_j\}_{j=1}^m$,

$$\left\|\sum_{j=1}^m a_j u_j\right\| \geqslant x^*\left(\sum_{j=1}^m a_j u_j\right) \geqslant \varepsilon \sum_{j=1}^m a_j .$$

It follows that, for every choice of scalars $\{a_j\}_{j=1}^m$, $\left\|\sum_{j=1}^m a_j u_j\right\| \geqslant \varepsilon \sum_{j=1}^m |a_j|/K$, where K is the unconditional constant of $\{x_n\}_{n=1}^\infty$. The $\{u_j\}_{j=1}^\infty$ are thus equivalent to the unit vector basis of l_1. □

Theorem 1.c.10. *Let X be a Banach space with an unconditional basis $\{x_n\}_{n=1}^\infty$. Then, the following three assertions are equivalent.*

(i) *The basis is boundedly complete.*

(ii) *X is weakly sequentially complete (i.e. if $\{y_i\}_{i=1}^\infty \subset X$ are such that $\lim_i x^*(y_i)$ exists for every $x^* \in X^*$ then there is a $y \in X$ such that $x^*(y)=\lim_i x^*(y_i)$ for every $x^* \in X^*$).*

(iii) *X does not have a subspace isomorphic to c_0.*

Proof. We assume, as we clearly may, that the unconditional constant of $\{x_n\}_{n=1}^\infty$ is 1. Since c_0 is not w sequentially complete it is clear that (ii) ⇒ (iii). We shall prove now that (iii) ⇒ (i). Assume that the basis is not boundedly complete. Then there exist scalars $\{a_n\}_{n=1}^\infty$ such that $\left\|\sum_{i=1}^n a_i x_i\right\| \leqslant 1$, for every n, but $\sum_{n=1}^\infty a_n x_n$ does not converge. It follows that there is an $\varepsilon>0$ and a sequence of integers $p_1<q_1<p_2<q_2\ldots$ so that if $u_j=\sum_{i=p_j}^{q_j} a_i x_i$ then $\|u_j\|\geqslant\varepsilon$, for every j. It follows from 1.c.7 that for every choice of $\{\lambda_j\}_{j=1}^m$,

$$\left\|\sum_{j=1}^m \lambda_j u_j\right\| \leqslant 2 \sup_j |\lambda_j| \left\|\sum_{j=1}^m u_j\right\| \leqslant 2 \sup_j |\lambda_j| .$$

On the other hand, $\left\|\sum_{j=1}^m \lambda_j u_j\right\| \geqslant \varepsilon \sup_j |\lambda_j|$ and hence, $\{u_j\}_{j=1}^\infty$ is equivalent to the unit vector basis in c_0.

In order to prove the remaining implication in 1.c.10 (i.e. that (i) ⇒ (ii)) we need the following lemma.

Lemma 1.c.11. *Let $\{x_n\}_{n=1}^\infty$ be an unconditional basis of a Banach space X with biorthogonal functionals $\{x_n^*\}_{n=1}^\infty$. Let $\{y_i\}_{i=1}^\infty$ be a bounded sequence in X such that $\lim_i x^*(y_i)$ exists for every $x^* \in X^*$, and such that $\lim_i x_n^*(y_i)=0$ for every n. Then, $\lim_i x^*(y_i)=0$ for every $x^* \in X^*$.*

Proof. Assume that, for some $x^* \in X^*$ and some $\varepsilon > 0$, $x^*(y_i) \geq \varepsilon$ for all i. Since $\lim_i x_n^*(y_i) = 0$ for every n, there is a subsequence $\{y_{i_k}\}_{k=1}^\infty$ of $\{y_i\}_{i=1}^\infty$ so that $\|y_{i_k} - u_k\| < 2^{-k}$, for some block basis $\{u_k\}_{k=1}^\infty$ of $\{x_n\}_{n=1}^\infty$. Since $x^*(u_k) > \varepsilon/2$, for sufficiently large k, the proof of 1.c.9 shows that $\{u_k\}_{k=1}^\infty$, and therefore also $\{y_{i_k}\}_{k=1}^\infty$, are equivalent to the unit vector basis of l_1. Thus, there is an element $y^* \in X^*$ such that $y^*(y_{i_k}) = (-1)^k$, for every k. This contradicts the assumption that $\lim y^*(y_i)$ exists. $\square$

We now prove (i) ⇒ (ii) of 1.c.10. Assume that $\{x_n\}_{n=1}^\infty$ is boundedly complete and that $\lim x^*(y_i)$ exists for every $x^* \in X^*$. Put $a_n = \lim_i x_n^*(y_i)$, $n = 1, 2, \ldots$. For every integer m, $\left\|\sum_{n=1}^m a_n x_n\right\| = \lim_i \|P_m y_i\| \leq \sup_i \|y_i\|$. Hence, $\sum_{n=1}^\infty a_n x_n$ converges to an element $y \in X$. By applying 1.c.11 to $\{y_i - y\}_{i=1}^\infty$ we get that $y_i \xrightarrow{w} y$. $\square$

The following theorem is an immediate consequence of 1.c.9, 1.c.10 and the results 1.b.4, 1.b.5 of the previous section.

Theorem 1.c.12. (a) *A Banach space X with an unconditional basis which does not have subspaces isomorphic to c_0 or l_1 must be reflexive. In particular, if X has an unconditional basis and X^{**} is separable then X is reflexive.*

(b) *A weakly sequentially complete Banach space with an unconditional basis is isomorphic to a conjugate space.*

(c) *If X has an unconditional basis and X^* is separable then X^* has an unconditional basis.*

In connection with assertion (c) above and Theorem 1.b.6 it is worthwhile to remark that the existence of an unconditional basis in X^* does not imply that X has an unconditional basis. For example, let K be the set of ordinals $\leq \omega^\omega$, endowed with their usual order topology. We shall prove in Vol. III that $C(K)$ does not have an unconditional basis. On the other hand $C(K)^*$, which is isometric to l_1, has an unconditional basis.

The preceding results were generalized by Bessaga and Pelczynski [10, 11], as follows

Theorem 1.c.13. *Let Y be a closed subspace of a Banach space X with an unconditional basis. Then,*

(a) *Y is weakly sequentially complete if and only if Y contains no subspace isomorphic to c_0.*

(b) *Each norm bounded set in Y is weakly conditionally compact (i.e. each bounded sequence in Y contains a subsequence which is Cauchy in the weak sense) if and only if Y contains no subspace isomorphic to l_1.*

(c) *Y is reflexive if and only if Y contains no subspace isomorphic to c_0 or l_1.*

We shall prove 1.c.13 in Vol. II in the more general setting of spaces Y which are subspaces of suitable Banach lattices. H. P. Rosenthal has proved that part

(b) above holds without any assumption on Y (i.e. for this we do not have to assume the existence of $X \supset Y$ with an unconditional basis). This is a deeper result and we shall discuss it in detail in Section 2.e below.

d. Examples of Spaces Without an Unconditional Basis

Since the notion of an unconditional basis is much stronger than that of a basis it is naturally much easier to exhibit examples of separable spaces which fail to have an unconditional basis than to exhibit spaces which fail to have a basis altogether. The most common non-reflexive classical function spaces, i.e. $C(0, 1)$ (cf. [68]) and $L_1(0, 1)$, fail to have an unconditional basis. That $L_1(0, 1)$ fails to have an unconditional basis can be deduced from 1.c.12(b) since $L_1(0, 1)$ is w sequentially complete and not isomorphic to a conjugate space (this latter fact will be proved in Vol. III). We present here a short proof of the fact, due to Pelczynski [115], that $L_1(0, 1)$ is not even isomorphic to a subspace of a space with an unconditional basis. The proof we give is due to V. D. Milman [107].

Proposition 1.d.1. *The space $L_1(0, 1)$ is not isomorphic to a subspace of a space with an unconditional basis.*

Proof. Let $\{r_n(t)\}_{n=1}^{\infty}$ be the Rademacher functions on $[0, 1]$ defined by $r_n(t) = \operatorname{sign} \sin 2^n \pi t$. Then, for every $x \in L_1(0, 1)$, we have

$$x(t)r_n(t) \xrightarrow{w} 0; \qquad \|x(t)+x(t)r_n(t)\| \to \|x(t)\|$$

(to check, e.g., the second statement observe that if x is the characteristic function of an interval $[k2^{-n}, (k+1)2^{-n}]$ then $\|x+r_m x\| = \|x\|$ for $m > n$).

Assume now that $L_1(0, 1) \subset Y$ and that Y has an unconditional basis $\{y_i\}_{i=1}^{\infty}$. Pick any $x_1 \in L_1(0, 1)$ with $\|x_1\| = 1$. By the observation above we can define inductively vectors $\{x_n\}_{n=2}^{\infty}$ in $L_1(0, 1)$ of the form

$$x_2 = x_1 \cdot r_{k_1}, \qquad x_3 = (x_1 + x_2) r_{k_2}, \ldots, \; x_n = \left(\sum_{j=1}^{n-1} x_j\right) r_{k_{n-1}}, \ldots$$

so that $\frac{1}{2} \leq \|x_n\| = \|x_1 + x_2 + \cdots + x_{n-1}\| \leq 2$, for all n, and so that $\|x_n - u_n\| \leq 2^{-n}$, where $\{u_n\}_{n=1}^{\infty}$ is a suitable block basis of $\{y_i\}_{i=1}^{\infty}$. It follows from these relations that $\{u_n\}_{n=1}^{\infty}$, and therefore $\{x_n\}_{n=1}^{\infty}$, is equivalent to the unit vector basis of c_0. This is impossible since $L_1(0, 1)$ is w-sequentially complete. □

Since every separable Banach space is isometric to a subspace of $C(0, 1)$ it follows from 1.d.1 that also $C(0, 1)$ cannot be embedded in a space with an unconditional basis.

We shall consider now another example of a separable Banach space which, among its other interesting properties, cannot be embedded in a space with an unconditional basis. This example, which is due to R. C. James [53, 54], had an important role in the development of Banach space theory and is still now a source of inspiration to many constructions in the theory.

Example 1.d.2. *The space J: A Banach space with a Schauder basis whose canonical image is of codimension* 1 *in its second dual J^{**} and is also isometric to J^{**}.*

Observe that, in spite of the fact that J is isometric to J^{**}, the space J is not reflexive. Since J^{**} is separable it cannot have a subspace isomorphic to c_0 or l_1. By 1.c.13, J is not isomorphic to a subspace of a space with an unconditional basis.

The space J consists of all sequences of scalars $x=(a_1, a_2, \ldots, a_n, \ldots)$ for which

$$\text{(i)} \qquad \|x\|=\sup \frac{1}{\sqrt{2}}\left[(a_{p_1}-a_{p_2})^2+(a_{p_2}-a_{p_3})^2+\cdots+(a_{p_{m-1}}-a_{p_m})^2+ \right.$$
$$\left. +(a_{p_m}-a_{p_1})^2\right]^{1/2}<\infty$$

and

$$\text{(ii)} \qquad \lim_n a_n=0\,.$$

The supremum in (i) is taken over all choices of m and $p_1<p_2<\cdots<p_m$. It is useful to note that

$$|||x|||=\sup\left[(a_{p_1}-a_{p_2})^2+(a_{p_2}-a_{p_3})^2+\cdots+(a_{p_{m-1}}-a_{p_m})^2\right]^{1/2}$$

is an equivalent norm on J (the special form of $\|\cdot\|$ is of importance only for proving that J^{**} is isometric and not merely isomorphic to J).

It is easy to verify that J is a Banach space and that the unit vectors $\{e_n\}_{n=1}^\infty$ form a monotone basis with respect to both norms. The vectors $\{e_1+e_2+\cdots+e_n\}_{n=1}^\infty$ have all norm 1 but they have no w limit point in J; therefore, J is not reflexive.

The unit vector basis is a shrinking basis of J. Indeed, assume that for some $x^*\in J^*$, some $\varepsilon>0$ and a normalized block basis $\{u_k\}_{k=1}^\infty$ of $\{e_n\}_{n=1}^\infty$ we have $x^*(u_k)\geq\varepsilon$, for all k. It is easily verified (by using $|||\cdot|||$) that $\sum_{k=1}^\infty u_k/k$ converges in J. Since $\sum_{k=1}^\infty x^*(u_k)/k$ does not converge we arrived at a contradiction. By 1.b.2, the space J^{**} consists of all sequences $(a_1, a_2, \ldots, a_n, \ldots)$ for which $\sup_n \left\|\sum_{i=1}^n a_i e_i\right\|<\infty$, i.e. all sequences for which (i) in the definition of J holds. Since (i) implies the existence of $\lim_n a_n$ we infer that J^{**} is the linear span of J (or, more precisely, of the canonical image of J in J^{**}) and the functional x_0^{**} defined by $x_0^{**}(e_n^*)=1$ for all n (i.e. the functional which corresponds to the sequence $(1, 1, 1, \ldots)$). The map $U: J^{**}\to J$ defined by

$$Ux^{**}=(-\lambda,\ x^{**}(e_1^*)-\lambda,\ x^{**}(e_2^*)-\lambda, \ldots)\,,$$

where $\lambda = \lim_n x^{**}(e_n^*)$, is an isometry of J^{**} onto J. (Recall that the norm in J^{**} is given by $\|x^{**}\| = \sup_n \left\| \sum_{i=1}^{n} x^{**}(e_i^*)e_i \right\|$.)

It is noteworthy to remark that the basis $\{e_n^*\}_{n=1}^\infty$ of J^* has the property that $\left\| \sum_{n=1}^\infty c_n e_n^* \right\| = \sum_{n=1}^\infty c_n$, whenever all the c_n are non-negative. Since J^* does not contain a subspace isomorphic to l_1 no subsequence of $\{e_n^*\}_{n=1}^\infty$ can be unconditional.

A useful variant of 1.d.2 shows that not only the one-dimensional space but every separable Banach space X can be realized as Z^{**}/Z for a suitable Z. This was proved in [55] under some restrictions on X and in the general form in [85].

Theorem 1.d.3. *Let X be a separable Banach space. Then there is a separable Banach space Z such that Z^{**} has a boundedly complete basis and Z^{**}/Z is isomorphic to X.*

For the proof of 1.d.3 we refer to [85]; we just mention here the definition of Z. Let $\{x_n\}_{n=1}^\infty$ be a sequence which is dense on the boundary of the unit ball of X. The space Z consists of all the sequences $z = (a_1, a_2, \ldots)$ of scalars for which

$$\text{(i)} \qquad \|z\| = \sup \left(\sum_{j=1}^{m} \left\| \sum_{i=p_{j-1}+1}^{p_j} a_i x_i \right\|^2 \right)^{1/2} < \infty$$

and

$$\text{(ii)} \qquad \sum_{i=1}^{\infty} a_i x_i = 0 .$$

The supremum in (i) is taken over all choices of integers m and $0 = p_0 < p_1 < \cdots < p_m$. The main point in the proof is to show that Z^{**} can be identified with the space of all sequences $(a_1, a_2, \ldots)$ for which condition (i) above holds. Observe that condition (i) implies that $\sum_{i=1}^\infty a_i x_i$ converges. The operator T from Z^{**} to X defined by $T(a_1, a_2, \ldots) = \sum_{i=1}^\infty a_i x_i$ is bounded and, by the density of $\{x_i\}_{i=1}^\infty$, it is actually a quotient map. The kernel of T is, by definition, the canonical image of Z in Z^{**} and thus Z^{**}/Z is isomorphic to X. From the description of Z^{**} it is obvious that the unit vectors $\{e_n\}_{n=1}^\infty$ form a boundedly complete basis of Z^{**}. From this fact and 1.b.6 it follows that Z has a shrinking basis.

In view of 1.b.2 it is possible to phrase 1.d.3 as follows. Let X be a Banach space. Then, there is a norm $\|\cdot\|$ on the space of the sequences of scalars which are eventually 0 so that X is isometric to Z_1/Z_2, where Z_1 consists of all sequences $(a_1, a_2, \ldots)$ such that $\sup_n \left\| \sum_{i=1}^n a_i e_i \right\| < \infty$ while Z_2 consists of all those sequences for which $\left\{ \sum_{i=1}^n a_i e_i \right\}_{n=1}^\infty$ is a Cauchy sequence.

Another construction of a space Z, having the properties required in 1.d.3, is given in [24]. This paper contains a construction of spaces Z satisfying $Z^{**}/Z \approx X$

also for a large natural class of non-separable spaces X. It seems to be unknown whether an arbitrary non-separable X can be represented as Z^{**}/Z, for a suitable Z.

Every space Z obtained in 1.d.3 is non-reflexive but has a separable second dual. Thus, by 1.c.13, it is not isomorphic to a subspace of a space with an unconditional basis. In all examples considered thus far the non-reflexivity of the space played a crucial role in establishing the non-existence of an unconditional basis. The first example of a separable reflexive space which cannot be embedded in a space with an unconditional basis was given in [78]. Later it was shown in [88] that no space with an unconditional basis can contain isometric copies of $L_p(0, 1)$, with p arbitrarily close to 1 or to ∞ (we shall present the proof of this fact in Vol. II). Thus, for every sequence $\{p_n\}_{n=1}^{\infty}$ of numbers so that $1<p_n<\infty$ for every n and either $\inf_n p_n=1$ or $\sup_n p_n=\infty$, the reflexive space $\left(\sum_{n=1}^{\infty} \oplus L_{p_n}(0, 1)\right)_2$ is not isomorphic to a subspace of a space with an unconditional basis.

The previous examples cannot be embedded isomorphically into a Banach space having an unconditional basis. There are also examples of spaces which fail to have an unconditional basis but do embed into a space having such a basis. The first and perhaps simplest example of this kind (cf. [82]) is the subspace D of l_1, spanned by $x_n=e_n-(e_{2n}+e_{2n+1})/2$, $n=1, 2,\ldots$, where $\{e_n\}_{n=1}^{\infty}$ are the unit vectors in l_1. The sequence $\{x_n\}_{n=1}^{\infty}$ forms a monotone basis of D. The fact that D does not have an unconditional basis follows from 1.c.12 since D is weakly sequentially complete without being isomorphic to a conjugate space (this latter fact will be proved in Vol. IV).

More complicated but also more interesting examples are obtained by using Enflo's solution to the basis problem (and its modification in [20] and [40]); there exist subspaces of c_0 and l_p, $2<p<\infty$, which fail even to have a basis. We shall discuss this in detail in the next chapter.

We mention now two questions related to the existence of unconditional bases which are still open.

Problem 1.d.4. *Let X be a Banach space with an unconditional basis and let Y be a complemented subspace of X. Does Y have an unconditional basis?*

The second question asks whether Banach's theorem on the existence of basic sequences can be strengthened in the following sense.

Problem 1.d.5. *Does every infinite dimensional Banach space X contain an unconditional basic sequence?*

The answer to 1.d.5 is positive whenever X is isomorphic to a subspace of a space with an unconditional basis (use 1.a.11), or more generally, to a subspace of a nice Banach lattice (this will be made precise and proved in Vol. II). In a Banach space which is a subspace of a space with an unconditional basis an unconditional basic sequence can be obtained from every sequence of vectors $\{u_n\}_{n=1}^{\infty}$ with $\|u_n\|=1$ and $u_n \xrightarrow{w} 0$, by passing to a subsequence. A natural approach to 1.d.5 is therefore to investigate whether in an arbitrary Banach space a sequence $\{u_n\}_{n=1}^{\infty}$, for which $||u_n||=1$ for all n and $u_n \xrightarrow{w} 0$, must have an unconditional basic subsequence. B. Maurey and H. P. Rosenthal [104] proved recently that this is not the case.

Example 1.d.6. *There is a Banach space having a Schauder basis $\{e_n\}_{n=1}^\infty$ which converges weakly to 0 but which has no unconditional subsequence.*

Let $1>\varepsilon>0$ and let $M=\{m_n\}_{n=1}^\infty$ be an increasing sequence of integers with $m_1=1$ so that

$$(*) \qquad \sum_{i=1}^\infty \sum_{j\neq i} \inf\left(\sqrt{\frac{m_i}{m_j}}, \sqrt{\frac{m_j}{m_i}}\right) \leq \varepsilon/2 .$$

We shall construct a collection Δ of sequences $\delta=\{\sigma_n\}_{n=1}^\infty$ of finite subsets σ_n of the integers N whose main properties are:

1. If $\delta=\{\sigma_n\}_{n=1}^\infty \in \Delta$ then, for every n, the largest integer in σ_n is smaller than the smallest integer in σ_{n+1} (i.e. $\max \sigma_n < \min \sigma_{n+1}$).

2. For $\delta=\{\sigma_n\}_{n=1}^\infty \in \Delta$ the cardinalities $\bar{\bar{\sigma}}_n$ of the sets σ_n satisfy $1=\bar{\bar{\sigma}}_1<\bar{\bar{\sigma}}_2<\bar{\bar{\sigma}}_3<\cdots$ and $\bar{\bar{\sigma}}_n \in M$, $n=1, 2,\ldots$.

3. If, for a pair of sequences $\delta^1=\{\sigma_n^1\}_{n=1}^\infty$ and $\delta^2=\{\sigma_n^2\}_{n=1}^\infty$ in Δ, we have $\bar{\bar{\sigma}}_{j+1}^1=\bar{\bar{\sigma}}_{k+1}^2$, for some integers j and k, then necessarily $j=k$ and $\sigma_i^1=\sigma_i^2$ for $1\leq i\leq k$.

4. For every infinite subset N_1 of the integers there is a $\delta=\{\sigma_n\}_{n=1}^\infty$ in Δ so that $\sigma_n\subset N_1$ for all n.

The family Δ is defined as follows: Pick a one to one mapping ψ from the set of all finite subsets of the integers into M which satisfies $\psi(\sigma)>\bar{\bar{\sigma}}$, for every σ, and take as Δ the family of all $\delta=\{\sigma_n\}_{n=1}^\infty$ which satisfy 1 above and for which $\bar{\bar{\sigma}}_1=1$ and $\bar{\bar{\sigma}}_{n+1}=\psi(\sigma_1 \cup \sigma_2 \cup \cdots \cup \sigma_n)$, $n=1, 2,\ldots$. With this definition of Δ it is obvious that 1, 2 and 4 hold. Let us verify that also 3 holds. We observe first that the growth condition (*) on $\{m_n\}_{n=1}^\infty$ is such that the following is true. If

$$m_{n_1}+m_{n_2}+\cdots+m_{n_k}=m_{s_1}+m_{s_2}+\cdots+m_{s_h}$$

for some choice of $n_1<n_2<\cdots<n_k$ and $s_1<s_2<\cdots<s_h$ then $k=h$ and $n_i=s_i$ for $1\leq i\leq k$. Assume now that, as in 3, $\bar{\bar{\sigma}}_{j+1}^1=\bar{\bar{\sigma}}_{k+1}^2$. By the definition of Δ and the fact that ψ is one to one it follows that $\bigcup_{i=1}^j \sigma_i^1=\bigcup_{i=1}^k \sigma_i^2$ and hence, $\sum_{i=1}^j \bar{\bar{\sigma}}_i^1=\sum_{i=1}^k \bar{\bar{\sigma}}_i^2$. By 2 and the remark made above on M it follows that $j=k$ and $\bar{\bar{\sigma}}_i^1=\bar{\bar{\sigma}}_i^2$ for $1\leq i\leq k$. From this and the fact that ψ is one to one we get that $\sigma_i^1=\sigma_i^2$ for $1\leq i\leq k$, as desired.

We pass now to the construction of the space E of Maurey and Rosenthal. It is the completion of the space of sequences $x=(a_1, a_2,\ldots)$ of scalars, which are eventually 0, with respect to the norm

$$||x||=\sup\left|\sum_{n=1}^\infty \left(\sum_{i\in\sigma_n} a_i\right)\bar{\bar{\sigma}}_n^{-1/2}\right|$$

where the supremum is taken over all sequences $\delta=\{\sigma_n\}_{n=1}^\infty$ in Δ. It is clear that the unit vectors $\{e_n\}_{n=1}^\infty$ form a monotone basis of E. Let $\delta=\{\sigma_j\}_{j=1}^\infty \in \Delta$ and put $u_j^{(\delta)}=\left(\sum_{i\in\sigma_j} e_i\right)/\bar{\sigma}_j^{1/2}$, $j=1, 2,\ldots$. Let $\{c_j\}_{j=1}^\infty$ be scalars and let n be an integer. By using in the definition of $\|\cdot\|$ elements $\eta^k=\{\tau_j^k\}_{j=1}^\infty \in \Delta$, $1\leqslant k\leqslant n$, with $\tau_j^k=\sigma_j$ if $1\leqslant j\leqslant k$ and $\min \tau_{k+1}^k>\max \sigma_n$, it follows that

$$\left\|\sum_{j=1}^n c_j u_j^{(\delta)}\right\|\geqslant \sup_{1\leqslant k\leqslant n}\left|\sum_{j=1}^k c_j\right|.$$

It follows from (*) and condition 3 on Δ, by an easy computation, that

$$\left\|\sum_{j=1}^n c_j u_j^{(\delta)}\right\|\leqslant(1+\varepsilon)\sup_{1\leqslant k\leqslant n}\left|\sum_{j=1}^k c_j\right|.$$

Hence, $\{u_j^{(\delta)}\}_{j=1}^\infty$ is equivalent to the summing basis in c. Consequently, by 4, every subsequence of $\{e_n\}_{n=1}^\infty$ has a block basis which is equivalent to the summing basis and thus no subsequence of $\{e_n\}_{n=1}^\infty$ can be unconditional.

It remains to show that $\{e_n\}_{n=1}^\infty$ tends weakly to 0. This follows from the fact that, for every increasing sequence $\{n_k\}_{k=1}^\infty$ of integers, we have

$$\lim_{k\to\infty}\|e_{n_1}+e_{n_2}+\cdots+e_{n_k}\|/k=0.$$

Indeed, by 2 and the definition of $\|\cdot\|$, we obtain that $\|e_{n_1}+e_{n_2}+\cdots+e_{n_k}\|\leqslant\sum_{i=1}^k \alpha_i$, where $\alpha_i=m_j^{-1/2}$ whenever $\sum_{h=1}^{j-1} m_h<i\leqslant\sum_{h=1}^{j} m_h$. Obviously, $\lim_k\left(\sum_{i=1}^k \alpha_i\right)/k=0$.

The space E described here is not reflexive. In Vol. II we shall show how to modify the construction of E in order to get even a uniformly convex space with a basis which tends weakly to 0 but has no unconditional subsequence. Thus, there is also a uniformly convex space which is not isomorphic to a subspace of a space with an unconditional basis.

e. The Approximation Property

A result which goes back to the beginnings of functional analysis asserts that the compact operators on a Hilbert space are exactly those operators which are limits in norm of operators of finite rank. One part of this assertion, namely that every $T\in L(X, Y)$ for which $\|T-T_n\|\to 0$ for suitable $\{T_n\}_{n=1}^\infty \in L(X, Y)$ with $\dim T_nX<\infty$ is compact, is trivially true for every pair of Banach spaces X and Y. It was realized long ago that the converse assertion is also true for many examples of spaces X and Y besides Hilbert spaces. For example, if Y has a Schauder basis $\{y_n\}_{n=1}^\infty$ then, for every compact $T\in L(X, Y)$, $\|T-P_nT\|\to 0$, where the $\{P_n\}_{n=1}^\infty$ are the projections associated to the basis $\{y_n\}_{n=1}^\infty$. The question whether the converse assertion is true for arbitrary Banach spaces X and Y (which was called for obvious reasons the approximation problem) was open for a long time. This problem was solved (in the negative) by P. Enflo [37]. The observation above shows that this solution provides also a negative solution to the basis problem. We shall present Enflo's solution or, more precisely, a simplified version of it, due to Davie [20], in the next chapter. Our purpose in this section is to investigate those Banach spaces which have one of the many variants of the "approximation property", i.e. those spaces X for which every compact T in $L(X, Y)$, or $L(Y, X)$, (Y arbitrary) is a limit in norm of a suitable sequence of finite rank operators. The investigation of the various variants of the approximation property and the relations between them was initiated by Grothendieck [48]. Many of the results presented here are taken from Grothendieck's memoir. The real impetus to the investigation of the approximation properties was however given by Enflo's result which ensured that the class of spaces which fail to have any of the approximation properties is not void. Much progress has been done in this study but, generally speaking, the situation is still very far from being clear. We shall mention below several of the many natural open problems concerning the approximation property. What is clear by now is that there are many examples of spaces which fail to have the approximation property even among spaces which are "nice" in other respects (there is, for example, a Banach lattice which fails to have the approximation property [138]. We shall present this example in Vol. II). It is also clear that the study of the approximation property is important in many contexts in Banach space theory and even in some areas of analysis outside the framework of this theory.

Definition 1.e.1. A Banach space X is said to have the *approximation property* (A.P. in short) if, for every compact set K in X and every $\varepsilon>0$, there is an operator $T\colon X\to X$ of finite rank (i.e. $Tx=\sum_{i=1}^{n} x_i^*(x)x_i$, for some $\{x_i\}_{i=1}^n\subset X$ and $\{x_i^*\}_{i=1}^n\subset X^*$) so that $\|Tx-x\|\leq\varepsilon$, for every $x\in K$.

Every space with a Schauder basis has the A.P. Indeed, for every compact K and every $\varepsilon>0$, we have $\|P_nx-x\|<\varepsilon$ for every $x\in K$ provided $n>n(\varepsilon, K)$, where $\{P_n\}_{n=1}^\infty$ are the projections associated to the basis.

In order to study the A.P. we need two general facts—one concerns the structure of compact sets in a Banach space and the other gives a concrete representation of the dual of some space of operators.

Proposition 1.e.2. *A closed subset K of a Banach space X is compact if and only if there is a sequence $\{x_n\}_{n=1}^\infty$ in X such that $\|x_n\|\to 0$ and $K\subset\overline{\operatorname{conv}}\,\{x_n\}_{n=1}^\infty$.*

Proof. It is easily checked that if $\|x_n\|\to 0$ then $\left\{\sum_{n=1}^\infty \lambda_n x_n;\ \lambda_n\geq 0,\ \sum_{n=1}^\infty \lambda_n\leq 1\right\}$ is compact and coincides with $\overline{\operatorname{conv}}\,\{x_n\}_{n=1}^\infty$. This proves the "if" part. We prove now the "only if" part. Let K be compact. Let $\{x_{i,1}\}_{i=1}^{n_1}$ be a finite set of elements of X so that $2K\subset\bigcup_{i=1}^{n_1} B(x_{i,1}, 1/4)$. Put

$$K_2=\bigcup_{i=1}^{n_1}\{(B(x_{i,1}, 1/4)\cap 2K)-x_{i,1}\}\,.$$

Then K_2 is a compact subset of $B(0, 1/4)$. Pick next $\{x_{i,2}\}_{i=1}^{n_2}$ in $B(0, 1/2)$ so that $2K_2 \subset \bigcup_{i=1}^{n_2} B(x_{i,2}, 1/4^2)$ and put

$$K_3 = \bigcup_{i=1}^{n_2} \{(B(x_{i,2}, 1/4^2) \cap 2K_2) - x_{i,2}\}.$$

We continue the inductive construction of $\{x_{i,j}\}_{i=1}^{n_j}$, $j=1, 2, 3, \ldots$ in an obvious way.

For every $x \in K$ there is an $1 \leqslant i_1 \leqslant n_1$ so that $2x - x_{i_1,1} \in K_2$; hence an $1 \leqslant i_2 \leqslant n_2$ so that $4x - 2x_{i_1,1} - x_{i_2,2} \in K_3$ and, in general,

$$x - (x_{i_1,1}/2 + x_{i_2,2}/2^2 + \cdots + x_{i_k,k}/2^k) \in 2^{-k} K_{k+1}.$$

It follows that $x \in \overline{\text{conv}}\,\{x_{i,j};\ 1 \leqslant i \leqslant n_j, j=1, 2, \ldots\}$. Since $\|x_{i,j}\| \leqslant 2 \cdot 4^{-j+1}$, for $j>1$ and every $i \leqslant n_j$, our assertion is proved. □

Proposition 1.e.3. *Let X and Y be Banach spaces and put on $L(X, Y)$ the topology τ of uniform convergence on compact sets in X (this is the locally convex topology generated by the seminorms of the form $\|T\|_K = \sup\{\|Tx\|, x \in K\}$, where K ranges over the compact subsets of X). Then, the continuous linear functionals on $(L(X, Y), \tau)$ consist of all functionals φ of the form*

$$\varphi(T) = \sum_{i=1}^{\infty} y_i^*(Tx_i), \quad \{x_i\}_{i=1}^{\infty} \subset X, \quad \{y_i^*\}_{i=1}^{\infty} \subset Y^*, \quad \sum_{i=1}^{\infty} \|x_i\|\,\|y_i^*\| < \infty.$$

Proof. Assume that φ has such a representation. We may clearly assume that $x_i \neq 0$ for every i. Let $\{\eta_i\}_{i=1}^{\infty}$ be a sequence of positive scalars tending to ∞ so that $\sum_{i=1}^{\infty} \eta_i \|x_i\|\,\|y_i^*\| = C < \infty$. Put $K = \{x_i/\|x_i\|\eta_i\}_{i=1}^{\infty} \cup \{0\}$. Then K is compact and

$$|\varphi(T)| \leqslant \sum_{i=1}^{\infty} \|y_i^*\|\,\|x_i\|\eta_i \|T(x_i/\|x_i\|\eta_i)\| \leqslant C\|T\|_K.$$

Conversely, assume that φ is a linear functional on $L(X, Y)$ so that $|\varphi(T)| \leqslant C\|T\|_K$, for some constant C and some compact set $K \subset X$. By 1.e.2 we may assume without loss of generality that $K = \overline{\text{conv}}\,\{x_n\}_{n=1}^{\infty}$, where $\|x_n\| \to 0$. Let $S: L(X, Y) \to (Y \oplus Y \oplus \cdots)_0$ be defined by $S(T) = (Tx_1, Tx_2, \ldots)$. Since $|\varphi(T)| \leqslant C\|S(T)\|$ it follows that there is a linear functional ψ defined on the closure of $SL(X, Y)$ so that $\varphi(T) = \psi(S(T))$. By the Hahn–Banach theorem we may extend ψ to a continuous linear functional on $(Y \oplus Y \oplus \cdots)_0$, i.e. to an element of $(Y^* \oplus Y^* \oplus \cdots)_1$. In other words there exist $\{y_n^*\}_{n=1}^{\infty}$ in Y^* so that $\sum_{n=1}^{\infty} \|y_n^*\| < \infty$ and $\varphi(T) = \sum_{n=1}^{\infty} y_n^* T(x_n)$. □

The next theorem, due to Grothendieck [48], clarifies the relation between the approximation property and the question of approximating compact operators, with which we started this section.

Theorem 1.e.4. *Let X be a Banach space. The following five assertions are equivalent.*

(i) *X has the approximation property.*

(ii) *For every Banach space Y the finite rank operators are dense in $L(Y, X)$, in the topology τ of uniform convergence on compact sets.*

(iii) *For every Banach space Y the finite rank operators are dense in $L(X, Y)$, in the topology τ of uniform convergence on compact sets.*

(iv) *For every choice of $\{x_n\}_{n=1}^\infty \subset X$, $\{x_n^*\}_{n=1}^\infty \subset X^*$ such that $\sum_{n=1}^\infty \|x_n^*\|\,\|x_n\| < \infty$ and $\sum_{n=1}^\infty x_n^*(x)x_n = 0$, for all $x \in X$, we have $\sum_{n=1}^\infty x_n^*(x_n) = 0$.*

(v) *For every Banach space Y, every compact $T \in L(Y, X)$ and every $\varepsilon > 0$ there is a finite rank operator $T_1 \in L(Y, X)$ with $\|T - T_1\| < \varepsilon$.*

Proof. The equivalence of (i) and (iv) is a consequence of 1.e.3. Indeed, by definition, (i) means that the identity operator is in the τ closure of the space of finite rank operators in $L(X, X)$. This happens if and only if every τ continuous linear functional φ on $L(X, X)$, which vanishes on operators of rank 1, vanishes also on the identity operator. By 1.e.3 this is exactly what (iv) means.

It is clear that (ii) or (iii) (with $Y = X$) imply (i). We shall show that (i) implies (ii) and (iii). Let $T \in L(Y, X)$. For every compact set $K \subset Y$ the set TK is compact in X. Hence, given $\varepsilon > 0$, we have by (i) a finite rank operator T_1 on X so that $\|T_1 Ty - Ty\| \leq \varepsilon$, for $y \in K$. Since $T_1 T$ is of finite rank we proved (ii). Let now $0 \neq T \in L(X, Y)$, let K be a compact set in X and let $\varepsilon > 0$. By (i) there is a finite rank operator T_1 on X so that $\|T_1 x - x\| \leq \varepsilon/\|T\|$, for $x \in K$. Then, $\|TT_1 x - Tx\| \leq \varepsilon$ for $x \in K$ and this proves (iii).

It remains to prove the equivalence of (i) and (v). Assume that (i) holds and let $T \in L(Y, X)$ be a compact operator. The set $K = \overline{TB_Y(0, 1)}$ is compact and hence, for every $\varepsilon > 0$, there is a finite rank operator T_1 on X so that $\|T_1 x - x\| \leq \varepsilon$ for $x \in K$. Then, $\|T_1 T - T\| \leq \varepsilon$ and thus (v) holds.

Assume that (v) holds and let K be a compact subset of X and $\varepsilon > 0$. By 1.e.2 we may assume without loss of generality that $K = \overline{\text{conv}}\,\{x_n\}_{n=1}^\infty$, with $\|x_n\| \downarrow 0$ and $\|x_1\| \leq 1$. Put $U = \overline{\text{conv}}\,\{\pm x_n/\|x_n\|^{1/2}\}_{n=1}^\infty$. Clearly, U is a compact convex set in X which is symmetric with respect to the origin. Let Y be the linear span of U in X, i.e. $Y = \bigcup_{n=1}^\infty nU$, and introduce in Y a norm $|||\cdot|||$ which makes U its unit ball (i.e. $|||y||| = \inf\{\lambda > 0;\ y/\lambda \in U\}$). A routine argument shows that $(Y, |||\cdot|||)$ is a Banach space (in particular, it is complete). The formal identity map from Y to X is compact and hence, by (v), there are $\{y_i^*\}_{i=1}^m \subset Y^*$ and $\{u_i\}_{i=1}^m \subset X$ so that $\left\|\sum_{i=1}^m y_i^*(x)u_i - x\right\| \leq \varepsilon/2$, for every $x \in U$, and hence $x \in K$. The $\{y_i^*\}_{i=1}^m$ are continuous with respect to $|||\cdot|||$ but need not be continuous with respect to $\|\cdot\|$ (and thus are not in general restrictions of elements of X^* to Y). In order to conclude the proof it is enough to verify the following statement. Given any $y^* \in Y^*$ and $\delta > 0$ (in our case we take $\delta = \varepsilon/2m \cdot \max_i \|u_i\|$) there is an $x^* \in X^*$ such that $|y^*(x) - x^*(x)| < \delta$ for $x \in K$ i.e. $|y^*(x_n) - x^*(x_n)| < \delta$, for every n.

Observe that, since $x_n/\|x_n\|^{1/2} \in U$, we have $|||x_n||| \leq \|x_n\|^{1/2}$, for every n, and thus $|||x_n|||\to 0$. For $n\geqslant n_0$ we have therefore $|y^*(x_n)|<\delta/2$. Put $K_0=2\delta^{-1}\overline{\text{conv}}\{\pm x_n\}_{n=n_0+1}^{\infty}$ (notice that the closures in $\|\cdot\|$ and $|||\cdot|||$ are the same) and $F=\{x;\ x\in \text{span}\,\{x_n\}_{n=1}^{n_0},\ y^*(x)=1\}$. Then, F is $\|\cdot\|$ closed, K_0 is $\|\cdot\|$ compact and $K_0\cap F=\varnothing$. By the geometric version of the Hahn–Banach theorem there is a $\|\cdot\|$ closed hyperplane $\hat{F}$ in X so that $F\subset\hat{F}$ and $\hat{F}\cap K_0=\varnothing$. Let $x^*\in X^*$ be such that $\hat{F}=\{x;\ x^*(x)=1\}$. Then, $x^*(x_n)=y^*(x_n)$ for $n\leqslant n_0$ and $|x^*(x_n)|<\delta/2$ for $n>n_0$. Consequently, $|x^*(x_n)-y^*(x_n)|<\delta$ for every n, as desired. □

In view of (ii), (iii) and (v) of 1.e.4 it is natural to ask what is the situation if we reverse the roles of X and Y in (v). The answer is given by the following result which is also due to Grothendieck [48].

Theorem 1.e.5. *Let X be a Banach space. Then, X^* has the approximation property if and only if, for every Banach space Y, every $\varepsilon>0$ and every compact $T\in L(X, Y)$, there is a finite rank operator $T_1\in L(X, Y)$ such that $\|T-T_1\|\leqslant\varepsilon$.*

Proof. Assume that, for every Y, every compact $T\in L(X, Y)$ can be approximated as above. Let Z be any Banach space, let $T\in L(Z, X^*)$ be compact and let $\varepsilon>0$. By considering the compact operator $T^*_{|X}\colon X\to Z^*$ it follows from our assumption that there are $\{x_i^*\}_{i=1}^n\subset X^*$, $\{z_i^*\}_{i=1}^n\subset Z^*$ so that, for every x with $\|x\|\leqslant 1$,

$$\Big\|T^*x-\sum_{i=1}^n x_i^*(x)z_i^*\Big\|\leqslant\varepsilon\,.$$

Hence, for every $z\in Z$ with $\|z\|\leqslant 1$,

$$\Big\|Tz(x)-\sum_{i=1}^n z_i^*(z)x_i^*(x)\Big\|\leqslant\varepsilon\,,$$

i.e. $\Big\|Tz-\sum_{i=1}^n z_i^*(z)x_i^*\Big\|\leqslant\varepsilon$. By 1.e.4, X^* has the A.P.

Assume, conversely, that X^* has the A.P. Let $T\in L(X, Y)$ be compact and $1/2>\varepsilon>0$. The operator $T^*\colon Y^*\to X^*$ is also compact and hence there are $\{y_i^{**}\}_{i=1}^n$ in Y^{**} and $\{x_i^*\}_{i=1}^n$ in X^* so that $\Big\|T^*y^*-\sum_{i=1}^n y_i^{**}(y^*)x_i^*\Big\|\leqslant\varepsilon$, whenever $\|y^*\|\leqslant 1$. It follows that $\Big\|Tx-\sum_{i=1}^n x_i^*(x)y_i^{**}\Big\|\leqslant\varepsilon$, whenever $\|x\|\leqslant 1$. This does not conclude the proof since the $\{y_i^{**}\}_{i=1}^n$ are not necessarily contained in Y. We have to "push" the $\{y_i^{**}\}_{i=1}^n$ into Y. This is done by using the following lemma.

Lemma 1.e.6 [90]. *Let X be a Banach space, let D be a finite-dimensional subspace of X^{**} and let $\varepsilon>0$. Then, there is an operator $S\colon D\to X$ such that $\|S\|\leqslant 1+\varepsilon$ and $S_{|D\cap X}$ is the identity.*

Let us first see how 1.e.6 is used to conclude the proof of 1.e.5. By the compactness of T there are $\{x_j\}_{j=1}^m$ in the unit ball $B_X(0, 1)$ of X so that $TB_X(0,1)\subset\bigcup_{j=1}^m B_Y(Tx_j, \varepsilon)$. Apply 1.e.6 to $D=\text{span}\,\{Tx_j\}_{j=1}^m\cup\{y_i^{**}\}_{i=1}^n$. We claim that

$\left\|Tx-\sum_{i=1}^{n} x_i^*(x)Sy_i^{**}\right\| \leqslant 4\varepsilon$, for every $x \in B_X(0, 1)$. Indeed, fix $x \in B_X(0, 1)$ and pick a j such that $\|Tx_j - Tx\| \leqslant \varepsilon$. Then, $\left\|Tx_j-\sum_{i=1}^{n} x_i^*(x)y_i^{**}\right\| \leqslant 2\varepsilon$. By applying S we get that $\left\|Tx_j-\sum_{i=1}^{n} x_i^*(x)Sy_i^{**}\right\| \leqslant 2\varepsilon(1+\varepsilon) \leqslant 3\varepsilon$ and this proves our assertion and concludes the proof of 1.e.5. □

Lemma 1.e.6 is a special instance of a result which plays a central role in the local theory of Banach spaces and which will be discussed in detail in Vol. IV. For the sake of completeness we present its proof (cf. [29]) also here.

Proof of 1.e.6. First we notice that $L(l_1^n, X^{**}) = L(l_1^n, X)^{**}$. This follows from the fact that the correspondence $T \to \{y_i = Te_i\}_{i=1}^n$ (where $\{e_i\}_{i=1}^n$ are the unit vectors of l_1^n) is an isometry from $L(l_1^n, X)$ onto $(\overbrace{X \oplus \cdots \oplus X}^{n \text{ times}})_\infty$.

Let $I: D \to X^{**}$ be the identity mapping and let $\varepsilon > 0$. There exist an n and vectors $\{u_j\}_{j=1}^n$ in D of norm $\leqslant 1+\varepsilon$ so that conv $\{u_j\}_{j=1}^n \supset B_D(0, 1)$. Hence, there is an operator $V: l_1^n \to D$, for which $\|V\| \leqslant 1+\varepsilon$ and $VB_{l_1^n}(0, 1) \supset B_D(0, 1)$. Since $IV \in L(l_1^n, X)^{**}$ there is a net $\{S_\alpha\} \subset L(l_1^n, X)$ with $\|S_\alpha\| \leqslant \|IV\| \leqslant 1+\varepsilon$ for all α and $\{S_\alpha\}$ converges to IV in the w^* topology of $L(l_1^n, X)^{**}$. Any pair $e \in l_1^n$, $x^* \in X^*$ defines a functional $(e, x^*) \in L(l_1^n, X)^*$, by setting $(e, x^*)(S) = x^*Se$. This implies that, for every $e \in l_1^n$, $S_\alpha e \xrightarrow{w^*} IVe$. Put $B = \{e \in l_1^n;\ Ve \in D \cap X\}$ and observe that $S_\alpha e \xrightarrow{w} IVe$ for all $e \in B$. Thus, by taking a suitable convex combination of S_α's and by using a standard perturbation argument, we can construct an operator $T: l_1^n \to X$ such that $T_{|B} = IV_{|B}$ and $\|T\| < 1+2\varepsilon$. If $D \ni d = Vv_1 = Vv_2$, for some $v_1, v_2 \in l_1^n$, then $v_1 - v_2 \in B$ and therefore $Tv_1 = Tv_2$. Hence, by setting $Sd = Tv$, where $v \in l_1^n$ is any vector satisfying $Vv = d$, we define an operator $S \in L(D, X)$ for which $\|S\| < 1+2\varepsilon$ and $S_{|D \cap X} = I_{|D \cap X}$. This completes the proof. □

The relation between the properties appearing in 1.e.4 and 1.e.5 is clarified in the following result.

Theorem 1.e.7. (a) *Let X be a Banach space. If X^* has the* A.P. *then X has the* A.P. *In particular, if X is reflexive then X has the* A.P. *if and only if X^* has the* A.P.

(b) *There is a separable Banach space having a Schauder basis whose dual is separable but fails to have the* A.P.

Proof. Assertion (a) follows immediately from the equivalence (i) ⇔ (iv) of 1.e.4. If $\{x_n\}_{n=1}^\infty \subset X$ and $\{x_n^*\}_{n=1}^\infty \subset X^*$ are such that $\sum_{n=1}^\infty \|x_n\|\,\|x_n^*\| < \infty$ and $\sum_{n=1}^\infty x_n^*(x)x_n = 0$, for every $x \in X$, then also $\sum_{n=1}^\infty x^*(x_n)x_n^* = 0$, for every $x^* \in X^*$. In other words $\sum_{n=1}^\infty Jx_n(x^*)x_n^* = 0$, where $J: X \to X^{**}$ is the canonical embedding. Since X^* has the A.P. $\sum_{n=1}^\infty Jx_n(x_n^*) = \sum_{n=1}^\infty x_n^*(x_n) = 0$.

The proof of assertion (b) uses of course the fact that there is a Banach space which fails to have the A.P. (this will be proved in Section 2.d below). Let X be a separable Banach space which does not have the A.P. By 1.d.3 there is a space Z so that Z^{**} has a basis and Z^{**}/Z is isomorphic to X. By passing to the duals we get that $Z^{***} \approx Z^* \oplus X^*$ (observe that for every Banach space Z there is a projection of norm 1 from Z^{***} onto Z^*. The projection is the map which assigns to every functional on Z^{**} its restriction to Z). Since X fails to have the A.P. the same is true for X^*, by assertion (a). It is trivial to verify that a complemented subspace of a space having the A.P. has also the A.P. Hence Z^{***} (which is a dual of a space Z^{**} with a basis) fails to have the A.P. If X is such that X^* is separable (e.g. if X is isomorphic to a subspace of c_0; see the remark following 1.e.8) then Z^{***} is separable. □

While investigating the approximation problem Grothendieck found many nice equivalent formulations of this problem. We give two of those formulations here. As stated, this result is of course only of historical interest. However, from a different point of view it is still useful since it shows the connection between the A.P. and some problems arising in classical analysis. Moreover, the simple and explicit proof of the result enables us to transfer each counterexample to one of the versions of the approximation problem to a counterexample to the others.

Proposition 1.e.8. *The following three assertions are equivalent.*

(i) *Every Banach space has the* A.P.

(ii) *Every matrix $A=(a_{i,j})_{i,j=1}^{\infty}$ of scalars, for which* $\lim_j a_{i,j}=0$, $i=1, 2,\ldots,$

$$\sum_{i=1}^{\infty} \max_j |a_{i,j}|<\infty \textit{ and } A^2=0, \textit{ satisfies } \operatorname{trace} A=\sum_{n=1}^{\infty} a_{nn}=0.$$

(iii) *Every continuous function $K(s, t)$ on $[0, 1)\times[0, 1]$, for which $\int_0^1 K(s, t)K(t, u)\,dt=0$ for every s and u, satisfies $\int_0^1 K(t, t)\,dt=0$.*

Proof. (ii) ⇒ (i). Let X be a Banach space which fails to have the A.P. Then, by 1.e.4, there are $\{x_n\}_{n=1}^{\infty}\subset X$, $\{x_n^*\}_{n=1}^{\infty}\subset X^*$ with $\sum_{n=1}^{\infty}\|x_n\|\,\|x_n^*\|<\infty$ and $\sum_{n=1}^{\infty} x_n^*(x)x_n=0$ for every $x\in X$ but $\sum_{n=1}^{\infty} x_n^*(x_n)\neq 0$. There is clearly no loss of generality to assume that $\|x_n\|\to 0$ and $\sum_{n=1}^{\infty}\|x_n^*\|<\infty$. The matrix $A=(x_i^*(x_j))_{i,j=1}^{\infty}$ satisfies $\lim_j x_i^*(x_j)=0$, $\sum_{i=1}^{\infty}\max_j |x_i^*(x_j)|<\infty$ and also $A^2=0$ (the entries of A^2 are expressions of the form $\sum_{n=1}^{\infty} x_n^*(x_i)x_k^*(x_n)$). However, trace $A=\sum_{n=1}^{\infty} x_n^*(x_n)\neq 0$.

(iii) ⇒ (ii). Assume that (ii) fails and that $A=(a_{i,j})$ is a counterexample. Put $\alpha_i=\max_j |a_{i,j}|$ and choose a sequence of positive numbers $\{\eta_i\}_{i=1}^{\infty}$ such that $\eta_i\to\infty$ and $\sum_{i=1}^{\infty}\alpha_i\eta_i<1$. Put $b_{i,j}=a_{i,j}/\alpha_i\eta_i$; then $\lim_{i,j} b_{i,j}=0$ (in the sense that, for every

$\varepsilon>0$, there are only finitely many pairs (i,j) for which $|b_{i,j}|>\varepsilon$). Let $1=t_1>s_1>t_2>s_2\dots$ be numbers such that $\lim_n t_n=0$ and $t_i-s_i>\alpha_i\eta_i$, for every i (this is possible since $\sum_{i=1}^{\infty}\alpha_i\eta_i<1$). Let $\{\varphi_i\}_{i=1}^{\infty}$ be continuous functions on $[0,1]$ such that $0\leqslant\varphi_i\leqslant 1$, φ_i vanishes outside $[s_i,t_i]$ and $\int_{s_i}^{t_i}\varphi_i(t)\,dt=\alpha_i\eta_i$. It is easy to verify that $K(s,t)=\sum_{i=1}^{\infty}\sum_{j=1}^{\infty}\sqrt{\varphi_i(s)\varphi_j(t)}\,b_{i,j}$ is continuous on $[0,1]\times[0,1]$ (at each point the sum consists of at most one summand). The fact that $A^2=0$ implies that $\int_0^1 K(s,t)K(t,u)\,dt=0$ for every s and u while

$$\int_0^1 K(t,t)\,dt=\sum_{n=1}^{\infty}\int_0^1\varphi_n(t)\,dt\cdot b_{n,n}=\sum_{n=1}^{\infty}\alpha_n\eta_n b_{nn}=\sum_{n=1}^{\infty}a_{n,n}\neq 0\,.$$

(i) $\Rightarrow$ (iii). Assume that there is a counterexample $K(s,t)$ to (iii). For every $s\in[0,1]$, let $f_s(t)\in C(0,1)$ be defined by $f_s(t)=K(s,t)$. The continuity of $K(s,t)$ implies that $K_0=\{f_s\}_{0\leqslant s\leqslant 1}$ is a compact subset of $C(0,1)$. We claim that $X_0=\overline{\operatorname{span}}\,K_0$ is a subspace of $C(0,1)$ which does not have the A.P. Indeed, assume that $\varepsilon>0$ is such that there is an operator T of finite rank $Tg=\sum_{i=1}^{n}x_i^*(g)f_i$ on X_0 such that $\|Tg-g\|<\varepsilon$ for all $g\in K_0$. Since span K_0 is dense in X_0 there is no loss of generality to assume (by increasing n) that $f_i\in K_0$ for all i, i.e. $f_i=f_{s_i}$, for suitable $s_i\in[0,1]$. The functionals x_i^* extend to functionals on $C(0,1)$ and can thus be considered as measures on $[0,1]$. Since the measures with finite support are w^* dense in the set of all measures it follows easily that we may assume that each x_i^* is a finitely supported measure and thus (by increasing n again) that each x_i^* is of the form $x_i^*(g)=\lambda_i g(u_i)$, for suitable $u_i\in[0,1]$ and scalars $\{\lambda_i\}_{i=1}^{n}$. In conclusion we get that, for every $s\in[0,1]$, $t\in[0,1]$,

$$\left|K(s,t)-\sum_{i=1}^{n}\lambda_i K(s,u_i)K(s_i,t)\right|\leqslant\varepsilon$$

and, in particular,

$$\left|K(t,t)-\sum_{i=1}^{n}\lambda_i K(s_i,t)K(t,u_i)\right|\leqslant\varepsilon\,.$$

This however is impossible for small enough ε since $\int_0^1 K(t,t)\,dt\neq 0$ while

$$\int_0^1 K(s_i,t)K(t,u_i)\,dt=0,\quad\text{for every } i.\quad\square$$

It is also instructive to note that it is very simple to verify directly that (i) $\Rightarrow$ (ii) in 1.e.8. Indeed, let $A=(a_{i,j})$ be a matrix which fails (ii) and put $x_i=(a_{i,1},a_{i,2},\dots)\in c_0$, for $i=1,2,\dots$. Then, the closed linear span X of $\{x_i\}_{i=1}^{\infty}$ in c_0 is a space which fails the A.P. (apply 1.e.4 to the vectors $\{x_i\}_{i=1}^{\infty} \in X$ and $\{e_i\}_{i=1}^{\infty} \in X^*$ where e_i is the restriction to X of the i'th unit vector in l_1). Thus, the analysis of Grothendieck of the approximation property shows that if there is a Banach space which does not have the A.P. then there is also a subspace of c_0 which does not have the A.P. In the next chapter we shall show that not only c_0 but also the sequence spaces l_p, for $p>2$, have subspaces which fail to have the A.P.

We mention now some of the open problems concerning the A.P.

Problem 1.e.9. *Let X be a Banach space such that every compact $T: X \to X$ is a limit in norm of finite rank operators from X into itself. Does X have the* A.P.*?*

Problem 1.e.10. *Does the space $L(l_2, l_2)$, of all bounded operators on l_2 with the usual operator norm, have the* A.P.*? Does the space H_∞, of all the bounded analytic functions on $\{z; |z|<1\}$ with the supremum norm, have the* A.P.*?*

Observe that the two concrete spaces appearing in 1.e.10, for which the A.P. has not yet been verified, are non-separable. The common separable spaces which appear in analysis have the A.P. and, as a matter of fact, as mentioned already in Section a they even have a Schauder basis. We would like to point out that it is usually much easier to verify that a given space has the A.P. than to construct a basis in this space. Let us illustrate this by considering the disc algebra A. As mentioned in Section a it is known by now that A has a basis. However, it is not easy to construct such a basis (and its existence was open for a long time). On the other hand it is very easy to verify that A has the A.P. Indeed, for $f(x)=a_0+a_1z+a_2z^2+\cdots \in A$ put $S_n f=a_0+a_1z+\cdots+a_nz^n$ and $\sigma_n f=(S_1 f+S_2 f+\cdots+S_n f)/n$, $n=1, 2, \ldots$. The classical result of Fejer states that $\|\sigma_n\| \leqslant 1$ for all n and that $\|\sigma_n f-f\| \to 0$, for every $f \in A$. This shows that the disc algebra has the A.P. (even the M.A.P. defined below).

In the definition 1.e.1 of the A.P. we imposed no requirement on the norm of the operator T. For a Banach space with a basis the operator T required in 1.e.1 can be chosen to be bounded by a constant independent of the compact set K (namely, by the basis constant). We shall study now this stronger version of the A.P.

Definition 1.e.11. Let X be a Banach space and let $1 \leqslant \lambda < \infty$. We say that X has the *λ-approximation property* (λ-A.P. in short) if, for every $\varepsilon>0$ and every compact set K in X, there is a finite rank operator T in X so that $\|Tx-x\| \leqslant \varepsilon$, for every $x \in K$, and $\|T\| \leqslant \lambda$. A Banach space is said to have the *bounded approximation property* (B.A.P. in short) if it has the λ-A.P., for some λ. A Banach space is said to have the *metric approximation property* (M.A.P. in short) if it has the 1-A.P.

Observe that in 1.e.11 it is enough to take instead of a general compact set K finite sets only. Indeed, given K and ε we find $\{x_i\}_{i=1}^{n}$ so that $K \subset \bigcup_{i=1}^{n} B(x_i, \varepsilon/3\lambda)$. If $\|T\| \leqslant \lambda$ and $\|Tx_i-x_i\| \leqslant \varepsilon/3$, for every i, then $\|Tx-x\| \leqslant \varepsilon$, for every $x \in K$ (in 1.e.1 it is of course essential that K is infinite; for finite sets K a T satisfying the requirements in 1.e.1 always exists trivially).

As we already observed, a space with a basis has the B.A.P. (and a space with a monotone basis has the M.A.P.). It is not known whether the converse is true.

Problem 1.e.12. *Does there exist a separable Banach space which has the* B.A.P. *but fails to have a basis?*

It is likely that the answer to 1.e.12 is negative and that the "right" relation between bases and the B.A.P. is the one given by the following result of [118] and [61].

Theorem 1.e.13. *A separable Banach space X has the* B.A.P. *if and only if X is isomorphic to a complemented subspace of a space with a basis.*

Proof [118]. The "if" part is trivial and so we have merely to prove the "only if" part. We start by making two observations.

1. Assume that X is separable and has the λ-A.P. Then there exists a sequence of finite rank operators $\{S_n\}_{n=1}^\infty$ on X so that $x=\sum_{n=1}^\infty S_n x$, for every $x \in X$, and $\left\|\sum_{i=1}^n S_i\right\| \leq \lambda$, for every n. Indeed, let $\{y_i\}_{i=1}^\infty$ be a dense sequence in X. There exist, for $n=1, 2,\ldots$, $T_n \in L(X, X)$ of finite rank such that $\|T_n\| \leq \lambda$ and $\|T_n y_i - y_i\| \leq n^{-1}$ for $1 \leq i \leq n$. The operators $\{S_n\}_{n=1}^\infty$ defined by $S_1 = T_1$ and $S_n = T_n - T_{n-1}$, for $n>1$, have the desired property.

2. Let B be a Banach space with $\dim B=n$. Then, there are operators $\{U_i\}_{i=1}^{n^2}$ in $L(B, B)$ such that $\dim U_k B=1$, $\left\|\sum_{i=1}^k U_i\right\| \leq 2$, for every $1 \leq k \leq n^2$, and $\sum_{i=1}^{n^2} U_i x = x$ for every $x \in B$. Indeed, let $\{x_j\}_{j=1}^n$ and $\{x_j^*\}_{j=1}^n$ be an Auerbach system for X (see 1.c.3). For $i=rn+j$, $0 \leq r < n$, $1 \leq j \leq n$ put $U_i x = x_j^*(x) x_j / n$. Then, for every $k=rn+j$, we get

$$\left\|\sum_{i=1}^k U_i\right\| \leq \left\|\sum_{i=1}^{rn} U_i\right\| + \sum_{i=1}^j \|U_{rn+i}\| = \|rI/n\| + \sum_{i=1}^j n^{-1} \leq 2$$

(I denotes the identity operator on B).

Let now X be a separable space having the B.A.P. and choose $\{S_n\}_{n=1}^\infty$ as in observation 1. Since every space $S_n X$ is finite dimensional we can construct, for each n, operators $\{U_{i,n}\}_{i=1}^{m_n}$ on $S_n X$, as in observation 2, where $m_n = (\dim S_n X)^2$. Put $V_j = U_{i,n} S_n$ if $j = m_1 + m_2 + \cdots + m_{n-1} + i$, $1 \leq i \leq m_n$, $n=1, 2,\ldots$. Then, for every $x \in X$, $x = \sum_{j=1}^\infty V_j x$, $\left\|\sum_{j=1}^k V_j\right\| \leq 5\lambda$ for every k and $\dim V_j X = 1$ for every j. Let v_j be a vector of norm 1 in $V_j X$, $j=1, 2,\ldots$. Let Y be the space consisting of all sequences of scalars $y=(a_1, a_2,\ldots)$ such that $\sum_{j=1}^\infty a_j v_j$ converges and put

$$\|y\| = \sup_k \left\|\sum_{j=1}^k a_j v_j\right\| .$$

The unit vectors form clearly a monotone basis of Y. Define $V: X \to Y$ by $Vx=(a_1, a_2, \dots)$, where a_j is the scalar determined by $V_j x = a_j v_j$, $j=1, 2, \dots$. Clearly, $\|V\| \leq 5\lambda$ and $\|V^{-1}\| \leq 1$, i.e. V is an isomorphism (into). Let $U: Y \to X$ be the operator defined by $U(a_1, a_2, \dots) = \sum_{j=1}^{\infty} a_j v_j$. Clearly, UV is the identity operator of X and hence VX is complemented in Y. □

A part of Theorem 1.e.4 can be generalized to the setting of B.A.P. or M.A.P. without any change in the proof. We state this for the M.A.P.

Proposition 1.e.14. *Let X be a Banach space. The following four assertions are equivalent.*

(i) *X has the* M.A.P.

(ii) *For every Banach space Y the finite rank operators of norm ≤ 1 are dense in the unit ball of $L(Y, X)$ in the topology τ.*

(iii) *For every Banach space Y the finite rank operators of norm ≤ 1 are dense in the unit ball of $L(X, Y)$ in the topology τ.*

(iv) *For every choice of $\{x_n\}_{n=1}^{\infty} \subset X$, $\{x_n^*\}_{n=1}^{\infty} \subset X^*$ such that $\sum_{n=1}^{\infty} \|x_n\| \|x_n^*\| < \infty$ and $\left|\sum_{n=1}^{\infty} x_n^*(Tx_n)\right| \leq \|T\|$, for every operator T of finite rank in $L(X, X)$, we have $\left|\sum_{n=1}^{\infty} x_n^*(x_n)\right| \leq 1$.*

Condition (v) of 1.e.4 does not generalize to this setting since the T_1 given there satisfies automatically $\|T_1\| \leq \|T\| + \varepsilon$ and it can actually be chosen always so that $\|T_1\| = \|T\|$.

We conclude this section with a discussion of the relation between the A.P. and the M.A.P. Grothendieck [48] proved the surprising result that in many cases the A.P. implies the M.A.P.

Theorem 1.e.15. *Let X be a separable space which is isometric to a dual space and which has the* A.P. *Then X has the* M.A.P.

For the proof of 1.e.15 we need two simple lemmas.

Lemma 1.e.16. *Let X be separable and let $\varepsilon > 0$. Then there exists a sequence of functions $\{f_i\}_{i=1}^{\infty}$ on the unit ball B_X of X so that $x = \sum_{i=1}^{\infty} f_i(x)$, for every x in B_X, each $f_i(x)$ is of the form $\sum_{j=1}^{\infty} \chi_{E_{i,j}}(x) x_{i,j}$, where $\{E_{i,j}\}_{j=1}^{\infty}$ are disjoint Borel sets of B_X, $\{x_{i,j}\}_{j=1}^{\infty} \subset B_X$ and $\sum_{i=1}^{\infty} \|f_i\|_\infty < 1 + \varepsilon$ where $\|f_i\|_\infty = \sup_x \|f_i(x)\| = \sup_j \|x_{i,j}\|$.*

Proof. We construct the $\{f_i\}_{i=1}^{\infty}$ inductively. Choose first an f_1 of the suitable form so that $\|f_1\|_\infty \leq 1$ and $\|x - f_1(x)\|_\infty \leq \varepsilon/2$, then an f_2 so that $\|f_2\|_\infty \leq \varepsilon/2$ and $\|x - f_1(x) - f_2(x)\|_\infty \leq \varepsilon/4$ and continue in an obvious manner. □

Lemma 1.e.17. *Let $X=Y^*$. The space of all operators T of the form*

$$(*)\qquad Tx=\sum_{i=1}^{n} x(y_i)x_i, \quad \text{with } \{x_i\}_{i=1}^{n}\subset X \quad \text{and} \quad \{y_i\}_{i=1}^{n}\subset Y,$$

is τ-dense in the space of all finite rank operators from X into itself.

Proof. It is enough to note that every $x^*\in X^*=Y^{**}$ is a limit (in the sense of uniform convergence on compact sets of X) of elements from $JY\subset Y^{**}$. □

Proof of 1.e.15. Let $X=Y^*$ be a space having the A.P. By 1.e.14 we have to show that if φ is a τ-continuous linear functional on $L(X, X)$ such that $|\varphi(T)|\leq||T||$, for finite rank operators, then $|\varphi(T)|\leq||T||$, for every $T\in L(X, X)$. We shall prove this in the following manner. For every $\varepsilon>0$ we shall construct a τ continuous linear functional ψ_ε on $L(X, X)$ such that $\psi_\varepsilon(T)=\varphi(T)$, for T of the form (*), and for which it will be evident that $|\psi_\varepsilon(T)|\leq(1+\varepsilon)||T||$ for every $T\in L(X, X)$. By the assumption that X has the A.P. and 1.e.17 it follows from this that $\psi_\varepsilon(T)=\varphi(T)$, for all $T\in L(X, X)$, and thus $|\varphi(T)|\leq(1+\varepsilon)||T||$ for every T. Since $\varepsilon>0$ is arbitrary this gives the desired result.

For the construction of ψ_ε we use 1.e.16. First we let $K=B_X\times B_{X^*}$. This is a compact metric space if we endow B_X with the w^* topology induced by Y and B_{X^*} with the w^* topology induced by X. To every T of the form (*) we assign a function $g_T\in C(K)$ by $g_T(x, x^*)=x^*(Tx)$. The special form of T ensures that g_T is continuous. The map $T\to g_T$ is an isometry. By the Hahn–Banach and the Riesz representation theorems it follows that there is a measure μ of norm 1 on K so that

$$\varphi(T)=\int_K x^*(Tx)\,d\mu, \quad T \text{ of the form } (*).$$

Apply now 1.e.16. Then for T of the form (*), we get

$$(\ast\ast)\qquad \varphi(T)=\sum_{i=1}^{\infty}\int_K x^*T(f_i(x))d\mu=\sum_{i=1}^{\infty}\sum_{j=1}^{\infty} x^*_{i,j}Tx_{i,j}$$

where $x^*_{i,j}$ is the functional on X defined by $x^*_{i,j}(x)=\int_{E_{i,j}\times B_{X^*}} x^*(x)d\mu$. Clearly $||x^*_{i,j}||\leq|\mu|(E_{i,j}\times B_{X^*})$ and hence $\sum_{j=1}^{\infty}||x^*_{i,j}||\leq||\mu||\leq1$, for every i. Also $\sum_{i=1}^{\infty}\sup_j||x_{i,j}||\leq1+\varepsilon$. It follows that the right-hand side of $(\ast\ast)$, which we denote by $\psi_\varepsilon(T)$, is a τ continuous functional on $L(X, X)$ which satisfies

$$|\psi_\varepsilon(T)|\leq||T||\cdot\sum_{i,j=1}^{\infty}||x^*_{i,j}||\,||x_{i,j}||\leq(1+\varepsilon)||T||. \quad □$$

It follows from 1.e.15 that, for separable reflexive spaces, the A.P. implies the M.A.P. The same is true for nonseparable reflexive spaces. This follows e.g. from the fact (cf. [83]) that if X is reflexive and X_0 is a separable subspace of X then there is a separable space Z, $X_0 \subset Z \subset X$ so that there is a projection of norm 1 from X onto Z.

In general, the A.P. does not imply the B.A.P. and the B.A.P. does not imply the M.A.P. This was shown by T. Figiel and W. B. Johnson [41]. In order to present their example it is convenient to use the following variant of the λ-A.P. A Banach space X is said to satisfy the (ε, λ)-A.P. if, for every finite dimensional subspace B of X and every $\delta > 0$, there is a finite rank operator T on X so that $\|Tx - x\| \leqslant (\varepsilon + \delta)\|x\|$ for $x \in B$ and $\|T\| \leqslant \lambda + \delta$. It is easily checked that the λ-A.P. is the same property as the $(0, \lambda)$-A.P. The (ε, λ)-A.P. implies the λ'-A.P, for some $\lambda' = \lambda'(\lambda, \varepsilon)$. More precisely,

Lemma 1.e.18. *A Banach space which has the* (ε, λ)*-A.P. with* $0 < \varepsilon < 1$ *has also the* $(1-\varepsilon)^{-1}\lambda$*-A.P.*

Proof. Let $\delta > 0$ be such that $\varepsilon + \delta < 1$ and let $B \subset X$ with $\dim B < \infty$. By our assumption we can find inductively a sequence $\{T_n\}_{n=1}^{\infty}$ of finite rank operators of norm $\leqslant \lambda + \delta$ on X so that $\|T_1 x - x\| \leqslant (\varepsilon + \delta)\|x\|$, for $x \in B$, and

$$\|T_{n+1}x - x\| \leqslant (\varepsilon + \delta)\|x\|, \quad \text{for } x \in \operatorname{span}\left\{B \cup \bigcup_{i=1}^{n} T_i X\right\}.$$

For $n \geqslant 1$ let $S_n \in L(X, X)$ be defined by $(I - S_n) = (I - T_n)(I - T_{n-1}) \ldots (I - T_1)$. Then. for $x \in B$, $\|(I - S_n)x\| \leqslant (\varepsilon + \delta)^n \|x\|$. Also,

$$S_n = (I - T_n)(I - T_{n-1}) \ldots (I - T_2)T_1 + (I - T_n) \ldots (I - T_3)T_2 + \cdots + T_n\,.$$

Hence, $\|S_n\| = (\lambda + \delta)((\varepsilon + \delta)^{n-1} + (\varepsilon + \delta)^{n-2} + \cdots + (\varepsilon + \delta) + 1) \leqslant (\lambda + \delta)/(1 - \varepsilon - \delta)$. Since δ can be taken arbitrarily small this proves our assertion. □

The main step in the construction of Figiel and Johnson is contained in the next lemma.

Lemma 1.e.19 *Let* X *be a Banach space and* $\lambda \geqslant 1$ *be a number such that* X *has the* λ*-A.P. in every equivalent norm. Then, for every* $\varepsilon > 0$*, the dual* X^* *of* X *has the* $(\varepsilon, \lambda(1 + 2\varepsilon^{-1}\lambda))$*-A.P.*

Proof. Let B be a finite dimensional subspace of X^*, let $\delta > 0$ and $\beta = \lambda + \delta$. Let C be a finite-dimensional subspace of X such that, for every $x^* \in B$, $\|x^*\| \leqslant (1 + \delta) \sup\{|x^*(x)|;\, x \in C,\, \|x\| = 1\}$. We fix $\varepsilon > 0$ and introduce a new norm $|||\cdot|||$ on X^* by $|||x^*||| = \|x^*\| + 2\varepsilon^{-1}\beta d(x^*, B)$. Since the unit ball of $|||\cdot|||$ is w^* compact this norm is induced by an equivalent norm in X which is also denoted by $|||\cdot|||$.

Since, by our assumption, $(X, |||\cdot|||)$ has the λ-A.P. there is a finite rank operator T on X so that $\|Tx - x\| \leqslant \delta\|x\|$, for $x \in C$, and $|||T||| \leqslant \beta$. Passing to the dual we get, for $x^* \in X^*$

$$(\dagger) \qquad \|T^*x^*\| + 2\varepsilon^{-1}\beta d(T^*x^*, B) \leqslant \beta(\|x^*\| + 2\varepsilon^{-1}\beta d(x^*, B))\,.$$

It follows that $\|T^*x^*\| \leqslant \beta(1+2\varepsilon^{-1}\beta)\|x^*\|$ and hence, $\|T\| \leqslant \beta(1+2\varepsilon^{-1}\beta)$. For $x^* \in B$ we get from (†) that $d(T^*x^*, B) \leqslant \varepsilon\|x^*\|/2$, i.e. there is a $y^* \in B$ such that $\|T^*x^*-y^*\| \leqslant \varepsilon\|x^*\|/2$. Let $y \in C$ be any element of norm 1. Then, $|T^*x^*(y)-x^*(y)| = |x^*(Ty-y)| \leqslant \delta\|x^*\|$. Hence,

$$\begin{aligned}\|x^*-y^*\| &\leqslant (1+\delta) \sup\{|x^*(y)-y^*(y)|, y \in C, \|y\|=1\}\\ &\leqslant (1+\delta)(\delta+\varepsilon/2)\|x^*\|,\end{aligned}$$

and consequently, $\|T^*x^*-x^*\| \leqslant ((1+\delta)(\delta+\varepsilon/2)+\varepsilon/2)\|x^*\|$. Since $\delta>0$ is arbitrary the lemma is proved. □

Example 1.e.20. *There is a separable Banach space X (which has even a separable dual) such that X has the* A.P. *but not the* B.A.P.

Proof. Let Z be a space such that Z has the A.P. but Z^* fails to have the A.P. and is separable (see 1.e.7(b)). By 1.e.18 and 1.e.19 we can find, for every integer n, an equivalent norm $|||\cdot|||_n$ on Z so that $(Z, |||\cdot|||_n)$ fails to have the n-A.P. The space $\left(\sum_{n=1}^{\infty} \oplus (Z, |||\cdot|||_n)\right)_2 = X$ clearly has the A.P., fails to have the B.A.P. and has a separable dual. □

Note that if Z has the B.A.P. and Z^* fails to have the A.P. then $(Z, |||\cdot|||_n)$, for $n>1$, is an example of a space which has the B.A.P. but not the M.A.P.

In connection with 1.e.15 and 1.e.20 we mention the following open problem.

Problem 1.e.21. *Let X be a Banach space having the* B.A.P. *Does there exist an equivalent norm $|||\cdot|||$ on X so that $(X, |||\cdot|||)$ has the* M.A.P.*?*

The investigation of the approximation property will be continued in Section 2.d. This section (starting from Theorem 2.d.3) can be read directly after the present section.

f. Biorthogonal Systems

The existence of separable Banach spaces which fail to have a basis motivates the attempts to try to use some weaker forms of coordinate systems. One approach, which has been studied for a long time and for which strong existence theorems are now available, is that of using biorthogonal systems.

Definition 1.f.1. Let X be a Banach space. A pair of sequences $\{x_n\}_{n=1}^{\infty}$ in X and $\{x_n^*\}_{n=1}^{\infty}$ in X^* is called a *biorthogonal system* if $x_m^*(x_n) = \delta_n^m$. A sequence $\{x_n\}_{n=1}^{\infty}$ in X is called a *minimal system* if there exists a sequence $\{x_n^*\}_{n=1}^{\infty}$ in X such that $(\{x_n\}_{n=1}^{\infty}, \{x_n^*\}_{n=1}^{\infty})$ is a biorthogonal system.

It is clear that a sequence $\{x_n\}_{n=1}^{\infty}$ is minimal if and only if, for every integer n, $x_n \notin [x_i]_{i=1, i\neq n}^{\infty}$. Observe that if $(\{x_n\}_{n=1}^{\infty}, \{x_n^*\}_{n=1}^{\infty})$ forms a biorthogonal system then both $\{x_n\}_{n=1}^{\infty}$ and $\{x_n^*\}_{n=1}^{\infty}$ are minimal systems. Every basic sequence $\{x_n\}_{n=1}^{\infty}$ is a minimal system. The functionals $\{x_n^*\}_{n=1}^{\infty}$ are in this case the biorthogonal functionals discussed in Section b (more precisely, extensions of those functionals from $[x_n]_{n=1}^{\infty}$ to all of X). Observe that, unlike the notion of a basis, the notions of a minimal system and of a biorthogonal system involve no natural ordering and they should therefore be considered as countable sets of elements rather than sequences. There are important examples of minimal systems which do not form a basic sequence in any ordering. For example, take $x_n(t)=e^{int}$, $n=0, \pm 1, \pm 2, \ldots$ in $\tilde{C}(0, 2\pi)$ (=the subspace of $C(0, 2\pi)$ consisting of those functions f for which $f(0)=f(2\pi)$). The corresponding functionals x_n^* in $\tilde{C}(0, 2\pi)^*$ are in this case the measures given by $x_n^*=e^{-int}\,dt$, $n=0, \pm 1, \pm 2, \ldots$. The fact that $\{x_n\}_{n=-\infty}^{\infty}$ does not form a basis under any ordering follows e.g. from a result of P. Cohen [19].

In Section c above we proved the existence of a nice biorthogonal system in any finite dimensional space (called there an Auerbach system). We shall present now existence theorems in the infinite dimensional case. Since we want to construct a biorthogonal system which has to play a role similar to that of a basis rather than a basic sequence we introduce first a definition of a suitable notion of completeness of a biorthogonal system.

Definition 1.f.2. A minimal system $\{x_n\}_{n=1}^{\infty}$ is called *fundamental* if $[x_n]_{n=1}^{\infty}$ is all of X (i.e. $x^*(x_n)=0$ for all $n \Rightarrow x^*=0$). A minimal system $\{x_n^*\}_{n=1}^{\infty}$ in X^* is called *total* if $x_n^*(x)=0$ for all $n \Rightarrow x=0$ (i.e. X^* is the w^* closed linear span of $\{x_n^*\}_{n=1}^{\infty}$).

If $\{x_n\}_{n=1}^{\infty}$ is a basis in X then it is clearly fundamental and its biorthogonal functionals are total. The trigonometric system mentioned above is total and fundamental in $\tilde{C}(0, 2\pi)$. The following general existence theorem is known for a long time (it was proved first by Markushevich [101]).

Proposition 1.f.3. *Let X be a separable Banach space. Then X contains a fundamental minimal system whose biorthogonal functionals are total (or more briefly, with a slight abuse of language, X contains a total and fundamental biorthogonal system).*

Proof. Let $\{y_n\}_{n=1}^{\infty}$ be a sequence of non-zero elements in X such that $[y_n]_{n=1}^{\infty}=X$ and let $\{y_n^*\}_{n=1}^{\infty}$ be a sequence in X^* such that $y_n^*(x)=0$ for all n implies $x=0$. We shall construct inductively elements $\{x_n\}_{n=1}^{\infty}\subset X$, $\{x_n^*\}_{n=1}^{\infty}\subset X^*$ such that $x_m^*(x_n)=\delta_n^m$, span $\{x_n\}_{n=1}^{\infty}\supset$ span $\{y_n\}_{n=1}^{\infty}$ and span $\{x_n^*\}_{n=1}^{\infty}\supset$ span $\{y_n^*\}_{n=1}^{\infty}$.

We start by taking $x_1=y_1$ and put $x_1^*=y_{k_1}^*/y_{k_1}^*(y_1)$, where k_1 is any integer such that $y_{k_1}^*(y_1)\neq 0$. Next, we take the smallest integer h_2 such that $y_{h_2}^* \notin$ span x_1^*. Put $x_2^*=y_{h_2}^*-x_1^*\cdot y_{h_2}^*(x_1)$ and let $x_2=(x_{k_2}-x_1\cdot x_1^*(x_{k_2}))/x_2^*(x_{k_2})$, where k_2 is any index such $x_2^*(x_{k_2})\neq 0$. It is easily checked that with this choice $x_n^*(x_m)=\delta_n^m$ for $1\leqslant n, m\leqslant 2$. In the next step we let h_3 be the smallest integer such that $y_{h_3}\notin$ span $\{x_1, x_2\}$. We put

$$x_3=y_{h_3}-x_1\cdot x_1^*(y_{h_3})-x_2\cdot x_2^*(y_{h_3})$$

and

$$x_3^*=(y_{k_3}^*-x_1^*\cdot y_{k_3}^*(x_1)-x_2^*\cdot y_{k_3}^*(x_2))/y_{k_3}^*(x_3)\,,$$

where k_3 is such that $y^*_{k_3}(x_3) \neq 0$. We continue in an obvious way. In the step $2n$ we start in X^* and construct first the element x^*_{2n} while in the step $2n+1$ we start by constructing x_{2n+1}. It is clear that span $\{x_i\}_{i=1}^{2n} \supset$ span $\{y_i\}_{i=1}^{n}$ and span $\{x^*_i\}_{i=1}^{2n} \supset$ span $\{y^*_i\}_{i=1}^{n}$, for every n, and that $x^*_j(x_i) = \delta^j_i$. $\square$

Remark. The proof given above shows that we can assure that the $\{x^*_n\}_{n=1}^\infty$ are not only total but also that $[x^*_n]_{n=1}^\infty$ is norming, i.e. that for every $x \in X$, $\|x\| = \sup\{|x^*(x)|;\ \|x^*\| \leq 1,\ x^* \in [x^*_n]_{n=1}^\infty\}$. Indeed, we have simply to start the construction with a sequence $\{y^*_n\}_{n=1}^\infty$ so that $[y^*_n]_{n=1}^\infty$ is norming. Similarly, we can assure that $[x^*_n]_{n=1}^\infty = X^*$ if X^* is separable.

The simple proposition 1.f.3 can be used for replacing bases (which may not exist) in several situations. There is, however, one obvious drawback to the construction given in 1.f.3. We have no control on $\|x_n\|$ and $\|x^*_n\|$; in general, $\sup_n \|x_n\|\,\|x^*_n\| = \infty$ (of course we can always normalize the system so that, e.g. $\|x_n\| = 1$, for every n, but then $\sup_n \|x^*_n\|$ may be ∞). If $\{x_n\}_{n=1}^\infty$ is a basis then clearly $\sup_n \|x^*_n\|\,\|x_n\| < \infty$. Solving a problem which was open for a long time, R. Ovsepian and A. Pelczynski [112] proved that 1.f.3 can be strengthened to ensure that we get also $\sup_n \|x^*_n\|\,\|x_n\| < \infty$.

Theorem 1.f.4. *In every separable and infinite-dimensional Banach space X there is a fundamental and total biorthogonal system $(\{x_n\}_{n=1}^\infty, \{x^*_n\}_{n=1}^\infty)$ so that $\|x_n\| \cdot \|x^*_n\| \leq 20$, for every n. If X^* is separable the system may be chosen so that, in addition, $[x^*_n]_{n=1}^\infty = X^*$.*

Proof. The proof is divided into two steps. The first step is a slight refinement of the proof of 1.f.3 which shows that it is possible to choose a fundamental and total biorthogonal system $(\{u_n\}_{n=1}^\infty, \{u^*_n\}_{n=1}^\infty)$ so that, for some subsequence $\{n_k\}_{k=1}^\infty$ of the integers, $\|u_{n_k}\| \cdot \|u^*_{n_k}\| \leq 3$, $k = 1, 2, \ldots$. The second step shows how, by using this well behaved subsequence, it is possible to replace the biorthogonal system by another one in which the entire sequence behaves well.

We begin with the proof of the first step. As in the proof of 1.f.3 we construct $\{u_n\}_{n=1}^\infty$ and $\{u^*_n\}_{n=1}^\infty$ inductively. The inductive construction will depend on n (mod 3). If $n = 3j+1$, resp. $n = 3j+2$, we do exactly the same which we did in the proof of 1.f.3 for n odd, resp. n even. This, by itself, will ensure that span $\{u_n\}_{n=1}^\infty \supset$ span $\{y_n\}_{n=1}^\infty$ and span $\{u^*_n\}_{n=1}^\infty \supset$ span $\{y^*_n\}_{n=1}^\infty$. The inductive construction for $n = 3j$ will be such that $\|u^*_{3j}\|\,\|u_{3j}\| \leq 3$. Assume that the u_n and u^*_n have been chosen for $n < 3j$. By 1.a.6 there is a vector u_{3j} in X such that $\|u_{3j}\| = 1$, $u^*_n(u_{3j}) = 0$ if $n < 3j$ and $\|x\| \leq 2\|x + \lambda u_{3j}\|$, $x \in$ span $\{u_n\}_{n=1}^{3j-1}$, λ scalar. The functional on span $\{u_n\}_{n=1}^{3j}$, which assigns the value 0 to u_n with $n < 3j$ and the value 1 to u_{3j}, has norm ≤ 3. We take as u^*_{3j} any Hahn–Banach extension of it to all of X^*. This concludes the proof of the first step.

For the second step we need the following lemma

Lemma 1.f.5. *Let X be a Banach space and let $\{u_i\}_{i=1}^{2^n} \subset X$ and $\{u^*_i\}_{i=1}^{2^n} \subset X^*$ be such* *that* $u_j^*(u_i)=\delta_i^j$. *Then, there exists a real unitary matrix* $A=(a_{k,i})$ *of order* $2^n\times 2^n$ *so that if*

$$x_k=\sum_{i=1}^{2^n} a_{k,i}u_i, \qquad x_k^*=\sum_{i=1}^{2^n} a_{k,i}u_i^*, \quad k=1,2,\ldots,2^n$$

then

1. $\max\limits_{1\le k\le 2^n}\|x_k\|\le(1+\sqrt{2})\max\limits_{1\le i<2^n}\|u_i\|+2^{-n/2}\|u_{2^n}\|$

2. $\max\limits_{1\le k\le 2^n}\|x_k^*\|\le(1+\sqrt{2})\max\limits_{1\le i<2^n}\|u_i^*\|+2^{-n/2}\|u_{2^n}^*\|$

3. $x_j^*(x_i)=\delta_i^j,\quad 1\le i,j\le 2^n$

4. $\operatorname{span}\{x_k\}_{k=1}^{2^n}=\operatorname{span}\{u_k\}_{k=1}^{2^n};\qquad \operatorname{span}\{x_k^*\}_{k=1}^{2^n}=\operatorname{span}\{u_k^*\}_{k=1}^{2^n}$.

Proof. The relations 3 and 4 hold for any choice of a unitary matrix A. We get that also 1 and 2 hold if we choose $A=(a_{k,i})$ so that, for $k=1,\ldots,2^n$,

$$\sum_{i=1}^{2^n}|a_{k,i}|\le 1+\sqrt{2}\quad\text{and}\quad |a_{k,2^n}|\le 2^{-n/2}.$$

Such an orthogonal matrix A exists; put for $0\le s\le n-1$, $0\le r\le 2^s-1$

$$a_{k,2^s+r}=\begin{cases}2^{(s-n)/2}, & 2^{n-s-1}2r<k\le 2^{n-s-1}(2r+1)\\ -2^{(s-n)/2}, & 2^{n-s-1}(2r+1)<k\le 2^{n-s-1}(2r+2)\\ 0 & \text{otherwise}\end{cases}$$

and $a_{k,2^n}=2^{-n/2}$ for every k. □

We give now the proof of step 2 of 1.f.4. We take the biorthogonal sequence which was constructed in step 1 and reorder it in such a manner that in the new order "most" elements are nicely bounded. More precisely, we give this sequence new indices such that with the new indices the following holds. There is a sequence of integers $\{n_j\}_{j=1}^\infty$ so that if k is not of the form $2^{n_1}+2^{n_2}+\cdots+2^{n_i}$ for some i then $\|u_k\|\le 1$ and $\|u_k^*\|\le 3$ while, for $k=2^{n_1}+2^{n_2}+\cdots+2^{n_i}$, $\|u_k\|2^{-n_i/2}<1/20$, $\|u_k^*\|2^{-n_i/2}<1/20$. It is clear that such a choice of indices can be made. Indeed, pick the first element in the original sequence, say an element $u\in X$ with a corresponding $u^*\in X^*$. We find an n_1 so that $\|u\|2^{-n_1/2}<1/20$ and $\|u^*\|2^{-n_1/2}<1/20$ and give this element the index 2^{n_1}. For $1\le k<2^{n_1}$ we take out of the original sequence the first $2^{n_1}-1$ elements whose norms are 1 and whose biorthogonal functionals have norms ≤ 3. We consider now the first element in the original sequence which was not chosen so far and use it to determine n_2. We continue in an obvious manner.

For every integer i we apply 1.f.5 to construct $x_k\in X$ and $x_k^*\in X^*$ for $2^{n_1}+2^{n_2}+\cdots+2^{n_{i-1}}<k\le 2^{n_1}+2^{n_2}+\cdots+2^{n_i}$ so that the span of the x_k with these indices

(resp. of the x_k^* with these indices) coincides with that of the u_k (resp. the u_k^*) with the same indices, $x_k^*(x_j)=\delta_k^j$, and

$$\|x_k\| \leq (1+\sqrt{2})\cdot 1+1/20<2.5$$
$$\|x_k^*\| \leq (1+\sqrt{2})\cdot 3+1/20<7.5 . \quad \square$$

The proof gives clearly a constant which is better than the constant 20 appearing in the statement of 1.f.4. By modifying the proof, Pelczynski [119] was able to show that 1.f.4 remains true if 20 is replaced by any constant >1. It seems to be open whether it is actually possible to get $\|x_n\|\,\|x_n^*\|=1$, for every n (in the finite dimensional case this is possible; this is exactly the assertion of Auerbach's Lemma 1.c.3).

Many of the notions we encountered in Sections a and b can be defined in a meaningful way also in the context of general biorthogonal systems. A minimal system $\{x_n\}_{n=1}^\infty$ is said to be equivalent to a minimal system $\{y_n\}_{n=1}^\infty$ if there is an isomorphism T of $[x_n]_{n=1}^\infty$ onto $[y_n]_{n=1}^\infty$ such that $Tx_n=y_n$ for all n. In analogy to 1.a.9 it is easy to prove the following stability theorem. If $\{x_n\}_{n=1}^\infty \subset X$ and $\{x_n^*\}_{n=1}^\infty \subset X^*$ form a biorthogonal system and if $\{y_n\}_{n=1}^\infty \subset X$ satisfy $\sum_{n=1}^\infty \|x_n-y_n\|\,\|x_n^*\|<1$ then $\{y_n\}_{n=1}^\infty$ is a minimal system which is equivalent to $\{x_n\}_{n=1}^\infty$. A minimal system $\{x_n\}_{n=1}^\infty$ is said to be shrinking if the following relation holds $[x_n^*]_{n=1}^\infty=([x_n]_{n=1}^\infty)^*$ (here the x_n^* are considered as functionals on $[x_n]_{n=1}^\infty$ only). A minimal system $\{x_n\}_{n=1}^\infty$ is said to be boundedly complete if, for every bounded sequence $\{y_i\}_{i=1}^\infty$ in $[x_n]_{n=1}^\infty$, the existence of $\lim_i x_n^*(y_i)=a_n$ for every n implies the existence of a vector y in $[x_n]_{n=1}^\infty$ such that $x_n^*(y)=a_n$ for every n. Clearly, if $\{x_n\}_{n=1}^\infty$ is a basic sequence these notions agree with those defined in Section b. The natural generalization of 1.b.5 is true in the present setting (with the same proof); let $(\{x_n\}_{n=1}^\infty, \{x_n^*\}_{n=1}^\infty)$ be a fundamental and total biorthogonal system for a Banach space X. Then X is reflexive if and only if $\{x_n\}_{n=1}^\infty$ is both shrinking and boundedly complete.

A detailed discussion of minimal systems and their applications is given in [106]. In this paper V. D. Milman uses, e.g. biorthogonal systems for proving 1.b.14. (In the original proof of 1.b.14 given by Milman there is however a wrong statement. He stated that a block minimal system (in the obvious definition of this notion) of a boundedly complete minimal system is again boundedly complete. As observed in [26] this is not true even for bases. What is true however is that every block minimal system of a boundedly complete system has a boundedly complete subsequence. This is a direct consequence of 1.b.12.)

We conclude this section by showing that there is no non-trivial generalization of the notion of an unconditional basis to the setting of biorthogonal systems. This was proved by various authors in several degrees of generality (see, e.g. [7] and [23]). We present here, following [98], a very simple variant.

Proposition 1.f.6. *Let $\{x_n\}_{n=1}^\infty$ be a fundamental minimal system in a Banach space X such that $\{x_n^*\}_{n=1}^\infty$ is total. Assume that, for every $x \in X$ and every subset σ of the* *integers, there is an element x_σ in X such that $x_n^*(x_\sigma)=x_n^*(x)$ for $n\in\sigma$ and $x_n^*(x_\sigma)=0$ for $n\notin\sigma$. Then, $\{x_n\}_{n=1}^\infty$ is already an unconditional basis of X.*

Proof. It is clearly enough to show that $\{x_n\}_{n=1}^\infty$ is a basis of X (the assumptions are independent of the order). By 1.a.3 we have to show that the operators $\{P_n\}_{n=1}^\infty$, defined by $P_nx=\sum_{i=1}^n x_i^*(x)x_i$, are uniformly bounded. Observe first that it follows from our assumption and the closed graph theorem that, for every subset σ of the integers, there is a bounded linear operator P_σ on X defined by $P_\sigma x=x_\sigma$. If the $\{P_n\}_{n=1}^\infty$ are not uniformly bounded we can construct inductively a sequence of integers $1=p_1<q_1<p_2<q_2\cdots$ and vectors $\{u_j\}_{j=1}^\infty$ so that $\|u_j\|=2^{-j}$, $u_j\in\operatorname{span}\{x_i\}_{i=p_j}^{p_{j+1}-1}$ and $\|P_{q_j}u_j\|\geq 1$ for $j=1,2,\ldots$. Put $\sigma=\bigcup_{j=1}^\infty\{i;\,p_j\leq i\leq q_j\}$. Then $\sum_{j=1}^\infty u_j$ converges but $\sum_{j=1}^\infty P_\sigma u_j$ fails to converge and this contradicts the continuity of P_σ. □

g. Schauder Decompositions

A Schauder basis decomposes, in a sense, a Banach space into a sum of one-dimensional spaces. It is sometimes useful to consider cruder decompositions where the components into which we decompose a given Banach space are subspaces of dimension larger than 1.

Definition 1.g.1. Let X be a Banach space. A sequence $\{X_n\}_{n=1}^\infty$ of closed subspaces of X is called a *Schauder decomposition* of X if every $x\in X$ has a unique representation of the form $x=\sum_{n=1}^\infty x_n$, with $x_n\in X_n$ for every n.

Observe that if $\dim X_n=1$ for every n, i.e. $X_n=\operatorname{span}\{x_n\}$ then $\{X_n\}_{n=1}^\infty$ is a Schauder decomposition of X if and only if $\{x_n\}_{n=1}^\infty$ is a Schauder basis of X. Many results concerning bases generalize trivially to the setting of Schauder decompositions. Every Schauder decomposition $\{X_n\}_{n=1}^\infty$ of a Banach space X determines a sequence of projections $\{P_n\}_{n=1}^\infty$ on X by putting $P_n\sum_{i=1}^\infty x_i=\sum_{i=1}^n x_i$. These projections are bounded linear operators and $\sup_n\|P_n\|<\infty$. The number $\sup_n\|P_n\|$ is called the *decomposition constant* of $\{X_n\}_{n=1}^\infty$. Conversely, every sequence of bounded projections $\{P_n\}_{n=1}^\infty$ on X such that $P_nP_m=P_{\min(n,m)}$ and $\lim_n P_nx=x$ for every $x\in X$ determines a unique Schauder decomposition of X by putting $X_1=P_1X$ and $X_n=(P_n-P_{n-1})X$ for $n>1$. As in 1.a.3 it is easily seen that it is possible to replace the condition $\lim_n P_nx=x$ by the apparently weaker conditions $\sup_n\|P_n\|<\infty$ and $\overline{\bigcup_{n=1}^\infty P_nX}=X$.

A decomposition $\{X_n\}_{n=1}^\infty$ is called *boundedly complete* if, for every sequence $\{x_n\}_{n=1}^\infty$ with $x_n \in X_n$, $n=1, 2, 3,\ldots$ for which $\sup_n \left\|\sum_{i=1}^n x_i\right\| < \infty$, the series $\sum_{i=1}^\infty x_i$ converges. The decomposition is called *shrinking* if, for every $x^* \in X^*$, we have $\|P_n^* x^* - x^*\| \to 0$. If this is the case the sequence $\{P_n^*\}_{n=1}^\infty$ determines a boundedly complete Schauder decomposition of X^*. A decomposition $\{X_n\}_{n=1}^\infty$ is *unconditional* if, for every $x \in X$, the series $\sum_{n=1}^\infty x_n$, which represents x, converges unconditionally. In this case, for every sequence $\theta = (\theta_1, \theta_2, \ldots)$ of signs, the operator M_θ defined by $M_\theta \sum_{n=1}^\infty x_n = \sum_{n=1}^\infty \theta_n x_n$ is a bounded linear operator. The constant $\sup_\theta \|M_\theta\|$ is called the *unconditional constant* of the decomposition.

The decompositions, which are most useful in applications, are those in which $\dim X_n < \infty$ for all n ($\sup_n \dim X_n$ need not be finite). Such decompositions are called *finite dimensional Schauder decompositions* or F.D.D. in short. The same proof as that of 1.b.5 shows that if $\{X_n\}_{n=1}^\infty$ is an F.D.D. of a Banach space X then X is reflexive and only if $\{X_n\}_{n=1}^\infty$ is boundedly complete and shrinking.

Of course, the interest in Schauder decompositions does not stem from results which generalize trivially theorems on bases. Their importance stems from the fact that there are results on F.D.D.'s whose analogues for bases are not known (and perhaps false) or do not even have a meaningful analogue in terms of bases. We shall illustrate this by proving here two results on F.D.D.'s which will be applied in the next chapter. The first theorem shows that the answer to Problem 1.b.10 concerning bases has a positive answer in the setting of F.D.D.'s. The second result is a theorem which makes sense only in the setting of F.D.D.

Theorem 1.g.2 [60]. *Let X be a separable infinite-dimensional Banach space. Then there exists a subspace Y of X such that both Y and X/Y have a* F.D.D. *Moreover, if X^* is separable Y may be chosen so that Y and X/Y have a shrinking* F.D.D.

Proof. By 1.f.3 (and the remark following it) there is a biorthogonal system $\{x_n\}_{n=1}^\infty \subset X$ and $\{x_n^*\}_{n=1}^\infty \subset X^*$ so that $[x_n]_{n=1}^\infty = X$ and $[x_n^*]_{n=1}^\infty$ is norming over X. We can therefore choose inductively finite sets $\sigma_1 \subset \sigma_2 \subset \cdots$ and $\eta_1 \subset \eta_2 \subset \cdots$ so that $\sigma = \bigcup_{n=1}^\infty \sigma_n$ and $\eta = \bigcup_{n=1}^\infty \eta_n$ are complementary infinite subsets of the positive integers and, for $n=1, 2,\ldots$,

$$\|x\| \leqslant (1+n^{-1}) \sup\{|x^*(x)|;\ \|x^*\|=1,\ x^* \in [x_i^*]_{i\in\sigma_n\cup\eta_n}\}, \text{ for every } x \in [x_i]_{i\in\sigma_n}.$$
$$\|x^*\| \leqslant (1+n^{-1}) \sup\{|x^*(x)|,\ \|x\|=1,\ x \in [x_i]_{i\in\eta_n\cup\sigma_{n+1}}\}, \text{ for every } x^* \in [x_i^*]_{i\in\eta_n}.$$

For every n let S_n and T_n be the projections on X defined by $S_n x = \sum_{i\in\sigma_n} x_i^*(x) x_i$, respectively $T_n x = \sum_{i\in\eta_n} x_i^*(x) x_i$. We claim that

(i) $\|T_{n|[x_i]_{i\in\sigma_{n+1}}^\perp}^*\| \leqslant 1+n^{-1}$

(ii) $\|S_{n|[x_i^*]_{i\in\eta_n}^\top}\| \leqslant 1+n^{-1}$.

Indeed, let $x^* \in [x_i]^{\perp}_{i\in\sigma_{n+1}}$ and pick an $x \in [x_i]_{i\in\eta_n\cup\sigma_{n+1}}$ so that $\|x\|=1$ and $\|T_n^*x^*\| \leqslant (1+n^{-1})|T_n^*x^*(x)|$. Since $T_nx-x \in [x_i]_{i\in\sigma_{n+1}}$ and $|T_n^*x^*(x)|=|x^*(x)+x^*(T_nx-x)|=|x^*(x)|$ we get that $\|T_n^*x^*\| \leqslant (1+n^{-1})\|x^*\|$. This proves (i); the proof of (ii) is similar.

We show next that for $x^* \in [x_i]^{\perp}_{i\in\sigma}$, $T_n^*x^* \xrightarrow{w^*} x^*$. By (i) the sequence $\{T_n^*x^*\}_{n=1}^{\infty}$ is bounded. Let y^* be any w^* limit point of $\{T_n^*x^*\}_{n=1}^{\infty}$. Then, clearly $y^*(x_i)=x^*(x_i)=0$ for $i\in\sigma$ and $y^*(x_i)=x^*(x_i)$ for $i\in\eta$. Since $\sigma\cup\eta$ is the set of all positive integers we deduce that $y^*=x^*$ and thus, indeed, $T_n^*x^* \xrightarrow{w^*} x^*$. It follows also that $[x_i]^{\perp}_{i\in\sigma}$ is the w^* closure of $[x_i^*]_{i\in\eta}$. Put $Y=[x_i^*]^{\top}_{i\in\eta}=[x_i]_{i\in\sigma}$. By the analogue of 1.b.9 for F.D.D.'s it follows that X/Y has an F.D.D. From (ii) it follows that $\{S_{n|Y}\}_{n=1}^{\infty}$ determines an F.D.D. in Y. This concludes the proof of the first assertion of the theorem.

Assume now that X^* is separable. Then the $\{x_n\}_{n=1}^{\infty}$ can be chosen so that, in addition, $[x_n^*]_{n=1}^{\infty}=X^*$. Also, we may assume that the norm in X is such that $\{y_n^*\}_{n=1}^{\infty}\subset X^*$, $y_n^* \xrightarrow{w^*} y^*$ and $\|y_n^*\|\to\|y^*\| \Rightarrow \|y_n^*-y_n\|\to 0$ (use 1.b.11). The proof given above shows that in this case $\|T_n^*x^*-x^*\|\to 0$ for every $x^*\in[x_i]^{\perp}_{i\in\sigma}$ (it is here that we use the factor $1+n^{-1}$ in (i); for the proof of the first assertion of 1.g.2 it would have been enough to replace $1+n^{-1}$ by 2, say). This shows that $[x_i^*]_{i\in\eta}$ is w^* closed and that the decomposition of X/Y constructed above is shrinking. Similarly, it follows from $[x_n^*]_{n=1}^{\infty}=X^*$ that $\{S_{n|Y}\}_{n=1}^{\infty}$ determines a shrinking decomposition of Y. □

For stating the next result we need first a definition.

Definition 1.g.3. Let $\{X_n\}_{n=1}^{\infty}$ be a Schauder decomposition of X. Let $1=k_1<k_2<k_3<\cdots$ be an increasing sequence of integers and put $Y_i=X_{k_i}\oplus X_{k_i+1}\oplus\cdots\oplus X_{k_{i+1}-1}$, $i=1,2,\ldots$. Then, the decomposition $\{Y_i\}_{i=1}^{\infty}$ of X is said to be a *blocking* of the decomposition $\{X_n\}_{n=1}^{\infty}$.

If $\{P_n\}_{n=1}^{\infty}$ is the sequence of projection associated to an F.D.D. $\{X_n\}_{n=1}^{\infty}$ then the subsequences of $\{P_n\}_{n=1}^{\infty}$ are exactly the sequences of projections associated to the blockings of $\{X_n\}_{n=1}^{\infty}$. The decomposition (resp. the unconditional decomposition) constant of $\{Y_i\}_{i=1}^{\infty}$ is smaller or equal to the decomposition (resp. the unconditional decomposition) constant of $\{X_n\}_{n=1}^{\infty}$. One word of caution should be said concerning this definition. The notion of the blocking of a decomposition is not the obvious generalization of the notion of a block basic sequence. The direct generalization of the notion of a block basic sequence to the setting of Schauder decomposition would be to consider decompositions $\{Y_i\}_{i=1}^{\infty}$ of subspaces of X such that $Y_i\subset X_{k_i}\oplus X_{k_i+1}\oplus\cdots\oplus X_{k_{i+1}-1}$ for every i.

The following result concerning blockings of F.D.D.'s was proved in [63] and [59]. It turns out to be very useful in the study of the structure of subspaces of L_p spaces (and other spaces as well). In view of its general nature we state and prove it here; however, its significance will become clear only in view of its applications (for example, in Section 2.d below).

Proposition 1.g.4. (a) *Let T: $X\to Y$ be a bounded linear operator. Let $\{B_n\}_{n=1}^{\infty}$ be a shrinking* F.D.D. *of X and let $\{C_n\}_{n=1}^{\infty}$ be an* F.D.D. *of Y. Let $\{\varepsilon_i\}_{i=1}^{\infty}$ be a sequence* *of positive numbers tending to* 0. *Then there are blockings* $\{B'_i\}_{i=1}^\infty$ *of* $\{B_n\}_{n=1}^\infty$ *and* $\{C'_i\}_{i=1}^\infty$ *of* $\{C_n\}_{n=1}^\infty$ *so that, for every* $x \in B'_i$, *there is a* $y \in C'_{i-1} \oplus C'_i$ *so that* $||Tx-y|| \leqslant \varepsilon_i ||x||$.

(b) *Let* $T: X \to Y$ *be a quotient map. Let* $\{B_n\}_{n=1}^\infty$ *be an* F.D.D. *of X and* $\{C_n\}_{n=1}^\infty$ *a shrinking* F.D.D. *of a subspace of Y. Let* $\{\varepsilon_i\}_{i=1}^\infty$ *be a sequence of positive numbers tending to* 0. *Then there is a constant K and blockings* $\{B'_i\}_{i=1}^\infty$ *of* $\{B_n\}_{n=1}^\infty$ *and* $\{C'_i\}_{i=1}^\infty$ *of* $\{C_n\}_{n=1}^\infty$ *so that, for every* $y \in C'_i$, *there is an* $x \in B'_i \oplus B'_{i+1}$ *such that* $||Tx-y|| \leqslant \varepsilon_i ||y||$ *and* $||x|| \leqslant K||y||$.

Both parts of 1.g.4 (which are in a sense dual to each other) assert that after a suitable blocking the given operator is close to being diagonal with respect to the blockings; e.g. in (a) TB'_i is "essentially" contained in $C'_{i-1} \oplus C'_i$.

Proof of (a). Let $\{P_n\}_{n=1}^\infty$, resp. $\{Q_n\}_{n=1}^\infty$, be the projections associated to the given decomposition of X, resp. Y. We note first that, for every $\varepsilon > 0$ and integer n, there is an integer m such that if $x \in X$ with $P_m x = 0$ then $||Q_n Tx|| \leqslant \varepsilon ||x||$. Indeed, otherwise there would be an n, an $\varepsilon > 0$ and a sequence of vectors $\{x_k\}_{k=1}^\infty$ in X so that $||x_k|| = 1$, $||Q_n T x_k|| \geqslant \varepsilon$ for all k and $\lim_k P_m x_k = 0$ for every m. Since $Q_n Y$ is finite-dimensional we may assume (by passing to a subsequence if necessary) that, for some $y^* \in Y^*$ with $||y^*|| = 1$, $T^* y^*(x_k) \geqslant \varepsilon/2$ for every k. This however contradicts the assumption that $\{B_n\}_{n=1}^\infty$ is shrinking.

Using this observation we construct two sequences of integers $1 = m_1 < m_2 < m_3 < \cdots$ and $1 = k_1 < k_2 < k_3 \cdots$ as follows. We pick m_2 so that if $P_{m_2} x = 0$ then $||Q_{k_1} Tx|| \leqslant \varepsilon_1 ||x||/2$. Next, we let k_2 be such that, for every $x \in P_{m_2} X$, $||Tx - Q_{k_2} Tx|| \leqslant \varepsilon_1 ||x||/2$. Then, we pick m_3 so that if $P_{m_3} x = 0$ then $||Q_{k_2} Tx|| \leqslant \varepsilon_2 ||x||/2$ and k_3 so that $||Tx - Q_{k_3} Tx|| \leqslant \varepsilon_2 ||x||/2$ for every $x \in P_{m_3} X$. We continue in an obvious manner. The sequences $\{1, m_2+1, m_3+1, \ldots\}$ and $\{1, k_2+1, k_3+1, \ldots\}$ determine blockings with the desired properties.

Proof of (b). Let again $\{P_n\}_{n=1}^\infty$ and $\{Q_n\}_{n=1}^\infty$ be the projections which correspond to the given decompositions (note that the $\{Q_n\}_{n=1}^\infty$ are defined only on the subspace $Y_0 = [C_n]_{n=1}^\infty$ of Y). Let $K = 4 + 4 \sup_n ||P_n||$. As above, we shall define the suitable sequences of integers $1 = m_1 < m_2 < \cdots$ and $1 = k_1 < k_2 < \cdots$ inductively. It will be clear how to choose the $\{m_i\}_{i=1}^\infty$ and $\{k_i\}_{i=1}^\infty$ once we show the following. For every $\varepsilon > 0$ and every integer n there is an integer k so that if $y \in Y_0$ with $Q_k y = 0$ there is an $x \in X$ with $P_n x = 0$, $||x|| \leqslant K||y||$ and $||Tx - y|| \leqslant \varepsilon ||y||$. Suppose this were false for some $\varepsilon > 0$ and integer n. Then, there is an $\varepsilon > 0$ and a sequence $\{y_j\}_{j=1}^\infty$ of vectors of norm 1 in Y_0 so that $d(y_j, TU) \geqslant \varepsilon$, $j = 1, 2, \ldots$ (where $U = \{x \in X; P_n x = 0, ||x|| \leqslant K\}$) and $\lim_j Q_k y_j = 0$ for every k (and thus, since $\{Q_n\}_{n=1}^\infty$ is shrinking, $y_j \xrightarrow{w} 0$). Since T is a quotient map there are $\{x_j\}_{j=1}^\infty$ in X such that $||x_j|| \leqslant 2$ and $Tx_j = y_j$, $j = 1, 2, \ldots$. Put $v_j = P_n x_j$ and $u_j = x_j - v_j$. Clearly, $||u_j|| \leqslant K/2$ for every j. Since $P_n X$ is finite dimensional we may assume (by passing to a subsequence if necessary) that $||v_j - v_1|| \leqslant \varepsilon/2$ for every j. Hence,

$$d(y_1 - y_j, TU) \leqslant ||(y_1 - y_j) - T(u_1 - u_j)|| = ||T(v_1 - v_j)|| \leqslant \varepsilon/2 .$$

Since $y_j \xrightarrow{w} 0$ the point y_1 belongs to the closed convex hull of $\{y_1 - y_j\}_{j=2}^\infty$ and consequently, $d(y_1, TU) \leq \varepsilon/2$. This contradicts the choice of y_1 and concludes the proof. □

Remark. The proof of 1.g.4 ensures that not only the specified blockings have the desired property but that the same is true for suitable blockings of these blockings. More precisely, assume for simplicity that $\varepsilon_1 > \varepsilon_2 > \varepsilon_3 \ldots$. Then, the blockings chosen in (a) have the following property. For every blocking $\{B_j''\}_{j=1}^\infty$ of $\{B_i'\}_{i=1}^\infty$ there is a blocking $\{C_j''\}_{j=1}^\infty$ of $\{C_i'\}_{i=1}^\infty$ so that, for every $x \in B_j''$, there is a $y \in C_{j-1}'' \oplus C_j''$ with $\|Tx - y\| \leq \varepsilon_j \|x\|$. Similarly, the blockings chosen in (b) have the following property. For every blocking $\{C_j''\}_{j=1}^\infty$ of $\{C_i'\}_{i=1}^\infty$ there is a blocking $\{B_j''\}_{j=1}^\infty$ of $\{B_i'\}_{i=1}^\infty$ so that, for every $y \in C_j''$, there is an $x \in B_j'' \oplus B_{j+1}''$ with $\|Tx - y\| \leq \varepsilon_j \|y\|$ and $\|x\| \leq K\|y\|$.

We turn now to a discussion of the relation between the existence of F.D.D.'s and that of existence of bases.

Since obviously the existence of an F.D.D. of a Banach space X implies that X has the B.A.P. we note first that, by Enflo's example, there are separable Banach spaces which fail to have an F.D.D. It is not known whether the existence of an F.D.D. implies the existence of a basis (this is a special instance of problem 1.e.12). What is trivially true is the following fact: Let $\{B_n\}_{n=1}^\infty$ be an F.D.D. of a Banach space X. Assume that every B_n has a basis $\{x_{i,n}\}_{i=1}^{k_n}$ with basis constant K_n and $\sup_n K_n < \infty$. Then, the sequence

$$x_{1,1}, x_{2,1}, \ldots, x_{k_1,1}, x_{1,2}, \ldots, x_{k_2,2}, x_{1,3}, \ldots$$

forms a basis of X whose basis constant is $\leq K \cdot \sup_n K_n$, where K is the decomposition constant of $\{B_n\}_{n=1}^\infty$.

For unconditional bases the situation is more involved. Let, for example, X be the space of all compact operators T on l_2 which have a triangular representing matrix with respect to the unit vector basis (i.e. $Te_n = \sum_{m=1}^n a_{n,m} e_m$, $n = 1, 2, \ldots$). Let B_n be the subspace of X consisting of those $T \in X$ such that $Te_j = 0$ for $j \neq n$ (i.e. for which $a_{j,m} = 0$ unless $j = n$). It is clear that B_n is isometric to l_2^n, $n = 1, 2, \ldots$. Moreover, it is trivial to check that $\{B_n\}_{n=1}^\infty$ forms an unconditional decomposition of X. Nevertheless, it follows from the results of [47] that X does not have an unconditional basis and it is not even complemented in a space with an unconditional basis. This space does however embed in a space with an unconditional basis. This is a special case of the following general result.

Theorem 1.g.5. *Let X be a Banach space admitting an unconditional* F.D.D. $\{B_n\}_{n=1}^\infty$. *Then X is isomorphic to a subspace of a space with an unconditional basis.*

Proof. Without loss of generality we may assume that the unconditional constant of $\{B_n\}_{n=1}^\infty$ is 1. For each n we choose a set of non-zero elements $\{x_{i,n}^*\}_{i=1}^{k_n}$ in B_n^* such that $\|x_{i,n}^*\| \leq 1$ for all i and so that, for every x^* in the unit ball of B_n^*, there is an i such that $\|x^* - x_{i,n}^*\| \leq 4^{-n}$.

Define a map T from $X_0=\operatorname{span}\{B_n\}_{n=1}^\infty$ into the space Y_0 of sequences of scalars which are eventually 0 by

$$T\sum_{n=1}^m x_n=(x_{1,1}^*(x_1),\, x_{2,1}^*(x_1),\dots,\, x_{k_1,1}^*(x_1),\, x_{1,2}^*(x_2),\dots,\, x_{k_2,2}^*(x_2),\dots)\,.$$

Let U be the unit ball of X_0 and let V be the convex hull of $\bigcup_\theta M_\theta TU$, where the union is taken over all sequences of signs $\theta=(\theta_1,\theta_2,\dots)$ and M_θ is the operator on Y_0 defined by $M_\theta(a_1,a_2,\dots)=(\theta_1a_1,\theta_2a_2,\dots)$. We introduce a norm in Y_0, by putting $\|y\|=\inf\{\lambda>0;\ y/\lambda\in V\}$, and we let Y be the completion of $(Y_0,\|\cdot\|)$. It is clear that the unit vectors form an unconditional basis of Y and that T extends to an operator of norm 1 from X into Y. We have to show that T is an isomorphism into.

Let $u=\sum_{n=1}^m u_n$ be an element of norm 1 in X_0. By the Hahn–Banach theorem and the fact that the unconditional constant of the decomposition is 1 there is an $x^*=\sum_{n=1}^m x_n^*$ in X^* with $\|x^*\|=1$, $x_n^*\in B_n^*$ and $1=x^*(u)=\sum_{n=1}^m x_n^*(u_n)$ (we identify in an obvious manner B_n^* with a suitable subspace of X^*). For each $1\le n\le m$ let i_n be such that $\|x_n^*-x_{i_n,n}^*\|\le 4^{-n}$. Let φ be the linear functional on Y_0 defined by

$$\varphi(a_1,a_2,a_3,\dots)=\sum_{n=1}^m a_{j_n}$$

where j_n is the index assigned to $x_{i_n,n}^*$ by T (i.e. $j_n=k_1+k_2+\cdots+k_{n-1}+i_n$). By the definition of φ we get that $\varphi(Tu)=\sum_{n=1}^m x_{i_n,n}^*(u_n)\ge\sum_{n=1}^m x_n^*(u_n)-\sum_{n=1}^m 4^{-n}\ge 1/2$. Also, for every $x=\sum_{n=1}^\infty x_n\in X_0$ with $\|x\|\le 1$ and for every θ,

$$\varphi(M_\theta Tx)=\sum_{n=1}^m \theta_n x_{i_n,n}^*(x_n)\le\sum_{n=1}^m \theta_n x_n^*(x_n)+\sum_{n=1}^\infty 4^{-n}=$$
$$x^*\left(\sum_{n=1}^m \theta_n x_n\right)+\sum_{n=1}^\infty 4^{-n}\le 3/2\,.$$

Hence, by the definition of the norm in Y, $\varphi\in Y^*$ and $\|\varphi\|\le 3/2$. Consequently, $\|Tu\|\ge 2\varphi(Tu)/3\ge 1/3$ and this concludes the proof. □

2. The Spaces c_0 and l_p

a. Projections in c_0 and l_p and Characterizations of these Spaces

The simplest examples of infinite-dimensional Banach spaces are l_p, $1 \leqslant p \leqslant \infty$, and c_0. These spaces appeared in many problems in analysis much before a systematic theory of normed linear spaces was developed, and, as a result of continuous efforts, their geometry is quite well known today.

The unit vectors $\{e_n\}_{n=1}^{\infty}$ form an unconditional basis (with unconditional constant 1) of c_0 and l_p, $1 \leqslant p < \infty$. Some very simple but important properties of this basis are exhibited in the following proposition.

Proposition 2.a.1. *Let X be either c_0 or l_p, $1 \leqslant p < \infty$, and let $\{u_j\}_{j=1}^{\infty}$ be a normalized block basis of the unit vector basis $\{e_n\}_{n=1}^{\infty}$. Then,*

(i) $\{u_j\}_{j=1}^{\infty}$ is equivalent to $\{e_n\}_{n=1}^{\infty}$ and $[u_j]_{j=1}^{\infty}$ is isometric to X.
(ii) There is a projection of norm 1 from X onto $[u_j]_{j=1}^{\infty}$.

Proof. We present the proof in the case of l_p. The proof for c_0 is the same but the notation is somewhat different. Let $u_j = \sum_{i=m_j+1}^{m_{j+1}} \lambda_i e_i$ with $\sum_{i=m_j+1}^{m_{j+1}} |\lambda_i|^p = 1$, $j = 1, 2, \ldots$. For every sequence of scalars $\{a_j\}_{j=1}^{\infty}$ with $\sum_{j=1}^{\infty} |a_j|^p < \infty$ we have

$$\left\| \sum_{j=1}^{\infty} a_j u_j \right\| = \left(\sum_{j=1}^{\infty} |a_j|^p \sum_{i=m_j+1}^{m_{j+1}} |\lambda_i|^p \right)^{1/p} = \left(\sum_{j=1}^{\infty} |a_j|^p \right)^{1/p}$$

and this proves (i).

To prove (ii) choose, for every j, an element $u_j^* \in \operatorname{span}\{e_i\}_{i=m_j+1}^{m_{j+1}} \subset l_p^*$ so that $\|u_j^*\| = u_j^*(u_j) = 1$. Then $u_j^*(u_k) = 0$ for $k \neq j$ and the operator P defined by $Px = \sum_{j=1}^{\infty} u_j^*(x) u_j$ is a projection of norm 1 from X onto $[u_j]_{j=1}^{\infty}$. Indeed, if $x = \sum_{i=1}^{\infty} a_i e_i \in l_p$ then $|u_j^*(x)|^p \leqslant \sum_{i=m_j+1}^{m_{j+1}} |a_i|^p$ for every j and thus $\|Px\|^p = \sum_{j=1}^{\infty} |u_j^*(x)|^p \leqslant \|x\|^p$. $\square$

From 2.a.1 we get immediately (by using 1.a.9) the following result.

Proposition 2.a.2. *Let X be either c_0 or l_p, $1 \leqslant p < \infty$. Then every infinite dimensional subspace Y of X contains a subspace Z which is isomorphic to X and complemented in X (and therefore also in Y).*

Another interesting fact which follows immediately from 2.a.1 (and 1.a.12) is that *no space of the family* c_0 *and* l_p, $1 \leq p < \infty$, *is isomorphic to a subspace of another member of this family*. By using 2.a.2 A. Pelczynski gave in [114] a complete characterization of the complemented subspaces of c_0 and l_p, $1 \leq p < \infty$.

Theorem 2.a.3. *Let* X *be either* c_0 *or* l_p, $1 \leq p < \infty$. *Then every infinite dimensional complemented subspace of* X *is isomorphic to* X.

Proof. Let Y be an infinite-dimensional complemented subspace of X; then $X = Y \oplus X_1$ for some Banach space X_1. By 2.a.2, $Y = Z \oplus Y_1$ with $Z \approx X$ and a suitable Banach space Y_1. Then,

$$X \oplus Y \approx X \oplus (Z \oplus Y_1) \approx (X \oplus X) \oplus Y_1 \approx X \oplus Y_1 \approx Y,$$

since $X \oplus X \approx X$. Furthermore, observe that if $X = l_p$, $1 \leq p < \infty$ (resp. $X = c_0$) then X is isometric to $(X \oplus X \oplus \cdots)_p$ (resp. $(X \oplus X \oplus \cdots)_0$); in other words, we can write for every such X, $X = (X \oplus X \oplus \cdots)_X$. Consequently,

$$\begin{aligned} X \oplus Y = (X \oplus X \oplus \cdots)_X \oplus Y &\approx ((Y \oplus X_1) \oplus (Y \oplus X_1) \oplus \cdots)_X \oplus Y \\ &\approx (X_1 \oplus X_1 \oplus \cdots)_X \oplus (Y \oplus Y \oplus \cdots)_X \oplus Y \\ &\approx (X_1 \oplus X_1 \oplus \cdots)_X \oplus (Y \oplus Y \oplus \cdots)_X \\ &\approx ((Y \oplus X_1) \oplus (Y \oplus X_1) \oplus \cdots)_X = X. \end{aligned}$$

Hence $X \approx X \oplus Y \approx Y$ and this concludes the proof (the verification that all the computations done here with infinite direct sums are valid, is straightforward). □

The elegant method of proof of 2.a.3 is called Pelczynski's *decomposition method.* This method (as well as several variants of it) is very useful in many contexts. Observe that all that we used in the computations above was that X and Y are each isomorphic to a complemented subspace of the other space and that X is isomorphic to an infinite direct sum of itself with respect to a suitable norm (the fact that $X \oplus X$ is also isomorphic to X is a consequence of this assumption). Let us point out that it is unknown whether the assumption that X and Y are each isomorphic to a complemented subspace of the other space suffices to ensure that X is isomorphic to Y.

The decomposition method has however one drawback: it is hard to give an explicit form of an isomorphism whose existence is proved by the decomposition method. From the practical point of view (though not from the formal theoretical point of view) the decomposition method is basically an existence proof. In simple cases where other methods are available they usually give more information than the decomposition method. We shall illustrate this by considering projections of norm 1 in c_0 or l_p, $1 \leq p < \infty$. In this case it is possible to give explicit representations of the projections and their ranges. In particular, if X is c_0 or l_p, $1 \leq p < \infty$ and P is a projection of norm 1 on X with infinite dimensional range than PX is isometric to X (this is a special case of a result due essentially to Ando [5] cf. also [79]). We shall prove this here only in the case $1 < p < \infty$, $p \neq 2$. For $p = 2$ the result is of course trivial. The cases l_1 and c_0 are somewhat different; the proofs in these cases are simpler.

Theorem 2.a.4. *Let $X=l_p$, for some $1<p<\infty$, $p\neq 2$ and let P be a projection of norm 1 in X. Then, there exist vectors $\{u_j\}_{j=1}^m$ of norm 1 in X (where $m=\dim PX$ is either an integer or ∞) of the form*

$$(*)\qquad u_j=\sum_{i\in\sigma_j}\lambda_i e_i,\ 1\leqslant j\leqslant m,\quad \text{with } \sigma_j\cap\sigma_k=\varnothing \text{ for } k\neq j$$

so that $Px=\sum_{j=1}^m u_j^*(x)u_j$, where $\{u_j^*\}_{j=1}^m\in X^*$ satisfy $\|u_j^*\|=u_j^*(u_j)=1$, $j=1,\ldots,m$. In particular, $PX=[u_j]_{j=1}^m$ is isometric to l_p^m.

Observe that the proof of 2.a.1(ii) shows that, conversely, every P defined as above is a projection of norm 1 on X. Note also that the $\{u_j\}_{j=1}^m$ do not necessarily form a block basis of $\{e_i\}_{i=1}^\infty$ according to definition 1.a.10; first, since the sets $\{\sigma_j\}_{j=1}^m$ may be infinite and even if all the sets $\{\sigma_j\}_{j=1}^m$ are finite the $\{u_j\}_{j=1}^m$ are then only a block basis of a suitable permutation of $\{e_i\}_{i=1}^\infty$. What 2.a.4 asserts is that the obvious generalization of 2.a.1(ii) to the setting in which a block basis of $\{e_n\}_{n=1}^\infty$ is replaced by vectors satisfying (*) gives the most general projection of norm 1 on l_p, $1<p<\infty$, $p\neq 2$.

Proof. We introduce first some notations. For $0\neq x\in l_p$ the support $\operatorname{supp} x$ of x is the set of integers i such that $x(i)$ (the i'th coordinate of x) is $\neq 0$. The function $\operatorname{sign} x$ on the integers is defined by $\operatorname{sign} x(i)=1$ if $x(i)>0$, $=-1$ if $x(i)<0$ and $=0$ if $x(i)=0$ (we give the proof in the case where the scalars are real; the same argument with some slight changes works also in the complex case). For any $0\neq x\in l_p$ let $\psi_p(x)$ be the unique element in $l_q=l_p^*$ for which $\psi_p(x)(x)=\|\psi_p(x)\|_q\|x\|_p=\|x\|_p^p$. The element $\psi_p(x)$ is given explicitly by $\psi_p(x)=|x|^{p-1}\operatorname{sign} x$ (i.e. $\psi_p(x)(i)=|x(i)|^{p-1}\operatorname{sign} x(i)$ for $1\leqslant i<\infty$); for $x=0$ put $\psi_p(x)=0$. Observe that $\psi_q(\psi_p(x))=x$ for every $x\in l_p$.

Let now P be a projection of norm 1 in $X=l_p$ with $2<p<\infty$ (by duality it is enough to prove the theorem in this case) and let $Y=PX$. Note first that if $x=Px$ then $P^*\psi_p(x)(x)=\psi_p(x)(Px)=\psi_p(x)(x)$ and hence, by the uniqueness of $\psi_p(x)$, we get that $P^*\psi_p(x)=\psi_p(x)$. By duality we get conversely that $P^*\psi_p(x)=\psi_p(x)\Rightarrow Px=x$. Thus,

$$\ker P=\{x^*;\ P^*x^*=x^*\}^\top=\{\psi_p(y);\ y\in Y\}^\top=\psi_p(Y)^\top .$$

Hence, since Y determines both PX and $\ker P$, it follows that Y determines P uniquely, i.e. on a subspace of l_p there is at most one projection of norm 1. The preceding remarks show also that $\psi_p(Y)=P^*X^*$ is a linear subspace of l_q. We shall exploit this fact in proving the following lemma.

Lemma 2.a.5. *Let Y be a subspace of l_p, $2<p<\infty$, on which there is a projection of norm 1. Let $y, z\in Y$ be two non-zero vectors. Then $r_z(y)\in Y$, where $r_z(y)$ is the* *restriction of y to the support of z, i.e. $r_z(y)(i)=y(i)$ if $i\in\operatorname{supp} z$ and $r_z(y)(i)=0$ if $i\notin\operatorname{supp} z$.*

Proof. Since $\psi_p(Y)$ is a linear subspace of l_q we get that, for every real t, $\psi_p(z+ty)-\psi_p(z)\in\psi_p(Y)$. A simple computation shows that $\lim_{t\to\infty}(\psi_p(z+ty)-\psi_p(z))/t$ exists (in norm) and is equal to $(p-1)|z|^{p-2}y$ and thus $|z|^{p-2}y\in\psi_p(Y)$. Hence, $u_1=|z|^{1-(q-1)}|y|^{q-1}\operatorname{sign} y=\psi_q(|z|^{p-2}y)\in\psi_q(\psi_p(Y))=Y$.

By repeating the same argument we get that

$$u_2=|z|^{1-(q-1)}|u_1|^{q-1}\operatorname{sign} u_1=|z|^{1-(q-1)^2}|y|^{(q-1)^2}\operatorname{sign} y\in Y,$$

and, in general, $u_n=|z|^{1-(q-1)^n}|y|^{(q-1)^n}\operatorname{sign} y\in Y$, $n=1,2,\ldots$. Since $p>2$ it follows that $q-1<1$ and hence, $u=\lim_n u_n=|z|\operatorname{sign} y\in Y$. Consequently,

$$r_z(y)=|y|\operatorname{sign}(|z|\operatorname{sign} y)=|y|\operatorname{sign} u\in Y. \quad \square$$

We return to the proof of 2.a.4. We show first the following. Let i_0 be an integer which belongs to the support of some element in Y. Then, among all $0\neq y\in Y$ such that $i_0\in\operatorname{supp} y$, there is an element y_0 whose support is minimal. In order to show this it is enough to verify that if $\operatorname{supp} y_1\supset\operatorname{supp} y_2\supset\cdots\ni i_0$, with $\{y_j\}_{j=1}^\infty\subset Y$, then there is a $\hat{y}\in Y$ such that $\operatorname{supp}\hat{y}=\bigcap_{j=1}^\infty\operatorname{supp} y_j$. Such an element is $\lim_j r_{y_j}(y_1)$, which belongs to Y by 2.a.5.

The subspace of Y consisting of those elements y for which $\operatorname{supp} y\subset\operatorname{supp} y_0$ is one-dimensional. Otherwise, there would exist $y, z\in Y$ such that $\operatorname{supp} y\subset\operatorname{supp} y_0$, $\operatorname{supp} z\subset\operatorname{supp} y_0$, $z(i_0)=y(i_0)=1$ and $z(i)\neq y(i)$ for some $i\in\operatorname{supp} y_0\sim\{i_0\}$. Then, $i_0\in\operatorname{supp}(z(i)y-y(i)z)\subset\operatorname{supp} y_0\sim\{i\}$ and this contradicts the minimality of $\operatorname{supp} y_0$.

It follows from the preceding observation that there is a set $\{\sigma_j\}_{j=1}^m$ (m finite or infinite) of disjoint subsets of the integers such that every element of Y vanishes outside $\bigcup_{j=1}^m\sigma_j$ and elements $\{u_j\}_{j=1}^m$ of norm 1 in Y such that $\sigma_j=\operatorname{supp} u_j$ and $\{y;\ y\in Y, \operatorname{supp} y\subset\sigma_j\}$ is the one dimensional space spanned by u_j; $j=1,2,\ldots,m$. From this and 2.a.5 we deduce that $Y=[u_j]_{j=1}^m$. The fact that P has the desired form follows from the fact that P is determined uniquely by Y. $\square$

In terms of the Banach-Mazur distance, 2.a.4 (and its analogue for c_0 and l_1) states that if X is either c_0 or l_p, for $1\leqslant p<\infty$, and P is a projection of norm 1 on X with $\dim PX=\infty$ then $d(X, PX)=1$. From 2.a.3 it is easy to deduce that there is a function $f(\lambda)$ defined on $\lambda\geqslant 1$ (and independent of p) such that if P is a projection on X with $\dim PX=\infty$ then $d(X, PX)\leqslant f(\|P\|)$. Since $f(1)\neq 1$ this does not reduce to 2.a.4 if $\|P\|=1$ and leaves open the question whether the following "perturbed form" of 2.a.4 is valid. Does there exist a function $g(\lambda)$ defined on $\lambda\geqslant 1$ such that $\lim_{\lambda\to 1} g(\lambda)=1$ and such that, for every projection P on X with $\dim PX=\infty$, $d(X, PX)\leqslant g(\|P\|)$?

The property of c_0 and l_p, $1\leqslant p<\infty$ exhibited in 2.a.3 is of enough interest to justify a special terminology.

Definition 2.a.6. An infinite-dimensional Banach space X is said to be *prime* if every infinite-dimensional complemented subspace of X is isomorphic to X.

Besides c_0 and l_p, $1 \leqslant p < \infty$ the only known example of a prime space is l_∞. It is very likely that there are many other examples of prime spaces. In Section 4.c we shall give some examples of Orlicz sequences spaces which are conjectured to be prime. We present now the proof that l_∞ is prime. It is clear that this case requires a proof different from that of 2.a.3 since l_∞ is not separable and thus does not have a Schauder basis.

Theorem 2.a.7 [84]. *The space l_∞ is prime.*

Proof. Let Y be an infinite dimensional complemented subspace of l_∞. The main point in the proof is to show that Y has a subspace Z isomorphic to l_∞. Such a subspace Z is necessarily complemented in Y. Indeed, if $T: Z \to l_\infty$ is an isomorphism define an extension $\hat{T}$ of T, from Y into l_∞, by $\hat{T}y = (y_1^*(y), y_2^*(y), \ldots)$, where y_i^* is a Hahn–Banach extension of the functional $z \to Tz(i)$, $i = 1, 2, \ldots$, (for $u \in l_\infty$, $u(i)$ denotes the i'th coordinate of u). The operator $Q = T^{-1}\hat{T}$ is a projection from Y onto Z. We are now in a position to apply Pelczynski's decomposition method to conclude that $Y \approx l_\infty$.

The proof that Y contains a subspace isomorphic to l_∞ is based on the following two general facts concerning $C(K)$ spaces which will be proved in Vol. III.

(i) *Let T be an operator from a $C(K)$ space into a Banach space X which does not have a subspace isomorphic to c_0. Then, T is weakly compact.*
(ii) *If $T: C(K) \to C(K)$ is weakly compact then T^2 is compact.*

By applying (i) and (ii) to a projection operator P on a $C(K)$ space we see that if the range of P is infinite dimensional then P is not compact, hence not even weakly compact (note that $P = P^2$). Thus, the range of P contains a subspace isomorphic to c_0. Since l_∞ is a $C(K)$ space (K = the Stone–Cech compactification of the integers) it follows that the complemented subspace Y of l_∞ contains a subspace isomorphic to c_0. Thus, there exist vectors $\{y_n\}_{n=1}^\infty$ in Y and a constant M such that

$$(\dagger) \qquad \sup_n |a_n| \leqslant \left\| \sum_{n=1}^\infty a_n y_n \right\| \leqslant M \cdot \sup_n |a_n|, \quad \text{if } \lim_n a_n = 0 .$$

It follows from ($\dagger$) that, for every integer i, $\sum_{n=1}^\infty |y_n(i)| \leqslant M$. Hence, the series $\sum_{n=1}^\infty a_n y_n$ is w^* convergent for every bounded sequence $\{a_n\}_{n=1}^\infty$ of scalars $\Big($i.e. $\sum_{n=1}^\infty a_n y_n(i)$ converges for every $i\Big)$. The right-hand inequality of ($\dagger$) remains valid for every choice of a bounded sequence of scalars $\{a_n\}_{n=1}^\infty$. It is not clear however that the same is true for the left-hand inequality of ($\dagger$). Moreover, since Y is not necessarily w^* closed the expression $\sum_{n=1}^\infty a_n y_n$ need not be in Y if $\{a_n\}_{n=1}^\infty$ does not converge to 0. We shall see that by passing to a subsequence we can overcome these two difficulties.

Lemma 2.a.8. *Let $\{y_n\}_{n=1}^\infty$ be a sequence of vectors in l_∞ which satisfies (†). Then there is a subsequence $\{y_{n_k}\}_{k=1}^\infty$ of $\{y_n\}_{n=1}^\infty$ so that*

$$(\ddagger)\qquad \tfrac{1}{2}\sup_k |a_k| \leqslant \left\| \sum_{k=1}^\infty a_k y_{n_k} \right\| \leqslant M \sup_k |a_k|; \quad \text{if } \sup_k |a_k| < \infty .$$

Proof. For $x \in l_\infty$ and $\varepsilon > 0$ we put $N(x, \varepsilon) = \{i;\ |x(i)| < \varepsilon\}$. We observe first that, for every ε, there is an integer n_0 so that, for infinitely many indices n, the restriction of y_n to the subset $N(y_{n_0}, \varepsilon)$ of the integers N has norm $\geqslant 1$. This follows from the fact that $\sum_{n=1}^\infty |y_n(i)| \leqslant M$ and thus, if $r > M/\varepsilon$, $\bigcup_{n=1}^r N(y_n, \varepsilon) = N$. Hence, we can take as n_0 one of the integers from 1 to r.

Using this observation we can find an integer n_1 and an infinite sequence of integers N_1 so that if $n \in N_1$ the restriction of y_n to $N(y_{n_1}, 1/8)$ has norm $\geqslant 1$. Pick i_1 such that $|y_{n_1}(i_1)| \geqslant 3/4$. By passing to a subsequence of N_1, if necessary, we may assume also that $\sum_{n \in N_1} |y_n(i_1)| < 1/8$.

Next we pick an $n_2 \in N_1$ and an infinite subsequence $N_2 \subset N_1$ so that, for $n \in N_2$, the norm of the restriction of y_n to $N(y_{n_1}, 1/8) \cap N(y_{n_2}, 1/8^2)$ is $\geqslant 1$. By our assumption on N_1 there is an $i_2 \in N(y_{n_1}, 1/8)$ so that $|y_{n_2}(i_2)| > 3/4$. By passing to an infinite subsequence of N_2, if necessary, we may assume also that $\sum_{n \in N_2} |y_n(i_2)| < 1/8^2$.

We continue this inductive construction in an obvious manner and obtain subsequences $\{n_k\}_{k=1}^\infty$ and $\{i_k\}_{k=1}^\infty$ of the integers so that $|y_{n_k}(i_k)| > 3/4$ for every k and

$$\sum_{\substack{j=1 \\ j \neq k}}^\infty |y_{n_j}(i_k)| \leqslant 1/8 + 1/8^2 + \cdots = 1/7 .$$

The subsequence $\{y_{n_k}\}_{k=1}^\infty$ satisfies (‡). Indeed, if $\sup_k |a_k| = 1$ pick a k_0 such that $|a_{k_0}| > 9/10$. Then,

$$\begin{aligned} \left\| \sum_{k=1}^\infty a_k y_{n_k} \right\| &\geqslant \left| \sum_{k=1}^\infty a_k y_{n_k}(i_{k_0}) \right| \geqslant |a_{k_0}|\,|y_{n_{k_0}}(i_{k_0})| - \sum_{k \neq k_0} |y_{n_k}(i_{k_0})| \\ &\geqslant (3/4)\cdot(9/10) - 1/7 > 1/2 . \end{aligned}$$

This proves the left-hand side of (‡). We observed already that the right-hand side of (‡) is automatically true. □

We return to the proof of 2.a.7. For simplicity of notation we assume, as we may, that $\{y_n\}_{n=1}^\infty$ itself satisfies (‡) For every infinite subset N_0 of the integers we let X_{N_0} be the subspace of l_∞ consisting of all vectors of the form $\sum_{n=1}^\infty a_n y_n$, where $\{a_n\}_{n=1}^\infty$ is bounded and $a_n = 0$ if $n \notin N_0$. By (‡) each such X_{N_0} is isomorphic to l_∞. We shall show now that there is an N_0 such that $X_{N_0} \subset Y$ and this will conclude the proof. Let $\{N_\gamma\}_{\gamma \in \Gamma}$ be an uncountable collection of infinite subsets of the integers such that $N_{\gamma_1} \cap N_{\gamma_2}$ is finite for every $\gamma_1 \neq \gamma_2$ (to prove the existence of such a collection assign to each real number t a sequence of rational numbers converging to t. In this way we get an uncountable collection of subsets of a countable set (the rationals) such that the intersection of every two different sets in the collection is finite). If, for each $\gamma \in \Gamma$, the space X_{N_γ} is not contained in Y we can find, for each γ, an $x_\gamma \in X_{N_\gamma}$ with $\|x_\gamma\|=1$ and $Tx_\gamma \neq 0$, where $T: l_\infty \to l_\infty/Y$ is the quotient map. By our assumption on the collection $\{N_\gamma\}_{\gamma\in\Gamma}$ and the fact that all the $\{y_n\}_{n=1}^\infty$ belong to Y it follows that, for every choice of scalars $\{b_k\}_{k=1}^m$ and every choice of distinct indices $\{\gamma_k\}_{k=1}^m$ in Γ, $\left\|\sum_{k=1}^m b_k T x_{\gamma_k}\right\| \leq 2M \sup_k |b_k|$. Hence, for every $\varphi \in (l_\infty/Y)^*$ and every $\varepsilon>0$, there are only finitely many $\gamma\in\Gamma$ such that $|\varphi(Tx_\gamma)|>\varepsilon$ and thus, there are only countably many $\gamma\in\Gamma$ such that $\varphi(Tx_\gamma)\neq 0$.

Since Y is complemented in l_∞ the space l_∞/Y is isomorphic to a subspace of l_∞. Hence, there is a sequence of functionals $\{\varphi_j\}_{j=1}^\infty$ in $(l_\infty/Y)^*$ which is total (i.e, if $u\in l_\infty/Y$ is such that $\varphi_j(u)=0$, for every j, then $u=0$). Since Γ is uncountable we deduce that there is a $\gamma\in\Gamma$ such that $\varphi_j(Tx_\gamma)=0$ for $j=1, 2, \ldots$. This however contradicts the assumption that the $\{\varphi_j\}_{j=1}^\infty$ are total and $Tx_\gamma\neq 0$. □

We already noted above that it seems likely that the fact that c_0 and l_p, $1\leq p<\infty$ are prime does not characterize these spaces among the class of all separable spaces and probably not even among all spaces having an unconditional basis. The main ingredients used in the proof of 2.a.3, namely the two parts of the simple Proposition 2.a.1, do however characterize c_0 and l_p, $1\leq p<\infty$ up to isomorphism. This (and its analogues in the function space and "local" settings) explains why the structure of $L_p(\mu)$ and $C(K)$ spaces is far simpler than that of general function spaces. We present first a result of M. Zippin [146] which shows that 2.a.1(i) characterizes in a very strong sense the unit vectors of c_0 or l_p.

Theorem 2.a.9. *Let X be a Banach space with a normalized basis $\{x_n\}_{n=1}^\infty$. Assume that $\{x_n\}_{n=1}^\infty$ is equivalent to all its normalized block bases. Then, $\{x_n\}_{n=1}^\infty$ is equivalent to the unit vector basis in c_0 or in some l_p, $1\leq p<\infty$.*

Proof. First notice that since $\{x_n\}_{n=1}^\infty$ is equivalent to $\{\theta_n x_n\}_{n=1}^\infty$, for every choice of signs $\{\theta_n\}_{n=1}^\infty$, we get that $\{x_n\}_{n=1}^\infty$ is unconditional. Hence, we can assume without loss of generality that the unconditional constant of $\{x_n\}_{n=1}^\infty$ is 1. Next, using a uniform boundedness argument, the following is proved: there exists a constant M so that, for every normalized block basis $\{u_j\}_{j=1}^\infty$ of $\{x_n\}_{n=1}^\infty$, the operator T which exhibits the equivalence of these basic sequences satisfies $\|T\|, \|T^{-1}\|\leq M$ or, equivalently,

$$(*)\qquad M^{-1}\left\|\sum_{j=1}^\infty a_j x_j\right\| \leq \left\|\sum_{j=1}^\infty a_j u_j\right\| \leq M\left\|\sum_{j=1}^\infty a_j x_j\right\|$$

for all choices of scalars $\{a_j\}_{j=1}^\infty$ such that $\sum_{j=1}^\infty a_j x_j$ converges. Taking in (*) $u_j=x_{m_j}$, where $m_1<m_2<\cdots$, $a_j=1$ for $j=1,\ldots,n$ and $a_j=0$ for $j>n$, we get

$$M^{-1}\left\|\sum_{j=1}^n x_j\right\| \leq \left\|\sum_{j=1}^n x_{m_j}\right\| \leq M\left\|\sum_{j=1}^n x_j\right\|, \quad n=1, 2, \ldots.$$

Let n and k be integers and construct blocks $\{u_j\}_{j=1}^{\infty}$ as follows:

$$u_1=\sum_{i=1}^{n^{k-1}} x_i \Big/ \Big\|\sum_{i=1}^{n^{k-1}} x_i\Big\|, \quad u_2=\sum_{i=n^{k-1}+1}^{2n^{k-1}} x_i \Big/ \Big\|\sum_{i=n^{k-1}+1}^{2n^{k-1}} x_i\Big\|, \ldots .$$

By applying (*) to these $\{u_j\}_{j=1}^{\infty}$ and suitably chosen $\{a_j\}_{j=1}^{\infty}$ we get that

$$M^{-2}\lambda(n)\lambda(n^{k-1})\leqslant \lambda(n^k)\leqslant M^2\lambda(n)\lambda(n^{k-1}), \quad n, k=1, 2,\ldots,$$

where $\lambda(n)=\Big\|\sum_{i=1}^{n} x_i\Big\|$. It follows easily by induction that

$$M^{-2k}\lambda(n)^k\leqslant \lambda(n^k)\leqslant M^{2k}\lambda(n)^k .$$

Let m, n and k be integers and denote by $[x]$ the integer part of a positive real number x. By using the preceding inequality we get that

$$\begin{aligned} M^{-2[k\log m]}\lambda(n)^{[k\log m]} &\leqslant \lambda(n^{[k\log m]})\leqslant \lambda(m^{[k\log n]+1}) \\ &\leqslant M^{2[k\log n]+2}\lambda(m)^{[k\log n]+1} . \end{aligned}$$

It follows from this, after some easy computations and letting $k\to\infty$, that

$$\left|\frac{\log \lambda(n)}{\log n}-\frac{\log \lambda(m)}{\log m}\right|\leqslant 2\log M\left(\frac{1}{\log n}+\frac{1}{\log m}\right) .$$

Hence, $c=\lim_n \log \lambda(n)/\log n$ exists. Passing to the limit, as $m\to\infty$, in the preceding inequality we get

(**) $$M^{-2}n^c\leqslant \lambda(n)\leqslant M^2 n^c, \quad n=1, 2,\ldots .$$

Since $1\leqslant \lambda(n)\leqslant n$ it follows that $0\leqslant c\leqslant 1$. If $c=0$ we get that $\lambda(n)\leqslant M^2$ and thus

$$\sup_n |a_n|\leqslant \Big\|\sum_{n=1}^{\infty} a_n x_n\Big\|\leqslant M^2 \sup_n |a_n|$$

for every sequence of scalars $\{a_n\}_{n=1}^{\infty}$ which are eventually 0, i.e. $\{x_n\}_{n=1}^{\infty}$ is equivalent to the unit vector basis of c_0.

If $c>0$ we put $p=1/c$. To prove the equivalence of $\{x_n\}_{n=1}^{\infty}$ with the unit vector basis of l_p we let $\{r_j\}_{j=1}^{m}$ be positive rational numbers with $r_j=k_j/k$, where $\{k_j\}_{j=1}^{m}$ and k are positive integers. It follows from (**) that

$$\Big\|\sum_{j=1}^{m} r_j^{1/p}x_j\Big\|=k^{-1/p}\Big\|\sum_{j=1}^{m} k_j^{1/p}x_j\Big\|\geqslant M^{-2}k^{-1/p}\Big\|\sum_{j=1}^{m} \lambda(k_j)x_j\Big\| .$$

Using again (*) with $u_j = \sum_{i=s_j+1}^{s_j+1} x_i \Big/ \left\| \sum_{i=s_j+1}^{s_j+1} x_i \right\|$, $a_j = \left\| \sum_{i=s_j+1}^{s_j+1} x_i \right\|$, $j=1,\ldots, m$, where $s_j = k_1 + k_2 + \cdots + k_{j-1}$, we get that

$$\left\| \sum_{j=1}^{m} \lambda(k_j) x_j \right\| \geqslant M^{-1} \left\| \sum_{j=1}^{m} a_j x_j \right\| \geqslant M^{-2} \left\| \sum_{j=1}^{m} a_j u_j \right\| = M^{-2} \lambda\left(\sum_{j=1}^{m} k_j \right)$$
$$\geqslant M^{-4} \left(\sum_{j=1}^{m} k_j \right)^{1/p} = M^{-4} k^{1/p} \left(\sum_{j=1}^{m} r_j \right)^{1/p},$$

and thus, $\left\| \sum_{j=1}^{m} r_j^{1/p} x_j \right\| \geqslant M^{-6} \left(\sum_{j=1}^{m} r_j \right)^{1/p}$. A similar argument shows that

$$\left\| \sum_{j=1}^{m} r_j^{1/p} x_j \right\| \leqslant M^{6} \left(\sum_{j=1}^{m} r_j \right)^{1/p}$$

and this completes the proof. □

Remarks. 1. The following terminology is used in several places. A normalized basis $\{x_n\}_{n=1}^{\infty}$ is said to be *perfectly homogeneous* if it is equivalent to any of its normalized block bases. Since 2.a.9 states that the perfectly homogeneous bases are exactly those which are equivalent to the unit vector basis of c_0 or some l_p, $1 \leqslant p < \infty$ this terminology is no longer very useful; it is just an abbreviation of "being equivalent to the unit vector basis of c_0 or some l_p, $1 \leqslant p < \infty$".

2. The proof of 2.a.9 uses only the fact that $\{x_n\}_{n=1}^{\infty}$ is equivalent to each of its normalized block bases with constant coefficients, i.e. blocks having the form $u_j = \sum_{n=m_j+1}^{m_j+1} x_n \Big/ \left\| \sum_{n=m_j+1}^{m_j+1} x_n \right\|$, for some increasing sequence of integers $\{m_j\}_{j=1}^{\infty}$. The significance of this remark will become clear in the next chapter.

We show now that a modified version of 2.a.1(ii) also characterizes the unit vector basis of c_0 or l_p.

Theorem 2.a.10 [92]. *Let $\{x_n\}_{n=1}^{\infty}$ be a normalized unconditional basis of a Banach space X. Assume that, for every permutation π of the integers and for every block basis $\{u_j\}_{j=1}^{\infty}$ of $\{x_{\pi(n)}\}_{n=1}^{\infty}$, the subspace $[u_j]_{j=1}^{\infty}$ is complemented. Then $\{x_n\}_{n=1}^{\infty}$ is equivalent to the unit vector basis of c_0 or l_p for some $1 \leqslant p < \infty$.*

The proof of 2.a.10 is based on a lemma which will be useful also in other contexts.

Lemma 2.a.11. *Let X be a Banach space with an unconditional basis $\{x_n\}_{n=1}^{\infty}$. Let $\{\lambda_n\}_{n=1}^{\infty}$ be a sequence of scalars tending to 0 and let $v_j = \sum_{n \in \delta_j} a_n x_n$, $w_j = \sum_{n \in \sigma_j} a_n x_n$, $j = 1, 2, \ldots$ be normalized block bases of a permutation of $\{x_n\}_{n=1}^{\infty}$ such that $\delta_i \cap \sigma_j = \varnothing$ for every i and j. Assume that $[v_j + \lambda_j w_j]_{j=1}^{\infty}$ is complemented in X. Then, for every choice of scalars $\{\eta_j\}_{j=1}^{\infty}$ such that $\sum_{j=1}^{\infty} \eta_j v_j$ converges, the series $\sum_{j=1}^{\infty} \lambda_j \eta_j w_j$ converges too.*

Proof. Let $u_j=v_j+\lambda_j w_j$, $j=1, 2,\ldots$ and let Q be a projection from X onto $[u_j]_{j=1}^\infty$. Put

$$Qv_i=\sum_{j=1}^\infty b_{i,j}u_j, \qquad Qw_i=\sum_{j=1}^\infty c_{i,j}u_j, \quad i=1, 2,\ldots.$$

Then, for every i, $b_{i,i}+\lambda_i c_{i,i}=1$ and, since $\sup_i |c_{i,i}|<\infty$, we get that $\lim_i b_{i,i}=1$. Let P_σ be the projection associated to the unconditional basis $\{x_n\}_{n=1}^\infty$ and the set $\sigma=\bigcup_{j=1}^\infty \sigma_j$ (i.e. $P_\sigma x_n=x_n$ if $n\in\sigma$ and $P_\sigma x_n=0$ if $n\notin\sigma$). Then, $P_\sigma Qv_i=\sum_{j=1}^\infty b_{i,j}\lambda_j w_j$, $i=1, 2,\ldots$. We regard $P_\sigma Q$ as an operator from $V=[v_i]_{i=1}^\infty$ into $W=[w_j]_{j=1}^\infty$. The matrix corresponding to $P_\sigma Q$ with respect to the given bases is $(b_{i,j}\lambda_j)$. By 1.c.8 the operator $D: V\to W$, which corresponds to the diagonal of this matrix (i.e. defined by $Dv_j=\lambda_j b_{j,j}w_j$, $j=1, 2,\ldots$), is also bounded. Thus the convergence of $\sum_{j=1}^\infty \eta_j v_j$ implies that of $\sum_{j=1}^\infty \eta_j Dv_j$ and therefore also that of $\sum_{j=1}^\infty \lambda_j\eta_j w_j$ (recall that $\lim_j b_{j,j}=1$). □

Proof of 2.a,10. Let $\{v_j\}_{j=1}^\infty$ and $\{w_j\}_{j=1}^\infty$ be normalized block bases of $\{x_{2n}\}_{n=1}^\infty$, respectively $\{x_{2n+1}\}_{n=1}^\infty$. We shall show that $\{v_j\}_{j=1}^\infty$ is equivalent to $\{w_j\}_{j=1}^\infty$. This will imply that $\{x_{2n}\}_{n=1}^\infty$ and $\{x_{2n+1}\}_{n=1}^\infty$ are both perfectly homogeneous and equivalent to each other. An application of 2.a.9 will thus conclude the proof.

Let $\{\eta_j\}_{j=1}^\infty$ be a sequence of scalars such that $\sum_{j=1}^\infty \eta_j v_j$ converges. By the assumption in 2.a.10 and Lemma 2.a.11 we get that, for every sequence $\{\lambda_j\}_{j=1}^\infty$ tending to 0, the series $\sum_{j=1}^\infty \lambda_j\eta_j w_j$ converges. If $\sum_{j=1}^\infty \eta_j w_j$ fails to converge we can find disjoint finite sets of integers $\{\sigma_i\}_{i=1}^\infty$ and an $\varepsilon>0$ such that $\left\|\sum_{j\in\sigma_i}\eta_j w_j\right\|\geq\varepsilon$ for every i. Put $u_i=\sum_{j\in\sigma_i}\eta_j w_j$, $i=1, 2,\ldots$. Since $\sum_{i=1}^\infty \lambda_i u_i$ converges whenever $\lambda_i\to 0$ we get that $\{u_i\}_{i=1}^\infty$ is equivalent to the unit vector basis of c_0. By using again 2.a.11 it follows that in this case $\{v_j\}_{j=1}^\infty$ is also equivalent to the unit vector basis of c_0 and (apply 2.a.11 once more) so is $\{w_j\}_{j=1}^\infty$. Thus, $\sum_{j=1}^\infty \eta_j w_j$ must converge. □

The use of permutations in the statement of 2.a.10 is necessary. This follows from the next proposition.

Proposition 2.a.12 [15]. *Let X be a Banach space with an unconditional basis $\{x_n\}_{n=1}^\infty$. Let $\{m_j\}_{j=1}^\infty$ be an increasing sequence of integers with $m_1=0$ and put $B_j=\operatorname{span}\{x_n\}_{n=m_j+1}^{m_{j+1}}$, $j=1, 2,\ldots$. If every block basis of $\{x_n\}_{n=1}^\infty$ spans a complemented subspace of X then the same is true for the natural basis of $Y=\left(\sum_{j=1}^\infty\oplus B_j\right)_p$, where $p=0$ or $1\leq p<\infty$.*

Proof. We may assume without loss of generality that the unconditional constant of $\{x_n\}_{n=1}^\infty$ is 1. By the "natural basis" of Y we mean the vectors $\{y_n\}_{n=1}^\infty$ defined as follows: if $m_j<n\leq m_{j+1}$ we let y_n be the element in the direct sum whose only non-

zero component is x_n in the j'th place. It is obvious that $\{y_n\}_{n=1}^\infty$ is an unconditional basis of Y whose unconditional constant is 1. For the proof we need the following trivial observation. If $\{P_j\}_{j=1}^\infty$ are projections in $\{B_j\}_{j=1}^\infty$ such that $\sup_j \|P_j\|<\infty$ then $Q(z_1, z_2,\ldots)=(P_1z_1, P_2z_2,\ldots)$ is a projection on Y with $\|Q\|=\sup_j \|P_j\|$ (here $z_j \in B_j$ for every j). Q is called the projection determined by $\{P_j\}_{j=1}^\infty$.

Let now $u_k=\sum_{n=q_k+1}^{q_{k+1}} a_n y_n$, $k=1, 2,\ldots$ be a normalized block basis of $\{y_n\}_{n=1}^\infty$. Let N_1 be the set of those integers k for which there exists a j such that $u_k \in B_j$ (i.e. $m_j \leqslant q_k < q_{k+1} \leqslant m_{j+1}$). By our assumption on $\{x_n\}_{n=1}^\infty$ there is a bounded linear projection P from X onto $\left[\sum_{i=q_k+1}^{q_{k+1}} a_n x_n\right]_{k \in N_1}$. By 1.c.8 (and the remark thereafter) there is no loss of generality to assume that if $q_k<n\leqslant q_{k+1}$ then $Px_n=\lambda_n \sum_{i=q_k+1}^{q_{k+1}} a_i x_i$, where $\lambda_n=0$ whenever $k \notin N_1$. Hence, $PB_j \subset B_j$ for every j. Let Q_1 be the projection on Y determined by $\{P_{|B_j}\}_{j=1}^\infty$. Then $Q_1 Y=[u_k]_{k \in N_1}$ and $Q_1 u_k=0$ if $k \notin N_1$.

We divide now $N \sim N_1$ into two sets $N_2 \cup N_3$ by taking every second element in $N \sim N_1$ (i.e. the first, third,...) into N_2 and the rest (i.e. the second, fourth,...) into N_3. By this choice of N_2 we ensure that if $k<h$ are two integers in N_2 then there is a j such that $q_{k+1} \leqslant m_j < q_h$ and a similar statement holds for N_3. Thus, the sequence $\{u_k\}_{k \in N_2}$ is equivalent to the unit vector basis of l_p, resp. c_0 (if $p=0$). (If $N \sim N_1$ is finite then N_2 is finite and we get only the unit vector basis of l_p^r for some finite r.) Also, there exist $\{u_k^*\}_{k \in N_2}$ such that $\|u_k^*\|=u_k^*(u_k)=1$ for every $k \in N_2$, $u_k^*(u_h)=0$ for every $h \neq k$, and $\{u_k^*\}_{k \in N_2}$ is equivalent to the unit vector basis in l_p^*, resp. c_0^*. The operator Q_2 on Y defined by $Q_2 u=\sum_{k \in N_2} u_k^*(y) u_k$ is a projection of norm 1 onto $[u_k]_{k \in N_2}$ so that $Q_2 u_k=0$ if $k \notin N_2$. In a similar manner we define Q_3. Then $Q=Q_1+Q_2+Q_3$ is a bounded linear projection from Y onto $[u_k]_{k=1}^\infty$. □

It follows from 2.a.12 that, e.g. the natural basis of $X=\left(\sum_{n=1}^\infty \oplus l_s^n\right)_p$ has, for every choice of s and p, the property that all its block bases span a complemented subspace of X. For $p \neq s$ the natural basis of this space is not perfectly homogeneous.

b. Absolutely Summing Operators and Uniqueness of Unconditional Bases

In this section we shall present some of the basic facts concerning the class of p-absolutely summing operators. These operators are, by definition, closely connected to the spaces $L_p(\mu)$. They are of importance in the geometric theory of general Banach spaces and, in particular, in the study of the structure of the classical spaces. In this section we apply them in proving that c_0 and l_1 have up to equivalence, only one normalized unconditional basis (namely the unit vector basis). Further applications of p-absolutely summing operators to the study of classical spaces will be presented in Vol. II.

Definition 2.b.1. Let X and Y be Banach spaces and let $p \geqslant 1$. An operator

$T \in L(X, Y)$ is called *p-absolutely summing* if there is a constant K so that, for every choice of an integer n and vectors $\{x_i\}_{i=1}^n$ in X, we have

$$\left(\sum_{i=1}^n \|Tx_i\|^p\right)^{1/p} \leqslant K \sup_{\|x^*\|\leqslant 1} \left(\sum_{i=1}^n |x^*(x_i)|^p\right)^{1/p}$$

The smallest possible constant K is denoted by $\pi_p(T)$. The class of all *p*-absolutely summing operators in $L(X, Y)$ is denoted by $\Pi_p(X, Y)$.

For $T \notin \Pi_p(X, Y)$ we shall put $\pi_p(T) = \infty$. The 1-absolutely summing operators will be simply called absolutely summing operators. A straightforward verification shows that, for every p, $\Pi_p(X, Y)$ is a linear subspace of $L(X, Y)$ and $\pi_p(T)$ defines a norm on $\Pi_p(X, Y)$ in which this space is even complete (i.e. a Banach space). It is also trivial to verify that if S and T are bounded linear operators whose composition is defined then $\pi_p(ST) \leqslant \pi_p(S) \cdot \|T\|$ and $\pi_p(ST) \leqslant \|S\| \pi_p(T)$. The name "absolutely summing" will become clear if we consider the definition for $p = 1$. Observe that $\sup_{\|x^*\|=1} \sum_{i=1}^n |x^*(x_i)| = \sup\left\{\left\|\sum_{i=1}^n \theta_i x_i\right\|, \theta_i = \pm 1, 1 \leqslant i \leqslant n\right\}$. This implies that *an operator $T \in L(X, Y)$ is absolutely summing if and only if, for every sequence* $\{x_n\}_{n=1}^\infty$ *in X such that $\sum_{n=1}^\infty x_n$ converges unconditionally, the series $\sum_{n=1}^\infty Tx_n$ converges absolutely* $\left(\text{i.e. } \sum_{n=1}^\infty \|Tx_n\| < \infty\right)$.

The following factorization theorem, due to A. Pietsch [121], clarifies the notion of *p*-absolutely summing operators for a general p and is a basic tool in several applications.

Theorem 2.b.2. *An operator $T \in L(X, Y)$ is p-absolutely summing for some $1 \leqslant p < \infty$ if and only if there is a regular probability measure μ on the unit ball B_{X^*} of X^* (in its w^* topology) and a constant K so that $\|Tx\| \leqslant K \left(\int_{B_{X^*}} |x^*(x)|^p \, d\mu(x^*)\right)^{1/p}$. Moreover, the smallest possible constant K for which such a measure μ exists is equal to $\pi_p(T)$.*

This theorem can be interpreted as follows. Let $I: X \to C(B_{X^*})$ be the canonical isometry defined by $Ix(x^*) = x^*(x)$. For a probability measure μ on B_{X^*} let J_μ be the formal identity map from $C(B_{X^*})$ into $L_p(B_{X^*}, \mu)$. Then, $T: X \to Y$ is *p*-absolutely summing if and only if there is a measure μ and a bounded linear operator S from $\overline{J_\mu IX} = Z$ into Y such that the following diagram commutes

$$
\begin{array}{ccc}
C(B_{X^*}) & \xrightarrow{\ J_\mu\ } & L_p(B_{X^*}, \mu) \supseteq Z \\
{\scriptstyle I}\big\uparrow & \nearrow{\scriptstyle J_\mu I} & \big\downarrow{\scriptstyle S} \\
X & \xrightarrow[\ T\ ]{} & Y
\end{array}
$$

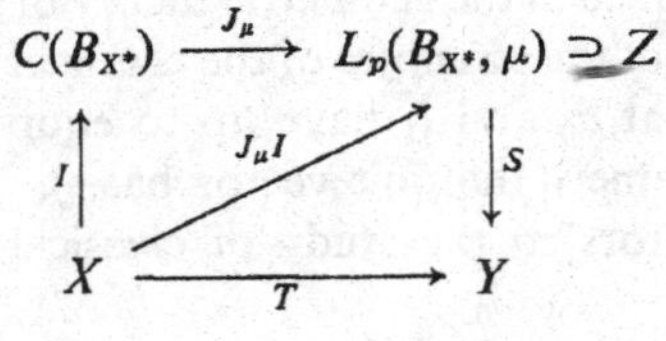

In general, S cannot be extended to an operator from $L_p(B_{X^*}, \mu)$ into Y (of course, if $p=2$ this is always possible since Z is then complemented in the Hilbert space $L_2(B_{X^*}, \mu)$).

Proof of 2.b.2. Assume first that such μ and K exist. Let $\{x_i\}_{i=1}^n$ be elements of X; then

$$\sum_{i=1}^n \|Tx_i\|^p \leqslant K^p \int_{B_{X^*}} \sum_{i=1}^n |x^*(x_i)|^p \, d\mu(x^*) \leqslant K^p \sup_{\|x^*\| \leqslant 1} \sum_{i=1}^n |x^*(x_i)|^p$$

and thus $\pi_p(T) \leqslant K < \infty$. Assume conversely, that $T \in \Pi_p(X, Y)$ and $\pi_p(T)=1$. Consider the following subsets of $C(B_{X^*})$

$$F_1 = \left\{ f \in C(B_{X^*});\ \sup_{x^* \in B_{X^*}} f(x^*) < 1 \right\},$$

$$F_2 = \operatorname{conv} \{ f; f(x^*) = |x^*(x)|^p,\ \|Tx\| = 1 \}.$$

The sets F_1 and F_2 are convex, F_1 is open and the assumption that $\pi_p(T)=1$ implies that $F_1 \cap F_2 = \emptyset$. By the Hahn–Banach and the Riesz representation theorems there exists a positive constant λ and a regular measure μ on B_{X^*} such that $f \in F_1 \Rightarrow \int_{B_{X^*}} f(x^*)\, d\mu(x^*) \leqslant \lambda$ and $f \in F_2 \Rightarrow \int_{B_{X^*}} f(x^*)\, d\mu(x^*) \geqslant \lambda$. Since F_1 contains all the negative functions the measure μ must be a positive measure and thus we may assume without loss of generality that it is a probability measure. Since F_1 contains the open unit ball of $C(B_{X^*})$ we get that $\lambda \geqslant 1$. Hence if $x \in X$ with $\|Tx\|=1$ then $\int_{B_{X^*}} |x^*(x)|^p \, d\mu(x^*) \geqslant 1$, i.e. for every $x \in X$, $\|Tx\|^p \leqslant \int_{B_{X^*}} |x^*(x)|^p \, d\mu(x^*)$. □

Several interesting facts can be read off directly from 2.b.2.

1. Since, for a probability measure μ, the norm of a function in $L_p(\mu)$ is always smaller than its norm in $L_r(\mu)$, if $p<r$, we get that for all Banach spaces X and Y, *$\Pi_p(X, Y) \subset \Pi_r(X, Y)$ whenever $1 \leqslant p < r < \infty$ and, moreover, $\pi_p(T) \geqslant \pi_r(T)$ for every $T \in L(X, Y)$.*

2. *Every p-absolutely summing operator is weakly compact.* For $p>1$ this is evident from the factorization diagram (since Z is reflexive). For $p=1$ use the preceding observation.

3. Theorem 2.b.2 provides a proof of a weak version of the Dvoretzky–Rogers theorem 1.c.2. In every infinite dimensional Banach space X there is a sequence $\{x_i\}_{i=1}^\infty$ such that $\sum_{i=1}^\infty x_i$ converges unconditionally but not absolutely. Indeed, assume that X is an infinite dimensional Banach space in which unconditional convergence implies absolute converges. Then the identity operator of X is 1-absolutely summing and hence also 2-absolutely summing. By 2.b.2 there is a Hilbert space H and operators $S_1: X \to H$, $S_2: H \to X$ such that $S_2 S_1$ is the identity of X. Thus, X is isomorphic to a subspace of H and is therefore isomorphic to a Hilbert space. It is however, obvious that in an infinite-dimensional Hilbert space there are unconditionally converging series which are not absolutely converging.

Before proceeding with the study of absolutely summing operators we prove a classical inequality of Khintchine which has very many applications in the study of the spaces l_p and $L_p(\mu)$, in general.

Theorem 2.b.3. *Let* $r_n(t)=\text{sign}\sin 2^n\pi t$, $n=0, 1, 2,\ldots$ *be the Rademacher functions on* $[0, 1]$. *For every* $1\leq p<\infty$ *there exist positive constants* A_p *and* B_p *so that*

$$A_p\Big(\sum_{n=1}^m |a_n|^2\Big)^{1/2} \leq \Big(\int_0^1 \Big|\sum_{n=1}^m a_n r_n(t)\Big|^p dt\Big)^{1/p} \leq B_p\Big(\sum_{n=1}^m |a_n|^2\Big)^{1/2}$$

for every choice of scalars $\{a_n\}_{n=1}^m$.

Proof. It is trivial that we can take $A_2=B_2=1$ and also $A_p=1$ if $p\geq 2$ and $B_p=1$ if $1\leq p<2$. It is clearly enough to show that we can find a suitable B_p if p is an even integer and a suitable A_p if $p=1$. By considering the real and purely imaginary parts of the $\{a_n\}_{n=1}^m$ separately we see immediately that it is enough to consider only real coefficients.

Observe that $\int_0^1 r_{n_1}^{k_1}(t)r_{n_2}^{k_2}(t)\ldots r_{n_s}^{k_s}(t)\,dt$, where $n_1<n_2<\cdots<n_s$, is 0 unless all the $\{k_i\}_{i=1}^s$ are even in which case the integral is equal to 1. A direct calculation shows therefore that $\int_0^1 \Big(\sum_{n=1}^m a_n r_n(t)\Big)^{2k} dt$ is equal to $\Sigma\gamma(2k_1, 2k_2,\ldots, 2k_s)a_{n_1}^{2k_1} a_{n_2}^{2k_2}\ldots a_{n_s}^{2k_s}$, where the sum is taken over all choices of $n_1, n_2,\ldots, n_s$ between 1 and m and all choices of positive integers $\{k_i\}_{i=1}^s$ such that $k=\sum_{i=1}^s k_i$. The explicit form of γ is given by $\gamma(2k_1, 2k_2,\ldots, 2k_s)=(2k_1+\cdots+2k_s)!/(2k_1)!(2k_2)!\ldots(2k_s)!$. By expanding $\Big(\sum_{n=1}^m a_n^2\Big)^k$ we get a very similar expression the only change being that $\gamma(2k_1, 2k_2,\ldots, 2k_s)$ is replaced by $\gamma(k_1, k_2,\ldots, k_s)$. Thus, the desired inequality for $p=2k$ holds if we take $B_{2k}^{2k}=\sup\gamma(2k_1, 2k_2,\ldots, 2k_s)/\gamma(k_1, k_2,\ldots, k_s)$. A short computation shows that this gives $B_{2k}\leq k^{1/2}$. It remains to verify the existence of A_1. Put $f(t)=\sum_{n=1}^m a_n r_n(t)$. Then, by Holder's inequality,

$$\int_0^1 |f(t)|^2\,dt=\int_0^1 |f(t)|^{2/3}|f(t)|^{4/3}\,dt\leq \Big(\int_0^1 |f(t)|\,dt\Big)^{2/3}\Big(\int_0^1 |f(t)|^4\,dt\Big)^{1/3}$$

$$\leq \Big(\int_0^1 |f(t)|\,dt\Big)^{2/3}\cdot B_4^{4/3}\Big(\int_0^1 |f(t)|^2\,dt\Big)^{2/3}.$$

This gives the desired result (with $A_1=B_4^{-2}$). □

We shall apply now Khintchine's inequality for identifying the class of p-absolutely summing operators in $L(X, X)$ for the simplest possible case, i.e. when $X=l_2$. It turns out that in this special case the space $\Pi_p(X, X)$ does not depend on p.

Theorem 2.b.4 [116]. *For every* $1 \leqslant p < \infty$ *the space* $\Pi_p(l_2, l_2)$ *consists exactly of the Hilbert–Schmidt operators.*

We recall that an operator T in $L(l_2, l_2)$ is called a Hilbert–Schmidt operator if $\sum_{n=1}^{\infty} \|Te_n\|^2 < \infty$, where $\{e_n\}_{n=1}^{\infty}$ is the unit vector basis in l_2. Also, if $\{y_n\}_{n=1}^{\infty}$ is any orthonormal basis in l_2 then

$$\sum_{n=1}^{\infty} \|Ty_n\|^2 = \sum_{n,m=1}^{\infty} |(Ty_n, e_m)|^2 = \sum_{m=1}^{\infty} \|T^* e_m\|^2,$$

hence, $\sum_{n=1}^{\infty} \|Ty_n\|^2 = \sum_{n=1}^{\infty} \|T^* y_n\|^2$ and this expression is independent of the particular choice of $\{y_n\}_{n=1}^{\infty}$ and is denoted by $\|T\|_{\mathrm{HS}}^2$.

Every Hilbert–Schmidt operator T is compact. Hence, as for any compact T in $L(l_2, l_2)$, there is an orthonormal basis $\{x_n\}_{n=1}^{\infty}$ of l_2, an orthonormal system $\{y_n\}_{n=1}^{\infty}$ in l_2 and scalars $\{\lambda_n\}_{n=1}^{\infty}$ such that $Tx_n = \lambda_n y_n$ for all n. An operator T given in this form is Hilbert–Schmidt if and only if $\sum_{n=1}^{\infty} |\lambda_n|^2 < \infty$.

Proof of 2.b.4. We show first that every Hilbert–Schmidt operator T is 1-absolutely summing. By observation 1 following 2.b.2 this will prove that T is p-absolutely summing for every $p > 1$.

Let T be given by $Tx_n = \lambda_n y_n$, $n = 1, 2, \ldots$ and let $\lambda = \left(\sum_{n=1}^{\infty} |\lambda_n|^2\right)^{1/2} < \infty$. Let $\{u_j\}_{j=1}^{m}$ be any m-tuple of vectors in l_2 and, for $0 \leqslant t \leqslant 1$, let $v(t) = \lambda^{-1} \sum_{n=1}^{\infty} r_n(t) \lambda_n x_n \in l_2$, where $\{r_n\}_{n=1}^{\infty}$ are the Rademacher functions. Then by 2.b.3,

$$\begin{aligned}
\sup_{\|x\| \leqslant 1} \sum_{j=1}^{m} |(u_j, x)| &\geqslant \sup_{0 \leqslant t \leqslant 1} \sum_{j=1}^{m} |(u_j, v(t))| \\
&\geqslant \lambda^{-1} \sum_{j=1}^{m} \int_0^1 \left| \sum_{n=1}^{\infty} r_n(t) \bar{\lambda}_n (u_j, x_n) \right| dt \\
&\geqslant A_1 \lambda^{-1} \sum_{j=1}^{m} \left(\sum_{n=1}^{\infty} |\lambda_n (u_j, x_n)|^2 \right)^{1/2} = A_1 \lambda^{-1} \sum_{j=1}^{m} \|Tu_j\|
\end{aligned}$$

and thus $\pi_1(T) \leqslant A_1^{-1} \cdot \lambda$.

Assume, conversely, that T is p-absolutely summing for some $p \geqslant 2$. It follows that, for every orthonormal basis $\{u_n\}_{n=1}^{\infty}$ of l_2,

$$\left(\sum_{n=1}^{n} \|Tu_n\|^p \right)^{1/p} \leqslant K \sup_{\|x\|=1} \left(\sum_{n=1}^{\infty} |(u_n, x)|^p \right)^{1/p} \leqslant K$$

and hence $\lim_n \|Tu_n\| = 0$. This implies easily that T is compact and thus there are orthonormal bases $\{x_n\}_{n=1}^{\infty}$ and $\{y_n\}_{n=1}^{\infty}$ in l_2 and scalars $\{\lambda_n\}_{n=1}^{\infty}$ such that $Tx_n = \lambda_n y_n$.

Let m be an integer and put $w(t)=\sum_{n=1}^m r_n(t)x_n$, $0\leq t\leq 1$, where $\{r_n\}_{n=1}^m$ are the Rademacher functions. Then, by 2.b.3,

$$\left(\sum_{n=1}^m |\lambda_n|^2\right)^{1/2}=\left(\int_0^1 \|Tw(t)\|^p\,dt\right)^{1/p}$$

$$\leq \pi_p(T)\sup_{\|x\|=1}\left(\int_0^1 |(w(t),x)|^p\,dt\right)^{1/p}$$

$$=\pi_p(T)\sup_{\|x\|=1}\left(\int_0^1 \left|\sum_{n=1}^m r_n(t)(x_n,x)\right|^p dt\right)^{1/p}$$

$$\leq \pi_p(T)B_p\sup_{\|x\|=1}\left(\sum_{n=1}^m |(x,x_n)|^2\right)^{1/2}\leq \pi_p(T)\cdot B_p\,.$$

Hence, $\sum_{n=1}^\infty |\lambda_n|^2<\infty$ and T is a Hilbert–Schmidt operator. □

Many interesting results concerning p-absolutely summing operators as well as several applications of these operators are based on the following inequality due to Grothendieck [49].

Theorem 2.b.5. *Let $(\alpha_{i,j})_{i,j=1}^n$ be a matrix of scalars such that $\left|\sum_{i,j=1}^n \alpha_{i,j}t_is_j\leq 1\right|$ for every choice of scalars $\{t_i\}_{i=1}^n$ and $\{s_j\}_{j=1}^n$ satisfying $|t_i|\leq 1$, $|s_j|\leq 1$. Then, for any choice of vectors $\{x_i\}_{i=1}^n$ and $\{y_j\}_{j=1}^n$ in a Hilbert space,*

$$\left|\sum_{i,j=1}^n \alpha_{i,j}(x_i,y_j)\right|\leq K_G\max_i\|x_i\|\max_j\|y_j\|\,,$$

where K_G is Grothendieck's universal constant (in case the scalars are real $K_G\leq (e^{\pi/2}-e^{-\pi/2})/2$).

Proof. It is clearly enough to prove the theorem for real scalars. Also, it is easily seen that it is enough to consider vectors $\{x_i\}_{i=1}^n$ and $\{y_j\}_{j=1}^n$ so that $\|x_i\|=\|y_i\|=1$ for every i. Since every finite dimensional subspace of a Hilbert space is isometric to l_2^k for some k we may assume that $\{x_i\}_{i=1}^n$, $\{y_j\}_{j=1}^n\subset l_2^k$ for some k. Let $S=\{u;\,u\in l_2^k,\ \|u\|=1\}$ and let μ be the unique probability measure on S which is rotation invariant. A simple two-dimensional computation shows that, for every choice of $x,y\in S$,

$$\int_S \operatorname{sign}(x,u)\operatorname{sign}(y,u)\,d\mu(u)=1-2\theta[x,y]/\pi$$

where $\theta[x,y]$ is the angle between x and y (i.e. $0\leq\theta[x,y]\leq\pi$ and $\cos\theta[x,y]=(x,y)$).

In view of our assumption on $(\alpha_{i,j})$ we have, for every $u \in S$ and for every $\{t_i\}_{i=1}^n$, $\{s_j\}_{j=1}^n$ with $|t_i| \leq 1$, $|s_j| \leq 1$, that

$$-1 \leq \sum_{i,j=1} \alpha_{i,j} s_i t_j \operatorname{sign}(x_i, u) \operatorname{sign}(y_j, u) \leq 1 .$$

By integrating with respect to μ we get that

$$-1 \leq \sum_{i,j=1}^{n} \alpha_{i,j} s_i t_j (1 - 2\theta[x_i, y_j]/\pi) \leq 1 .$$

Hence, the matrix $(\alpha_{i,j}(1-2\theta[x_i, y_j]/\pi)$ also satisfies the assumptions made on $(\alpha_{i,j})$. Thus, by iterating the same argument, we get for every integer m that

$$\left| \sum_{i,j=1}^{n} \alpha_{i,j} (1 - 2\theta[x_i, y_j]/\pi)^m \right| \leq 1 .$$

Since, for every $1 \leq i, j \leq n$,

$$(x_i, y_j) = \cos \theta[x_i, y_j] = \sin(\pi/2 - \theta[x_i, y_j])$$
$$= \sum_{m=0}^{\infty} (-1)^m (\pi/2 - \theta[x_i, y_j])^{2m+1}/(2m+1)!$$

we conclude that

$$\left| \sum_{i,j=1}^{n} \alpha_{i,j}(x_i, y_j) \right| \leq \sum_{m=0}^{\infty} (\pi/2)^{2m+1}/(2m+1)! = (e^{\pi/2} - e^{-\pi/2})/2 . \square$$

Several other proofs of 2.b.5 can be found in the literature (cf. [12, 103, 123]). Some give also a better estimate for K_G. The best possible value of K_G seems to be unknown.

The next result, due to Grothendieck [49], is actually a restatement of 2.b.5 in terms of absolutely summing operators.

Theorem 2.b.6. *Every bounded linear operator T from l_1 into l_2 is absolutely summing and $\pi_1(T) \leq K_G \|T\|$.*

Proof. Let $\{e_j\}_{j=1}^\infty$ be the unit vector basis of l_1 and $u_i = \sum_{j=1}^{m} \alpha_{i,j} e_j$, $i = 1, 2, \ldots, n$ be vectors in l_1^m, for some m, such that $\sum_{i=1}^{n} |x^*(u_i)| \leq \|x^*\|$, for every $x^* \in l_1^*$. Let $\{s_j\}_{j=1}^m$ be scalars of absolute value ≤ 1 and let $x_s^* \in l_1^*$ be defined by $x_s^*(e_j) = s_j$ if $1 \leq j \leq m$ and $x^*(e_j) = 0$ if $j > m$. For every choice of $\{t_i\}_{i=1}^n$ such that $|t_i| \leq 1$, $1 \leq i \leq m$,

$$\left| \sum_{j=1}^{m} \sum_{i=1}^{n} \alpha_{i,j} t_i s_j \right| \leq \sum_{i=1}^{n} |t_i| \left| \sum_{j=1}^{m} \alpha_{i,j} s_j \right| \leq \sum_{i=1}^{n} |x_s^*(u_i)| \leq 1 .$$

For every $1 \leqslant i \leqslant n$ let $y_i \in l_2$ be such that $||y_i|| = 1$ and $(Tu_i, y_i) = ||Tu_i||$. Then, by 2.b.5

$$\sum_{i=1}^{n} ||Tu_i|| = \sum_{i=1}^{n} (Tu_i, y_i) = \sum_{i=1}^{n} \sum_{j=1}^{m} \alpha_{i,j}(Te_j, y_i) \leqslant K_G ||T|| . \quad \square$$

Let us point out that 2.b.6 is in a sense a joint characterization of l_1 and l_2. It is proved in [87] that *if X has an unconditional basis and, for some Banach space Y, $\Pi_1(X, Y) = L(X, Y)$ then X is isomorphic to l_1 and Y is isomorphic to a Hilbert space.* If we do not assume that X has an unconditional basis then there are more spaces which satisfy $\Pi_1(X, l_2) = L(X, l_2)$. For example, $X = L_1(0, 1)$ has this property.

There are many examples of pairs of spaces X and Y such that $L(X, Y) = \Pi_p(X, Y)$, for some $p \geqslant 2$. For instance, we have the following:

Theorem 2.b.7 [87]. *Every bounded linear operator T from c_0 into l_p, with $1 \leqslant p \leqslant 2$, is 2-absolutely summing and $\pi_2(T) \leqslant K_G ||T||$.*

Proof. Let $\{e_i\}_{i=1}^\infty$ and $\{f_j\}_{j=1}^\infty$ be the unit vector bases of c_0, respectively l_p, and let $T \in L(c_0, l_p)$. Let $(\alpha_{i,j})_{i,j=1}^\infty$ be the matrix defined by $Te_i = \sum_{j=1}^{\infty} \alpha_{i,j} f_j$, $i = 1, 2, \ldots$. For every choice of $y^* = (a_1, a_2, \ldots) \in l_p^*$ with $||y^*|| = 1$ and every choice of scalars $\{t_i\}_{i=1}^\infty$, $\{s_j\}_{j=1}^\infty$ of absolute value $\leqslant 1$ such that $\lim_i t_i = 0$ we have

$$(*) \qquad \left|\sum_{i,j=1}^{\infty} \alpha_{i,j} a_j t_i s_j\right| = \left|y_s^*\left(\sum_{i=1}^{\infty} Tt_i e_i\right)\right| \leqslant ||T|| \left\|\sum_{i=1}^{\infty} t_i e_i\right\| \leqslant ||T|| ,$$

where $y_s^* = (s_1 a_1, s_2 a_2, \ldots) \in l_p^*$. Let $\{x_k\}_{k=1}^n$ be n vectors in c_0 of the form $x_k = \sum_{i=1}^{m} b_{k,i} e_i$, for some integer m, which satisfy $\sum_{k=1}^{n} |x^*(x_k)|^2 \leqslant 1$ whenever $x^* \in c_0^*$ with $||x^*|| \leqslant 1$. In particular, by taking as x^* the unit vectors in l_1 we get that, for $1 \leqslant i \leqslant m$, $\sum_{k=1}^{n} b_{k,i}^2 \leqslant 1$. By considering the m vectors $u_i = (b_{1,i}, b_{2,i}, \ldots, b_{n,i})$, $1 \leqslant i \leqslant m$ in l_2^n it follows from (*) and 2.b.5 that

$$\sum_{j=1}^{\infty} \left\|\sum_{i=1}^{m} a_j \alpha_{i,j} u_i\right\|_2 = \sum_{j=1}^{\infty} a_j \left(\sum_{k=1}^{n} \left(\sum_{i=1}^{m} b_{k,i} \alpha_{i,j}\right)^2\right)^{1/2} \leqslant K_G ||T|| .$$

Since this holds whenever $||(a_1, a_2, \ldots)|| = 1$ (in l_p^*) we get that

$$(\overset{*}{*}) \qquad \left(\sum_{j=1}^{\infty} \left(\sum_{k=1}^{n} \left(\sum_{i=1}^{m} b_{k,i} \alpha_{i,j}\right)^2\right)^{p/2}\right)^{1/p} \leqslant K_G ||T|| .$$

Put $c_{j,k} = \left|\sum_{i=1}^{m} b_{k,i} \alpha_{i,j}\right|^p$. By using the triangle inequality in $l_{2/p}$ (recall that $p \leqslant 2$), i.e

$$\left(\sum_{k=1}^{n} \left(\sum_{j=1}^{\infty} c_{j,k}\right)^{2/p}\right)^{p/2} \leqslant \sum_{j=1}^{\infty} \left(\sum_{k=1}^{n} c_{j,k}^{2/p}\right)^{p/2}$$

we deduce from $(\ast\ast)$ that

$$(\ast\ast\ast) \qquad \left(\sum_{k=1}^{n}\left(\sum_{j=1}^{\infty}\left|\sum_{i=1}^{m} b_{k,i}\alpha_{i,j}\right|^{p}\right)^{2/p}\right)^{1/2} \leq K_G\|T\|.$$

Since $Tx_k = \sum_{i=1}^{m} b_{k,i} Te_i = \sum_{j=1}^{\infty}\sum_{i=1}^{m} b_{k,i}\alpha_{i,j}f_j$, $k=1,2,\ldots,n$, $(\ast\ast\ast)$ states that

$$\left(\sum_{k=1}^{n}\|Tx_k\|^2\right)^{1/2} \leq K_G\|T\|$$

and thus $\pi_2(T) \leq K_G\|T\|$. □

We state now without proof a result of L. Schwartz [133] and S. Kwapien [77] which shows what happens if we take $p>2$ in 2.b.7.

Theorem 2.b.8. *Let $2<p<r<\infty$; then $L(c_0, l_p)=\Pi_r(c_0, l_p)$. There are however operators in $L(c_0, l_p)$ which are not p-absolutely summing.*

There are of course many other situations which can be investigated, e.g. the structure of $\Pi_r(l_p, l_s)$ for arbitrary r, p, and s. The only operators T which are relatively easy to investigate in the general case are the diagonal operators (i.e. operators of the form $Te_i=\lambda_i f_i$, $i=1,2,\ldots$, where $\{e_i\}_{i=1}^{\infty}$, resp. $\{f_i\}_{i=1}^{\infty}$, are the unit vector bases in l_p, resp. l_s. For a discussion of these results and several of their variants we refer to [122].

We turn now to the question of uniqueness of unconditional bases in the spaces c_0 and l_p, $1 \leq p < \infty$. We consider first the simplest case, i.e. $p=2$. It follows from the parallelogram identity in Hilbert space that the average of $\left\|\sum_{i=1}^{n}\theta_i x_i\right\|^2$, over all choices of signs $\{\theta_i\}_{i=1}^{n}$, is equal to $\sum_{i=1}^{n}\|x_i\|^2$. Thus, *if $\{u_n\}_{n=1}^{\infty}$ is a normalized unconditional basis in l_2 then $\sum_{n=1}^{\infty} a_n u_n$ converges if and only if $\sum_{n=1}^{\infty} a_n^2<\infty$.* In other words, every normalized unconditional basis of l_2 is equivalent to the unit vector basis (this observation goes back to G. Köthe [72] and E. R. Lorch [98]). The situation in c_0 and l_1 is similar to that in l_2 but the proof is more difficult (since it is based on 2.b.5).

Proposition 2.b.9 [87]. *Every normalized unconditional basis in l_1 or in c_0 is equivalent to the unit vector basis of the space.*

Proof. Let $\{x_n\}_{n=1}^{\infty}$ be a normalized unconditional basis of l_1. Let $\{a_n\}_{n=1}^{\infty}$ be a sequence of scalars such that $\sum_{n=1}^{\infty} a_n x_n$ converges. Let S be the operator from c_0 into l_1 defined by $S(\lambda_1, \lambda_2, \ldots)=\sum_{n=1}^{\infty}\lambda_n a_n x_n$. By 2.b.7 $\pi_2(S) \leq K_G\|S\| \leq K_G M\left\|\sum_{n=1}^{\infty} a_n x_n\right\|$, where M is the unconditional constant of $\{x_n\}_{n=1}^{\infty}$. Hence, $\left(\sum_{n=1}^{\infty}|a_n|^2\right)^{1/2} \leq K_G M\left\|\sum_{n=1}^{\infty} a_n x_n\right\|$.

It follows that the operator $T: l_1 \to l_2$, defined by $T\left(\sum_{n=1}^{\infty} a_n x_n\right)=(a_1, a_2, \ldots)$, is a bounded linear operator with $\|T\| \leqslant K_G M$. By 2.b.6, $\pi_1(T) \leqslant K_G \|T\| \leqslant K_G^2 M$. Hence, for every $x=\sum_{n=1}^{\infty} a_n x_n$ in l_1,

$$\sum_{n=1}^{\infty} |a_n| = \sum_{n=1}^{\infty} \|T a_n x_n\| \leqslant K_G^2 M \sup_{\theta_n=\pm 1} \left\|\sum_{n=1}^{\infty} \theta_n a_n x_n\right\| \leqslant K_G^2 M^2 \left\|\sum_{n=1}^{\infty} a_n x_n\right\| .$$

Since obviously $\left\|\sum_{n=1}^{\infty} a_n x_n\right\| \leqslant \sum_{n=1}^{\infty} |a_n|$ we deduce that $\{x_n\}_{n=1}^{\infty}$ is equivalent to the unit vector basis in l_1.

Let now $\{y_n\}_{n=1}^{\infty}$ be normalized unconditional basis in c_0. By 1.c.9 the basis $\{y_n\}_{n=1}^{\infty}$ is shrinking and thus the biorthogonal functionals $\{y_n^*\}_{n=1}^{\infty}$ associated to $\{y_n\}_{n=1}^{\infty}$ form an unconditional basis of l_1 such that $1 \leqslant \|y_n^*\| \leqslant M$, for some M and $n=1, 2, \ldots$. By the first part of the proposition the sequence $\{y_n^*\}_{n=1}^{\infty}$ is equivalent to the unit vector basis in l_1. This implies immediately that $\{y_n\}_{n=1}^{\infty}$ is equivalent to the unit vector basis of c_0. □

Remarks. 1. The same argument as that used in the proof of 2.b.9 shows that if $\{X_n\}_{n=1}^{\infty}$ is an unconditional Schauder decomposition of l_1, respectively c_0, then there is a constant M so that, for every choice of $x_n \in X_n$, $n=1, 2, \ldots$ with $\sum_{n=1}^{\infty} x_n$ converging, we have $\sum_{n=1}^{\infty} \|x_n\| \leqslant M \left\|\sum_{n=1}^{\infty} x_n\right\|$, respectively $\left\|\sum_{n=1}^{\infty} x_n\right\| \leqslant M \sup_n \|x_n\|$.

2. The arguments used in the beginning of the proofs of 2.b.9 and 2.b.7 show also the following. Let $\{u_n\}_{n=1}^{\infty}$ be a sequence of elements in l_p, $1 \leqslant p < 2$ such that $\sum_{n=1}^{\infty} u_n$ converges unconditionally; then $\sum_{n=1}^{\infty} \|u_n\|^2 < \infty$. This fact was first proved by Orlicz [110] and his proof is simpler than the one given here (which relies on 2.b.5). Similarly, we deduce from 2.b.8 that if $\sum_{n=1}^{\infty} u_n$ converges unconditionally in l_p, for $p>2$, then $\sum_{n=1}^{\infty} \|u_n\|^r < \infty$ for every $r>p$. This however is not the best possible result. We shall see in Vol. II that we can take also $r=p$.

In contrast to the cases of c_0, l_1 and l_2 there are normalized unconditional bases in l_p, $1<p<\infty$, $p \neq 2$ which are not equivalent to the unit vector basis [114]. In order to see this we observe that the Khintchine inequality 2.b.3 shows that the mapping $T: l_2 \to L_p(0, 1)$, defined by $T(a_1, a_2, \ldots) = \sum_{n=1}^{\infty} a_n r_n(t)$, is an isomorphism for every $1 \leqslant p < \infty$.

Let now $p \geqslant 2$ and let P be the orthogonal projection from $L_p(0, 1)$ (which is, under our assumption on p, a linear subspace of $L_2(0, 1)$) onto $[r_n]_{n=1}^{\infty}$. The projection P is defined by

$$Pf(t) = \sum_{n=1}^{\infty} \int_0^1 f(s) r_n(s)\, ds \cdot r_n(t) .$$

It is bounded in L_p since

$$\|Pf\|_p \leq B_p \|Pf\|_2 \leq B_p \|f\|_2 \leq B_p \|f\|_p .$$

For every n let F_n be the subspace of $L_p(0, 1)$ spanned by the characteristic functions of the intervals $[k/2^n, (k+1)/2^n]$, $k=0, 1, \ldots, 2^n-1$. Clearly, $F_n \supset E_n = [r_i]_{i=1}^n$, F_n is isometric to $l_p^{2^n}$ and $PF_n = E_n$ for every n. In other words, there is a constant K_p so that, for every n, there exist a subspace C_n of $l_p^{2^n}$ with $d(C_n, l_2^n) \leq K_p$ and a projection from $l_p^{2^n}$ onto C_n of norm $\leq K_p$. Consequently, the space $\left(\sum_{n=1}^{\infty} \oplus l_2^n\right)_p$ is, for $2 \leq p < \infty$, isomorphic to a complemented subspace of $\left(\sum_{n=1}^{\infty} \oplus l_p^{2^n}\right)_p = l_p$. By 2.a.3, $\left(\sum_{n=1}^{\infty} \oplus l_2^n\right)_p$ is isomorphic to l_p for $2 \leq p < \infty$ and, by duality, also for $1 < p < 2$. The obvious unit vector basis in $\left(\sum_{n=1}^{\infty} \oplus l_2^n\right)_p$ is therefore, for $1 < p < \infty, p \neq 2$, an example of a normalized unconditional basis in l_p (more precisely, in a space isomorphic to l_p) which is not equivalent to the unit vector basis of l_p. Observe that the basis we have just discussed has, by 2.a.12, the property that all its block bases span a complemented subspace of l_p. Observe also that by 2.b.9 the space l_1 is not isomorphic to $\left(\sum_{n=1}^{\infty} \oplus l_2^n\right)_1$ and, similarly, c_0 is not isomorphic to $\left(\sum_{n=1}^{\infty} \oplus l_2^n\right)_0$.

The preceding discussion concerning uniqueness of unconditional bases proves a part of the following theorem.

Theorem 2.b.10 [97]. *A Banach space has, up to equivalence, a unique unconditional basis if and only if it is isomorphic to one of the following three spaces*: c_0, l_1 *or* l_2.

We shall conclude the proof of 2.b.10 in Section 3.a below.

We would like to show now how the weak form of 1.a.8 (namely, that every infinite-dimensional Banach space with a basis has at least two non-equivalent normalized bases) can be deduced from 2.b.10. Let $\{x_n\}_{n=1}^{\infty}$ be a normalized basis of a Banach space X. If $\{x_n\}_{n=1}^{\infty}$ is not unconditional then, for some sequence of signs $\{\theta_n\}_{n=1}^{\infty}$, the normalized basis $\{\theta_n x_n\}_{n=1}^{\infty}$ of X is not equivalent to $\{x_n\}_{n=1}^{\infty}$. If $\{x_n\}_{n=1}^{\infty}$ is unconditional then, by 2.b.10, X has a normalized unconditional basis which is not equivalent to $\{x_n\}_{n=1}^{\infty}$ unless X is c_0, l_1 or l_2. Thus, we have only to show that these three spaces have conditional (i.e. not unconditional) bases. For l_1 and c_0 this assertion is trivial. The vectors $e_1, e_2-e_1, e_3-e_2, \ldots$ form a conditional basis of l_1 ($\{e_n\}_{n=1}^{\infty}$ denotes as usual the unit vector basis). The summing basis is a conditional basis of c (which is of course isomorphic to c_0 via the map $T(a_1, a_2, \ldots) = (\lim_n a_n, a_1 - \lim_n a_n, a_2 - \lim_n a_n, \ldots)$, from c onto c_0). For l_2 the assertion is by no means obvious.

Proposition 2.b.11. *The space* l_2 *has a conditional basis.*

This proposition was proved first by Babenko [6] who used harmonic analysis to construct concrete examples of conditional bases in Hilbert function spaces. For example, he showed that, for $0 < \alpha < 1/2$, the sequence $|t|^\alpha e^{int}$, $n=0, \pm 1, \pm 2, \ldots$

is a conditional basis of $L_2(-\pi, \pi)$. The proof we present here is completely different and is due to C. A. McCarthy and J. Schwartz [105].

Proof. It is clearly enough to construct, for every even integer n, a set $\{Q_i\}_{i=1}^n$ of projections in l_2^n so that $\dim Q_i l_2^n = 1$ for every i, $Q_i Q_j = 0$ if $i \neq j$, $\left\| \sum_{i=1}^k Q_i \right\| \leq 2$ for $1 \leq k \leq n$ and $\lim_n \|Q_1 + Q_3 + \cdots + Q_{n-1}\| = \infty$. Indeed, these assumptions on $\{Q_i\}_{i=1}^n$ imply that $\sum_{i=1}^n Q_i = I$ and thus if $0 \neq x_i \in Q_i l_2^n$, $1 \leq i \leq n$, then $\{x_i\}_{i=1}^n$ is a basis of l_2^n whose basis constant is ≤ 2 and whose unconditional constant tends to ∞ with n. From this it is obvious how to construct a conditional basis in $l_2 = \left(\sum_{n=1}^\infty \oplus l_2^n \right)_2$.

In order to construct suitable Q_i we pick a sequence $\{\alpha_k\}_{k=1}^\infty$ of positive numbers such that

$$\sum_{k=1}^\infty k\alpha_k^2 \leq 1 \quad \text{and} \quad \lambda_n = \frac{1}{n} \sum_{k=1}^n \left(\sum_{i=1}^k \alpha_i \right)^2 \to \infty, \quad \text{as } n \to \infty$$

(take, e.g. $\alpha_k = \delta / k \log k$, for $k > 1$ and a suitable $\delta > 0$).

Let now $n = 2m$ and let Q_i be the operator on l_2^n whose matrix with respect to the unit vector basis is defined as follows:

If $i = 2j - 1$, $j = 1, \ldots, m$ the matrix which represents Q_i has all its columns equal to 0 except the i'th column which is equal to $(0, 0, \ldots, 0, 1, \alpha_1, 0, \alpha_2, 0, \ldots, \alpha_{m-j+1})$ (for typographical reasons we write this column here as a row). If $i = 2j$, $j = 1, \ldots, m$ the matrix which represents Q_i has all rows equal to 0 except the i'th row which is equal to $(-\alpha_j, 0, -\alpha_{j-1}, 0, \ldots, -\alpha_1, \overset{i}{1}, 0, 0, \ldots, 0)$. It is clear that each Q_i is a rank one operator and also that $Q_i^2 = Q_i$ (observe that the matrices representing the Q_i are triangular with a diagonal which is equal to 0 except that i'th element which is equal to 1). It is also easy to verify that $Q_{i_1} \cdot Q_{i_2} = 0$ for $i_1 \neq i_2$.

In order to estimate the norm of $Q_1 + Q_2 + \cdots + Q_i$, $1 \leq i \leq n$, we write this operator in the form of $D_i + R_i$ where D_i is the operator which is represented by the diagonal of the matrix representing $Q_1 + Q_2 + \cdots + Q_i$. (Thus the matrix representing D_i is a diagonal matrix whose diagonal is $(1, 1, \ldots, \overset{i}{1}, 0, \ldots, 0)$.) Observe that the non-zero entries in the matrix representing R_i (i.e. the off diagonal entries in the matricial representation of $Q_1 + Q_2 + \cdots + Q_i$) consist of at most k times the term α_k, $k = 1, 2, \ldots, m$. Thus the Hilbert Schmidt norm of R_i is $\leq \left(\sum_{k=1}^m k\alpha_k^2 \right)^{1/2}$. Hence,

$$\|Q_1 + Q_2 + \cdots + Q_i\| \leq \|D_i\| + \|R_i\| \leq 1 + \left(\sum_{k=1}^m k\alpha_k^2 \right)^{1/2} \leq 2.$$

In order to estimate the norm of $Q_1 + Q_3 + \cdots + Q_{2m-1}$ let x be the vector $(1, 0, 1, \ldots, 1, 0)$ in l_2^n. Then $\|x\| = m^{1/2}$ and

$$(Q_1 + Q_3 + \cdots + Q_{2m-1})x = (1, \alpha_1, 1, \alpha_1 + \alpha_2, \ldots, 1, \alpha_1 + \alpha_2 + \cdots + \alpha_m).$$

Hence

$$\|Q_1+Q_3+\cdots+Q_{2m-1}\|^2 \geqslant \|(Q_1+Q_3+\cdots+Q_{2m-1})x\|^2/\|x\|^2$$

$$\geqslant m^{-1} \sum_{k=1}^{m} \left(\sum_{i=1}^{k} \alpha_i \right)^2 = \lambda_m . \quad \square$$

c. Fredholm Operators, Strictly Singular Operators and Complemented Subspaces of $l_p \oplus l_r$

We have proved in Section a that each infinite-dimensional complemented subspace of c_0 or l_p, $1<p<\infty$ is isomorphic to the entire space. In this section we are going to present a result of I. S. Edelstein and P. Wojtaszczyk [35] which shows that the only isomorphism types of infinite dimensional complemented subspaces of $l_p \oplus l_r$ are the obvious ones, i.e. l_p, l_r and $l_p \oplus l_r$. The proof of this result makes use of some notions (like strictly singular and Fredholm operators) which enter into many other contexts in functional analysis. A large part of this section will be devoted to proving the basic results concerning these important notions. Our presentation will go a little beyond the material which is strictly needed for the proof of the result of Edelstein and Wojtaszczyk. Concerning the proof itself let us point out already here one interesting feature. It uses spectral theory in an essential way. This topic had so far only few applications in the structure theory of Banach spaces. The section concludes with a study of the unconditional bases of the spaces $l_p \oplus l_r$.

The fact that makes the approach of Edelstein and Wojtaszczyk work is that, for $p \neq r$, the space l_p does not have an infinite-dimensional subspace isomorphic to a subspace of l_r (see the remark following 2.a.2). Let us give this property a formal name.

Definition 2.c.1. Two infinite-dimensional Banach spaces X and Y are called *totally incomparable* if there exists no infinite dimensional Banach space Z which is isomorphic to a subspace of X and to a subspace of Y.

As mentioned above, any two different spaces of the set $\{c_0\} \cup \{l_p;\ 1 \leqslant p < \infty\}$ are totally incomparable. The notion of "totally incomparable spaces" was introduced by H. P. Rosenthal [126]. In this paper it is proved that *X and Y are totally incomparable if and only if, whenever U is a Banach space with subspaces X_1 and Y_1 which are isomorphic to X, respectively Y, the algebraic sum $X_1 + Y_1$ is closed in U.*

Definition 2.c.2. An operator T: $X \to Y$ is called *strictly singular* if the restriction of T to any infinite-dimensional subspace of X is not an isomorphism.

If X and Y are totally incomparable any operator from X to Y is strictly singular. For general spaces X and Y, every compact operator is strictly singular. The formal identity map from l_p to l_r, if $r>p$, is an example of a strictly singular operator which is not compact. In this connection it is worthwhile to mention the following result of H. R. Pitt.

Proposition 2.c.3. *Let $1\leqslant p<r<\infty$. Then, every bounded linear operator from l_r into l_p is compact. The same is true for every linear operator from c_0 into l_p.*

Proof. Assume that T is a non-compact operator from l_r into l_p. Then there is a sequence $\{x_n\}_{n=1}^\infty$ in l_r so that $x_n \xrightarrow{w} 0$ and $\|Tx_n\|\geqslant \varepsilon$ for some $\varepsilon>0$ and all integers n. By passing to a subsequence we may assume by 1.a.9 and 2.a.1 that $\{x_n\}_{n=1}^\infty$ is equivalent to the unit vector basis in l_r and $\{Tx_n\}_{n=1}^\infty$ is equivalent to the unit vector basis in l_p. Since the formal identity map from l_r into l_p is not bounded (recall that $r>p$) we arrived at a contradiction. The proof in the case where c_0 replaces l_r is the same. □

Observe that the proof of 2.c.3 shows also that a $T\in L(l_p, l_p)$ is strictly singular if and only if it is compact.

We present now two simple propositions concerning strictly singular operators due to T. Kato [69].

Proposition 2.c.4. *Let X and Y be infinite-dimensional Banach spaces. Assume that $T\colon X\to Y$ is an operator such that the restriction of T to any subspace of X of finite codimension is not an isomorphism (this is the case in particular if TX is not closed in Y). Then, for every $\varepsilon>0$ there is an infinite-dimensional subspace Z of X so that $T_{|Z}$ is compact and $\|T_{|Z}\|\leqslant\varepsilon$.*

Proof. For every $\delta>0$ and every finite set $\{x_i^*\}_{i=1}^m$ of elements in X^* there is, by our assumption, an element $x\in X$ with $\|x\|=1$, $\|Tx\|<\delta$ and $x_i^*(x)=0$, $1\leqslant i\leqslant m$. Hence, for every $\varepsilon>0$ (see the proof of 1.a.5), there exists a normalized basic sequence $\{x_n\}_{n=1}^\infty$ in X, with basis constant $\leqslant 2$, so that $\|Tx_n\|\leqslant\varepsilon\cdot 8^{-n}$, $n=1, 2, \ldots$. The space $Z=[x_n]_{n=1}^\infty$ has the desired property. □

Observe that if, for some $X_0\subset X$ of finite codimension, $T_{|X_0}$ is an isomorphism then TX_0, and thus also TX, are closed in Y.

Proposition 2.c.5. (i) *The sum of two strictly singular operators is strictly singular.*
(ii) *The composition of a strictly singular and a bounded operator is strictly singular.*

Proof. (i) Let T and S be strictly singular operators from X to Y and let Z be an infinite-dimensional subspace of X. By 2.c.4 there is an infinite dimensional subspace Z_1 of Z such that $T_{|Z_1}$ is compact. By applying 2.c.4 again we get an infinite-dimensional subspace Z_2 of Z_1 such that $S_{|Z_2}$ is compact. Clearly, $(S+T)_{|Z_2}$ is compact and thus $S+T$ is not an isomorphism on Z. The verification of (ii) is straightforward. □

Another way of formulating 2.c.5 is to say that the strictly singular operators form an operator ideal.

It should be pointed out that, in contrast to the ideal of compact (or weakly compact) operators, the dual operator T^* of a strictly singular operator T need not be strictly singular. For example, let X be a separable Banach space which does not contain a subspace isomorphic to l_1 (e.g. $X=l_p$, for $p>1$, or $X=c_0$). Let T be a quotient map from l_1 onto X (cf. 2.f). By 2.a.2 T is strictly singular. The operator $T^*: X \to l_\infty$ is an isometry (into) and thus is not strictly singular.

We introduce now another important notion, namely that of the index of an operator.

Definition 2.c.6. Let $T: X \to Y$ be a bounded linear operator for which TX is closed. Put

$$\alpha(T)=\dim \ker T, \qquad \beta(T)=\dim Y/TX\,.$$

If either $\alpha(T)<\infty$ or $\beta(T)<\infty$ we define the *index* $i(T)$ of T by $i(T)=\alpha(T)-\beta(T)$. If $\alpha(T)$ and $\beta(T)$ are both finite (i.e. if $i(T)$ is defined and is finite) then T is called a *Fredholm operator*.

Observe that if the algebraic dimension of Y/TX is finite then, by the open mapping theorem, it follows that TX is closed in Y.

A typical example of a Fredholm operator of index $k>0$ is the operator, in c_0 or l_p, which sends $(a_1, a_2, \ldots)$ into $(a_{k+1}, a_{k+2}, \ldots)$.

We shall prove now several general results concerning Fredholm and strictly singular operators which are due to I. E. Gohberg and M. G. Krein [46] and to T. Kato [69]. Observe first that $T: X \to Y$ is a Fredholm operator if and only if there exist subspaces X_1 and B of X; Y_1 and C of Y, so that $X=X_1 \oplus B$, $Y=Y_1 \oplus C$, $T_{|B}=0$, $T_{|X_1}$ is an isomorphism onto Y_1 and $\dim B<\infty$, $\dim C<\infty$. With this notation we have $\alpha(T)=\dim B$ and $\beta(T)=\dim C$. This observation and simple finite-dimensional linear algebra prove the following proposition.

Proposition 2.c.7. (i) *Let $T: X \to Y$ be a Fredholm operator and let $S: X \to Y$ be a finite rank operator. Then, $T+S$ is a Fredholm operator and $i(T+S)=i(T)$.*

(ii) *Let $T_1: X \to Y$, $T_2: Y \to Z$ be Fredholm operators. Then, T_2T_1 is a Fredholm operator and $i(T_2T_1)=i(T_2)+i(T_1)$.*

(iii) *Let $T: X \to Y$ be a Fredholm operator. Then $T^*: Y^* \to X^*$ is also a Fredholm operator for which $\alpha(T^*)=\beta(T)$, $\beta(T^*)=\alpha(T)$ and therefore, $i(T^*)=-i(T)$.*

In the sequel we shall need the following finite dimensional lemma.

Lemma 2.c.8. *Let X be a Banach space and let B and C be subspaces of X with* $\dim B<\infty$ *and* $\dim C>\dim B$. *Then, there is an $x \in C$ such that $\|x\|=d(x, B)=1$.*

Proof. Assume first that X is a strictly convex Banach space. For every $x \in C$ with $\|x\|=1$ let $f(x)$ be the unique point in B for which $\|x-f(x)\|=d(x, B)$. Clearly, f is continuous and $f(-x)=-f(x)$. By the antipodal map theorem of

Borsuk [32, p. 347] and the fact that dim $C >$ dim B we infer that there is an x with $||x||=1$ such that $f(x)=0$, i.e. $d(x, B)=1$.

In the general case we observe that we may assume without loss of generality that dim $C=$ dim $B+1<\infty$ and that $X=$ span $\{B, C\}$. Hence, X is finite-dimensional and thus has an equivalent strictly convex norm $|||\cdot|||$. For every integer n let $||\cdot||_n$ be the strictly convex norm on X defined by $||x||_n=||x||+|||x|||/n$, where $||\cdot||$ is the given norm in X. By the first part of the proof we can find, for every integer n, an $x_n \in C$ with $||x_n||_n=1$ and $\inf\{||x_n-y||_n;\ y \in B\}=1$. If x is any limiting point of the sequence $\{x_n\}_{n=1}^\infty$ then $||x||=d(x, B)=1$, as desired. □

Proposition 2.c.9. *Let T: $X \to Y$ be an operator for which $i(T)$ is defined. There is a number $\lambda(T)>0$ so that if S: $X \to Y$ satisfies $||S||<\lambda(T)$ then*

(i) $\alpha(T+S) \leqslant \alpha(T)$.
(ii) *$T+S$ has a closed range and $\beta(T+S) \leqslant \beta(T)$.*
(iii) $i(T+S)=i(T)$.

Proof. Every operator T with a closed range admits a unique factorization of the form

$$X \xrightarrow{\pi_T} X/\ker T \xrightarrow{\tilde{T}} TX \xrightarrow{i_T} Y$$

where π_T is the natural quotient map, $\tilde{T}$ is an isomorphism and i_T is the inclusion map. Put $\lambda(T)=||\tilde{T}^{-1}||^{-1}$. Since $\lambda(T)=\lambda(T^*)$ we may assume without loss of generality that $\alpha(T)<\infty$ (otherwise, we work with T^*).

Let $||S||<\lambda(T)$ and suppose that $\alpha(T+S)>\alpha(T)$. Then, by 2.c.8, there is an $x \in \ker(T+S)$ such that $1=||x||=d(x, \ker T)=||\pi_T x||$. Hence,

$$||Tx||=||\tilde{T}\pi_T x|| \geqslant \lambda(T)\,.$$

On the other hand, $Sx=-Tx$ and thus $||Tx||=||Sx|| \leqslant ||S||$. This contradicts the assumption that $||S||<\lambda(T)$ and (i) is thus proved.

Assume that $T+S$ does not have a closed range. Let $0<\varepsilon<\lambda(T)-||S||$. By 2.c.4 there is an infinite dimensional subspace Z of X so that $||(T+S)_{|Z}||<\varepsilon$. By 2.c.8 there is an $z \in Z$ such that $1=||z||=||\pi_T z||$. Hence,

$$\lambda(T)>||S||+\varepsilon \geqslant ||Sz||+||(S+T)z|| \geqslant ||Tz||=||\tilde{T}\pi_T z|| \geqslant \lambda(T)\,,$$

and we arrived at a contradiction. If $\beta(T)=\infty$ this concludes the proof of (ii). If $\beta(T)<\infty$ we can apply (i) to T^* and thereby conclude the proof of (ii).

In order to prove (iii) it is enough to verify it with some $\varepsilon>0$ instead of the $\lambda(T)$ defined above. Indeed, once we show that $||U|| \leqslant \varepsilon_T$ implies $i(T+U)=i(T)$, we can apply this fact (by (i) and (ii)) to $T+tS$ for every $t \in [0, 1]$ provided of course $||S||<\lambda(T)$. The function $i(T+tS)$ is thus a continuous function of t which takes only discrete values; hence, it must be constant, i.e. $i(T)=i(T+S)$.

We prove the existence of such an $\varepsilon>0$ first in the case $\alpha(T)=0$, i.e. when T is an isomorphism. Let U: $X \to Y$ satisfy $||U||<\lambda(T)/2$ and let $x \in X$ with $||x||=1$.

Then $\|(T+U)x\| \geqslant \lambda(T) - \|Ux\| > \lambda(T)/2$ and hence, $\lambda(T+U) > \lambda(T)/2$. By (i) and (ii) we have $\alpha(T+U)=0$ and $\beta(T+U) \leqslant \beta(T)$. Since $\|U\| < \lambda(T+U)$ we may apply (ii) also to $T+U$ and deduce that $\beta(T) \leqslant \beta(T+U)$. Consequently, $\beta(T) = \beta(T+U)$ and thus also $i(T)=i(T+U)$.

The general case (i.e. $\alpha(T)>0$ but finite) reduces easily to the previous case. Indeed, let $X=X_1 \oplus \ker T$ and let $W=Y \oplus \ker T$. Let $T_1: X \to W$ be such that $T_{1|X_1}=T_{|X_1}$ and T_1 is the identity on $\ker T$. Then, $\alpha(T_1)=0$ and $\beta(T_1)=\beta(T)$, i.e. $i(T_1)=i(T)-\alpha(T)$. Let P be the natural projection from W onto Y and let $U: X \to Y$. Since $\alpha(T_1)=0$ we have that if $\|U\| < \lambda(T_1)/2$ then $i(T_1+U)=i(T_1)=i(T)-\alpha(T)$. By 2.c.7(i) the index of $T+U=P(T_1+U)$, as an operator from X to W, is the same as that of T_1+U. Hence, the index of $T+U$, as an operator from X to Y, is $i(T_1+U)+\alpha(T)$, i.e. is equal to $i(T)$. □

Proposition 2.c.10. *Let T: $X \to Y$ be an operator with closed range for which $\alpha(T) < \infty$. Let S: $X \to Y$ be strictly singular. Then, $\alpha(T+S) < \infty$, $T+S$ has a closed range and $i(T+S)=i(T)$.*

Proof. Let $X_1 \subset X$ be such that $X=X_1 \oplus \ker T$. Then, $T_{|X_1}$ is an isomorphism. If $\dim \ker (T+S)=\infty$ there would exist an infinite dimensional subspace Z of X_1 on which $T+S=0$, i.e. $S=-T$ is invertible. This contradicts the assumption that S is strictly singular. If $(T+S)X$ were not closed we could, by 2.c.4, find an infinite dimensional subspace Z of X_1 so that $\|(T+S)_{|Z}\| < \|T_{|X_1}^{-1}\|^{-1}$. But then $S_{|Z}$ is an isomorphism and we reached again a contradiction.

By what we have already proved the number $i(T+tS)$ is defined for all $t \in [0, 1]$. By 2.c.9, $i(T+tS)$ is a continuous function of t and thus a constant, i.e. $i(T)=i(T+S)$. □

Proposition 2.c.11. *Let T: $X \to X$ be a Fredholm operator. There is an $\varepsilon > 0$ so that $\alpha(T+\lambda I)$ is constant on the set $\{\lambda; 0 < |\lambda| \leqslant \varepsilon\}$.*

Proof. All the iterates of T have a closed range and thus $Y=\bigcap_{n=1}^{\infty} T^n X$ is a closed subspace of X (which may consist of the vector 0 only). Clearly, $TY \subset Y$; we shall prove that $TY=Y$. Observe first that since $\alpha(T) < \infty$ there is an integer k so that $T^k X \cap \ker T = T^n X \cap \ker T$, for every $n > k$. Let $y \in Y$ and let $u \in T^k X$ and $v \in T^n X$ (for some $n > k$) be such that $Tu=y=Tv$. Then, $u-v \in T^k X \cap \ker T \subset T^n X$, i.e. $u \in T^n X$. Since this is true for every $n > k$ we get that $u \in Y = \bigcap_{n=1}^{\infty} T^n X$.

Let $T_1 = T_{|Y}$. We have shown that $\beta(T_1)=0$ and thus, by 2.c.9, $\beta(T_1+\lambda I_Y)=0$ for λ sufficiently small. By using 2.c.9 again, we deduce that $i(T_1+\lambda I_Y)=i(T_1)$ and thus also $\alpha(T_1+\lambda I_Y)=\alpha(T_1)$ if λ is sufficiently small. To conclude the proof we have only to observe that, for $\lambda \neq 0$, $\ker (T+\lambda I) = \ker (T_1+\lambda I_Y)$. Indeed, if $Tx=-\lambda x$, with $\lambda \neq 0$, then $x \in \bigcap_{n=1}^{\infty} T^n X = Y$. □

The next lemma involves spectral properties of operators and it therefore requires that we work with a complex Banach space.

Lemma 2.c.12. *Let X be a complex Banach space and let P be a projection in X. Let $S: X \to X$ be a strictly singular operator and let $Q=P+S$. Then, the spectrum $\sigma(Q)$ of Q is countable and its only possible limit points are* 0 *and* 1.

Proof. Let C denote the complex plane. By 2.c.10 the operator $Q-\lambda I$ is a Fredholm operator of index 0 for every $\lambda \in C \sim \{0, 1\}$. By 2.c.11, any two points in $C \sim \{0, 1\}$ can be connected by a finite chain of discs such that $\alpha(Q-\lambda I)$ is constant on each disc (with the possible exception of the center of the disc). Consequently, $\alpha(Q-\lambda I)$ is constant on $C \sim \{0, 1\}$ except for a set of isolated points. This constant value is 0 since $Q-\lambda I$ is invertible for $|\lambda| > \|Q\|$. Since $i(Q-\lambda I)=0$ it follows that also $\beta(Q-\lambda I)=0$ except for a set of isolated points in $C \sim \{0, 1\}$. □

We are now ready to prove the theorem of I. S. Edelstein and P. Wojtaszczyk [35] and P. Wojtaszczyk [143].

Theorem 2.c.13. *Let X and Y be two Banach spaces so that every operator from Y into X in strictly singular. Let P be a projection of $X \oplus Y$ onto an infinite-dimensional subspace Z. Then there exist an automorphism τ_0 of $X \oplus Y$ and complemented subspaces X_0 of X and Y_0 of Y such that $\tau_0 Z = X_0 \oplus Y_0$.*

Proof. We shall first assume that X and Y are complex Banach spaces. As every operator from $X \oplus Y$ into itself, the projection P has a natural representation as a matrix $P=\begin{pmatrix} S_1 & S_3 \\ S_2 & S_4 \end{pmatrix}$, where $S_1: X \to X$, $S_2: X \to Y$, $S_3: Y \to X$ and $S_4: Y \to Y$.

Since S_3 is strictly singular the same is true for $P-Q$, where $Q=\begin{pmatrix} S_1 & 0 \\ S_2 & S_4 \end{pmatrix}$. We want to replace Q by a projection $\tilde{P}$ having a similar matricial representation (i.e. zero in the upper right corner). We achieve this by putting

$$\tilde{P}=\frac{1}{2\pi i} \int_\Gamma R(\lambda, Q)\, d\lambda\,,$$

where $R(\lambda, Q)$ is the resolvent of Q and Γ is a closed simple curve which does not intersect $\sigma(Q)$, has the number 1 in its interior and 0 and 2 in its exterior. Such a Γ exists by 2.c.12.

The operator $\tilde{P}$ is a projection in $X \oplus Y$ so that $\sigma(Q_{|\tilde{P}(X\oplus Y)}) \subset$ interior of Γ and $\sigma(Q_{|(I-\tilde{P})(X\oplus Y)}) \subset$ exterior of Γ (cf. [33, VII.3.11]). Moreover, it is easily checked that $\tilde{P}=\begin{pmatrix} P_1 & 0 \\ P_2 & P_4 \end{pmatrix}$, where P_1 is a projection in X, P_4 a projection in Y and P_2 an operator from X into Y.

The operator $I+\tilde{P}-Q$ on $X \oplus Y$ leaves $\tilde{P}(X \oplus Y)$ invariant and is invertible there (since $(I+\tilde{P}-Q)_{|\tilde{P}(X\oplus Y)}=(2I-Q)_{|\tilde{P}(X\oplus Y)}$ and 2 is in the exterior of Γ). This operator also leaves $(I-\tilde{P})(X \oplus Y)$ invariant and is invertible on this subspace (since 1 is in the interior of Γ). Hence, $I+\tilde{P}-Q$ is an invertible operator on $X \oplus Y$ and thus, by 2.c.10, $T_1=I+\tilde{P}-P=(I+\tilde{P}-Q)+(Q-P)$ is a Fredholm operator of index 0 on $X \oplus Y$.

We shall show next that $T_1 Z$ is a subspace of finite codimension in $\tilde{P}(X \oplus Y)$ Since $T_1 P = \tilde{P}P$ it follows that $T_1 Z = T_1 P(X \oplus Y) \subset \tilde{P}(X \oplus Y)$. To show that $T_1 Z$ is of finite codimension in $\tilde{P}(X \oplus Y)$ is the same as to show that Z is of finite codimension in $T_1^{-1}(\tilde{P}(X \oplus Y))$ (this follows from the fact that T_1 is a Fredholm operator; in particular, it is obvious that $T_1 Z$ is closed). If $\dim T_1^{-1}(\tilde{P}(X \oplus Y))/Z = \infty$ there would exist an infinite dimensional subspace W of $X \oplus Y$ so that $P_{|W} = 0$, $T_{1|W}$ is an isomorphism and $T_1 W \subset \tilde{P}(X \oplus Y)$. For $w \in W$ we have $T_1 w = w + \tilde{P}w$ and thus $w = T_1 w - \tilde{P}w \in \tilde{P}(X \oplus Y)$, i.e. $W \subset \tilde{P}(X \oplus Y)$. On $\tilde{P}(X \oplus Y)$ the operator Q is invertible (since 0 is in the exterior of Γ). But $Q_{|W} = (Q-P)_{|W}$ and, since $Q-P$ is strictly singular, we arrived at a contradiction.

Consider now the projection $\tilde{\tilde{P}} = \begin{pmatrix} P_1 & 0 \\ 0 & P_4 \end{pmatrix}$ on $X \oplus Y$. The next step in the proof is to construct an automorphism T_2 of $X \oplus Y$ so that $T_2 \tilde{P} T_2^{-1} = \tilde{\tilde{P}}$. We put $S = \tilde{P} - \tilde{\tilde{P}}$ and notice that $S^2 = S\tilde{P}S = \tilde{P}S\tilde{P} = 0$. Hence, $\tilde{S} = S\tilde{P} - \tilde{P}S$ satisfies

$$\tilde{S}^2 = S\tilde{P}S\tilde{P} - S\tilde{P}S - \tilde{P}S^2\tilde{P} + \tilde{P}S\tilde{P}S = 0\,.$$

It follows that $T_2 = I - \tilde{S}$ is an automorphism of $X \oplus Y$ with $T_2^{-1} = I + \tilde{S}$ and we have

$$T_2 \tilde{P} T_2^{-1} = (I - \tilde{S})\tilde{P}(I + \tilde{S}) = (\tilde{P} - S)\tilde{P}(\tilde{P} - S) = \tilde{\tilde{P}}\tilde{P}\tilde{\tilde{P}} = \tilde{\tilde{P}}\,.$$

The operator $T = T_2 T_1 T_2^{-1}$, which is also a Fredholm operator of index 0 in $X \oplus Y$, maps Z into a subspace of finite codimension of $\tilde{\tilde{P}}(X \oplus Y) = X_1 \oplus Y_1$, where $X_1 = P_1 X$ and $Y_1 = P_4 Y$. Since T is a Fredholm operator of index 0 there is a finite rank operator R on $X \oplus Y$ so that $\tau_1 = T + R$ is an automorphism of $X \oplus Y$. There are finite dimensional subspaces B and C of $X \oplus Y$ so that $\tau_1(Z) \oplus B = X_1 \oplus Y_1 \oplus C$. In particular, $\tau_1(Z) \subset X_2 \oplus Y_2$ with $X_2 = X_1 \oplus C_X$ and $Y_2 = Y_1 \oplus C_Y$ where C_X and C_Y are suitable finite dimensional subspaces of X, respectively Y.

We recall now the fact that if V_1 and V_2 are two closed subspaces of a Banach space U with $\dim U/V_1 = \dim U/V_2 < \infty$ then there is an automorphism of U which maps V_1 onto V_2 (the automorphism can be chosen to be the identity on $V_1 \cap V_2$ since $\dim V_1/(V_1 \cap V_2) = \dim V_2/(V_1 \cap V_2) < \infty$).

Hence, if X_0 and Y_0 are subspaces of finite codimension in X_2, respectively Y_2, so that $\dim(X_2 \oplus Y_2)/\tau_1(Z) = \dim(X_2 \oplus Y_2)/(X_0 \oplus Y_0)$ then there is an automorphism τ_2 of $X_2 \oplus Y_2$ onto itself which maps $\tau_1(Z)$ onto $X_0 \oplus Y_0$. Since $X_2 \oplus Y_2$ is complemented in $X \oplus Y$, τ_2 can be extended to an automorphism $\hat{\tau}_2$ of $X \oplus Y$. Consequently, $\tau_0 = \hat{\tau}_2 \tau_1$ is an automorphism of $X \oplus Y$ so that $\tau_0(Z) = X_0 \oplus Y_0$. Since X_0 is complemented in X and Y_0 is complemented in Y this concludes the proof in the complex case.

We point out now the modification needed in the case where the scalars are real. If X and Y are real Banach spaces we pass to their natural "complexifications" $\hat{X}$, respectively $\hat{Y}$. For example, $\hat{X} = \{(u, v);\, u, v \in X\}$, where we put $(a+ib)(u, v) = (au - bv, av + bu)$ and $\|(u, v)\| = \max\{(\|au - bv\|^2 + \|av + bu\|^2)^{1/2};\, a^2 + b^2 = 1\}$. The given projection P on $X \oplus Y$ induces in an obvious way a projection $\hat{P}$ on $\hat{X} \oplus \hat{Y}$.

The proof proceeds as in the complex case. We have only to ensure that the operator $(2\pi i)^{-1} \int_\Gamma R(\lambda, \hat{Q})\, d\lambda$, used in the proof, is of the form $\tilde{\tilde{P}}$, for a suitable operator $\tilde{P}$ on $X \oplus Y$. This is ensured if, e.g. we take as Γ a rectangle symmetric with respect to the real axis. □

Remarks. 1. Observe that the assumptions on X and Y in 2.c.13 are satisfied in particular if X and Y are totally incomparable. In this case the proof 2.c.13, can be simplified somewhat. Indeed, we could take here $Q=\begin{pmatrix} S_1 & 0 \\ 0 & S_4 \end{pmatrix}$ and get that $\tilde{P}=\begin{pmatrix} P_1 & 0 \\ 0 & P_4 \end{pmatrix}$. So the step of producing $\tilde{\tilde{P}}$ becomes unnecessary in this case.

2. It follows from 2.c.13 that if Y is an arbitrary Banach space and X is l_p, $1 \leq p < \infty$ or c_0 then any complemented subspace of $X \oplus Y$ is of the form Y_0 or $X \oplus Y_0$ for some complemented subspace Y_0 of Y. Indeed, if there is a non-strictly singular operator from Y into X then, by 2.a.2, Y has a complemented subspace isomorphic to X and thus $Y \oplus X \approx Y$.

From 2.c.13 and 2.a.3 we get, by an obvious induction argument, the following result

Theorem 2.c.14. *Let $\{X_i\}_{i=1}^m$ be distinct spaces out of the set $\{c_0\} \cup \{l_p;\, 1 \leq p < \infty\}$. Then any infinite-dimensional complemented subspace of $X_1 \oplus X_2 \oplus \cdots \oplus X_m$ is isomorphic to $X_{i_1} \oplus X_{i_2} \oplus \cdots \oplus X_{i_k}$, for some $1 \leq i_1 < i_2 < \cdots < i_k \leq m$.*

There is no analogue to 2.c.14 if we consider general subspaces instead of complemented subspaces. We shall discuss this situation in Section 4.d.

We shall apply now 2.c.14 to describe the structure of unconditional bases in spaces of the form $l_{p_1} \oplus l_{p_2} \oplus \cdots \oplus l_{p_m}$.

Theorem 2.c.15 [35]. *Let $\{X_i\}_{i=1}^m$ be distinct spaces out of the set $\{c_0\} \cup \{l_p;\, 1 \leq p < \infty\}$. Let $\{z_n\}_{n=1}^\infty$ be an unconditional basis of $X_1 \oplus X_2 \oplus \cdots \oplus X_m$. Then we can partition the integers N into m disjoint sets $\{N_i\}_{i=1}^m$ such that $[z_n]_{n \in N_i}$ is isomorphic to X_i, $i=1,\ldots, m$.*

This theorem reduces the question of describing the unconditional bases in the direct sum $\sum_{i=1}^m \oplus l_{p_i}$ to that of describing the unconditional bases in each component. In particular, we get by 2.c.15 and 2.b.9 (cf. also the remark preceding it) the following result.

Theorem 2.c.16. *Let Z be one of the spaces, $l_1 \oplus l_2$, $c_0 \oplus l_1$, $c_0 \oplus l_2$ or $c_0 \oplus l_1 \oplus l_2$. Let $\{z'_n\}_{n=1}^\infty$ and $\{z''_n\}_{n=1}^\infty$ be two normalized unconditional bases of Z. Then there is a permutation π of the positive integers such that $\{z'_n\}_{n=1}^\infty$ is equivalent to $\{z''_{\pi(n)}\}_{n=1}^\infty$.*

The proof of 2.c.15 is by induction on m. The argument needed in the general induction step is very similar to that used in the case $m=2$. We present here only the proof for $m=2$. Also, there is some difference between the reflexive and non-reflexive case; we shall treat here only the reflexive case. The main step of the proof consists of the following lemma.

Lemma 2.c.17. *Let $1<p<r<\infty$ and let $z_n=(x_n, y_n)$, $n=1, 2,\ldots$ be a normalized unconditional basis of $l_p \oplus l_r$ with unconditional constant K ($l_p \oplus l_r$ is normed by $\|(x, y)\|=\|x\|+\|y\|$ for $x \in l_p$, $y \in l_r$). Let $\{n_k\}_{k=1}^\infty$ be a subsequence of the integers so that $\alpha=\sup_k \|x_{n_k}\|<K^{-1}$. Then, $[z_{n_k}]_{k=1}^\infty$ is isomorphic to l_r.*

Proof. The proof we give here is very similar to that of 2.a.11. We claim first that, for every sequence of scalars $\{\lambda_k\}_{k=1}^\infty$ such that $\sum_{k=1}^\infty \lambda_k y_{n_k}$ converges unconditionally, the series $\sum_{k=1}^\infty \lambda_k z_{n_k}$ converges too. To prove this let $\{a_n^k\}_{n=1}^\infty$ and $\{b_n^k\}_{n=1}^\infty$ be scalars so that

$$(x_{n_k}, 0)=\sum_{n=1}^\infty a_n^k z_n, \qquad (0, y_{n_k})=\sum_{n=1}^\infty b_n^k z_n, \quad k=1, 2,\ldots.$$

Then, $a_{n_k}^k+b_{n_k}^k=1$ and, since $|a_{n_k}^k|\leqslant K\|x_{n_k}\|$, we get that $b_{n_k}^k \geqslant 1-K\alpha>0$ for every k. By Remark 2 following 1.c.8 we deduce that

$$\Big\|\sum_{k=1}^m \lambda_k b_{n_k}^k z_{n_k}\Big\| \leqslant K \sup_{\pm} \Big\|\sum_{j=1}^m \pm\lambda_j y_{n_j}\Big\|.$$

Thus, if $\{\lambda_k\}_{k=1}^\infty$ is such that $\sum_{k=1}^\infty \lambda_k y_{n_k}$ converges unconditionally the series $\sum_{k=1}^\infty \lambda_k b_{n_k}^k z_{n_k}$, and therefore also $\sum_{k=1}^\infty \lambda_k z_{n_k}$, converge too.

If $[z_{n_k}]_{k=1}^\infty$ is not isomorphic to l_r it would follow from 2.c.14 that $[z_{n_k}]_{k=1}^\infty$ contains a subspace isomorphic to l_p. Hence, by 1.a.9, there is a normalized block basis $w_j=\sum_{k\in\sigma_j} \alpha_k z_{n_k}$, $j=1, 2,\ldots$ of $\{z_{n_k}\}_{k=1}^\infty$ which is equivalent to the unit vector basis of l_p (here $\{\sigma_j\}_{j=1}^\infty$ denotes a sequence of disjoint finite subsets of the integers). Let Q be the natural projection from $l_p \oplus l_r$ onto l_r and let $\{P_m\}_{m=1}^\infty$ be the projections on l_r which are associated to the unit vector basis. Since $z_n \xrightarrow{w} 0$ we get that, for every m, $\lim_{n\to\infty} \|P_m Q_{|[z_k]_{k=n}^\infty}\|=0$. Hence, by passing to a subsequence if necessary, we may assume that there exists a sequence $\{m_j\}_{j=1}^\infty$ of integers so that $\|(P_{m_{j+1}}-P_{m_j})Qz-Qz\|\leqslant 2^{-j}\|z\|$, for every $z\in B_j=[z_{n_k}]_{k\in\sigma_j}$. In particular, if $u_j\in B_j$ for $j=1, 2,\ldots, s$ then

$$\Big\|\sum_{j=1}^s Qu_j\Big\|\leqslant \sum_{j=1}^s 2^{-j}\|u_j\|+\Big(\sum_{j=1}^s \|Qu_j\|^r\Big)^{1/r}.$$

Let $\{\beta_j\}_{j=1}^\infty$ be a sequence of scalars so that $\sum_{j=1}^\infty |\beta_j|^r<\infty$ and let $\{\theta_k\}_{k=1}^\infty$ be a sequence of signs. Put $u_j=\sum_{k\in\sigma_j}\theta_k\alpha_k z_{n_k}$, $j=1,2,\ldots$. Then, $\|u_j\|\leqslant K\|w_j\|\leqslant K$ and hence,

$$\left\|\sum_{j=1}^{\infty}\sum_{k\in\sigma_j}\beta_j\theta_k\alpha_k y_{n_k}\right\|=\left\|\sum_{j=1}^{\infty}\beta_j Qu_j\right\|\leqslant\sum_{j=1}^{\infty}K2^{-j}|\beta_j|+K\left(\sum_{j=1}^{\infty}|\beta_j|^r\right)^{1/r}.$$

In other words, $\sum_{j=1}^{\infty}\sum_{k\in\sigma_j}\beta_j\alpha_k y_{n_k}$ converges unconditionally. By the first part of the proof this implies that $\sum_{j=1}^{\infty}\beta_j w_j=\sum_{j=1}^{\infty}\sum_{k\in\sigma_j}\beta_j\alpha_k z_{n_k}$ converges. This however contradicts the assumption that $\{w_j\}_{j=1}^{\infty}$ is equivalent to the unit vector basis of l_p and that $p<r$. □

Lemma 2.c.18. *Let $1<p<r<\infty$ and let $z_n=(x_n,y_n)$, $n=1,2,\ldots$, be a normalized unconditional basis of $l_p\oplus l_r$. Let $\{n_k\}_{k=1}^{\infty}$ be a subsequence of the integers so that $\|x_{n_k}\|\geqslant\alpha$, for some $\alpha>0$ and every integer k. Then $[z_{n_k}]_{k=1}^{\infty}$ is isomorphic to l_p.*

Proof. Every subsequence $\{n'_k\}_{k=1}^{\infty}$ of $\{n_k\}_{k=1}^{\infty}$ has, by 1.a.12 and 2.a.1, a subsequence $\{n''_k\}_{k=1}^{\infty}$ such that $\{x_{n''_k}\}_{k=1}^{\infty}$ and $\{y_{n''_k}/\|y_{n''_k}\|\}_{k=1}^{\infty}$ are equivalent to the unit vector bases of l_p, respectively l_r (if $y_{n''_k}=0$ the situation is even simpler). For such a sequence $\{n''_k\}_{k=1}^{\infty}$ the basic sequence $\{z_{n''_k}\}_{k=1}^{\infty}$ is also equivalent to the unit vector basis of l_p. It follows from this remark and 2.c.14 that $[z_{n_k}]_{k=1}^{\infty}$ is isomorphic to either l_p or $l_p\oplus l_r$.

Assume that $[z_{n_k}]_{k=1}^{\infty}$ is isomorphic to $l_p\oplus l_r$ and let $z^*_{n_k}=(y^*_{n_k},x^*_{n_k})$, $k=1,2,\ldots$ be the functionals biorthogonal to z_{n_k} in $(l_p\oplus l_r)^*=l_r^*\oplus l_p^*$. Since $[z^*_{n_k}]_{k=1}^{\infty}\approx([z_{n_k}]_{k=1}^{\infty})^*\approx l_r^*\oplus l_p^*$ it follows by 2.c.17, that $\limsup_k\|y^*_{n_k}\|>0$ and hence, by the first part of this proof, there is a subsequence $\{n'_k\}_{k=1}^{\infty}$ of $\{n_k\}_{k=1}^{\infty}$ such that $[z^*_{n'_k}]_{k=1}^{\infty}$ is isomorphic to l_r^*. Hence, $[z_{n'_k}]_{k=1}^{\infty}$ is isomorphic to l_r. Using again the first part of this proof we arrive at a contradiction ($\{z_{n'_k}\}_{k=1}^{\infty}$ cannot have a subsequence equivalent to the unit vector basis of l_p). □

It is now obvious how to conclude the *proof of* 2.c.15 (for $l_p\oplus l_r$ with $1<p<r<\infty$). We simply take $N_1=\{n;\ \|x_n\|>1/2K\}$ and $N_2=\{n;\ \|x_n\|\leqslant 1/2K\}$. Then, $[z_n]_{n\in N_1}$ is isomorphic to l_p and $[z_n]_{n\in N_2}$ is isomorphic to l_r. □

A stronger version of 2.c.15 was proved recently by P. Wojtaszczyk [143].

We conclude this section by mentioning (an admittedly vaguely stated) problem arising from 2.c.16.

Problem 2.c.19. *Describe all the separable Banach spaces which have, up to equivalence and to a permutation, a unique normalized unconditional basis.*

d. Subspaces of c_0 and l_p and the Approximation Property, Complementably Universal Spaces

This section is devoted mainly to the study of subspaces of c_0 and l_p, $1<p<\infty$. We present first a result of W. B. Johnson and M. Zippin on the structure of subspaces of c_0 or l_p, $1<p<\infty$ having a shrinking basis. The main part of this section is devoted to Davie's proof of the fact that c_0 and l_p, $2<p<\infty$ have subspaces which fail to have the approximation property. The section concludes with some general results concerning universal spaces and the approximation property.

We start with the result of Johnson and Zippin [63]. This result gives actually more information than indicated in the preceding paragraph. It describes even the structure of subspaces of quotients of c_0 or l_p, $1<p<\infty$ and the assumptions concerning the existence of a basis can be somewhat relaxed. Before stating this result let us observe that the notions of a quotient of a subspace and that of a subspace of a quotient coincide. Let X be a subspace of a quotient space Y of Z and let $T: Z\to Y$ be the quotient map. Then X is a quotient of the subspace $T^{-1}X$ of Z. Conversely, assume that X is a quotient space of a subspace Y of Z and let $T\colon Y\to X$ be the quotient map. Then X is isometric to $Y/\ker T$ which is a subspace of the quotient space $Z/\ker T$ of Z.

Theorem 2.d.1. *Let $\{B_n\}_{n=1}^\infty$ be a sequence of finite-dimensional Banach spaces. Let X be an infinite dimensional subspace of a quotient space of $\left(\sum_{n=1}^\infty \oplus B_n\right)_p$, $1<p<\infty$ (resp. $\left(\sum_{n=1}^\infty \oplus B_n\right)_0$) having a shrinking* F.D.D. *Then X is isomorphic to $\left(\sum_{k=1}^\infty \oplus D_k\right)_p$ (resp. $\left(\sum_{k=1}^\infty \oplus D_k\right)_0$), for a suitable sequence of finite dimensional spaces $\{D_k\}_{k=1}^\infty$.*

Proof. We prove the theorem first in the case $1<p<\infty$. Let $\{C_k\}_{k=1}^\infty$ be an F.D.D. of X and let $T\colon \left(\sum_{n=1}^\infty \oplus B_n\right)_p \to Y$ be a quotient map, where $Y\supset X$. By 1.g.4(b) there exists a constant K, a blocking $\{B'_n\}_{n=1}^\infty$ of $\{B_n\}_{n=1}^\infty$ and a blocking $\{C'_k\}_{k=1}^\infty$ of $\{C_k\}_{k=1}^\infty$ so that, for every $x\in C'_k$, there is a $u\in B'_k\oplus B'_{k+1}$ with $\|u\|\leq K\|x\|$ so that $\|Tu-x\|\leq 2^{-k}\|x\|$. Let $x=\sum_{k=1}^\infty x_k$ be an element in X with $x_k\in C'_k$, for every k, and assume that $\sum_{k=1}^\infty \|x_k\|^p<\infty$. For $k=1, 2,\dots$ let $u_k\in B'_k\oplus B'_{k+1}$ satisfy $\|u_k\|\leq K\|x_k\|$ and $\|Tu_k-x_k\|\leq 2^{-k}\|x_k\|$. Put $u=u_1+u_3+u_5+\cdots$. Then, $\|u\|=\left(\sum_{i=1}^\infty \|u_{2i-1}\|^p\right)^{1/p}\leq K\left(\sum_{k=1}^\infty \|x_k\|^p\right)^{1/p}$. Since $\left\|Tu-\sum_{i=1}^\infty x_{2i-1}\right\|\leq \sum_{k=1}^\infty 2^{-k}\|x_k\|$ we get that $\left\|\sum_{i=1}^\infty x_{2i-1}\right\|\leq \|u\|+\sum_{k=1}^\infty 2^{-k}\|x_k\|$ and thus $\left\|\sum_{i=1}^\infty x_{2i-1}\right\|\leq M\left(\sum_{k=1}^\infty \|x_k\|^p\right)^{1/p}$, for a suitable constant M. The same calculation shows that this estimate holds also for $\left\|\sum_{i=1}^\infty x_{2i}\right\|$ and thus

$$(*)\qquad \left\|\sum_{k=1}^\infty x_k\right\|\leq 2M\left(\sum_{k=1}^\infty \|x_k\|^p\right)^{1/p}.$$

In order to obtain an estimate on $\|x\|$ from below we pass to the dual. The space X^* is a subspace of a quotient space of $\left(\sum_{k=1}^\infty \oplus (B'_n)^*\right)_q$, where $q^{-1}+p^{-1}=1$. By applying 1.g.4(b) once more and repeating the argument above it follows that there is a blocking $\{D_k\}_{k=1}^\infty$ of $\{C'_k\}_{k=1}^\infty$ so that, whenever $x^* = \sum_{k=1}^\infty x_k^*$ with $x_k^* \in D_k^*$, $k=1, 2,\ldots$, we have

$$(**) \qquad \|x^*\| \leq M'\left(\sum_{k=1}^\infty \|x_k^*\|^q\right)^{1/q}$$

for a suitable constant M'. Clearly, (*) remains true if we consider decompositions with respect to the blocking $\{D_k\}_{k=1}^\infty$, i.e. if $x = \sum_{k=1}^\infty x_k$ with $x_k \in D_k$, $k=1, 2,\ldots$ (see the remark following 1.g.4). From (*) and (**) it follows that X is isomorphic to $\left(\sum_{k=1}^\infty \oplus D_k\right)_p$.

The proof for the case $\left(\sum_{n=1}^\infty \oplus B_n\right)_0$ is simpler. In this case it is enough to do the first step. The inequality (*) reads $\left\|\sum_{k=1}^\infty x_k\right\| \leq 2M \sup_k \|x_k\|$ while the inequality $\left\|\sum_{k=1}^\infty x_k\right\| \geq \sup_k \|x_k\|/M'$, for some M', follows from the fact that $\{C_k\}_{k=1}^\infty$, and thus $\{C'_k\}_{k=1}^\infty$, are F.D.D.'s. We remark that only in this case is the assumption that $\{C_k\}_{k=1}^\infty$ is shrinking a real restriction. In the case $1<p<\infty$ this assumption holds automatically since X is reflexive. □

Remarks. 1. Theorem 2.d.1 is no longer true if $p=1$. This is completely obvious if we consider subspaces of quotients or even only quotients since every separable Banach space is a quotient space of l_1. If we consider only subspaces of l_1, Theorem 2.d.1 is valid for a trivial reason—no infinite dimensional subspace of l_1 has a shrinking F.D.D. Without the requirement that the F.D.D. be shrinking the theorem fails even for subspaces. We mentioned already in Section 1.d that $\{e_n-(e_{2n}+e_{2n+1})/2\}_{n=1}^\infty$ is a basic sequence in l_1 whose span is not isomorphic to a conjugate space and thus is not of the form $\left(\sum_{n=1}^\infty \oplus D_n\right)_1$.

2. It is unknown whether the assumption of the existence of a F.D.D. can be replaced by the weaker assumption that X has the approximation property. This is a special instance of Problem 1.e.12.

Theorem 2.d.1 does not answer of course all the natural questions on subspaces of l_p, $1<p<\infty$ or c_0, even on those subspaces which have a basis. For example, the answer to the following problem is unknown even under the assumption of the existence of a basis.

Problem 2.d.2. *Let X be an infinite-dimensional Banach space and let $1<p<\infty$ be such that X is a subspace, as well as a quotient space, of l_p. Is X isomorphic to l_p?*

We pass now to the construction of subspaces of c_0 and l_p, $2<p<\infty$ which do not have the approximation property. We reproduce here the proof due to A. M.

Davie [20, 21]. Our approach will be to construct an infinite matrix of the type appearing in 1.e.8 (this is the main step) and then deduce from the existence of such a matrix the existence of the desired subspaces.

Theorem 2.d.3. *There exists an infinite matrix $A=(a_{i,j})_{i,j=1}^{\infty}$ of scalars such that for every i, $a_{i,j}\neq 0$ only for finitely many indices j, $\sum_{i=1}^{\infty}(\max_j |a_{i,j}|)^r<\infty$ for every $r>2/3$, $A^2=0$ and* trace $A=\sum_{i=1}^{\infty} a_{i,i}\neq 0$.

Proof. We shall work with complex scalars since this is somewhat more convenient. For every $k=0, 1, 2,\ldots$ we let U_k be a unitary matrix of order $3\cdot 2^k$ (the specific choice of U_k will be made later). We partition U_k as follows $\begin{pmatrix} 2^{(k+1)/2}P_k \\ 2^{k/2}Q_k \end{pmatrix}=U_k$, where $2^{(k+1)/2}P_k$ is the $2^{k+1}\times 3\cdot 2^k$ matrix consisting of the first 2^{k+1} rows of U_k and $2^{k/2}Q_k$ is the $2^k\times 3\cdot 2^k$ matrix consisting of the last 2^k rows of U_k. Since $U_kU_k^*=I_{3\cdot 2^k}$ (where I_m denotes the identity matrix of order m) we get that

$$P_kP_k^*=2^{-(k+1)}I_{2^{k+1}},\qquad Q_kQ_k^*=2^{-k}I_{2^k},\qquad P_kQ_k^*=Q_kP_k^*=0,$$
$$k=0, 1, 2,\ldots.$$

Consider now the matrix

$$A=(a_{i,j})$$
$$=\begin{pmatrix} P_0^*P_0 & P_0^*Q_1 & 0 & 0 & 0 & \cdots \\ -Q_1^*P_0 & P_1^*P_1-Q_1^*Q_1 & P_1^*Q_2 & 0 & 0 & \cdots \\ 0 & -Q_2^*P_1 & P_2^*P_2-Q_2^*Q_2 & P_2^*Q_3 & 0 & \cdots \\ 0 & 0 & -Q_3^*P_2 & P_3^*P_3-Q_3^*Q_3 & P_3^*Q_4 & \cdots \\ \cdot & \cdot & \cdot & \cdot & \cdot & \cdot \end{pmatrix}$$

It is easily verified that $A^2=0$. Clearly, trace $(P_k^*P_k-Q_k^*Q_k)=0$ for $k=1, 2,\ldots$ and thus trace $A=$ trace $P_0^*P_0=1$. We shall construct the U_k's so that each element in the k'th block of rows in A (i.e. each element of the matrices $-Q_k^*P_{k-1}$, $P_k^*P_k-Q_k^*Q_k$ and $P_k^*Q_{k+1}$) is $\leq C(k+1)^{1/2}2^{-3k/2}$, where C is a constant. Since the k'th block contains $3\cdot 2^k$ rows this will imply that, for $r>2/3$,

$$\sum_{i=1}^{\infty}\left(\max_j |a_{i,j}|\right)^r\leq\sum_{k=0}^{\infty}3\cdot 2^kC^r(k+1)^{r/2}2^{-3rk/2}<\infty.$$

Observe that, since $P_k^*Q_{k+1}=(Q_{k+1}^*P_k)^*$, it is enough to examine only the matrices $Q_k^*P_{k-1}$ and $P_k^*P_k-Q_k^*Q_k$. For the construction of the $\{U_k\}_{k=0}^{\infty}$ we need two lemmas.

Lemma 2.d.4. (a) *Let $\{\alpha_j\}_{j=1}^{n}$ be complex numbers and let $\{\theta_j\}_{j=1}^{n}$ be independent* *random variables each taking the values* $+1$ *and* -1 *with probability* $1/2$. *Then there is an absolute constant* K *so that*

$$\text{Probability}\left\{\left|\sum_{j=1}^{n} \theta_j\alpha_j\right| > K\left(\sum_{j=1}^{n} |\alpha_j|^2 \log n\right)^{1/2}\right\} < Kn^{-3}.$$

(b) *The same assertion as* (a) *with the only difference that each* θ_j *takes the value* 2 *with probability* $1/3$ *and* -1 *with probability* $2/3$.

Proof. We shall prove assertion (a) only; the proof of (b) is very similar. By considering separately the real and the imaginary parts it follows that it is enough to prove (a) for real $\{\alpha_j\}_{j=1}^n$. Also, there is no loss of generality to assume that $\sum_{j=1}^n \alpha_j^2=1$. We write $\theta=(\theta_1, \theta_2,\ldots, \theta_n)$ and $f(\theta)=\left|\sum_{j=1}^n \theta_j\alpha_j\right|$. We denote by E the expectation with respect to the probability distribution of θ (i.e. the average over all 2^n possible choices of θ). Then, for every $\lambda>0$,

$$E(e^{\lambda f(\theta)}) \leq E(e^{\lambda \sum_j \theta_j\alpha_j} + e^{-\lambda \sum_j \theta_j\alpha_j}) = 2\prod_{j=1}^{n} (e^{\lambda\alpha_j}+e^{-\lambda\alpha_j})/2\,.$$

Since $(e^x+e^{-x})/2 \leq e^{x^2}$ for real x we get that

$$\text{Probability}\{\lambda f(\theta)-\lambda^2-3\log n>0\} \leq E(e^{\lambda f(\theta)-\lambda^2-3\log n}) \leq 2n^{-3}\,.$$

The desired result follows by taking, e.g. $\lambda=(3\log n)^{1/2}$. □

Lemma 2.d.5. *Let* G *be an Abelian group of order* $3\cdot 2^k$ *for some integer* k. *Then it is possible to divide the characters of* G *into two sets: one consisting of* 2^{k+1} *elements, denoted by* $\{\tau_j\}_{j=1}^{2^{k+1}}$, *and another consisting of* 2^k *elements, denoted by* $\{\sigma_j\}_{j=1}^{2^k}$, *so that for every* $g\in G$

$$\left|2\sum_{j=1}^{2^k} \sigma_j(g) - \sum_{j=1}^{2^{k+1}} \tau_j(g)\right| \leq L(k+1)^{1/2}2^{k/2}$$

where L *is an absolute constant.*

Recall that a character γ of an Abelian group G is a homomorphism from G into the multiplicative group $\{z;\ |z|=1\}$ in the plane. An Abelian group of order m has exactly m characters and any two different characters are orthogonal, i.e. $\sum_{g\in G} \gamma(g)\gamma'(g)=0$ if $\gamma\neq\gamma'$. For our purpose it is enough to consider a cyclic group G of order m. In this case the characters of G are given by $\gamma_k(g_0^l)=e^{-2\pi ikl/m}$, $0\leq k<m$, $0\leq l<m$, where g_0 is a generator of G.

Proof. Let $\{\gamma_j\}_{j=1}^{3\cdot 2^k}$ be an enumeration of all the characters of G. By 2.d.4(b) there is a choice of $\{\theta_j\}_{j=1}^{3\cdot 2^k}$ such that θ_j is either 2 or -1, for every j, and so that, for some constant L,

$$\left|\sum_{j=1}^{3\cdot 2^k} \theta_j \gamma_j(g)\right| \leq L(k+1)^{1/2} 2^{k/2}, \quad \text{for every } g \in G.$$

The number of these inequalities is $n = 3\cdot 2^k$ and thus such $\{\theta_j\}_{j=1}^n$ exist, by 2.d.4(b), whenever $n < n^3/K$, i.e. for $n > n_0 = \sqrt{K}$. (For $n < n_0$ there is nothing to prove.) By taking, in particular, g to be the identity element of G we get that $\left|\sum_{j=1}^{3\cdot 2^k} \theta_j\right| \leq L(k+1)^{1/2} 2^{k/2}$. Thus, by changing if necessary at most $2L(k+1)^{1/2}2^{k/2}$ of the θ_j and replacing L by $2L$, we may assume that $\sum_{j=1}^{3\cdot 2^k} \theta_j = 0$, i.e. exactly 2^{k+1} of the θ_j are equal to -1 and 2^k of the θ_j are equal to 2. Those γ_j for which the corresponding θ_j are equal to -1 are denoted by $\{\tau_j\}_{j=1}^{2^{k+1}}$ and the rest by $\{\sigma_j\}_{j=1}^{2^k}$. From the choice of $\{\theta_j\}_{j=1}^{3\cdot 2^k}$ it is clear that this partition of the characters into two sets has the required properties. □

We return to the proof of 2.d.3. We make the set $\{1, 2, \ldots, 3\cdot 2^k\}$ into an Abelian group G_k and define the τ_j^k and σ_j^k as in 2.d.5. We let the rows of P_k be $3^{-1/2}2^{-(2k+1)/2}\tau_j^k$, $1 \leq j \leq 2^{k+1}$ (i.e. the entries of P_k are $3^{-1/2}2^{-(2k+1)/2}\tau_j^k(i)$, where $i \in G_k$, i.e. $1 \leq i \leq 3\cdot 2^k$) and let the rows of Q_k be $3^{-1/2}2^{-k}\theta_j^k\sigma_j^k$, $1 \leq j \leq 2^k$, where the θ_j^k are either $+1$ or -1 and will be determined below. Whatever is the choice of $\{\theta_j^k\}_{j=1}^{2^k}$ the matrix U_k defined in this manner is unitary (this follows from the orthogonality property of the characters).

We now write down explicitly the elements of $Q_k^* P_{k-1}$ and $P_k^* P_k - Q_k^* Q_k$. They are

(1) $$3^{-1}\cdot 2^{1/2-2k} \sum_{j=1}^{2^k} \theta_j^k \overline{\sigma_j^k(h)} \tau_j^{k-1}(g), \quad h \in G_k,\ g \in G_{k-1}$$

(2) $$3^{-1}\cdot 2^{-2k}\left(\frac{1}{2}\sum_{j=1}^{2^{k+1}} \tau_j^k(h) - \sum_{j=1}^{2^k} \sigma_j^k(h)\right), \quad h \in G_k.$$

In the derivation of (2) we used the fact that the τ_j^k and σ_j^k are characters, i.e. that $\tau_j^k(h_1 \circ h_2) = \tau_j^k(h_1)\tau_j^k(h_2)$, where $h_1 \circ h_2$ denotes the multiplication in G_k. We have to show that, for suitable choices of $\{\theta_j^k\}_{j=1}^{2^k}$, $k = 0, 1, 2, \ldots$ all these terms are in absolute value $\leq C(k+1)^{1/2}2^{-3k/2}$. The expressions appearing in (2) are independent of θ_j^k and they are of the right order of magnitude by Lemma 2.d.5. Observe that the number of terms written in (1) is $3\cdot 2^{k-1}\cdot 3\cdot 2^k \leq 5\cdot n^2$, where $n = 2^k$. Hence, by Lemma 2.d.4(a), we infer that, whenever $5\cdot 2^{2k} < 2^{3k}/K$, the $\{\theta_j^k\}_{j=1}^{2^k}$ can be chosen so that all terms in (1) are in absolute value of the required order of magnitude (for the finitely many k for which $5\cdot 2^{2k} > 2^{3k}/K$ there is of course nothing to prove). This concludes the construction of $\{U_k\}_{k=1}^{\infty}$ and thus also the proof of 2.d.3. □

The matrix A, constructed in 2.d.3, satisfies in particular $\sum_i \max_j |a_{i,j}| < \infty$. Thus, by 1.e.8, the rows of A span a subspace of c_0 which does not have the A.P. The fact that the matrix of 2.d.3 has a stronger property enables us to show, by the same argument which was used in the proof of 1.e.8, that also the spaces l_p, for $2<p<\infty$ have subspaces which fail to have the A.P.

Theorem 2.d.6. *The spaces c_0 and l_p, $2<p<\infty$, have subspaces which do not have the approximation property.*

Proof. We have to prove the theorem only in the case of l_p. Let A be a matrix satisfying the requirements of 2.d.3 (the assumption that for every i only finitely many j satisfy $a_{i,j}\neq 0$ will not be used here). Put $\lambda_i=\max_j |a_{i,j}|$, $1\leq i<\infty$ and $b_{i,j}=(\lambda_j/\lambda_i)^{1/(p+1)}a_{i,j}$, $1\leq i,j<\infty$. The matrix $B=(b_{i,j})_{i,j=1}^{\infty}$ satisfies $B^2=0$ and trace $B=$ trace $A\neq 0$. Let $y_i=(b_{i,1}, b_{i,2},\ldots)$. Then, since $p/(p+1)>2/3$, $y_i\in l_p$ and

$$\|y_i\|=\left(\sum_{j=1}^{\infty}|b_{i,j}|^p\right)^{1/p}\leq\lambda_i^{p/(p+1)}\left(\sum_{j=1}^{\infty}\lambda_j^{p/(p+1)}\right)^{1/p}\leq L\lambda_i^{p/(p+1)},$$

for some constant L. Consequently, $\sum_{i=1}^{\infty}\|y_i\|<\infty$. Denote by $\{e_i\}_{i=1}^{\infty}$ the unit unit vector basis of l_q. For every y in $[y_i]_{i=1}^{\infty}\subset l_p$ we have $\sum_{i=1}^{\infty} y_ie_i(y)=0$, but $\sum_{i=1}^{\infty} e_i(y_i)=$ trace $B\neq 0$. Hence, by 1.e.4, the space $[y_i]_{i=1}^{\infty}$ does not have the A.P. □

Remarks. 1. Since every subspace of l_2 has the A.P. the argument used in the preceding proof shows that if $A=(a_{i,j})_{i,j=1}^{\infty}$ is a matrix such that $A^2=0$ and $\sum_{i=1}^{\infty}(\sup_j|a_{i,j}|)^{2/3}<\infty$ then trace $A=0$. Thus, the exponent 2/3 appearing in 2.d.3, is the smallest possible one.

2. In vol II we shall present a proof of the fact that the spaces l_p for $1\leq p<2$ also have subspaces which fail to have the A.P.

3. In 1.e.8 we presented a result of Grothendieck which shows that the approximation problem has also an equivalent formulation in terms of continuous functions $K(s,t)$ on $[0,1]\times[0,1]$. We can now give a precise answer also to this problem. By using the proof of (iii) ⇒ (ii) of 1.e.8 and the matrix given in 2.d.3 we obtain a function $K(s,t)$ on $[0,1]\times[0,1]$ which satisfies a Lipschitz condition of every order $<1/2$ and so that $\int_0^1 K(s,t)K(t,u)\,dt=0$, for every s and u, while $\int_0^1 K(t,t)\,dt\neq 0$. On the other hand it can be shown that there is no function which satisfies a Lipschitz condition of order 1/2 and has these properties.

It is not easy to construct an infinite dimensional subspace X of l_p, for $2<p<\infty$, which is not isomorphic to l_p. In fact, prior to the work of Davie [20] and Figiel [40] who exhibited such X which fail to have the A.P., no such examples were known. Once we know of the existence of such X it is possible to show that there are "many" such examples. We show first that, for $2<p<\infty$ (and, in fact, for every

$1 \leqslant p < \infty, p \neq 2$), there is an infinite dimensional subspace X of l_p so that X has the A.P. but it is not isomorphic to l_p.

Proposition 2.d.7. (a) *Let* $1 < p < \infty$ *and let* Y *be a Banach space which is not isomorphic to a complemented subspace of* $L_p(0, 1)$. *Assume that* $Y = \overline{\bigcup_{n=1}^{\infty} B_n}$ *with* $B_n \subset B_{n+1}$ *and* $\dim B_n < \infty$ *for every* n. *Then,* $\left(\sum_{n=1}^{\infty} \oplus B_n\right)_p$ *is not isomorphic to* l_p.

(b) *For every* $1 \leqslant p < \infty, p \neq 2$ *there is a subspace of* l_p *which is isomorphic to* $\left(\sum_{n=1}^{\infty} \oplus B_n\right)_p$, *for suitable finite dimensional* $\{B_n\}_{n=1}^{\infty}$, *and which is not isomorphic to* l_p *itself.*

Outline of proof. The proof uses arguments whose natural setting is the local theory of Banach spaces. These arguments will be discussed in detail in Vol. IV; here we only outline them. If $\left(\sum_{n=1}^{\infty} \oplus B_n\right)_p$ is isomorphic to l_p then there is a λ and operators $U_n: B_n \to l_p$, $V_n: l_p \to B_n$, $n = 1, 2, \ldots$ so that $V_n U_n = I_{B_n}$ and $\|U_n\| \|V_n\| \leqslant \lambda$, for every n. Using this fact and a compactness argument we get that Y must be isomorphic to a complemented subspace of $L_p(0, 1)$, contradicting the assumption in (a).

To prove part (b), let first $2 < p < \infty$ and let Y be a subspace of l_p which does not have the A.P. Let $\{y_n\}_{n=1}^{\infty}$ be a sequence which spans Y and put $B_n = [y_i]_{i=1}^{n}$, $n = 1, 2, \ldots$. It is easily seen that $\left(\sum_{n=1}^{\infty} \oplus B_n\right)_p$ is isomorphic to a subspace of l_p which, by part (a), is not isomorphic to l_p itself.

For $1 \leqslant p < 2$ we take $p < r < 2$ and apply part (a) to $Y = l_r$. It will be shown in Vol. II that $\left(\sum_{n=1}^{\infty} \oplus l_r^n\right)_p$ is isomorphic to a subspace of l_p but that l_r is not isomorphic to a complemented subspace of $L_p(0, 1)$. Thus, $\left(\sum_{n=1}^{\infty} \oplus l_r^n\right)_p$ is not isomorphic to l_p. □

The examples constructed for $1 \leqslant p < 2$ have an unconditional basis. Also for $2 < p < \infty$, l_p has a subspace with an unconditional basis which is not isomorphic to l_p itself. This is a consequence of Szankowski's result [138] that a certain lattice does not have the A.P. We shall discuss this fact in detail in Vol. IV.

We conclude this section with some results involving universal spaces. These results show that there are "many" spaces which fail to have the A.P. even if we consider only subspaces of l_p, for some fixed $2 < p < \infty$. It is well known that $C(0, 1)$ is a universal Banach space in the sense that every separable Banach space Y is isometric to a subspace of $C(0, 1)$. In general, Y is not isomorphic to a complemented subspace of $C(0, 1)$. The question whether there is a separable Banach space X which contains isomorphic copies of all separable spaces as complemented subspaces turns out to be closely related to the approximation property. The following two theorems answer this question.

Theorem 2.d.8 [118, 66]. *There is a Banach space* X *having a basis such that every* *separable Banach space having the* B.A.P. *is isomorphic to a complemented subspace of X.*

Theorem 2.d.9 [62]. *There is no separable Banach space X so that every separable Banach space Y is isomorphic to a complemented subspace of X. Moreover, such an X fails to exist even if we consider only those separable spaces Y which have the* A.P. *or, alternatively, all the subspaces Y of l_p which fail to have the* A.P. *for any fixed p, $2<p<\infty$.*

Since for every sequence $\{Y_n\}_{n=1}^{\infty}$ of separable Banach spaces there is trivially a separable Banach space X which contains all the Y_n as complemented subspaces $\left(\text{e.g. } X=\left(\sum_{n=1}^{\infty}\oplus Y_n\right)_p\right)$, the last statement in 2.d.9 asserts in particular that, for every $2<p<\infty$, there are uncountably many mutually non-isomorphic subspaces of l_p all of which fail to have the A.P.

We prove first Theorem 2.d.8. This theorem is an immediate consequence of Theorem 1.e.13 and the second assertion of the following theorem due essentially to Pelczynski [117].

Theorem 2.d.10. (a) *There exists a separable Banach space U_1 having an unconditional basis $\{x_i\}_{i=1}^{\infty}$ such that every unconditional basic sequence (in an arbitrary separable Banach space) is equivalent to a subsequence of $\{x_i\}_{i=1}^{\infty}$.*

(b) *There exists a separable Banach space U_2 having a Schauder basis $\{x_n\}_{n=1}^{\infty}$ such that, for any basic sequence $\{y_k\}_{k=1}^{\infty}$, there is a subsequence $\{n_k\}_{k=1}^{\infty}$ of the integers such that $\{y_k\}_{k=1}^{\infty}$ is equivalent to $\{x_{n_k}\}_{k=1}^{\infty}$ and the natural projection P on $[x_{n_k}]_{k=1}^{\infty}$ (defined by $Px_{n_k}=x_{n_k}$, $k=1, 2,\ldots$ and $Px_n=0$ if $n\notin\{n_k\}_{k=1}^{\infty}$) is bounded.*

The spaces U_1 and U_2 are determined uniquely, up to isomorphism, by the properties appearing in (a), *resp.* (b).

Proof. We shall present a proof due to Schechtman [131] which is considerably shorter than the original proof. We start with the construction of U_1. Let $\{u_n\}_{n=1}^{\infty}$ be a sequence which is dense in $C(0, 1)$. We introduce a norm $|||\cdot|||_1$ in the space U_0 of all sequences $x=(a_1, a_2, a_3,\ldots)$ of scalars which are eventually 0, by putting,

$$|||x|||_1=\sup\left\{\left\|\sum_{n=1}^{\infty}\theta_n a_n u_n\right\|;\ \theta_n=\pm 1,\ n=1, 2,\ldots\right\}.$$

Let $x_n=(0, 0,\ldots, \overset{n}{1}, 0,\ldots)$, $n=1,\ldots$. The sequence $\{x_n\}_{n=1}^{\infty}$ is an unconditional basis of the completion U_1 of U_0 with respect to $|||\cdot|||_1$. Let $\{y_k\}_{k=1}^{\infty}$ be any unconditional basis of a separable Banach space. By the universality of $C(0, 1)$ we may assume that $y_k\in C(0, 1)$, for every k. Let K be the unconditional constant of $\{y_k\}_{k=1}^{\infty}$. Choose integers $\{n_k\}_{k=1}^{\infty}$ so that $\|y_k-u_{n_k}\|\leq\|y_k\|/K\cdot 2^{k+2}$ for every k. By 1.a.9 the sequence $\{u_{n_k}\}_{k=1}^{\infty}$ is an unconditional basic sequence equivalent to $\{y_k\}_{k=1}^{\infty}$. From the definition of $|||\cdot|||_1$ it follows that $\{x_{n_k}\}_{k=1}^{\infty}$ is equivalent to $\{u_{n_k}\}_{k=1}^{\infty}$ and thus to $\{y_k\}_{k=1}^{\infty}$.

We pass now to the proof of the existence of U_2. Without the requirement of the existence of P we could have proceeded in a way very similar to that used in the preceding case. In the case of an unconditional basis the natural projection is of course defined and bounded for every subsequence $\{n_k\}_{k=1}^{\infty}$. The main point in our construction now will be that, in spite of the fact that we deal with a conditional basis, the natural projection on a large collection of subsequences of the integers (called "branches") will be bounded.

Let $\{u_n\}_{n=1}^{\infty}$ be a sequence which is dense in $C(0, 1)$. Let φ be a one to one function from the set of all finite sequences of positive integers onto the positive integers so that $\varphi(i_1, i_2, \ldots, i_k) < \varphi(i_1, i_2, \ldots, i_k, i_{k+1})$ for every choice of k and $\{i_j\}_{j=1}^{k+1}$. A subsequence of the integers is called a branch if it has the form

$$\varphi(i_1), \varphi(i_1, i_2), \varphi(i_1, i_2, i_3), \ldots,$$

for a suitable choice of integers $\{i_j\}_{j=1}^{\infty}$. The set of all branches will be denoted by $\mathscr{B}$. For $n = \varphi(i_1, i_2, \ldots, i_k)$ we put $v_n = u_{i_k}$. We define now a norm $|||\cdot|||_2$ on the space U_0 of sequences of scalars $x = (a_1, a_2, a_3, \ldots)$ which are eventually 0, by putting,

$$|||x|||_2 = \sup\left\{\left\|\sum_{j=1}^{n} \chi_B(j) a_j v_j\right\|; n = 1, 2, \ldots, B \in \mathscr{B}\right\}$$

where $\chi_B(j)$ is 1 if $j \in B$ and is 0 if $j \notin B$. Let $x_n = (0, 0, \ldots, 0, \overset{n}{1}, 0, \ldots)$, $n = 1, 2, \ldots$. Then $\{x_n\}_{n=1}^{\infty}$ forms a basis of the completion U_2 of U_0 with respect to $|||\cdot|||_2$. It is clear that if $B = \{n_k\}_{k=1}^{\infty} \in \mathscr{B}$ then the natural projection from U_2 onto $[x_{n_k}]_{k=1}^{\infty}$ is bounded (in fact, it has norm 1).

Let $\{y_k\}_{k=1}^{\infty}$ be a basic sequence in $C(0, 1)$ and let K be its basis constant. Choose integers $\{i_k\}_{k=1}^{\infty}$ so that $\|u_{i_k} - y_k\| \leq \|y_k\|/2^{k+2}K$, and let $n_k = \varphi(i_1, i_2, \ldots, i_k)$, $k = 1, 2, \ldots$. Then, $\{n_k\}_{k=1}^{\infty} \in \mathscr{B}$ and, by 1.a.9, $\{y_k\}_{k=1}^{\infty}$ is equivalent to $\{u_{i_k}\}_{k=1}^{\infty}$, i.e. to $\{v_{n_k}\}_{k=1}^{\infty}$. It follows immediately from the definition of $|||\cdot|||_2$ and the fact that the intersection of two different branches is an initial segment of these branches that $\{x_{n_k}\}_{k=1}^{\infty}$ is equivalent to the basic sequence $\{v_{n_k}\}_{k=1}^{\infty}$ and thus to $\{y_k\}_{k=1}^{\infty}$. This proves that U_2 has the desired properties.

We prove now the uniqueness of U_1 and U_2. We actually prove the uniqueness in the following stronger sense: every separable Banach space which has an unconditional basis (respectively, a basis) and which contains isomorphic copies of all separable spaces with an unconditional basis (resp. with a basis) as complemented subspaces must be isomorphic to U_1 (resp. U_2). We present the proof in the case of U_1 (the proof for U_2 is identical). Let V be another space which has the above mentioned property of U_1. Then, there exist Banach spaces X and Y such that $V \approx U_1 \oplus X$ and $U_1 \approx V \oplus Y$. Also, there is a Banach space Z so that $U_1 \approx (U_1 \oplus U_1 \oplus \cdots)_2 \oplus Z$ and thus,

$$U_1 \oplus U_1 \approx U_1 \oplus (U_1 \oplus U_1 \oplus \cdots)_2 \oplus Z \approx (U_1 \oplus U_1 \oplus \cdots)_2 \oplus Z \approx U_1 .$$

Similarly, $V \oplus V \approx V$. Thus, we get that

$$U_1 \approx V \oplus Y \approx V \oplus V \oplus Y \approx V \oplus U_1 \approx U_1 \oplus U_1 \oplus X \approx U_1 \oplus X \approx V. \quad \square$$

Remarks. 1. Some further interesting properties of the space U_1 will be presented in Section 3.b.

2. In connection with 2.d.10 (a) it is worthwhile to mention the following result of Schechtman [131] concerning unconditional basic sequences in l_p. *Let* $1<p<\infty$; *then, there exists an unconditional basic sequence* $\{x_n\}_{n=1}^\infty$ *in* l_p *so that every unconditional basic sequence in* l_p *is equivalent to a subsequence* $\{x_{n_k}\}_{k=1}^\infty$ *of* $\{x_n\}_{n=1}^\infty$. The proof of this fact is based on 2.d.1.

We shall now give a partial *proof of* 2.d.9. We have seen that for every $2<p<\infty$ there is a subspace Y_p of l_p which does not have the A.P. Actually, the spaces obtained in Davie's construction fail to have an apparently weaker property namely, the *compact approximation property* (C.A.P. in short). A Banach space Y is said to have the C.A.P. if the identity operator on Y is in the closure of the set of compact operators from Y into itself with respect to the topology τ of uniform convergence on compact subsets of Y. That the spaces Y_p fail to have the C.A.P. is evident from the argument in [20] but it is not as apparent from the approach presented in this section via infinite matrices. In [20] Davie defines a sequence of τ continuous linear functionals $\beta_n(T)$, $n=1,\ldots$ so that $\beta(T)=\lim_n \beta_n(T)$ exists for every T and is a τ continuous linear functional. The $\{\beta_n\}_{n=1}^\infty$ satisfy $\beta_n(I)=1$ for every n while $\lim_n \beta_n(T)=0$ for every compact operator $T\colon Y_p\to Y_p$. Thus, β is a τ continuous functional which vanishes on the compact operators and is equal to 1 on the identity operator.

Assume now that X is a separable space such that, for every $2<p<\infty$, there is a subspace Z_p of X which is isomorphic to Y_p and so that there is a bounded linear projection Q_p from X onto Z_p. There is an uncountable subset A of $(2,\infty)$ and a $\lambda<\infty$ so that $\|Q_p\|\leqslant\lambda$ for every $p\in A$.

Since each Z_p does not have the C.A.P. it follows that there are finite sets $\{z_{i,p}\}_{i=1}^{n(p)}$ of unit vectors in Z_p and an $\varepsilon_p>0$ so that, whenever T is a compact operator on Z_p for which $\|z_{i,p}-Tz_{i,p}\|<\varepsilon_p$ for $1\leqslant i\leqslant n(p)$, then $\|T\|>\lambda^2$. Let B be an uncountable subset of A so that $n(p)$ is constant (say $=n$) on B and $\inf_{p\in B}\varepsilon_p=\varepsilon>0$. Since B is uncountable and X is separable there exist $p<r$ in B so that $\|z_{i,p}-z_{i,r}\|<(\lambda+\lambda^2)^{-1}\varepsilon$ for $1\leqslant i\leqslant n$.

The proof of 2.c.3 shows also that every operator from a subspace of l_r into l_p is compact. Thus, every operator from Z_r to Z_p is compact. In particular, $T=Q_rQ_{p|Z_r}$ is a compact operator from Z_r into itself with $\|T\|\leqslant\lambda^2$. Note that, for $1\leqslant i\leqslant n$,

$$\begin{aligned}\|z_{i,r}-Tz_{i,r}\|&=\|z_{i,r}-Q_rQ_pz_{i,r}\|\leqslant\|z_{i,r}-Q_rz_{i,p}\|+\|Q_rz_{i,p}-Q_rQ_pz_{i,r}\|\\&=\|Q_r(z_{i,r}-z_{i,p})\|+\|Q_rQ_p(z_{i,p}-z_{i,r})\|\\&\leqslant(\lambda+\lambda^2)\|z_{i,r}-z_{i,p}\|<\varepsilon,\end{aligned}$$

but this contradicts the choice of $\{z_{i,r}\}_{i=1}^n$. This proves the first assertion in the statement of 2.d.9. The proofs of the other two statements of 2.d.9 are more complicated and we do not reproduce them here (we refer the reader to [62]). We just remark that the proof of the fact that there is no separable space which contains isomorphic copies of all separable spaces having the A.P. as complemented subspaces is based on the construction presented in 1.e.20 of a space which has the A.P. but fails to have the B.A.P. The proof of the fact that there is no separable space which contains isomorphic copies of all subspaces of l_p (for a fixed $2<p<\infty$) which fail to have the A.P., as complemented subspaces is obtained by modifying Davie's construction. (Instead of constructing one space the same method is used to construct a suitable uncountable family of spaces.)

e. Banach Spaces Containing l_p or c_0

A long standing open problem going back to Banach's book was the following: *Does every infinite-dimensional Banach space have a subspace isomorphic to either c_0 or l_p, for some $1\leq p<\infty$?* For the common examples of Banach spaces it was easy to show that the answer is positive while, in general, the problem seemed to be quite difficult. Therefore, it came as a surprise when a rather simple counterexample was constructed by B. S. Tsirelson [140]. Tsirelson constructed an example of a reflexive space with an unconditional basis which contains no isomorphic copy of any l_p space, $1<p<\infty$ (actually, his example contains no uniformly convexifiable subspace). We shall present here the dual of Tsirelson's original example which also solves the question stated above. In our presentation we follow T. Figiel and W. B. Johnson [42].

Example 2.e.1. *There is a reflexive Banach space T with an unconditional basis which contains no isomorphic copy of any l_p space, $1\leq p<\infty$.*

Proof. We start by defining a sequence of norms on T_0, the space of all sequences of scalars which are eventually zero. We denote by $\{t_n\}_{n=1}^\infty$ the unit vector basis of T_0 and set, for $x=\sum_{n=1}^\infty a_n t_n \in T_0$, $\|x\|_0=\max_n |a_n|$, and for $m\geq 0$

$$\|x\|_{m+1}=\max\left\{\|x\|_m,\ 2^{-1}\max \sum_{j=1}^k \left\|\sum_{n=p_j+1}^{p_{j+1}} a_n t_n\right\|_m\right\},$$

where the inner max is taken over all choices of $k\leq p_1<p_2<\cdots<p_{k+1}$, $k=1, 2,\ldots$. Obviously, $\|x\|=\lim_{n\to\infty}\|x\|_m$ exists for all $x\in T_0$ and defines a norm on T_0. The unit vectors $\{t_n\}_{n=1}^\infty$ form a normalized unconditional basis of the completion T of T_0. It is also clear that

$$\|x\|=\max\left\{\max_n |a_n|,\ 2^{-1}\sup\left(\sum_{j=1}^k\left\|\sum_{n=p_j+1}^{p_{j+1}} a_n t_n\right\|,\right.\right.$$
$$\left.\left. k\leq p_1<p_2<\cdots<p_{k+1},\ k=1, 2,\ldots\right)\right\},$$

for every $x=\sum_{n=1}^{\infty} a_n t_n \in T$. Consequently, for any k and any sequence of k normalized blocks $u_j=\sum_{n=p_j+1}^{p_{j+1}} a_n t_n$, $1\leqslant j\leqslant k$ with $k\leqslant p_1<p_2<\cdots<p_{k+1}$, we have

$$(*)\qquad \sum_{j=1}^{k}|c_j|\geqslant\left\|\sum_{j=1}^{k}c_j u_j\right\|\geqslant 2^{-1}\sum_{j=1}^{k}|c_j|\,,$$

for every choice of scalars $\{c_j\}_{j=1}^{k}$. This fact and 1.a.12 show that T contains no subspace isomorphic to c_0 or to some l_p, $1<p<\infty$.

To show that T contains no subspace isomorphic to l_1 is more difficult. To this end we have to apply a result of R. C. James (cf. Proposition 2.e.3 below) which ensures that if T has a subspace isomorphic to l_1 then there is a normalized block basis $\{v_j\}_{j=0}^{\infty}$ of $\{t_n\}_{n=1}^{\infty}$ so that, for every choice of scalars $\{b_j\}_{j=0}^{\infty}$,

$$\sum_{j=0}^{\infty}|b_j|\geqslant\left\|\sum_{j=0}^{\infty}b_j v_j\right\|\geqslant(8/9)\sum_{j=0}^{\infty}|b_j|\,,$$

and, in particular,

$$(\substack{*\\*})\qquad \|v_0+r^{-1}(v_1+v_2+\cdots+v_r)\|\geqslant 16/9,\quad r=1,2,\ldots.$$

Consider now integers $k\leqslant p_1<p_2<\cdots<p_{k+1}$ and let $\{P_j\}_{j=1}^{k}$ be the projections associated to the basis $\{t_n\}_{n=1}^{\infty}$ so that $P_j t_n=t_n$ if $p_j<n\leqslant p_{j+1}$ and $P_j t_n=0$, otherwise. Let n_0 be the largest integer for which t_n belongs to the support of v_0. If $k\geqslant n_0$ then

$$\sum_{j=1}^{k}\|P_j(v_0+r^{-1}(v_1+\cdots+v_r))\|=\sum_{j=1}^{k}\|P_j r^{-1}(v_1+\cdots+v_r)\|\leqslant 2\,.$$

If $k<n_0$ we set

$$\delta=\{i;\ \|P_j v_i\|\neq 0 \text{ for at least two values of } j\}\,,$$
$$\sigma=\{i;\ \|P_j v_i\|\neq 0 \text{ for at most one value of } j\}\,.$$

Then, since δ has at most $k-1$ elements, we get that

$$\begin{aligned}&\sum_{j=1}^{k}\|P_j(v_0+r^{-1}(v_1+\cdots+v_r))\|\\&\leqslant\sum_{j=1}^{k}\|P_j v_0\|+r^{-1}\Big(\sum_{i\in\delta}\sum_{j=1}^{k}\|P_j v_i\|+\sum_{i\in\sigma}\sum_{j=1}^{k}\|P_j v_i\|\Big)\\&\leqslant 2\|v_0\|+r^{-1}\Big(2\sum_{i\in\delta}\|v_i\|+\sum_{i\in\sigma}\|v_i\|\Big)\\&\leqslant 2+r^{-1}(2(k-1)+r-k+1)\\&\leqslant 3+(k-1)r^{-1}\leqslant 3+(n_0-1)r^{-1}\,.\end{aligned}$$

Therefore, by taking $r \geqslant 2n_0$, we get that

$$\sum_{j=1}^{k} \|P_j(v_0+r^{-1}(v_1+\cdots+v_r))\| \leqslant 7/2 .$$

It follows from the definition of the norm in T that $\|v_0+r^{-1}(v_1+\cdots+v_r)\| \leqslant 7/4$ and this contradicts (⁑).

Since T has an unconditional basis but does not contain subspaces isomorphic to c_0 or l_1 we deduce from 1.c.12 that T is reflexive. □

Remark. The relation (*) actually shows that T has no uniformly convexifiable infinite-dimensional subspace. Figiel and Johnson show in [42] how to modify T so as to obtain an example of a uniformly convex space with an unconditional basis containing no isomorphic copy of any l_p; $1 \leqslant p < \infty$ (we shall discuss this matter in Vol. II).

Closely related to the problem with which we started this section and which was solved by Tsirelson's example is the problem whether a space which contains an isomorphic copy of some l_p must actually contain almost isometric copies of this space. More precisely,

Problem 2.e.2. *Let X be the space l_p for some $1<p<\infty$, with the usual norm $\|\cdot\|$. Let $|||\cdot|||$ be an equivalent norm on X. Given $\varepsilon>0$, does there exist a subspace Y of X so that $d((Y, |||\cdot|||), (X, \|\cdot\|))<1+\varepsilon$?*

This problem, which is called for obvious reasons the "distortion problem", is still open. Some partial positive answers to it will be described in Vol. III and Vol. IV. Some negative results concerning the analogous distortion problem in $L_p(0, 1)$, $p \neq 2$, are given in [88]. The most interesting case in 2.e.2 is the case $p=2$.

Notice that in the statement of 2.e.2 we excluded the cases $X=l_1$ and $X=c_0$. In these cases the answer to the problem is known to be positive. This is a result of R. C. James [56] which has already been used in the proof of 2.e.1.

Proposition 2.e.3. *Let $(X, \|\cdot\|)$ be the space l_1 or c_0 with its usual norm. Let $|||\cdot|||$ be an equivalent norm on X. Then, for every $\varepsilon>0$, there is a subspace Y of X with $d((Y, |||\cdot|||), (X, \|\cdot\|))<1+\varepsilon$.*

Proof. Assume that $X=l_1$ and that $\alpha|||x||| \leqslant \|x\| \leqslant |||x|||$, for some $\alpha>0$ and all $x \in X$. Let $\varepsilon>0$ and let $\{P_n\}_{n=1}^{\infty}$ be the natural projections induced by the unit vector basis. For every n put $\lambda_n = \sup\{\|x\|;\ |||x|||=1, P_n x=0\}$. Clearly, $\lambda_n \downarrow \lambda$, for some $1 \geqslant \lambda \geqslant \alpha$. Let n_0 be such that $\lambda_{n_0} < \lambda(1+\varepsilon)$. By the definition of the $\{\lambda_n\}_{n=1}^{\infty}$ there is a block basis $\{y_k\}_{k=1}^{\infty}$ of the unit vector basis so that, for all k, $|||y_k|||=1$, $P_{n_0} y_k=0$ and $\|y_k\|>\lambda/(1+\varepsilon)$. For every choice of scalars $\{a_k\}_{k=1}^{\infty}$ we have $P_{n_0}\left(\sum_{k=1}^{\infty} a_k y_k\right)=0$ and hence

$$\left|\left|\left|\sum_{k=1}^{\infty} a_k y_k\right|\right|\right| \geqslant \lambda_{n_0}^{-1} \left\|\sum_{k=1}^{\infty} a_k y_k\right\| = \lambda_{n_0}^{-1} \sum_{k=1}^{\infty} |a_k|\,\|y_k\|$$

$$\geqslant \lambda_{n_0}^{-1}(1+\varepsilon)^{-1}\lambda \sum_{k=1}^{\infty} |a_k| \geqslant (1+\varepsilon)^{-2} \sum_{k=1}^{\infty} |a_k| .$$

On the other hand, by the triangle inequality, $\left|\left|\left|\sum_{k=1}^{\infty} a_k y_k\right|\right|\right| \leqslant \sum_{k=1}^{\infty} |a_k| \, |||y_k||| \leqslant \sum_{k=1}^{\infty} |a_k|$ and thus $d(([y_k]_{k=1}^{\infty}, |||\cdot|||), l_1) \leqslant (1+\varepsilon)^2$.

The proof for c_0 is similar. By replacing the "sup" in the definition of λ_n, by "inf", we get that, for some constant λ, there is a block basis $\{y_k\}_{k=1}^{\infty}$ of the unit vector basis of c_0 with $|||y_k|||=1$, $\|y_k\| < \lambda(1+\varepsilon)$ for every k, and $\left|\left|\left|\sum_{k=1}^{\infty} a_k y_k\right|\right|\right| \leqslant (1+\varepsilon)^2 \cdot \max_k |a_k|$ for every sequence of scalars $\{a_k\}_{k=1}^{\infty}$ tending to 0. An estimate from below can again be deduced from the triangle inequality. Indeed, assume that $|a_{k_0}| = \max_k |a_k|$; then

$$\left|\left|\left|\sum_{k=1}^{\infty} a_k y_k\right|\right|\right| \geqslant |||2a_{k_0} y_{k_0}||| - \left|\left|\left|\sum_{k=1}^{\infty} a_k y_k - 2a_{k_0} y_{k_0}\right|\right|\right|$$

$$\geqslant 2|a_{k_0}| - (1+\varepsilon)^2 \cdot \max_k |a_k| = (1-2\varepsilon-\varepsilon^2) \max_k |a_k| . \ \square$$

The preceding proof does not work for $1<p<\infty$ since the triangle inequality cannot be used to get one of the desired inequalities automatically. The only thing which can obviously be done is to choose a block basis $\{y_k\}_{k=1}^{\infty}$ of the unit vector basis so that $|||y_k|||=1$ for every k and $\left|\left|\left|\sum_{k=1}^{\infty} a_k y_k\right|\right|\right| \geqslant \left(\sum_{k=1}^{\infty} |a_k|^p\right)^{1/p} \Big/ (1+\varepsilon)$ for every choice of scalars $\{a_k\}_{k=1}^{\infty}$, and a block basis $\{z_k\}_{k=1}^{\infty}$ of the unit vector basis so that $|||z_k|||=1$ for every k and $\left|\left|\left|\sum_{k=1}^{\infty} a_k z_k\right|\right|\right| \leqslant (1+\varepsilon)\left(\sum_{k=1}^{\infty} |a_k|^p\right)^{1/p}$ for every choice of scalars $\{a_k\}_{k=1}^{\infty}$.

We return to the discussion centered around the question with which we started this section. While there are no known good criteria for a space to contain subspaces isomorphic to l_p, for some $1<p<\infty$, the situation is different for c_0 and especially for l_1. We shall first present a simple characterization for spaces containing c_0 and then present a deep result characterizing spaces containing l_1.

Proposition 2.e.4 [9]. *A Banach space X has a subspace isomorphic to c_0 if and only if there is a sequence $\{x_n\}_{n=1}^{\infty}$ in X so that $\sum_{n=1}^{\infty} |x^*(x_n)| < \infty$ for every $x^* \in X^*$ but $\sum_{n=1}^{\infty} x_n$ fails to converge.*

Proof. The "only if" part is trivial: we simply take as $\{x_n\}_{n=1}^{\infty}$ a basic sequence which is equivalent to the unit vector basis of c_0. To prove the "if" part let $\{x_n\}_{n=1}^{\infty}$ be such that $\sum_{n=1}^{\infty} |x^*(x_n)| < \infty$ for every $x^* \in X^*$ and $\sum_{n=1}^{\infty} x_n$ diverges. It follows from the uniform boundedness principle that there is a constant M so that $\sum_{n=1}^{\infty} |x^*(x_n)| \leqslant M\|x^*\|$ for every $x^* \in X^*$. Since $\sum_{n=1}^{\infty} x_n$ diverges there is an $\varepsilon>0$ and integers $p_1<q_1<p_2<q_2<\cdots$ so that $\left\|\sum_{n=p_k}^{q_k} x_n\right\| \geqslant \varepsilon$ for every k. Put $y_k = \sum_{n=p_k}^{q_k} x_n$, $k=1, 2, \ldots$.

Since $\sum_{k=1}^{\infty} |x^*(y_k)| < \infty$ for every $x^* \in X^*$ it follows that $y_k \xrightarrow{w} 0$. By 1.a.12 (and the remark following it) we may assume without loss of generality that $\{y_k\}_{k=1}^{\infty}$ forms a basic sequence with basis constant K (otherwise, pass to a subsequence). For every finite sequence of scalars $\{a_k\}_{k=1}^{m}$ we have $\left\|\sum_{k=1}^{m} a_k y_k\right\| \geq \varepsilon \max_k |a_k|/2K$ and also

$$\left\|\sum_{k=1}^{m} a_k y_k\right\| = \sup\left\{\left|\sum_{k=1}^{\infty} a_k x^*(y_k)\right|;\ \|x^*\| \leq 1\right\} \leq M \max_k |a_k|.$$

Thus, $\{y_k\}_{k=1}^{\infty}$ is equivalent to the unit vector basis of c_0. □

Remark. A series $\sum_{n=1}^{\infty} x_n$ for which $\sum_{n=1}^{\infty} |x^*(x_n)| < \infty$ for every $x^* \in X^*$ is said to be *weakly unconditionally convergent* (w.u.c.). It is easily seen that $\sum_{n=1}^{\infty} x_n$ is a w.u.c. series in an arbitrary Banach space X if and only if $\sum_{n=1}^{\infty} a_n x_n$ converges unconditionally whenever $a_n \to 0$.

We pass now to a fundamental result due to H. P. Rosenthal [129].

Theorem 2.e.5. *Let $\{x_n\}_{n=1}^{\infty}$ be a bounded sequence in a Banach space X. Then, $\{x_n\}_{n=1}^{\infty}$ has a subsequence $\{x_{n_i}\}_{i=1}^{\infty}$ satisfying one of the two mutually exclusive alternatives:*

(i) *$\{x_{n_i}\}_{i=1}^{\infty}$ is equivalent to the unit vector basis of l_1.*
(ii) *$\{x_{n_i}\}_{i=1}^{\infty}$ is a weak Cauchy sequence.*

Consequently, the unit ball of X is weakly conditionally compact if and only if no closed subspace of X is isomorphic to l_1.

The proof given in [129] is valid only for real spaces. L. Dor [31] adapted the proof to the complex case. We shall reproduce here a simplified proof due to J. Farahat [38].

Proof. By using the canonical embedding of X into X^{**} we can consider each x_n as an affine continuous function f_n on the unit ball S of X^*. In this setting we have to prove that if $\{f_n\}_{n=1}^{\infty}$ does not have a subsequence which converges pointwise then it has a subsequence which is equivalent, in the sup norm on S, to the unit vector basis of l_1.

Suppose that no subsequence of $\{f_n\}_{n=1}^{\infty}$ converges pointwise on S. Let $\mathscr{D} = \{D_k^1, D_k^2\}_{k=1}^{\infty}$ be the countable family of all pairs of open discs in the complex plane for which both centers and radii are rational and such that

$$\operatorname{diam} D_k^1 = \operatorname{diam} D_k^2 < d(D_k^1, D_k^2)/2, \quad k = 1, 2, \ldots.$$

Then there exists an index k_0 and an infinite subsequence $\{f_n\}_{n \in M}$, $M \subset N$, such that for every subsequence $\{f_n\}_{n \in L}$ with $L \subset M$ there is an $s_L \in S$ for which the sequence of scalars $\{f_n(s_L)\}_{n\in L}$ has points of accumulation in both $D^1_{k_0}$ and $D^2_{k_0}$. Indeed, if this were false we could construct a sequence of infinite subsequences of the integers $N\supset M_1\supset M_2\cdots\supset M_k\supset\cdots$ such that, for every k and every $s\in S$, the sequence $\{f_n(s)\}_{n\in M_k}$ does not have points of accumulation in both discs D^1_k and D^2_k. Let $L=\{m_k\}_{k=1}^\infty$ be a subsequence of the integers so that $m_k\in M_k$ for every k. In view of our assumption, $\{f_n\}_{n\in L}$ does not converge pointwise on S and thus there is an $s_0\in S$ for which $\{f_n(s_0)\}_{n\in L}$ has at least two distinct points of accumulation d_1 and d_2. This, however, contradicts our assumption concerning M_k, for k chosen so that $d_1\in D^1_k$ $d_2\in D^2_k$.

Let α be the center of $D^1_{k_0}$ and β the center of $D^2_{k_0}$. There is no loss of generality to assume that $\beta-\alpha$ is real and positive (otherwise, replace $\{f_n\}_{n\in M}$, $D^1_{k_0}$ and $D^2_{k_0}$ by $\{\gamma f_n\}_{n\in M}$, $\gamma D^1_{k_0}$ and $\gamma D^2_{k_0}$, respectively, where $\gamma=|\beta-\alpha|/(\beta-\alpha)$).

To continue the proof we need the following lemma.

Lemma 2.e.6. *Let $\{A_j, B_j\}_{j=1}^\infty$ be a sequence of pairs of subsets of a set S so that $A_j\cap B_j=\varnothing$, $j=1, 2,\ldots$. Assume that there is no subsequence $\{A_{j_h}, B_{j_h}\}_{h=1}^\infty$ so that, for every $s\in S$, either $\lim_h \chi_{A_{j_h}}(s)=0$ or $\lim_h \chi_{B_{j_h}}(s)=0$ (here χ_A denotes the characteristic function of A). Then there exists an infinite subsequence $\{A_j, B_j\}_{j\in J}$ so that*

$$\Big(\bigcap_{j\in\delta} A_j\Big)\cap\Big(\bigcap_{j\in\sigma} B_j\Big)\neq\varnothing$$

for any pair of disjoint finite sets δ, $\sigma\subset J$ (such a sequence $\{A_j, B_j\}_{j\in J}$ is called Boolean independent).

Once 2.e.6 is proved, the proof of 2.e.5 can be completed as follows.

Let k_0 and $M=\{n_j\}_{j=1}^\infty$ be as above. Set

$$A_j=\{s; f_{n_j}(s)\in D^1_{k_0}\},\qquad B_j=\{s; f_{n_j}(s)\in D^2_{k_0}\},\quad j=1, 2,\ldots.$$

The properties of k_0 and M show that the assumptions in 2.e.6 are satisfied. Thus, there exists a Boolean independent infinite subsequence $\{A_j, B_j\}_{j\in J}$. We shall show that if $d=d(D^1_{k_0}, D^2_{k_0})$ and $c_j=a_j+ib_j$, $j\in J$ are arbitrary complex scalars then

$$\Big\|\sum_{j\in J} c_j f_{n_j}\Big\|\geq(d/8)\sum_{j\in J}|c_j|\,,$$

i.e. that $\{f_{n_j}\}_{j\in J}$ is equivalent to the unit vector basis of l_1. Let σ be any finite subset of J and assume, for simplicity, that $\sum_{j\in\sigma}|a_j|\geq\sum_{j\in\sigma}|b_j|$. Set $\sigma_+=\{j; j\in\sigma, a_j\geq 0\}$, $\sigma_-=\sigma\sim\sigma_+$ and choose

$$s_1\in\Big(\Big(\bigcap_{j\in\sigma_+} B_j\Big)\cap\Big(\bigcap_{j\in\sigma_-} A_j\Big)\Big);\qquad s_2\in\Big(\Big(\bigcap_{j\in\sigma_+} A_j\Big)\cap\Big(\bigcap_{j\in\sigma_-} B_j\Big)\Big).$$

Since, for $z_1\in D^1_{k_0}$ and $z_2\in D^2_{k_0}$, we have

$$\operatorname{Re}(z_2-z_1)\geq d\quad\text{and}\quad\operatorname{Im}(z_2-z_1)\leq\operatorname{diam} D^1_{k_0}<d/2,$$

it follows that

$$\Big\|\sum_{j\in\sigma} c_j f_{n_j}\Big\| = \sup_{s\in S}\Big|\sum_{j\in\sigma} c_j f_{n_j}(s)\Big| \geqslant \operatorname{Re}\sum_{j\in\sigma} c_j f_{n_j}((s_1-s_2)/2)$$

$$\geqslant \frac{1}{2}\sum_{j\in\sigma} a_j \operatorname{Re}(f_{n_j}(s_1)-f_{n_j}(s_2)) - \frac{1}{2}\sum_{j\in\sigma} |b_j \operatorname{Im}(f_{n_j}(s_1)-f_{n_j}(s_2))|$$

$$\geqslant \frac{d}{2}\sum_{j\in\sigma}|a_j| - \frac{d}{4}\sum_{j\in\sigma}|b_j| \geqslant \frac{d}{4}\sum_{j\in\sigma}|a_j| \geqslant \frac{d}{8}\sum_{j\in\sigma}|c_j|\,. \quad \square$$

Proof of 2.e.6. The proof is based on the following combinatorial result, due to C. St. J. A. Nash-Williams [108] (for a simpler proof, as well as a more general result, see [36]): *Let $\mathscr{P}_\infty(L)$ denote the set of all infinite subsets of a countable set L and let $\mathscr{P}\subset\mathscr{P}_\infty(N)$ be a closed subset ($\mathscr{P}_\infty(N)$ is identified with $\{0,1\}^N$, endowed with the product topology). Then $\mathscr{P}$ is a Ramsey set, i.e. for every $M\in\mathscr{P}_\infty(N)$, there exists an $L\in\mathscr{P}_\infty(M)$ such that either $\mathscr{P}_\infty(L)\subset\mathscr{P}$ or $\mathscr{P}_\infty(L)\subset\mathscr{P}_\infty(N)\sim\mathscr{P}$.*

Let the sequence $\{A_j, B_j\}_{j=1}^\infty$ satisfy the assumptions of 2.e.6 and, for notational convenience, denote B_j by $-A_j$, $j=1,2,\ldots$. For each integer k consider the subset $\mathscr{P}_k\subset\mathscr{P}_\infty(N)$ consisting of all $M=\{n_h\}_{h=1}^\infty$ for which $\bigcap_{h=1}^{k}(-1)^h A_{n_h}\neq\varnothing$. Each of the sets $\mathscr{P}_k$ is closed and so is $\mathscr{P}=\bigcap_{k=1}^{\infty}\mathscr{P}_k$. Thus $\mathscr{P}$ is a Ramsey set; this means that there exists an $L=\{m_p\}_{p=1}^\infty\in\mathscr{P}_\infty(N)$ so that either $\mathscr{P}_\infty(L)\subset\mathscr{P}$ or $\mathscr{P}_\infty(L)\subset\mathscr{P}_\infty(N)\sim\mathscr{P}$. In our case we must have $\mathscr{P}_\infty(L)\subset\mathscr{P}$. Indeed, in view of our assumption, there is an $s_0\in S$ so that both sets $\{n\in L;\ s_0\in A_n\}$ and $\{n\in L;\ s_0\in B_n\}$ are infinite which implies the existence of an infinite subsequence $L_0=\{n_h\}_{h=1}^\infty$ of L so that $s_0\in(-1)^h A_{n_h}$, $h=1,2,\ldots$. Thus, $L_0\in\mathscr{P}$ and therefore $\mathscr{P}_\infty(L)\subset\mathscr{P}$.

Let $J=\{m_{2p}\}_{p=1}^\infty$; we claim that $\{A_j, B_j\}_{j\in J}$ is Boolean independent. Indeed, let $\{\theta_p\}_{p=1}^k$ be a finite sequence of signs. Construct a subset $L_1=\{n_h\}_{h=1}^\infty$ of L which contains the integers $m_2, m_4,\ldots, m_{2k}$, scattered among $n_1, n_2,\ldots, n_{2k}$, so that if $n_h=m_{2p}$ then $\theta_p=(-1)^h$. We have $\bigcap_{p=1}^{k}\theta_p A_{m_{2p}}\supset\bigcap_{h=1}^{2k}(-1)^h A_{n_h}\neq\varnothing$. $\square$

We pass now to another result which characterizes Banach spaces containing l_1. This result is due to Odell and Rosenthal [109] and Rosenthal [130].

Theorem 2.e.7. *Let X be a separable Banach space. Then, the following assertions are equivalent*

(i) *X does not contain a subspace isomorphic to l_1.*
(ii) *Every element in X^{**} is the w^*-limit of a sequence of elements in X (i.e. in the canonical image of X in X^{**}).*
(iii) *The cardinality of X^{**} is equal to that of X (i.e. the cardinality of the continuum).*
(iv) *Every bounded sequence in X^{**} has a w^* convergent subsequence.*

We shall outline the proof of a part of this theorem; namely of the equivalence of (i), (ii) and (iii). Assume that every element of X^{**} is a w^* limit of a sequence of elements in X. This sequence can obviously be taken out of any fixed countable dense set in X. Hence, the cardinality of X^{**} is at most the cardinality of the set of all subsets of the integers, i.e. that of the continuum. Thus (ii) $\Rightarrow$ (iii). It is also trivial that (iii) $\Rightarrow$ (i) since l_1^{**} has a cardinality larger than that of the continuum (observe, e.g. that if Γ is a set of the cardinality of the continuum then $l_1(\Gamma)\subset C(0, 1)^*\subset l_\infty = l_1^*$ and thus l_1^{**} has $l_\infty(\Gamma)$ as a quotient space).

Outline of the proof of (i) $\Rightarrow$ (ii). Assume that there is an $x_0^{**} \in X^{**}$ with $\|x_0^{**}\|=1$ which is not a w^* limit of a sequence of elements in X. Let S be the unit ball of X^* with the w^* topology. Since X is separable S is a compact metric space. The function $x_0^{**}(x^*)$ on S is not the pointwise limit of a sequence of continuous functions on S, i.e. does not belong to the first Baire class on S. This fact is however not obvious; our assumption on x_0^{**} implies immediately only that it is not the pointwise limit of a sequence of affine continuous functions on S. A direct proof of this fact is given in [109]; it can also be deduced from a general result of Choquet concerning functions of Baire class 1 (cf. [1, p. 16]).

Once we know that $x_0^{**}(x^*)$ is not in Baire class 1 on S it follows from the classical characterization of Baire class 1 functions (cf. [51, p. 288]) that there is a closed non-empty set K in S so that the restriction of x_0^{**} to K has no points of continuity. Let $\mathscr{D}=\{D_k^1, D_k^2\}_{k=1}^\infty$ be the family of discs used in the proof of 2.e.5 and let, for $k=1, 2,\ldots,$

$$F_k=\{x^* \in K;\ \text{in any } w^* \text{ neighbourhood } G \text{ of } x^* \text{ there are } y^* \in K\cap G \text{ and } z^* \in K\cap G \text{ so that } x_0^{**}(y^*) \in D_k^1,\ x_0^{**}(z^*) \in D_k^2\}\,.$$

Then, clearly, each F_k is a closed subset of K and $K=\bigcup_{k=1}^\infty F_k$. By the Baire category theorem there is an integer k_0 so that F_{k_0} has a non-empty interior, say G_0, relative to K. Thus, for every non-empty open subset G of G_0, there are $y^*, z^* \in G$ so that $x_0^{**}(y^*) \in D_{k_0}^1$ and $x_0^{**}(z^*) \in D_{k_0}^2$.

We shall prove that there is a sequence $\{x_n\}_{n=1}^\infty$ in the unit ball of X so that the sequence $\{A_n, B_n\}_{n=1}^\infty$, where

$$A_n=\{x^* \in G_0;\ x^*(x_n) \in D_{k_0}^1\}, \qquad B_n=\{x^* \in G_0;\ x^*(x_n) \in D_{k_0}^2\}\,,$$

is Boolean independent. Once this is proved it follows, as in the proof of 2.e.5, that $\{x_n\}_{n=1}^\infty$ i.e. equivalent to the unit vector basis in l_1.

We choose the $\{x_n\}_{n=1}^\infty$ inductively. By the definition of G_0 there are $y^*, z^* \in G_0$ so that $x_0^{**}(y^*) \in D_{k_0}^1$ and $x_0^{**}(z^*) \in D_{k_0}^2$. By the w^* density of the unit ball of X in the unit ball of X^{**} there is an $x_1 \in X$ with $\|x_1\|\leq 1$ so that $y^*(x_1) \in D_{k_0}^1$ and $z^*(x_1) \in D_{k_0}^2$. With this choice of x_1 we have that $A_1\neq\emptyset$, $B_1\neq\emptyset$ and obviously $A_1\cap B_1=\emptyset$. Assume that $\{x_n\}_{n=1}^m$ have already been chosen in the unit ball of X so that, for every choice of signs $\theta=(\theta_1, \theta_2,\ldots, \theta_m)$, $\bigcap_{n=1}^m \theta_n A_n\neq\emptyset$ (where for notational convenience we put $-A_n=B_n$). Since $\bigcap_{n=1}^m \theta_n A_n$ is an open subset of G_0 there are, for every such θ, elements y_θ^* and z_θ^* in $\bigcap_{n=1}^m \theta_n A_n$ so that $x_0^{**}(y_\theta^*) \in D_{k_0}^1$ and

$x_0^{**}(z_\theta^*) \in D_{k_0}^2$. By the w^* density of the unit ball of X in that of X^{**} we get that there is an x_{m+1} with $\|x_{m+1}\| \leq 1$ so that $y_\theta^*(x_{m+1}) \in D_{k_0}^1$ and $z_\theta^*(x_{m+1}) \in D_{k_0}^2$ for all the 2^m possible choices of $(\theta_1, \theta_2, \ldots, \theta_m)$. With this choice of x_{m+1} it is clear that $\bigcap_{n=1}^{m+1} \theta_n A_n \neq \emptyset$ for all $(m+1)$-tuples of signs $(\theta_1, \theta_2, \ldots, \theta_{m+1})$. □

There are also other interesting theorems characterizing Banach spaces containing l_1. These theorems involve the function spaces $C(0, 1)$ and $L_1(0, 1)$ and we shall discuss them in Vol. III.

A question which goes back to Banach is whether every separable Banach space X whose dual is non-separable must contain a subspace isomorphic to l_1. This question was answered negatively by two, independently constructed, counterexamples. One (cf. [57] and also [91]), denoted by JT and called the *James tree*, is obtained from the space J (cf. Example 1.d.2) by replacing its index set (i.e. the integers) by an infinite tree. The second example (cf. [91]), denoted by JF and called the *James function space*, is the continuous analogue of J. This space is easier to define (but more difficult to analyse) than JT. The space JF is the completion of the linear span of characteristic functions of subintervals of $[0, 1]$ with respect to the norm

$$\|f\| = \sup \left(\sum_{i=0}^{n-1} \left(\int_{t_i}^{t_{i+1}} f(t)\, dt \right)^2 \right)^{1/2}$$

where the supremum is taken over all partitions $0 = t_0 < t_1 < \cdots < t_n = 1$ of $[0, 1]$. For both spaces the verification that their duals are non-separable is trivial while the proof that they do not contain l_1 is more difficult. Since the discovery of these two examples several other counterexamples have been found. We mention, in particular, an example due to Hagler [50] of a space X which, among its other interesting properties, satisfies the following: X is separable, X^* is non-separable, every infinite-dimensional subspace Y of X has a subspace isomorphic to c_0 and every infinite-dimensional subspace Z of X^* has a subspace isomorphic to l_1. We shall present in Vol. IV still another counterexample to Banach's question which is obtained by using a counterexample of James [58] to an important question in the local theory of Banach spaces.

We pass now to some questions involving duality which concern Banach spaces containing c_0 or l_1.

Proposition 2.e.8 [9]. *Let X be a Banach space such that X^* contains a subspace isomorphic to c_0. Then, X has a complemented subspace isomorphic to l_1. Consequently, X^* has a subspace isomorphic to l_∞.*

Proof. Let $T: c_0 \to X^*$ be an isomorphism into and let $\{e_n\}_{n=1}^\infty$ denote the unit vector basis of c_0. The map $x \to Sx = (Te_1(x), Te_2(x), \ldots)$ is the restriction of T^* to $X \subset X^{**}$ and thus it maps X into l_1. Since T^* maps X^{**} onto l_1 and the unit ball of X is w^* dense in the unit ball of X^{**} there exists a constant K such that, for every n, there is an $x_n \in X$ with $\|x_n\| \leq K$, $Te_n(x_n) = 1$ and $\sum_{i=1}^{n-1} |Te_i(x_n)| < 1/n$. The sequence $\{Sx_n\}_{n=1}^\infty$ has, by 1.a.9, 1.a.12 and 2.a.1, a subsequence $\{Sx_{n_k}\}_{k=1}^\infty$ which is equivalent to the unit vector basis of l_1 and whose span is complemented in l_1 by a projection P. Hence, for some constant M and every choice of scalars $\{a_k\}_{k=1}^\infty$, we have

$$\left\|\sum_{k=1}^\infty a_k x_{n_k}\right\| \leq K \sum_{k=1}^\infty |a_k| \leq KM \left\|\sum_{k=1}^\infty a_k S x_{n_k}\right\| \leq KM\|S\| \left\|\sum_{k=1}^\infty a_k x_{n_k}\right\| .$$

Thus, S is invertible on $Y=[x_{n_k}]_{k=1}^\infty$ which implies that Y is isomorphic to l_1 and $S^{-1}PS$ is a projection from X onto Y. □

Proposition 2.e.8 shows in particular that c_0 *is not isomorphic to a subspace of a separable conjugate space.*

We cannot interchange the roles of c_0 and l_1 in 2.e.8. If, e.g. $X=l_1$, then X^* contains l_1 as a subspace without X having c_0 as a subspace. It is also possible for X^* to contain l_1 as a subspace without X having c_0 as a quotient space. This is e.g. the case for $X=l_\infty$ (it will be proved in Vol. II that every separable quotient space of l_∞ is reflexive (cf. Proposition 2.f.4 below)). However, for separable spaces, we have the following result [60].

Proposition 2.e.9. *Let X be a separable space such that X^* contains a subspace isomorphic to l_1. Then c_0 is isomorphic to a quotient space of X.*

Proof. Let $\{x_n^*\}_{n=1}^\infty$ be a sequence in X^* which is equivalent to the unit vector basis of l_1. Since X is separable there is a subsequence $\{x_{n_k}^*\}_{k=1}^\infty$ which converges w^*. The sequence $y_k^*=x_{n_{2k+1}}^*-x_{n_{2k}}^*$, $k=1, 2,\dots$ is also equivalent to the unit vector basis of l_1 and converges w^* to 0. By the proof of 1.b.7 there is a subsequence $\{y_{k_j}^*\}_{j=1}^\infty$ of $\{y_k^*\}_{k=1}^\infty$ which is a w^* basic sequence. Hence, by 1.b.9, the space $X/([y_{k_j}^*]_{j=1}^\infty)^\top$ is isomorphic to c_0. □

We conclude this section with an open problem which is closely related to the results discussed above.

Problem 2.e.10. *Does every infinite-dimensional Banach space X contain an infinite dimensional subspace which is either reflexive or isomorphic to c_0 or to l_1?*

By 1.c.12 the answer to 2.e.10 is positive if X has a subspace with an unconditional basis (thus, 2.e.10 is a weak version of Problem 1.d.5). Another partial answer to 2.e.10 is given in 1.b.14.

f. Extension and Lifting Properties, Automorphisms of l_∞, c_0 and l_1

The spaces l_∞ and c_0 have important "extension properties" which characterize them while l_1 is characterized by a "lifting property". Some of these properties were mentioned briefly in the previous sections. Here we shall treat them in detail. The main part of this section will be devoted to the study of automorphisms (i.e. invertible linear operators) of l_∞, c_0 and l_1. It turns out that these spaces are surprisingly rich in automorphisms.

We start by considering the extension property of l_∞.

Definition 2.f.1. A Banach space X is said to be *injective* if, for every Banach space Y containing X as a subspace, there is a bounded linear projection from Y onto X.

Injective spaces can be characterized by extension properties for operators, as shown in the following simple proposition.

Proposition 2.f.2. *The following three assertions concerning a Banach space X are equivalent.*

(i) *X is injective.*
(ii) *For every Banach space $Y \supset X$, every Banach space Z and every $T \in L(X, Z)$ there is a $\hat{T} \in L(Y, Z)$ which extends T.*
(iii) *For every pair of Banach spaces $Z \supset Y$ and every $T \in L(Y, X)$ there is a $\hat{T} \in L(Z, X)$ which extends T.*

Proof. Assertion (i) is a special case of both (ii) and (iii) (take e.g. in (ii) $Z = X$ and T the identity operator) and thus (ii) $\Rightarrow$ (i) and (iii) $\Rightarrow$ (i). Assume now that (i) holds and let Y, Z and T be given as in (ii). Let P be a projection from Y onto X. The operator $\hat{T} = TP$ has the desired property. It remains to prove that (i) $\Rightarrow$ (iii). Let $\{x_\gamma^*\}_{\gamma \in \Gamma}$ be a set of functionals of norm 1 on X such that $\|x\| = \sup_\gamma |x_\gamma^*(x)|$ for every $x \in X$ (if X is separable Γ can of course be taken to be countable). Define an isometry S, from X into $l_\infty(\Gamma)$, by $Sx(\gamma) = x_\gamma^*(x)$, $\gamma \in \Gamma$. For every $\gamma \in \Gamma$ let $z_\gamma^* \in Z^*$ be a norm preserving extension of $T^* x_\gamma^*$ from Y to Z and let $\hat{T}_0 \in L(Z, l_\infty(\Gamma))$ be defined by $\hat{T}_0 z(\gamma) = z_\gamma^*(z)$, $\gamma \in \Gamma$. Let P be a projection from $l_\infty(\Gamma)$ onto SX (here we use (i)). Then, $\hat{T} = S^{-1} P \hat{T}_0$ has the desired property. □

The proof of (i) $\Rightarrow$ (iii) above shows in particular that, for every set Γ, *the space $l_\infty(\Gamma)$ is injective* (there is even a projection of norm 1 from any $Z \supset l_\infty(\Gamma)$ onto $l_\infty(\Gamma)$). The spaces $l_\infty(\Gamma)$ are not the only injective spaces. We shall discuss in detail the question of the structure of injective spaces in Vol. III. We shall show here only that l_∞ is characterized by the fact that it is the "smallest" injective space. Theorem 2.a.7 states that every injective infinite-dimensional subspace of l_∞ is isomorphic to l_∞ and, in particular, that there are no separable infinite-dimensional injective spaces. Essentially the same argument as that used in 2.a.7 proves however the following stronger version of 2.a.7.

Theorem 2.f.3 [125]. *Every infinite-dimensional injective Banach space has a subspace isomorphic to l_∞.*

This theorem is a consequence of

Proposition 2.f.4. *Let Γ be a set and T a non-weakly compact operator from $l_\infty(\Gamma)$ into some Banach space Z. Then, there exists a subspace U of $l_\infty(\Gamma)$ which is isomorphic to l_∞ so that $T_{|U}$ is an isomorphism.*

We show first that 2.f.4 implies 2.f.3. Let X be an infinite-dimensional injective space and let Γ be such that X is isometric to a subspace X_0 of $l_\infty(\Gamma)$. Let P be a projection from $l_\infty(\Gamma)$ onto X_0. By fact (ii) stated in the proof of 2.a.7 P is not w compact and thus, by 2.f.4, X_0 has a subspace isomorphic to l_∞.

Proof of 2.f.4. The starting point of the proof is a stronger variant of fact (i) which was used in the proof of 2.a.7: every bounded non-w compact operator T, defined on a $C(K)$ space, is an isomorphism on a suitable subspace which is isomorphic to c_0. This result is due to Pelczynski [114] and will be proved in Vol. III.

Let $T: l_\infty(\Gamma) \to Z$ be non-w compact. Then, by the preceding remark, there are vectors $\{y_i\}_{i=1}^\infty$ in $l_\infty(\Gamma)$ which are equivalent to the unit vector basis of c_0 and so that $T_1 = T_{|[y_i]_{i=1}^\infty}$ is an isomorphism. Since $[y_i]_{i=1}^\infty$ is separable we may consider it as a subspace of l_∞. Since l_∞ and $l_\infty(\Gamma)$ are injective there are bounded linear operators $S_1: l_\infty \to l_\infty(\Gamma)$ and $S_2: Z \to l_\infty$ so that $S_{1|[y_i]_{i=1}^\infty} = \text{identity}$ and $S_{2|[Ty_i]_{i=1}^\infty} = T_1^{-1}$, i.e. the following diagram commutes

$$\begin{array}{ccccccc}
l_\infty & \xrightarrow{S_1} & l_\infty(\Gamma) & \xrightarrow{T} & Z & \xrightarrow{S_2} & l_\infty \\
\bigcup & & \bigcup & & \bigcup & & \bigcup \\
[y_i]_{i=1}^\infty & \xrightarrow{id} & [y_i]_{i=1}^\infty & \xrightarrow{T_1} & [Ty_i]_{i=1}^\infty & \xrightarrow{T_1^{-1}} & [y_i]_{i=1}^\infty
\end{array}$$

Let I be the identity operator of l_∞. The operator $I - S_2TS_1: l_\infty \to l_\infty$ vanishes on $[y_i]_{i=1}^\infty$. By the proof of 2.a.7, $I - S_2TS_1$ must vanish on a subspace V of l_∞ which is isomorphic to l_∞. Hence, S_2TS_1 is the identity on V and therefore $U = S_1V$ is isomorphic to l_∞ and $T_{|U}$ is an isomorphism. □

We pass now to the extension property of c_0. The space c_0 is, by 2.a.7 (as every other separable space), not injective. If we consider however only separable spaces containing c_0 then c_0 behaves as an injective space. More precisely, we have the following theorem due to Sobczyk [136].

Theorem 2.f.5. *Let Y be a separable Banach space containing c_0. Then there is a projection of norm ≤ 2 from Y onto c_0.*

Proof (due to Veech [142]). Let $\{e_n^*\}_{n=1}^\infty$ denote the unit vector basis in $l_1 = c_0^*$. For every n let $y_n^* \in Y^*$ be a norm preserving extension of e_n^* from c_0 to Y. Let $F = B_{Y^*} \cap (c_0^\perp)$ and let d be a translation invariant metric on Y^* which induces on B_{Y^*} the w^* topology (here, the separability of Y is used). Since every w^* limit point of the set $\{y_n^*\}_{n=1}^\infty$ belongs to F it follows that $d(y_n^*, F) \to 0$. Let $\{z_n^*\}_{n=1}^\infty$ be a sequence of elements in F such that $\lim_n d(y_n^*, z_n^*) = 0$, i.e. $w^* \lim (y_n^* - z_n^*) = 0$. The operator $P: Y \to c_0$, defined by $Py = (y_1^*(y) - z_1^*(y), y_2^*(y) - z_2^*(y), \ldots)$, is a projection of norm ≤ 2. □

The property of c_0, which was exhibited in 2.f.5, actually characterizes c_0 among the separable Banach spaces. More precisely, a separable infinite-dimensional Banach space which is complemented in every separable space containing it must be isomorphic to c_0. This is a difficult result, due to M. Zippin [149], and will be presented in Vol. III.

We shall present now a result on quotient spaces of c_0 whose proof uses 2.f.5.

Theorem 2.f.6 [64]. *Every quotient space of c_0 is isomorphic to a subspace of c_0.*

Proof. Let X be a quotient space of c_0. Since X^* is clearly separable we deduce from 1.g.2 that there is a closed subspace Y of X so that both Y and X/Y have shrinking F.D.D.'s. By 2.d.1 there exist sequences $\{B_n\}_{n=1}^\infty$ and $\{C_n\}_{n=1}^\infty$ of finite dimensional Banach spaces so that Y is isomorphic to $\left(\sum_{n=1}^\infty \oplus B_n\right)_0$ and X/Y is isomorphic to $\left(\sum_{n=1}^\infty \oplus C_n\right)_0$. Since $\left(\sum_{n=1}^\infty \oplus B_n\right)_0$ is clearly isomorphic to a subspace of $(c_0 \oplus c_0 \oplus \cdots)_0 = c_0$ we get that Y, and similarly X/Y, are isomorphic to subspaces of c_0. Let $T\colon Y \to c_0$ and $S\colon X/Y \to c_0$ be isomorphisms into and let $Q\colon X \to X/Y$ be the quotient map. The operator T can be extended to a bounded linear operator $\hat{T}\colon X \to c_0$. Indeed, by 2.f.2, T can be extended to an operator $\hat{T}_0\colon X \to l_\infty$. Since $\hat{T}_0 X$ is separable there is, by 2.f.5, a projection P from span $\{\hat{T}_0 X, c_0\}$ onto c_0. The operator $\hat{T} = P\hat{T}_0$ maps X into c_0 and extends T. It is easily verified that the operator $R\colon X \to c_0 \oplus c_0$, defined by $Rx = (\hat{T}x, SQx)$, is an isomorphism into. □

We turn now to the study of the space l_1 which has a property dual to the extension property, namely the "lifting property".

Proposition 2.f.7. *Let X and Y be Banach spaces so that there is an operator S from Y onto X. Then, for every $T \in L(l_1, X)$, there is a $\hat{T} \in L(l_1, Y)$ for which $S\hat{T} = T$. Moreover, if S is a quotient map then, for every $\varepsilon > 0$, $\hat{T}$ may be chosen so that $\|\hat{T}\| \leq (1+\varepsilon)\|T\|$.*

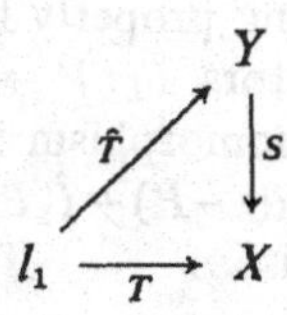

Proof. Let $\{e_n\}_{n=1}^\infty$ be the unit vector basis of l_1 and put $x_n = Te_n$, $n = 1, 2, \ldots$. Then, $\sup_n \|x_n\| < \infty$. It follows from the open mapping theorem that there is a bounded sequence $\{y_n\}_{n=1}^\infty$ in Y such that $Sy_n = x_n$, $n = 1, 2, \ldots$. The operator $\hat{T}$, defined by $\hat{T}(a_1, a_2, \ldots) = \sum_{n=1}^\infty a_n y_n$, has the desired property. If S is a quotient map then, for every $\varepsilon > 0$, we can choose the $\{y_n\}_{n=1}^\infty$ so that $\|y_n\| \leq \|x_n\|(1+\varepsilon)$, for all n. This proves the second assertion. □

The property of l_1 exhibited in 2.f.7 characterizes l_1. We say that a Banach space Z has the *lifting property* if, for every operator S from a Banach space Y onto a space X and for every $T \in L(Z, X)$, there is a $\hat{T} \in L(Z, Y)$ so that $T = S\hat{T}$. Every separable infinite dimensional Banach space with the lifting property is isomorphic to l_1. Indeed, if we take $Y = l_1$, $X = Z$, S a quotient map from l_1 onto Z and T=identity of Z, in the definition of the lifting property we deduce that there is a $\hat{T} \in L(Z, l_1)$ so that $S\hat{T} = I_Z$. Consequently, $\hat{T}$ is an isomorphism into and $\hat{T}S$ is a projection from l_1 onto $\hat{T}Z$. Thus, by 2.a.3, $Z \approx l_1$.

The spaces having the lifting property have been characterized also in the non-separable case. It is clear from the proof of 2.f.7 that, for every set Γ, the space $l_1(\Gamma)$ has the lifting property. Köthe [73] generalized 2.a.3 to the non-separable case and showed that, conversely, *every space with the lifting property is isomorphic to* $l_1(\Gamma)$, *for some set* Γ.

We used in the preceding paragraph (and in several places in the preceding sections) the fact that *every separable Banach space X is a quotient space of l_1*. Let us recall the proof of this simple fact. We let $\{x_n\}_{n=1}^\infty$ be a dense sequence in the unit ball of X and define $T: l_1 \to X$ by $T(a_1, a_2 \ldots) = \sum_{n=1}^\infty a_n x_n$. It is clear that $\|T\| \leq 1$ and, since the image under T of the unit ball of l_1 is dense in the unit ball of X, we get that T is a quotient map. This proof shows that T is highly non-unique; it depends on the choice of the sequence $\{x_n\}_{n=1}^\infty$ (actually the $\{x_n\}_{n=1}^\infty$ need not even be dense in the unit ball of X. It is easily seen that a necessary and sufficient condition for T to be a quotient map is that $B_X = \overline{\text{conv}}\,\{x_n\}_{n=1}^\infty$). It is therefore somewhat surprising that, up to an automorphism of l_1, T is actually unique.

Theorem 2.f.8 [89]. *Let T_1 and T_2 be two linear operators mapping l_1 onto the same Banach space X which is not isomorphic to l_1. Then there exists an automorphism τ of l_1 (i.e. an invertible linear operator from l_1 onto l_1) so that $T_1 = T_2\tau$. In particular,* ker T_1 *is isomorphic to* ker T_2.

Proof. By 2.a.2 and 2.a.3 there are subspaces U and V of l_1 so that $l_1 = U \oplus V$, $U \subset \ker T_1$ and $U \approx V \approx l_1$. Let P be the projection of l_1 onto V which maps U to 0. Then, $T_1 = T_1 P$ and thus $T_{1|V}$ maps V onto X. Using twice the lifting property for $T_{1|V}$ (acting on $V \approx l_1$) and T_2 we conclude that there are operators $\hat{T}_1: V \to l_1$ and $\hat{T}_2: l_1 \to V$ so that $T_2\hat{T}_1 = T_{1|V}$ and $T_1\hat{T}_2 = T_2$. Let R be an isomorphism from U onto l_1 normalized so that $\|R^{-1}\| > 1$ and put $S = (I - \hat{T}_1\hat{T}_2)R(I-P) + \hat{T}_1 P$, where I is the identity on l_1. The operator S maps l_1 into l_1 and satisfies

$$T_2 S = (T_2 - T_2\hat{T}_1\hat{T}_2)R(I-P) + T_2\hat{T}_1 P = T_1 P = T_1\,.$$

Passing to the duals we get that $S^* = (I-P^*)R^*(I - \hat{T}_2^*\hat{T}_1^*) + P^*\hat{T}_1^*$ and therefore,

$$\|S^*x^*\| \geq \tfrac{1}{2}\|P\|^{-1} \max\left(\|(I-P^*)R^*(I-\hat{T}_2^*\hat{T}_1^*)x^*\|,\ \|P^*\hat{T}_1^*x^*\|\right)$$

for every $x^* \in l_1^*$. Since $\|P^*\hat{T}_1^*x^*\| \geq \|\hat{T}_1^*x^*\|$ and, similarly,

$$\|(I-P^*)R^*(I-\hat{T}_2^*\hat{T}_1^*)x^*\| \geq \|R^{-1}\|^{-1}(\|x^*\| - \|\hat{T}_2\|\,\|\hat{T}_1^*x^*\|)$$

we get that

$$\begin{aligned}\|S^*x^*\| &\geqslant \tfrac{1}{2}\|P\|^{-1}\|R^{-1}\|^{-1}\max(\|x^*\|-\|\hat{T}_2\|\,\|\hat{T}_1^*x^*\|,\ \|\hat{T}_1^*x^*\|)\\ &\geqslant (4\|P\|\,\|R^{-1}\|(1+\|\hat{T}_2\|))^{-1}\|x^*\|\,.\end{aligned}$$

Hence, S^* is an isomorphism and thus S is an operator from l_1 onto l_1. By using again the lifting property of l_1 we deduce that there is an $\tilde{I}\in L(l_1, l_1)$ such that $S\tilde{I}=I$. The operator $I-\tilde{I}S$ is a projection from l_1 onto $\ker S$.

Let Q be a projection in l_1 such that $Ql_1\subset\ker T_2$ and $Ql_1\approx(I-Q)l_1\approx l_1$. Then l_1 can be decomposed into the direct sum $l_1=\ker S\oplus W_1\oplus W_2$ so that $SW_1=Ql_1$ and $SW_2=(I-Q)l_1$. Let $\tau\in L(l_1, l_1)$ be an operator which maps $\ker S\oplus W_1$ isomorphically onto Ql_1 and whose restriction to W_2 is equal to $S_{|W_2}$. Then, τ is an automorphism and $T_1=T_2\tau$ since $\ker S\oplus W_1\subset\ker T_1$ and $T_1=T_2S$. □

Remarks. 1. The requirement that X is not isomorphic to l_1 was used in the proof only to ensure that $\ker T_1$ and $\ker T_2$ are both infinite-dimensional. If $X=l_1$ and if $\dim\ker T_1\neq\dim\ker T_2$ (i.e. if one of them is finite and the second infinite or if both are finite but different) then, clearly, there does not exist an automorphism τ so that $T_1=T_2\tau$.

2. Theorem 2.f.8 shows, in particular, that to every separable Banach space X (with $X\not\approx l_1$) there corresponds an, up to isomorphism unique, subspace of l_1, namely the kernel of any quotient map from l_1 onto X. This correspondence is not however, one to one. For example, if T is a quotient map from l_1 onto some Banach space X then $\tilde{T}: l_1\oplus l_1\to X\oplus l_1$, defined by $\tilde{T}(u, v)=(Tu, v)$, is also a quotient map. Clearly, $\ker T=\ker\tilde{T}$ though, in general, X is not isomorphic to $X\oplus l_1$.

It is possible that 2.f.8 or even some weaker versions of it characterize the space l_1. For example, the following problem is open.

Problem 2.f.9. *Let X be an infinite-dimensional separable Banach space. Assume that, for every pair of quotient maps T_1 and T_2 from X onto the same space Y (with $Y\not\approx X$), there is an automorphism τ of X so that $T_1=T_2\tau$. Is X isomorphic to either l_1 or l_2?*

Notice that, for $X=l_2$, the assumptions in 2.f.9 are trivially satisfied.

Theorem 2.f.8 can be dualized to theorems concerning extension of isomorphisms in c_0 and l_∞. We state first the result for c_0.

Theorem 2.f.10 [89]. *Let Y be a closed subspace of c_0 and let T be an isomorphism from Y into c_0 so that* $\dim c_0/Y=\dim c_0/TY=\infty$. *Then, there is an automorphism τ of c_0 so that $\tau_{|Y}=T$. In particular, $c_0/Y\approx c_0/TY$.*

The proof of 2.f.10 is very similar to the proof of the first part of the corresponding result for l_∞ (Theorem 2.f.12 below) and therefore we omit it. Theorem 2.f.10 enables us to associate to every Banach space X isomorphic to a subspace of c_0 (with $X\not\approx c_0$) a, unique up to isomorphism, quotient space of c_0 (denoted by c_0/X). In analogy to 2.f.9 we have the following problem.

Problem 2.f.11. *Let X be a separable infinite-dimensional Banach space. Assume that, for every pair Y, Z of isomorphic subspaces of X of infinite-codimension, there is an automorphism τ of X so that $\tau Y=Z$. Is X isomorphic to c_0 or l_2?*

We pass now to the space l_∞.

Theorem 2.f.12 [89]. *Let Y be a subspace of l_∞ and let T be an isomorphism from Y into l_∞ so that* $\dim l_\infty/Y=\dim l_\infty/TY=\infty$. *Then,*

(i) *If l_∞/Y and l_∞/TY are both non-reflexive there is an automorphism τ of l_∞ which extends T (i.e. $\tau_{|Y}=T$).*
(ii) *If only one of the spaces l_∞/Y and l_∞/TY is reflexive T cannot be extended to an automorphism of l_∞.*
(iii) *If l_∞/Y and l_∞/TY are both reflexive then every extension $\hat{T}$ of T to an operator from l_∞ into itself is a Fredholm operator. The index $i(\hat{T})$ of $\hat{T}$ does not aepend on the particular choice of $\hat{T}$ and thus defines an integer valued invariant of T denoted by $\hat{\imath}(T)$. The operator T can be extended to an automorphism of l_∞ if and only if $\hat{\imath}(T)=0$.*

Since assertion (ii) is trivial only (i) and (iii) require a proof.

Proof of 2.f.12(i). We show first that if Y is a subspace of l_∞ with l_∞/Y non-reflexive then there is a projection P on l_∞ with $PY=\{0\}$ and $\dim Pl_\infty=\infty$ (i.e. $Pl_\infty\approx l_\infty$). Indeed, since the quotient map $\varphi\colon l_\infty\to l_\infty/Y$ is not w compact there is, by 2.f.4, a subspace U of l_∞ with $U\approx l_\infty$ so that $\varphi_{|U}$ is an isomorphism. Since l_∞ is injective there is a projection P_0 from l_∞/Y onto φU. The projection $P=(\varphi_{|U})^{-1}P_0\varphi$ has the desired properties.

Let $T\colon Y\to l_\infty$ be an isomorphism such that l_∞/Y and l_∞/TY are both non-reflexive. By the preceding remark there are projections P and Q on l_∞ so that $PY=QTY=\{0\}$ and $Pl_\infty\approx Ql_\infty\approx l_\infty$. Let I denote the identity operator of l_∞. The subspace $(I-Q)l_\infty$ contains TY and hence, by the injectivity of l_∞, there is an $S_1\colon l_\infty\to(I-Q)l_\infty$ such that $S_{1|Y}=T$. Similarly, there is an $S_2\colon l_\infty\to(I-P)l_\infty$ so that $S_{2|TY}=T^{-1}$. Let R be an isomorphism from l_∞ onto Ql_∞ normalized so that $\|R^{-1}\|>1$ and define $\hat{T}\colon l_\infty\to l_\infty$ by $\hat{T}=S_1+R(I-S_2S_1)$. Since $(I-S_2S_1)_{|Y}=0$ it follows that $\hat{T}_{|Y}=T$. We show next that $\hat{T}$ is an isomorphism into l_∞. Indeed, let $x\in l_\infty$; then

$$\begin{aligned}\|\hat{T}x\| &\geq \max(\|Q\hat{T}x\|/\|Q\|,\ \|(I-Q)\hat{T}x\|/\|I-Q\|)\\ &\geq \tfrac{1}{2}\|Q\|^{-1}\max(\|R(I-S_2S_1)x\|,\ \|S_1x\|)\\ &\geq \tfrac{1}{2}\|Q\|^{-1}\|R^{-1}\|^{-1}\max(\|x\|-\|S_2\|\,\|S_1x\|,\ \|S_1x\|)\\ &\geq \|x\|/2\|Q\|\,\|R^{-1}\|(\|S_2\|+1)\,.\end{aligned}$$

The operator $\hat{T}$ is not necessarily an automorphism of l_∞ since $\hat{T}$ need not map l_∞ onto l_∞. In order to replace $\hat{T}$ by an automorphism consider the subspace $\hat{T}(I-P)l_\infty$ of l_∞. This is a complemented subspace of infinite-codimension in l_∞ (since $(I-P)l_\infty$ is of infinite codimension in l_∞). Hence, by 2.a.7, there is a subspace W of l_∞ so that $l_\infty=W\oplus\hat{T}(I-P)l_\infty$ and $W\approx l_\infty$. Let R_0 be an isomorphism from

Pl_∞ onto W. The operator $\tau = R_0P + \hat{T}(I-P)$ is an automorphism of l_∞ and $\tau_{|Y} = R_0P_{|Y} + \hat{T}(I-P)_{|Y} = \hat{T}_{|Y} = T$. □

Proof of 2.f.12(iii). We remark first that if K is a w compact operator on l_∞ then $I+K$ is a Fredholm operator of index 0. This follows from the fact that K^2 is a compact operator (cf. fact (ii) in the proof of 2.a.7) and that the classical Riesz theory for compact operators applies also to operators which have a compact power (cf. [33, VII.4.6]). Alternatively, we could apply 2.c.10 and use the fact that K is strictly singular. Indeed, let $V \subset l_\infty$ be such that $K_{|V}$ is an isomorphism. By the injectivity of l_∞ there is a $K_0: l_\infty \to l_\infty$ such that $K_{0|KV} = K^{-1}$. Then $K_0K_{|V}$ = identity and since $(K_0K)^2$ is compact it follows that dim $V < \infty$.

Let now R and S be operators from l_∞ into itself such that $R_{|Y} = T$, $S_{|TY} = T^{-1}$ (R and S exist since l_∞ is injective). The operator $I - SR$ vanishes on Y and thus it factors through the reflexive space l_∞/Y. Consequently, $I - SR$ is w compact and hence SR is a Fredholm operator of index 0. Similarly, since l_∞/TY is reflexive, RS is a Fredholm operator of index 0. Hence, both R and S are Fredholm operators and $i(R) + i(S) = 0$. This proves that $i(R)$ does not depend on the choice of R and, thus, $\hat{\imath}(T)$ is well defined. If there is an automorphism τ which extends T then $\hat{\imath}(T) = i(\tau) = 0$. Conversely, if $\hat{\imath}(T) = 0$ then T can be extended to an $\acute{R}: l_\infty \to l_\infty$ with index 0. By adding to R a finite rank operator (which maps ker R isomorphically on a subspace U of l_∞ such that $U \oplus Rl_\infty = l_\infty$) we get an automorphism τ of l_∞ which extends T. □

Remarks. There are subspaces Y of l_∞ such that l_∞/Y is reflexive and infinite-dimensional. For example, *l_2 is isomorphic to a quotient space of l_∞*. This follows from the following facts. By 2.b.3 $L_1(0, 1)$ has a subspace isomorphic to l_2 and thus l_2 is a quotient space of $L_\infty(0, 1)$. *The space $L_\infty(0, 1)$ is isomorphic to l_∞* [113]. Indeed, $L_\infty(0, 1)$ as a dual of a separable space is isometric to a subspace of l_∞. The usual proof of the Hahn–Banach theorem can be used to show that, for every $Y \supset X$ and every $T \in L(X, L_\infty(0, 1))$, there is a $\hat{T} \in L(Y, L_\infty(0, 1))$ with $\|\hat{T}\| = \|T\|$ and $\hat{T}_{|X} = T$. Thus, $L_\infty(0, 1)$ is injective and its isomorphism with l_∞ follows by using the decomposition method or, alternatively, from 2.a.7.

Let us verify now that cases (ii) and (iii) of 2.f.12 can actually occur. Let l_∞/Y be isomorphic to l_2. Since $l_\infty \oplus l_\infty = l_\infty$ there is a subspace Y_0 of l_∞ so that Y_0 is isometric to Y and $l_\infty/Y_0 \approx l_\infty/Y \oplus l_\infty$, i.e. l_∞/Y_0 is non-reflexive. By letting T be an isometry from Y to Y_0 we get an instance where case (ii) occurs. Let now $S: l_\infty \to l_\infty$ be the shift operator defined by $S(a_1, a_2, \ldots) = (0, a_1, a_2, \ldots)$. For a positive integer k let $T_k = S^k_{|Y}$. Then, l_∞/Y and l_∞/T_kY are both reflexive and $\hat{\imath}(T_k) = k$, and $\hat{\imath}(T_k^{-1}) = -k$.

In connection with 2.f.12 it is of interest to study further the subspaces Y of l_∞ so that l_∞/Y is reflexive. In this direction we have the following.

Proposition 2.f.13 [89]. *Let Y be a subspace of l_∞ such that l_∞/Y is reflexive. Then Y has a subspace isomorphic to l_∞.*

Proof. We show first that Y has a subspace isomorphic to c_0. Otherwise, Y and c_0

would be totally incomparable and thus (cf. the result quoted following 2.c.1) $Y+c_0$ would be a closed subspace of l_∞ with $\dim(Y\cap c_0)<\infty$. Consequently, the quotient map $l_\infty \to l_\infty/Y$ would be an isomorphism on an infinite-dimensional subspace of c_0 which contradicts the reflexivity to l_∞/Y.

By using 2.a.8 we get that there is a sequence $\{y_n\}_{n=1}^\infty$ in Y which is equivalent to the unit vector basis of c_0 so that, for every infinite subset M of the integers, the w^* closure X_M of $[y_n]_{n\in M}$ is isomorphic to l_∞. Let $\{N_\gamma\}_{\gamma\in\Gamma}$ be an uncountable set of subsets of the integers so that $N_{\gamma_1}\cap N_{\gamma_2}$ is finite for every $\gamma_1\neq\gamma_2$. We claim that, for some $\gamma\in\Gamma$, $X_{N_\gamma}\subset Y$. Assume that, for every γ, there is an $x_\gamma\in X_{N_\gamma}$ with $\|x_\gamma\|=1$ and $x_\gamma\notin Y$. Let $T\colon l_\infty\to l_\infty/Y$ be the quotient map. Since Γ is uncountable there is a sequence $\{\gamma_i\}_{i=1}^\infty$ of elements in Γ and a $\delta>0$ such that $\|Tx_{\gamma_i}\|\geqslant\delta$ for every i. Since $[y_n]_{n=1}^\infty\subset Y$ it follows that, for every choice of scalars $\{a_i\}_{i=1}^j$, $\left\|\sum_{i=1}^j a_iTx_{\gamma_i}\right\|\leqslant\max_{1\leqslant i\leqslant j}|a_i|$ and thus $S\colon c_0\to l_\infty/Y$, defined by $S(a_1, a_2,\ldots)=\sum_{i=1}^\infty a_iTx_{\gamma_i}$, is a bounded operator, and $Tx_{\gamma_i}\to 0$ weakly in l_∞/Y. Since l_∞/Y is reflexive S is weakly compact and therefore also compact (recall that in l_1 a sequence converges w if and only if it converges in norm and that an operator is compact, resp. w compact, if and only if its conjugate is such an operator). Thus, $\{Tx_{\gamma_i}\}_{i=1}^\infty$ has a subsequence which converges in norm. The limit of this subsequence must be 0 but this contradicts the fact that $\|Tx_{\gamma_i}\|\geqslant\delta$ for all i. □

It follows from 2.f.12 and 2.f.13 that if Y is separable then, for any pair of isomorphisms T_1 and T_2 of Y into l_∞, there is an automorphism τ of l_∞ so that $\tau T_1=T_2$ and, in particular, $l_\infty/T_1Y\approx l_\infty/T_2Y$. It makes sense therefore to talk of the space l_∞/Y.

Theorems 2.f.8, 2.f.10 and 2.f.12 show that there are many automorphisms of l_1, c_0 and l_∞. The automorphisms constructed in these theorems are in general not isometries. The isometries form a rather restricted class and it is easy to determine their most general form.

Proposition 2.f.14. *Let τ be an isometry of c_0 or l_p, $1\leqslant p\leqslant\infty$, $p\neq 2$ onto itself. Then there is a sequence $\{\theta_i\}_{i=1}^\infty$ of signs and a permutation π of the integers so that $\tau(a_1, a_2,\ldots)=(\theta_1a_{\pi(1)}, \theta_2a_{\pi(2)}, \theta_3a_{\pi(3)},\ldots)$.*

Proof. Let us first consider the spaces l_p, $1\leqslant p<\infty$, $p\neq 2$. It is not hard to verify (and for $p=1$ it is trivial) that if $x=(a_1, a_2,\ldots)$ and $y=(b_1, b_2,\ldots)$ then $\|x+y\|^p=\|x-y\|^p=\|x\|^p+\|y\|^p$ if and only if x and y have disjoint supports (i.e. $a_ib_i=0$ for every i). Hence, if τ is an isometry of l_p into itself and $\{e_i\}_{i=1}^\infty$ denotes the unit vector basis of l_p then $\{\tau e_i\}_{i=1}^\infty$ have mutually disjoint supports. Consequently, if τ is, in addition, onto then there must exist a permutation π of the integers and signs $\{\theta_i\}_{i=1}^\infty$ so that $\tau e_i=\theta_ie_{\pi(i)}$ for every i. The case of c_0 can be proved similarly or by using the fact that if τ is an isometry of c_0 onto c_0 then τ^* is an isometry of l_1 onto l_1. In the case of l_∞ the simplest way to prove 2.f.14 is to deduce it as a special case of the general result on the structure of isometries of $C(K)$ spaces (the classical Banach–Stone theorem which will be discussed in Vol. III). □

3. Symmetric Bases

a. Properties of Symmetric Bases, Examples and Special Block Bases

It is a trivial observation that the unit vector basis $\{e_n\}_{n=1}^\infty$ of c_0 or l_p, $p\geq 1$, besides being unconditional, is equivalent to any of its permutations. Moreover, the unit ball of each of these spaces is a symmetric body in the sense that it contains a sequence $(a_1, a_2, \ldots)$ if and only if it contains the sequence $(a_{\pi(1)}, a_{\pi(2)}, \ldots)$, for every permutation π of the integers. The purpose of this chapter is to study those Banach spaces which share this property with c_0 and l_p.

Definition 3.a.1 [134]. A basis $\{x_n\}_{n=1}^\infty$ of a Banach space X is said to be *symmetric* if, for any permutation π of the integers, $\{x_{\pi(n)}\}_{n=1}^\infty$ is equivalent to $\{x_n\}_{n=1}^\infty$.

It is interesting to compare this definition with 1.c.6 where it is shown that a basis $\{x_n\}_{n=1}^\infty$ is unconditional if and only if, for each permutation π, $\{x_{\pi(n)}\}_{n=1}^\infty$ is a basic sequence. For $\{x_n\}_{n=1}^\infty$ to be a symmetric basis one has to require that, in addition, the basic sequence $\{x_{\pi(n)}\}_{n=1}^\infty$ be equivalent to $\{x_n\}_{n=1}^\infty$ for each such π. This remark shows that *every symmetric basis is also unconditional.*

Let $\{x_n\}_{n=1}^\infty$ be a symmetric basis of a Banach space X. Then, for each permutation π of the integers, the operator $V_\pi\colon X\to X$ defined by $V_\pi\left(\sum_n a_n x_n\right)=\sum_n a_n x_{\pi(n)}$ is evidently an automorphism of X. Moreover, the family $\{V_\pi\}$ of all these automorphisms induced by permutations forms a uniformly bounded group. Indeed, if this were not true then, by the uniform boundedness principle, we could produce a vector $x=\sum_{n=1}^\infty a_n x_n \in X$, a sequence $\{\pi_j\}_{j=1}^\infty$ of permutations of the integers and a sequence $\{\sigma_j\}_{j=1}^\infty$ of finite subsets of the integers with

$$\sigma_j \cap \sigma_k = \emptyset \quad \text{and} \quad \pi_j(\sigma_j)\cap \pi_k(\sigma_k)=\emptyset\,,$$

for $j\neq k$, such that $\left\|\sum_{n\in\sigma_j} a_n x_{\pi_j(n)}\right\|\geq 1$ for all j. Then, for any permutation π_0 such that $\pi_0(n)=\pi_{2j}(n)$ for $n\in\sigma_{2j}$, $j=1, 2,\ldots$, it would follow that $\sum_{n=1}^\infty a_n x_{\pi_0(n)}$ diverges, contrary to our assumption.

Since, as remarked above, $\{x_n\}_{n=1}^\infty$ is also unconditional, we get that $K=\sup_{\theta,\pi}\|M_\theta V_\pi\|<\infty$ (recall that, for every choice of signs $\theta=\{\theta_n\}_{n=1}^\infty$, the operator M_θ is defined by $M_\theta\left(\sum_{n=1}^\infty a_n x_n\right)=\sum_{n=1}^\infty a_n\theta_n x_n$). The number K, thus defined, is called the *symmetric constant* of $\{x_n\}_{n=1}^\infty$. Notice that the symmetric constant of any symmetric basis is always larger than or equal to the unconditional constant of that basis. If, for $x=\sum_{n=1}^\infty a_n x_n \in X$, we put

$$\|x\|_0=\sup_{\theta_n=\pm 1}\sup_{\pi}\left\|\sum_{n=1}^\infty a_n\theta_n x_{\pi(n)}\right\|$$

then $\|\cdot\|_0$ is a new norm on X such that $\|x\|\leqslant\|x\|_0\leqslant K\|x\|$ for all $x\in X$. This new norm satisfies $\left\|\sum_{n=1}^\infty a_n\theta_n x_{\pi(n)}\right\|_0=\left\|\sum_{n=1}^\infty a_n x_n\right\|_0$ for every permutation π of the integers and every choice of signs $\theta_n=\pm 1$. If X is equipped with such a norm (called sometimes a symmetric norm) then its unit ball is a symmetric body with respect to all permutations of the coordinates and all changes of signs.

We introduce now a notion which is slightly weaker than that of a symmetric basis.

Definition 3.a.2. A basis $\{x_n\}_{n=1}^\infty$ of a Banach space X is called *subsymmetric* if it is unconditional and, for every increasing sequence of integers $\{n_i\}_{i=1}^\infty$, $\{x_{n_i}\}_{i=1}^\infty$ is equivalent to $\{x_n\}_{n=1}^\infty$.

The requirement that $\{x_n\}_{n=1}^\infty$ be unconditional is not redundant: a basis which is equivalent to each of its subsequences need not be unconditional. For instance, the summing basis of the space c, introduced in Section 1.c, is subsequence equivalent but not unconditional.

Let $\{x_n\}_{n=1}^\infty$ be a subsymmetric basis of a Banach space X. It is easily checked that, for any increasing sequence of integers $\{n_i\}_{i=1}^\infty$, the operator $S_{\{n_i\}}; X\to X$, defined by $S_{\{n_i\}}\left(\sum_{n=1}^\infty a_n x_n\right)=\sum_{i=1}^\infty a_i x_{n_i}$, is an isomorphism from X onto $[x_{n_i}]_{i=1}^\infty$. Moreover, the family of all these isomorphisms is uniformly bounded; hence, we can define the *subsymmetric constant* of $\{x_n\}_{n=1}^\infty$ as the number

$$K_1=\sup_{\theta,\{n_i\}}\|M_\theta S_{\{n_i\}}\|<+\infty.$$

If, for $x=\sum_{n=1}^\infty a_n x_n\in X$, we put $\|x\|_0=\sup_{\theta_i=\pm 1}\sup_{\{n_i\}}\left\|\sum_{i=1}^\infty a_i\theta_i x_{n_i}\right\|$ then $\|\cdot\|_0$ is a new norm on X which satisfies $\|x\|\leqslant\|x\|_0\leqslant K_1\|x\|$ and $\left\|\sum_{i=1}^\infty a_i\theta_i x_{n_i}\right\|_0=\left\|\sum_{n=1}^\infty a_n x_n\right\|_0$ for all $x\in X$. Such a norm is called a subsymmetric norm.

The relationship between symmetric and subsymmetric bases is described in the following simple result.

Proposition 3.a.3 [134]. *Every symmetric basis is subsymmetric.*

Proof. Let X be a Banach space with a symmetric basis $\{x_n\}_{n=1}^\infty$ whose symmetric constant is 1. For every increasing sequence of integers $\{n_i\}_{i=1}^\infty$ and every integer k there is a permutation π such that $\pi(i)=n_i$ for $1\leqslant i\leqslant k$. Then, for any choice of scalars $(a_1,\ldots,a_k)$, we have $\left\|\sum_{i=1}^{k} a_i x_i\right\| = \left\|\sum_{i=1}^{k} a_i x_{\pi(i)}\right\| = \left\|\sum_{i=1}^{k} a_i x_{n_i}\right\|$, which completes the proof since k is arbitrary. □

It follows easily from 3.a.3 and its proof that, for any symmetric basis, the symmetric constant is larger than or equal to its subsymmetric constant.

That the notions of a symmetric and of a subsymmetric basis do not coincide is shown by the following example, due to D. J. H. Garling [44]. Let Y be the space of all sequences of scalars $y=(a_1, a_2,\ldots)$ for which

$$\|y\| = \sup \sum_{i=1}^{\infty} |a_{n_i}| i^{-1/2} < \infty,$$

where the supremum is taken over all increasing sequences of integers $\{n_i\}_{i=1}^{\infty}$. It is easily checked that Y, endowed with the norm defined above, is a Banach space whose unit vectors $\{e_n\}_{n=1}^{\infty}$ form a subsymmetric basis. However, the unit vectors do not form a symmetric basis in Y. Indeed, for each fixed k, the vector $y^{(k)} = (1, 2^{-1/2},\ldots, k^{-1/2}, 0, 0,\ldots)$ is obtained from $z^{(k)} = (k^{-1/2}, (k-1)^{-1/2},\ldots, 1, 0, 0,\ldots)$ by a suitable permutation of the integers but, in spite of this, $\|y^{(k)}\| = \sum_{n=1}^{k} n^{-1}$ and $\|z^{(k)}\| = \sum_{n=1}^{k} (k-n+1)^{-1/2} n^{-1/2}$, i.e. $\sup_k \|y^{(k)}\| = \infty$ while $\sup_k \|z^{(k)}\| < \infty$.

In the past, spaces with a subsymmetric basis were studied only very little. However, quite recently, there has been a growing interest in the notion of a subsymmetric basis especially since it has turned out that such bases arise naturally in the local theory of Banach spaces. We shall discuss this in more detail in Vol. III.

The most interesting class of spaces with a symmetric basis is perhaps that of Orlicz sequence spaces which were originally introduced by W. Orlicz [111]. An Orlicz function M is a convex, non-decreasing continuous function on $[0,\infty)$ such that $M(0)=0$ and $\lim_{t\to\infty} M(t)=\infty$. To any Orlicz function M one can associate the space l_M of all the sequences $x=(a_1, a_2,\ldots)$ of scalars such that $\sum_{n=1}^{\infty} M(|a_n|/\rho) < \infty$ for some $\rho>0$. The conditions imposed on M make l_M into a Banach space when it is equipped with the norm

$$\|x\| = \inf\left\{\rho>0;\ \sum_{n=1}^{\infty} M(|a_n|/\rho) \leqslant 1\right\}.$$

The space l_M is called an Orlicz sequence space and it can be easily seen that, for $M(t)=t^p$ with $p\geqslant 1$, l_M coincides with the classical space l_p. The unit vectors $\{e_n\}_{n=1}^{\infty}$ form clearly a symmetric basis of $h_M = [e_n]_{n=1}^{\infty}$ and, in general, h_M does not coincide with l_M. We shall discuss this matter, as well as many other questions regarding the structure of Orlicz spaces, in the next chapter.

Another class of spaces with a symmetric basis related to the l_p spaces is that of Lorentz sequence spaces. For every $p\geqslant 1$ and every non-increasing sequence of positive numbers $w=\{w_n\}_{n=1}^{\infty}$ we consider the space $d(w,p)$ of all sequences of scalars $x=(a_1, a_2, \ldots)$ for which $\|x\|=\sup\left(\sum_{n=1}^{\infty} |a_{\pi(n)}|^p w_n\right)^{1/p}<\infty$, the supremum being taken over all permutations π of the integers. It is easily checked that $d(w, p)$, endowed with the norm $\|\cdot\|$ defined above, is a Banach space. If $\inf_n w_n>0$ then $d(w, p)$ is isomorphic to l_p. Another trivial case occurs when $\sum_{n=1}^{\infty} w_n<\infty$: then, $d(w, p)\approx l_\infty$. We shall exclude these two trivial cases, i.e. we shall assume that $\lim_{n\to\infty} w_n=0$, $\sum_{n=1}^{\infty} w_n=\infty$ and $w_1=1$ (to normalize the unit vectors). When these conditions are satisfied $d(w, p)$ is called a Lorentz sequence space. We shall discuss these spaces in Section 4.e.

There is also a general procedure to construct symmetric bases starting with an arbitrary sequence of vectors. Let X be a Banach space and let $\{x_n\}_{n=1}^{\infty}$ be an arbitrary sequence of non-zero vectors in X. Denote by Z the completion of the space of all sequences of scalars $z=(a_1, a_2, \ldots)$ which are eventually equal to zero, with respect to the norm $\|z\|_Z=\sup_{\theta_n=\pm 1}\sup_{\pi}\left\|\sum_{n=1}^{\infty} a_n\theta_n x_{\pi(n)}\right\|$. The space Z, equipped with the norm $\|\cdot\|_Z$, is a Banach space in which the unit vector basis is a symmetric basis.

New spaces with a symmetric basis can be constructed from old ones by some averaging processes. Sometimes, we obtain in this way new interesting spaces. For instance, let μ be a probability measure on the interval $[1, 2]$ and let X_μ be the space of all sequences of scalars $x=(a_1, a_2, \ldots)$ with the norm $\|x\|_\mu=\int_1^2\left(\sum_{n=1}^{\infty}|a_n|^p\right)^{1/p}d\mu(p)<\infty$. We shall see in Vol. II that X_μ is a subspace of $L_1(0, 1)$. These spaces, as well as spaces obtained by averaging suitable Orlicz norms, play a role in the investigation of the structure of $L_1(0, 1)$.

Let X be a Banach space with a normalized symmetric basis $\{x_n\}_{n=1}^{\infty}$ whose symmetric constant is 1. In the study of subspaces of X a special role is played by block bases with constant coefficients (with respect to $\{x_n\}_{n=1}^{\infty}$). Let $\sigma=\{\sigma_j\}_{j=1}^{\infty}$ be a sequence of consecutive disjoint finite subsets of the integers (i.e. the greatest integer in σ_j precedes the smallest integer in σ_{j+1}, for all j), let $\bar{\bar{\sigma}}_j$ denote the number of elements in σ_j and define the operator P_σ on X, as follows.

$$P_\sigma(x)=\sum_{j=1}^{\infty}\left(\left(\sum_{n\in\sigma_j} a_n\right)\Big/\bar{\bar{\sigma}}_j\right)\left(\sum_{n\in\sigma_j} x_n\right),\quad x=\sum_{n=1}^{\infty} a_n x_n\in X\,.$$

The operator P_σ, thus defined, is called *the averaging projection* or *the conditional expectation with respect to* σ and has the following property.

Proposition 3.a.4. *For every X, $\{x_n\}_{n=1}^{\infty}$ and σ as above, the averaging projection P_σ is a norm-one linear projection in X whose range is the closed linear span of the vectors $u_j=\sum_{n\in\sigma} x_n$, $j=1, 2, \ldots$. In other words, the closed linear span of any block basis with constant coefficients is complemented.*

If $\{x_n\}_{n=1}^{\infty}$ is only a normalized subsymmetric basis whose subsymmetric constant is 1, the same holds with the exception that $\|P_\sigma\|\leq 2$.

Proof. Fix an integer k and let $\{\pi_i\}_{i=1}^{m(k)}$, with $m(k)=\bar{\bar{\sigma}}_1\cdot\bar{\bar{\sigma}}_2\ldots\bar{\bar{\sigma}}_k$, be the set of all distinct permutations π such that π leaves invariant all the integers outside $\sigma_1\cup\sigma_2\cup\cdots\cup\sigma_k$ and, for $1\leqslant j\leqslant k$, $\pi(\sigma_j)=\sigma_j$ and π restricted to σ_j acts as a cyclic permutation on σ_j (by cyclic permutation τ on a set $\{n_i\}_{i=1}^k$, where $n_1<n_2<\cdots<n_k$, we mean a permutation of the form $\tau_j(n_i)=n_{i+j\,(\mathrm{mod}\,k)}$, $1\leqslant i\leqslant k$, $1\leqslant j\leqslant k$). For every vector $y=\sum_{n=1}^\infty a_n x_n\in X$ and every $1\leqslant i\leqslant m(k)$ we have $\left\|\sum_{n=1}^\infty a_{\pi_i(n)}x_n\right\|=\|y\|$, in the symmetric case, and $\left\|\sum_{n=1}^\infty a_{\pi_i(n)}x_n\right\|\leqslant 2\|y\|$, in the subsymmetric case.

By averaging over $i=1, 2,\ldots, m(k)$ we get that

$$\left\|\sum_{j=1}^k\left(\left(\sum_{n\in\sigma_j}a_n\right)\Big/\bar{\bar{\sigma}}_j\right)u_j+\sum_{n\notin\bigcup_{j=1}^k\sigma_j}a_nx_n\right\|,$$

and therefore also $\left\|\sum_{j=1}^k\left(\left(\sum_{n\in\sigma_j}a_n\right)\Big/\bar{\bar{\sigma}}_j\right)u_j\right\|$, are $\leqslant\|y\|$, in the symmetric case, and $\leqslant 2\|y\|$, in the subsymmetric case. Since k is arbitrary it follows that P_σ is a well-defined operator on X whose norm is $\leqslant 1$, respectively $\leqslant 2$. Obviously, P_σ is a projection whose range is $[u_j]_{j=1}^\infty$. □

In the case $\{x_n\}_{n=1}^\infty$ is a symmetric basis 3.a.4 clearly remains true even if the u_j's are taken to be a block basis with constant coefficients of some permutation of $\{x_n\}_{n=1}^\infty$. As pointed out by J. T. Woo [145], there exist also non-symmetric unconditional bases which have this property, i.e. in which every block basis with constant coefficients of any permutation of the basis spans a complemented subspace. Such a basis is, for example, the natural basis of the space $X_{\infty,r}$, $r\geqslant 1$ (see the remark following 4.d.7 below).

As a first application of 3.a.4, we bring the following.

Proposition 3.a.5. *Let X be a Banach space with a symmetric basis $\{x_n\}_{n=1}^\infty$. For every $\sigma=\{\sigma_j\}_{j=1}^\infty$ denote by U the closed linear span of the vectors $u_j=\sum_{n\in\sigma_j}x_n$, $j=1, 2,\ldots$. Then X is isomorphic to $X\oplus U$.*

Proof. We shall use a simplified argument due to W. B. Johnson. Let $\{v_n\}_{n=1}^\infty$ be a normalized block basis with constant coefficients of $\{x_n\}_{n=1}^\infty$ so that every possible normalized block with constant coefficients of $\{x_n\}_{n=1}^\infty$ appears infinitely many times in the sequence $\{v_n\}_{n=1}^\infty$. This property of $\{v_n\}_{n=1}^\infty$ ensures that $V=[v_n]_{n=1}^\infty$ satisfies $V\oplus V\approx V$ and $V\oplus U\approx V$ for every U as above. It follows that we can apply the decomposition method for the spaces $X\oplus V$ and X since $X\oplus X\approx X$, $X\oplus V\approx(X\oplus V)\oplus(X\oplus V)$, and, by 3.a.4, $X\oplus V$ is a complemented subspace of X. We conclude that

$$X\approx X\oplus V\approx X\oplus V\oplus U\approx X\oplus U\,.\ \square$$

We are now prepared to complete the proof of the fact that c_0, l_1 and l_2 are the only Banach spaces which have, up to equivalence, a unique unconditional basis.

Proof of 2.b.10. Assume that a Banach space X has, up to equivalence, a unique normalized unconditional basis $\{x_n\}_{n=1}^\infty$. Then, for each permutation π of the integers, $\{x_{\pi(n)}\}_{n=1}^\infty$ is equivalent to $\{x_n\}_{n=1}^\infty$ and therefore $\{x_n\}_{n=1}^\infty$ is a symmetric basis of X. Let $\{u_j\}_{j=1}^\infty$ be any normalized block basis with constant coefficients of $\{x_n\}_{n=1}^\infty$ and let $U=[u_j]_{j=1}^\infty$. By 3.a.5, $X\approx X\oplus U$, which implies that $\{x_1, u_1, x_2, u_2, \ldots, x_n, u_n, \ldots\}$ is a normalized unconditional basis of X. Thus, $\{u_j\}_{j=1}^\infty$ is equivalent to $\{x_n\}_{n=1}^\infty$ and, by the second remark following 2.a.9, it follows that $\{x_n\}_{n=1}^\infty$ is a perfectly homogeneous basis, i.e. $\{x_n\}_{n=1}^\infty$ is equivalent to the unit vector basis of c_0 or l_p for some $p\geqslant 1$. Since each of the spaces l_p with $p\neq 1, 2$ has an unconditional basis which is not equivalent to the unit vector basis (see the discussion preceding 2.b.10), the only remaining possibilities are c_0, l_1 and l_2. □

It is interesting to point out that *a Banach space with an unconditional basis can have either a unique normalized unconditional basis, as in the case of l_1, l_2 and c_0, or uncountably many non-equivalent unconditional bases.* The construction of infinitely many unconditional bases is quite simple (cf. [52]). Let X be a space having a least two non-equivalent normalized unconditional bases. Then, we can easily check that X admits a normalized unconditional basis, say $\{x_n\}_{n=1}^\infty$, which is not symmetric. For such a basis there is a partition of the integers into disjoint infinite subsets $\{N_j\}_{j=1}^\infty$ so that, for each j, $\{x_n\}_{n\in N_j}$ is not a symmetric basis sequence. Let $\pi_{j,1}$ be a permutation of N_j such that $\{x_n\}_{n\in N_j}$ and $\{x_{\pi_{j,1}(n)}\}_{n\in N_j}$ are not equivalent. For each sequence $\{\eta(j)\}_{j=1}^\infty$ of zeros and ones, we consider the basis $\bigcup_{j=1}^\infty \{x_{\pi_{j,\eta(j)}(n)}\}_{n\in N_j}$, where $\pi_{j,0}$ denotes simply the identity on N_j. It is easily seen that distinct sequences of zeros and ones give rise to non-equivalent unconditional bases of X.

Proposition 3.a.4 is also useful if we desire to study the behavior of the sums of vectors of a normalized symmetric or subsymmetric basis.

Proposition 3.a.6. *Let $\{x_n\}_{n=1}^\infty$ be a normalized symmetric (subsymmetric) basis of a Banach space X with symmetric (subsymmetric) constant equal to 1. Let $\{x_n^*\}_{n=1}^\infty$ be the sequence of the functionals biorthogonal to $\{x_n\}_{n=1}^\infty$ and, for each n, put $\lambda(n)=\|x_1+x_2+\cdots+x_n\|$ and $\mu(n)=\|x_1^*+x_2^*+\cdots+x_n^*\|$. Then,*

$$\lambda(n)\mu(n)=n \qquad (n\leqslant\lambda(n)\mu(n)\leqslant 2n)\,,$$

for all integers n.

Proof. Since $\left(\sum_{k=1}^n x_k^*\right)\left(\sum_{k=1}^n x_k\right)=n$ the inequality $\lambda(n)\mu(n)\geqslant n$ is true for every space with a basis. By 3.a.4, applied to a single block σ, we obtain the reverse inequality (i.e. $\lambda(n)\mu(n)\leqslant n$ in the symmetric case or $\lambda(n)\mu(n)\leqslant 2n$ in the subsymmetric case). Indeed,

$$\mu(n)=\left\|\sum_{k=1}^n x_k^*\right\|=\sup_{\{a_k\}}\left(\sum_{k=1}^n a_k\right)\Big/\left\|\sum_{k=1}^n a_k x_k\right\|\leqslant n/\lambda(n),$$

(respectively, $2n/\lambda(n)$). □

For every space with a normalized symmetric basis $\{x_n\}_{n=1}^\infty$ with symmetric constant equal to 1, $\{\lambda(n)\}_{n=1}^\infty$ is a non-decreasing sequence which tends to ∞ unless $\{x_n\}_{n=1}^\infty$ is equivalent to the unit vector basis of c_0 (see e.g. the proof of 2.a.9). In most of the known examples $\lambda(\cdot)$ is even a "concave" function on the integers in the sense that $\{\lambda(n+1)-\lambda(n)\}_{n=1}^\infty$ is a non-increasing sequence. This fact is not true in general but we have the following result (cf. [148]).

Proposition 3.a.7. *Let $(X, \|\cdot\|)$ be a Banach space with a symmetric basis $\{x_n\}_{n=1}^\infty$ whose symmetric constant is equal to* 1. *Then there exists a new norm $\|\cdot\|_0$ on X such that:*

(i) $\|x\| \leq \|x\|_0 \leq 2\|x\|$ *for all* $x \in X$,

(ii) *The symmetric constant of $\{x_n\}_{n=1}^\infty$ with respect to $\|\cdot\|_0$ is equal to* 1,

(iii) *if we put* $\lambda_0(n)=\left\|\sum_{i=1}^n x_i\right\|_0$, $n=1, 2,\ldots$ *then* $\{\lambda_0(n+1)-\lambda_0(n)\}_{n=1}^\infty$ *is a non-increasing sequence, i.e. $\lambda_0(\cdot)$ is a concave function on the integers.*

Proof. Let $\{x_n^*\}_{n=1}^\infty$ denote again the sequence of the functionals biorthogonal to $\{x_n\}_{n=1}^\infty$ and let $\lambda(n)=\left\|\sum_{i=1}^n x_i\right\|$, $\mu(n)=\left\|\sum_{i=1}^n x_i^*\right\|$ for $n\geq 1$ and $\lambda(0)=0$. For each integer n, we put further $\lambda_0(n)=\sup \sum_{i=1}^n [\lambda(k_i)-\lambda(k_i-1)]$, where the supremum is taken over all n-tuples of integers $\{k_1<k_2<\cdots<k_n\}$. By 3.a.6 we get, for every $n\geq 1$,

$$\lambda(n)\leq \lambda_0(n)=\sup\Big\{\sum_{k_i\leq n} (\lambda(k_i)-\lambda(k_i-1))+\sum_{k_i>n} (k_i/\mu(k_i)-(k_i-1)/\mu(k_i-1))\Big\}$$
$$\leq \lambda(n)+n/\mu(n)=2\lambda(n)\,.$$

Now, fix $\varepsilon>0$ and n and choose suitable integers so that

$$\lambda_0(n-1)<\varepsilon/2+\sum_{i=1}^{n-1}(\lambda(h_i)-\lambda(h_i-1)),$$

$$\lambda_0(n+1)<\varepsilon/2+\sum_{i=1}^{n+1}(\lambda(j_i)-\lambda(j_i-1))\,.$$

It follows immediately that $\lambda_0(n-1)+\lambda_0(n+1)\leq \varepsilon+2\lambda_0(n)$ which shows that $\lambda_0(n+1)-\lambda_0(n)\leq\lambda_0(n)-\lambda_0(n-1)$ since $\varepsilon>0$ is arbitrary. The sequence $\lambda_0(n)$ is therefore concave. Observe that a similar argument shows also that $k\lambda_0(n)\leq n\lambda_0(k)$ for all $1\leq k\leq n$. In order to complete the proof it suffices to construct a new norm $\|\cdot\|_0$ satisfying (i), (ii) and such that $\left\|\sum_{i=1}^n x_i\right\|_0=\lambda_0(n)$ for all n. This is achieved by setting,

$$\|x\|_0=\max\Big\{\|x\|, \sup_n \sup_\pi \lambda_0(n) \sum_{i=1}^n |a_{\pi(i)}|/n\Big\} \quad \text{for } x=\sum_{n=1}^\infty a_n x_n \in X\,,$$

where the inner supremum ranges over all the permutations of the integers. The norm $\|\cdot\|_0$ has indeed all the desired properties since, by 3.a.4,

$$\lambda(n) \sum_{i=1}^{n} |a_{\pi(i)}|/n \leq \left\| \sum_{i=1}^{n} |a_{\pi(i)}| x_i \right\| \leq \|x\|$$

for all x, n and π as above. □

The converse to 3.a.7 is also true in the sense that, for every concave non-decreasing sequence of positive numbers $\{\lambda_n\}_{n=1}^{\infty}$, there exists at least one Banach space X having a symmetric basis $\{x_n\}_{n=1}^{\infty}$ with symmetric constant equal to 1 such that $\left\|\sum_{i=1}^{n} x_i\right\| = \lambda_n$ for every n. For instance, we can take the space $d(w, 1)$ with the sequence $w = \{w_n\}_{n=1}^{\infty}$ defined as follows: $w_1 = \lambda_1$ and, for $n > 1$, $w_n = \lambda_n - \lambda_{n-1}$. The sequence $\{\lambda_n\}_{n=1}^{\infty}$ does not determine the equivalence class of $\{x_n\}_{n=1}^{\infty}$ except for the following two extreme cases: if $\sup_n \lambda_n < \infty$ then $\{x_n\}_{n=1}^{\infty}$ must be equivalent to the unit vector basis of c_0 and, if $\limsup_n \lambda(n)/n > 0$ then $\{x_n\}_{n=1}^{\infty}$ is necessarily equivalent to the unit vector basis of l_1 (this last fact is an immediate consequence of 3.a.4 or of 3.a.6). If we exclude these two cases there are always many mutually non-isomorphic spaces with a symmetric basis $\{x_n\}_{n=1}^{\infty}$ such that $\left\|\sum_{i=1}^{n} x_i\right\| = \lambda_n$, for all n. For example, in addition to the space $d(w, 1)$ considered above, we can take the closed linear span of the functionals biorthogonal to the unit vector basis of $d(w', 1)$, where $w' = \{w'_n\}_{n=1}^{\infty}$ is defined as follows: $w'_1 = 1/\lambda_1$ and, for $n > 1$, $w'_n = n/\lambda_n - (n-1)/\lambda_{n-1}$ (see the discussion following 4.e.4 below).

Besides block bases with constant coefficients there is another type of special block bases which is of interest in the study of symmetric bases.

Definition 3.a.8 [4]. Let X be a Banach space with a symmetric basis $\{x_n\}_{n=1}^{\infty}$ and let $N_i = \{n_{i,1} < n_{i,2} < \cdots\}$, $i = 1, 2, \ldots$ be any sequence of disjoint infinite subsets of the integers. For every $0 \neq \alpha = \sum_{n=1}^{\infty} a_n x_n \in X$, the sequence $\{u_i^{(\alpha)}\}_{i=1}^{\infty}$, defined by $u_i^{(\alpha)} = \sum_{j=1}^{\infty} a_j x_{n_{i,j}}$, $i = 1, 2, \ldots$, is called a *block basis generated by the vector* α.

Clearly, $\{u_i^{(\alpha)}\}_{i=1}^{\infty}$ is a symmetric basic sequence and its equivalence type does not depend on the particular choice of the sets $\{N_i\}_{i=1}^{\infty}$. Observe also that we use the term "block basis" though, strictly speaking, the $u_i^{(\alpha)}$'s do not form a block basis of $\{x_n\}_{n=1}^{\infty}$.

The following elegant result concerning block bases generated by one vector was proved in [15].

Theorem 3.a.9. *Let X be a Banach space with a symmetric basis $\{x_n\}_{n=1}^{\infty}$. Let $\alpha = \sum_{n=1}^{\infty} a_n x_n \neq 0$ and let $\{u_i^{(\alpha)}\}_{i=1}^{\infty}$ be a block basis generated by the vector α. Then $\{u_i^{(\alpha)}\}_{i=1}^{\infty}$ is equivalent to $\{x_n\}_{n=1}^{\infty}$ if and only if $[u_i^{(\alpha)}]_{i=1}^{\infty}$ is a complemented subspace of X.*

Proof. Fix an integer h so that $a_h \neq 0$. Then, for every $x \in X$, put

$$Px = \sum_{i=1}^{\infty} (b_{i,h}/a_h) u_i^{(\alpha)},$$

where $b_{i,h}$ is the coefficient of $x_{n_{i,h}}$ in the expansion of x with respect to the basis $\{x_n\}_{n=1}^{\infty}$. If $\{u_i^{(\alpha)}\}_{i=1}^{\infty}$ is equivalent to $\{x_n\}_{n=1}^{\infty}$ then P is a bounded linear projection from X onto $[u_i^{(\alpha)}]_{i=1}^{\infty}$. Indeed, if K_α is chosen so that $\left\|\sum_{i=1}^{\infty} c_i u_i^{(\alpha)}\right\| \leq K_\alpha \left\|\sum_{i=1}^{\infty} c_i x_i\right\|$ for every choice of $\{c_i\}_{i=1}^{\infty}$ then $\|P\| \leq K_\alpha/|a_h|$.

To prove the converse we first notice that whenever $\sum_{i=1}^{\infty} c_i u_i^{(\alpha)}$ converges so does $\sum_{i=1}^{\infty} c_i a_h x_{n_{i,h}}$, and thus also $\sum_{i=1}^{\infty} c_i x_i$. Assume now that $[u_i^{(\alpha)}]_{i=1}^{\infty}$ is a complemented subspace of X and consider a series $\sum_{i=1}^{\infty} c_i x_i$ which converges in X. Choose an increasing sequence of integers $\{k_m\}_{m=1}^{\infty}$ so that $\left\|\sum_{i=k_m}^{\infty} c_i x_i\right\| \leq 2^{-m}$ for all m and, for $k_m \leq i < k_{m+1}$, set $v_i^{(\alpha)} = \sum_{j=1}^{m} a_j x_{n_{i,j}}$, $\lambda_i = \|u_i^{(\alpha)} - v_i^{(\alpha)}\|$ and $w_i^{(\alpha)} = (u_i^{(\alpha)} - v_i^{(\alpha)})/\lambda_i$. Then $u_i^{(\alpha)} = v_i^{(\alpha)} + \lambda_i w_i^{(\alpha)}$ for all i (notice that this is true even if $\lambda_i = 0$ in which case $w_i^{(\alpha)}$ is not well-defined and can be taken to be an arbitrary unit vector). We have,

$$\left\|\sum_{i=k_1}^{\infty} c_i v_i^{(\alpha)}\right\| = \left\|\sum_{m=1}^{\infty} \sum_{i=k_m}^{k_{m+1}-1} c_i \sum_{j=1}^{m} a_j x_{n_{i,j}}\right\| = \left\|\sum_{j=1}^{\infty} a_j \sum_{m=j}^{\infty} \sum_{i=k_m}^{k_{m+1}-1} c_i x_{n_{i,j}}\right\| \leq \sum_{j=1}^{\infty} |a_j|/2^{j-1},$$

which shows that $\sum_{i=1}^{\infty} c_i v_i^{(\alpha)}$ converges. In order to conclude the proof we have to prove that $\sum_{i=1}^{\infty} c_i \lambda_i w_i^{(\alpha)}$ converges too. This however is an immediate consequence of 2.a.11. □

In the spaces c_0 and l_p, $p \geq 1$ all the block bases generated by one vector are equivalent to the unit vector basis. This can also happen in other spaces with a symmetric basis, for example, in the Lorentz sequence spaces $d(w, p)$ with $w = \{n^{-1}\}_{n=1}^{\infty}$ and $p \geq 1$ arbitrary (see 4.e.5). However, if this situation occurs both in the space and in its dual then the underlying space has to be isomorphic to c_0 or to l_p for some $p \geq 1$. This was proved by Z. Altshuler [2].

Theorem 3.a.10 . *Let $\{x_n\}_{n=1}^{\infty}$ be a symmetric basis of a Banach space X and let $\{x_n^*\}_{n=1}^{\infty}$ be the functionals biorthogonal to $\{x_n\}_{n=1}^{\infty}$. Then, X is isomorphic to either c_0 or to l_p for some $p \geq 1$ if (and only if) all the block bases of $\{x_n\}_{n=1}^{\infty}$ and of $\{x_n^*\}_{n=1}^{\infty}$, which are generated by one vector, are equivalent to $\{x_n\}_{n=1}^{\infty}$, respectively $\{x_n^*\}_{n=1}^{\infty}$.*

Proof. We assume, as we may without loss of generality, that the symmetric constant of $\{x_n\}_{n=1}^{\infty}$ is equal to 1. Next, we observe that if $\alpha = \sum_{n=1}^{\infty} a_n x_n$ and $\beta = \sum_{n=1}^{\infty} b_n x_n$

are two non-zero vectors in X then $\left\|\sum_{i=1}^{\infty} a_i u_i^{(\beta)}\right\| = \left\|\sum_{i=1}^{\infty} b_i u_i^{(\alpha)}\right\|$. Using this fact and the uniform boundedness principle it follows immediately that if all block bases of $\{x_n\}_{n=1}^{\infty}$, which are generated by one-vector, are equivalent to $\{x_n\}_{n=1}^{\infty}$ then there exists a constant K so that $\left\|\sum_{i=1}^{\infty} b_i u_i^{(\alpha)}\right\| \leq K \cdot \|\alpha\| \cdot \|\beta\|$ for all $\alpha, \beta \in X$. Using the similar inequality for $\{x_n^*\}_{n=1}^{\infty}$ and a standard duality argument we get that $K^{-1}\|\alpha\| \cdot \|\beta\| \leq \left\|\sum_{i=1}^{\infty} b_i u_i^{(\alpha)}\right\|$.

Put $\lambda(n) = \left\|\sum_{i=1}^{n} x_i\right\|$ and use the preceding inequalities for $\alpha = \sum_{i=1}^{n} x_i$ and $\beta = \sum_{i=1}^{k} x_i$, with n and k being arbitrary integers. We get that

$$K^{-1} \leq \lambda(nk)/\lambda(n) \cdot \lambda(k) \leq K, \quad n, k = 1, 2, \ldots .$$

Hence (see the proof of 2.a.9), either $\sup_n \lambda(n) < \infty$ or $K^{-1} \leq \lambda(n)/n^{1/p} \leq K$ for some $p \geq 1$. If $\sup_n \lambda(n) < \infty$ then X is isomorphic to c_0. If $p = 1$ then X is isomorphic to l_1, by the remark following 3.a.7. We can therefore assume that $p > 1$. By duality and 3.a.6 it suffices to prove that, for every choice of scalars $\{b_i\}_{i=1}^{n}$, we have

$$\left\|\sum_{i=1}^{n} b_i x_i\right\| \leq K\left(\sum_{i=1}^{n} |b_i|^p\right)^{1/p} ; \quad n = 1, 2, \ldots .$$

We shall prove this fact only for $n = 2$; it will become clear from the proof how to generalize the argument for a general n.

Put $\beta_1 = b_1 x_1 + b_2 x_2$. By the inequalities established above we have

$$\|\beta_1\|^2 \leq K\|b_1 u_1^{(\beta_1)} + b_2 u_2^{(\beta_1)}\| = K\|\beta_2\|,$$

where $\beta_2 = b_1^2 x_1 + b_1 b_2 (x_2 + x_3) + b_2^2 x_4$. Similarly, we get that

$$\|\beta_1\| \cdot \|\beta_2\| \leq K\|b_1 u_1^{(\beta_2)} + b_2 u_2^{(\beta_2)}\| = K\|\beta_3\| ,$$

where $\beta_3 = b_1^3 x_1 + b_1^2 b_2 (x_2 + x_3 + x_4) + b_1 b_2^2 (x_5 + x_6 + x_7) + b_2^3 x_8$. Continuing inductively we get that

$$\|\beta_1\| \cdot \|\beta_{m-1}\| \leq K\|\beta_m\|, \quad m > 1 ,$$

where $\beta_m = \sum_{j=0}^{m} b_1^{m-j} b_2^j \sum_{i \in \sigma_j} x_i$ with $\{\sigma_j\}_{j=0}^{m}$ being disjoint subsets of integers for which $\bar{\bar{\sigma}}_j = \binom{m}{j}$. Hence,

$$\|\beta_1\|^m \leq K^{m-1}\|\beta_m\| \leq K^{m-1} \sum_{j=0}^{m} |b_1|^{m-j} |b_2|^j \lambda(\bar{\bar{\sigma}}_j) \leq K^m \sum_{j=0}^{m} |b_1|^{m-j} |b_2|^j (\bar{\bar{\sigma}}_j)^{1/p}$$

$$= K^m \sum_{j=0}^{m} \binom{m}{j}^{1/p} |b_1|^{m-j} |b_2|^j .$$

By Holder's inequality we deduce that

$$\|\beta_1\|^m \leq K^m(m+1)^{(p-1)/p}(|b_1|^p+|b_2|^p)^{m/p}, \quad m=1, 2,\ldots$$

and thus, by taking the m'th root and letting $m \to \infty$, we obtain the desired inequality, i.e. $\|\beta_1\| \leq K(|b_1|^p+|b_2|^p)^{1/p}$. $\square$

b. Subspaces of Spaces with a Symmetric Basis

The properties of the spaces with a symmetric basis might create the impression that their subspaces, or perhaps their complemented subspaces, should inherit some of the symmetric structure. The next result (cf. [86]) shows that this is not the case.

Theorem 3.b.1. *Every Banach space with an unconditional basis is isomorphic to a complemented subspace of a space with a symmetric basis.*

Proof. We may assume without loss of generality that X is a Banach space with a normalized unconditional basis $\{x_n\}_{n=1}^\infty$ whose unconditional constant is equal to 1. We denote by X_0 the algebraic span of $\{x_n\}_{n=1}^\infty$ and by Y_0 the vector space of all sequences of scalars which are eventually equal to zero.

Let $\{\alpha_i\}_{i=1}^\infty$, $\{\beta_i\}_{i=1}^\infty$ be decreasing sequences of positive numbers and $\{n_i\}_{i=1}^\infty$ an increasing sequence of integers such that $\alpha_1=\beta_1=n_1=1$ and, for $i>1$, $\alpha_i/\alpha_{i-1}<2^{-i}$, $\beta_i N_{i-1}<2^{-i-1}$ and $\alpha_i\beta_i n_i=1$, where $N_i=\sum_{j=1}^{i} n_j$. We consider now the linear operator T from X_0 into Y_0 which maps a vector $x=\sum_{i=1}^{k} \lambda_i x_i \in X_0$ into the sequence $y=Tx=(a_1, a_2,\ldots) \in Y_0$ so that $a_1=\lambda_1\alpha_1$, $a_j=\lambda_i\alpha_i$ for $N_{i-1}<j\leq N_i$, $i=2, 3,\ldots, k$ and $a_j=0$ for $j>N_k$.

Let K be the convex hull of all sequences $z=(c_1, c_2,\ldots) \in Y_0$ such that $|c_n| \leq |a_{\pi(n)}|$, $n=1, 2,\ldots$, for some permutation π of the integers and some sequence $y=(a_1, a_2,\ldots)$ which belongs to the image, under T, of the unit ball of X_0. With the aid of K we introduce a norm on Y_0 by putting

$$\|z\|=\inf\{\rho>0;\ \rho^{-1}z \in K\}, \quad z \in Y_0\,.$$

Then the closure of K in the normed space Y_0 coincides with its unit ball and therefore T becomes a norm-one operator. It follows that T has a unique continuous extension to a norm-one operator from X into the norm completion Y of Y_0. We shall show that T is an isomorphism. Fix an $x=\sum_{i=1}^{k} \lambda_i x_i \in X_0$ with $\|x\|=1$. By the Hahn–Banach theorem there exist scalars $\{t_i\}_{i=1}^k$ such that $\sum_{i=1}^{k} t_i\lambda_i=1$ and $\sum_{i=1}^{k} |t_i\mu_i| \leq 1$

whenever $\left\|\sum_{i=1}^{k} \mu_i x_i\right\| \leqslant 1$ (in particular, $|t_i| \leqslant 1$ for $i=1, 2, \ldots, k$). We define now a linear functional y^* on Y_0 by putting

$$y^*(z) = \sum_{i=1}^{k} \sum_{j=N_{i-1}+1}^{N_i} t_i \beta_i c_j, \quad z=(c_1, c_2, \ldots) \in Y_0 .$$

Since $\alpha_i \beta_i n_i = 1$ for all i we get that

$$y^*(Tx) = \sum_{i=1}^{k} \sum_{j=N_{i-1}+1}^{N_i} t_i \beta_i \lambda_i \alpha_i = \sum_{i=1}^{k} t_i \lambda_i = 1 ,$$

i.e. $\|Tx\| \geqslant 1/\|y^*\|$. Hence, T would be an isomorphism provided we could show that $\|y^*\|$, which a priori depends on x, is bounded by a number independent of the choice of x.

Let π be a permutation of the integers, let $u = \sum_{i=1}^{\infty} \mu_i x_i$ be a vector in X_0 with norm $\leqslant 1$ and put $Tu = (b_1, b_2, \ldots) \in Y_0$. For every $1 \leqslant i \leqslant k$ we split the set $\sigma_i = \{j;\, N_{i-1} < j \leqslant N_i\}$ into three disjoint (possibly empty) subsets: $\sigma_{i,1} = \{j \in \sigma_i;\, \pi(j) \in \sigma_i\}$, $\sigma_{i,2} = \{j \in \sigma_i;\, \pi(j) \leqslant N_{i-1}\}$ and $\sigma_{i,3} = \{j \in \sigma_i;\, \pi(j) > N_i\}$. Then, by the choice of $\{\alpha_i\}_{i=1}^{\infty}$, $\{\beta_i\}_{i=1}^{\infty}$ and $\{n_i\}_{i=1}^{\infty}$, we have for every $i \leqslant k$

$$\sum_{j \in \sigma_{i,1}} |t_i \beta_i b_{\pi(j)}| \leqslant n_i \beta_i \alpha_i |t_i \mu_i| = |t_i \mu_i| ,$$

$$\sum_{j \in \sigma_{i,2}} |t_i \beta_i b_{\pi(j)}| \leqslant N_{i-1} \beta_i \leqslant 2^{-i-1} ,$$

$$\sum_{j \in \sigma_{i,3}} |t_i \beta_i b_{\pi(j)}| \leqslant n_i \beta_i \alpha_{i+1} \leqslant 2^{-i-1} .$$

Hence, if $\hat{\pi} Tu$ denotes the vector $(b_{\pi(1)}, b_{\pi(2)}, \ldots) \in Y_0$ then

$$|y^*(\hat{\pi} Tu)| \leqslant \sum_{i=1}^{k} \sum_{j=N_{i-1}+1}^{N_i} |t_i \beta_i b_{\pi(j)}| \leqslant \sum_{i=1}^{k} (|t_i \mu_i| + 2 \cdot 2^{-i-1}) \leqslant 2 ,$$

i.e. $\|y^*\| \leqslant 2$. This implies that $\|Tx\| \geqslant 1/2$, for all $x \in X$ with $\|x\| = 1$, i.e. T is an isomorphism from X into Y. This completes the proof since the unit vectors of Y_0 form a symmetric basis in Y and TX is complemented in Y in view of 3.a.4 (TX is clearly the closed linear span of a block basis with constant coefficients). □

Two variants of 3.b.1 are presented in the following theorem.

Theorem 3.b.2. *Every reflexive (uniformly convex) Banach space with an unconditional basis is isomorphic to a complemented subspace of some reflexive (uniformly convex) space with a symmetric basis.*

The reflexive case has been originally proved by A. Szankowski [137] and the uniformly convex one by W. J. Davis [22]. We present here the approach of Davis which, besides proving the uniformly convex case, provides an alternative proof of 3.b.1, as well as of the reflexive case. The approach of Davis, which in essence is an interpolation method (based on some ideas from [24]) proceeds as follows.

Let E and F be two Banach sequence spaces so that the unit vectors $\{e_k\}_{k=1}^\infty$ form a symmetric basis with symmetric constant equal to 1 in both E and F. We assume that

(i) For each sequence of scalars $\alpha=(a_1, a_2,\ldots)\in F$, $\|\alpha\|_E\leq\|\alpha\|_F$ and

(ii) $\lim_{n\to\infty}\lambda_E(n)/\lambda_F(n)=0$, where, for $n\geq 1$, we set $\lambda_E(n)=\left\|\sum_{k=1}^n e_k\right\|_E$ and $\lambda_F(n)=\left\|\sum_{k=1}^n e_k\right\|_F$.

For every number $m\geq 1$ we define a new norm on E by putting

$$\|\alpha\|_m=\inf\{(\|\beta\|_E^2+\|\gamma\|_F^2)^{1/2};\ \alpha=m^{-1}\beta+m\gamma \text{ with } \beta\in E \text{ and } \gamma\in F\}.$$

Let $\alpha=m^{-1}\beta+m\gamma$ be as above; then $\|\alpha\|_E\leq(m^{-2}+m^2)^{1/2}(\|\beta\|_E^2+\|\gamma\|_F^2)^{1/2}\leq 2m(\|\beta\|_E^2+\|\gamma\|_F^2)^{1/2}$, which implies that $\|\alpha\|_E\leq 2m\|\alpha\|_m$ for every $\alpha\in E$. On the other hand, we clearly have $\|\alpha\|_m\leq m\|\alpha\|_E$, $\alpha\in E$. A better estimate can be obtained for $\gamma\in F$, namely $\|\gamma\|_m\leq m^{-1}\|\gamma\|_F$.

It follows from this discussion that, for every $m\geq 1$, the space $E_m=(E,\|\cdot\|_m)$ is isomorphic to E and the identity mapping of F into E_m has norm $\leq m^{-1}$.

We compute now the value of $\lambda_{E_m}(n)=\left\|\sum_{k=1}^n e_k\right\|_m$, $n\geq 1$.

Lemma 3.b.3. *With λ_E, λ_F and λ_{E_m} as above we have*

$$\lambda_{E_m}(n)=(m^{-2}\lambda_E^{-2}(n)+m^2\lambda_F^{-2}(n))^{-1/2}$$

for every integer n and every $m\geq 1$. For a fixed n the maximal value of $\lambda_{E_m}(n)$ is equal to

$$\max_{m\geq 1}\lambda_{E_m}(n)=(2^{-1}\lambda_E(n)\cdot\lambda_F(n))^{1/2}$$

and is attained at $m=(\lambda_F(n)/\lambda_E(n))^{1/2}$.

Proof. Since the norm of any sequence in E or in F does not increase when the sequence is replaced by its restriction to the first n components or by its average over all possible permutations of the integers $\{1, 2,\ldots, n\}$ it follows that

$$\lambda_{E_m}(n)=\inf\{(B^2\lambda_E^2(n)+C^2\lambda_F^2(n))^{1/2};\ Bm^{-1}+Cm=1\}.$$

However, for any constants B and C so that $Bm^{-1}+Cm=1$, we have

$$1\leq(m^{-2}\lambda_E^{-2}(n)+m^2\lambda_F^{-2}(n))^{1/2}\cdot(B^2\lambda_E^2(n)+C^2\lambda_F^2(n))^{1/2}.$$

This proves the first assertion of the lemma. The second assertion is an immediate consequence of the first. □

Let now X be a Banach space with a normalized unconditional basis $\{x_n\}_{n=1}^\infty$ whose unconditional constant is equal to 1. For every increasing sequence of numbers $\{1 \leq m_1 < m_2 < \cdots\}$ so that $\sum_{n=1}^\infty m_n^{-1} < \infty$ we let $Y = Y(E, F, X, \{m_n\}_{n=1}^\infty)$ be the space of all sequences of scalars $\alpha \in E$ for which

$$\|\alpha\|_Y = \left\| \sum_{n=1}^\infty \|\alpha\|_{m_n} x_n \right\|_X < \infty .$$

The condition imposed on the sequence $\{m_n\}_{n=1}^\infty$ ensures that the identity mapping of F into Y is a bounded operator. Indeed, for $\gamma \in F$, we have

$$\|\gamma\|_Y = \left\| \sum_{n=1}^\infty \|\gamma\|_{m_n} x_n \right\|_X \leq \|\gamma\|_F \cdot \left\| \sum_{n=1}^\infty m_n^{-1} x_n \right\|_X \leq \|\gamma\|_F \cdot \sum_{n=1}^\infty m_n^{-1} .$$

In particular, it follows that the unit vectors belong to Y and, as easily checked, they form there a symmetric basis with symmetric constant equal to 1. We can prove now the embedding result (cf. [22]).

Proposition 3.b.4. *For every E, F and X as above there exists an increasing sequence of numbers $\{m_n\}_{n=1}^\infty$ with $\sum_{n=1}^\infty m_n^{-1} < \infty$ such that $Y = Y(E, F, X, \{m_n\}_{n=1}^\infty)$ contains a complemented subspace isomorphic to X. Moreover, every infinite dimensional subspace of Y contains an infinite dimensional subspace which is isomorphic to a subspace of E or to a subspace of X.*

Proof. We recall that by 3.b.3, for each fixed n, the maximal value of $\lambda_{E_m}(n)$ is attained at $m(n) = (\lambda_F(n)/\lambda_E(n))^{1/2}$. The condition (ii) imposed on E and F therefore implies that $m(n) \to \infty$ as $n \to \infty$. Thus, we can construct an increasing sequence of integers $\{n_j\}_{j=1}^\infty$ so that the corresponding numbers $m_j = m(n_j)$, $j = 1, 2, \ldots$ satisfy

$$m_j^{-1} \sum_{i=1}^{j-1} m_i + m_j \cdot \sum_{i=j+1}^\infty m_i^{-1} < 2^{-j-1} \quad \text{for all } j .$$

Next, we choose a sequence $\{\sigma_j\}_{j=1}^\infty$ of disjoint subsets of integers so that $\bar{\sigma}_j = n_j$ and put

$$u_j = \Big(\sum_{k \in \sigma_j} e_k \Big) / (2^{-1} \lambda_E(n_j) \cdot \lambda_F(n_j))^{1/2}, \quad j = 1, 2, \ldots .$$

Then, by 3.b.3 and the definition of the m_j's, we have $\|u_j\|_{m_j} = 1$ for all j and, for $i \neq j$,

$$\begin{aligned} \|u_j\|_{m_i} &= \lambda_{E_{m_i}}(n_j) / (2^{-1} \lambda_E(n_j) \cdot \lambda_F(n_j))^{1/2} \\ &= 2^{1/2} / (m_i^{-2} \cdot \lambda_E^{-1}(n_j) \cdot \lambda_F(n_j) + m_i^2 \cdot \lambda_E(n_j) \cdot \lambda_F^{-1}(n_j))^{1/2} \\ &= 2^{1/2} / (m_i^{-2} m_j^2 + m_i^2 m_j^{-2})^{1/2} \leq 2^{1/2} \min(m_i/m_j, m_j/m_i) . \end{aligned}$$

The lacunarity condition imposed on the sequence $\{m_j\}_{j=1}^\infty$ implies that $\sum_{j\neq i} \|u_j\|_{m_i} \leq 2^{1/2} \cdot 2^{-i-1}$ for all i. Hence, for any choice of $\{c_j\}_{j=1}^\infty$, we have

$$0 \leq \left\|\sum_{j=1}^\infty c_j u_j\right\|_{m_i} - |c_i| \leq \sum_{j\neq i} |c_j|\,\|u_j\|_{m_i} \leq 2^{1/2} \cdot 2^{-i-1} \max_j |c_j|\,,$$

which further implies that in $Y = Y(E, F, X, \{m_i\}_{i=1}^\infty)$ we have

$$\left\|\sum_{j=1}^\infty c_j u_j\right\|_Y = \left\|\sum_{i=1}^\infty \left\|\sum_{j=1}^\infty c_j u_j\right\|_{m_i} x_i\right\|_X \leq \left\|\sum_{i=1}^\infty c_i x_i\right\|_X + 2^{-1/2} \max_i |c_i|\,,$$

i.e. $\left\|\sum_{j=1}^\infty c_j u_j\right\|_Y \leq 2\left\|\sum_{i=1}^\infty c_i x_i\right\|_X$. On the other hand, we clearly have

$$\left\|\sum_{j=1}^\infty c_j u_j\right\|_Y \geq \left\|\sum_{i=1}^\infty |c_i| x_i\right\|_X = \left\|\sum_{i=1}^\infty c_i x_i\right\|_X .$$

This shows that $\{x_n\}_{n=1}^\infty$ is equivalent to the block basis $\{u_j\}_{j=1}^\infty$ in Y and, since the u_j's have constant coefficients, $[u_j]_{j=1}^\infty$ is complemented there, by 3.a.4.

Let now V be an infinite dimensional subspace of Y and suppose that the formal identity mapping of V into E is a strictly singular operator. Then, there exists a basic sequence $\{w_j\}_{j=1}^\infty$ in V with $\|w_j\|_Y = 1$ for all j and $\|w_j\|_E \to 0$. Since each of the norms $\|\cdot\|_{m_i}$, $i = 1, 2, \ldots$ used in the construction of Y is equivalent to the norm of E we can find an increasing sequence of integers $\{q_k\}_{k=1}^\infty$ and a subsequence $\{w_{j_k}\}_{k=1}^\infty$ of $\{w_j\}_{j=1}^\infty$ so that $\left\|\sum_{i=q_k}^\infty \|w_{j_k}\|_{m_i} x_i\right\|_X < 2^{-k-1}$, $k \geq 1$ and

$$\left\|\sum_{i=1}^{q_{k-1}} \|w_{j_k}\|_{m_i} x_i\right\|_X < 2^{-k-1}, \quad k > 1 .$$

The basic sequence $\{w_{j_k}\}_{k=1}^\infty$ is easily seen to be equivalent to the block basis $y_k = \sum_{i=q_{k-1}}^{q_k - 1} \|w_{j_k}\|_{m_i} x_i$, $k = 1, 2, \ldots$ of $\{x_n\}_{n=1}^\infty$. □

We note that 3.b.4 already provides an alternative proof for 3.b.1. We can also give now the

Proof of 3.b.2. Let X be a space with an unconditional basis $\{x_n\}_{n=1}^\infty$. We may assume without loss of generality that $\{x_n\}_{n=1}^\infty$ is normalized and has unconditional constant equal to 1. Take $E = l_q$ and $F = l_p$ with $1 < p < q < \infty$ and observe that conditions (i) and (ii) are satisfied with this choice of E and F. Using 3.b.4 we construct the space $Y = Y(E, F, X, \{m_n\}_{n=1}^\infty)$ so that it contains a complemented subspace isomorphic to X. We recall that the unit vectors $\{e_n\}_{n=1}^\infty$ form a symmetric basis in Y.

Assume now that X is reflexive. Then, by the last part of 3.b.4, Y contains no subspace isomorphic to l_1 or to c_0. Consequently, by 1.c.12, Y is a reflexive space too.

We consider now the case when X is a uniformly convex space. Uniform convexity will be studied quite in detail in Vol. II. For the present proof we just recall that the *modulus of convexity* δ_U of a Banach space U is defined by

$$\delta_U(\varepsilon)=\inf\{1-\|u+v\|/2;\ \|u\|=\|v\|=1,\ \|u-v\|\geqslant\varepsilon\},\quad 0<\varepsilon<2.$$

A Banach space U is *uniformly convex* provided $\delta_U(\varepsilon)>0$ for every $0<\varepsilon<2$.

We note that in our case the spaces E_m, $m\geqslant 1$ are uniformly uniformly convex in the sense that their moduli of convexity δ_{E_m} satisfy $\inf_m \delta_{E_m}(\varepsilon)>0$ for every $0<\varepsilon<2$. Indeed, the mapping $Q_m\colon (E\oplus F)_2\to E_m$, defined by $Q_m(\beta,\gamma)=m^{-1}\beta+m\gamma$ for $\beta\in E$ and $\gamma\in F$, maps the unit ball of $(E\oplus F)_2$ onto the unit ball of E_m. Thus, for any $m\geqslant 1$ and $0<\varepsilon<2$, we have $\delta_{E_m}(\varepsilon)\geqslant\delta_{(E\oplus F)_2}(\varepsilon)$. By using an argument of M. M. Day [27] we shall show that, under the present assumptions, the direct sum $Z=\left(\sum_{n=1}^\infty\oplus E_{m_n}\right)_X$ is also a uniformly convex space. This direct sum is defined as the space of all the sequences $\{\alpha^{(n)}\}_{n=1}^\infty$ with $\alpha^{(n)}\in E_{m_n}$ for all n for which $\|\{\alpha^{(n)}\}_{n=1}^\infty\|_Z=\left\|\sum_{n=1}^\infty \|\alpha^{(n)}\|_{m_n}x_n\right\|_X<\infty$. Since Y is clearly a subspace of Z (more precisely, the "diagonal" of Z) the uniform convexity of Z would imply that Y is also a uniformly convex space, thus completing the proof.

Let $\{\alpha^{(n)}\}_{n=1}^\infty$ and $\{\beta^{(n)}\}_{n=1}^\infty$ be two elements of Z so that

$$\|\{\alpha^{(n)}\}_{n=1}^\infty\|_Z=\|\{\beta^{(n)}\}_{n=1}^\infty\|_Z=1\quad\text{and}\quad\|\{\alpha^{(n)}-\beta^{(n)}\}_{n=1}^\infty\|_Z\geqslant\varepsilon$$

for some $0<\varepsilon<2$. We first consider the case where $\|\alpha^{(n)}\|_{m_n}=\|\beta^{(n)}\|_{m_n}$ for all n. Put $a_n=\|\alpha^{(n)}\|_{m_n}$, $c_n=\|\alpha^{(n)}-\beta^{(n)}\|_{m_n}$, $n=1,2,\ldots$, $\sigma=\{n;\ c_n/a_n>\varepsilon/2\}$ and $\delta(\varepsilon)=\inf_m\delta_{E_m}(\varepsilon)$. Since $\|\alpha^{(n)}+\beta^{(n)}\|_{m_n}\leqslant 2a_n(1-\delta(c_n/a_n))$ we get that

$$\begin{aligned}\|\{\alpha^{(n)}+\beta^{(n)}\}_{n=1}^\infty\|_Z&\leqslant 2\left\|\sum_{n=1}^\infty a_n(1-\delta(c_n/a_n))x_n\right\|_X\\&\leqslant 2\left\|(1-\delta(\varepsilon/2))\sum_{n\in\sigma}a_nx_n+\sum_{n\notin\sigma}a_nx_n\right\|_X.\end{aligned}$$

We also have $1\geqslant\left\|\sum_{n\notin\sigma}a_nx_n\right\|_X\geqslant(2/\varepsilon)\left\|\sum_{n\notin\sigma}c_nx_n\right\|_X$, which implies that $\left\|\sum_{n\in\sigma}a_nx_n\right\|_X\geqslant\frac12\left\|\sum_{n\in\sigma}c_nx_n\right\|_X\geqslant\varepsilon/4$. Thus, by the uniform convexity of X, we get that

$$\|\{\alpha^{(n)}+\beta^{(n)}\}_{n=1}^\infty\|_Z\leqslant 2(1-\eta(\varepsilon)),$$

where $\eta(\varepsilon)=\delta_X(\varepsilon\delta(\varepsilon/2)/8)$.

In the general case put $\gamma^{(n)}=\|\alpha^{(n)}\|_{m_n}\cdot\beta^{(n)}/\|\beta^{(n)}\|_{m_n}$, $n=1,2,\ldots$. Since $\|\gamma^{(n)}\|_{m_n}=\|\alpha^{(n)}\|_{m_n}$ for all n we have $\|\{\gamma^{(n)}\}_{n=1}^\infty\|_Z=1$. We observe that

$$\begin{aligned}\|\{\alpha^{(n)}-\gamma^{(n)}\}_{n=1}^\infty\|_Z&\geqslant\|\{\alpha^{(n)}-\beta^{(n)}\}_{n=1}^\infty\|_Z-\|\{\beta^{(n)}-\gamma^{(n)}\}_{n=1}^\infty\|_Z\\&\geqslant\varepsilon-\left\|\sum_{n=1}^\infty\|\beta^{(n)}-\gamma^{(n)}\|_{m_n}x_n\right\|_X\\&=\varepsilon-\left\|\sum_{n=1}^\infty\big|\|\alpha^{(n)}\|_{m_n}-\|\beta^{(n)}\|_{m_n}\big|x_n\right\|_X.\end{aligned}$$

If $\left\|\sum_{n=1}^{\infty} |\,\|\alpha^{(n)}\|_{m_n} - \|\beta^{(n)}\|_{m_n}|x_n\right\|_X \geqslant \eta(\varepsilon/2)$ then, by the uniform convexity of X, we get that

$$\|\{\alpha^{(n)}+\beta^{(n)}\}_{n=1}^{\infty}\|_Z \leqslant \left\|\sum_{n=1}^{\infty} (\|\alpha^{(n)}\|_{m_n} + \|\beta^{(n)}\|_{m_n})x_n\right\|_X \leqslant 2(1-\delta_X(\eta(\varepsilon/2))) .$$

Otherwise, $\|\{\alpha^{(n)}-\gamma^{(n)}\}_{n=1}^{\infty}\|_Z \geqslant \varepsilon - \eta(\varepsilon/2) \geqslant \varepsilon/2$ which, by the first part of this proof, implies that

$$\begin{aligned}\|\{\alpha^{(n)}+\beta^{(n)}\}_{n=1}^{\infty}\|_Z &\leqslant \|\{\alpha^{(n)}+\gamma^{(n)}\}_{n=1}^{\infty}\|_Z + \|\{\beta^{(n)}-\gamma^{(n)}\}_{n=1}^{\infty}\|_Z \\ &\leqslant 2(1-\eta(\varepsilon/2)) + \eta(\varepsilon/2) \leqslant 2(1-\eta(\varepsilon/2)/2) .\end{aligned}$$

Hence, Z and therefore also Y, are uniformly convex spaces. □

As an immediate application of 3.b.2 we get that each of the spaces $L_p(0, 1)$, $1<p<\infty$, $p \neq 2$ is isomorphic to a complemented subspace of a uniformly convex space with a symmetric basis. It will be shown in Vol. III that none of these spaces themselves has a symmetric basis.

An interesting application of 3.b.1 is connected with the universal space U_1 of Pelczynski, which was introduced in 2.d.10(a). We recall that U_1 is a space with an unconditional basis $\{u_n\}_{n=1}^{\infty}$, which is universal in the sense that every other space with an unconditional basis is isomorphic to a complemented subspace of U_1. The space U_1 is determined uniquely, up to isomorphism, by this universality property.

It turns out that U_1 *has a symmetric basis*. Indeed, by 3.b.1, this universal space U_1 is isomorphic to a complemented subspace of a space Y with a symmetric basis. It follows that Y itself is a universal space for all spaces with an unconditional basis and therefore, by the uniqueness of U_1, we get that Y is isomorphic to U_1. Hence, U_1 has a symmetric basis. This fact is somewhat surprising if we recall that, by its construction, the natural basis $\{u_n\}_{n=1}^{\infty}$ of U_1 has the property that every other unconditional basis is equivalent to a subsequence $\{u_{n_j}\}_{j=1}^{\infty}$ of $\{u_n\}_{n=1}^{\infty}$.

We pass now to the question of uniqueness, up to equivalence, of a symmetric basis. This property, which is shared by a relatively large class of spaces with a symmetric basis, is quite useful in applications and it is applied mostly in the following typical situation. Let X and Y be two Banach spaces with symmetric bases $\{x_n\}_{n=1}^{\infty}$, respectively $\{y_n\}_{n=1}^{\infty}$, and assume that $\{x_n\}_{n=1}^{\infty}$ is, up to equivalence, the unique symmetric basis of X. In this case, if we wish to check whether X is isomorphic to Y or not it suffices to check whether $\{x_n\}_{n=1}^{\infty}$ and $\{y_n\}_{n=1}^{\infty}$ are equivalent which, in general, is much easier.

Proposition 3.b.5. *The spaces c_0 and l_p with $1 \leqslant p<\infty$ have, up to equivalence, a unique symmetric basis.*

Proof. Let X be either c_0 or l_p for some $1 \leqslant p<\infty$ and let $\{x_n\}_{n=1}^{\infty}$ be a symmetric basis of X. Let $\{e_{n_j}^*\}_{n=1}^{\infty}$ denote the sequence of the functionals biorthogonal to the unit vector basis $\{e_n\}_{n=1}^{\infty}$ of X. Choose a subsequence $\{x_{n_i}\}_{i=1}^{\infty}$ of $\{x_n\}_{n=1}^{\infty}$ such that

$\lim_{i\to\infty} e_m^* x_{n_i}$ exists for each m. If all these limits are equal to 0 it follows from 1.a.12 and 2.a.1(ii) that $\{x_{n_i}\}_{i=1}^\infty$, and thus also $\{x_n\}_{n=1}^\infty$, are equivalent to the unit vector basis of X. If $\lim_{i\to\infty} e_m^* x_{n_i} \neq 0$ for some m then a subsequence of $\{x_{n_i}\}_{i=1}^\infty$, and therefore also $\{x_n\}_{n=1}^\infty$, are equivalent to the unit vector basis of l_1. □

Remark. The proof actually shows that every symmetric basic sequence in X is equivalent to the unit vector basis.

In contrast to the case of c_0 and l_p, the universal space U_1 (for all the spaces with an unconditional basis) *has uncountably many mutually non-equivalent symmetric bases.* Moreover, for every $p \geqslant 1$, the space U_1 has a symmetric basis $\{u_n^{(p)}\}_{n=1}^\infty$ such that, for every $\varepsilon > 0$, there is a constant $K_\varepsilon > 0$ with the property that

$$K_\varepsilon \Big(\sum_n |a_n|^{p+\varepsilon}\Big)^{1/(p+\varepsilon)} \leqslant \Big\|\sum_n a_n u_n^{(p)}\Big\| \leqslant \Big(\sum_n |a_n|^p\Big)^{1/p},$$

for every choice of $\{a_n\}$. Indeed, take $F = l_p$ and $E = l_{M_p}$, where $M_p(t) = t^p/(1 + |\log t|)$, and construct the space $Y_p = Y(E, F, U_1, \{m_n\}_{n=1}^\infty)$ so that Y_p contains a complemented subspace isomorphic to U_1 and $\sum_{n=1}^\infty m_n^{-1} < 1$. This is possible in view of 3.b.4 and it implies that $U_1 \approx Y_p$, by the uniqueness of the universal space U_1. To complete the proof we observe that

$$\|\alpha\|_p \geqslant \Big(\sum_{n=1}^\infty m_n^{-1}\Big) \|\alpha\|_F \geqslant \|\alpha\|_{Y_p} \geqslant \|\alpha\|_{m_1} \geqslant K\|\alpha\|_{l_{M_p}},$$

for some constant $K > 0$ and every sequence $\alpha \in l_p$. This means that the unit vector basis of Y_p is mapped, by the isomorphism between Y_p and U_1, into a symmetric basis $\{u_n^{(p)}\}_{n=1}^\infty$ of U_1 having all the desired properties.

Unlike the case of uniqueness of unconditional bases (cf. 2.b.10) it does not seem likely that there is a way to describe those spaces having a unique symmetric basis. It is of interest to point out that from the isometric point of view a symmetric basis is unique. More precisely, we have the following result due to A. Pelczynski and S. Rolewicz (cf. [124, Th. IX.8.3]). Let X and Y be Banach spaces having normalized bases $\{x_n\}_{n=1}^\infty$, resp. $\{y_n\}_{n=1}^\infty$, whose symmetric constants are equal to 1. Then, X is isometric to Y if and only if the basis $\{x_n\}_{n=1}^\infty$ is isometrically equivalent to $\{y_n\}_{n=1}^\infty$, i.e. the map $T: X \to Y$ defined by $T\Big(\sum_{n=1}^\infty a_n x_n\Big) = \sum_{n=1}^\infty a_n y_n$ is an isometry.

Another problem related to the uniqueness of symmetric bases involves the possible number of mutually non-equivalent symmetric bases of one space. In all the examples studied so far (see also the Section 4.b, 4.c and 4.e for the cases of Orlicz and Lorentz sequences spaces) either there is a unique symmetric basis or there are uncountably many non-equivalent symmetric bases. Hence, we have the following question.

Problem 3.b.6. *Is there any Banach space with exactly* 2 (*or any other finite number, or* $\aleph_0$) *non-equivalent symmetric bases?*

In 2.a.6 we have introduced the notion of a prime space, i.e. a Banach space in which all infinite-dimensional complemented subspaces are isomorphic to each other. There are relatively few prime spaces (the only known examples are c_0 and l_p, $1 \leq p \leq \infty$). We introduce now a related class of spaces which is however much larger.

Definition 3.b.7. A Banach space X is said to be *primary* if, for every bounded projection Q on X, either QX or $(I-Q)X$ is isomorphic to X.

In Vol. II we shall prove that the common classical function spaces $C(K)$ (with K being a compact metric topological space) and $L_p(0, 1)$, $1 \leq p \leq \infty$ are all primary. We shall show now that the universal space U_1 (of 2.d.10(a)) is primary. This fact is proved by using the decomposition method of Pelczynski described in the proof of 2.a.3 (observe that this method can be applied since

$$U_1 \approx (U_1 \oplus U_1 \oplus \cdots \oplus U_1 \oplus \cdots)_2)$$

and the following result of P. G. Casazza and Bor-Luh Lin [16].

Proposition 3.b.8. *Let X be a Banach space with a subsymmetric basis $\{x_n\}_{n=1}^\infty$. Then, for every bounded projection Q on X, either QX or $(I-Q)X$ contains a subspace isomorphic to X which is complemented in X.*

Proof. We may assume that $\{x_n\}_{n=1}^\infty$ is a normalized basis with subsymmetric constant equal to 1. We can also assume that $x_n \xrightarrow{w} 0$ since, otherwise, X is isomorphic to l_1 which is prime. By putting $Qx_n = \sum_{k=1}^\infty \lambda_{n,k} x_k$, $n=1, 2, \ldots$ we can find an infinite subset N_1 of the integers so that either $|\lambda_{n,n}| \geq 1/2$ or $|1-\lambda_{n,n}| \geq 1/2$ for all $n \in N_1$. If, for instance, the first alternative holds then, by 1.a.12, there are an infinite subset $N_2 = \{n_1 < n_2 < \cdots < n_j < \cdots\}$ of N_1 and a block basis $\{u_j\}_{j=1}^\infty$ of $\{x_n\}_{n=1}^\infty$ such that $\{Qx_{n_j}\}_{j=1}^\infty$ satisfies $\|Qx_{n_j} - u_j\| < 1/5 \cdot 2^{j+3} \cdot \|Q\|$ for all j.

Let $\{x_n^*\}_{n=1}^\infty$ be the sequence of the functionals biorthogonal to $\{x_n\}_{n=1}^\infty$. Then, $|x_{n_j}^* u_j| \geq |x_{n_j}^* Q x_{n_j}| - 1/2^{j+3} \geq |\lambda_{n_j, n_j}| - 1/2^4 \geq 1/4$ for all j. It follows that the convergence of a series $\sum_{j=1}^\infty a_j Q x_{n_j}$, which is equivalent to that of $\sum_{j=1}^\infty a_j u_j$, implies the convergence of $\sum_{j=1}^\infty a_j x_{n_j}$. On the other hand, it is obvious that $\sum_{j=1}^\infty a_j Q x_{n_j}$ converges whenever $\sum_{j=1}^\infty a_j x_{n_j}$ is a convergent series. Hence, since $\{x_n\}_{n=1}^\infty$ is subsymmetric, we get that $[Qx_{n_j}]_{j=1}^\infty$ is a subspace of QX which is isomorphic to X. Using the equivalence between $\{u_j\}_{j=1}^\infty$ and $\{x_n\}_{n=1}^\infty$ it is easy to show that the operator $P: X \to X$, defined by

$$P\left(\sum_{n=1}^\infty a_n x_n\right) = \sum_{j=1}^\infty (a_{n_j} / x_{n_j}^* u_j) u_j, \quad \sum_{n=1}^\infty a_n x_n \in X,$$

is a bounded projection onto $[u_j]_{j=1}^\infty$ with $\|P\| \leq 5\|Q\|$. Thus, by 1.a.9(ii), $[Qx_{n_j}]_{j=1}^\infty$ is also complemented in X.

If the second alternative holds, i.e. if $|1-\lambda_{n,n}| \geq 1/2$ for all $n \in N_1$ we use $I-Q$ instead of Q. □

Proposition 3.b.8 suggests the following

Problem 3.b.9. *Is every Banach space X, with a symmetric or even subsymmetric basis, primary?*

Observe that, by the decomposition method, this problem would have a positive answer provided we could show that the factor containing a complemented subspace isomorphic to X, say QX, satisfies $QX \approx QX \oplus QX$. In general, however, a space with a symmetric basis might have complemented subspaces which are not isomorphic to their square (this follows from T. Figiel [39] and 3.b.1).

Let us mention another space which is primary; the space J of James, introduced in 1.d.2 (cf. P. G. Casazza [14]).

We conclude this section by discussing some examples of spaces with a symmetric basis which have some special properties. T. Figiel and W. B. Johnson [42] have constructed a Banach space with a symmetric basis which contains no subspace isomorphic to c_0 or to l_p, $p \geq 1$. Their example disproved the feeling that, by some fixed point argument, it might be possible to come up with a positive answer to the question whether every space with a symmetric basis contains c_0 or l_p (such a method works for Orlicz sequence spaces, as shown in 4.a.9 below). The construction of Figiel and Johnson is based on the procedure described in 3.b.4 and the proof of 3.b.2. More precisely, in the notation used there, their example is just the space $Y(c_0, l_1, T, \{m_n\}_{n=1}^\infty)$, where $\{m_n\}_{n=1}^\infty$ is any increasing sequence satisfying $m_n \geq 1$ and $\sum_{n=1}^\infty m_n^{-1} < \infty$, and T is the dual of the space of Tsirelson introduced in 2.e.1. We recall that T is a reflexive space with an unconditional basis $\{t_n\}_{n=1}^\infty$. The proof of 2.e.1. actually shows that T *contains no subsymmetric basic sequence.* Instead of discussing here in detail the example of Figiel and Johnson we present a modified example, due to Z. Altshuler [3], which has some additional interesting properties.

Example 3.b.10. *A Banach space Y with a symmetric basis $\{e_n\}_{n=1}^\infty$ in which all symmetric basic sequences are equivalent to $\{e_n\}_{n=1}^\infty$ but which contains no subspace isomorphic to c_0 or l_p, $p \geq 1$.*

The additional interest in this example stems from the fact, proved in 3.a.10, that a space X with a symmetric basis $\{x_n\}_{n=1}^\infty$ has to be isomorphic to c_0 or to l_p for some $p \geq 1$ if all symmetric basic sequences in X are equivalent to each other and the same holds in X^*. The example 3.b.10 shows that it is not enough to assume that this property holds only in X.

We first define a sequence of symmetric norms on c_0 as follows. For any integer n and any $\alpha = (a_1, a_2, \ldots) \in c_0$ we put

$$\|\alpha\|_n = \sup_k \sup_\pi \left(\sum_{j=1}^k |a_{\pi(j)}| j^{-1} \right) / (2^n + 2^{-n} s_k),$$

where $s_k = \sum_{j=1}^{k} j^{-1}$, $k=1, 2,\ldots$ and the inner supremum ranges over all permutations of the integers. It is easily checked that

$$2^{-n-1}\|\alpha\|_0 \leqslant \|\alpha\|_n \leqslant 2^n\|\alpha\|_0\,,$$

for all n and all $\alpha \in c_0$, where $\|\cdot\|_0$ denotes the norm in c_0.

Let Y be the space of all sequences $\alpha \in c_0$ for which $\|\alpha\|_Y = \left\| \sum_{n=1}^{\infty} \|\alpha\|_n t_n \right\|_T < \infty$, where $\{t_n\}_{n=1}^{\infty}$ is the unconditional basis of the space T introduced in 2.e.1. The unit vectors $\{e_i\}_{i=1}^{\infty}$ belong to Y since, for each i,

$$\|e_i\|_Y = \left\| \sum_{n=1}^{\infty} \|e_i\|_n t_n \right\|_T = \left\| \sum_{n=1}^{\infty} t_n/(2^n+2^{-n}) \right\|_T \leqslant 1\,.$$

Moreover, they form a symmetric basis of Y with symmetric constant equal to 1. We also note that $\lim_{k\to\infty} \left\| \sum_{i=1}^{k} e_i \right\|_Y = \infty$. Indeed, let n and k be such that $s_k/2 < 2^{2n} \leqslant 2s_k$. Then, $\left\| \sum_{i=1}^{k} e_i \right\|_Y \geqslant \left\| \sum_{i=1}^{k} e_i \right\|_n \geqslant s_k/(2^n+2^{-n}s_k) \geqslant s_k^{1/2}/3$, which proves our assertion since $\lim_{k\to\infty} s_k = \infty$. It follows from 3.b.5 that Y is not isomorphic to c_0.

In order to prove that Y has the desired properties we first remark that every symmetric basic sequence in Y is equivalent to a (symmetric) block basis of $\{e_i\}_{i=1}^{\infty}$ (apply the argument used in the proof of 3.b.5). We begin by treating some particular cases of such block bases.

Lemma 3.b.11. *Let* $u_m = \sum_{i=q_m+1}^{q_{m+1}} c_i e_i$, $m=1, 2,\ldots$ *be a normalized block basis of the unit vector basis* $\{e_i\}_{i=1}^{\infty}$ *of* Y. *If* $\lim_{i\to\infty} c_i = 0$ *then there exists a subsequence* $\{u_{m_j}\}_{j=1}^{\infty}$ *of* $\{u_m\}_{m=1}^{\infty}$ *which is equivalent to a block basis* $\{x_j\}_{j=1}^{\infty}$ *of* $\{t_n\}_{n=1}^{\infty}$, *the natural basis of* T.

Proof. The relation between the norms $\|\cdot\|_n$ and $\|\cdot\|_0$ of c_0 shows that, for each fixed m and N, we have

$$\sum_{n=1}^{N-1} \|u_m\|_n \leqslant \sum_{n=1}^{N-1} 2^n\|u_m\|_0 \leqslant 2^N \max\{|c_i|\,;\, q_m < i \leqslant q_{m+1}\}\,.$$

Therefore, we can construct inductively two increasing sequences of integers, $\{m_j\}_{j=1}^{\infty}$ and $\{N_j\}_{j=1}^{\infty}$, such that

$$\left\| \sum_{n=1}^{N_{j-1}-1} \|u_{m_j}\|_n t_n + \sum_{n=N_j}^{\infty} \|u_{m_j}\|_n t_n \right\| < 2^{-j-1} \quad \text{for all } j \text{ (we take } N_0=1)\,.$$

As easily checked, this implies that the basic sequence $\{u_{m_j}\}_{j=1}^{\infty}$ is equivalent to the block basis $x_j = \sum_{n=N_{j-1}}^{N_j-1} \|u_{m_j}\|_n t_n$, $j=1, 2,\ldots$ of $\{t_n\}_{n=1}^{\infty}$. $\square$

It follows from 3.b.11 and the fact that T contains no subsymmetric basic sequences that there is no symmetric block basis $u_m=\sum_{i=q_m+1}^{q_{m+1}} c_i e_i$, $m=1, 2, \ldots$ of $\{e_i\}_{i=1}^{\infty}$ so that the sequence $\{c_i\}_{i=1}^{\infty}$ tends to zero. This implies that $\{e_i\}_{i=1}^{\infty}$ is not equivalent to the unit vector basis of l_p, $p\geqslant 1$ and thus, by 3.b.5, Y is not isomorphic to any l_p with $p\geqslant 1$.

We consider next block bases of the unit vector basis of Y which are generated by one vector (defined in 3.a.8).

Proposition 3.b.12. *Every block basis $\{u_i^{(\alpha)}\}_{i=1}^{\infty}$ of $\{e_i\}_{i=1}^{\infty}$, which is generated by a vector $\alpha\in Y$, is equivalent to $\{e_i\}_{i=1}^{\infty}$.*

Proof. We first recall that it suffices to show that, for any fixed $0\neq\alpha=\sum_{i=1}^{\infty} a_i e_i\in Y$, a series $\sum_{i=1}^{\infty} b_i u_i^{(\alpha)}$ converges whenever $\beta=\sum_{i=1}^{\infty} b_i e_i\in Y$. We further observe that it is actually enough to prove that $\sum_{i=1}^{\infty} a_i u_i^{(\alpha)}$ is a convergent series for every $0\neq\alpha=\sum_{i=1}^{\infty} a_i e_i\in Y$. Indeed, if this were the case then, for α and β as above with $a_i\geqslant 0$ and $b_i\geqslant 0$ for all i, we would get that $\sum_{i=1}^{\infty} (a_i+b_i)u_i^{(\alpha+\beta)}$, and therefore also $\sum_{i=1}^{\infty} b_i u_i^{(\alpha)}$, is a convergent series.

Fix $\alpha=\sum_{i=1}^{\infty} a_i e_i\in Y$ with $1\geqslant a_1\geqslant a_2\geqslant\cdots\geqslant a_i\geqslant\cdots\geqslant 0$ and notice that in order to check whether $\sum_{i=1}^{\infty} a_i u_i^{(\alpha)}$ converges in Y we have to compute the $\|\cdot\|_n$-norms of the double sequence $\{a_i a_j\}_{i,j=1}^{\infty}$ (the numbers $a_i a_j$, $i, j=1, 2, \ldots$ are the coefficients in the expansion of $\sum_{i=1}^{\infty} a_i u_i^{(\alpha)}$ with respect to $\{e_i\}_{i=1}^{\infty}$). Let $a(t)$ be a non-increasing function on $[1,\infty)$ such that $a(i)=a_i$ for all integers i. If, for some integer m, $i\cdot j=m$ then at least one of the integers i or j is $\geqslant m^{1/2}$ and therefore $a_i a_j\leqslant a(m^{1/2})$. It follows that the non-increasing rearrangement of $\{a_i a_j\}_{i,j=1}^{\infty}$ (as a single sequence) is majorated by the sequence $\beta=(b_1, b_2, \ldots)$ whose explicit form is

$$\beta=(\overbrace{a(1^{1/2})}^{\tau(1)\text{—times}}, \overbrace{a(2^{1/2}), a(2^{1/2})}^{\tau(2)\text{—times}}, \ldots, \overbrace{a(m^{1/2}), \ldots, a(m^{1/2})}^{\tau(m)\text{—times}}, \ldots),$$

where $\tau(m)$ is the number of distinct divisors of m. Thus, for every n, we have $\left\|\sum_{i=1}^{\infty} a_i u_i^{(\alpha)}\right\|_n\leqslant\|\beta\|_n$.

For each integer m let $\varphi(m)$ be the first place where $a(m)$ appears in the sequence β. Then, for $\varphi(m)\leqslant k<\varphi(m+1)$, we have

$$\begin{aligned}\Big(\sum_{i=1}^{k} b_i i^{-1}\Big)/(2^n+2^{-n}s_k)&\leqslant\Big(\sum_{j=1}^{m}\sum_{i=\varphi(j)}^{\varphi(j+1)-1} b_i i^{-1}\Big)/(2^n+2^{-n}s_k)\\&\leqslant\Big(\sum_{j=1}^{m} b_{\varphi(j)}\varphi(j)^{-1}(\varphi(j+1)-\varphi(j))\Big)/(2^n+2^{-n}s_{\varphi(m)})\\&\leqslant\Big(\sum_{j=1}^{m} a_j\varphi(j)^{-1}(\varphi(j+1)-\varphi(j))\Big)/(2^n+2^{-n}s_{\varphi(m)})\,.\end{aligned}$$

Since $s_{\varphi(m)} \geqslant \log \varphi(m)$ we get that

$$\|\beta\|_n \leqslant \sup_m \left(\sum_{j=1}^m a_j \varphi(j)^{-1}(\varphi(j+1)-\varphi(j)) \right) / (2^n + 2^{-n} \log \varphi(m)),$$
$$n=1, 2, \ldots .$$

To estimate further the norm of β we use the fact that $\sum_{i=1}^k \tau(i) = k \log k + (2\gamma-1)k + O(k^{1/2})$, where $\gamma = 0.57721\ldots$ is the constant of Euler (this formula is proved in many books on number theory; e.g. see [80]). It follows that there are constants C_1 and C_2 so that, for $j \geqslant 1$, we have

$$\varphi(j) = 1 + \sum_{i=1}^{j^2-1} \tau(i) \geqslant 1 + C_1 j^2 \log j,$$

and $\varphi(j+1)-\varphi(j) \leqslant C_2(1+j \log j)$. Since s_m behaves asymptotically as $\log m$ we get, by substituting the estimates of $\varphi(j)$ and $\varphi(j+1)-\varphi(j)$ in that of $\|\beta\|_n$, that

$$\|\beta\|_n \leqslant C_3 \sup_m \left(\sum_{j=1}^m a_j j^{-1} \right) / (2^n + 2^{-n} s_m) = C_3 \|\alpha\|_n ,$$

for all n and for some constant $C_3 < \infty$. This implies that $\|\beta\|_Y \leqslant C_3 \|\alpha\|_Y$, i.e. $\left\| \sum_{i=1}^\infty a_i u_i^{(\alpha)} \right\|_Y \leqslant C_3 \|\alpha\|_Y$ for all $\alpha \in Y$. □

We pass now to the study of general symmetric block bases of $\{e_i\}_{i=1}^\infty$. Let $u_m = \sum_{i=q_m+1}^{q_{m+1}} c_i e_i$, $m=1, 2, \ldots$ be a normalized symmetric block basis of $\{e_i\}_{i=1}^\infty$ in Y. We can assume without loss of generality that in each block u_m the coefficients c_i are arranged in non-increasing order and that they are non-negative.

Suppose first that, for every $\varepsilon > 0$, there exists an integer $r = r(\varepsilon)$ such that $\left\| \sum_{i=q_m+r}^{q_{m+1}} c_i e_i \right\|_Y < \varepsilon$ for all m for which $q_{m+1} - q_m \geqslant r$. In this case $\{u_m\}_{m=1}^\infty$ is equivalent to a block basis generated by one vector and thus, by 3.b.12, it is equivalent to $\{e_i\}_{i=1}^\infty$. Indeed, by putting $v_m = \sum_{i=1}^{q_{m+1}-q_m} c_{i+q_m} e_i$, $m=1, 2, \ldots$ and using a standard diagonal argument we can find a subsequence $\{v_{m_j}\}_{j=1}^\infty$ of $\{v_m\}_{m=1}^\infty$ such that $a_i = \lim_{j \to \infty} c_{i+q_{m_j}}$ exists for every i. For $\varepsilon > 0$ and $r = r(\varepsilon)$ we have $\left\| \sum_{i=r}^{q_{m_j+1}-q_{m_j}} c_{i+q_{m_j}} e_i \right\|_Y < \varepsilon$ for all j which implies that $\left\| \sum_{i=r}^s a_i e_i \right\|_Y \leqslant \varepsilon$, for every $s > r$. Hence, $\alpha = \sum_{i=1}^\infty a_i e_i \in Y$ and

$$\|\alpha - v_{m_j}\| \leqslant 2\varepsilon + \left\| \sum_{i=1}^{r-1} (a_i - c_{i+q_{m_j}}) e_i \right\|_Y \quad \text{for all } j .$$

It follows that a suitable subsequence of $\{u_{m_j}\}_{j=1}^\infty$ is equivalent to a block basis generated by α (recall that each v_{m_j} is a "translation" of u_{m_j}).

Finally, suppose that there exist an $\varepsilon>0$ and an increasing sequence of integers $\{m_h\}_{h=1}^\infty$ such that $q_{m_h+1}-q_{m_h}>h$ and $\left\|\sum_{i=q_{m_h}+h+1}^{q_{m_h+1}} c_i e_i\right\|_Y \geq \varepsilon$ for all h. Put $v_h=\sum_{i=q_{m_h}+1}^{q_{m_h}+h} c_i e_i$ and $w_h=\sum_{i=q_{m_h}+h+1}^{q_{m_h+1}} c_i e_i$. Then, $u_{m_h}=v_h+w_h$ and $\|w_h\|_Y \geq \varepsilon$ for every integer h. Notice also that $c_{q_{m_h}+h} \geq c$, for some constant $c>0$ and every h, would imply $1 \geq \|v_h\|_Y \geq c\left\|\sum_{i=1}^h e_i\right\|_Y$, $h=1, 2, \ldots$, i.e. $c=0$. Thus, $\lim_{h\to\infty} c_{q_{m_h}+h}=0$ which means that $\{w_h\}_{h=1}^\infty$ is a block of $\{e_i\}_{i=1}^\infty$ with coefficients tending to zero.

By using 3.b.11 and passing to a subsequence if necessary we can assume that $\{w_h\}_{h=1}^\infty$ is equivalent to a block basis $\{x_h\}_{h=1}^\infty$ of $\{t_n\}_{n=1}^\infty$, the unit vector basis of T. The definition of the norm in T implies the existence of a constant $A_1>0$ such that, for every k, we have $\left\|\sum_{h=k+1}^{2k} x_h\right\|_T \geq A_1 k$. It follows that, for every integer k and some constant A_2,

$$\left\|\sum_{h=1}^{2k} u_{m_h}\right\|_Y \geq \left\|\sum_{h=1}^{2k} w_h\right\|_Y \geq A_2 \left\|\sum_{h=k+1}^{2k} x_h\right\|_T \geq A_1 A_2 k\,.$$

Since $\{u_{m_h}\}_{h=1}^\infty$ is a symmetric basic sequence we get, by the discussion following 3.a.7, that $\{u_{m_h}\}_{h=1}^\infty$, and thus also $\{u_m\}_{m=1}^\infty$, are equivalent to the unit vector basis of l_1. This would imply that there exists in Y a block basis of $\{e_i\}_{i=1}^\infty$ with coefficients tending to zero which is also equivalent to the unit vector basis of l_1. This however is impossible in view of 3.b.11 and the discussion thereafter. □

Remark. It is interesting to compare the construction of Y in 3.b.10 with the general method of constructing spaces with a symmetric basis which has been described in the proof of 3.b.2. It can be shown that the space Y of 3.b.10 coincides with the space denoted, in the notation of 3.b.4, by $Y(c_0, d(w, 1), T, \{2^n\}_{n=1}^\infty)$, where $w=\{n^{-1}\}_{n=1}^\infty$.

4. Orlicz Sequence Spaces

a. Subspaces of Orlicz Sequence Spaces which have a Symmetric Basis

Most of this chapter is devoted to a quite detailed study of Orlicz sequence spaces, of particular interest being the relation between these spaces and the l_p spaces. Later on in this chapter we study the subspaces and the quotients of subspaces of the direct sum $l_p \oplus l_r$. It turns out that this topic is closely connected to the study of Orlicz sequence spaces. In the last section of this chapter we present some results on another class of spaces with a symmetric basis, namely Lorentz sequence spaces.

The introduction of Orlicz functions has been inspired by the obvious role played by the functions t^p in the definition of the spaces l_p or, more generally $L_p(\mu)$. It is quite natural to try to replace t^p by a more general function M and then to consider the set of all sequences of scalars $\{a_n\}_{n=1}^\infty$ for which the series $\sum_{n=1}^\infty M(|a_n|)$ converges. W. Orlicz [111] has checked the restrictions which have to be imposed on the function M in order to make this set of sequences into a suitable Banach space. His study led to the following definition of the so-called Orlicz functions and Orlicz sequence spaces (for basic material on Orlicz spaces the reader is referred also to [75]).

Definition 4.a.1. An *Orlicz function* M is a continuous non-decreasing and convex function defined for $t \geq 0$ such that $M(0)=0$ and $\lim_{t\to\infty} M(t)=\infty$. If $M(t)=0$ for some $t>0$, M is said to be a *degenerate Orlicz function.*

To any Orlicz function M we associate the space l_M of all sequences of scalars $x=(a_1, a_2, \ldots)$ such that $\sum_{n=1}^\infty M(|a_n|/\rho)<\infty$ for some $\rho>0$. The space l_M equipped with the norm

$$\|x\|=\inf\left\{\rho>0;\ \sum_{n=1}^\infty M(|a_n|/\rho)\leq 1\right\}$$

is a Banach space usually called an Orlicz sequence space.

Of particular interest is the subspace h_M of l_M consisting of those sequences $x=(a_1, a_2, \ldots)\in l_M$ for which $\sum_{n=1}^\infty M(|a_n|/\rho)<\infty$ *for every* $\rho>0$. Some basic properties of h_M are collected in the following proposition.

Proposition 4.a.2. *Let M be an Orlicz function. Then h_M is a closed subspace of l_M and the unit vectors $\{e_n\}_{n=1}^\infty$ form a symmetric basis of h_M.*

Proof. It is clear that the unit vectors form a symmetric basic sequence in l_M. Therefore, both assertions of the proposition will be proved if we show that h_M coincides with $[e_n]_{n=1}^\infty$. An element $x=(a_1, a_2,\ldots)$ belongs to $[e_n]_{n=1}^\infty$ if and only if, for every $\rho>0$, there exists an integer $N=N(\rho)$ such that $\left\|\sum_{n=N}^\infty a_ne_n\right\|\leqslant\rho$, i.e. if and only if $\sum_{n=N}^\infty M(|a_n|/\rho)\leqslant 1$. □

It is easily verified that if M is a degenerate Orlicz function then $l_M\approx l_\infty$ and $h_M\approx c_0$. Since this case is not interesting in the present context *we shall assume from now on that all the Orlicz functions considered in the sequel are non-degenerate unless specified otherwise.*

In general, the spaces l_M and h_M are distinct. In order to give conditions for l_M to coincide with h_M we need the following definition.

Definition 4.a.3. An Orlicz function M is said to satisfy the *Δ_2-condition* at zero if $\limsup_{t\to 0} M(2t)/M(t)<\infty$.

It is easily checked that the Δ_2-condition at 0 implies that, for every positive number Q, $\limsup_{t\to 0} M(Qt)/M(t)<\infty$ (this condition is sometimes called the Δ_Q-condition). The importance of the Δ_2-condition is illustrated by the following result.

Proposition 4.a.4. *For an Orlicz function M the following conditions are equivalent.*

(i) *M satisfies the Δ_2-condition at 0.*
(ii) *$l_M=h_M$.*
(iii) *The unit vectors form a boundedly complete symmetric basis of l_M.*
(iv) *l_M is separable.*
(v) *l_M contains no subspace isomorphic to l_∞.*

Proof. The fact that the convergence of a series $\sum_{n=1}^\infty M(|a_n|/\rho)$ implies that of $\sum_{n=1}^\infty M(|a_n|/\eta)$ follows easily from the Δ_Q-condition at zero with $Q=\rho/\eta$. This proves the implication (i) ⇒ (ii). In order to prove that (ii) ⇒ (iii) we use 4.a.2 and the fact that $\sup_n\left\|\sum_{i=1}^n a_ie_i\right\|\leqslant 1$, for some sequence $\{a_i\}_{i=1}^\infty$, implies that $\sum_{i=1}^\infty M(|a_i|)\leqslant 1$, i.e. $(a_1, a_2,\ldots)\in l_M=h_M$ and thus $\sum_{i=1}^\infty a_ie_i$ converges. It is obvious that (iii) ⇒ (iv) ⇒ (v). Assume now that an Orlicz function M does not satisfy the Δ_2-condition at zero. Then, we can find a sequence $\{t_n\}_{n=1}^\infty$ such that $M(2t_n)/M(t_n)>2^{n+1}$ and $M(t_n)\leqslant 2^{-n}$. Let k_n be integers chosen so that $2^{-(n+1)}<$

$k_n M(t_n) \leqslant 2^{-n}$ for all n. Then $\sum_{n=1}^{\infty} k_n M(t_n) \leqslant 1$ while $k_n M(2t_n) > 1$. Thus, for any choice of scalars $\{a_n\}_{n=1}^{\infty}$, we have that

$$2^{-1} \sup_n |a_n| \leqslant \|(\overbrace{a_1 t_1, \ldots, a_1 t_1}^{k_1 \text{ times}}, \overbrace{a_2 t_2, \ldots, a_2 t_2}^{k_2 \text{ times}}, \ldots, \overbrace{a_n t_n, \ldots, a_n t_n}^{k_n \text{ times}}, \ldots)\| \leqslant \sup_n |a_n|$$

and, therefore, (v) $\Rightarrow$ (i). □

The definitions of l_M and of h_M show that, up to an isomorphism, what really matters is the behavior of M in the neighborhood of $t=0$: if two Orlicz functions M_1 and M_2 coincide on an interval $0 \leqslant t \leqslant t_0$ then l_{M_1} and l_{M_2} consist of the same sequences and the norms induced by M_1 and M_2 are equivalent. The same is true for h_{M_1} and h_{M_2}. More generally, we have the following result.

Proposition 4.a.5. *Let M_1 and M_2 be two Orlicz functions. Then, the following assertions are equivalent.*

(i) $l_{M_1} = l_{M_2}$ *(i.e. both spaces consist of the same sequences) and the identity mapping is an isomorphism between l_{M_1} and l_{M_2}.*
(ii) *The unit vector bases of h_{M_1} and h_{M_2} are equivalent.*
(iii) *M_1 and M_2 are equivalent at zero, i.e. there exist constants $k>0$, $K>0$ and $t_0>0$ such that, for all $0 \leqslant t \leqslant t_0$, we have*

$$K^{-1} M_2(k^{-1} t) \leqslant M_1(t) \leqslant K M_2(kt)\,.$$

The proof of this proposition is very simple. The implication (ii) $\Rightarrow$ (iii), for instance, is proved by comparing the norms of $\{e_1 + \cdots + e_n\}_{n=1}^{\infty}$ in l_{M_1} and l_{M_2}.

If at least one of the functions satisfies the Δ_2-condition at zero then the equivalence at zero of M_1 and M_2 can be expressed in a simpler form: there exist constants K and $t_0>0$ such that $K^{-1} \leqslant M_1(t)/M_2(t) \leqslant K$ for all $0 < t \leqslant t_0$.

There are many instances where an Orlicz function M is defined only in a neighborhood of zero. In this situation the function M can be extended for $t > t_0$ so that it becomes an Orlicz function on the entire positive line. By 4.a.5 the corresponding spaces l_M and h_M will be the same regardless of the way we have extended M. The norms associated to two distinct extensions might be different but always equivalent.

Every Orlicz function M, being non-decreasing and convex, has a right-derivative $p(t)$ for every $t>0$ and $M(t) = \int_0^t p(s)\, ds$. Since the function p is non-negative and non-decreasing it follows that $1 \leqslant tp(t)/M(t)$ for all $t>0$. In particular, this implies by differentiation that $M(t)/t$ is a non-decreasing function. This fact can be also obtained directly from the convexity of M since, for $0<s<t$, we have $M(s) \leqslant (s/t)M(t) + (1-s/t)M(0) = (s/t)M(t)$.

For every Orlicz function M there exists an Orlicz function M_0 which is equivalent to M and has a continuous derivative. We simply put $M_0(t) = \int_0^t (M(s)/s)\, ds$ and get $M(t) \geqslant M_0(t) \geqslant \int_{t/2}^t (M(s)/s)\, ds \geqslant M(t/2)$ for all $t>0$.

The ratio $tp(t)/M(t)$ is also related to the Δ_2-condition. More precisely, *an Orlicz function M satisfies the Δ_2-condition at zero if and only if*

$$\limsup_{t\to 0} tp(t)/M(t) < \infty .$$

Indeed, if $M(2t) \leqslant KM(t)$, for some constant K and for $0 \leqslant t \leqslant t_0$, then

$$tp(t) \leqslant \int_t^{2t} p(s)\, ds = M(2t) - M(t) \leqslant KM(t) .$$

Conversely, if $tp(t)/M(t) \leqslant K_1$, $0 < t \leqslant t_0$ then

$$\log (M(2t)/M(t)) = \int_t^{2t} (p(s)/M(s))\, ds \leqslant K_1 \log 2,\ 0 < t \leqslant t_0/2 ,$$

i.e. $M(2t) \leqslant 2^{K_1} M(t)$.

The fact that Orlicz sequence spaces are a natural generalization of l_p spaces might suggest that they have a structure almost as simple as that of l_p spaces. Already the study of subspaces of Orlicz sequence spaces shows that this is not the case. For instance, there is no simple description of a general subspace with a basis of an Orlicz sequence space similar to that for l_p spaces (cf. 2.d.1). As we shall see in the next section, the structure of complemented subspaces of l_M is much more involved. It is however possible to describe quite satisfactorily those subspaces of an h_M space which themselves possess a symmetric basis. To present this approach we need the following lemma which was proved in [81] and [93] under the assumption that M satisfies the Δ_2-condition; the fact that this lemma, as well as 4.a.7 and 4.a.8 below, are valid without the Δ_2-condition was noticed in [70].

Lemma 4.a.6. *Let M be an Orlicz function and consider the following subsets of* $C(0, \tfrac{1}{2})$

$$E_{M,\Lambda} = \overline{\{M(\lambda t)/M(\lambda);\ 0 < \lambda < \Lambda\}}, \qquad 0 < \Lambda \leqslant \infty, \qquad E_M = \bigcap_{\Lambda > 0} E_{M,\Lambda},$$

$$C_{M,\Lambda} = \overline{\operatorname{conv} E_{M,\Lambda}}, \qquad C_M = \bigcap_{\Lambda > 0} C_{M,\Lambda} .$$

where the closure is taken in the norm topology of $C(0, \tfrac{1}{2})$. *Then* $E_{M,\Lambda}$, E_M, $C_{M,\Lambda}$ *and* C_M *are non-void norm compact subsets of* $C(0, \tfrac{1}{2})$ *consisting entirely of Orlicz functions which might be degenerate.*

Proof. Let p be the right-derivative of M. Then, for $\lambda > 0$, we have

$$M(\lambda) = \int_0^{\lambda} p(s)\, ds \geqslant \int_{\lambda/2}^{\lambda} p(s)\, ds \geqslant \tfrac{1}{2}\lambda p(\lambda/2) .$$

Hence, for $0 \leqslant t_1, t_2 \leqslant 1/2$ and any $\lambda > 0$, we get that

$$\left| \frac{M(\lambda t_1)}{M(\lambda)} - \frac{M(\lambda t_2)}{M(\lambda)} \right| \leqslant |t_1 - t_2| \frac{\lambda p(\lambda/2)}{M(\lambda)} \leqslant 2|t_1 - t_2|$$

which shows that the functions in $E_{M,\infty}$, considered as elements of $C(0, \frac{1}{2})$, are equi-continuous. In addition, these functions are uniformly bounded by 1. Thus, $E_{M,\infty}$ is a norm compact subset of $C(0, \frac{1}{2})$ and so are all the other sets defined above. It is quite clear that every $N \in C_{M,\infty}$ is an Orlicz function on $[0, \frac{1}{2}]$ (as a uniform limit of Orlicz functions) though it is even possible that $N(t)=0$ for every $t \in [0, \frac{1}{2}]$. The functions of $C_{M,\infty}$ will be extended, for convenience, to Orlicz functions defined on $[0, \infty)$. $\square$

Remark. It is easily checked that the statement of 4.a.6 remains valid if, instead of $[0, \frac{1}{2}]$, we use any other interval $[0, t_0]$ with $0<t_0<1$. If the Δ_2-condition at zero does hold for M then, for any $0<\Lambda<\infty$, $\sup_{0<\lambda\leq\Lambda} (\lambda p(\lambda)/M(\lambda))<\infty$ which implies that the set $C_{M,\Lambda}$ is compact also when considered as a subset of $C(0, 1)$, the closure being subsequently taken also in $C(0, 1)$. We shall use this observation later on in this section.

Let M be an Orlicz function, $\{e_n\}_{n=1}^\infty$ the unit vectors in l_M and $u_j = \sum_{i=n_{j-1}+1}^{n_j} c_i e_i$, $j=1,2,\ldots$ $(n_0=0)$ any normalized block basis of $\{e_n\}_{n=1}^\infty$. To every vector u_j we associate the function $M_j(t) = \sum_{i=n_{j-1}+1}^{n_j} M(|c_i|t)$. Since $\sum_{i=n_{j-1}+1}^{n_j} M(|c_i|)=1$ it follows immediately that the functions $\{M_j\}_{j=1}^\infty$, as elements of $C(0, \frac{1}{2})$, belong to the set $C_{M,1}$. By 4.a.6 there exists a subsequence $\{M_{j_k}\}_{k=1}^\infty$ of $\{M_j\}_{j=1}^\infty$ and an Orlicz function $N \in C_{M,1}$, which might be degenerate, so that $|M_{j_k}(t)-N(t)| \leq 2^{-k}$, $0 \leq t \leq \frac{1}{2}$, $k=1, 2, \ldots$. Assume that N is not degenerate; then, we get that $\sum_{k=1}^\infty M_{j_k}(|a_k|)<\infty$ if and only if $\sum_{k=1}^\infty N(|a_k|)<\infty$, i.e. the subsequence $\{u_{j_k}\}_{k=1}^\infty$ is equivalent to the unit vector basis of h_N and $[u_{j_k}]_{k=1}^\infty \approx h_N$. Moreover, the map $(a_1, a_2, \ldots) \to (a_1c_1, \ldots, a_1c_{n_1}, a_2c_{n_1+1}, \ldots, a_2c_{n_2}, \ldots)$ is an isomorphism from l_N into l_M. If $N(t)=0$ for some $t>0$ then $\{u_{j_k}\}_{k=1}^\infty$ is equivalent to the unit vector basis of c_0 which, in this case, is isomorphic to h_N.

We also note that every subsymmetric basic sequence in h_M is equivalent to some normalized block basis of $\{e_n\}_{n=1}^\infty$ (see the proof of 3.b.5 and the remark thereafter). These facts and 1.a.11 prove the following result (cf. [81]).

Proposition 4.a.7. *For every Orlicz function M the following assertions are true.*

(i) *Every infinite-dimensional subspace Y of h_M contains a closed subspace Z which is isomorphic to some Orlicz sequence space h_N.*
(ii) *Let X be a subspace of h_M which has a subsymmetric basis $\{x_n\}_{n=1}^\infty$. Then X is isomorphic to some Orlicz sequence space h_N and $\{x_n\}_{n=1}^\infty$ is equivalent to the unit vector basis of h_N.*

The functions N appearing in (i) and (ii) might be degenerate. The discussion above shows that the function N appearing in 4.a.7 belongs to the compact convex set $C_{M,1}$ introduced in 4.a.6. The next result proved in [93, 94] shows that $C_{M,1}$ actually "coincides" with the collection of all subspaces of h_M which have a subsymmetric (or a symmetric) basis.

Theorem 4.a.8. *Let M be any Orlicz function. An Orlicz sequence space h_N, where N might be a degenerate Orlicz function, is isomorphic to a subspace of h_M if and only if N is equivalent to some function in $C_{M,1}$.*

Proof. We have to prove only the "if" part. Let $N \in C_{M,1}$ and observe that the extreme points of $C_{M,1}$ are contained in the compact set $E_{M,1}$. The correspondence $\lambda \to M(\lambda t)/M(\lambda)$ is a continuous map from the interval $I_0=(0, 1]$ into $E_{M,1}$ and, therefore, it may be extended uniquely to a map $\omega \to M_\omega$ from βI_0, the Stone–Čech compactification of I_0, onto $E_{M,1}$. By the Krein–Milman theorem there exists a probability measure μ on βI_0 so that

$$N(t)=\int_{\beta I_0} M_\omega(t)\,d\mu(\omega), \quad 0 \leqslant t \leqslant \tfrac{1}{2}.$$

For every integer n put $\Lambda_n=1$ if $\mu(I_0)>0$ and $\Lambda_n=1/2^{n+1}$ if $\mu(I_0)=0$. Then, choose a sequence of probability measures $\{\mu_n\}_{n=1}^\infty$ on βI_0 so that μ_n is supported by the interval $(0, \Lambda_n)$ and

$$\left|N(t)-\int_0^{\Lambda_n} (M(\lambda t)/M(\lambda))\,d\mu_n(\lambda)\right| < 1/2^{n+1}$$

for all n and for all $t \in [0, \frac{1}{2}]$. Fix $0<\tau<1$ and, for every n and j, set $\alpha_{j,n}=\int_{\tau^j\Lambda_n}^{\tau^{j-1}\Lambda_n} d\mu_n(\lambda)/M(\lambda)$. Then,

$$\sum_{j=1}^\infty [\alpha_{j,n}]M(\tau^j\Lambda_n t)-1/2^{n+1} \leqslant N(t) \leqslant \sum_{j=1}^\infty [\alpha_{j,n}]M(\tau^{j-1}\Lambda_n t) + M(t)\Lambda_n/(1-\tau)+1/2^{n+1}$$

for all n and all $t \in [0, \frac{1}{2}]$. Choosing integers k_n so that

$$\sum_{j=k_n+1}^\infty [\alpha_{j,n}]M(\tau^{j-1}\Lambda_n/2) \leqslant 1/2^{n+1}, \quad n=1, 2, \ldots$$

we get that

$$(*) \qquad F_n(\tau t)-1/2^{n+1} \leqslant N(t) \leqslant F_n(t)+M(t)\Lambda_n/(1-\tau)+1/2^n$$

for all n and all $t \in [0, \frac{1}{2}]$, where $F_n(t)=\sum_{j=1}^{k_n} [\alpha_{j,n}]M(\tau^{j-1}\Lambda_n t)$. We also observe that, in the case $\mu(I_0)>0$, $N(t) \geqslant \gamma M(\lambda_0 t)$ for some constants $1>\gamma$, $\lambda_0>0$ and for every $t \in [0, \frac{1}{2}]$. Thus, in this case

$$(\overset{*}{*}) \qquad \tfrac{1}{2}(F_n(\tau t)+M(\lambda_0 t)\gamma-1/2^n) \leqslant N(t) \leqslant F_n(t)+M(t)/(1-\tau)+1/2^n.$$

Let $\{\eta_n\}_{n=1}^\infty$ be disjoint subsets of integers and let $\{\eta_{j,n}\}_{j=0}^\infty$ be a disjoint splitting of η_n so that $\eta_{0,n}$ consists only of one element m_n and $\eta_{j,n}$ has $[\alpha_{j,n}]$ elements for $j=1, 2,\ldots, k_n$ and is void for $j>k_n$. Then, the vectors

$$u_n=\Delta_n\left[\sum_{j=1}^{k_n} \tau^j \sum_{i\in\eta_{j,n}} e_i+e_{m_n}\right]\in h_M;\quad n=1, 2,\ldots$$

form a basic sequence which, by the inequalities (*) and ($\overset{*}{*}$), is equivalent to the unit vector basis of h_N ($\{e_i\}_{i=1}^\infty$ denotes here the unit vector basis of h_M). □

We have presented in 2.e.1 the negative solution of Tsirelson to the question whether every Banach space contains a subspace isomorphic to c_0 or to l_p for some $1\leqslant p<\infty$. Nevertheless, there are some important classes of Banach spaces which are not connected a priori to some l_p or to c_0 but for which this problem has a positive answer. The next theorem gives a positive answer in the case of Orlicz sequences spaces (and in view of 4.a.7(i) also for their subspaces) in very precise terms (cf. [93] and [95]).

Theorem 4.a.9. *The space l_p, or c_0 if $p=\infty$, is isomorphic to a subspace of an Orlicz sequence space h_M if and only if $p\in[\alpha_M, \beta_M]$ where*

$$\alpha_M=\sup\Big\{q;\ \sup_{0<\lambda,t\leqslant 1} M(\lambda t)/M(\lambda)t^q<\infty\Big\}\quad \textit{and}$$

$$\beta_M=\inf\Big\{q;\ \inf_{0<\lambda,t\leqslant 1} M(\lambda t)/M(\lambda)t^q>0\Big\}.$$

Proof. It is easily checked that we always have $1\leqslant\alpha_M\leqslant\beta_M\leqslant\infty$ and $\beta_M<\infty$ if and only if M satisfies the Δ_2-condition at zero. It follows from 4.a.4 that the Δ_2-condition holds for M if and only if the unit vectors of l_M form a boundedly complete basis. Hence, by 1.e.10, $\beta_M=\infty$ if and only if c_0 is isomorphic to a subspace of h_M. From now on we consider only finite values of p. For $p\notin[\alpha_M, \beta_M]$ it is easily seen that the function t^p is not equivalent to any function in $C_{M,1}$ and therefore, by 4.a.8, l_p is not isomorphic to a subspace of h_M. If $\alpha_M=\beta_M<\infty$ the Δ_2-condition holds for M and using the remark following 4.a.6 we can consider $C_{M,1}$ as a convex compact subset of $C(0, 1)$. Let $0<\tau<1$ and consider the map T_τ on $C_{M,1}$ defined by $T_\tau N(t)=N(\tau t)/N(\tau)$. Since $C_{M,1}\subset C(0, 1)$ T_τ is well-defined, continuous and it maps $C_{M,\Lambda}$ into $C_{M,\Lambda\tau}$ for every $0<\Lambda\leqslant 1$. Hence, by the Schauder–Tychonoff fixed point theorem [33, V.10.5] T_τ has a fixed point $N_\tau\in C_M$ which, by definition, satisfies $N_\tau(\tau t)=N_\tau(\tau)N_\tau(t)$, $0\leqslant t\leqslant 1$. Putting $q_\tau=\log N_\tau(\tau)/\log\tau$, we get that $N_\tau(\tau^n)=\tau^{nq_\tau}$ for all n. Let $t\in(0, 1)$ and choose an integer n such that $\tau^n<t\leqslant\tau^{n-1}$. Then,

$$\begin{aligned}|N_\tau(t)-t^{q_\tau}|&\leqslant|N_\tau(t)-N_\tau(\tau^n)|+|\tau^{nq_\tau}-t^{q_\tau}|\leqslant 2(\tau^{(n-1)q_\tau}-\tau^{nq_\tau})\\&\leqslant 2q_\tau(1-\tau),\quad 0\leqslant t\leqslant 1\,.\end{aligned}$$

By letting $\tau\to 1$ and using the compactness of C_M we get that $t^q\in C_M$ for some $q\geqslant 1$ (observe that $\sup_\tau q_\tau<\infty$ because of the Δ_2-condition). As remarked above, this q must be equal to $\alpha_M=\beta_M$.

In the case $\alpha_M<\beta_M$ we choose $p\in(\alpha_M,\beta_M)$. By the definition of α_M and β_M there are numbers $0<u_n<v_n<w_n\leqslant 1$ with $w_n\to 0$, $u_n/v_n\to 0$ such that $n\varphi(u_n)<\varphi(v_n/2)$, $n\varphi(w_n)<\varphi(v_n/2)$, where $\varphi(t)=M(t)/t^p$. Put

$$M_n(t)=A_n^{-1}\int_{u_n/w_n}^{1} M(tsw_n)s^{-p-1}\,ds\,,$$

where $A_n=\int_{u_n/w_n}^{1} M(sw_n)s^{-p-1}\,ds$. Clearly, $M_n\in C_{M,w_n}$ for all n. Let $a_n=u_n/w_n$ and $b_n=v_n/w_n$; by substituting $y=ts$ we get that $M_n(t)=A_n^{-1}t^p\int_{a_nt}^{t} M(yw_n)y^{-p-1}\,dy$. Since $\int_{a_nt}^{t}=\int_{a_n}^{1}+\int_{a_nt}^{a_n}-\int_{t}^{1}$ it follows that $M_n(t)=t^p+f_n(t)-g_n(t)$, where

$$f_n(t)=A_n^{-1}t^p\int_{a_nt}^{a_n} M(yw_n)y^{-p-1}\,dy\leqslant A_n^{-1}a_n^{-p}M(u_n)$$

and

$$g_n(t)=A_n^{-1}t^p\int_{t}^{1} M(yw_n)y^{-p-1}\,dy\leqslant A_n^{-1}M(w_n)\,.$$

On the other hand, since $b_n/a_n=v_n/u_n\to\infty$, we get, for n sufficiently large, that

$$A_n\geqslant\int_{b_n/2}^{b_n} M(sw_n)s^{-p-1}\,ds\geqslant 2^{-1}M(v_n/2)b_n^{-p}\,.$$

It follows that

$$f_n(t)\leqslant 2^{p+1}\varphi(u_n)/\varphi(v_n/2)\leqslant 2^{p+1}/n \quad\text{and}\quad g_n(t)\leqslant 2^{p+1}\varphi(w_n)/\varphi(v_n/2)\leqslant 2^{p+1}/n\,,$$

i.e. $M_n(t)\to t^p$ uniformly on $[0,\frac{1}{2}]$. Hence, $t^p\in C_M$ and this concludes the proof in view of 4.a.8 and the fact that C_M is closed. □

Let us make a few comments on 4.a.9. The proof above shows that, for any finite value of p in $[\alpha_M,\beta_M]$, we have $t^p\in C_M$; hence, as easily checked in the proof of 4.a.8, we get that h_M actually contains almost isometric copies of l_p. In the case $p=\infty$ the same is true for c_0 (use 2.e.3).

For the actual computation of the interval $[\alpha_M,\beta_M]$ it is of some interest to point out (cf. [95 p. 374]) that $\alpha_M=\sup a_N$ and $\beta_M=\inf b_N$ if $\beta_M<\infty$, where the supremum, respectively the infimum, is taken over all Orlicz functions N which are equivalent to M at zero, and

$$a_N=\liminf_{t\to 0} tN'(t)/N(t),\qquad b_N=\limsup_{t\to 0} tN'(t)/N(t)\,.$$

Theorems 4.a.8 and 4.a.9 give a complete description of those Orlicz sequence spaces and, in particular, of those l_p spaces which embed isomorphically into a given Orlicz sequence space. There are, however, some interesting questions of a more specific nature on mappings between Orlicz sequence spaces. We conclude this section by presenting some results, due to N. J. Kalton [67], which give necessary and sufficient conditions for the identity mapping between two Orlicz sequence spaces to be an isomorphism on some infinite dimensional subspace.

Theorem 4.a.10. *Let M and $\tilde{M}$ be two Orlicz functions satisfying the Δ_2-condition at zero so that the formal identity mapping $T: l_M \to l_{\tilde{M}}$ is bounded (i.e. $M(t) \geq A\tilde{M}(t)$ for some constant $A>0$ and for every $t \in [0, 1]$). Then T is strictly singular if and only if, for every constant $B<\infty$, there exists a finite sequence $\{\tau_i\}_{i=1}^m \subset (0, 1]$ so that*

$$\sum_{i=1}^m M(\tau_i t) \geq B \sum_{i=1}^m \tilde{M}(\tau_i t)$$

for every $t \in [0, 1]$.

Proof. We shall prove here only the necessity since this is the part which will be used in Section 2.c below. Fix $B<\infty$ and suppose that the formal identity mapping T from l_M into $l_{\tilde{M}}$ is a strictly singular operator. By using the Δ_2-condition and the fact that $M(t) \geq A\tilde{M}(t)$ for all $t \in [0, 1]$, it is easily checked that the set

$$D = D(M, \tilde{M}, B) = \overline{\text{conv}}\,\{(M(\lambda t) - B\tilde{M}(\lambda t))/M(\lambda);\ 0<\lambda\leq 1\}$$

is a norm compact subset of $C(0, 1)$.

We shall show now that the set D contains no non-positive function. Assume the contrary, i.e. that there is $f \in D$ with $f(t) \leq 0$ for all $t \in [0, 1]$. Let $\omega \to M_\omega$ and $\omega \to \tilde{M}_\omega$ denote the continuous mappings from βI_0, the Stone–Čech compactification of the interval $I_0=(0, 1]$, onto $E_{M,1}$, respectively $E_{\tilde{M},1}$, such that $M_\lambda(t) = M(\lambda t)/M(\lambda)$ and $\tilde{M}_\lambda(t) = \tilde{M}(\lambda t)/\tilde{M}(\lambda)$ for $\lambda \in I_0$. Let R_ω be the unique extension to βI_0 of the (bounded) function $\tilde{M}(\lambda)/M(\lambda)$, $\lambda \in I_0$. Then, by the definition of D, there is a probability measure μ on βI_0 so that $f(t) = \int_{\beta I_0} (M_\omega(t) - BR_\omega\tilde{M}_\omega(t))\,d\mu(\omega)$, $0 \leq t \leq 1$. It follows that

$$A \int_{\beta I_0} \tilde{M}_\omega(t) R_\omega\, d\mu(\omega) \leq \int_{\beta I_0} M_\omega(t)\, d\mu(\omega) \leq B \int_{\beta I_0} \tilde{M}_\omega(t) R_\omega\, d\mu(\omega), \quad 0 \leq t \leq 1\,.$$

Put $N(t) = \int_{\beta I_0} M_\omega(t)\, d\mu(\omega)$, $\tilde{N}(t) = \int_{\beta I_0} \tilde{M}_\omega(t) R_\omega\, d\mu(\omega)$ and notice that $N \in C_{M,1}$ and $\tilde{N}(t)/\tilde{N}(1) \in C_{\tilde{M},1}$. Hence, by 4.a.8, the unit vectors in l_N (respectively in $l_{\tilde{N}}$) form a basis equivalent to a block basis $\{u_n\}_{n=1}^\infty$ (respectively $\{\tilde{u}_n\}_{n=1}^\infty$) of the unit vectors in l_M (respectively in $l_{\tilde{M}}$). Reviewing the proof of 4.a.8 it is easy to check that the blocks $\tilde{u}_n$, $n=1, 2, \ldots$ can be constructed as to actually coincide with u_n, $n=1, 2, \ldots$. Indeed, if we keep the notation of 4.a.8 and approximate μ by the same sequence of measures $\{\mu_n\}_{n=1}^\infty$ on $(0, \Lambda_n)$, for both N and $\tilde{N}$, the forms of u_n and $\tilde{u}_n$ depend only on the numbers

$$\alpha_{j,n} = \int_{\tau^j \Lambda_n}^{\tau^{j-1}\Lambda_n} d\mu_n(\lambda)/M(\lambda), \quad \text{respectively } \tilde{\alpha}_{j,n} = \int_{\tau^j \Lambda_n}^{\tau^{j-1}\Lambda_n} (R_\lambda/\tilde{M}(\lambda))\, d\mu_n(\lambda)\,.$$

But $\alpha_{j,n}=\tilde{\alpha}_{j,n}$ for all j and n which proves that $u_n=\tilde{u}_n$. Hence, $T_{|[u_n]_{n=1}^{\infty}}$ is an isomorphism and this contradicts our assumption.

Since the compact set D is disjoint from $\{f; f(t)\leqslant 0\}$, by the geometric form of the Hahn–Banach theorem and by the Riesz representation theorem for functionals in $C(0, 1)$, we get that there exists a probability measure ν on $[0, 1]$ such that $\inf\left\{\int\limits_0^1 f(t)\, d\nu(t); f\in D\right\}>0$. This measure ν can be approximated by a convex combination, with rational coefficients, of point-mass measures. Thus, there exist integers $\{n_j\}_{j=1}^k$ and reals $\{t_j\}_{j=1}^k$, with $0<t_j\leqslant 1$ for all j, so that $\sum\limits_{j=1}^k n_j(M(\lambda t_j)-B\tilde{M}(\lambda t_j))>0$, $0<\lambda\leqslant 1$. To complete the proof of the necessity we just take as $\{\tau_i\}_{i=1}^m$ the points $\{t_j\}_{j=1}^k$, each t_j being repeated n_j times. □

The condition appearing in 4.a.10 takes a particularly simple form if $M(t)=t^p$.

Corollary 4.a.11 [67]. *Let M be an Orlicz function satisfying the Δ_2-condition at zero and such that $M(t)\geqslant At^p$ for some $A>0$, $1\leqslant p<\infty$ and for every $0\leqslant t\leqslant 1$. Then the identity mapping from l_M into l_p is a strictly singular operator if and only if*

$$\liminf_{\varepsilon\to 0}\ \inf_{0<s\leqslant 1} \frac{1}{\log 1/\varepsilon}\int\limits_{\varepsilon}^{1} \frac{M(st)}{s^p t^{p+1}}\, dt=\infty .$$

Proof. Again, as in 4.a.10, we shall prove only the necessity so we assume that the identity mapping from l_M into l_p is strictly singular. By 4.a.10, for every $B>0$, there are $\{\tau_i\}_{i=1}^m\subset(0, 1]$ so that $\sum\limits_{i=1}^m M(\tau_i su)\geqslant Bs^p u^p \sum\limits_{i=1}^m \tau_i^p$ for every $s, u\in[0, 1]$. Put $\tau=\min\limits_{1\leqslant i\leqslant m} \tau_i$. Then, for $0<\varepsilon<\tau^2$, we have

$$B\sum_{i=1}^m \tau_i^p \log\frac{\tau}{\varepsilon}=B\sum_{i=1}^m \tau_i^p \int\limits_{\varepsilon/\tau}^{1}\frac{du}{u}\leqslant\sum_{i=1}^m \int\limits_{\varepsilon/\tau}^{1} \frac{M(\tau_i su)}{s^p u^{p+1}}\, du$$

$$=\sum_{i=1}^m \tau_i^p \int\limits_{\tau_i\varepsilon/\tau}^{\tau_i} \frac{M(st)}{s^p t^{p+1}}\, dt\leqslant \sum_{i=1}^m \tau_i^p \int\limits_{\varepsilon}^{1} \frac{M(st)}{s^p t^{p+1}}\, dt .$$

Thus, for $0<\varepsilon<\tau^2$ and for every $s\in(0, 1]$, we get that

$$\frac{1}{\log 1/\varepsilon}\int\limits_{\varepsilon}^{1} \frac{M(st)}{s^p t^{p+1}}\, dt\geqslant B\frac{\log \tau/\varepsilon}{\log 1/\varepsilon}\geqslant\frac{B}{2}$$

which proves our assertion since B is arbitrary. □

Using 4.a.11 it is easy to check that the identity operator from l_{M_p}, with $M_p(t)=t^p(1+|\log t|)$, into l_p is strictly singular in spite of the fact that l_{M_p} contains complemented subspaces isomorphic to l_p (cf. 4.c.1 below).

b. Duality and Complemented Subspaces

Let M be a non-degenerate Orlicz function whose right-derivative p satisfies $p(0)=0$ and $\lim_{t\to\infty} p(t)=\infty$. These restrictions exclude only the case when $M(t)$ is equivalent to t, i.e. $l_M \approx l_1$. Consider the right-inverse q of p which is defined by $q(u)=\sup\{t; p(t)\leqslant u\}$, $u\geqslant 0$. It is easily verified that q is a right-continuous non-decreasing function such that $q(0)=0$ and $q(u)>0$ whenever $u>0$. Put $M^*(u)=\int_0^u q(v)\,dv$ for $u\geqslant 0$. Then, M^* is also a non-degenerate Orlicz function and q is its right-derivative. The function M^*, defined in this way, is called *the function complementary to M*. It is clear that M is the function complementary to M^*, i.e. $M^{**}=M$.

A quick glance at the graph of p shows that, for any t and $u\geqslant 0$, we have the so-called *Young inequality*, namely

$$tu\leqslant M(t)+M^*(u),$$

with equality holding if $u=p(t)$ (or $t=q(u)$). In other words, for any $u\geqslant 0$, we have

$$uq(u)=M(q(u))+M^*(u),$$

i.e. M^* satisfies

$$M^*(u)=\max\{tu-M(t); 0<t<\infty\}.$$

With the aid of the complementary function M^* we can introduce a new norm on l_M by putting,

$$|||x|||_M=\sup\left\{\sum_{n=1}^{\infty} a_n b_n; \sum_{n=1}^{\infty} M^*(|b_n|)\leqslant 1\right\}$$

for $x=(a_1, a_2, \ldots)\in l_M$. The norm $|||\cdot|||_M$ satisfies

$$||x||_M\leqslant |||x|||_M\leqslant 2||x||_M$$

for every $x\in l_M$. The right-hand side inequality follows directly from Young's inequality. In order to prove the left-hand side inequality let $x=(a_1, a_2, \ldots)\in l_M$ be such that $|||x|||_M=1$ and take $b_n=p(|a_n|)$. Then $|a_n|b_n=M(|a_n|)+M^*(b_n)$ for all n. If $\sum_{i=1}^{\infty} M^*(b_i)>1$ it would follow by the convexity of M^* that $M^*\left(b_n\Big/\sum_{i=1}^{\infty} M^*(b_i)\right)\leqslant M^*(b_n)\Big/\sum_{i=1}^{\infty} M^*(b_i)$ and thus, $\sum_{n=1}^{\infty} M^*\left(b_n\Big/\sum_{i=1}^{\infty} M^*(b_i)\right)\leqslant 1$. Using the fact that $|||x|||_M=1$ we would get that

$$\sum_{i=1}^{\infty} M^*(b_i)\geqslant \sum_{n=1}^{\infty} |a_n|b_n=\sum_{n=1}^{\infty} M(|a_n|)+\sum_{n=1}^{\infty} M^*(b_n)$$

and this is a contradiction. Thus, $\sum_{i=1}^{\infty} M^*(b_i) \leq 1$ and therefore $1 \geq \sum_{n=1}^{\infty} |a_n| b_n \geq \sum_{n=1}^{\infty} M(|a_n|)$, i.e. $||x||_M \leq 1$.

The complementary function can be used to describe the dual space of an Orlicz sequence space.

Proposition 4.b.1. *Let M and M^* be complementary Orlicz functions. Then $h_M^* \approx l_{M^*}$, $l_M^* \approx h_{M^*}^{**}$ and if, in addition, M^* satisfies the Δ_2-condition at zero then $h_M^{**} \approx l_M$.*

Proof. Let $x^* \in h_M^*$ and put $c_n = x^*(e_n)$, $n = 1, 2, \ldots$. Then,

$$||(c_1, c_2, \ldots)||_{M^*} = \sup\left\{\sum_{n=1}^{\infty} a_n c_n;\ \sum_{n=1}^{\infty} M(|a_n|) \leq 1\right\}$$

$$= \sup\left\{x^*\left(\sum_{n=1}^{\infty} a_n e_n\right);\ ||(a_1, a_2, \ldots)||_M \leq 1\right\} = ||x^*||\ .$$

As easily checked, this implies that the map $x^* \to (x^*(e_1), x^*(e_2), \ldots)$ defines an isometry from h_M^* onto l_{M^*}, endowed with $|||\cdot|||_{M^*}$, and therefore an isomorphism from h_M^* onto l_{M^*}. The other two assertions follow immediately from the existence of this isomorphism and 4.a.4. □

Combining 4.b.1 with 4.a.4 we get the following criterion for reflexivity of Orlicz sequence spaces.

Proposition 4.b.2. *Let M and M^* be complementary Orlicz function. Then h_M (or l_M) is reflexive if and only if both M and M^* satisfy the Δ_2-condition at zero.*

We have already remarked that M satisfies the condition Δ_2 at zero if and only if $b_M = \limsup_{t \to 0} tp(t)/M(t) < \infty$. If we assume, in addition, that both p and q are continuous functions then M^* satisfies the Δ_2-condition if and only if $1 < a_M = \liminf_{t \to 0} tp(t)/M(t)$. More precisely, we have $a_M = b_{M^*}/(b_{M^*} - 1)$. Indeed, if $uq(u)/M^*(u) \leq b_{M^*} + \varepsilon$ for some $\varepsilon > 0$ and $0 < u \leq u_0$ then

$$uq(u)/M(q(u)) \geq (b_{M^*} + \varepsilon)/(b_{M^*} + \varepsilon - 1)\ .$$

Because of the continuity of q we further get $tp(t)/M(t) \geq (b_{M^*} + \varepsilon)/(b_{M^*} + \varepsilon - 1)$ in some neighborhood of $t = 0$. This shows that $a_M \geq b_{M^*}/(b_{M^*} - 1)$. The opposite inequality is proved in a similar manner.

Sometimes the following terminology is used: an Orlicz function M is said to satisfy the *Δ_2^*-condition* at zero if $\liminf_{t \to 0} tp(t)/M(t) > 1$. We can thus restate 4.b.2 in the following form.

Proposition 4.b.2'. *Let M be an Orlicz function with a continuous strictly-increasing* *derivative. Then l_M the (or h_M) is reflexive if and only if $1<a_M\leqslant b_M<\infty$ (i.e. both the Δ_2- and the Δ_2^*-conditions hold for M).*

Observe that M' is strictly increasing if and only if q, the derivative of M^*, is continuous.

For reflexive Orlicz sequence spaces the duality between subspaces and quotient spaces is reflected by the following result.

Theorem 4.b.3. *Let l_M be a reflexive Orlicz sequence space and let M^* be the function complementary to M. Then the following assertions hold.*

(i) *An Orlicz sequence space l_N is isomorphic to a quotient space of l_M if and only if N^* is equivalent to a function in $C_{M^*,1}$.*
(ii) *An Orlicz function N is equivalent at zero to a function in $E_{M,1}$ if and only if N^* is equivalent to a function in $E_{M^*,1}$.*
(iii) $\alpha_M^{-1}+\beta_{M^*}^{-1}=1$, $\alpha_{M^*}^{-1}+\beta_M^{-1}=1$.
(iv) *l_M contains a subspace isomorphic to l_p for some $p\geqslant 1$ if and only if l_M has a quotient space isomorphic to l_p.*

Before proving the theorem let us mention that Examples 4.c.1 and 4.c.2, to be presented in the sequel, show that, in general, property (ii) is not shared by the sets $C_{M,1}$ and $C_{M^*,1}$ or by C_M and C_{M^*}. In other words, condition (iv) above is not necessarily valid when the space l_p is replaced by a general Orlicz sequence space.

Proof of 4.b.3. The assertion (i) is just a restatement of 4.a.8. To prove (ii) we notice that for any $0<\lambda, u$

$$\begin{aligned}(M(\lambda t)/M(\lambda))^*(u)&=\max\{tu-M(\lambda t)/M(\lambda);\,0<t<\infty\}\\&=M^*(\lambda^{-1}M(\lambda)u)/M(\lambda)\\&=(M^*(\mu u)/M^*(\mu))\cdot(M^*(\mu)/M(\lambda))\,,\end{aligned}$$

where $\mu=\lambda^{-1}M(\lambda)$. It is easily checked that the reflexivity of l_M and 4.b.2′ imply that $\mu\to 0$, while the ratio $M^*(\mu)/M(\lambda)$ remains bounded and bounded away from zero, as $\lambda\to 0$. This completes the proof of (ii), and therefore also that of (iii), which is an immediate consequence of (ii). Finally, (iv) follows from (iii) and 4.a.9. □

A simple application of 4.a.9 and 4.b.3 is the following generalization of 2.c.3. *Let M and N be two Orlicz functions satisfying the Δ_2-condition at zero. Then, every bounded linear operator from l_M into l_N is compact if and only if $\alpha_M>\beta_N$.* The "if" part of this assertion is proved by using arguments similar to those used in 2.c.3. In order to prove the "only if" part assume that $\alpha_M\leqslant\beta_N$. Let T_1 be a quotient map from l_M onto l_{α_M}, let I be the formal identity map from l_{α_M} into l_{β_N} and let T_2 be an isomorphism from l_{β_N} into l_N. Then, the operator $T=T_2IT_1$ is a non-compact operator from l_M into l_N.

We turn now to the study of complemented subspaces of an Orlicz sequence space. The results proved so far yield immediately a necessary condition as well as a sufficient condition on an Orlicz function N, for l_N to be isomorphic to a complemented subspace of l_M. Assume that l_M is reflexive; then, by 4.b.3, a necessary condition is

(*) N is equivalent to a function in $C_{M,1}$ and N^* is equivalent to a function in $C_{M^*,1}$.

In order to derive a sufficient condition observe that in an Orlicz space l_M normalized block bases of the unit vector basis with constant coefficients correspond (via the general correspondence between blocks and functions in $C_{M,1}$, which was described in 4.a.8) to functions in $E_{M,1}$. Hence, by 3.a.4, a sufficient condition is

($\overset{*}{*}$) N is equivalent to a function in $E_{M,1}$.

We shall present in the next section examples which show that (*) is not a sufficient condition and that ($\overset{*}{*}$) is not a necessary condition (in both cases l_M will be reflexive and $N(t)$ equivalent to t^p for some p).

There is, however, a weaker version of ($\overset{*}{*}$) which is already a necessary condition for l_N to be isomorphic to a complemented subspace of l_M. To explain the definition below we first write down the negation of ($\overset{*}{*}$) in an explicit manner: N is not equivalent to any function in $E_{M,1}$ if and only if

(†) For every $K \geq 1$ there exist m_K points $t_i \in (0, 1/2)$ such that, for every $\lambda \in (0, 1)$, there is at least one index i, $1 \leq i \leq m_K$ for which

$$M(\lambda t_i)/M(\lambda)N(t_i) \notin [K^{-1}, K] .$$

Definition 4.b.4. Let M be an Orlicz function. A function N is said to be *strongly non-equivalent* to $E_{M,1}$ if (†) holds with the additional requirement that m_K can be chosen so that $m_K = o(K^\alpha)$ as $K \to \infty$, for every $\alpha > 0$.

Theorem 4.b.5 [94]. *Let l_M be a separable Orlicz sequence space and N an Orlicz function which is strongly non-equivalent to $E_{M,1}$. Then l_N is not isomorphic to a complemented subspace of l_M.*

Proof. Suppose that l_N is isomorphic to a complemented subspace of l_M and let $\{e_n\}_{n=1}^\infty$ denote the unit vector basis of l_M. By 1.a.12 and 1.a.9(ii) there exists a normalized block basis $w_j = \sum_{i \in \sigma_j} a_i e_i, j = 1, 2, \ldots$ of $\{e_n\}_{n=1}^\infty$ such that $\{w_j\}_{j=1}^\infty$ is equivalent to the unit vector basis of l_N and there exists a projection P from l_M onto $[w_j]_{j=1}^\infty$. By passing to a subsequence and changing the signs of the coefficients, if necessary, we may assume that $a_i > 0$ for all $i \in \sigma_j$ and that the functions $N_j(t) = \sum_{i \in \sigma_j} M(a_i t)$ satisfy $|N_j(t) - \tilde{N}(t)| \leq 2^{-j}$, $j = 1, 2, \ldots$ for some Orlicz function $\tilde{N} \in C_{M,1}$ equivalent to N and for every $t \in [0, 1]$. We also observe that, since M satisfies the Δ_2-condition at zero, there exists a $p < \infty$ so that $M(st) \leq s^p M(t)$ for $s > 1$ and every $t > 0$.

Assume now that N is strongly non-equivalent to $E_{M,1}$. Then, there are a number K and m_K points $t_h \in (0, 1)$, $h = 1, 2, \ldots, m_K$ so that

$$m_K/K^{1/p} \leq \min(2^{-1} \cdot 4^{-1/p} \|P\|^{-1}, 2^{-p-1} \|P\|^{-p})$$

and, for every $\lambda \in (0, 1)$, there exists at least one h, $1 \leq h \leq m_K$ for which $M(\lambda t_h)/M(\lambda)\tilde{N}(t_h) \notin [K^{-1}, K]$. By passing to a subsequence of $\{N_j\}_{j=1}^\infty$, if necessary, we may assume without loss of generality that $|N_j(t) - \tilde{N}(t)| \leq \min\{\tilde{N}(t_h);\ 1 \leq h \leq m_K\}$ for all $t \in [0, 1]$.

We split now each of the sets σ_j into $2m_K$ disjoint subsets of integers δ_j^h and η_j^h so that, for every $1 \leq h \leq m_K$, we have $M(a_i t_h)/M(a_i)\tilde{N}(t_h) < K^{-1}$ if $i \in \delta_j^h$ and $M(a_i t_h)/M(a_i)\tilde{N}(t_h) > K$ for $i \in \eta_j^h$. Then, for $j \geq 1$ and $1 \leq h \leq m_K$, we have

$$K\tilde{N}(t_h) \sum_{i \in \eta_j^h} M(a_i) < \sum_{i \in \eta_j^h} M(a_i t_h) \leq N_j(t_h) \leq 2\tilde{N}(t_h)$$

which implies that $\sum_{i \in \eta_j^h} M(a_i) \leq 2/K$. Thus, $\sum_{h=1}^{m_K} \sum_{i \in \eta_j^h} M(a_i) \leq 2m_K/K$ for $j \geq 1$.

Every function $F \in C_{M,1}$ satisfies $F(st)/F(t) \leq s^p$ for all $s > 1$ and $t > 0$. In particular, if we put $F_j(t) = \sum_{h=1}^{m_K} \sum_{i \in \eta_j^h} M(a_i t)$ then $F_j(t)/F_j(1) \in C_{M,1}$ and therefore

$$F_j(2\|P\|) \leq 2^p \|P\|^p F_j(1) \leq 2^{p+1} \|P\|^p m_K/K, \quad j \geq 1 .$$

The condition imposed on the ratio m_K/K implies that $F_j(2\|P\|) \leq 1$, i.e. the vectors $v_j = \sum_{h=1}^{m_K} \sum_{i \in \eta_j^h} a_i e_i$, $j \geq 1$, have norms $\leq 1/2\|P\|$.

Put, for $1 \leq h \leq m_K$ and $j = 1, 2, \ldots$, $u_j^h = \sum_{i \in \delta_j^h} a_i e_i$ and let Q_j be the norm one projection from l_M onto $[e_i]_{i \in \sigma_j}$. Then, $w_j = Q_j P w_j = \sum_{h=1}^{m_K} Q_j P u_j^h + Q_j P v_j$, $j = 1, 2, \ldots$ which implies that

$$\sum_{h=1}^{m_K} \|Q_j P u_j^h\| \geq \|w_j\| - \|Q_j P v_j\| \geq 1 - \|P\| \cdot \|v_j\| \geq \tfrac{1}{2} .$$

Hence, for every $j \geq 1$, there exists at least one index $1 \leq h_j \leq m_K$ such that $\|Q_j P u_j^{h_j}\| \geq 1/2m_K$. It follows that if we put $P u_j^{h_j} = \sum_{i=1}^{\infty} d_{i,j} w_i$ then

$$|d_{j,j}| = \|Q_j P u_j^{h_j}\| \geq 1/2m_K$$

for all $j \geq 1$. By 1.c.8 the linear operator $D: [u_j^{h_j}]_{j=1}^\infty \to [w_j]_{j=1}^\infty$, defined by $D u_j^{h_j} = d_{j,j} w_j$, $j \geq 1$ (i.e. the "diagonal" of P), is bounded and $\|D\| \leq \|P\|$. Consequently, for any set of coefficients $\{b_j\}_{j=1}^J$, we have

$$\left\| \sum_{j=1}^J b_j w_j \right\| \leq 2m_K \left\| \sum_{j=1}^J b_j d_{j,j} w_j \right\| \leq 2m_K \|P\| \cdot \left\| \sum_{j=1}^J b_j u_j^{h_j} \right\| .$$

Choosing an integer J so that $2 \leq \sum_{j=1}^J \tilde{N}(t_{h_j}) \leq 3$ we get that $1 \leq \sum_{j=1}^J N_j(t_{h_j}) \leq 4$. Thus, in view of the correspondence between the functions $\{N_j\}_{j=1}^J$ and the blocks $\{w_j\}_{j=1}^J$, we have

$$1 \leqslant \left\| \sum_{j=1}^{J} t_{h_j} w_j \right\| \leqslant 2m_K \|P\| \cdot \left\| \sum_{j=1}^{J} t_{h_j} u_j^{h_j} \right\| .$$

On the other hand,

$$\sum_{j=1}^{J} \sum_{i \in \sigma_j^{h_j}} M(a_i t_{h_j}) \leqslant K^{-1} \sum_{j=1}^{J} \tilde{N}(t_{h_j}) \sum_{i \in \sigma_j^{h_j}} M(a_i) \leqslant 3K^{-1} .$$

Therefore, by the fact that $F(st)/F(t) \leqslant s^p$, for all $F \in C_{M,1}$ and $t>0$ we conclude that $\left\| \sum_{j=1}^{J} t_{h_j} u_j^{h_j} \right\| \leqslant 3^{1/p} K^{-1/p}$. This implies that $1/2m_K \|P\| \leqslant 3^{1/p} K^{-1/p}$, which contradicts the choice of K and m_K. $\square$

The results on subspaces of Orlicz sequence spaces (and especially the method used to prove 4.a.9) put in evidence the mappings $T_\lambda \colon C_{M,1} \to C_{M,1}$ defined by $(T_\lambda N)(t) = N(\lambda t)/N(\lambda)$. The pair $(C_{M,1}, \{T_\lambda\})$ forms what is called a flow in topological dynamics. Some standard notions and reasonings from topological dynamics yield interesting facts on Orlicz sequence spaces if we consider this particular flow. For instance, the notion of a minimal set from topological dynamics has some applications in our context. The notion of a minimal Orlicz function M will be defined only for those M which satisfy the Δ_2-condition. Therefore, using the remark following 4.a.6, we can and shall consider the sets $E_{M,1}$, $C_{M,1}$, etc. as compact subsets of $C(0,1)$ (rather than subsets of $C(0,\frac{1}{2})$). In this way the mapping T_λ is well-defined for every $0<\lambda<1$.

Definition 4.b.6. An Orlicz function M satisfying the Δ_2-condition at zero is called *minimal* if the set $E_{M,1}$ has no proper closed subsets which are invariant under the flow $(T_\lambda; 0<\lambda<1)$. In other words, if $E_{N,1}=E_{M,1}$ for every $N \in E_{M,1}$.

Let M be any Orlicz function satisfying the Δ_2-condition at zero. A standard application of Zorn's lemma to the set $E_{M,1}$, endowed with the order $F \prec G \Leftrightarrow F \in E_{G,1}$, shows that $E_{M,1}$ *contains at least one minimal Orlicz function.*

Minimal Orlicz sequence spaces have the following property.

Proposition 4.b.7 [94]. *Let M be a minimal Orlicz function. Then every block basis with constant coefficients of the unit vector basis of l_M spans a subspace which is isomorphic to l_M itself.*

Proof. It follows from 3.a.5 that if U is a subspace of l_M which is spanned by a block basis with constant coefficients then $l_M \approx l_M \oplus U$. On the other hand, by 4.a.7 and its proof, $U \approx l_N \oplus V$ for some Orlicz function $N \in E_{M,1}$ and some Banach space V. Since M is minimal we have $M \in E_{N,1}$ and thus, by (*), $U \approx l_M \oplus W$ for some space W. Consequently,

$$l_M \approx l_M \oplus U \approx l_M \oplus l_M \oplus W \approx l_M \oplus W \approx U . \quad \square$$

It is worthwhile to compare 4.b.7 with the second remark following 2.a.9: for a minimal Orlicz function M which is not equivalent to any t^p the isomorphism between U and l_M cannot be always induced by mapping the n'th block to the n'th unit vector.

We shall present later on in this chapter some examples of minimal Orlicz functions which are not equivalent to any t^p. Actually, we will see that for any interval $[\alpha, \beta]$ there is a minimal Orlicz function M with $\alpha_M=\alpha$ and $\beta_M=\beta$; the only restriction being that $\alpha>1$ (it is easily verified that if M is minimal and $\alpha_M=1$ then $M(t)=ct$).

The proposition above states that for minimal Orlicz sequence spaces the "obvious" complemented subspaces of l_M (i.e. those spanned by block bases with constant coefficients) are necessarily isomorphic to l_M. It is possible that this is also true for any other complemented subspaces. We thus formulate.

Problem 4.b.8. *Assume that M is a minimal Orlicz function. Is then l_M a prime Banach space?*

It would be actually of interest to decide whether l_M is prime even for a single example of a minimal function M other than t^p.

In Section 3.b we have seen that the universal space U_1 of Pelczynski has uncountably many mutually non-equivalent symmetric bases. Among the Orlicz sequence spaces there are also many examples of spaces having at least two non-equivalent symmetric bases. The construction of such spaces is based on the following remark: *if l_M is isomorphic to l_N but M is not equivalent to N then l_M has at least two non-equivalent symmetric bases, namely the unit vector bases of l_M and of l_N.*

This observation can be used in the case of minimal Orlicz functions in order to prove the next result.

Theorem 4.b.9 [95]. *Let M be a minimal Orlicz function which is not equivalent to any t^p, $1\leq p<\infty$. Then l_M has uncountably many mutually non-equivalent symmetric bases.*

Proof. It follows from the definition of minimality, condition (⁑) and the decomposition method of Pelczynski that, for every $N\in E_{M,1}$, the space l_N is isomorphic to l_M. Therefore, in view of the preceding remark, it suffices to show that $E_{M,1}$ contains uncountably many mutually non-equivalent functions.

Assume that there are only countably many equivalence classes in $E_{M,1}$ and denote by $\{M_n\}_{n=1}^{\infty}$ their representatives. Then, by Baire's category theorem, one of the sets

$$F_{n,k}=\{N\in E_{M,1}; k^{-1}\leq N(t)/M_n(t)\leq k, 0<t\leq 1\},$$

whose union covers entirely the set $E_{M,1}$, contains a relatively open set G. By minimality there exists, for every $N\in E_{M,1}$, a $\lambda\in(0,1)$ such that $N(\lambda t)/N(\lambda)\in G$

and thus, $E_{M,1}$ consists of exactly one equivalence class. In order to complete the proof it suffices to show that all the functions in $E_{M,1}$ are uniformly equivalent to M. For $0<\lambda<1$ we put $G_\lambda=\{N\in E_{M,1};\ N(\lambda t)/N(\lambda)\in G\}$. The sets G_λ are open and, as remarked above, $E_{M,1}=\bigcup\{G_\lambda;\ 0<\lambda\leq 1\}$. Hence, by the compactness of $E_{M,1}$, there is a $\lambda_0>0$ such that $E_{M,1}=\bigcup\{G_\lambda;\ \lambda_0\leq\lambda\leq 1\}$. It follows that, for every $0<\mu\leq 1$, there is a λ with $\lambda_0\leq\lambda\leq 1$ such that $M(\lambda\mu t)/M(\lambda\mu)\in G$, i.e. $k^{-1}\leq M(\lambda\mu t)/M(\lambda\mu)M(t)\leq k$ (if we assume, as we may, that $G\subset F_{n,k}$ and $M_n=M$). Since $\lambda\geq\lambda_0$ the Δ_2-condition implies the existence of a constant $A>0$ so that, for all $0<\mu,\ t\leq 1$,

$$A^{-1}\leq M(\mu t)/M(\mu)M(t)\leq A\,.$$

This implies that M is equivalent to t^p for some $1\leq p<\infty$, contrary to the assumption. □

The concept of a minimal Orlicz sequence space can be extended in a natural manner to the more general setting of spaces with a symmetric basis. A symmetric basis $\{x_n\}_{n=1}^\infty$ of a Banach space X is said to be *minimal symmetric* provided every block basis of $\{x_n\}_{n=1}^\infty$ with constant coefficients spans a subspace which is isomorphic to the whole space X.

In view of 4.b.9 it is natural to ask whether the only spaces with a minimal symmetric basis which is unique, up to equivalence, are c_0 and l_p, $1\leq p<\infty$.

There are many interesting examples of Orlicz sequence spaces which do have, up to equivalence, a unique symmetric basis. A sufficient condition for this to happen is given in the next proposition (cf. [93]).

Proposition 4.b.10. *Let M be an Orlicz function for which the set C_M contains no Orlicz function equivalent to M itself. Then the unit vector basis is, up to equivalence, the unique symmetric basis of h_M.*

Proof. If h_M has in addition to the unit vector basis $\{e_n\}_{n=1}^\infty$ another symmetric basis $\{f_n\}_{n=1}^\infty$ then each of these two bases is equivalent to a block basis of the other. It is easily seen that if, in both block basis representations, the coefficients do not tend to zero then $\{f_n\}_{n=1}^\infty$ is equivalent to $\{e_n\}_{n=1}^\infty$. If this is not the case then $\{e_n\}_{n=1}^\infty$ is equivalent to a block basis of itself with coefficients tending to zero. In view of 4.a.8 and its proof this implies that C_M contains a function which is equivalent to M itself. □

A simple consequences of 4.b.10 is that *Orlicz sequence spaces h_M for which $\lim_{t\to 0} tM'(t)/M(t)$ exists (i.e. $a_M=b_M$) have, up to equivalence, a unique symmetric basis*. This follows from the fact that, in this case, C_M consists only of one Orlicz function, namely t^p, provided $p=\lim_{t\to 0} tM'(t)/M(t)$ is finite or $f(t)=0$ in $[0, 1/2]$ if $p=\infty$.

Besides minimal Orlicz functions it is of interest to consider some maximal ones. More precisely, we shall construct Orlicz functions $U_{\alpha,\beta}$ which are universal for the class of all Orlicz functions M with $[\alpha_M, \beta_M] \subset (c, d)$. We need first the following lemma.

Lemma 4.b.11. *Let F and G be two continuous non-decreasing convex functions, defined on an interval $[\tau, 1]$ with $0<\tau<1$. Assume that*

(i) *$F(1)=G(1)=1$, $0<F(\tau)<1$, $0<G(\tau)<1$ and,*
(ii) *for some numbers c and d such that $1<c<d$, $F'(1)=G'(1)=c$ and $c \leqslant tF'(t)/F(t) \leqslant d$, $c \leqslant tG'(t)/G(t) \leqslant d$ for every $t \in [\tau, 1]$ (here F' and G' stand for the right-derivatives except for $t=1$ where they mean the left-derivatives).*

Then the function

$$H(t)=\begin{cases} F(t), & \tau \leqslant t \leqslant 1 \\ F(\tau)G(t/\tau), & \tau^2 \leqslant t < \tau \end{cases}$$

is continuous, convex, non-decreasing and $H'(1)=c \leqslant tH'(t)/H(t) \leqslant d$ for every $t \in [\tau^2, 1]$.

The proof is straightforward.

Theorem 4.b.12. *For every $1 \leqslant c<d<\infty$ there exists an Orlicz function $U=U_{c,d}$ such that*

(i) *$c \leqslant tU'(t)/U(t) \leqslant d$ for all $t \in [0, 1]$,*
(ii) *for every Orlicz function M with $c \leqslant tM'(t)/M(t) \leqslant d$ for all $t \in (0, 1]$ there exists in E_U a function equivalent to M,*
(iii) *there exists a constant $K_{c,d}$ such that, for every M with $c \leqslant tM'(t)/M(t) \leqslant d$, there is a norm-one projection P_M in l_U such that $d(P_M l_U, l_M) \leqslant K_{c,d}$.*

Proof. Assume first that $c>1$ and choose a sequence $\{N_n(t)\}_{n=1}^{\infty}$ of Orlicz functions which is dense in the set $\mathscr{F}$ of all Orlicz functions N satisfying $N(1)=1$, $N'(1)=c$ and $c \leqslant tN'(t)/N(t) \leqslant d$ for all $t \in (0, 1]$. Put $\tau_n=2^{-2^{n-1}}$, $n=1, 2, \ldots$ and define

$$U(t)=\begin{cases} N_1(t), & \tau_1 \leqslant t \leqslant 1 \\ N_n(t/\tau_n)U(\tau_n), & \tau_{n+1} \leqslant t < \tau_n,\ n=1, 2, \ldots \\ 0, & t=0\,. \end{cases}$$

In view of 4.b.11, U is an Orlicz function defined on the entire interval $[0, 1]$ and such that $c \leqslant tU'(t)/U(t) \leqslant d$ for all $0<t \leqslant 1$, Moreover, for every n and every $\tau_n \leqslant t \leqslant 1$, we get that $U(\tau_n t)/U(\tau_n)=N_n(t)$ which implies that E_U contains all the functions of $\mathscr{F}$.

Let now M be any Orlicz function such that $M(1)=1$ and $c \leqslant tM'(t)/M(t) \leqslant d$ for all $0<t \leqslant 1$. Choose $t_1=t_{c,d}$ so that $t_1 c(d-1)/(c-1)d=1/2$ and t_2 so that $c(M'(t_1)(t_2-t_1)+M(t_1))=M'(t_1)t_2$. It is easily verified that

$$t_1 \leqslant t_2 \leqslant t_1 c(d-1)/(c-1)d=1/2\,.$$

Define

$$M_0(t)=\begin{cases} M(t), & 0\leqslant t\leqslant t_1\\ M(t_1)+M'(t_1)(t-t_1), & t_1<t\leqslant t_2\\ (M(t_1)+M'(t_1)(t_2-t_1))(t/t_2)^c, & t_2<t\leqslant 1\,. \end{cases}$$

Then the function $M_1(t)=M_0(t)/M_0(1)\in\mathscr{F}$ and $K^{-1}M_1(t)\leqslant M(t)\leqslant KM_1(t)$ for all $t\in[0,1]$ and for some constant $K=K_{c,d}$, independent of M. This concludes the proof in the case $c>1$.

The case $c=1$ can be reduced to $c>1$ in the following manner. Fix $d>1$ and let $U=U_{2,d+1}$ be a universal function corresponding to the numbers 2 and $d+1$ (whose existence follows from the previous case). Put $U_0(t)=U(t)/t$. In general, U_0 need not be convex so, instead, we consider the function $U_1(t)=\int\limits_0^t(U_0(s)/s)\,ds$. It is easily checked that U_1 is an Orlicz function satisfying $1\leqslant tU_1'(t)/U_1(t)\leqslant d$ and $U_1(t)\leqslant U_0(t)\leqslant dU_1(t)$ for all $0<t\leqslant 1$. We shall show that $U_1=U_{1,d}$ has all the desired properties. Let N be any Orlicz function such that $N(1)=1$ and $1\leqslant tN'(t)/N(t)\leqslant d$ for all $0<t\leqslant 1$. Then, $M(t)=tN(t)$ satisfies $M(1)=1$ and $2\leqslant tM'(t)/M(t)\leqslant d+1$ for all $0<t\leqslant 1$; hence, by the first part of the proof, E_U contains a function M_1 such that $K^{-1}M_1(t)\leqslant M(t)\leqslant KM_1(t)$, $0\leqslant t\leqslant 1$, with K being a constant independent of M. This means that, for some sequence $\lambda_n\to 0$, $U(\lambda_n t)/U(\lambda_n)\to M_1(t)$ for all $0<t\leqslant 1$. By passing to a subsequence of $\{\lambda_n\}_{n=1}^\infty$, if needed, we can assume with no loss of generality that, for some $N_1\in E_{U_1}$, $U_1(\lambda_n t)/U_1(\lambda_n)\to N_1(t)$ uniformly for $0\leqslant t\leqslant 1$. It is easily verified that $(dK)^{-1}N(t)\leqslant N_1(t)\leqslant dKN(t)$ for all $0\leqslant t\leqslant 1$. □

Remarks. It follows immediately from ($\ast$) and Pelczynski's decomposition method that $l_{U_{c,d}}$ is determined uniquely, up to an isomorphism, by c and d. If we choose c and d so that $c^{-1}+d^{-1}=1$ then $U_{c,d}^*$ is also universal for the same c and d. Consequently, $l_{U_{c,d}}$ is isomorphic to $l_{U_{c,d}}^*$. This constitutes a non-trivial example, i.e. different from l_2, of *a Banach space with a symmetric basis which is isomorphic to its own conjugate*. The fact that all the universal spaces corresponding to the same c and d are isomorphic to each other shows that $l_{U_{c,d}}$ is another example of a space with infinitely many mutually non-equivalent symmetric bases. Indeed, it suffices to observe that, by proper rearrangements of the sequence $\{N_n\}_{n=1}^\infty$ in the construction of $U_{c,d}$, we obtain uncountably many universal functions $\{U_{c,d}^{(\alpha)}\}_\alpha$ such that, for $\alpha\neq\beta$, $U_{c,d}^{(\alpha)}$ is not equivalent at zero to $U_{c,d}^{(\beta)}$.

c. Examples of Orlicz Sequence Spaces

The difference between the structure of Orlicz sequence spaces and that of l_p spaces is best illustrated by considering suitable examples. In the beginning of this section we present some examples of Orlicz functions which are given by concrete formulas. In addition to these examples we shall describe a general method of constructing Orlicz functions by using sequences of zeros and ones.

Example 4.c.1. *A reflexive Orlicz sequence space l_M with a unique symmetric basis such that the sets $C_{M^*,1}$ and $C^*_{M,1}=\{N^*; N\in C_{M,1}\}$ have a totally different structure (recall that, up to equivalence at zero, the set $E_{M^*,1}$ coincides with $E^*_{M,1}$ in view of 4.b.3.*(ii)).

Let $M(t)=t^p |\log t|^\alpha$ with $1<p<\infty$ and $\alpha>0$. It is easily checked that M is an Orlicz function on some interval $[0, t_0]$ with $t_0>0$. Thus, M can be extended to an Orlicz function on $[0,\infty)$ but for the present discussion the values of M outside a neighborhood of $t=0$ are of no importance. A trivial computation shows that $\lim_{t\to 0} tM'(t)/M(t)=p$. Hence, by 4.b.2′, l_M is reflexive and, by 4.b.10, the unit vectors are, up to equivalence, the unique symmetric basis of l_M. We also have

$$\lim_{\lambda\to 0} M(\lambda t)/M(\lambda)=\lim_{\lambda\to 0} t^p(1+\log t/\log \lambda)^\alpha=t^p, \quad 0<t\leqslant t_0 .$$

Therefore, the sets E_M and C_M consist both of only one function, namely t^p. It follows that the set $E_{M,1}$ has exactly two equivalence classes: t^p and functions equivalent to M itself. Also $C_{M,1}$ consists of these two equivalent classes. Indeed, let $N(t)=\int_0^{t_0} (M(\lambda t)/M(\lambda))\, d\mu(\lambda)$, for some probability measure μ on $[0, t_0]$ (where $M(0t)/M(0)$ stands for t^p). Observe that, for a fixed $0<t\leqslant t_0$, $M(\lambda t)/M(\lambda)$ is an increasing function of λ. Hence, for every $0<t_1<t_0$,

$$M(t_0t)/M(t_0)\geqslant N(t)\geqslant \mu([t_1, t_0])M(t_1t)/M(t_1)$$

and, unless μ is concentrated in the origin, N is equivalent to M. We conclude that, up to an isomorphism, the only Orlicz sequence subspaces of l_M are l_p and l_M itself, and both are also complemented subspaces of l_M.

In order to study the quotient spaces of l_M we have to compute the complementary function M^*. The exact computation of M^* is quite complicated but, for our purposes, it suffices to find a function which is equivalent to M^*. Observe first that M is equivalent at zero to $M_1(t)=\int_0^t f(s)\, ds$, where $f(s)=s^{p-1} |\log s|^\alpha$. Take q so that $p^{-1}+q^{-1}=1$ and put $g(t)=t^{q-1} |\log t|^{\alpha(1-q)}$. Since

$$f(g(t))=g^{p-1}(t) |\log g(t)|^\alpha=t |\log t|^{-\alpha}|(q-1) \log t+\alpha(1-q) \log |\log t|\,|^\alpha$$

we get that $\lim_{t\to 0} f(g(t))/t=(q-1)^\alpha$. This implies that g is equivalent at zero to the inverse function of f and thus, M^* is equivalent to $\int_0^t g(s)\, ds$ which, in turn, is equivalent to the function $t^q |\log t|^{\alpha(1-q)}$. By abuse of notation we shall put $M^*(t)=t^q |\log t|^{\alpha(1-q)}$.

It is easily verified that E_{M^*} and C_{M^*} consist only of one function, namely t^q. However, the set $C_{M^*,1}$ turns out to contain infinitely many equivalence classes: for example, for every $0<\varepsilon<\alpha(q-1)$, $C_{M^*,1}$ contains a function equivalent to

$t^q |\log t|^{-\varepsilon}$. Indeed, the function $N_\varepsilon(t)=\int_0^{e^{-1}} (M^*(\lambda t)/M^*(\lambda))\lambda^{-1} |\log \lambda|^{-1-\varepsilon}\, d\lambda$ clearly satisfies $N_\varepsilon(t)/N_\varepsilon(1) \in C_{M^*,1}$ and

$$\begin{aligned} N_\varepsilon(t) &= t^q \int_0^{e^{-1}} \left(\frac{\log \lambda}{\log \lambda + \log t}\right)^{\alpha(q-1)} \lambda^{-1} |\log \lambda|^{-1-\varepsilon} d\lambda \\ &= t^q \int_1^\infty \left(\frac{u}{u+|\log t|}\right)^{\alpha(q-1)} u^{-1-\varepsilon}\, du \\ &= t^q |\log t|^{-\varepsilon} \int_{|\log t|^{-1}}^\infty \left(\frac{v}{1+v}\right)^{\alpha(q-1)} v^{-1-\varepsilon}\, dv \\ &= Ct^q |\log t|^{-\varepsilon} + t^q \cdot O(|\log t|^{\alpha(1-q)}) \quad \text{as } t \to 0, \end{aligned}$$

for some constant $C>0$. It follows that N_ε is equivalent to $t^q |\log t|^{-\varepsilon}$.

It is clear from the discussion in Section b above that, for any $N \in C_{M^*,1}$ such that N is not equivalent to a function in $E_{M^*,1}$ (i.e. to M^* itself or to t^q), l_N is isomorphic to a subspace of l_{M^*} (or, equivalently, l_{N^*} is isomorphic to a quotient space of l_M) but not to a complemented subspace of l_{M^*}. □

We consider next an example introduced in [81] and investigated in [81] and [94].

Example 4.c.2. *A reflexive Orlicz sequence space l_M with a unique symmetric basis for which $\alpha_M < \beta_M$ and C_{M^*} is different from $C_M^* = \{N^*;\, N \in C_M\}$ (recall that, in 4.c.1, $C_{M^*} = C_M^* = \{t^q\}$).*

Let $M(t)=t^{p+\sin(\log|\log t|)}$; it is easily checked that, for $p-\sqrt{2}>1$, M is an Orlicz function in some neighborhood of $t=0$. Put

$$U(\lambda)=\lambda M'(\lambda)/M(\lambda)=p+\sin(\log|\log \lambda|)+\cos(\log|\log \lambda|).$$

Then, for any $1>\delta>0$, $\lim_{\lambda\to 0}(U(\lambda s)-U(\lambda))=0$ uniformly for $s\in[\delta,1]$. Suppose now that $\lim_{n\to\infty} U(\lambda_n)=r$ for some $r\in[a_M, b_M]$ and some sequence $\lambda_n \to 0$. By using the uniformity of the limit above and by passing to a subsequence, if needed, we can assume that

$$r-n^{-1} \leqslant U(\lambda_n s) \leqslant r+n^{-1} \quad \text{for } 2^{-n} \leqslant s \leqslant 1.$$

Integrating this inequality between t and 1 we get that

$$t^{r+n^{-1}} \leqslant M(\lambda_n t)/M(\lambda_n) \leqslant t^{r-n^{-1}} \quad \text{for } 2^{-n} \leqslant t \leqslant 1,$$

i.e. $M(\lambda_n t)/M(\lambda_n)$ tends to t^r. It follows that $E_M=\{t^r;\, p-\sqrt{2}\leqslant r\leqslant p+\sqrt{2}\}$ while C_M consists of all the functions N which can be represented as

$$N(t)=\int_{p-\sqrt{2}}^{p+\sqrt{2}} t^r\, d\mu(r)$$

for some probability measure μ on $[p-\sqrt{2}, p+\sqrt{2}]$. By taking, for instance, μ to be uniformly distributed on the interval $[r, p+\sqrt{2}]$ we get a function equivalent to $t^r/|\log t|$. A simple computation shows that if $N \in C_M$ is represented by a measure μ and r is the smallest number in the support of μ then $\lim_{\lambda \to 0} N(\lambda t)/N(\lambda) = t^r$. Consequently, C_M contains no function equivalent to M itself and, by 4.b.10, l_M has, up to equivalence, a unique symmetric basis.

In order to prove that the sets C_M^* and C_{M^*} are different we shall show that every Orlicz function, which is simultaneously equivalent to a function in C_M and to a function in $C_{M^*}^*$, is already equivalent to a function in E_M, i.e. to t^{r_0} for some $p-\sqrt{2} \leq r_0 \leq p+\sqrt{2}$. Indeed, let $N \in C_M$ and let r_0 be the smallest number in the support of the measure μ representing N. Then, as remarked above, $E_N = \{t^{r_0}\}$ and

$$N(t)/t^{r_0} = \int_{r_0}^{p+\sqrt{2}} t^{r-r_0}\, d\mu(r) \leq \mu([r_0, r]) + t^{r-r_0} \quad \text{for } r_0 < r \leq p+\sqrt{2}\,.$$

If N is not equivalent to t^{r_0} then $\mu(\{r_0\}) = 0$ and hence, $\lim_{t \to 0} N(t)/t^{r_0} = 0$. It follows that $\lim_{t \to 0} N^*(t)/t^{q_0} = \infty$, where $q_0^{-1} + r_0^{-1} = 1$.

On the other hand, if N^* is equivalent at zero to a function in

$$C_{M^*} = \overline{\text{conv}}\,\{t^s;\ s_1 \leq s \leq s_2\}\,,$$

where $s_1^{-1} + (p+\sqrt{2})^{-1} = 1$ and $s_2^{-1} + (p-\sqrt{2})^{-1} = 1$, then, up to equivalence,

$$N^*(t) = \int_{s_1}^{s_2} t^s\, d\nu(s)$$

for some probability measure ν on $[s_1, s_2]$. Since $E_{N^*} = \{t^{q_0}\}$ the smallest number in the support of ν must be q_0. This however is a contradiction since it implies that $\limsup_{t \to 0} N^*(t)/t^{q_0} \leq 1$. □

The following example, constructed by N. J. Kalton [67], shows that Theorem 4.b.5 is no longer valid if we replace "strong non-equivalence" by "non-equivalence".

Example 4.c.3. *A separable Orlicz sequence space l_M which contains complemented subspaces isomorphic to l_p for some $1 \leq p < \infty$ but t^p is not equivalent to any function in $E_{M,1}$.*

We first define a sequence of functions $\{f_n\}_{n=1}^\infty$ on $[0, \infty)$ in the following way: for each integer n, f_n is the function of period $P_n = 2^{2^{2n}}$ such that

$$f_n(t) = \begin{cases} 0, & 0 \leq t \leq P_n - 4 \cdot 2^n \\ \frac{1}{2}(t-P_n) + 2 \cdot 2^n, & P_n - 4 \cdot 2^n < t \leq P_n - 2 \cdot 2^n \\ \frac{1}{2}(P_n - t), & P_n - 2 \cdot 2^n < t \leq P_n\,. \end{cases}$$

It is easily seen that f_n is a continuous function on $[0,\infty)$ whose maximal value is equal to 2^n. Let $f(t)=\max\{f_n(t); n=1,2,\ldots\}$ and observe that $|f(t_1)-f(t_2)|\leq \frac{1}{2}|t_1-t_2|$ for any $t_1, t_2>0$. Fix $p>3/2$ and put $M(t)=t^p e^{f(-\log t)}$, for $0<t\leq 1$, and $M(0)=0$. The function M is continuous but not necessarily convex on $[0,1]$. However, it is easily checked that, with M' standing for the right-derivative of M, we have

$$p-\tfrac{1}{2}\leq tM'(t)/M(t)=p-f'(-\log t)\leq p+\tfrac{1}{2}$$

for every $t\in(0,1]$. By the condition imposed on p we get that $M(t)/t$ is an increasing function and thus, M is equivalent at zero to an Orlicz function M_1, defined on $[0,\infty)$ (take, e.g. $M_1(t)=\int_0^t (M(u)/u)\,du$). Moreover, M_1 satisfies the Δ_2-condition at zero since $tM'(t)/M(t)\leq p+\frac{1}{2}$.

It is evident from the definition of M that the formal identity mapping T from l_{M_1} into l_p is a bounded operator. We shall show that T is not strictly singular. This would imply that l_{M_1} contains a complemented subspace isomorphic to l_p. Indeed, if W is an infinite-dimensional subspace of l_{M_1} for which $T_1=T_{|W}$ is an isomorphism then, by 2.a.2, TW contains a subspace $V\approx l_p$ so that there exists a bounded projection P from l_p onto V. It is easily checked that $Q=T_1^{-1}PT$ is a bounded projection from l_{M_1} onto its subspace $T_1^{-1}V$, which is clearly isomorphic to l_p.

To prove that T is not a strictly singular operator we use 4.a.11. Put $\varepsilon_n=e^{-P_n}$, $n=1,2,\ldots$. Then,

$$\frac{1}{\log 1/\varepsilon_n}\int_{\varepsilon_n}^{1}\frac{M(t)}{t^{p+1}}\,dt=\frac{1}{P_n}\int_{\varepsilon_n}^{1}\frac{e^{f(-\log t)}}{t}\,dt=\frac{1}{P_n}\int_0^{P_n} e^{f(u)}\,du\,.$$

Notice that, for $t\in[0,P_n]$, $f(t)=\max\{f_i(t); 1\leq i\leq n\}$. Hence, by using the definition of f_i, we get that

$$\frac{1}{\log 1/\varepsilon_n}\int_{\varepsilon_n}^{1}\frac{M(t)}{t^{p+1}}\,dt\leq\frac{1}{P_n}\sum_{i=1}^{n}\int_0^{P_n} e^{f_i(u)}\,du\leq\frac{1}{P_n}\sum_{i=1}^{n}\frac{P_n}{P_i}\,4\cdot 2^i\cdot e^{2^i}$$

$$\leq 4\sum_{i=1}^{\infty}2^{-2^{2i}}\cdot 2^i\cdot e^{2^i}<\infty\,,$$

for every integer n, and this proves that T is strictly singular.

It remains to show that t^p is not equivalent to any function in $E_{M_1,1}$. Assume to the contrary that there exists a constant $K>1$ such that, for any $\tau>0$, there is a $u=u(\tau)$ for which $e^{-K}t^p\leq M(e^{-u}t)/M(e^{-u})\leq e^K t^p$, $e^{-\tau}\leq t\leq 1$ or, equivalently $|f(u+v)-f(u)|\leq K$, $0\leq v\leq\tau$. Thus, for $u\leq t_1, t_2\leq u+\tau$, we have

$$|f(t_2)-f(t_1)|\leq 2K\,.$$

For each $t>0$ let $n(t)$ be the least integer for which $f_n(t)=f(t)$. Choose $\tau>3\cdot 2^{64K^2}$ and observe that if $0<f(t)<2^{n(t)}-2K$ for some $u+\tau/3\leq t\leq u+2\tau/3$ then there exists a t_1 so that $f(t)+3K>f_{n(t)}(t_1)>f(t)+2K$ and $|t_1-t|=2(f_{n(t)}(t_1)-f(t))<6K<\tau/3$. It follows that $u<t_1<u+\tau$ and this contradicts the fact that $f(t_1)-f(t)>2K$. Thus, for every $u+\tau/3\leqslant t\leqslant u+2\tau/3$, either $f(t)=0$ or $f(t)\geqslant 2^{n(t)}-2K$. Since $\tau/3\geqslant P_2$ there exists a $t\in[u+\tau/3, u+2\tau/3]$ so that $n(t)>1$. We also notice that $n(t)$ is not a constant on $[u+\tau/3, u+2\tau/3]$ since this would imply that the variation of f on this interval exceeds $\tau/12>2K$. Hence, if $t_1, t\in[u+\tau/3, u+2\tau/3]$ are chosen so that $n(t)-n(t_1)\geqslant 1$ then $f(t)\neq 0$ (otherwise $n(t)=1$) and therefore $f(t)\geqslant 2^{n(t)}-2K\geqslant 2\cdot 2^{n(t_1)}-2K\geqslant 2f(t_1)-2K$, i.e. $f(t_1)\leqslant 2K+f(t)-f(t_1)\leqslant 4K$. Consequently,

$$2^{n(t)}\leqslant 2K+f(t)\leqslant 2K+f(t)-f(t_1)+f(t_1)\leqslant 8K\,.$$

It follows that, for every $t\in[u+\tau/3, u+2\tau/3]$, we have $f(t)=\max\{f_i(t);\,1\leqslant i\leqslant n_0\}$, where n_0 is the largest integer satisfying $2^{n_0}\leqslant 8K$. Hence, on this interval, f has period $P_{n_0}=2^{2^{2n_0}}\leqslant 2^{64K^2}<\tau/3$ which means that f takes both values 0 and 2^{n_0} there. This however implies that the variation of f on $[u+\tau/3, u+2\tau/3]$ is equal to $2^{n_0}\geqslant 4K$, which leads to a contradiction. □

We present now a general procedure of constructing (or representing) Orlicz functions M in a form in which the set $E_{M,1}$ can be easily described (cf. [95]).

Fix $0<\tau<1$ and let F and G be two strictly increasing continuous convex functions on the interval $[\tau, 1]$ such that

(i) $F(1)=G(1)=1$, $0<F(\tau)<1$, $0<G(\tau)<1$.
(ii) $F'(1)=G'(1)$, $F'(1)\leqslant tF'(t)/F(t)$ and $G'(1)\leqslant tG'(t)/G(t)$ for all $t\in[\tau, 1]$.
(iii) $F(\tau)=\tau^{p_1}$ and $G(\tau)=\tau^{p_2}$ for some $1<p_1<p_2$.

For every sequence of digits $\eta=\{\eta(n)\}_{n=1}^\infty$ with $\eta(n)$ equal to 0 or to 1 for each n we define a function M_η on $[0, 1]$ in the following way. We put $M_\eta(1)=1$, $M_\eta(0)=0$ and, for $\tau^n\leqslant t<\tau^{n-1}$, $n=1, 2,\ldots$

$$M_\eta(t)=\begin{cases}M_\eta(\tau^{n-1})F(t/\tau^{n-1}) & \text{if } \eta(n)=0\\ M_\eta(\tau^{n-1})G(t/\tau^{n-1}) & \text{if } \eta(n)=1\,.\end{cases}$$

Using 4.b.11 it is easily verified that M_η is indeed an Orlicz function on $[0, 1]$ which satisfies the Δ_2-condition at zero.

The function M_η has the following quite obvious properties.

(a) $M_\eta(\tau^k)=\tau^{p_1k+(p_2-p_1)\sum_{n=1}^k\eta(n)}$, $k=1, 2,\ldots$ which implies that, up to equivalence, M_η is solely determined by p_1, p_2 and η and does not depend on the particular choice of F and G.

(b) For the same τ, p_1 and p_2 and for two different sequences of digits $\eta=\{\eta(n)\}_{n=1}^\infty$ and $\rho=\{\rho(n)\}_{n=1}^\infty$, the functions M_η and M_ρ are equivalent if and only if

$$\sup_k\left|\sum_{n=1}^k\eta(n)-\sum_{n=1}^k\rho(n)\right|<\infty.$$

(c) For fixed τ, p_1 and p_2, the set of all the functions of the form M_η, with η being a sequence of zeros and ones, is a norm compact subset of $C(0, 1)$ and the map $\eta \to M_\eta$ is a homeomorphism from $\{0, 1\}^{\aleph_0}$, equipped with the product topology, into $C(0, 1)$.

(d) Consider the map T defined by $(TN)(t)=N(\tau t)/N(\tau)$ and let Φ be the shift by one to the left, i.e. $(\Phi\eta)(n)=\eta(n+1)$. Then $TM_\eta=M_{\Phi\eta}$.

(e) Up to equivalence, $E_{M_\eta,1}$ consists of functions of the form M_ρ (for the same τ, p_1 and p_2) with ρ being a pointwise limit of sequences having the form $\{\Phi^{k_j}\eta\}_{j=1}^\infty$.

The following proposition describes the interval $[\alpha_{M_\eta}, \beta_{M_\eta}]$ corresponding to an Orlicz function of the type M_η (see 4.a.9).

Proposition 4.c.4. *Let τ, p_1, p_2, η and M_η be as above. Then,*

$$\alpha_{M_\eta}=p_1+(p_2-p_1)\liminf_{k\to\infty}\inf_n \delta(n+1, n+k)\,,$$

$$\beta_{M_\eta}=p_1+(p_2-p_1)\limsup_{k\to\infty}\sup_n \delta(n+1, n+k)\,,$$

where $\delta(n+1, n+k)$ denotes the density of ones between the numbers $n+1$ and $n+k$, i.e. $\delta(n+1, n+k)=\sum_{i=n+1}^{n+k}\eta(i)/k$.

Proof. Since M_η satisfies the Δ_2-condition at zero it suffices to consider in the definition of α_{M_η} only expressions of the form

$$M_\eta(\tau^n\cdot\tau^k)/M_\eta(\tau^n)\tau^{pk}=\tau^{(p_1-p)k+(p_2-p_1)\sum_{i=n+1}^{n+k}\eta(i)}\,.$$

Hence, α_{M_η} is the supremum of all numbers p for which

$$\inf_{n,k}\left((p_1-p)k+(p_2-p_1)\sum_{i=n+1}^{n+k}\eta(i)\right)>-\infty\,.$$

A simple argument shows that α_{M_η} is therefore equal to the expression given in the statement. The proof for β_{M_η} is similar. □

We present now a characterization of minimal Orlicz function of the form M_η.

Proposition 4.c.5. *Let M_η be as above. The function M_η is equivalent to a minimal Orlicz function if and only if there exists a constant K such that, for every integer k, there is an integer $n=n(k)$ with the following property: for every integer h there is an $m\leq n$ such that*

$$\left|\sum_{i=h+m+1}^{h+m+j}\eta(i)-\sum_{i=1}^{j}\eta(i)\right|\leq K \quad \text{for } j=1, 2,\dots, k\,.$$

Proof. Assume first that such K and $n(k)$ do exist. Let $\{h_l\}_{l=1}^\infty$ be an increasing sequence of integers for which the pointwise limit of $\{\Phi^{h_l}\eta\}_{l=1}^{\infty}$ exists. Denote this limit sequence by ρ and fix k. Then there exists an index h_l so that

$$\rho(i)=(\Phi^{h_l}\eta)(i)=\eta(h_l+i) \quad \text{for } 1\leqslant i\leqslant k+n(k).$$

By our assumption, there is an $1\leqslant m\leqslant n(k)$ so that

$$\left|\sum_{i=m+1}^{m+j}\rho(i)-\sum_{i=1}^{j}\eta(i)\right|=\left|\sum_{i=h_l+m+1}^{h_l+m+j}\eta(i)-\sum_{i=1}^{j}\eta(i)\right|\leqslant K, \quad j=1,2,\ldots,h.$$

Hence, by using the observations (e) and (b) above, it follows that $E_{M_\rho,1}$ contains a function equivalent to M_η. By the remark preceding 4.b.7 it follows that M_η is equivalent to a minimal Orlicz function.

In order to prove the converse we suppose that there exists a minimal function N such that $A^{-1}\leqslant N(t)/M_\eta(t)\leqslant A$ for some constant $A>0$ and every $0<t\leqslant 1$. Assume now that the condition described in the statement is not satisfied for any K and $n(k)$. This means that, for every integer K, there exist an integer $k(K)$ and sequences $\{h_1(n,K)\}_{n=1}^{\infty}$, $\{h_2(n,K)\}_{n=1}^{\infty}$ for which $h_2(n,K)-h_1(n,K)=n$, $n=1,2,\ldots$ and such that, for every $h_1(n,K)\leqslant h<h_2(n,K)$, there is a $j\leqslant k(K)$ with

$$\left|\sum_{i=h+1}^{h+j}\eta(i)-\sum_{i=1}^{j}\eta(i)\right|>K.$$

For a fixed K let η_K be any limit point of the sequence $\{\Phi^{h_1(n,K)}\eta\}_{n=1}^{\infty}$. Then $M_{\eta_K}\in E_{M_\eta,1}$ and, for every integer m, there is a $j\leqslant k(K)$ so that

$$\left|\sum_{i=m+1}^{m+j}\eta_K(i)-\sum_{i=1}^{j}\eta(i)\right|>K.$$

It follows from the observations (a), (b) and (e) that, for any $N_0\in E_{M_{\eta_K}}$, there exists $0<t\leqslant 1$ for which the ratio $N_0(t)/M_\eta(t)$ is not in the interval $[B^{-1}\tau^{K(p_2-p_1)}, B\tau^{-K(p_2-p_1)}]$, where B is a constant depending only on F and G. This however contradicts the minimality of N when K is chosen so that $B\tau^{-K(p_2-p_1)}>A^4$. □

The previous propositions will now be used to study some concrete minimal Orlicz sequence spaces.

Example 4.c.6 [94, 95]. *A minimal Orlicz sequence space l_M whose interval $[\alpha_M,\beta_M]$ reduces to a single point p and which does not have any complemented subspace isomorphic to l_p.*

Let $0<\tau<1$ and $1<p_1<p_2$ be arbitrary. We construct simultaneously two sequences of zeros and ones, $\eta=\{\eta(i)\}_{i=1}^{\infty}$ and $\rho=\{\rho(i)\}_{i=1}^{\infty}$, as follows. Put $\eta(1)=0$, $\rho(1)=1$ and, for $n=0,1,2,\ldots$,

$$\begin{aligned}
&\eta(2^{3n}+i)=\rho(i) &&\text{and}\quad \rho(2^{3n}+i)=\rho(i) &&\text{for } 1\leqslant i\leqslant 2^{3n},\\
&\eta(2^{3n+1}+i)=\eta(i) &&\text{and}\quad \rho(2^{3n+1}+i)=\rho(i) &&\text{for } 1\leqslant i\leqslant 2^{3n+1},\\
&\eta(2^{3n+2}+i)=\eta(i) &&\text{and}\quad \rho(2^{3n+2}+i)=\eta(i) &&\text{for } 1\leqslant i\leqslant 2^{3n+2}.
\end{aligned}$$

Thus, these two sequences begin as follows

$$\eta=(0, \overbrace{1}, \overbrace{0, 1}, \overbrace{0, 1, 0, 1}, \overbrace{1, 1, 1, 1, 0, 1, 0, 1},\dots),$$
$$\rho=(1, \overbrace{1}, \overbrace{1, 1}, \overbrace{0, 1, 0, 1}, \overbrace{1, 1, 1, 1, 0, 1, 0, 1},\dots).$$

We shall prove first that M_η (and also M_ρ) is equivalent to a minimal Orlicz function. For every n let A_n (respectively B_n) be the block consisting of the first 2^{3n} digits in η (respectively ρ). By the inductive definition of η and ρ both A_{n+1} and B_{n+1} contain a block equal to A_n and a block equal to B_n. Since ρ can be written as a succession of blocks $C_1C_2C_3\dots$, where each block C_j is equal either to A_{n+1} or to B_{n+1}, it follows that every block of ρ of length $\geqslant 3\cdot 2^{3(n+1)}$ contains in it either A_{n+1} or B_{n+1} and thus, both A_n and B_n. This implies that the condition 4.c.5 holds for M_η and also for M_ρ, with $K=0$ and $n(k)=3\cdot 2^{3(n+1)}$ whenever $k\leqslant 2^{3n}$. This proves that both M_η and M_ρ are equivalent to minimal functions.

The functions M_η and M_ρ are not equivalent at zero because

$$\sum_{i=1}^{2^{3n}} (\rho(i)-\eta(i))=2^n \quad \text{for all } n\,.$$

On the other hand, η is the pointwise limit of the sequence $\{\Phi^{2^{3n+2}}\rho\}_{n=1}^\infty$ and ρ that of $\{\Phi^{2^{3n}}\eta\}_{n=1}^\infty$. Hence, $M_\eta\in E_{M_\rho,1}$ and $M_\rho\in E_{M_\eta,1}$ which, by condition $\binom{*}{*}$ of the preceding section, implies that $l_{M_\rho}\approx l_{M_\eta}$. We therefore conclude that neither M_η nor M_ρ is equivalent to t^p for some $p\geqslant 1$. We note in passing that, by 4.b.9, l_{M_η} has uncountably many mutually non-equivalent symmetric bases.

In order to determine the interval associated to M_η let us denote by a_n (respectively b_n) the number of times the digit 1 appears in the block A_n (respectively B_n). Then, $a_0=0$, $b_0=1$ and

$$a_{n+1}=4a_n+4b_n, \qquad b_{n+1}=2a_n+6b_n, \quad n=0, 1, 2,\dots.$$

The density of ones in the block A_n (respectively B_n) is equal to $a_n/2^{3n}$ (respectively $b_n/2^{3n}$). Easy computations show that $(b_n-a_n)/2^{3n}=1/4^n$ and $a_n/2^{3n}=\sum_{i=0}^{n-1} 1/2\cdot 4^i$ for all n. It follows immediately that $\lim_{n\to\infty} a_n/2^{3n}=\lim_{n\to\infty} b_n/2^{3n}=2/3$. Let $\varepsilon>0$ be given and choose an integer n so that $|a_n/2^{3n}-2/3|<\varepsilon$ and $|b_n/2^{3n}-2/3|<\varepsilon$. It is easily checked that the density of ones in any block of ρ of length $2^{3n}\cdot l$ is between $(l-2)(2/3-\varepsilon)/l$ and $((l-2)(2/3+\varepsilon)+2)/l$. Therefore, for l large enough, the density is between $2/3-2\varepsilon$ and $2/3+2\varepsilon$. By 4.c.3 it follows that $\alpha_{M_\rho}=\beta_{M_\rho}=p_1+2(p_2-p_1)/3$. Since l_{M_ρ} is isomorphic to l_{M_η} we get that M_η has the same interval.

It remains to show that, for $p=p_1+2(p_2-p_1)/3$, l_{M_ρ} contains no complemented subspace isomorphic to l_p. In order to prove this we use 4.b.5 and show that the function t^p is strongly non-equivalent to $E_{M_\rho,1}$. For an integer n put $m(n)=3\cdot 2^{3(n+1)}+2^{3n}$ and assume the existence of an integer k and of a constant $K>0$ so that

$$K^{-1}\tau^{pi}\leqslant M_\rho(\tau^k\cdot\tau^i)/M_\rho(\tau^k)\leqslant K\tau^{pi}, \quad i=1, 2,\dots, m(n)\,.$$

Let $1 \leqslant j \leqslant 3 \cdot 2^{3(n+1)}$; by using the above inequality with $i=j$ and $i=j+2^{3n}$ we get

$$K^{-2}\tau^{p\cdot 2^{3n}} \leqslant M_\rho(\tau^{k+j}\tau^{2^{3n}})/M_\rho(\tau^{k+j}) \leqslant K^2\tau^{p\cdot 2^{3n}}, \quad 1 \leqslant j \leqslant 3\cdot 2^{3(n+1)}.$$

Recall now that every block of ρ of length $\geqslant 3\cdot 2^{3(n+1)}$ contains a block equal to A_n and a block equal to B_n. This implies that there are integers j_1, j_2, $1 \leqslant j_1, j_2 < 3\cdot 2^{3(n+1)}$ for which

$$M_\rho(\tau^{k+j_1}\tau^{2^{3n}})/M_\rho(\tau^{k+j_1}) = M_\rho(\tau^{2^{3n}}), \quad M_\rho(\tau^{k+j_2}\tau^{2^{3n}})/M_\rho(\tau^{k+j_2}) = M_\eta(\tau^{2^{3n}}).$$

It follows that $K^{-4} \leqslant M_\rho(\tau^{2^{3n}})/M_\eta(\tau^{2^{3n}}) \leqslant K^4$. On the other hand,

$$M_\rho(\tau^{2^{3n}})/M_\eta(\tau^{2^{3n}}) = \tau^{(p_2-p_1)\left(\sum_{i=1}^{2^{3n}}\rho(i) - \sum_{i=1}^{2^{3n}}\eta(i)\right)} = \tau^{(p_2-p_1)2^n}.$$

This means that for $\tau^{-(p_2-p_1)2^{n-1}} \leqslant K < \tau^{-(p_2-p_1)2^{n-2}}$ it suffices to take $m_K = 3\cdot 2^{3(n+1)} + 2^{3n}$ points in order to prove that condition (†) of 4.b.4 holds. Since, for any $\alpha > 0$, $m_K = o(K^\alpha)$ we get that t^p is strongly non-equivalent to $E_{M_\rho,1}$. □

The next example is similar to 4.c.6 with the exception that its interval can be chosen arbitrarily.

Example 4.c.7 [95]. *A minimal Orlicz sequence space l_M with $\alpha_M < \beta_M$ which contains no complemented subspace isomorphic to l_p for $p \geqslant 1$.*

Let $0 < \tau < 1$ and $1 < p_1 < p_2$ be arbitrarily chosen. For every $0 < \alpha < \beta < 1$ construct a sequence of positive integers $\{n_j\}_{j=1}^\infty$ such that $\sum_{j=1}^\infty 1/n_j \leqslant \alpha$ and $\prod_{j=1}^\infty (1 - 1/n_j) \geqslant \beta$. We define two sequences ρ and η of zeros and ones, as follows. Let $m_j = n_1 \cdot n_2 \ldots n_{j-1}$ $(m_1 = 1)$ and let A_j (respectively B_j) denote the block of the first m_j digits of ρ (respectively η). A_1 consists of the digit 1 and B_1 of the digit 0 while, for $j > 1$, A_j and B_j are defined inductively by

$$A_{j+1} = \overbrace{A_jA_j\ldots A_j}^{n_j-1 \text{ times}}B_j, \qquad B_{j+1} = \overbrace{B_jB_j\ldots B_j}^{n_j-1 \text{ times}}A_j.$$

The same argument as in 4.c.6 shows that, for this ρ, the Orlicz function M_ρ is minimal (the condition in the statement of 4.c.5 is satisfied with $K=0$ and $n(m_{j-1}) = 3m_j$). To estimate the size of the interval of M_ρ we remark that the density of ones in A_j is larger than $\prod_{i=1}^{j-1} (1 - 1/n_i) \geqslant \beta$ while the density of ones in B_j is less than $\sum_{i=1}^{j-1} 1/n_i \leqslant \alpha$. It follows from 4.c.4 that $\alpha_{M_\rho} \leqslant p_1 + \alpha(p_2 - p_1)$ and $\beta_{M_\rho} \geqslant p_1 + \beta(p_2 - p_1)$, i.e. $\beta_{M_\rho} - \alpha_{M_\rho} = (\beta - \alpha)(p_2 - p_1) > 0$.

In spite of the fact that l_{M_ρ} has subspaces isomorphic to l_p for an entire interval of p's it does not admit any l_p as a complemented subspace. Since the proof is similar to that presented in 4.c.6 we do not reproduce it here. □

We conclude the study of functions of the form M_η by proving that they actually represent all reflexive Orlicz sequence spaces.

Proposition 4.c.8. *For every Orlicz function M such that l_M is reflexive there exist $1<p_1<p_2$, $0<\tau<1$ and a sequence $\eta=\{\eta(n)\}_{n=1}^{\infty}$ of zeros and ones such that the corresponding function M_η is equivalent to M.*

Proof. We may assume that $M(1)=1$. Since l_M is reflexive it follows from 4.b.2′ that, for some $1<p_1<p_2$ and all $0<t\leqslant 1$, $p_1\leqslant tM'(t)/M(t)\leqslant p_2$. Choose $0<\tau<1$ so that $p_1\tau-p_1+1=\tau^{p_2}$. Then, the functions $F(t)=t^{p_1}$ and $G(t)=p_1t-p_1+1$ satisfy in the interval $[\tau, 1]$ all the assumptions appearing in the definition of functions of the type M_η.

Now, we construct inductively a sequence $\eta=\{\eta(n)\}_{n=1}^{\infty}$ as follows. We set $\eta(1)=1$ and if $M_\eta(\tau^n)\tau^{p_1}\leqslant M(\tau^{n+1})$ we put $\eta(n+1)=0$; otherwise, $\eta(n+1)=1$. It is easily verified that

$$M_\eta(\tau^n)\leqslant M(\tau^n)\leqslant \tau^{p_1-p_2}M_\eta(\tau^n) \quad \text{for all } n\geqslant 1\,,$$

i.e. M_η is equivalent to M. □

Remark. The construction above does not work for a non-reflexive space l_M satisfying the Δ_2-condition because, in this case, p_1 is necessarily equal to 1 and thus, 4.b.11 cannot be used. However, the same construction can be still performed with $F(t)=t$, $G(t)=t^{p_2}$ and $0<\tau<1$ arbitrarily chosen. The function $M_\eta(t)$ obtained in this way is equivalent to $M(t)$ and $M_\eta(t)/t$ in an increasing function but, in general, $M_\eta(t)$ need not be convex.

d. Modular Sequence Spaces and Subspaces of $l_p \oplus l_r$

In the study of Orlicz sequence spaces we have already encountered spaces which are generated by a sequence of Orlicz functions. More precisely, we have seen that, for any block basis $\{u_n\}_{n=1}^{\infty}$ of the unit vector basis of an Orlicz sequence space l_M, there exists a sequence of Orlicz functions $\{M_n\}_{n=1}^{\infty}$ such that a series $\sum_{n=1}^{\infty} a_n u_n$ converges if and only if $\sum_{n=1}^{\infty} M_n(|a_n|/\rho)<\infty$ for some $\rho>0$. It is therefore natural to consider the following class of sequence spaces.

Definition 4.d.1. Let $\{M_n\}_{n=1}^{\infty}$ be a sequence of Orlicz functions. The space $l_{(M_n)}$, is the Banach space of all sequences $x=(a_1, a_2,\ldots)$ with $\sum_{n=1}^{\infty} M_n(|a_n|/\rho)<\infty$ for some $\rho>0$, equipped with the norm

$$\|x\|=\inf\left\{\rho>0;\ \sum_{n=1}^{\infty} M_n(|a_n|/\rho)\leqslant 1\right\}.$$

The space $l_{(M_n)}$ is called a *modular sequence space.*

An important subspace of $l_{(M_n)}$ is $h_{(M_n)}$ which consists of those sequences $x=(a_1, a_2,\ldots)\in l_{(M_n)}$ such that $\sum_{n=1}^{\infty} M_n(|a_n|/\rho)<\infty$ *for every* $\rho>0$.

The notion of equivalence between sequences of Orlicz functions is defined in the following way: *two sequences of Orlicz functions* $\{M_n\}_{n=1}^{\infty}$ *and* $\{N_n\}_{n=1}^{\infty}$ *are said to be equivalent if* $l_{\{M_n\}}$ *and* $l_{\{N_n\}}$ *are equal as sets, i.e. they consist of the same sequences.* It is easily checked, using the closed graph theorem, that if $\{M_n\}_{n=1}^{\infty}$ and $\{N_n\}_{n=1}^{\infty}$ are equivalent then the identity map from $l_{\{M_n\}}$ onto $l_{\{N_n\}}$ is an isomorphism. Analytic conditions for the equivalence of two sequences of Orlicz functions are in general quite unnatural. A sufficient condition, used often in the sequel, is the following: $\{M_n\}_{n=1}^{\infty}$ *and* $\{N_n\}_{n=1}^{\infty}$ *are equivalent provided there exist numbers* $K>0$, $t_n \geq 0$, $n=1, 2,\ldots$ *and an integer* n_0 *so that*

(a) $K^{-1}M_n(t) \leq N_n(t) \leq KM_n(t)$ *for all* $n \geq n_0$ *and* $t \geq t_n$

(b) $\sum_{n=1}^{\infty} M_n(t_n) < \infty$.

The proof is straightforward.

In order to avoid technical difficulties which arise in some non-interesting cases we assume, unless stated otherwise, that the functions $\{M_n\}_{n=1}^{\infty}$, $\{N_n\}_{n=1}^{\infty}$ etc., are nondegenerate, strictly increasing and have a derivative for every $t \geq 0$. A sequence of Orlicz functions $\{M_n\}_{n=1}^{\infty}$ is said to be normalized if $M_n(1)=1$ for all n. If $\{M_n\}_{n=1}^{\infty}$ is not normalized we put $N_n(t)=M_n(\tau_n t)$, where $M_n(\tau_n)=1$. Then, $\{N_n\}_{n=1}^{\infty}$ is normalized and $l_{\{N_n\}}$ is isomorphic to $l_{\{M_n\}}$. Since we consider here only linear topological properties of modular sequence spaces we can and shall assume that $\{M_n\}_{n=1}^{\infty}$ is a normalized sequence.

In the study of Orlicz sequence spaces a crucial role is played by the Δ_2-condition at zero. What we need in the present case is a "uniform" Δ_2-condition.

Definition 4.d.2. A sequence of Orlicz functions $\{M_n\}_{n=1}^{\infty}$ is said to satisfy the *uniform* Δ_2*-condition* (Δ_2^**-condition*) at zero if there exist a number $p>1$ ($r>1$) and an integer n_0 such that, for all $t \in (0, 1)$ and $n \geq n_0$, we have $tM_n'(t)/M_n(t) \leq p$ ($tM_n'(t)/M_n(t) > r$).

An equivalent definition of the uniform Δ_2-condition at zero is obtained by requiring the existence of a constant $K<\infty$ and of an integer n_0 such that $M_n(2t)/M_n(t) \leq K$ for all $n \geq n_0$ and $t \in (0, 1/2]$. The uniform Δ_2-condition is not preserved by equivalence. For example, let $\{M_n\}_{n=1}^{\infty}$ be any sequence of Orlicz functions for which the uniform Δ_2-condition at zero holds. Choose real numbers $t_n>0$, $n=1, 2,\ldots$ so that $\sum_{n=1}^{\infty} M_n(t_n)<\infty$; then, we can easily construct Orlicz functions N_n, $n=1, 2,\ldots$ which do not satisfy the Δ_2-condition (even individually) but such that $N_n(t)=M_n(t)$ for all n and $t \geq t_n$. As observed above, $\{M_n\}_{n=1}^{\infty}$ and $\{N_n\}_{n=1}^{\infty}$ are equivalent.

The importance of the uniform Δ_2-condition is illustrated by the following result.

Proposition 4.d.3 [144]. *For any sequence of Orlicz functions* $\{M_n\}_{n=1}^{\infty}$ *the following conditions are equivalent.*

(i) *The sequence* $\{M_n\}_{n=1}^{\infty}$ *is equivalent to a sequence* $\{N_n\}_{n=1}^{\infty}$ *which satisfies the uniform* Δ_2*-condition at zero.*

(ii) $l_{(M_n)}=h_{(M_n)}$.
(iii) *The unit vectors form a boundedly complete normalized unconditional basis of* $l_{(M_n)}$.
(iv) $l_{(M_n)}$ *is separable.*
(v) $l_{(M_n)}$ *contains no subspace isomorphic to* l_∞.

Notice that the unit vectors are normalized for it is assumed that $M_n(1)=1$ for all n. The proof of 4.d.3 is quite similar to that of 4.a.4 and we do not reproduce it here.

We pass now to another routine matter, namely that of duality of modular spaces. Let $\{M_n\}_{n=1}^\infty$ be a sequence of Orlicz functions such that none of which is equivalent to t. Let $\{M_n^*\}_{n=1}^\infty$ be the corresponding sequence of complementary functions. In general, the functions M_n^* do not satisfy $M_n^*(1)=1$ for all n but we still can consider the space $l_{(M_n^*)}$. For every $y=(a_1, a_2, \ldots)\in l_{(M_n^*)}$ we put

$$|||y|||=\sup\left\{\sum_{n=1}^\infty a_n b_n;\ \sum_{n=1}^\infty M(|b_n|)\leq 1\right\}.$$

It follows from Young's inequality that $\|y\|\leq|||y|||\leq 2\|y\|$ for every $y\in l_{(M_n^*)}$.

Let $x^*\in h_{(M_n)}^*$ and put $c_n=x^*(e_n)$, $n=1, 2, \ldots$, where $\{e_n\}_{n=1}^\infty$ denotes, as usual, the sequence of the unit vectors. Then, we have

$$|||(c_1, c_2, \ldots)|||=\sup\left\{x^*\left(\sum_{n=1}^\infty b_n e_n\right);\ \sum_{n=1}^\infty M(|b_n|)\leq 1\right\}=\|x^*\|,$$

i.e. the mapping $x^*\to(x^*(e_1), x^*(e_2), \ldots)$ defines an isomorphism from $h_{(M_n)}^*$ into $l_{(M_n^*)}$. It is easily verified that this isomorphism is onto, i.e. $h_{(M_n)}^*\approx l_{(M_n^*)}$.

We present now without proof a result which generalizes 4.b.2′ to the case of modular sequence spaces.

Proposition 4.d.4 [144]. *A modular sequence space* $l_{(M_n)}$ *is reflexive if and only if* $\{M_n\}_{n=1}^\infty$ *is equivalent to a sequence of Orlicz functions* $\{N_n\}_{n=1}^\infty$ *for which the uniform* Δ_2*- and* Δ_2^**-conditions hold.*

In the sequel we shall be interested only in modular sequence spaces satisfying the uniform Δ_2-condition at zero. These spaces are related to Orlicz sequence spaces in a very simple way.

Theorem 4.d.5. *Let* $1\leq q<s<\infty$. *A Banach space is a modular sequence space* $l_{(M_n)}$, *with* M_n *satisfying* $q\leq tM_n'(t)/M_n(t)\leq s$ *for all* n *and* $t\in(0, 1)$, *if and only if it is isomorphic to the closed linear span of a block basis of the unit vectors in some Orlicz sequence space* l_M, *with* M *satisfying* $q\leq tM'(t)/M(t)\leq s$ *for every* $t>0$.

Proof. The "if" part is immediate (use, e.g. the argument preceding 4.a.7). Conversely, let $\{M_n\}_{n=1}^\infty$ be a sequence of Orlicz functions satisfying our hypotheses for some $1\leq q<s<\infty$. Let $U=U_{q,s}$ be the universal Orlicz function whose existence is ensured by 4.b.13. The function U has the property that there are a constant $K=K_{q,s}$ and functions $N_n\in E_U$, $n=1, 2, \ldots$ such that $K^{-1}\leq N_n(t)/M_n(t)\leq K$ for all n and

$t \in (0, 1)$. The arguments already used in the proof of 4.a.8 show that there is a block basis $\{u_n\}_{n=1}^\infty$ (with constant coefficients) of the unit vector basis of l_U such that $[u_n]_{n=1}^\infty \approx l_{(N_n)} \approx l_{(M_n)}$. We remark that $[u_n]_{n=1}^\infty$ is actually a complemented subspace of l_U. □

We turn now our attention to a special class of modular sequence spaces. Fix $1 \leq r < p < \infty$ and let $\{f_n\}_{n=1}^\infty$ and $\{g_n\}_{n=1}^\infty$ be the unit vector bases of l_p, respectively l_r. For a sequence $w = \{w_n\}_{n=1}^\infty$ of positive reals put $e_n = f_n + w_n g_n$, $n = 1, 2, \ldots$ and let $X_{p,r,w}$ be the closed linear span of $\{e_n\}_{n=1}^\infty$ in $(l_p \oplus l_r)_\infty$. In many cases, the space $X_{p,r,w}$ is isomorphic to either l_p, l_r or to $l_p \oplus l_r$. A non-trivial case which is of importance occurs when the sequence $w = \{w_n\}_{n=1}^\infty$ satisfies the condition

$$(*) \qquad \sum_{n=1}^\infty w_n^{pr/(p-r)} = \infty, \quad w_n \to 0 \text{ and } w_n < 1 \text{ for all } n.$$

The spaces $X_{p,2,w}$ were introduced by H. P. Rosenthal [127]. These spaces play an important role in the study of complemented subspaces of $L_p(0, 1)$ and they will be further investigated in Vol. II. Some of the results, to be presented in this section, have been originally proved by probabilistic methods in the case when $r = 2$ (cf. [128]). The probabilistic methods do not work for other values of r and therefore we present here an approach based on modular and Orlicz sequence spaces. This method is due to J. T. Woo [145].

Before describing these results we prove that $X_{p,r,w}$ does not really depend on the sequence w provided condition (*) holds.

Proposition 4.d.6 [127, 145]. *Let $w = \{w_n\}_{n=1}^\infty$ and $w' = \{w'_n\}_{n=1}^\infty$ be two sequences both satisfying the condition (*). Then the spaces $X_{p,r,w}$ and $X_{p,r,w'}$ are isomorphic.*

Proof. Since w satisfies (*) there are disjoint finite subsets $\{\sigma_j\}_{j=1}^\infty$ of integers so that the numbers $v_j = \left(\sum_{n \in \sigma_j} w_n^{pr/(p-r)}\right)^{(p-r)/pr}$ satisfy $w'_j \leq v_j \leq 2w'_j$ for all j.

Let $\{e_n\}_{n=1}^\infty$ be the natural basis of $X_{p,r,w}$ and put

$$h_j = \left(\sum_{n \in \sigma_j} w_n^{r/(p-r)} e_n\right) \Big/ \left(\sum_{n \in \sigma_j} w_n^{pr/(p-r)}\right)^{1/p}, \quad j = 1, 2, \ldots.$$

Then, for every choice of scalars $\{a_j\}_{j=1}^\infty$, we have

$$\begin{aligned}
\left\|\sum_{j=1}^\infty a_j h_j\right\| &= \max\left\{\left[\sum_{j=1}^\infty \left(\sum_{n \in \sigma_j} |a_j|^p w_n^{pr/(p-r)}\right) \Big/ \left(\sum_{n \in \sigma_j} w_n^{pr/(p-r)}\right)\right]^{1/p},\right.\\
&\qquad \left.\left[\sum_{j=1}^\infty \left(\sum_{n \in \sigma_j} |a_j|^r w_n^{r^2/(p-r)} w_n^r\right) \Big/ \left(\sum_{n \in \sigma_j} w_n^{pr/(p-r)}\right)^{r/p}\right]^{1/r}\right\}\\
&= \max\left\{\left(\sum_{j=1}^\infty |a_j|^p\right)^{1/p}, \left(\sum_{j=1}^\infty |a_j|^r \left(\sum_{n \in \sigma_j} w_n^{pr/(p-r)}\right)^{(p-r)/p}\right)^{1/r}\right\}\\
&= \max\left\{\left(\sum_{j=1}^\infty |a_j|^p\right)^{1/p}, \left(\sum_{j=1}^\infty |a_j|^r v_j^r\right)^{1/r}\right\}.
\end{aligned}$$

This shows that $\{h_j\}_{j=1}^\infty$ is equivalent to the natural basis of $X_{p,r,w'}$. We shall prove now that $[h_j]_{j=1}^\infty$ is a complemented subspace of $X_{p,r,w}$. For any sequence of scalars $\{a_n\}_{n=1}^\infty$ which are eventually equal to zero we set

$$P\Big(\sum_{n=1}^\infty a_n e_n\Big)=\sum_{j=1}^\infty\Big[\Big(\sum_{n\in\sigma_j} a_n w_n^{(pr-r)/(p-r)}\Big)\Big/\Big(\sum_{n\in\sigma_j} w_n^{pr/(p-r)}\Big)^{(p-1)/p}\Big]h_j\,.$$

We have

$$\Big\|P\Big(\sum_{n=1}^\infty a_n e_n\Big)\Big\|=\max\Big\{\Big[\sum_{j=1}^\infty\Big|\sum_{n\in\sigma_j} a_n w_n^{(pr-r)/(p-r)}\Big|^p\Big/\Big(\sum_{n\in\sigma_j} w_n^{pr/(p-r)}\Big)^{p-1}\Big]^{1/p},$$

$$\Big[\sum_{j=1}^\infty\Big|\sum_{n\in\sigma_j} a_n w_n^{(pr-r)/(p-r)}\Big|^r\Big/\Big(\sum_{n\in\sigma_j} w_n^{pr/(p-r)}\Big)^{r-1}\Big]^{1/r}\Big\}$$

and, using twice Holder's inequality, we obtain

$$\Big\|P\Big(\sum_{n=1}^\infty a_n e_n\Big)\Big\|\leq\Big\|\sum_{j=1}^\infty\sum_{n\in\sigma_j} a_n e_n\Big\|\leq\Big\|\sum_{n=1}^\infty a_n e_n\Big\|\,.$$

Hence, the operator P extends to a norm-one projection from $X_{p,r,w}$ onto $[h_j]_{j=1}^\infty$. We remark that the proof that $[h_j]_{j=1}^\infty$ is complemented in $X_{p,r,w}$ is valid for any choice of $\{\sigma_j\}_{j=1}^\infty$ (i.e. it does not depend on the particular relation between v_j and w'_j which was taken into account when the sets $\{\sigma_j\}_{j=1}^\infty$ were constructed). The existence of the projection P shows that $X_{p,r,w'}$ is isomorphic to a complemented subspace of $X_{p,r,w}$ and, by symmetry, also that $X_{p,r,w}$ is isomorphic to a complemented subspace of $X_{p,r,w'}$. Thus, by Pelczynski's decomposition method (see the proof of 2.d.10), it would follow that $X_{p,r,w}\approx X_{p,r,w'}$ provided we prove that each of the spaces $X_{p,r,w}$ and $X_{p,r,w'}$ is isomorphic to its own square. This means that in order to complete the proof it suffices to prove the following lemma.

Lemma 4.d.7. *For each sequence* $w=\{w_n\}_{n=1}^\infty$ *satisfying the condition* (*) *we have* $X_{p,r,w}\oplus X_{p,r,w}\approx X_{p,r,w}$.

Proof. We first split the integers N into disjoint infinite subsets $N_1, N_2,\ldots, N_k,\ldots$ so that, for each k, $\sum_{n\in N_k} w_n^{pr/(p-r)}=\infty$. Put $w^{(k)}=\{w_n\}_{n\in N_k}$. By the first part of the proof of 4.d.6 there exists a constant $A<\infty$ such that, for each k, there is a block basis $\{u_j^{(k)}\}_{j\in N_k}$ of the natural basis of $X_{p,r,w^{(k)}}$, considered as a subspace of $X_{p,r,w}$, which is equivalent to the natural basis of $X_{p,r,w}$, $d([u_j^{(k)}]_{j\in N_k}, X_{p,r,w})\leq A$, and whose span $[u_j^{(k)}]_{j\in N_k}$ is the range of a norm-one projection in $X_{p,r,w^{(k)}}$.

Using the remark made in the proof of 4.d.6 we get that the space $Y=[u_j^{(k)}]_{j\in N_k, k\in N}$ is also complemented in $X_{p,r,w}$, i.e. $X_{p,r,w}=Y\oplus Z$ for some Banach space Z. Every $y\in Y$ can be represented uniquely as $y=\sum_{k=1}^\infty y^{(k)}$ with

$y^{(k)} = \sum_{j \in N_k} c_j^{(k)} u_j^{(k)} \in [u_j^{(k)}]_{j \in N_k}$ for all k. Moreover, for some constant C depending only on A, we have for every $y \in Y$ that

$$C^{-1}\|y\| \leqslant \max\left\{\left(\sum_{k=1}^{\infty} \sum_{j \in N_k} |c_j^{(k)}|^p\right)^{1/p}, \left(\sum_{k=1}^{\infty} \sum_{j \in N_k} |c_j^{(k)}|^r w_j^r\right)^{1/r}\right\} \leqslant C\|y\| .$$

This implies that $Y \approx \left\{y = \sum_{k=1}^{\infty} y^{(k)};\ y^{(k)} = 0 \text{ for } k > 1\right\} \oplus \left\{y = \sum_{k=1}^{\infty} y^{(k)};\ y^{(1)} = 0\right\}$, i.e. $Y \approx X_{p,r,w} \oplus Y$. It follows that $X_{p,r,w} \oplus X_{p,r,w} \approx X_{p,r,w} \oplus Y \oplus Z \approx Y \oplus Z = X_{p,r,w}$ and this completes the proof. □

Until now we have considered only the spaces $X_{p,r,w}$ with $1 \leqslant r < p < \infty$. If, in the definition of $X_{p,r,w}$, we replace the space l_p by c_0 we obtain a subspace of $(c_0 \oplus l_r)_\infty$, denoted by $X_{\infty,r,w}$. The condition (*) is replaced in this case by

$$(*)_\infty \qquad \sum_{n=1}^{\infty} w_n^r = \infty, \quad w_n \to 0 \text{ and } w_n < 1 \text{ for all } n .$$

By the same method as in 4.d.6 it can be shown that, for any two sequences $w = \{w_n\}_{n=1}^\infty$ and $w_n' = \{w_n'\}_{n=1}^\infty$ which both satisfy the condition $(*)_\infty$, the spaces $X_{\infty,r,w}$ and $X_{\infty,r,w'}$ are isomorphic. The blocks $\{h_j\}_{j=1}^\infty$ used in the proof are in this case blocks with constant coefficients. The proof that P is a norm-one projection shows in this case that every block basis with constant coefficients of any permutation of the natural basis of $X_{\infty,r,w}$ spans a complemented subspace.

In view of 4.d.6 and the preceding remark it is justified to use the notation $X_{p,r}$, $1 \leqslant r < p \leqslant \infty$ instead of $X_{p,r,w}$ without reference to the particular sequence w (satisfying the condition (*), respectively $(*)_\infty$).

We shall show now that, for any fixed $1 \leqslant r < p \leqslant \infty$ and $w = \{w_n\}_{n=1}^\infty$ satisfying (*), respectively $(*)_\infty$, the natural basis of $X_{p,r,w}$ is equivalent to the unit vector basis of some modular space.

We first consider the case $1 \leqslant r < p < \infty$. For every n we define the function

$$M_{w_n}(t) = \max\{t^p, w_n^r t^r\} = \begin{cases} w_n^r t^r, & 0 \leqslant t \leqslant w_n^{r/(p-r)} \\ t^p, & w_n^{r/(p-r)} < t < \infty . \end{cases}$$

It is clear that each M_{w_n} is an Orlicz function satisfying $r \leqslant tM'_{w_n}(t)/M_{w_n}(t) \leqslant p$ for every $t > 0$ for which the derivative exists. Moreover, it is easily checked that a series $\sum_{n=1}^{\infty} M_{w_n}(|a_n|)$ converges if and only if $\max\left\{\left(\sum_{n=1}^{\infty} |a_n|^p\right)^{1/p}, \left(\sum_{n=1}^{\infty} |a_n|^r w_n^r\right)^{1/r}\right\} < \infty$. This implies that the natural basis of $X_{p,r,w}$ is equivalent to the unit vector basis of $l_{(M_{w_n})}$. Consequently, $X_{p,r} \approx l_{(M_{w_n})}$.

In the case when $p = \infty$ we put

$$M_{w_n}(t) = \max\{t^n, w_n^r t^r\} .$$

Then, for any $\rho>0$, a series $\sum_{n=1}^{\infty} M_{w_n}(|a_n|/\rho)$ converges if and only if

$$\max\left\{\sum_{n=1}^{\infty}|a_n|^p\rho^{-p},\ \left(\sum_{n=1}^{\infty}|a_n|^r w_n^r\rho^{-r}\right)^{1/r}\right\}<\infty.$$

It follows from this fact that the natural basis of $X_{\infty,r,w}$ is equivalent to the unit vector basis of $h_{(M_{w_n})}$. Obviously, in this case the sequence $\{M_{w_n}\}_{n=1}^{\infty}$ is not equivalent to any sequence satisfying the uniform Δ_2-condition at zero.

In order to determine the dual of $X_{p,r}$ we restrict ourselves to the case $1<r<p\leq\infty$ (for $r=1$, the functions complementary to M_{w_n} are degenerate). We take q and s so that $p^{-1}+q^{-1}=1$ and $r^{-1}+s^{-1}=1$ ($q=1$ if $p=\infty$). For $p<\infty$ the function $M_{w_n}^*$, complementary to M_{w_n}, can be computed directly from the definition and we get

$$M_{w_n}^*(t)=\begin{cases} t^s/sr^{s-1}w_n^s, & 0\leq t\leq rw_n^{s/(s-q)}\\ tw_n^{(sq-s)/(s-q)}-w_n^{sq/(s-q)}, & rw_n^{s/(s-q)}<t\leq pw_n^{s/(s-q)}\\ t^q/qp^{q-1}, & pw_n^{s/(s-q)}<t<\infty. \end{cases}$$

For computations it is however more convenient to use the non-convex function N_{w_n}, defined by

$$N_{w_n}(t)=\min\{t^q, t^s/w_n^s\},$$

which has the property that $A^{-1}\leq N_{w_n}(t)/M_{w_n}^*(t)\leq A$ for every $t>0$ and for some constant $A>0$ which depends only on p and r but not on t and n. This formula is valid also for $p=\infty$, i.e. $q=1$.

It follows from this discussion that the dual $Y_{q,s}$, $1\leq q<s<\infty$, of $X_{p,r}$, consists of all sequences $y=(a_1, a_2,\ldots)$ for which $\sum_{n=1}^{\infty} N_{w_n}(|a_n|)<\infty$.

Before we state the main result of this section we prove an elementary technical lemma which will be used in the sequel.

Lemma 4.d.8. [102] *Let Q_0 be a continuous function on an interval* $[0, t_0]$ *so that*

(i) $Q_0(0)=0$ *and* $Q_0(t)>0$ for $0<t\leq t_0$,
(ii) Q_0 *has a right derivative which satisfies* $0\leq tQ_0'(t)/Q_0(t)\leq 1$ *for* $0<t<t_0$.

Then there exists a concave increasing twice continuously differentiable function Q so that $Q_0(t)/8\leq Q(t)\leq Q_0(t)$ for $0\leq t\leq t_0$.

Proof. Let $\varphi(t)$ be a continuously differentiable function on $[0, t_0]$ so that $1/4\leq\varphi(t)\leq 1/2$, $\varphi'(t)\geq 0$, $4t\varphi'(t)+tQ_0'(t)/Q_0(t)\leq 1$ for $0<t<t_0$ and so that $\varphi'(t)>0$ whenever t is an interior point of $\{t;\ Q_0'(t)=0\}$. These properties of φ ensure that $Q_1=Q_0\varphi$ is a continuous strictly increasing function on $[0, t_0]$ satisfying $tQ_1'(t)/Q_1(t)\leq 1$ on that interval. Therefore, the inverse function N_1 of Q_1 is also strictly increasing and satisfies $tN_1'(t)/N(t)\geq 1$, in a certain neighbourhood of 0. It follows that $N_2(t)=\int_0^t(N_1(u)/u)du$ is a continuously differentiable convex strictly increasing function such that $N_1(t/2) \leqslant N_2(t) \leqslant N_1(t)$ in some neighborhood of 0. In order to get a twice continuously differentiable function we take $N(t) = \int_0^t (N_2(u)/u)\,du$ which satisfies $N_1(t/4) \leqslant N(t) \leqslant N_1(t)$. The inverse function Q of N is a twice continuously differentiable concave function on $[0, t_0]$ which satisfies $Q(t)/4 \leqslant Q_1(t) \leqslant Q(t)$ and thus also $Q_0(t)/8 \leqslant Q(t) \leqslant Q_0(t)$ for $t \in [0, t_0]$. □

We are now ready to prove the theorem of J. T. Woo [145].

Theorem 4.d.9. *Let $1 \leqslant q < s < \infty$. A Banach space X with a symmetric basis is isomorphic to a subspace of $Y_{q,s}$ if and only if X is isomorphic to an Orlicz sequence space l_M with M satisfying $q \leqslant tM'(t)/M(t) \leqslant s$ for every $t > 0$.*

Proof. Let X be a Banach space with a symmetric basis $\{x_n\}_{n=1}^\infty$ which is isomorphic to a subspace of $Y_{q,s}$. Since $Y_{q,s} \approx l_{(M^*_{w_n})}$ we can assume with no loss of generality that $\{x_n\}_{n=1}^\infty$ is equivalent to a normalized block basis $\{u_j\}_{j=1}^\infty$ of the unit vector basis of $l_{(M^*_{w_n})}$. As in the case of Orlicz sequence spaces (see the discussion preceding 4.a.7), there exists a sequence of Orlicz functions $\{N_j\}_{j=1}^\infty$, with each N_j being a normalized convex combination of the $M^*_{w_n}$'s, such that a series $\sum_{j=1}^\infty a_j u_j$ converges if and only if $\sum_{j=1}^\infty N_j(|a_j|) < \infty$. The functions $\{M^*_{w_n}\}_{n=1}^\infty$, and thus also $\{N_j\}_{j=1}^\infty$, satisfy the uniform Δ_2-condition at zero; more precisely, we have $q \leqslant tN_j'(t)/N_j(t) \leqslant s$ for all j and $t > 0$. Hence, for any N which is a uniform limit of a subsequence of $\{N_j\}_{j=1}^\infty$, $\{u_j\}_{j=1}^\infty$ is equivalent to the unit vector basis of l_N. Thus, $X \approx l_N$ and $q \leqslant tN'(t)/N(t) \leqslant s$ for every $t > 0$. (Instead of repeating the argument used in 4.a.7 we could have also applied 4.d.5.)

In order to prove the converse we consider an Orlicz function $M(t)$ which satisfies the assumptions of the theorem, is not equivalent to t^q and for which $M(1) = 1$. We can exclude the case when $M(t)$ is equivalent at zero to t^q since $Y_{q,s}$ contains even a complemented subspace isomorphic to l_q. Indeed, by taking a subsequence of integers $\{n_i\}_{i=1}^\infty$ for which $\sum_{i=1}^\infty w_{n_i}^{pr/(p-r)} < \infty$ if $p < \infty$, or $\sum_{i=1}^\infty w_{n_i}^r < \infty$ if $p = \infty$, we get that $l_{(M^*_{w_{n_i}})}$ is a complemented subspace of $l_{(M^*_{w_n})} \approx Y_{q,s}$ which is isomorphic to l_q.

Put $Q_0(t) = M(t^{1/(s-q)})/t^{q/(s-q)}$. In view of the conditions imposed on M we get that Q_0 satisfies all the requirements of 4.d.8. Hence, there is a concave increasing function Q with two continuous derivatives and $Q(0) = 0$ which is equivalent to Q_0. Let $w = \{w_n\}_{n=1}^\infty$ be a sequence satisfying (*) if $p < \infty$ or $(*)_\infty$ if $p = \infty$, let $a_n = w_n^{s/(s-q)}$ and $b_n = a_n^{2(s-q)/(2s-q)}$. Put

$$G_n(t) = C_n^{-1} \int_{a_n}^1 N_{w_n}(b_n t u)\left[-(s-q)u^{q-2s-1} Q''(a_n^{s-q} b_n^q u^{q-s})\right] du\,,$$

where

$$C_n = \int_{a_n}^1 N_{w_n}(b_n u)\left[-(s-q)u^{q-2s-1} Q''(a_n^{s-q} b_n^q u^{q-s})\right] du\,.$$

This formula is inspired by the method used in the proof of 4.a.9. By substituting the explicit value of N_{w_n} and putting $v=a_n^{s-q}b_n^{q-s}u^{q-s}$ we get that

$$\begin{aligned}
G_n(t)=C_n^{-1}\Big[t^s\int_{a_n}^{a_n/b_nt} -(s-q)a_n^{q-s}b_n^s u^{q-s-1}Q''(a_n^{s-q}b_n^{q-s}u^{q-s})\,du \\
+t^q\int_{a_n/b_nt}^{1} -(s-q)b_n^q u^{2q-2s-1}Q''(a_n^{s-q}b_n^{q-s}u^{q-s})\,du\Big] \\
=C_n^{-1}\Big[t^s\int_{b_n^{q-s}}^{t^{s-q}} Q''(v)\,dv+t^q\int_{t^{s-q}}^{(a_n/b_n)^{s-q}} vQ''(v)\,dv\Big] \\
=C_n^{-1}\Big\{t^s[Q'(t^{s-q})-Q'(b_n^{q-s})]+t^q[(a_n/b_n)^{s-q}Q'((a_n/b_n)^{s-q}) \\
-t^{s-q}Q'(t^{s-q})]+t^q\int_{(a_n/b_n)^{s-q}}^{t^{s-q}} Q'(v)\,dv\Big\} \\
=C_n^{-1}\{-t^sQ'(b_n^{q-s})+t^q(a_n/b_n)^{s-q}Q'((a_n/b_n)^{s-q}) \\
+t^qQ(t^{s-q})-t^qQ((a_n/b_n)^{s-q})\}\,.
\end{aligned}$$

Since the behavior of M at ∞ is not important we may suppose that $\lim_{t\to\infty} Q'(t)=0$. This implies that $Q'(b_n^{q-s})\to 0$ as $n\to\infty$. Moreover, from the fact that $\lim_{t\to 0} tQ'(t)=\lim_{t\to 0} Q(t)=0$ it follows that also the second term and the fourth term in the parenthesis tend to zero since $a_n/b_n=a_n^{q/(2s-q)}\to 0$ as $n\to\infty$. Thus, for $t=1$, we get that $C_nG_n(1)=C_n\to Q(1)$. Consequently, the following limit exists uniformly for $t\in[0,1]$

$$G(t)=\lim_{n\to\infty} G_n(t)=t^qQ(t^{s-q})/Q(1),$$

and, since Q is equivalent to Q_0, we get that G is equivalent to M. From the definition of G_n it follows that, for every integer k, there is an integer $n(k)$ and real numbers $\{u_{i,k}\}_{i=1}^{j(k)}$ and $\{\lambda_{i,k}\}_{i=1}^{j(k)}$ in $[0,1]$ with $\sum_{i=1}^{j(k)}\lambda_{i,k}=1$ such that

$$\Big|G(t)-\sum_{i=1}^{j(k)}\lambda_{i,k}N_{w_{n(k)}}(b_{n(k)}tu_{i,k})\Big|\leq 2^{-k},\quad t\in[0,1]\,.$$

Since $\lambda N_{w_n}(bt)=N_{v_n}(ct)$ for suitable v_n and c we can write the previous inequality also in the form

$$\Big|G(t)-\sum_{i=1}^{j(k)}N_{v_{i,k}}(c_{i,k}t)\Big|\leq 2^{-k},\quad t\in[0,1]$$

for some $v_{i,k}$ and $c_{i,k}$. This clearly shows that l_G, or equivalently l_M, is isomorphic to a subspace of the modular sequence space $l_{(N_{v_{i,k}})}$ (recall that the non-convex functions $N_{v_{i,k}}$ can be always replaced by proper Orlicz functions). In order to conclude the proof we remark that the sequence $\{v_{i,k}\}_{i=1}^{j(k)}{}_{k=1}^{\infty}$ tends to zero and if

$\sum_{k=1}^{\infty}\sum_{i=1}^{j(k)} v_{i,k}^{pr/(p-r)}=\infty$ in the case when $p<\infty$, or $\sum_{k=1}^{\infty}\sum_{i=1}^{j(k)} v_{i,k}^{r}=\infty$ in the case when $p=\infty$, then $l_{(N_{v_{i,k}})}\approx Y_{q,s}$. If, on the other hand, $\sum_{k=1}^{\infty}\sum_{i=1}^{j(k)} v_{i,k}^{pr/(p-r)}<\infty$, respectively $\sum_{k=1}^{\infty}\sum_{i=1}^{j(k)} v_{i,k}^{r}<\infty$, then by applying Holder's inequality we get that the unit vectors of $l_{(N_{v_{i,k}})}$ form a basis equivalent to the unit vector basis of l_q. Thus, $l_M\approx l_q$ and we have already proved that $Y_{q,s}$ contains a subspace isomorphic to l_q. □

Using 4.d.9 together with 4.d.5 we obtain the following corollary.

Theorem 4.d.10. *Let $1\leqslant q<s<\infty$ and let $\{M_n\}_{n=1}^{\infty}$ be a sequence of Orlicz functions satisfying $q\leqslant tM_n'(t)/M_n(t)\leqslant s$ for all n and $t\in(0,1)$. Then $Y_{q,s}$ contains a subspace isomorphic to $l_{(M_n)}$.*

By its construction the space $Y_{q,s}$ is a quotient of $l_q\oplus l_s$. Thus, every Orlicz sequence space "between" q and s is isomorphic to a subspace of a quotient of $l_q\oplus l_s$. Comparing 4.d.9 and 4.d.10 with 2.c.14 and 2.d.1 we note the marked difference between the behavior of subspaces of quotients of $l_q\oplus l_s$ from that of complemented subspaces of $l_q\oplus l_s$ and also from that of subspaces of quotients of l_p.

e. Lorentz Sequence Spaces

The Lorentz sequence spaces $d(w,p)$ which have already been mentioned in Section 3.a, as well as the Lorentz function spaces (cf. [99]), were introduced in connection with some problems of harmonic analysis and interpolation theory. We do not study here this aspect; instead, we present briefly some results regarding their geometric structure. Let us first recall the definition of a Lorentz sequence space.

Definition 4.e.1. Let $1\leqslant p<\infty$ and let $w=\{w_n\}_{n=1}^{\infty}$ be a non-increasing sequence of positive numbers such that $w_1=1$, $\lim_{n\to\infty} w_n=0$ and $\sum_{n=1}^{\infty} w_n=\infty$. The Banach space of all sequences of scalars $x=(a_1,a_2,\ldots)$ for which

$$\|x\|=\sup_{\pi}\left(\sum_{n=1}^{\infty}|a_{\pi(n)}|^p w_n\right)^{1/p}<\infty,$$

where π ranges over all the permutations of the integers, is denoted by $d(w,p)$ and it is called a *Lorentz sequence space.*

If $\{a_n^*\}_{n=1}^{\infty}$ is a non-increasing rearrangement of the sequence $\{a_n\}_{n=1}^{\infty}$, i.e. $\{a_n^*\}_{n=1}^{\infty}$

is a non-increasing sequence obtained from $\{|a_n|\}_{n=1}^\infty$ by a suitable permutation of the integers then $\|x\|=\left(\sum_{n=1}^\infty a_n^{*p}w_n\right)^{1/p}$ for $x=(a_1, a_2,\ldots)\in d(w,p)$.

Lorentz sequence spaces are, in general, different from the previously defined classes of spaces with a symmetric basis. For instance, the uniqueness of the symmetric basis in l_p (see 3.b.5) easily implies that no Lorentz sequence space is isomorphic to an l_p space. In some cases a Lorentz sequence space is isomorphic to an Orlicz sequence space. An easy computation shows that this is the case, for example, if $w_n=1/(1+\log n)$ for $n\geqslant 1$ and $p\geqslant 1$ arbitrary: the corresponding space $d(w,p)$ is isomorphic to l_M, where $M(t)=t^p/(1+|\log t|)$. By the uniqueness of the symmetric basis in any Lorentz sequence space (see 4.e.4 below), a space $d(w,p)$ is isomorphic to an Orlicz sequence space l_M if and only if they are identical, i.e. they consist of the same sequences. In other words, $d(w,p)\approx l_M$ if and only if

$$(\dagger) \qquad \sum_{n=1}^\infty \lambda_n^p w_n<\infty \Leftrightarrow \sum_{n=1}^\infty M(\lambda_n)<\infty,$$

whenever $\{\lambda_n\}_{n=1}^\infty$ is a non-increasing sequence of reals tending to zero.

G. G. Lorentz [100] has found necessary and sufficient conditions on w for the existence of an Orlicz function M for which (†) holds and also necessary and sufficient conditions on M so that there exists a sequence $w=\{w_n\}_{n=1}^\infty$ for which (†) is satisfied. The conditions given by G. G. Lorentz were actually stated in the more general context of function spaces but they can be easily translated into the sequence spaces language, as follows.

Let $w=\{w_n\}_{n=1}^\infty$ be a strictly decreasing sequence of reals such that $w_1=1$, $\lim_{n\to\infty} w_n=0$ and $\sum_{n=1}^\infty w_n=\infty$ (the requirement that w be strictly decreasing sequence is not really a restriction since every $d(w,p)$ space is isomorphic to a space $d(w',p)$ for which w' is a strictly decreasing sequence). Construct two continuous functions $W(t)$ and $S(t)$, defined both on $[1,\infty)$, such that W is strictly decreasing, S is concave and strictly increasing, $W(n)=w_n$ and $S(n)=s_n=\sum_{i=1}^n w_i$ for $n=1,2,\ldots$. Since S^{-1} is a strictly increasing convex function on $[1,\infty)$ it follows easily that $F(t)=1/S^{-1}(1/t)$ satisfies $1\leqslant tF'(t)/F(t)$ for all $t\in(0,1)$ for which the derivative exists. This implies that $F(t)/t$ is non-decreasing and therefore, F is equivalent at zero to the Orlicz function $N(t)=\int_0^t (F(u)/u)\,du$. With these notations we are prepared to state the results of G. G. Lorentz [100].

Theorem 4.e.2. *A Lorentz sequence space $d(w,1)$ is isomorphic to an Orlicz sequence space l_M (i.e. $d(w,1)=l_M$ as sets) if and only if*

(i) *there exists a constant $\gamma>0$ such that $\sum_{n=1}^\infty 1/W^{-1}(\gamma w_n)<\infty$.*

In this case M is equivalent to the Orlicz function N, defined above.

We also have

Theorem 4.e.2′. *An Orlicz sequence space l_M is isomorphic to a Lorentz sequence space $d(w, 1)$ if and only if*

(i′) *there exists a constant $\delta>0$ so that $\int_1^\infty M^*(\delta M^{*-1}(1/t))\,dt<\infty$.*

If this is the case the sequence w may be taken to be $w_n=M^{*-1}(1/n)$, $n=1, 2,\ldots$.

Since these two theorems are not used in the sequel we do not reproduce their proofs here. The fact that both 4.e.2 and 4.e.2′ apply only in the case $p=1$ is not really a restriction. It is relatively easy to check that, for any $p>1$, we have $d(w, p)\approx l_{M_p}$ if and only if $d(w, 1)\approx l_{M_1}$, where the Orlicz functions M_p and M_1 are connected by the relation $M_p(t)=M_1(t^p)$.

Proposition 4.e.3. *Let $\{e_n\}_{n=1}^\infty$ be the unit vector basis of a Lorentz sequence space $d(w, p)$ with $p\geqslant 1$. Then every normalized block basis $u_n=\sum_{i=q_n+1}^{q_{n+1}} a_i e_i$, $n=1, 2,\ldots$ such that $\lim_{i\to\infty} a_i=0$ contains, for every $\varepsilon>0$, a subsequence $\{u_{n_j}\}_{j=1}^\infty$ which is $1+\varepsilon$-equivalent to the unit vector basis of l_p and so that $[u_{n_j}]_{j=1}^\infty$ is complemented in $d(w, p)$.*

Consequently, every infinite dimensional subspace of $d(w, p)$ contains complemented subspaces which are nearly isometric to l_p.

Proof. Since every change of signs and every permutation of the integers induces an isometry in $d(w, p)$ we may assume, by switching to a subsequence if necessary, that $\{a_i\}_{i=1}^\infty$ is a non-increasing sequence of positive numbers.

Fix $\varepsilon>0$ and construct by induction two increasing sequences of integers $\{n_j\}_{j=1}^\infty$ and $\{r_j\}_{j=1}^\infty$ such that $q_{n_j}<r_j<q_{n_j+1}$, $Q_{j-1}=\sum_{k=1}^{j-1}(q_{n_k+1}-q_{n_k})\leqslant r_j-q_{n_j}$ and $\left(\sum_{i=q_{n_j}+1}^{r_j} a_i^p w_{i-q_{n_j}}\right)^{1/p}<\varepsilon/2^{j+1}$ for all j. Then, for any set of coefficients $\{\lambda_j\}_{j=1}^\infty$, we have

$$\begin{aligned}\left\|\sum_{j=1}^\infty \lambda_j u_{n_j}\right\| &\geqslant \left\|\sum_{j=1}^\infty \lambda_j \sum_{i=r_j+1}^{q_{n_j+1}} a_i e_i\right\| - \sum_{j=1}^\infty |\lambda_j| \left\|\sum_{i=q_{n_j}+1}^{r_j} a_i e_i\right\|\\ &\geqslant \left(\sum_{j=1}^\infty |\lambda_j|^p \sum_{i=r_j+1}^{q_{n_j+1}} |a_i|^p w_{Q_{j-1}-r_j+i}\right)^{1/p} - \frac{\varepsilon}{2}\max_j |\lambda_j|\\ &\geqslant \left(\sum_{j=1}^\infty |\lambda_j|^p \sum_{i=r_j+1}^{q_{n_j+1}} |a_i|^p w_{i-q_{n_j}}\right)^{1/p} - \frac{\varepsilon}{2}\max_j |\lambda_j|\\ &\geqslant (1-\varepsilon)\left(\sum_{j=1}^\infty |\lambda_j|^p\right)^{1/p}.\end{aligned}$$

On the other hand, it is easily checked that, for every normalized block basis and therefore in particular for $\{u_{n_j}\}_{j=1}^\infty$, we have $\left\|\sum_{j=1}^\infty \lambda_j u_{n_j}\right\|\leqslant\left(\sum_{j=1}^\infty |\lambda_j|^p\right)^{1/p}$.

In order to prove that $[u_{n_j}]_{j=1}^\infty$ is complemented in $d(w,p)$ we set

$$P\Big(\sum_{n=1}^\infty c_n e_n\Big)=\sum_{j=1}^\infty\Big\{\Big(\sum_{i=r_j+1}^{q_{n_j+1}} c_i a_i^{p-1} w_{i-q_{n_j}}\Big)\Big/\Big(\sum_{i=r_j+1}^{q_{n_j+1}} a_i^p w_{i-q_{n_j}}\Big)\Big\}u_{n_j}.$$

Then, using Holder's inequality we get

$$\begin{aligned}\Big\|P\Big(\sum_{n=1}^\infty c_n e_n\Big)\Big\|^p &\leq \sum_{j=1}^\infty\Big|\sum_{i=r_j+1}^{q_{n_j+1}} c_i a_i^{p-1} w_{i-q_{n_j}}\Big|^p\Big/\Big(\sum_{i=r_j+1}^{q_{n_j+1}} a_i^p w_{i-q_{n_j}}\Big)^p\\ &\leq \sum_{j=1}^\infty\Big(\sum_{i=r_j+1}^{q_{n_j+1}} |c_i|^p w_{i-q_{n_j}}\Big)\Big/\Big(\sum_{i=r_j+1}^{q_{n_j+1}} a_i^p w_{i-q_{n_j}}\Big)\\ &\leq (1+\varepsilon)^p\Big\|\sum_{n=1}^\infty c_n e_n\Big\|^p,\end{aligned}$$

i.e. P is a bounded linear projection from $d(w,p)$ onto $[u_{n_j}]_{j=1}^\infty$. The last assertion follows easily from 1.a.11. $\square$

The following result from [4] is an immediate corollary of 4.e.3.

Theorem 4.e.4. *Let X be a subspace of a Lorentz sequence space $d(w,p)$, $p\geq 1$ which has a symmetric basis $\{x_n\}_{n=1}^\infty$. Then, up to equivalence, $\{x_n\}_{n=1}^\infty$ is the unique symmetric basis of this subspace.*

Proof. Assume that $\{y_n\}_{n=1}^\infty$ is another symmetric basis of X. If $X\approx l_p$ then, by 3.b.5, $\{x_n\}_{n=1}^\infty$ and $\{y_n\}_{n=1}^\infty$ are equivalent. Otherwise, by 1.a.12, $\{x_n\}_{n=1}^\infty$ is equivalent to a block basis $u_n=\sum_{i=q_n+1}^{q_{n+1}} a_i e_i$, $n=1,2,\ldots$ of $\{e_n\}_{n=1}^\infty$, the unit vector basis of $d(w,p)$, and $\{y_m\}_{m=1}^\infty$ is equivalent to a block basis $v_m=\sum_{n=r_m+1}^{r_{m+1}} b_n x_n$, $m=1,2,\ldots$ of $\{x_n\}_{n=1}^\infty$, with $\limsup_{n\to\infty} b_n\neq 0$ (for, by 4.e.3, $\lim_{n\to\infty} b_n=0$ implies that $\{y_n\}_{n=1}^\infty$ is equivalent to the unit vector basis of l_p, contrary to our assumption). Hence, by the symmetry of $\{y_n\}_{n=1}^\infty$, we may assume without loss of generality that there is an $\varepsilon>0$ so that, for every m, $|b_n|\geq\varepsilon$ for some $r_m<n<r_{m+1}$. Consequently, if a series $\sum_{n=1}^\infty \lambda_n y_n$ converges so does $\sum_{n=1}^\infty \lambda_n x_n$. By interchanging the roles of $\{x_n\}_{n=1}^\infty$ and $\{y_n\}_{n=1}^\infty$ we deduce the equivalence of these two bases. $\square$

It follows from 4.e.3 and 1.c.12(a) that a Lorentz sequence space $d(w,p)$ is reflexive if and only if $p>1$. The dual $d^*(w,p)$ of $d(w,p)$ is never isomorphic to a Lorentz sequence space. Indeed, assume that $d^*(w,p)=d(w',q)$ for some sequence w' and for some $q\geq 1$. By 4.e.3 we must have $p^{-1}+q^{-1}=1$ (if $p=1$ the assertion is entirely obvious so we assume that $p>1$). Let $\{e_n\}_{n=1}^\infty$ and $\{f_n\}_{n=1}^\infty$ be the unit vector bases of $d(w',q)$, respectively $d^*(w,p)$. Then, a series $\sum_{n=1}^\infty \lambda_n e_n$ converges whenever $\lambda=(\lambda_1, \lambda_2,...)\in l_q$ while there is a $\lambda \notin l_q$ for which $\sum_{n=1}^{\infty} \lambda_n e_n$ converges. On the other hand, the convergence of $\sum_{n=1}^{\infty} \lambda_n f_n$ implies that $\lambda \in l_q$ while there is a $\lambda \in l_q$ for which $\sum_{n=1}^{\infty} \lambda_n f_n$ fails to converge. This contradicts 4.e.4.

It is easy to give an explicit representation of $d^*(w, p)$. In the case $p>1$ this space consists of all sequences $x=(a_1, a_2,...)$ for which

$$\|x\|=\inf \sup_n \left(\sum_{i=1}^{n} a_i^*\right)\Big/\left(\sum_{i=1}^{n} |y_i| w_i^{1/p}\right)<\infty,$$

where the infimum is taken over all $y=(y_1, y_2,...)\in l_q$ with $\|y\|=1$, where $p^{-1}+q^{-1}=1$ and $\{a_i^*\}_{i=1}^{\infty}$ denotes a non-increasing rearrangement of $\{|a_i|\}_{i=1}^{\infty}$ (cf. [45]).

In the remainder of this section we present, without proofs, some results concerning symmetric basis sequences in $d(w, p)$ spaces. These sequences have been studied in [4]. It has been proved there that *every symmetric basic sequence in a Lorentz sequence space $d(w, p)$ is equivalent either to the unit vector basis of l_p or to a block basis $\{u_n^{(\alpha)}\}_{n=1}^{\infty}$ generated by a vector $0\neq\alpha=\sum_{n=1}^{\infty} a_n e_n \in d(w, p)$* (recall definition 3.a.8).

In many cases, block bases generated by one vector are themselves equivalent to the unit vector basis of some Lorentz sequence space but this is not true in general. For example, it has been shown in [4] that in $d(w, p)$, with $w_n=1/n^{1/2}(1+\log n)^2$, $n=1, 2,...$ and $p\geq 1$ arbitrary, the block basis generated by the vector $\alpha=\sum_{n=1}^{\infty} e_n/n^{1/2p}$ spans a subspace (with a symmetric basis) which is not isomorphic to any Lorentz sequence space or to any l_p space. On the other hand, there are cases where all block bases generated by one vector are equivalent to each other. These Lorentz sequence spaces have been characterized in [4].

Theorem 4.e.5. *In a Lorentz sequence space $d(w, p)$ there are exactly two non-equivalent symmetric basic sequences (namely, the unit vector basis of l_p and that of $d(w, p)$ itself) if and only if $\sup_{n,k} s_{nk}/s_n s_k<\infty$, where $s_n=\sum_{i=1}^{n} w_i$.*

References

1. Alfsen, E. M.: Compact convex sets and boundary integrals. Berlin-Heidelberg-New York: Springer, 1971.
2. Altshuler, Z.: Characterization of c_0 and l_p among Banach spaces with a symmetric basis. Israel J. Math. **24**, 39–44 (1976).
3. Altshuler, Z.: A Banach space with a symmetric basis which contains no l_p or c_0, and all its symmetric basic sequences are equivalent. Compositio Math. (1977).
4. Altshuler, Z., Casazza, P. G., Lin, B. L.: On symmetric basic sequences in Lorentz sequence spaces. Israel J. Math. **15**, 140–155 (1973).
5. Ando, T.: Contractive projections in L_p-spaces. Pacific J. Math. **17**, 391–405 (1966).
6. Babenko, K. I.: On conjugate functions. Dokl. Akad. Nauk. SSSR **62**, 157–160 (1948) [Russian].
7. Bachelis, G. F., Rosenthal, H. P.: On unconditionally converging series and biorthogonal systems in a Banach space. Pacific J. Math. **37**, 1–5 (1971).
8. Banach. S.: Théorie des opérations linéaires. Warszawa, 1932.
9. Bessaga, C., Pelczynski, A.: On bases and unconditional convergence of series in Banach spaces. Studia Math. **17**, 151–164 (1958).
10. Bessaga, C., Pelczynski, A.: A generalization of results of R. C. James concerning absolute bases in Banach spaces. Studia Math. **17**, 165–174 (1958).
11. Bessaga, C., Pelczynski, A.: On subspaces of a space with an absolute basis. Bull. Acad. Sci. Pol. **6**, 313–314 (1958).
12. Blei, R. C.: A uniformity property for $\Lambda(2)$ sets and Grothendieck inequality. to appear.
13. Botschkariev, S. V.: Existence of a basis in the space of analytic functions, and some properties of the Franklin system. Matem. Sbornik **24**, 1–16 (1974), [translated from Russian].
14. Casazza, P. G.: James' quasi-reflexive space is primary. Israel J. Math. **26**, 294–305 (1977).
15. Casazza, P. G., Lin, B. L.: On symmetric basic sequences in Lorentz sequence spaces II. Israel J. Math. **17**, 191–218 (1974).
16. Casazza, P. G., Lin, B. L.: Projections on Banach spaces with symmetric bases. Studia Math. **52**, 189–193 (1974).
17. Ciesielski, Z.: Properties of the orthonormal Franklin system. Studia Math. **23**, 141–157 (1963).
18. Ciesielski, Z., Domsta, J.: Construction of an orthonormal basis in $C^m(I^d)$ and $W_p^m(I^d)$. Studia Math. **41**, 211–224 (1972).
19. Cohen, P. J.: On a conjecture of Littlewood and idempotent measures. Amer. J. Math. **82**, 191–212 (1960).
20. Davie, A. M.: The approximation problem for Banach spaces. Bull. London Math. Soc. **5**, 261–266 (1973).
21. Davie, A. M.: The Banach approximation problem. J. Approx. Theory **13**, 392–394 (1975).
22. Davis, W. J. Embedding spaces with unconditional bases. Israel J. Math. **20**, 189–191 (1975).
23. Davis, W. J., Dean, D. W., Singer, I.: Multipliers and unconditional convergence of biorthogonal expansions. Pacific J. Math. **37**, 35–39 (1971).
24. Davis, W. J., Figiel, T., Johnson, W. B., Pelczynski, A.: Factoring weakly compact operators. J. Funct. Anal. **17**, 311–327 (1974).
25. Davis, W. J., Johnson, W. B.: A renorming of nonreflexive Banach spaces. Proc. Amer. Math. Soc. **37**, 486–488 (1973).

26. Davis, W. J., Singer, I.: Boundedly complete *M* bases and complemented subspaces in Banach spaces. Trans. Amer. Math. Soc. **175**, 299–326 (1973).
27. Day, M. M.: Some more uniformly convex spaces. Bull. Amer. Math. Soc. **47**, 504–507 (1941).
28. Day, M. M.: Normed linear spaces, Third Edition. Berlin-Heidelberg-New York: Springer 1973.
29. Dean, D. W.: The equation $L(E, X^{**})=L(E, X)^{**}$ and the principle of local reflexivity. Proc. Amer. Math. Soc. **40**, 146–148 (1973).
30. Dean, D. W., Singer, I, Sternbach, L.: On shrinking basic sequences in Banach spaces. Studia Math. **40**, 23–33 (1971).
31. Dor, L. E.: On sequences spanning a complex l_1 space. Proc. Amer. Math. Soc. **47**, 515–516 (1975).
32. Dugundji, J.: Topology. Boston: Allyn and Bacon Inc. 1966.
33. Dunford, N., Schwartz, J.: Linear operators, Vol. I. New York: Interscience 1958
34. Dvoretzky, A., Rogers, C. A.: Absolute and unconditional convergence in normed linear spaces. Proc. Nat. Acad. Sci. (U.S.A.) **36**, 192–197 (1950).
35. Edelstein, I. S., Wojtaszczyk, P.: On projections and unconditional bases in direct sums of Banach spaces. Studia Math. 56, 263–276 (1976).
36. Ellentuck, E.: A new proof that analytic sets are Ramsey. J. Symbolic Logic **39**, 163–165 (1974).
37. Enflo, P.: A counterexample to the approximation property in Banach spaces. Acta Math. **130**, 309–317 (1973).
38. Farahat, J.: Espaces de Banach contenant l_1 d'apres H. P. Rosenthal. Seminaire Maurey-Schwartz, Ecole Polytechnique, 1973–74.
39. Figiel, T.: An example of an infinite dimensional Banach space non-isomorphic to its Cartesian square. Studia Math. **42**, 295–306 (1972).
40. Figiel, T.: Further counterexamples to the approximation problem, dittoed notes.
41. Figiel, T., Johnson, W. B.: The approximation property does not imply the bounded approximation property. Proc. Amer. Math. Soc. **41**, 197–200 (1973).
42. Figiel, T., Johnson, W. B.: A uniformly convex Banach space which contains no l_p. Compositio Math. **29**, 179–190 (1974).
43. Figiel, T., Lindenstrauss, J., Milman, V.: The dimension of almost spherical sections of convex sets. Acta Math. (1977).
44. Garling, D. J. H.: Symmetric bases of locally convex spaces. Studia Math. **30**, 163–181 (1968).
45. Garling, D. J. H.: A class of reflexive symmetric *BK* spaces. Canad. J. Math. **21**, 602–608 (1969).
46. Gohberg, I. C., Krein, M. G.: Fundamental theorems on deficiency numbers, root numbers and indices of linear operators [Russian]. Usp. Math. Nauk **12**, 43–118 (1957) [Translated in Amer. Math. Soc. Translations, vol. 13].
47. Gordon, Y., Lewis, D. R.: Absolutely summing operators and local unconditional structures. Acta Math. **133**, 27–48 (1974).
48. Grothendieck, A.: Produits tensoriels topologiques et especas nucleaires. Memo. Amer. Math. Soc. **16** (1955).
49. Grothendieck, A.: Resume de la theorie metrique des produits tensoriels topologiques. Bol. Soc. Mat. Sao Pãulo **8**, 1–79 (1956).
50. Hagler, J.: A counterexample to several questions about Banach spaces. Studia Math. to appear.
51. Hausdorff, F.: Set theory. New York: Chelsea, 1962.
52. Hennefeld, J.: On nonequivalent normalized unconditional bases for Banach spaces. Proc. Amer. Math. Soc. **41**, 156–158 (1973).
53. James, R. C.: Bases and reflexivity of Banach spaces. Ann. of Math. **52**, 518–527 (1950).
54. James, R. C.: A non-reflexive Banach space isometric with its second conjugate. Proc. Nat. Acad. Sci. (U.S.A.) 37, 174–177 (1951).
55. James, R. C.: Separable conjugate spaces. Pacific J. Math. **10**, 563–571 (1960).

56. James, R. C.: Uniformly non-square Banach spaces. Ann. of Math. **80**, 542–550 (1964).
57. James, R. C.: A separable somewhat reflexive Banach space with non-separable dual. Bull. Amer. Math. Soc. **80**, 738–743 (1974).
58. James, R. C.: A nonreflexive Banach space that is uniformly nonoctahedral. Israel J. Math. **18**, 145–155 (1974).
59. Johnson, W. B.: On quotients of L_p which are quotients of l_p. Compositio Math. 33, (1976).
60. Johnson, W. B., Rosenthal, H. P.: On w^*-basic sequences and their applications to the study of Banach spaces. Studia Math. **43**, 77–92 (1972).
61. Johnson, W. B., Rosenthal, H. P., Zippin, M.: On bases, finite-dimensional decompositions and weaker structures in Banach spaces. Israel J. Math. **9**, 488–506 (1971).
62. Johnson, W. B. Szankowski, A.: Complementably universal Banach spaces. Studia Math. 58, 91–97 (1976).
63. Johnson, W. B., Zippin, M.: On subspaces of quotients of $(\sum G_n)_{l_p}$and $(\sum G_n)_{c_0}$. Israel J. Math. **13**, 311–316 (1972).
64. Johnson, W. B., Zippin, M.: Subspaces and quotient spaces of $(\sum G_n)_{l_p}$ and $(\sum G_n)_0$. Israel J. Math. **17**, 50–55 (1974).
65. Kadec, M. I.: On the connection between weak and strong convergence. Dopovidi Akad. Nauk Ukrain 9, 949–952 (1959), [Ukrainian].
66. Kadec, M. I.: On complementably universal Banach spaces. Studia Math. **40**, 85–89 (1971).
67. Kalton, N. J.: Orlicz sequence spaces without local convexity. to appear.
68. Karlin, S.: Bases in Banach spaces. Duke Math. J. **15**, 971–985 (1948).
69. Kato, T.: Perturbation theory for nullity deficiency and other quantities of linear operators. J. Analyse Math. **6** 273–322 (1958).
70. Kircev, K. P., Troianskii, S. L.: On Orlicz spaces associated to Orlicz functions not satisfying the Δ_2-condition [Russian]. Serdica **1**, 88–95 (1975).
71. Klee, V. L.: Mappings into normed linear spaces. Fund. Math. **49**, 25–34 (1960/61).
72. Köthe, G.: Das Trägheitsgesetz der quadratischen Formen im Hilbertschen Raum. Math. Z. **41**, 137–152 (1936).
73. Köthe, G.: Hebbare lokalkonvexe Räume. Math. Ann. **165**, 181–195 (1966).
74. Köthe, G.: Topological vector spaces I. Berlin-Heidelberg-New York: Springer 1969.
75. Krasnoselskii, M. A., Rutickii, Ya. B.: Convex functions and Orlicz spaces. Groningen, Netherlands 1961 [Translated from Russian].
76. Krein, M. G., Milman, D. P., Rutman, M. A.: On a property of the basis in Banach space. Zapiski Mat. T. (Kharkov) **16**, 106–108 (1940) [Russian].
77. Kwapien, S.: On a theorem of L. Schwartz and its applications to absolutely summing operators. Studia Math. **38**, 193–201 (1970).
78. Kwapien, S., Pelczynski, A.: The main triangle projection in matrix spaces and its application. Studia Math. **34**, 43–68 (1970).
79. Lacey, E.: The isometric theory of classical Banach spaces. Berlin-Heidelberg-New York: Springer 1974.
80. LeVeque, W. J.: Elementary theory of numbers. London: Addison-Wesley 1962.
81. Lindberg, K. J.: On subspaces of Orlicz sequence spaces. Studia Math. **45**, 119–146 (1793).
82. Lindenstrauss, J.: On a certain subspace of l_1. Bull. Acad. Polon. Sci. **12**, 539–542 (1964).
83. Lindenstrauss, J.: On non-separable reflexive Banach spaces. Bull. Amer. Math. Soc. **72**, 967–970 (1966).
84. Lindenstrauss, J.: On complemented subspaces of m. Israel J. Math. **5**, 153–156 (1967).
85. Lindenstrauss, J.: On James' paper "separable conjugate spaces". Israel J. Math. **9**, 279–284 (1971).
86. Lindenstrauss, J.: A remark on symmetric bases. Israel J. Math. **13**, 317–320 (1972).
87. Lindenstrauss, J., Pelczynski, A.: Absolutely summing operators in $\mathscr{L}_p$ spaces and their applications. Studia Math. **29**, 275–326 (1968).
88. Lindenstrauss, J., Pelczynski, A.: Contributions to the theory of the classical Banach spaces. J. Funct. Anal. **8**, 225–249 (1971).
89. Lindenstrauss, J., Rosenthal, H. P.: Automorphisms in c_0, l_1 and m. Israel J. Math. **7**, 227–239 (1969).

90. Lindenstrauss, J., Rosenthal, H. P.: The $\mathscr{L}_p$ spaces. Israel J. Math. **7**, 325–349 (1969).
91. Lindenstrauss, J., Stegall, C.: Examples of separable spaces which do not contain l_1 and whose duals are non-separable. Studia Math. **54**, 81–105 (1975).
92. Lindenstrauss, J., Tzafriri, L.: On the complemented subspaces problem. Israel J. Math. **9**, 263–269 (1971).
93. Lindenstrauss, J., Tzafriri, L.: On Orlicz sequence spaces. Israel J. Math. **10**, 379–390 (1971).
94. Lindenstrauss, J., Tzafriri, L.: On Orlicz sequence spaces II. Israel J. Math. **11**, 355–379 (1972).
95. Lindenstrauss, J., Tzafriri, L.: On Orlicz sequence spaces III. Israel J. Math. **14**, 368–389 (1973).
96. Lindenstrauss, J., Tzafriri, L.: Classical Banach spaces. Lecture Notes in Math. Berlin-Heidelberg-New York: Springer 1973.
97. Lindenstrauss, J., Zippin, M.: Banach spaces with a unique unconditional basis. J. Funct. Anal. 3, 115–125 (1969).
98. Lorch, E. R.: Bicontinuous linear transformations in certain vector spaces. Bull. Amer. Math. Soc. **45**, 564–569 (1939).
99. Lorentz, C. G.: Some new functional spaces. Ann. of Math. **51**, 37–55 (1950).
100. Lorentz, G. G.: Relations between function spaces. Proc. Amer. Math. Soc. **12**, 127–132 (1961).
101. Markushevich, A. I.: On a basis in the wide sense for linear spaces. Dokl. Akad. Nauk **41**, 241–244 (1943).
102. Matuszewska, W.: Regularly increasing functions in connection with the theory of $L^*\varphi$-spaces. Studia Math. **21**, 317–344 (1962).
103. Maurey, B.: Une nouvelle demonstration d'un théorème de Grothendieck. Seminaire Maurey-Schwartz, 1972–1973.
104. Maurey, B., Rosenthal, H. P.: Normalized weakly null sequences with no unconditional subsequence. Studia Math. to appear.
105. McCarthy, C. A., Schwartz, J.: On the norm of a finite Boolean algebra of projections and applications to theorems of Kreiss and Morton. Comm. Pure Appl. Math. **18**, 191–201 (1965).
106. Milman, V. D.: The geometric theory of Banach spaces, Part I. Usp. Math. Nauk. **25**, 113–173 (1970) [Russian]. English translation in Russian Math. Surveys **25**, 111–170 (1970).
107. Milman, V. D.: The geometric theory of Banach spaces, Part II. Usp. Math. Nauk **26**, 73–149 (1971) [Russian]. English translation in Russian Math. Surveys **26**, 79–163 (1971).
108. Nash-Williams, C. St. J. A.: On well-quasi-ordering transfinite sequences. Proc. Cambridge Phil. Soc. **61**, 33–39 (1965).
109. Odell, E., Rosenthal, H. P.: A double dual characterization of separable Banach spaces containing l_1. Israel J. Math. **20**, 375–384 (1975).
110. Orlicz, W.: Über unbedingte Konvergenz in Funktionräumen. Studia Math. **1**, 83–85 (1930).
111. Orlicz, W.: Über eine gewisse Klasse von Räumen vom Typus B. Bull. Intern. Acad. Pol. 8, 207–220 (1932).
112. Ovsepian, R. I., Pelczynski, A.: The existence in every separable Banach space of a fundamental total and bounded biorthogonal sequence and related constructions of uniformly bounded orthonormal systems in L^2. Studia Math. **54**, 149–159 (1975).
113. Pelczynski, A.: On the isomorphism of the spaces m and M. Bull. Acad. Pol. Sci. **6**, 695–696 (1958).
114. Pelczynski, A.: Projections in certain Banach spaces. Studia Math. **19**, 209–228 (1960).
115. Pelczynski, A.: On the impossibility of embedding of the space L in certain Banach spaces. Coll. Math. **8**, 199–203 (1961).
116. Pelczynski, A.: A characterization of Hilbert Schmidt operators. Studia Math. **28**, 355–360 (1967).
117. Pelczynski, A.: Universal bases. Studia Math. **32**, 247–268 (1969).
118. Pelczynski, A.: Any separable Banach space with the bounded approximation property is a complemented subspace of a Banach space with a basis. Studia Math. **40**, 239–242 (1971).
119. Pelczynski, A.: All separable Banach spaces admit for every $\varepsilon > 0$ fundamental and total biorthogonal sequences bounded by $1+\varepsilon$. Studia Math. 55, 295–304 (1976).

120. Pelczynski, A., Singer, I.: On non-equivalent bases and conditional bases in Banach spaces. Studia Math. **25**, 5–25 (1964).
121. Pietsch, A.: Absolute p-summierende Abbildugen in normierten Räumen. Studia Math. **28**, 333–353 (1967).
122. Pietsch, A.: Theorie der Operatorenideale. A forthcoming book.
123. Reitz, R. E.: A proof of the Grothendieck inequality. Israel J. Math. **19**, 271–276 (1974).
124. Rolewicz, S.: Metric Linear Spaces. Monografie Matematyczne Warsaw 1972.
125. Rosenthal, H. P.: On complemented and quasi-complemented subspaces of quotients of $C(S)$ for Stonian S. Proc. Nat. Acad. Sci. (U.S.A.) **60**, 1165–1169 (1968).
126. Rosenthal, H. P.: On totally incomparable Banach spaces. J. Funct. Anal. **4**, 167–175 (1969).
127. Rosenthal, H. P.: On the subspaces of L^p $(p>2)$ spanned by sequences of independent random variables. Israel. J. Math. **8**, 273–303 (1970).
128. Rosenthal, H. P.: On the span in L^p of sequences of independent random variables (II). Proc. of the 6th Berkeley Symp. on Prob. and Stat., Berkeley, Calif. 1971.
129. Rosenthal, H. P.: A characterization of Banach spaces containing l_1. Proc. Nat. Acad. Sci. (U.S.A.) **71**, 2411–2413 (1974).
130. Rosenthal, H. P.: Point-wise compact subsets of the first Baire class. Amer. J. Math. to appear.
131. Schechtman, G.: On Pelczynski's paper "Universal bases". Israel J. Math. **22**, 181–184 (1975).
132. Schonefeld, S.: Schauder bases in the Banach spaces $C^k(T^q)$. Trans Amer. Math. Soc. **165**, 309–318 (1972).
133. Schwartz, L.: Probabilités cylindriques et applications radonifiantes. C. R. Acad. Paris **268**, 646–648 (1969).
134. Singer, I.: On Banach spaces with a symmetric basis. Rev. Math. Pure Appl. 6, 159–166 (1961).
135. Singer, I.: Bases in Banach spaces I. Berlin-Heidelberg-New York: Springer 1970.
136. Sobczyk, A.: Projection of the space m on its subspace c_0. Bull. Amer. Math. Soc. **47**, 938–947 (1941).
137. Szankowski, A.: Embedding Banach spaces with unconditional bases into spaces with symmetric bases. Israel J. Math. **15**, 53–59 (1973).
138. Szankowski, A.: A Banach lattice without the approximation property. Israel J. Math. 24, 329–337 (1976).
139. Tong, A. E.: Diagonal submatrices of matrix maps. Pacific J. Math. **32**, 551–559 (1970).
140. Tsirelson, B. S.: Not every Banach space contains l_p or c_0. Functional Anal. Appl. **8**, 138–141 (1974) [translated from Russian].
141. Urbanik, K.: Some prediction problems for strictly stationary processes. Proc. of the 5th Berkeley Symp. II (part I), pp. 235–258.
142. Veech, W. A.: Short proof of Sobczyk's theorem. Proc. Amer. Math. Soc. **28**, 627–628 (1971).
143. Wojtaszczyk, P.: On projections and unconditional bases in direct sums of Banach spaces II. Studia Math. to appear.
144. Woo, J. Y. T.: On modular sequence spaces. Studia Math. **48**, 271–289 (1973).
145. Woo, J. Y. T.: On a class of universal modular sequence spaces. Israel J. Math. **20**, 193–215 (1975).
146. Zippin, M.: On perfectly homogeneous bases in Banach spaces. Israel J. Math. **4**, 265–272 (1966).
147. Zippin, M.: A remark on bases and reflexivity in Banach spaces. Israel J. Math. **6**, 74–79 (1968).
148. Zippin, M.: Interpolation of operators of weak type between rearrangement invariant function spaces. J. Funct. Anal. **7**, 267–284 (1971).
149. Zippin, M.: The separable extension problem. Israel J. Math. **26**, 372–387 (1977).

Subject Index

absolutely summing operators (see operators)
absolute convergence 16, 64, 65
approximation problem 29
— property (A.P.) 29, 30, 32, 35, 40–42, 84, 86, 90, 94, 95
— —, bounded (B.A.P.) 37, 38, 41, 42, 51, 92, 95
— —, compact (C.A.P.) 94
— —, (ε, λ) 41
— — in subspaces of c_0 and l_p 90, 91
— — in X^* 33, 34, 41, 42
— —, λ 37
— —, metric (M.A.P.) 37–42
Auerbach system 16, 17, 38, 43, 46
averaging projections (see projections)

Baire class 102
basic sequence 1, 2, 4, 5–7, 92
— —, boundedly complete 13
— —, block 6
— — in X^* 8, 10
— —, shrinking 14
— —, weak* 10, 11
basis 1, 7, 30, 34, 37, 38, 43, 44, 47, 48, 91
—, biorthogonal functionals associated to 7, 11, 22, 118
—, block 6, 7, 19, 49, 53, 59, 61, 62, 73, 133, 135, 141
—, boundedly complete 9, 10, 12, 14, 22, 26, 138, 143, 168
—, conditional (see conditional bases)
—, complemented subspace of a space with a 38
—, constant 2, 19
—, duality of 7, 10
—, monotone 2, 38
—, normalized 2
— of X^* 8–10
—, perfectly homogeneous 61, 62
—, perturbation of a 5
—, problem 4, 27, 30, 34, 51
—, projections associated to 2, 7, 30
basis, shrinking 8, 10, 14, 21, 22, 85
—, subsymmetric (see subsymmetric basis)
—, symmetric (see symmetric basis)
—, unconditional (see unconditional basis)
—, uniqueness of 5
—, weak null 28
biorthogonal functionals (see basis)
— system 42, 44, 46
Boolean independence 100, 101
Borsuk's antipodal map theorem 77
bounded approximation property (see approximation property)
boundedly complete basic sequence (see basic sequence)
— — basis (see basis)
— — minimal system (see minimal system)

compact sets in Banach spaces 30
— —, uniform convergence on 31, 32
complementably universal spaces (see universal spaces)
complexification 81
condition Δ_2 at zero 138–140, 143, 145, 146, 148, 149, 152, 166
— — — —, uniform 167, 168, 172, 173
— Δ_2^* — —, 148, 149
— — — —, uniform 167, 168
— Δ_Q 138
conditional basis 74
— expectation 116

Davie's construction 87, 90, 94, 95
decomposition method (see Pelczynski's decomposition method)
disc algebra, basis in 4, 37
distortion problem 97
Dvoretzky–Rogers theorem 16, 65

equivalence of bases 5, 59, 71, 73, 129
— — Orlicz functions 139, 153, 159, 166

equivalence of sequences of Orlicz functions 167
— — subsequences of a basis 114
extension of operators (see operators)
— property 106, 107

Fredholm operators (see operators)
fundamental system 43, 44, 46
Franklin system 4

Grothendieck's inequality 68
— universal constant 68, 70

Haar system 3, 19, 20

independent random variables 87
index of an operator (see operators)
injective spaces 105, 106, 111
interpolation 125

James' function space 103
— tree space 103
— space J (see space J)

Khintchine's inequality 66, 72

lifting property 104, 107–109
Lorentz sequence spaces 115, 116, 130, 132, 137, 175–179
— — —, unit vector basis of 177

matrix 35, 68, 87, 89
—, block diagonal 21
—, diagonal of 20, 21, 151
—, representation of operators 20, 83
—, trace of 35, 90
—, unitary 45, 87, 89
Maurey–Rosenthal example 28
metric approximation property (see approximation property)
minimal system 42, 43, 46
— —, block 46
— —, boundedly complete 46
— —, shrinking 46
modular sequence spaces 166–169
modulus of convexity 128

norming functionals 44

operator ideal 77
operators, absolutely summing 63, 64, 67
—, between Orlicz sequence spaces 149
—, compact 29, 32, 33, 51, 57, 67, 76, 94, 149
—, diagonal 20, 21, 71, 151
—, diagonal with respect to blocking 50
—, extension of 105, 110
—, Fredholm 77–79, 110, 111
—, from c_0 into l_p 70, 71, 76
— from l_1 into l_2 69
— from l_r into l_p 76
—, Hilbert-Schmidt 67, 74
—, index of 77, 79, 80, 110
—, matrix representation of (see matrix)
—, p-absolutely summing 63–65, 67, 69–71
— of finite rank 29, 32, 33, 39
—, space of 37
—, spectrum of 80
—, strictly singular 75, 77, 79, 80, 145, 146, 160
—, weakly compact 57, 65, 106, 111
Orlicz functions 115, 137, 138, 140, 141, 145, 146, 150, 173, 176
— —, complementary functions of 147–149
— —, degenerate 137, 138, 140, 141, 147, 167
— —, equivalence of (see equivalence)
— —, interval associated to 143, 144, 149, 153–155, 158, 162, 163, 165, 168, 175
— —, minimal 152–154, 162, 163, 165
— —, right derivative of 139, 140, 143, 147, 149, 155, 172, 173
— sequence spaces 115, 116, 130, 137, 141, 143, 145, 150, 166, 169, 173, 176, 177
— — —, dual of 147
— — —, quotient spaces of 149
— — —, unit vector basis of 141, 150, 152, 166, 168

Pelczynski's decomposition method 54, 57, 93, 111, 117, 131, 132, 156, 170
perturbation of a basis (see basis)
Pietsch's factorization theorem 64
primary spaces 131
prime spaces 57, 131, 153
projections, averaging 116
—, associated to a basis (see basis)

quotient map 11, 50, 78, 86, 108, 109

quotient map of an Orlicz sequence space (see Orlicz sequence spaces)
— space 10, 12, 14, 26, 35, 85, 86, 104

Rademacher functions 24, 66, 67
Ramsey sets 101
rearrangement of a sequence 175
reflexivity 9, 14, 23, 27, 34, 40, 46, 95, 104, 110, 124, 127
—, local 33
— of Lorentz sequence spaces 178
— of modular sequence spaces 168
— of Orlicz sequence spaces 148–150, 157, 158, 165, 166
— of quotient spaces 110, 111
representation of second dual 8

Schauder basis (see basis)
— decomposition 47, 49
— —, blocking of 49
— —, boundedly complete 48
— —, constant 47, 49
— —, finite dimensional (F.D.D.) 48, 49, 85, 86
— —, shrinking 48, 49, 50
— —, unconditional 48, 72
— —, unconditional constant 49
— system 3, 20
— Tychonoff fixed point theorem 143
sequences of zeros and ones 156, 161, 164, 166
set C_M 140, 143, 144, 158
— C_{MA} 140, 143, 149, 150, 157
— E_M 140
— E_{MA} 140, 149, 157, 161
shrinking basic sequence (see basic sequence)
— basis (see basis)
— minimal system (see minimal system)
space $C(0, 1)$ 24
— —, basis of 3, 43
— —, compact sets in 141, 152, 162
— $C(K)$ 57, 131
— $C^k(I^n)$, basis of 4
— c, basis of 3, 20, 29 (see also summing basis)
— c_0 22, 23, 27, 37, 53, 54, 70–73, 84, 86, 90, 97, 98, 103, 104, 106, 110, 118, 130, 132, 143, 144
— —, automorphisms of 109
— —, basis of 3, 19, 53, 59, 61, 120, 121, 129
— —, isometries of 112
— —, projections in 56
— —, quotient spaces of 107, 109
— $c_0 \oplus l_r$ 171
space $d(w, p)$ (see Lorentz sequence spaces)
— h_M (see Orlicz sequences spaces)
— H_∞ 37
— J 25, 103, 132
— JF (see James's function space)
— JT (see James's tree space)
— $L_1(0, 1)$ 24
— $L_p(0, 1)$ 27, 63, 65, 72, 91, 97, 129, 131
— —, basis of 3, 19, 20
— $L_\infty(0, 1)$ 111
— l_1 21, 23, 27, 69, 71–73, 97, 99, 101, 103, 104, 118
— —, automorphisms of 108
— —, quotient spaces of 108
— $l_1(\Gamma)$ 108
— l_2 18, 71, 73, 110, 111, 118, 153
— l_M (see Orlicz sequence spaces)
— l_p 27, 53, 54, 70, 73, 84, 86, 90, 91, 95, 131, 132, 143, 144, 146, 149, 159, 163
— —, basis of 3, 19, 53, 59, 61, 73, 94, 99, 121, 129, 177
— —, isometries of 112
— —, projections in 55, 56
— —, quotient spaces of 86, 92, 175
— $l_p \oplus l_r$ 75, 82, 137, 166, 169
— —, unconditional bases of 82, 83
— —, quotient spaces of 175
— l_∞ 57, 58, 103, 105, 111, 138, 168
— —, automorphisms of 110
— —, quotient spaces of 111, 112
— $l_\infty(\Gamma)$ 105, 106
— $X_{p,r}$ 171, 172
— $X_{p,r,w}$ 169
— $X_{p,2,w}$ 169
— $X_{\infty,r}$ 171
— $X_{\infty,r,w}$ 171
— $Y_{q,s}$ 172, 173, 175
strong non-equivalence of Orlicz functions 150, 159, 164
subsymmetric basis 114, 116, 131, 132, 141
— —, block basis with constant coefficients of a 116
— —, sums of vectors of a 118
— constant 114, 115
— norm 114
summing basis 20, 29, 73
symmetric basis 113, 114, 116, 117, 123, 124, 129, 132, 136–138, 140, 153, 178, 179
— —, block bases generated by one vector of a 120, 121, 134, 135, 179
— —, block bases with constant coefficients of a 116, 117, 127, 150, 152, 171
— —, functionals biorthogonal to 118, 120, 121
— —, minimal 154

symmetric basis, sums of vectors of a 118, 125, 139
— —, uniqueness of 129, 130, 153, 154, 157, 158, 164, 176, 178
— constant 114, 115, 119
— norm 114

Total system 43, 44, 46
totally incomparable spaces 75, 82, 112
trigonometric system 43
Tsirelson's example 95, 97, 132, 133, 136, 143

unconditional basic sequence 18, 28, 92
— basis 15, 18, 19, 21–23, 27, 46, 47, 51, 61, 62, 70, 91, 92, 95, 113, 123, 124, 126, 168
— — in X^* 23
— —, projections associated to 18
— —, spaces without an 24

unconditional basis, uniqueness of 63, 71, 73, 84, 117, 118, 130
— constant 18, 19
— convergence 15, 64, 65, 72, 99
— F.D.D. (see Schauder decomposition)
uniform convexifiability 97
— convexity 97, 124, 127–129
universal Orlicz sequence spaces 154, 155, 156
— spaces 84, 91, 92, 129–131
— — in l_p 94

weak Cauchy sequences 99
— conditional compactness 23
— sequential completeness 23, 27
— unconditional convergence (w.u.c.) 99
weak* convergence 101
— limit 101

Young's inequality 147

Joram Lindenstrauss   Lior Tzafriri

Classical Banach Spaces II

Function Spaces

Springer-Verlag
Berlin Heidelberg New York 1979

Joram Lindenstrauss
Lior Tzafriri

Department of Mathematics, The Hebrew University of Jerusalem
Jerusalem, Israel

AMS Subject Classification (1970): 46-02, 46 A40, 46Bxx, 46Jxx

ISBN 3-540-08888-1 Springer-Verlag Berlin Heidelberg New York
ISBN 0-387-08888-1 Springer-Verlag New York Heidelberg Berlin

Library of Congress Cataloging in Publication Data. Lindenstrauss, Joram, 1936-. Classical Banach spaces. (Ergebnisse der Mathematik und ihrer Grenzgebiete; 92, 97). Bibliography: v. 1, p.; v. 2, p. Includes index. CONTENTS: 1. Sequence spaces. 2. Function spaces. 1. Banach spaces. 2. Sequence spaces. 3. Function spaces. I. Tzafriri, Lior, 1936- joint author. II. Title. III. Series QA322.2.L56. 1977-515'.73. 77-23131.

Typesetting: William Clowes & Sons Ltd., Beccles and London.
Printing and bookbinding: Konrad Triltsch, Würzburg.
2141/3140-543210

To Naomi and Marianne

Preface

This second volume of our book on classical Banach spaces is devoted to the study of Banach lattices. The writing of an entire volume on this subject within the framework of Banach space theory became possible only recently due to the substantial progress made in the seventies.

The structure of Banach lattices is much simpler than that of general Banach spaces and their theory is therefore more complete and satisfactory. Many of the results concerning Banach lattices are not valid (and sometimes even do not make sense) for general Banach spaces. Naturally, the theory of Banach lattices has many tools which are specific to this theory. We would like to draw attention in particular to the notions of p-convexity and p-concavity and their variants which seem to be especially useful in studying Banach lattices. We are convinced that these notions, which play a central role in the present volume, will continue to dominate the theory of Banach lattices and will be also useful in the various applications of lattice theory to other branches of analysis.

The table of contents is quite detailed and should give a clear idea of the material discussed in each section. We would like to make here only a few comments on the contents of this volume. The basic standard theory of Banach lattices is contained in Section 1.a and in a part of Section 1.b. The theory of p-convexity and p-concavity in Banach spaces is presented in detail in Sections 1.d and 1.f. Chapter 2 is devoted to a detailed study of the structure of rearrangement invariant function spaces on $[0, 1]$ and $[0, \infty)$. The usefulness of the notions of p-convexity and p-concavity will become apparent from their various applications in Chapter 2. Three of the sections in this volume are concerned with the general theory of Banach spaces rather than with Banach lattices. Section 1.e contains (part of) the theory of uniform convexity in general Banach spaces. Section 1.g deals with the approximation property. It complements (but is independent of) the discussion of this property in Vol. I. Section 2.g deals with geometric aspects of interpolation theory in general Banach spaces.

The various sections of this volume vary as far as their degree of difficulty is concerned. The first four sections in Chapter 1 and the first three sections of Chapter 2 are easier than the rest of the volume. The technically most difficult sections are Sections 1.g and 2.e. The results of Section 1.g are not used elsewhere in this volume and Sections 2.f and 2.g can be read without being acquainted with 2.e.

The prerequisites for the reading of this volume include besides standard material from functional analysis and measure theory only a superficial knowledge of the material presented in Vol. I of this book [79]. The (rather infrequent) references to Vol. I are marked here as follows: I.1.d.6 means for example item 6 in Section 1.d of Vol. I. In the present volume a much more extensive use is made of ideas and results from probability theory than in Vol. I. For the convenience of the reader without a probabilistic background we tried to discuss briefly in the appropriate places the notions and results from probability theory which we apply. The notation used in this volume is essentially the standard one, which is explained for example in the beginning of Vol. I. A few notations will be introduced and explained throughout the text.

The overlap between this volume and existing books on lattice theory is small and consists mostly of the standard material presented in Sections 1.a and (partially) 1.b. The books of W. A. J. Luxemburg and A. C. Zaanen [90] and H. H. Schaefer [118] contain much additional material on vector lattices. We do not treat here the theory of positive operators presented in [118]. Further information on isometric aspects of Banach lattice theory can be found in E. Lacey [71]. Sections 2.e and 2.f are based almost entirely on material taken from the memoir [58]. This memoir contains more results and details on the subject matter of 2.e and 2.f. The lecture notes of B. Beauzamy [6] contain further material on interpolation spaces in the spirit of the discussion in Section 2.g.

In the writing of this volume we benefited very much from long discussions with W. B. Johnson, B. Maurey and G. Pisier. We are very grateful to them for many valuable suggestions. We are also very grateful to J. Arazy who read the entire manuscript of this volume and made many corrections and suggestions. We wish also to thank Z. Altshuler and G. Schechtman for their help in the preparation of the manuscript.

The main part of this volume was written while we both were members of the Institute for Advanced Studies of the Hebrew University. The volume was completed while the first-named author visited the University of Texas at Austin and the second-named author visited the University of Copenhagen (supported in part by the Danish National Science Research Council). In these respective institutions we both gave lectures based on a preliminary version of this volume. We benefited much from comments made by those who attended these lectures. We wish to express our thanks to all these institutions as well as to the U.S. National Science Foundation (which supported us during the summers of 1977 and 1978 while we stayed at the Ohio State University) for providing us with excellent working conditions.

Finally, we express our indebtedness to Susan Brink and Nita Goldrick who very patiently and expertly typed various versions of the manuscript of this volume.

Joram Lindenstrauss
August 1978 Lior Tzafriri

Table of Contents

1. Banach Lattices 1

a. Basic Definitions and Results 1

Characterizations of σ-completeness and σ-order continuity. Ideals, bands and band projections. Boolean algebras of projections and cyclic spaces.

b. Concrete Representation of Banach Lattices 14

Abstract L_p and M spaces. Joint characterizations of $L_p(\mu)$ and $c_0(\Gamma)$ spaces. The functional representation theorem for order continuous Banach lattices. Köthe function spaces. The Fatou property.

c. The Structure of Banach Lattices and their Subspaces . . . 31

Property (u). Weak completeness and reflexivity. Existence of unconditional basic sequences.

d. p-Convexity in Banach Lattices 40

Functional calculus in general Banach lattices. The definition and basic properties of p-convexity and p-concavity. Duality. The p-convexification and p-concavification procedures. Some connections with p-absolutely summing operators. Factorization through L_p spaces.

e. Uniform Convexity in General Banach Spaces and Related Notions. 59

The definition of the moduli of convexity and smoothness. Duality. Asymptotic behavior of the moduli. The moduli of $L_2(X)$. Convergence of series in uniformly convex and uniformly smooth spaces. The notions of type and cotype. Kahane's theorem. Connections between the moduli and type and cotype.

f. Uniform Convexity in Banach Lattices and Related Notions . . 79

Uniformly convex and smooth renormings of a $(p>1)$-convex and $(q<\infty)$-concave Banach lattice. The concepts of upper and lower estimates for disjoint elements. The relations between these notions and those of type, cotype, p-convexity, q-concavity, p-absolutely summing operators, etc. Examples. Two diagrams summarizing the various connections.

g. The Approximation Property and Banach Lattices. 102

Examples of Banach lattices and of subspaces of l_p, $p\neq 2$, without the B.A.P. The connection between the type and the cotype of a space and the existence of subspaces without the B.A.P. A space different from l_2 all of whose subspaces have the B.A.P.

2. Rearrangement Invariant Function Spaces 114

a. Basic Definitions, Examples and Results 114

The definition of r.i. function spaces on $[0,1]$ and $[0,\infty)$. Conditional expectations. Interpolation between L_1 and L_∞.

b. The Boyd Indices 129

The definition of Boyd indices. Duality. The Rademacher functions in an r.i. function space on [0, 1]. The characterization of Boyd indices in terms of existence of l_p^n's for all n on disjoint vectors having the same distribution. Operators of weak type (p, q). Boyd's interpolation theorem.

c. The Haar and the Trigonometric Systems 150

Basic results on martingales. The unconditionality of the Haar system in $L_p(0, 1)$ spaces $(1<p<\infty)$ and in more general r.i. function spaces on [0, 1]. Reproducibility of bases. The boundedness of the Riesz projection and applications.

d. Some Results on Complemented Subspaces 168

The isomorphism between an r.i. function space X on [0, 1] with non-trivial Boyd indices and the spaces $X(l_2)$ and Rad X. Complemented subspaces of X with an unconditional basis. Subspaces spanned by a subsequence of the Haar system. R.i. spaces with non-trivial Boyd indices are primary.

e. Isomorphisms Between r.i. Function Spaces; Uniqueness of the r.i. Structure 181

Isomorphic embeddings. Classification of symmetric basic sequences in r.i. function spaces of type 2. Uniqueness of the r.i. structure for $L_p(0, 1)$ spaces and for $(q<2)$-concave r.i. spaces. Other applications.

f. Applications of the Poisson Process to r.i. Function Spaces . . . 202

The isomorphism between $L_p(0, 1)$ and $L_p(0, \infty)\cap L_2(0, \infty)$ for $p>2$. R.i. function spaces in $[0, \infty)$ isomorphic to a given r.i. function space on [0, 1]. Isometric embeddings of $L_r(0,1)$ into $L_p(0,1)$ for $1\leq p<r<2$ and in other r.i. function spaces. The complementation of the spaces $X_{p,2}$ in $L_p(0,1)$, $p>2$ and generalizations.

g. Interpolation Spaces and their Applications 215

Interpolation pairs. General interpolation spaces and applications to the construction of r.i. function spaces without unique r.i. structure. The Lions–Peetre method of interpolation. Uniform convexity and type in interpolation spaces.

References 233

Subject Index 239

1. Banach Lattices

a. Basic Definitions and Results

The function spaces which appear in real analysis are usually ordered in a natural way. This order is related to the norm and is important in the study of the space as a Banach space. In this volume we study partially ordered Banach spaces whose order and norm are related by the following axioms.

Definition 1.a.1. A partially ordered Banach space X over the reals is called a *Banach lattice* provided

(i) $x \leqslant y$ implies $x+z \leqslant y+z$, for every $x, y, z \in X$,
(ii) $ax \geqslant 0$, for every $x \geqslant 0$ in X and every non negative real a.
(iii) for all $x, y \in X$ there exists a least upper bound (l.u.b.) $x \vee y$ and a greatest lower bound (g.l.b.) $x \wedge y$,
(iv) $\|x\| \leqslant \|y\|$ whenever $|x| \leqslant |y|$, where the absolute value $|x|$ of $x \in X$ is defined by $|x| = x \vee (-x)$.

Observe that in (iii) above it is enough e.g. to require the existence of the l.u.b. The greatest lower bound can then be defined by $x \wedge y = -((-x) \vee (-y))$ (or by $x \wedge y = x + y - x \vee y$). It follows from (i), (ii) and (iii) that, for every $x, y, z \in X$,

$$|x-y| = |x \vee z - y \vee z| + |x \wedge z - y \wedge z| ,$$

and thus, by (iv), the lattice operations are norm continuous. It is perhaps worthwhile to make a comment concerning the proof of the preceding identity. Its deduction from (i), (ii) and (iii), while definitely not hard, is not completely straightforward. On the other hand, it is trivial to check the validity of this identity if x, y, and z are real numbers. We shall prove below (cf. 1.d.1 and the discussion preceding it) a general result which asserts, in particular, that any inequality (and thus also any identity) which involves lattice operations and algebraic operations (i.e. sums and multiplication by scalars) is valid in an arbitrary Banach lattice if it is valid in the real line.

The continuity of lattice operations implies, in particular, that the set $C = \{x;\, x \in X,\, x \geqslant 0\}$ is norm closed. The set C, which is a convex cone, is called the positive cone of X. For an element x in a Banach lattice X we put $x_+=x\vee 0$ and $x_-=-(x\wedge 0)$. Obviously, $x=x_+-x_-$ (and thus $X=C-C$) and $|x|=x_++x_-$. Two elements $x, y\in X$ for which $|x|\wedge|y|=0$ are said to be *disjoint*.

Every space with a basis $\{x_n\}_{n=1}^\infty$, whose unconditional constant is equal to one, is a Banach lattice when the order is defined by $\sum_{n=1}^\infty a_n x_n \geqslant 0$ if and only if $a_n\geqslant 0$, for all n. This order is called the order induced by the unconditional basis. In the sequel, whenever we consider an abstract space with an unconditional basis as a Banach lattice, the order will be defined as above unless stated otherwise. For a general space with an unconditional basis endowed with the order defined above, axioms (i), (ii) and (iii) of 1.a.1 always hold but (iv) has to be replaced by

(iv′) there exists a constant M such that $\|x\|\leqslant M\|y\|$ whenever $|x|\leqslant|y|$.

As in the case of a space with an unconditional basis, every partially ordered Banach space satisfying (i), (ii), (iii) and (iv′) can be renormed, by putting $\|x\|_0=\sup\{\|y\|;\ |y|\leqslant|x|\}$, so that it becomes a Banach lattice.

There are many important lattices which are not induced by an unconditional basis. Clearly, every $L_p(\mu)$ space, $1\leqslant p\leqslant\infty$ and every $C(K)$ space is a Banach lattice with the pointwise order. Unless μ is purely atomic (and σ-finite), respectively, K is finite, these lattices are not induced by an unconditional basis. The separable Banach lattices $L_1(0,1)$ and $C(0,1)$ do not have an unconditional basis (in fact, they do not even embed in a space with an unconditional basis, cf. I.1.d.1). The spaces $L_p(0,1)$, $1<p<\infty$, have an unconditional basis, namely the Haar basis (cf. 2.c.5 below), but the natural order in $L_p(0,1)$ (i.e. the pointwise order) is completely different from the order induced by the basis.

Every Banach lattice X has the so-called *decomposition property*: if x_1, x_2 and y are positive elements in X and $y\leqslant x_1+x_2$ then there are $0\leqslant y_1\leqslant x_1$ and $0\leqslant y_2\leqslant x_2$ such that $y=y_1+y_2$. This property is easily checked if we take $y_1=x_1\wedge y$ and $y_2=y-y_1$. The converse is not true in general: there exist partially ordered Banach spaces having the decomposition property which are not lattices.

A linear operator T from a *vector lattice* X (i.e. a linear space satisfying (i), (ii) and (iii) of 1.a.1) into a vector lattice Y is called *positive* if $Tx\geqslant 0$ for every $x\geqslant 0$ in X. It is clear that a positive operator T from X to Y, which is one to one and onto, and whose inverse is also positive, preserves the lattice structure, i.e.

$$T(x_1\vee x_2)=Tx_1\vee Tx_2 \quad\text{and}\quad T(x_1\wedge x_2)=Tx_1\wedge Tx_2\,,$$

for all $x_1, x_2\in X$. Such an operator is called an *order preserving* operator or an *order isomorphism*. Two vector lattices X and Y are said to be *order isomorphic* if there is an order isomorphism from X onto Y. For example, a normalized unconditional basis $\{x_n\}_{n=1}^\infty$ in a Banach space X is equivalent to a permutation of a normalized unconditional basis $\{y_n\}_{n=1}^\infty$ in a Banach space Y if and only if X, with the order induced by $\{x_n\}_{n=1}^\infty$, is order isomorphic to Y, with the order induced by $\{y_n\}_{n=1}^\infty$. Observe that a positive linear map T between Banach lattices is automatically continuous. Indeed, otherwise there would exist a sequence $\{x_n\}_{n=1}^\infty$ such that $\|x_n\|=2^{-n}$ and $\|Tx_n\|\geqslant 2^n$, for all n, but this contradicts the fact that

$\|Tx_n\| \leqslant \left\| T \sum_{j=1}^{\infty} |x_j| \right\|$, for $n=1, 2, \ldots$ In particular, an order isomorphism between Banach lattices is also an isomorphism from the linear topological point of view. The Banach lattices X and Y are said to be *order isometric* if there exists a linear isometry T from X onto Y which is also an order isomorphism.

By a *sublattice* of a Banach lattice X we mean a linear subspace Y of X so that $x \vee y$ (and thus also $x \wedge y = x + y - x \vee y$) belongs to Y whenever $x, y \in Y$. Unless stated explicitly otherwise, we shall assume that a sublattice is also norm closed. Among the sublattices of a Banach lattice X we single out the ideals. An *ideal* in X is a linear subspace Y for which $y \in Y$ whenever $|y| \leqslant |x|$ for some $x \in Y$. (Again, unless stated otherwise, we assume that it is also norm closed.) If Y is an ideal in X then the quotient space X/Y becomes a Banach lattice if we take as its positive cone the image of the positive cone of X. It is easily checked that $Tx_1 \vee Tx_2 = T(x_1 \vee x_2)$ for every $x_1, x_2 \in X$, where $T: X \to X/Y$ denotes the quotient map. In order to verify that (iv) of 1.a.1 holds in X/Y we have to show that $\inf \{\|x_1 - y\|; y \in Y\} \leqslant \inf \{\|x_2 - y\|; y \in Y\}$, whenever $0 \leqslant x_1 \leqslant x_2$. This is done as follows. Let $y \in Y$ and observe that $x_1 - y \leqslant (x_1 - y)_+ \leqslant x_1 + y_-$. Consequently, since Y is an ideal, $(x_1 - y)_+ = x_1 - z$, for some $z \in Y$. Since $0 \leqslant x_1 - z \leqslant (x_2 - y)_+$ we deduce that $\|x_1 - z\| \leqslant \|x_2 - y\|$.

If $\{x_\alpha\}_{\alpha \in A}$ is a set in a Banach lattice we denote by $\bigvee_{\alpha \in A} x_\alpha$ or by $\text{l.u.b.}\{x_\alpha\}_{\alpha \in A}$ the (unique) element $x \in X$ which has the following properties: (1) $x \geqslant x_\alpha$ for all $\alpha \in A$ and (2) whenever $z \in X$ satisfies $z \geqslant x_\alpha$ for all $\alpha \in A$ then $z \geqslant x$. Unless the set A is finite, $\bigvee_{\alpha \in A} x_\alpha$ need not always exist in a Banach lattice. An ideal Y in a Banach lattice X is called a *band* if, for every subset $\{y_\alpha\}_{\alpha \in A}$ of Y such that $\bigvee_{\alpha \in A} y_\alpha$ exists in X, this element belongs already to Y.

The dual X^* of a Banach lattice X is also a Banach lattice provided that its positive cone is defined by $x^* \geqslant 0$ in X^* if and only if $x^*(x) \geqslant 0$, for every $x \geqslant 0$ in X. It is easily verified that, for any $x^*, y^* \in X^*$ and every $x \geqslant 0$ in X, we have

$$(x^* \vee y^*)(x) = \sup \{x^*(u) + y^*(x-u); 0 \leqslant u \leqslant x\}$$

and

$$(x^* \wedge y^*)(x) = \inf \{x^*(v) + y^*(x-v); 0 \leqslant v \leqslant x\}.$$

The Banach lattice X^* has the property that *every non-empty order bounded set $\mathscr{F}$ in X^* has a l.u.b.* In order to prove this fact we first replace $\mathscr{F}$ by the family $\mathscr{G}$ of all suprema of finite subsets of $\mathscr{F}$. The set $\mathscr{G}$ is upward directed, order bounded and has a l.u.b. if and only if $\mathscr{F}$ has a l.u.b. For every $x \geqslant 0$ in X we put $f(x) = \sup \{x^*(x); x^* \in \mathscr{G}\}$. It is easily checked that f is an additive and positively homogeneous functional on the positive cone of X and thus it extends uniquely to an element of X^*. Clearly, this element is the l.u.b. of $\mathscr{G}$.

Since every $x^* \in X^*$ can be decomposed as a difference of two non-negative elements, it follows that every norm bounded monotone sequence $\{x_n\}_{n=1}^{\infty}$ in X is weak Cauchy. If, in addition, $x_n \xrightarrow{w} x$ for some $x \in X$ then $\|x_n - x\| \to 0$ as $n \to \infty$. This is a consequence of the fact that weak convergence to x implies the existence of convex combinations of the x_n's which tend strongly to x.

Proposition 1.a.2. *The canonical embedding i of a Banach lattice X into its second dual X^{**} is an order isometry from X onto a sublattice of X^{**}.*

Proof. It is obvious that i is a positive operator. What we have to show is that $ix \vee iy = i(x \vee y)$, for all $x, y \in X$. We prove this first under the assumption that $x \wedge y = 0$. For every $u^* \geq 0$ in X^*, we have

$$\begin{aligned}(ix \vee iy)(u^*) &= \sup\{ix(v^*) + iy(u^* - v^*);\ 0 \leq v^* \leq u^*\}\\ &= \sup\{u^*(y) + v^*(x - y);\ 0 \leq v^* \leq u^*\}.\end{aligned}$$

By putting $w^*(z) = \sup_n u^*(z \wedge nx)$, for each $z \geq 0$ in X, we define a bounded linear functional $w^* \in X^*$ (the linearity of w^* is a consequence of the identity $(a+b) \wedge c \leq a \wedge c + b \wedge c \leq (a+b) \wedge 2c$, which holds for all $a, b, c \geq 0$ in X). The functional w^* satisfies $0 \leq w^* \leq u^*$, $w^*(x) = u^*(x)$ and (since $x \wedge y = 0$) $w^*(y) = 0$. It follows that $(ix \vee iy)(u^*) \geq u^*(y) + w^*(x - y) = u^*(x + y) = u^*(x \vee y) = i(x \vee y)(u^*)$, for every positive $u^* \in X^*$. Hence, $ix \vee iy \geq i(x \vee y)$ and, by the positivity of i, we deduce that $ix \vee iy = i(x \vee y)$.

Assume now that x, y are arbitrary elements in X. Put $u = x - x \wedge y$, $v = y - x \wedge y$. Then $u \wedge v = 0$ and hence, $iu \vee iv = iu + iv$. Consequently, $iu \wedge iv = 0$ and thus $ix \wedge iy = i(x \wedge y)$, which concludes the proof. □

In general, iX is not an ideal of X^{**}. We shall present in 1.b.16 below a necessary and sufficient condition for iX to be an ideal of X^{**}.

Definition 1.a.3. A Banach lattice X is said to be *conditionally order complete* (*σ-order complete*) or, briefly, *complete* (*σ-complete*) if every order bounded set (sequence) in X has a l.u.b.

The discussion preceding 1.a.2 shows that every Banach lattice X, which is the dual of another Banach lattice, is complete. In particular, every reflexive lattice is complete. The simplest examples of concrete complete Banach lattices are the $L_p(\mu)$ spaces with $1 \leq p \leq \infty$ (though $L_1(0, 1)$ is not a conjugate space). Banach lattices generated by unconditional bases are also complete; the supremum can be taken coordinatewise. On the other hand, $C(0, 1)$ is not σ-complete. In fact, we have the following result, due to H. Nakano [103] and M. H. Stone [122].

Proposition 1.a.4. (i) *The space $C(K)$ of all continuous functions on a compact Hausdorff topological space K is a σ-complete Banach lattice if and only if K is basically disconnected, i.e. the closure of every open F_σ-set in K is open.*

(ii) *The space $C(K)$ is a complete Banach lattice if and only if K is extremally disconnected, i.e. the closure of every open set in K is open.*

Proof. The proof of both assertions is similar. We shall present here only the proof of (i).

Assume that $C(K)$ is σ-complete. Let $\{E_n\}_{n=1}^{\infty}$ be a sequence of closed subsets of K so that $E=\bigcup_{n=1}^{\infty} E_n$ is open. For every integer n we construct a function $f_n \in C(K)$ such that $f_n(t)=1$ for $t \in E_n, f_n(t)=0$ for $t \notin E$ and $0 \leqslant f_n(t) \leqslant 1$ whenever $t \in K$. Since the sequence $\{f_n\}_{n=1}^{\infty}$ is order bounded by the function identically equal to 1 on K there exists $f=\bigvee_{n=1}^{\infty} f_n \in C(K)$. It is clear that $f(t)=1$ for $t \in E$ and $f(t)=0$ for $t \notin \bar{E}$. Hence, the set $\bar{E}$ is both open and closed.

Conversely, suppose that K is basically disconnected. For every $f \in C(K)$ put $E_f(\lambda)=\{t; f(t)<\lambda\}$ and observe that $E_f(\lambda)$ is an open F_σ-set since $E_f(\lambda)= \bigcup_{n=1}^{\infty} \{t; f(t) \leqslant \lambda - 1/n\}$. Let $\{g_n\}_{n=1}^{\infty}$ be a bounded sequence of elements of $C(K)$. By our assumption on K the set $\bigcap_{n=1}^{\infty} \overline{(E_{g_n}(\lambda))}$ is a closed G_δ-set. Hence, its complement is an open F_σ-set whose closure must be open. It follows that the set $E(\lambda)= \operatorname{int} \bigcap_{n=1}^{\infty} \overline{(E_{g_n}(\lambda))}$ is both open and closed.

Put

$$g_0(t)=\sup\{\lambda;\, t \notin E(\lambda)\}.$$

Notice that the above supremum exists since $\{g_n\}_{n=1}^{\infty}$ is bounded and that g_0 is continuous on K for both sets

$$\{t;\, g_0(t)<\lambda\}=\bigcup_{\mu<\lambda} E(\mu)$$

and

$$\{t;\, g_0(t)>\lambda\}=\bigcup_{\mu>\lambda} (K \sim E(\mu))$$

are open. The function g_0 is the l.u.b. of $\{g_n\}_{n=1}^{\infty}$. Indeed, since $E(\lambda) \subset \overline{E_{g_n}(\lambda)}$, $n=1, 2, \ldots$ we get that $g_n \leqslant g_0$ for all n and if $g_n \leqslant h, n=1,2,\ldots$ for some $h \in C(K)$ then

$$E_h(\lambda) \subset \bigcap_{n=1}^{\infty} E_{g_n}(\lambda) \subset \bigcap_{n=1}^{\infty} \overline{E_{g_n}(\lambda)}\,.$$

In view of the fact that $E_h(\lambda)$ is open it follows that $E_h(\lambda) \subset E(\lambda)$ for every real λ i.e. $h \geqslant g_0$. □

It should be pointed out that no infinite compact metric space K is basically disconnected. The simplest example of an extremally disconnected space is βN, the Stone–Čech compactification of the integers. It is, of course, easy to check directly that $l_\infty = C(\beta N)$ is indeed a complete Banach lattice. A simple example of a σ-complete $C(K)$ space, which is not complete, is the subspace of $l_\infty(\Gamma)$, with Γ uncountable, spanned by the constant function and the functions with countable support.

The following fact, due to Meyer-Nieberg [98], is very useful in applications.

Theorem 1.a.5. *A Banach lattice which is not σ-complete contains a sequence of mutually disjoint elements equivalent to the unit vector basis of c_0.*

Proof. Let $\{x_n\}_{n=1}^\infty \subset X$ be an order bounded sequence which does not have a l.u.b. By replacing $\{x_n\}_{n=1}^\infty$ by the sequence $\left\{\bigvee_{j=1}^n x_j\right\}_{n=1}^\infty$ we can assume with no loss of generality that $0 \leqslant x_1 \leqslant x_2 \leqslant \cdots \leqslant x_n \leqslant \cdots \leqslant x$, for some $x \in X$. If $\{x_n\}_{n=1}^\infty$ converges in norm to an element of X then, obviously, this element is also the l.u.b. of $\{x_n\}_{n=1}^\infty$. Otherwise, there is an $\alpha > 0$ and a subsequence $\{x_{n_j}\}_{j=1}^\infty$ of $\{x_n\}_{n=1}^\infty$ so that the vectors $u_j = x_{n_{j+1}} - x_{n_j}$ satisfy $\|u_j\| \geqslant \alpha$, $u_j \geqslant 0$ and $\sum_{k=1}^j u_k \leqslant x$ for all j.

We claim now that, for every $\varepsilon > 0$ and every $\beta > 0$, there exists a subsequence $\{v_k\}_{k=1}^\infty$ of $\{u_j\}_{j=1}^\infty$ so that $\|(v_k - \beta v_1)_+\| \geqslant \alpha - \varepsilon$ for all $k > 1$. Indeed, if this is not true then there is a subsequence $\{w_k\}_{k=1}^\infty$ of $\{u_j\}_{j=1}^\infty$ such that $\|(w_k - \beta w_j)_+\| < \alpha - \varepsilon$ for all $k > j$. It follows that, for any k, we have

$$\|x\| \geqslant \left\|\sum_{i=1}^k w_i\right\| = \beta^{-1}\left\|k w_{k+1} - \sum_{i=1}^k (w_{k+1} - \beta w_i)\right\|$$

$$= \beta^{-1}\left\|k w_{k+1} - \sum_{i=1}^k (w_{k+1} - \beta w_i)_+ + \sum_{i=1}^k (w_{k+1} - \beta w_i)_-\right\|.$$

Since $k w_{k+1} \geqslant \sum_{i=1}^k (w_{k+1} - \beta w_i)_+$ we get that

$$\|x\| \geqslant \beta^{-1}\left\|k w_{k+1} - \sum_{i=1}^k (w_{k+1} - \beta w_i)_+\right\| \geqslant \beta^{-1}(k\alpha - k(\alpha - \varepsilon)) = \beta^{-1} k\varepsilon$$

and this is contradictory for large values of k.

Now, fix $0 < \varepsilon < \alpha/2$ and construct a subsequence $\{v_k\}_{k=1}^\infty$ of $\{u_j\}_{j=1}^\infty$ so that $\|(v_k - \beta v_1)_+\| \geqslant \alpha - \varepsilon$ for all $k > 1$, where $\beta = 2\|x\|/\varepsilon$. Put $y_1 = \beta^{-1}(\beta v_1 - x)_+$ and $y_k = (v_k - \beta v_1)_+$ for $k > 1$. It is clear that $y_1 \wedge y_k = 0$ for every $k > 1$. By the choice of the sequence $\{v_k\}_{k=1}^\infty$ we also get that $y_k \leqslant v_k \leqslant x$, $\|y_k\| \geqslant \alpha - \varepsilon$ for $k > 1$, and $\|y_1\| = \|(v_1 - \beta^{-1}x)_+\| \geqslant \|v_1\| - \beta^{-1}\|x\| - \|(v_1 - \beta^{-1}x)_-\| \geqslant \alpha - \varepsilon$.

Applying again this argument to the sequence $\{y_k\}_{k=2}^\infty$, instead of $\{u_j\}_{j=1}^\infty$, and with $\varepsilon/2$, instead of ε, we can produce a new subsequence for which the norms of its elements are $\geqslant \alpha - \varepsilon - \varepsilon/2$, each element is $\leqslant x$ and the first two elements are mutually disjoint and also disjoint from the rest of the sequence. Continuing by induction we obtain a sequence $\{z_k\}_{k=1}^\infty$, of mutually disjoint elements of X, so that $\|z_k\| \geqslant \alpha - 2\varepsilon$ and $z_k \leqslant x$ for all k. This sequence is clearly equivalent to the unit vector basis of c_0. □

The converse of 1.a.5 is evidently false since e.g. c_0 itself is σ-complete.

Definition 1.a.6. A Banach lattice X is said to have *an order continuous norm* (*σ-order continuous norm*) or, briefly, to be *order continuous* (*σ-order continuous*) if, for every downward directed set (sequence) $\{x_\alpha\}_{\alpha \in A}$ in X with $\bigwedge_{\alpha \in A} x_\alpha = 0$, $\lim_\alpha \|x_\alpha\| = 0$.

A simple example of a σ-order continuous Banach lattice, which is not order continuous, is the subspace of $l_\infty(\Gamma)$ spanned by $c_0(\Gamma)$ and the function identically equal to one, where Γ is an uncountable set. Typical examples of order complete Banach lattices, which are not σ-order continuous, are l_∞ and $L_\infty(0, 1)$.

Proposition 1.a.7 [85]. *A σ-complete Banach lattice X, which is not σ-order continuous, contains a subspace isomorphic to l_∞. Moreover, the unit vectors of l_∞ correspond, under this isomorphism, to mutually disjoint elements of X.*

Proof. Assume that $\{x_n\}_{n=1}^\infty$ is a non-convergent decreasing sequence in X with $\bigwedge_{n=1}^\infty x_n = 0$. The sequence $\{x_1 - x_n\}_{n=1}^\infty$ is increasing, order bounded and not strong Cauchy. It follows from the proof of 1.a.5 that there exists a sequence $\{z_k\}_{k=1}^\infty$ of mutually disjoint elements in X, which is equivalent to the unit vector basis of c_0 and for which $0 < z_k \leqslant x_1$, $k \geqslant 1$. For $a = \{a_k\}_{k=1}^\infty \in l_\infty$, with $a_k \geqslant 0$ for every k, we put $Ta = \bigvee_{k=1}^\infty a_k z_k$ (the supremum exists since X is σ-complete and $a_k z_k \leqslant x_1 \sup_{1 \leqslant m < \infty} |a_m|$ for every k). It is easily checked that T is an isomorphism from the positive cone of l_∞ into that of X which extends uniquely to an isomorphism from l_∞ into X. □

The result 1.a.7 implies, in particular, that a separable σ-complete Banach lattice is σ-order continuous. It is easily seen from the definition that a separable σ-order continuous Banach lattice is already order continuous. Thus, every separable σ-complete Banach lattice is order continuous. The converse to this assertion is also true even without the separability assumption. Every order continuous Banach lattice is also order complete. This is the main assertion of the following proposition.

Proposition 1.a.8. *Let X be a Banach lattice. Then the following assertions are equivalent.*

(i) *X is σ-complete and σ-order continuous.*

(ii) *Every order bounded increasing sequence in X converges in the norm topology of X.*

(iii) *X is order continuous.*
(iv) *X is (order) complete and order continuous.*

Proof. The equivalence (i) $\Leftrightarrow$ (ii) follows directly from the definitions (that (ii) $\Leftrightarrow$ (i) was used already in the proofs of 1.a.5 and 1.a.7). We have just to prove that (ii) $\Rightarrow$ (iii) and that (iii) $\Rightarrow$ (iv).

(ii) $\Rightarrow$ (iii): Let $\{x_\alpha\}_{\alpha\in A}$ be a downward directed set satisfying $\bigwedge_{\alpha\in A} x_\alpha=0$. If the net $\{x_\alpha\}_{\alpha\in A}$ does not converge to 0 there are $\delta>0$ and a decreasing sequence $\{x_{\alpha_j}\}_{j=1}^\infty$ in this net so that $\|x_{\alpha_j}-x_{\alpha_{j+1}}\|\geqslant\delta$, for every j, and this contradicts (ii).

(iii) $\Rightarrow$ (iv). Let U be an order bounded set in X. By adding to U the finite suprema of its elements we may assume that $U=\{x_\alpha\}_{\alpha\in A}$ is upward directed. Let V by the set of all upper bounds of U i.e. $V=\{y;\, y\in X,\, y\geqslant x \text{ for all } x\in U\}$. Clearly, V is downward directed and so is also $V-U=\{y-x;\, y\in V,\, x\in U\}$. We claim that 0 is the g.l.b. of $V-U$. Indeed, let $z\geqslant 0$ satisfy $z\leqslant y-x$, for every $y\in V$ and $x\in U$. Then $y\in V \Rightarrow y-z\in V$ and thus, by induction, $y-nz\in V$ for every integer n. Consequently, $\|z\|=0$. Hence, by (iii), there are, for every $\varepsilon>0$, an $\alpha\in A$ and a $y\in V$ so that $\|y-x_\alpha\|\leqslant\varepsilon$ and therefore $\|x_\beta-x_\alpha\|\leqslant\varepsilon$ for every $\beta>\alpha$ in A. Thus, the net $\{x_\alpha\}_{\alpha\in A}$ converges to a limit x which is the l.u.b. of U. □

We shall encounter below several less trivial and more interesting characterizations of order continuous Banach lattices (see 1.a.11, 1.b.14 and 1.b.16). Condition (i) of 1.a.8 is however the easiest to check characterization of order continuity in concrete examples.

In the study of spaces with an unconditional basis $\{x_n\}_{n=1}^\infty$ the projections P_σ (with σ being a subset of the integers), defined by,

$$P_\sigma\left(\sum_{n=1}^\infty a_n x_n\right)=\sum_{n\in\sigma} a_n x_n$$

play a fundamental role. A natural generalization of these projections is possible in every σ-complete Banach lattice.

Let X be a σ-complete Banach lattice; to every $x\geqslant 0$ we associate a projection P_x in the following way. For $z\geqslant 0$ in X we put

$$P_x(z)=\bigvee_{n=1}^\infty (nx\wedge z)$$

and, for a general $y=y_+-y_-\in X$, we set

$$P_x(y)=P_x(y_+)-P_x(y_-).$$

It is easily verified that, for every x in the positive cone of X, P_x is a norm one positive linear projection. Note also that, for every $x, y\geqslant 0$ in X, we have

$$x\wedge(y-P_x(y))=0\,.$$

To understand better the definition above we should point out that in the case when X is a σ-complete Banach lattice of functions with the pointwise order, the projections $\{P_x\}_{x\geqslant 0}$ are just "multiplications" by characteristic functions. For instance, in $X=L_p(\mu)$, $1\leqslant p\leqslant\infty$, the projection P_x is the multiplication operator by the characteristic function of the support of the function $x\in L_p(\mu)$, i.e. $\{t; x(t)\neq 0\}$.

Using these projections we can decompose Banach lattices into a direct sum of ideals with a weak unit. An element $e\geqslant 0$ of a Banach lattice X is said to be a *weak unit* of X if $e\wedge x=0$ for $x\in X$ implies $x=0$ (there is also a notion of strong unit which will be mentioned in the next section in connection to the study of M spaces). In a σ-complete Banach lattice X an element e is a weak unit if and only if $P_e(x)=x$ for every $x\in X$. Indeed, for every $x\geqslant 0$ in X, we have $e\wedge(x-P_e(x))=0$. Therefore, if e is a weak unit then $x-P_e(x)=0$. Conversely, $e\wedge x=0$ clearly implies $x=0$ if we assume that $x=P_e(x)$ for every $x\in X$.

Proposition 1.a.9 [63]. *Any order continuous Banach lattice X can be decomposed into an unconditional direct sum of a (generally uncountable) family of mutually disjoint ideals $\{X_\alpha\}_{\alpha\in A}$, each X_α having a weak unit $x_\alpha>0$. More precisely, every $y\in X$ has a unique representation of the form $y=\sum_{\alpha\in A} y_\alpha$ with $y_\alpha\in X_\alpha$, only countably many $y_\alpha\neq 0$ and the series converging unconditionally. Moreover, if Z is a separable subspace of X then one of the indices α, say α_0, can be chosen so that $Z\subset X_{\alpha_0}$.*

Proof. By Zorn's lemma there exists a maximal family $\{x_\alpha\}_{\alpha\in A}$ of mutually disjoint positive elements of X. Let X_α be the set of all $x\in X$ such that $|x|\wedge y=0$ whenever $x_\alpha\wedge y=0$. It is easily checked that $X_\alpha=P_{x_\alpha}X$, that it is an ideal (even a band) of X and that x_α is a weak unit for X_α.

Fix $y\geqslant 0$ in X and consider the set $\{P_{x_\alpha}(y)\}_{\alpha\in A}$. Then, by 1.a.8, for any countable subset $A_0=\{\alpha_1,\alpha_2,\ldots,\alpha_n,\ldots\}$ of A, the series $\sum_{n=1}^{\infty} P_{x_{\alpha_n}}(y)$ converges strongly to $\bigvee_{n=1}^{\infty} P_{x_{\alpha_n}}(y)$. This implies that only countably many of the $\{P_{x_\alpha}(y)\}_{\alpha\in A}$ are different from zero and that $\sum_{\alpha\in A} P_{x_\alpha}(y)$ converges unconditionally to an element y_0 of X which clearly satisfies $y_0\leqslant y$. If $y-y_0>0$ then, by the maximality of the family $\{x_\alpha\}_{\alpha\in A}$, there is at least one index $\alpha\in A$ such that $x_\alpha\wedge(y-y_0)\neq 0$. This is however a contradiction since $0\leqslant x_\alpha\wedge(y-y_0)\leqslant x_\alpha\wedge(y-P_{x_\alpha}(y))=0$.

If Z is a separable subspace of X then Z is contained in the direct sum of at most a countable number of the X_α's, say $\{X_{\alpha_1}, X_{\alpha_2},\ldots\}$. By taking $x_{\alpha_0}=\sum_{n=1}^{\infty} x_{\alpha_n}/2^n\|x_{\alpha_n}\|$ we can replace the indices $\{\alpha_n\}_{n=1}^{\infty}$ by the single index α_0 and get that Z is contained in the ideal X_{α_0} generated by x_{α_0}. □

Proposition 1.a.9 shows that, for each of the bands X_α, we have $X=X_\alpha\oplus X_\alpha^\perp$, where $X_\alpha^\perp$ is the set of all $x\in X$ which are disjoint from every $y\in X_\alpha$. A band Y of a Banach lattice X is called a *projection band* if

$$X=Y\oplus Y^\perp,$$

where $Y^{\perp}=\{x\in X;\ |x|\wedge|y|=0$ whenever $y\in Y\}$. The set $Y^{\perp}$ is also a band of X, called the *polar* of Y. The (positive) projection P_Y from X onto Y, which vanishes on $Y^{\perp}$, is called a *band projection*. It is easily seen that in the case when X is a σ-complete Banach lattice and $\{P_x\}_{x\geq 0}$ are the projections associated to X as above then P_x is a band projection whose range is the band generated by x i.e. the set of all $u\in X$ for which $|u|\wedge y=0$ whenever $x\wedge y=0$. In general, there exist band projections other than the $\{P_x\}_{x\geq 0}$ (e.g. in $L_p(\mu)$, $1\leq p<\infty$, with μ being a non σ-finite measure). There are also bands which are not projection bands (e.g. the subspace of $C(0,1)$ consisting of all the functions which vanish on $[0,1/2]$). A simple characterization of those bands which are projection bands is given by the following result.

Proposition 1.a.10. *A band Y of a Banach lattice X is a projection band if and only if, for every $x\geq 0$ in X,*

$$P_Y(x)=\bigvee\{y\in Y;\ 0\leq y\leq x\}$$

exists in X. In this case, $x=P_Y(x)+P_{Y^{\perp}}(x)$, where $P_Y(x)\in Y$, $P_{Y^{\perp}}(x)\in Y^{\perp}$ and $P_{Y^{\perp}}(x)=\bigvee\{z\in Y^{\perp};\ 0\leq z\leq x\}$.

Proof. If Y is a projection band and $0\leq x=u+v$ with $u\in Y$ and $v\in Y^{\perp}$ then, for every $y\in Y$ satisfying $0\leq y\leq x$, we have $x-y=(u-y)+v$. Since $x-y\geq 0$ we get that $u-y\geq 0$, i.e. $u=\bigvee\{y\in Y;\ 0\leq y\leq x\}=P_Y(x)$. Similarly, we also get that $v=\bigvee\{z\in Y^{\perp};\ 0\leq z\leq x\}=P_{Y^{\perp}}(x)$.

Conversely, if $P_Y(x)$ exists in X for every $x\geq 0$ in X then, since Y is a band, we obtain that $P_Y(x)\in Y$. Put $w=x-P_Y(x)$ and take any $y\geq 0$ in Y. The element $w\wedge y$ also belongs to Y and $0\leq w\wedge y\leq x-P_Y(x)$. Therefore, $P_Y(x)+w\wedge y\leq x$ which, in view of the definition of $P_Y(x)$ as a supremum, implies that $w\wedge y=0$, i.e. $w\in Y^{\perp}$. □

It follows, in particular, that in a complete Banach lattice every band is the range of a positive contractive (i.e. of norm one) projection. If we require that the same holds for every ideal then we get a property which characterizes order continuous Banach lattices.

Proposition 1.a.11 (Ando [3]). *A Banach lattice X is order continuous if and only if every ideal of X is the range of a positive projection from X.*

Proof. Assume first that X is order continuous and thus also order complete by 1.a.8. Since every ideal in X is, by definition, a closed subspace we get that it is also a band and, by 1.a.10, there is a positive contractive projection on it.

Conversely, let X be a Banach lattice in which every ideal is the range of a positive projection. We prove first that every ideal Y in X is a projection band. Let P be a positive projection from X onto such an ideal Y and let $x\geq 0$ be an element in X. It is enough to show that $(x-Px)_+ \in Y^\perp$ since then

$$x = Px - (Px-x)_+ + (x-Px)_+ \in Y + Y^\perp.$$

If $(x-Px)_+ \notin Y^\perp$ then, since Y is an ideal, there would exist a $y \in Y$ with $0 < y \leqslant (x-Px)_+$. Then $y = Py \leqslant Px$ and hence $y \leqslant (x-Px)_+ \leqslant x - y$, i.e. $2y \leqslant x$. An easy induction argument shows that $ny \leqslant x$, for every integer n, and this clearly leads to a contradiction.

Let now $\{x_\alpha\}_{\alpha \in A}$ be a downward directed set in X with $\bigwedge_{\alpha \in A} x_\alpha = 0$. We may clearly assume that $x_\alpha \leqslant x$, for some $x \in X$ and all $\alpha \in A$. Let $\varepsilon > 0$ and, for $\alpha \in A$, let P_α be the band projection on the ideal generated by $(x_\alpha - \varepsilon x)_+$ (which is a projection band by the first part of the proof). Since $(I-P_\alpha)(x_\alpha - \varepsilon x) = -(x_\alpha - \varepsilon x)_- \leqslant 0$ it follows that

$$\|x_\alpha\| \leqslant \|P_\alpha x_\alpha\| + \|(I-P_\alpha)x_\alpha\| \leqslant \|P_\alpha x\| + \varepsilon\|x\|, \quad \alpha \in A.$$

Thus, in order to prove that $\|x_\alpha\| \downarrow 0$, it suffices to show that $\|P_{\alpha_0} x\| \leqslant \varepsilon$ for some α_0. Put $y_\alpha = (I-P_\alpha)x$, $\alpha \in A$. Since $x_\alpha \geqslant P_\alpha x_\alpha \geqslant \varepsilon P_\alpha x$ it follows that $\bigwedge_{\alpha \in A} P_\alpha x$ exists and is equal to $0 = \bigwedge_{\alpha \in A} x_\alpha$ and hence, $x = \bigvee_{\alpha \in A} y_\alpha$. Consequently, by the first part of the proof, x belongs to the ideal generated by $\{y_\alpha\}_{\alpha \in A}$. Using the fact that this set is directed upward it follows that there is a $z \in X$, a constant M and an $\alpha_0 \in A$ so that $\|x - z\| \leqslant \varepsilon$ and $0 \leqslant z \leqslant M y_{\alpha_0}$. Since $x \wedge z \leqslant x \wedge M y_{\alpha_0} = (P_{\alpha_0} x + y_{\alpha_0}) \wedge M y_{\alpha_0} \leqslant y_{\alpha_0}$ we get that

$$\|P_{\alpha_0} x\| = \|x - y_{\alpha_0}\| \leqslant \|x - x \wedge z\| \leqslant \|x - z\| \leqslant \varepsilon. \quad \square$$

We introduce next a notion which was originally considered in connection with multiplicity theory for spectral operators.

Definition 1.a.12. (i) A family $\mathscr{B}$ of commuting bounded linear projections on a Banach space X is called a *Boolean algebra* (B.A.) *of projections* if

$$P, Q \in \mathscr{B} \Rightarrow PQ,\ P+Q-PQ \in \mathscr{B}.$$

A B.A. of projections $\mathscr{B}$ will be ordered by $P \leqslant Q \Leftrightarrow P = PQ$. With this order $\mathscr{B}$ is a lattice with $P \wedge Q = PQ$ and $P \vee Q = P + Q - PQ$.

(ii) A B.A. of projections $\mathscr{B}$ is said to be *σ-complete* if, for every sequence $\{P_n\}_{n=1}^\infty$ in $\mathscr{B}$, $\bigvee_{n=1}^\infty P_n$ and $\bigwedge_{n=1}^\infty P_n$ exist in $\mathscr{B}$ and, moreover,

$$\left(\bigvee_{n=1}^\infty P_n\right) X = \overline{\operatorname{span}}\left\{\bigcup_{n=1}^\infty P_n X\right\}, \quad \left(\bigwedge_{n=1}^\infty P_n\right) X = \bigcap_{n=1}^\infty P_n X.$$

(iii) With X and $\mathscr{B}$ as in (i) and $x \in X$ the subspace $\mathcal{M}(x)=\overline{\text{span}}\{Px;\, P \in \mathscr{B}\}$ of X is called the *cyclic space* generated by x. It is the smallest closed linear subspace of X which contains the vector x and is invariant under $\mathscr{B}$. If there exists a vector x such that $X=\mathcal{M}(x)$ then x is called a *cyclic vector* and X a cyclic space (with respect to $\mathscr{B}$).

A simple example of a B.A. of projections is the family $\mathscr{E}$ of all multiplication operators on $X=L_p(\mu)$, $1 \leqslant p \leqslant \infty$, by characteristic functions of measurable sets. If p is finite this B.A. is σ-complete. It is also evident that the space X is cyclic with respect to $\mathscr{E}$ if and only if μ is a σ-finite measure. In this case one can choose as a cyclic vector in X any function $f \in L_p(\mu)$ for which $\mu(\{t; f(t)=0\})=0$. If X is a space with an unconditional basis $\{x_n\}_{n=1}^\infty$ the set of all natural projections P_σ; σ being a subset of the integers, is again a σ-complete B.A. of projections and X is always a cyclic space. As a cyclic vector in this case we can take any vector $x=\sum_{n=1}^\infty a_n x_n$, with $a_n \neq 0$ for all n. The previous examples are just two particular cases of the following result of general character [44], [63].

Theorem 1.a.13. *Let X be an order continuous Banach lattice with a weak unit $e>0$ (in particular, X can be any separable σ-complete Banach lattice). Then the family $\{P_x\}_{x \geqslant 0}$ of projections on X forms a σ-complete B.A. of projections and $X=\mathcal{M}(e)$ is a cyclic space with respect to this B.A.*

Proof. The first part of the statement is proved by direct verification if we observe that (i) P_e acts as the identity operator on X, (ii) for $x, y \geqslant 0$, $P_x P_y = P_{x \wedge y}$ (iii) for $x \geqslant y \geqslant 0$ we have $P_x - P_y = P_{x - P_y(x)}$ and (iv) the σ-completeness of $\{P_x\}_{x \geqslant 0}$ follows from the σ-completeness and σ-order continuity of X. In order to prove that $X=\mathcal{M}(e)$, fix $x \geqslant 0$ in X and, for any real $\lambda \geqslant 0$, put $x(\lambda)=P_{(\lambda e - x)_+}(e)$. If $\lambda \leqslant \eta$ then $(\lambda e - x)_+ \leqslant (\eta e - x)_+$ which implies that $x(\lambda) \leqslant x(\eta)$. This means that $x(\cdot)$ is a non-decreasing map from $[0, \infty)$ into the positive cone of X. As in the case of the scalar Riemann–Stieltjes integral we can consider the integral $\int_0^\infty \lambda\, dx(\lambda)$ with respect to the X-valued measure $dx(\cdot)$. This integral should be understood as the limit in norm, if it exists, of sums of the form $\sum_{i=1}^n \lambda_i (x(\lambda_i) - x(\lambda_{i-i}))$. We shall prove that $x=\int_0^\infty \lambda\, dx(\lambda)$ and this will imply that $x \in \mathcal{M}(e)$. For this we need the following lemma.

Lemma 1.a.14. *For every element $0 \leqslant z \in X$ which satisfies $z=P_z(e)$ (or equivalently, $z \wedge (e-z)=0$) the following two assertions are true:*

(i) *If $z \leqslant x(\lambda)$ for some $\lambda \geqslant 0$ then $P_z(x) \leqslant \lambda z$.*
(ii) *If $z \leqslant e - x(\lambda)$ for some $\lambda \geqslant 0$ then $\lambda z \leqslant P_z(x)$.*

Proof. Observe that $z \leqslant x(\lambda)$ implies that

$$0 \leqslant (x-\lambda e)_+ \wedge z \leqslant (x-\lambda e)_+ \wedge P_{(\lambda e - x)_+}(e) = (x-\lambda e)_+ \wedge \bigvee_{n=1}^\infty (n(\lambda e - x)_+ \wedge e) = 0.$$

It follows that $P_z((x-\lambda e)_+)=0$ and, since $x \leqslant (x-\lambda e)_+ + \lambda e$, we also get that $P_z(x) \leqslant P_z((x-\lambda e)_+) + \lambda P_z(e) = \lambda z$. The proof of (ii) is similar. $\square$

In order to complete the *proof of* 1.a.13 we fix $\varepsilon>0$ and $\Lambda<\infty$, and let $\Delta=\{0=\lambda_0<\lambda_1<\ldots<\lambda_n=\Lambda\}$ be a partition of the interval $[0, \Lambda]$ such that $\max_{1\leqslant i\leqslant n}(\lambda_i-\lambda_{i-1})<\varepsilon$. Put $z_i=x(\lambda_i)-x(\lambda_{i-1})$ and observe that $z_i=P_{z_i}(e)$ and $z_i \leqslant e-x(\lambda_{i-1})$. Since we also have $z_i \wedge z_j=0$, whenever $i\neq j$, it follows from 1.a.14(ii) that

$$s(\Delta)=\sum_{i=1}^{n}\lambda_{i-1}z_i \leqslant \sum_{i=1}^{n} P_{z_i}(x)=P_{\sum_{i=1}^{n} z_i}(x)=P_{x(\Lambda)}(x) .$$

On the other hand, $z_i \leqslant x(\lambda_i)$ and therefore, by 1.a.14(i),

$$P_{x(\Lambda)}(x)=\sum_{i=1}^{n} P_{z_i}(x) \leqslant \sum_{i=1}^{n} \lambda_i z_i = S(\Delta) .$$

However, $S(\Delta)-s(\Delta)\leqslant \varepsilon x(\Lambda) \leqslant \varepsilon e$ which implies that $\|S(\Delta)-s(\Delta)\| \leqslant \varepsilon\|e\|$. Thus, by fixing Λ and letting $\varepsilon\to 0$, we get that $\int_0^{\Lambda} \lambda\, dx(\lambda)=P_{x(\Lambda)}(x)$ for every Λ. Since X is σ-complete and $x(\Lambda)\leqslant e$ for every Λ we obtain that $x(\infty)=\lim_{\Lambda\to\infty} x(\Lambda)$ exists in X. But $e-x(\infty)$ satisfies the condition of 1.a.14 and, clearly, $e-x(\infty)\leqslant e-x(\Lambda)$ for every Λ. Thus, again by 1.a.14(ii), it follows that $\Lambda(e-x(\infty)) \leqslant P_{(e-x(\infty))}(x)$, i.e. $\Lambda\|e-x(\infty)\| \leqslant \|P_{(e-x(\infty))}(x)\|$. Since Λ is arbitrary we get that $x(\infty)=e$ i.e. $x(\Lambda)\to e$ as $\Lambda\to\infty$. Hence, by the σ-completeness of the B.A. of projections $\{P_x\}_{x\geqslant 0}$, we conclude that $P_{x(\Lambda)}(x)\to P_e(x)=x$ as $\Lambda\to\infty$. Consequently, $\int_0^{\infty}\lambda\, dx(\lambda)=x$. $\square$

Remarks. 1. A variant of Theorem 1.a.13 holds for Banach lattices without a weak unit. By combining 1.a.13 with 1.a.9, we get that, for any order continuous Banach lattice X, the projections $\{P_x\}_{x\geqslant 0}$ form a σ-complete B.A. of projections on every cyclic subspace of X and X is an (not necessarily countable) unconditional direct sum of cyclic subspaces. In general, X itself need not be a cyclic space and $\{P_x\}_{x\geqslant 0}$ need not be a B.A. on X because it could lack the property of complementation (take e.g. an $L_p(\mu)$ space, $1\leqslant p\leqslant\infty$ where μ is not a σ-finite measure).

2. If X is complete but not order continuous it is not true in general that $\{P_x\}_{x\geqslant 0}$ is a σ-complete B.A. of projections. For instance, let P_n be the natural projection in l_∞ whose range is the closed linear span of the first n unit vectors. Then $\bigvee_{n=1}^{\infty} P_n$ exists in $\{P_x\}_{x\geqslant 0}$ and is equal to the identity operator. However,

$$\left(\bigvee_{n=1}^{\infty} P_n\right) l_\infty = l_\infty \neq c_0 = \overline{\operatorname{span}}\left\{\bigcup_{n=1}^{\infty} P_n l_\infty\right\} .$$

To conclude this section we would like to mention without giving the proof that there is also a converse result to 1.a.13. *Every cyclic space with respect to a σ-complete B.A. of projections can be ordered and given an equivalent norm with which it becomes an order continuous Banach lattice with a weak unit.* This fact is due essentially to Bade [4] (see H. H. Schaefer [118] Section V.3 for a concise presentation.) In this paper Bade showed that if $\mathcal{M}(x)$ is a cyclic space then, to every element y in $\mathcal{M}(x)$, there corresponds a function $f(\omega)$ on some fixed measure space so that y is the integral of f with respect to some $\mathcal{M}(x)$-valued measure. The space of all the functions which correspond in this manner to vectors in $\mathcal{M}(x)$ is a vector lattice which satisfies (with respect to the norm induced by $\mathcal{M}(x)$) axiom (iv′) of the beginning of this section. Thus, after a suitable renorming, Y becomes a Banach lattice. That the space Y in Bade's representation is order continuous (i.e. σ-order complete and σ-order continuous) is easily verified by using standard results from measure theory. As a weak unit for Y we can take, of course, the function which corresponds to x (namely, the function identically equal to one). In the next section (cf. 1.b.14), we shall prove a representation theorem for order continuous lattices with a weak unit which is essentially equivalent to Bade's representation.

b. Concrete Representation of Banach Lattices

Many Banach lattices which appear in the literature are, in fact, spaces of functions; e.g. spaces of continuous functions on some compact Hausdorff space or spaces of measurable functions on a measure space (Ω, Σ, μ), with the order defined by $f \leqslant g \Leftrightarrow f(\omega) \leqslant g(\omega)$, for μ-a.e. $\omega \in \Omega$. There are many cases when abstract Banach lattices can be represented as concrete lattices of functions. Such representation theorems are very convenient since they facilitate, e.g., the application of many results of measure theory to the study of Banach lattices.

The best known theorems in this direction are those of S. Kakutani [63], [64] on the concrete representation of the so-called abstract L_p and M spaces. We present first these results. The representation theorems of Kakutani are followed by several results (of an isometric as well as of an isomorphic nature) which give joint characterizations of the abstract L_p spaces and M spaces among general Banach lattices. We then prove a functional representation theorem for general order continuous Banach lattices with a weak unit. Such lattices can be represented as suitable spaces of measurable functions on a measure space (Köthe function spaces). The section ends with a brief discussion of general properties of Köthe function spaces.

Definition 1.b.1. (i) Let $1 \leqslant p < \infty$. A Banach lattice X for which $\|x+y\|^p = \|x\|^p + \|y\|^p$, whenever $x, y \in X$ and $x \wedge y = 0$, is called an *abstract L_p space.* (ii) A Banach lattice X for which $\|x+y\| = \max(\|x\|, \|y\|)$, whenever $x, y \in X$ and $x \wedge y = 0$, is called an *abstract M space.*

It is obvious that every $L_p(\mu)$ space is an abstract L_p space if $p<\infty$ or an abstract M space if $p=\infty$. The converse is also true if $p<\infty$ (cf. S. Kakutani [63]).

Theorem 1.b.2. *An abstract L_p space X, $1\leqslant p<\infty$, is order isometric to an $L_p(\mu)$ space over some measure space (Ω, Σ, μ). If X has a weak unit then μ can be chosen to be a finite measure.*

Proof. By 1.a.5 and 1.a.7, X is σ-complete and σ-order continuous. Thus, by 1.a.13 and 1.a.9 (see also Remark 1 following 1.a.13), X is an unconditional direct sum of mutually disjoint cyclic spaces X_α, $\alpha\in A$ (with respect to the family $\mathscr{B}$ of projections $\{P_x\}_{x\geqslant 0}$). By the p-additivity of the norm we get that X is actually a direct sum in the l_p sense of cyclic spaces, i.e. $X=\left(\sum_{\alpha\in A}\oplus X_\alpha\right)_p$. It is therefore sufficient to show that each X_α is order isometric to some $L_p(\mu_\alpha)$ space.

For simplicity of notation we shall assume that X has a weak unit $e>0$ with $\|e\|=1$, i.e. that X itself is a cyclic space $\mathscr{M}(e)$ with respect to the B.A. of projections $\mathscr{B}=\{P_x\}_{x\geqslant 0}$. We want to construct a probability space (Ω, Σ, μ) and an order isometry T of X onto $L_p(\Omega, \Sigma, \mu)$ such that Te is the function identically equal to one. This is achieved by using the following well-known representation theorem, due to M. H. Stone [121].

Theorem 1.b.3. *Every Boolean algebra with a unit is isomorphic to the Boolean algebra of all simultaneously open and closed subsets of a totally disconnected compact Hausdorff space.*

A proof of this classical result can be found, e.g. in [32] I.2.1.

In view of 1.b.3 we shall identify $\mathscr{B}$ with the B.A. Σ_0 of all simultaneously open and closed subsets of a totally disconnected space and, if a projection $P_x\in\mathscr{B}$ corresponds to a set $\sigma\in\Sigma_0$, we shall write P_σ instead of P_x. By putting, $\mu(\sigma)=\|P_\sigma(e)\|^p$ for $\sigma\in\Sigma_0$ we clearly define an additive measure on (Ω, Σ_0). Actually, the measure μ is vacuously σ-additive on (Ω, Σ_0) since, by the fact that every set in Σ_0 is both open and compact, no set in Σ_0 can be expressed as a union of infinitely many mutually disjoint non-void sets from Σ_0. Thus, by the Caratheodory extension theorem (e.g. cf. [32] III.5.8), μ has a σ-additive extension, still denoted by μ, to the σ-field Σ of subsets of Ω generated by Σ_0. This extension has the additional property that every $\delta\in\Sigma$ differs from some set $\sigma\in\Sigma_0$ by a set of μ-measure zero (i.e. that, up to sets of μ-measure zero, Σ_0 is a σ-field). Indeed, suppose that some set $\delta\in\Sigma$ is the union of an increasing sequence $\{\sigma_n\}_{n=1}^\infty$ of elements of Σ_0. Then, by the σ-completeness and σ-order continuity of X and 1.a.13, $\bigvee_{n=1}^\infty P_{\sigma_n}=P_\sigma$ exists in $\mathscr{B}$ and $P_\sigma(e)=\lim_{n\to\infty} P_{\sigma_n}(e)$. It follows that $\sigma\supset\delta$ (since $\sigma\supset\sigma_n$ for all n) and

$$\mu(\sigma\sim\delta)=\lim_{n\to\infty}\mu(\sigma\sim\sigma_n)=\lim_{n\to\infty}\|P_\sigma(e)-P_{\sigma_n}(e)\|^p=0\,.$$

In order to complete the proof of 1.b.2 we define an order isometry T from X onto $L_p(\Omega, \Sigma, \mu)$ in the following way: if $\{\sigma_j\}_{j=1}^m$ are arbitrary disjoint sets in Σ_0 and $\{a_j\}_{j=1}^m$ are arbitrary scalars then we put

$$T\left(\sum_{j=1}^m a_j P_{\sigma_j}(e)\right)=\sum_{j=1}^m a_j \chi_{\sigma_j}.$$

Since $\{P_{\sigma_j}(e)\}_{j=1}^m$ are mutually disjoint elements of X it follows that

$$\left\|\sum_{j=1}^m a_j P_{\sigma_j}(e)\right\|^p=\sum_{j=1}^m |a_j|^p \|P_{\sigma_j}(e)\|^p=\sum_{j=1}^m |a_j|^p \mu(\sigma_j),$$

i.e. T extends uniquely to an order isometry of $X=\mathcal{M}(e)$ onto $L_p(\Omega, \Sigma, \mu)$. (Here we have used the fact that elements of the form $\sum_{j=1}^k b_j P_{x_j}(e)$ are dense in X since X is a cyclic space and each expression of the form $\sum_{j=1}^k b_j P_{x_j}(e)$ can be written as $\sum_{j=1}^m a_j P_{y_j}(e)$, with $\{P_{y_j}\}_{j=1}^m$ being mutually disjoint projections in $\mathscr{B}$). □

Corollary 1.b.4. *Any closed sublattice of an $L_p(\mu)$ space, $1 \leqslant p < \infty$, is order isometric to an $L_p(\nu)$ space, for a suitable measure ν.*

Before stating the result on the representation of the abstract M spaces we recall the following classical result of S. Kakutani (and also M. H. Stone) on the structure of sublattices of $C(K)$ spaces (see e.g. the proof of [32] IV.6.16).

Theorem 1.b.5. *Let X be a closed linear subspace of a $C(K)$ space. Let $\mathscr{F}$ be the collection of all the triples (k_1, k_2, λ) with $k_1, k_2 \in K$ and $\lambda \geqslant 0$ so that*

(*) $\qquad f(k_1)=\lambda f(k_2)$,

for all $f \in X$. Then X is a sublattice of $C(K)$ if and only if X contains every function $f \in C(K)$ which satisfies (*) for all triples (k_1, k_2, λ) in $\mathscr{F}$.

We also point out that in $C(K)$ spaces the function $f \equiv 1$ plays a special role which is clarified by the following definition. An element $e > 0$ of a Banach lattice X is said to be a *strong unit* of X provided that $\|x\| \leqslant 1$ if and only if $|x| \leqslant e$. The space c_0 is an example of an abstract M space without a strong unit which has, however, a weak unit.

Theorem 1.b.6. *Any abstract M space X is order isometric to a sublattice of a $C(K)$ space, for some compact Hausdorff space K. If, in addition, X has a strong unit then X is order isometric to a $C(K)$ space.*

Proof. We show first that X^* is an abstract L_1 space. Let x_1^* and x_2^* be two positive elements in X^* with $x_1^* \wedge x_2^* = 0$. Fix $\varepsilon > 0$ and, for $i = 1, 2$, let x_i be positive elements of norm one in X so that $x_i^*(x_i) \geq (1-\varepsilon)\|x_i^*\|$. Since $x_1^* \wedge x_2^* = 0$ there exist $u_i, v_i \geq 0$ in X so that

$$x_i = u_i + v_i, \qquad x_1^*(u_i) \leq \varepsilon\|x_1^*\|, \qquad x_2^*(v_i) \leq \varepsilon\|x_2^*\|, \quad i = 1, 2\,.$$

Clearly, $x_1^*(v_1) \geq (1-2\varepsilon)\|x_1^*\|$, $x_2^*(u_2) \geq (1-2\varepsilon)\|x_2^*\|$. Put $w_1 = (v_1 - u_2)_+$, $w_2 = (u_2 - v_1)_+$ and notice that $w_1 \wedge w_2 = 0$, $\|w_i\| \leq 1$ and $x_i^*(w_i) \geq (1-3\varepsilon)\|x_i^*\|$, $i = 1, 2$. Since X is an M space it follows that $\|w_1 + w_2\| \leq 1$, and thus

$$\|x_1^* + x_2^*\| \geq (x_1^* + x_2^*)(w_1 + w_2) \geq x_1^*(w_1) + x_2^*(w_2) \geq (1-3\varepsilon)(\|x_1^*\| + \|x_2^*\|)\,.$$

Since $\varepsilon > 0$ was arbitrary we get that X^* is an L_1 space. Hence, by 1.b.2, X^* is order isometric to $L_1(\mu)$, for some μ, and X^{**} is therefore isometric to $L_\infty(\mu)$. *The space $L_\infty(\mu)$ is, in turn, order isometric to a $C(K)$ space, for some compact Hausdorff K.* This well known fact can be proved in various ways; the most common approach is by using that $L_\infty(\mu)$ is a commutative C^* algebra (cf. e.g. [33] IX.3.7). Since, by 1.a.2, the canonical image of X in X^{**} is a sublattice of X^{**} the first part of the theorem is already proved.

Suppose now that X has a strong unit e. Then, for every $0 \leq x^* \in X^*$, we have $\|x^*\| = \sup\{x^*(x);\ 0 \leq x \in X,\ \|x\| \leq 1\} = x^*(e)$. This shows that the function $f \in L_1(\mu)$, which corresponds to x^* (under the order isometry between X^* and $L_1(\mu)$), satisfies $\int f\, d\mu = x^*(e)$, i.e. e, considered as an element of $X^{**} = L_\infty(\mu) = C(K)$, is the function identically equal to one. It follows from 1.b.5 that the sublattice X of $C(K)$, which contains the function identically equal to one, is obtained in the following way: the set K is divided into a certain family of equivalence classes $\{K_\beta\}_{\beta \in H}$ so that X consists exactly of all those functions in $C(K)$ which are constant on each equivalence class. Thus, if H is the compact Hausdorff space obtained from K by identifying each class K_β to a point $\beta \in H$, X is order isometric to $C(H)$. □

Remarks. 1. The original definition of an abstract L_1 or M space given by S. Kakutani was slightly different from that presented in 1.b.1 in the sense that he required that $\|x+y\| = \|x\| + \|y\|$, respectively $\|x \vee y\| = \max(\|x\|, \|y\|)$, be valid *for all x and y in the positive cone of X.*

2. Y. Benyamini has proved in [8] that *every separable abstract M space is isomorphic to a $C(K)$ space* though, obviously, not necessarily isometric to a $C(K)$ space. He has also shown that there exist *non-separable* abstract M spaces which are not isomorphic to any $C(K)$ space [9].

The classes of lattices introduced in 1.b.1, and concretely represented in 1.b.2 and 1.b.6, are the most important lattices which appear in analysis. From the point of view of Banach space theory itself their significance stems also from the fact that these lattices are the only ones which admit an abstract characterization similar in spirit to 1.b.1. We shall present next several results of an isometric as well as of an isomorphic nature which clarify this point. The first result proved in this direction is due to Bohnenblust [13]. This theorem, which was proved by Bohnenblust at about the same time in which Kakutani proved the representation theorems 1.b.2 and 1.b.6, inspired all the subsequent results in this direction as well as the analogous results in the setting of spaces with an unconditional basis (cf. section I.2.a).

Theorem 1.b.7. *Let X be a Banach lattice of dimension at least 3 for which there exists a function $F(s, t)$ (defined on $\{(s, t); s \geq 0, t \geq 0\}$) such that, for all $x, y \in X$ with $|x| \wedge |y| = 0$, we have $\|x+y\| = F(\|x\|, \|y\|)$. Then X is either an abstract L_p space, for some $1 \leq p < \infty$, or an abstract M space.*

(*Another way to express the assumption on X in 1.b.7 is the following: If $x_1, x_2, y_1, y_2 \in X$ satisfy $|x_1| \wedge |y_1| = |x_2| \wedge |y_2| = 0$, $\|x_1\| = \|x_2\|$ and $\|y_1\| = \|y_2\|$ then $\|x_1 + y_1\| = \|x_2 + y_2\|$.*)

Proof. It is easily verified that the axioms on the norm in a Banach lattice force the function F to have the following properties.

(1) $F(0, 1) = 1$
(2) $F(s, t) = F(t, s), \quad s, t \geq 0,$
(3) $F(rs, rt) = rF(s, t), \quad r, s, t \geq 0,$
(4) $F(r, F(s, t)) = F(F(r, s), t), \quad r, s, t \geq 0,$
(5) $F(s_1, t_1) \leq F(s_2, t_2), \quad 0 \leq s_1 \leq s_2, \quad 0 \leq t_1 \leq t_2,$
(6) $F(t, 1-t) \leq 1, \quad 0 \leq t \leq 1.$

We shall show that (1)–(6) imply that $F(s, t)$ is either $(s^p + t^p)^{1/p}$, for some $1 \leq p < \infty$, or $\max\{s, t\}$. Define numbers λ_n, $n = 1, 2, \ldots$ inductively by $\lambda_1 = 1$ and $\lambda_{n+1} = F(1, \lambda_n)$. Evidently, $\{\lambda_n\}_{n=1}^{\infty}$ is a non-decreasing sequence. By induction on m we get that $\lambda_{n+m} = F(\lambda_n, \lambda_m)$, $n, m = 1, 2, \ldots$ Indeed, by (2) and (4),

$$\lambda_{n+m+1} = F(1, \lambda_{n+m}) = F(1, F(\lambda_n, \lambda_m)) = F(F(1, \lambda_m), \lambda_n) = F(\lambda_n, \lambda_{m+1}).$$

Another simple induction on m proves that $\lambda_n \lambda_m = \lambda_{nm}$, $n, m = 1, 2, \ldots$ Indeed, by (3) and the identity proved above,

$$\lambda_n \lambda_{m+1} = \lambda_n F(1, \lambda_m) = F(\lambda_n, \lambda_n \lambda_m) = F(\lambda_n, \lambda_{nm}) = \lambda_{n(m+1)}.$$

If $\lambda_2 = F(1, 1) = 1$ then, by (1), (3) and (5), $F(s, t) = \max\{s, t\}$. If $\lambda_2 > 1$ we observe first that the monotonicity of $\{\lambda_n\}_{n=1}^{\infty}$ and the relation $\lambda_n \lambda_m = \lambda_{m+n}$ imply that $\log \lambda_n / \log n$ is independent of n (use the fact that if h and $k(h)$ are such that $m^k \leq n^h \leq m^{k+1}$ then $\lambda_m^k \leq \lambda_n^h \leq \lambda_m^{k+1}$, and let $h \to \infty$). Hence, $\lambda_n = n^{1/p}$ for some p which, by (6), must satisfy $p \geq 1$. Thus,

$$F(n^{1/p}, m^{1/p}) = F(\lambda_n, \lambda_n) = \lambda_{n+m} = (n+m)^{1/p}, \quad n, m = 1, 2, \ldots,$$

and consequently, by (3) and (5), $F(s, t) = (s^p + t^p)^{1/p}$ for $s, t \geq 0$. □

The next theorem, due to Ando [3], characterizes the L_p spaces and some M spaces by an intrinsic Banach space property.

Theorem 1.b.8. *Let X be a Banach lattice of dimension ≥ 3. Then X is order isometric to $L_p(\mu)$, for some $1 \leq p < \infty$ and measure μ, or to $c_0(\Gamma)$, for some index set Γ, if and only if there is a contractive positive projection from X onto any (closed) sublattice of it.*

The proof of 1.b.8 will be broken up into three lemmas.

Lemma 1.b.9. *In an $L_p(\mu)$ space, $1 \leq p < \infty$, there exists a contractive positive projection onto every sublattice.*

Lemma 1.b.10. *An abstract M space is an order continuous lattice if and only if it is order isometric to $c_0(\Gamma)$, for some index set Γ.*

Lemma 1.b.11. *Let X be a Banach lattice of dimension 3 so that there is a contractive projection from X onto any of its sublattices of dimension 2. Then X is order isometric to l_p^3, for some $1 \leq p \leq \infty$.*

Let us first show that 1.b.8 is indeed a consequence of 1.b.9, 1.b.10 and 1.b.11.

Proof of 1.b.8. The "only if" part of 1.b.8 for $1 \leq p < \infty$ is 1.b.9. The "only if" part for $c_0(\Gamma)$ is proved as follows. By 1.b.10, every sublattice Y of $c_0(\Gamma)$ is the closed linear span of a set $\{y_\lambda\}_{\lambda \in \Lambda}$ of vectors in $c_0(\Gamma)$ having the form

$$y_\lambda = \sum_{\gamma \in \Gamma_\lambda} a_\gamma e_\gamma, \quad \lambda \in \Lambda,$$

where $a_\gamma > 0$, $\{e_\gamma\}_{\gamma \in \Gamma}$ are the unit vector basis of $c_0(\Gamma)$ and $\{\Gamma_\lambda\}_{\lambda \in \Lambda}$ are mutually disjoint subsets of Γ. For every λ, let γ_λ be such that $a_{\gamma_\lambda} = \|y_\lambda\|$. A positive contractive projection from X onto Y is given by the formula

$$P\Big(\sum_{\gamma \in \Gamma} b_\gamma e_\gamma\Big) = \sum_{\lambda \in \Lambda} b_{\gamma_\lambda} y_\lambda / \|y_\lambda\|,$$

whenever $\sum_{\gamma \in \Gamma} b_\gamma e_\gamma \in c_0(\Gamma)$.

For the proof of the "if" part of 1.b.8 we note first that, by 1.a.11, X must be an order continuous lattice. By 1.b.11, there is, for any three disjoint positive vectors $\{x_1, x_2, x_3\}$ of norm one in X, a $1 \leq p \leq \infty$ so that $\left\|\sum_{i=1}^3 a_i x_i\right\| = \left(\sum_{i=1}^3 |a_i|^p\right)^{1/p}$ if $p < \infty$ (respectively, $\max_{1 \leq i \leq 3} |a_i|$ if $p = \infty$). We show that this p does not depend on the triple. To this end we notice first, by an easy induction on n, that, for every n-tuple $\{y_i\}_{i=1}^n$ of disjoint positive vectors of norm one in X, there is a $1 \leq p \leq \infty$ such that $\left\|\sum_{i=1}^n a_i y_i\right\| = \left(\sum_{i=1}^n |a_i|^p\right)^{1/p}$ (prove first that all the triples $y_{i_1}, y_{i_2}, y_{i_3}$, $1 \leq i_1 \leq i_2 \leq i_3 \leq n$, have a common p, and then use the induction hypothesis). If now $\{x_j\}_{j=1}^3$ and $\{u_j\}_{j=1}^3$ are any two triples of disjoint positive vectors of norm one in X then, for every $\varepsilon>0$, there is a n-tuple $\{y_i\}_{i=1}^n$ of disjoint positive vectors in X so that the distance of each of the six given vectors from $[y_i]_{i=1}^n$ is less than ε. (By 1.a.13, we may assume that X is a cyclic space $\mathscr{M}(e)$ and we can thus take $y_i=P_ie$, $1\leqslant i\leqslant n$, where the $\{P_i\}_{i=1}^n$ are suitable disjoint projections in the underlying Boolean algebra.) This approximation argument clearly shows that the p associated to $\{x_j\}_{j=1}^3$ is equal to the one associated to $\{u_j\}_{j=1}^3$. If the p, which is common to all triples in X, is finite then, by 1.b.2, X is order isometric to $L_p(\mu)$. If this common p is ∞ then, by 1.b.6 and 1.b.10, X is order isometric to $c_0(\Gamma)$. □

Proof of 1.b.9. Let Y be a sublattice of $X=L_p(\Omega, \Sigma, \mu)$, $1\leqslant p<\infty$. For every finite set $B=\{f_i\}_{i=1}^n$ of disjoint positive vectors of norm one in Y, there is a positive contractive projection P_B from X onto $[f_i]_{i=1}^n=l_p^n$, defined by

$$P_B f=\sum_{i=1}^n\left(\int_\Omega f(\omega)f_i(\omega)^{p-1}\,d\mu\right)f_i,\quad f\in X.$$

We partially order the set $\mathscr{B}$ of finite sets of disjoint positive vectors of norm one in Y by $\{y_i\}_{i=1}^n<\{z_j\}_{j=1}^m$ if $[y_i]_{i=1}^n\subset[z_j]_{j=1}^m$. Assume first that $1<p<\infty$. For every $f\in X$ and every $B\in\mathscr{B}$, the vector P_Bf belongs to the w compact subset $\{y;\ \|y\|\leqslant\|f\|\}$ in Y. Hence, by Tychonoff's theorem, the net $\{P_B\}_{B\in\mathscr{B}}$ of operators from X to Y, has a subnet which converges to some limit point P (in the topology of pointwise convergence on X taking in Y the w topology). It is clear that P is a positive contractive projection from X onto Y.

If $p=1$ we note first that, by 1.b.2, Y is an $L_1(\nu)$ space for a suitable positive measure ν on some compact Hausdorff space K(K is the one point compactification of the union of a set $\{K_\gamma\}_{\gamma\in\Gamma}$ of disjoint compact sets so that $\nu_{|K_\gamma}$ is finite for every γ). We can thus consider Y in a canonical way as a subspace of $C(K)^*$. There is a positive contractive projection P_0 from $C(K)^*$ onto Y; we simply take as $P_0\eta$, where η is a finite Borel measure on K, its absolutely continuous part with respect to ν. We return to the argument given above for $p>1$. We consider now each P_B as an operator from X into $C(K)^*$. Since the unit ball in $C(K)^*$ is w^* compact there is a subnet of $\{P_B\}_{B\in\mathscr{B}}$ which converges pointwise on X (taking in $C(K)^*$ the w^* topology) to a map P from X into $C(K)^*$. Clearly, P is a positive operator of norm one whose restriction to Y is the identity. Hence, P_0P is a positive contractive projection from X onto Y. □

Before proving 1.b.10 let us introduce the following notion. An element $x>0$ is called an *atom* of a Banach lattice X if $\{y\in X;\ 0\leqslant y\leqslant x\}=\{\lambda x;\ 0\leqslant\lambda\leqslant 1\}$. It is easily verified that in a σ-complete Banach lattice an element $x>0$ is an atom if and only if x cannot be written as $x=y+z$ with $y, z\neq 0$ and $y\wedge z=0$.

Proof of 1.b.10. Let X be an order continuous M space. Let $\{x_\gamma\}_{\gamma\in\Gamma}$ be the set of all the atoms of X of norm one. By 1.a.8, X is order complete and thus it is clear that $Y=[x_\gamma]_{\gamma\in\Gamma}$ is a band of X which is order isometric to $c_0(\Gamma)$. We have to prove that $Y=X$ i.e. that $Y^\perp$ consists of the 0 element only. Assume that $y>0$ is an element of norm one in $Y^\perp$. Since no element of $Y^\perp$ is an atom of X and X is an M

space, every $z>0$ in $Y^{\perp}$ can be written as $z=u+v$ with $u \wedge v=0$, $\|u\|=\|z\|$, $v\neq 0$. Hence, if F is a maximal (with respect to inclusion) downward directed chain of elements $\{z_\alpha\}_{\alpha\in A}$ satisfying $0\leqslant z_\alpha\leqslant y$, $\|z_\alpha\|=1$ then F does not have a g.l.b. This contradicts the order continuity of X. □

Proof of 1.b.11 ([3], cf. also [71]). We start with the trivial observation that, by the convexity of the norm in a Banach space, whenever v and w are two vectors with $\|v\|<1$ and $w\neq 0$, there is a unique positive t so that $\|v+tw\|=1$.

Let x, y, z be the three disjoint positive vectors of norm one which span X. For every $0\leqslant\alpha<1$ and $0\leqslant\theta<2\pi$, let $u_\theta=(y\cos\theta+z\sin\theta)/\|y\cos\theta+z\sin\theta\|$ and let $r(\alpha,\theta)$ be the unique positive number so that $\|\alpha x+r(\alpha,\theta)u_\theta\|=1$. Let P_θ be a contractive projection from X onto span $\{x, u_\theta\}$. Since the basis $\{x, y, z\}$ of X has an unconditional constant equal to one we may assume without loss of generality that $P_\theta y$ and $P_\theta z$ are multiples of u_θ (otherwise, pass to the "diagonal" of P_θ cf. I.1.c.8). Thus, there exists a unit vector $v_\theta\in$ span $\{y, z\}$ in the kernel of P_θ. We have that $\|\alpha x+r(\alpha,\theta)u_\theta+tv_\theta\|\geqslant 1$, for every α and θ as above and every real t. Geometrically, this means that the line in the y, z plane through $r(\alpha,\theta)u_\theta$ in the direction of v_θ is a line of support to the convex planar set which is defined (in polar coordinates) by $0\leqslant r\leqslant r(\alpha,\theta)$, $0\leqslant\theta<2\pi$. In particular, for every θ in which the curve $K_\alpha=\{(\theta, r);\ r=r(\alpha,\theta)\}$ has a tangent (and thus, for all θ except possibly for countably many values) this tangent is in the direction of v_θ. Since v_θ does not depend on α the curves K_α are all homothetic to each other, i.e. $r(\alpha,\theta)/r(\beta,\theta)$ is independent of θ (speaking analytically, the derivative of the absolutely continuous function $g(\theta)=r(\alpha,\theta)/r(\beta,\theta)$ vanishes whenever it exists and thus this function is a constant). Since $r(0,\theta)\equiv 1$ it follows that $r(\alpha,\theta)=r(\alpha)$ is independent of θ, i.e. $\|\alpha x+r(\alpha)u\|=1$, for every $0\leqslant\alpha<1$ and $u\in$ span $\{y, z\}$ of norm one. In other words, there is a function $F(s,t)$ so that $\|\alpha x+u\|=F(\alpha,\|u\|)$, for every $\alpha\geqslant 0$ and every $u\in$ span $\{y, z\}$. Similarly, there exist functions $G(s,t)$ and $H(s,t)$ so that, for every $\alpha\geqslant 0$,

$$\|\alpha y+v\|=G(\alpha,\|v\|),\ v\in \text{span}\,\{x, z\} \text{ and } \|\alpha z+w\|=H(\alpha,\|w\|),\ w\in\text{span}\,\{x, y\}\,.$$

Since

$$F(s,t)=\|sx+ty\|=G(t,s)=\|ty+sz\|=H(s,t)=\|sz+tx\|=F(t,s)\,.$$

we get that $F\equiv H\equiv G$. The desired result follows now by using 1.b.7. □

In the paper [3] Ando proved also a dual version of 1.b.8 which has the esthetical advantage that it characterizes all the abstract L_p and M spaces simultaneously (and does not single out a special subclass of the M spaces). In order to state this result, let us introduce the following notion. Let X be a Banach space and Y a closed subspace of X. An operator $T\colon Y^*\to X^*$ is called a *simultaneous extension operator* if $Ty^*{}_{|Y}=y^*$, for every $y^*\in Y^*$. If P is a projection from X onto Y then P^* is a simultaneous extension operator. The converse need not be true unless X is reflexive (i.e. not every simultaneous extension operator is necessarily the adjoint of an operator from X to Y).

Theorem 1.b.8′. *A Banach lattice X of dimension $\geqslant 3$ is an abstract L_p space, for some $1 \leqslant p < \infty$, or an abstract M space if and only if, for every sublattice Y of X, there is a positive simultaneous extension operator of norm one from Y^* to X^*.*

For a reflexive X, 1.b.8′ is completely equivalent to 1.b.8 by duality. For a non-reflexive X, the derivation of 1.b.8′ from 1.b.8 requires some quite simple arguments which we omit however.

We pass now to the isomorphic versions of 1.b.7 and 1.b.8 which we state and prove only in the order continuous case (cf. [127], [128], [77]).

Theorem 1.b.12. *Let X be an order continuous Banach lattice. Then the following assertions are equivalent.*

(1) *X is order isomorphic to either $L_p(\mu)$, for some $1 \leqslant p < \infty$ and some measure μ, or to $c_0(\Gamma)$, for some set Γ.*

(2) *There exist a non-negative valued function $F(t_1, t_2, \ldots)$ (of infinitely many real variables) and a constant A so that, for every choice of a sequence $\{x_n\}_{n=1}^{\infty}$ of disjoint elements in X such that $\sum_{n=1}^{\infty} x_n$ converges, we have*

$$A^{-1}F(\|x_1\|, \|x_2\|, \ldots) \leqslant \left\| \sum_{n=1}^{\infty} x_n \right\| \leqslant AF(\|x_1\|, \|x_2\|, \ldots)\,.$$

(3) *Every sublattice of X is complemented.*

For the proof of 1.b.12 we need the following lemma.

Lemma 1.b.13. *Let X be an order continuous Banach lattice and let $1 \leqslant p \leqslant \infty$. Assume that every sequence of disjoint elements of X of norm one is equivalent to the unit vector basis in l_p (in c_0 if $p=\infty$). Then X is order isomorphic to $L_p(\mu)$, for some measure μ (to $c_0(\Gamma)$, for some Γ, if $p=\infty$).*

Proof. Assume first that $1 \leqslant p < \infty$ and put

$$|||x||| = \sup \left(\sum_{i=1}^{n} \|x_i\|^p \right)^{1/p},$$

where the supremum is taken over all finite sequences $\{x_i\}_{i=1}^n$ of disjoint elements such that $|x| = \sum_{i=1}^{n} x_i$. The supremum may a-priori be infinite for some x but, by the decomposition property, it is easily seen that $|||\cdot|||$ satisfies the triangle inequality. Clearly, $\|x\| \leqslant |||x|||$, for every $x \in X$. We claim that there is a constant $K < \infty$ so that $|||x||| \leqslant K\|x\|$, for every $x \in X$. Assume to the contrary that there exist positive $x^m \in X$, $m = 1, 2, \ldots$ so that $|||x^m||| > m$ and $\|x^m\| = 1$. Let $m_1 = 2$ and let

$\{x_i^{m_1}\}_{i=1}^{n_1}$ be disjoint vectors in X so that $x^{m_1}=\sum_{i=1}^{n_1} x_i^{m_1}$ and $\sum_{i=1}^{n_1} \|x_i^{m_1}\|^p \geq 2^p$. Let $\{P_{1,i}\}_{i=1}^{n_1}$ be the band projections on the bands generated by $x_i^{m_1}$, $i=1,2,\ldots,n_1$ and let $P_{1,0}=I-\sum_{i=1}^{n_1} P_{1,i}$. Since

$$|||x^m||| \leq \sum_{i=0}^{n_1} |||P_{1,i}x^m|||$$

we may assume without loss of generality that $\lim_m |||P_{1,n_1}x^m||| \to \infty$. Hence, there are an $m_2>m_1$ and disjoint vectors $\{x_i^{m_2}\}_{i=1}^{n_2}$, whose sum is $P_{1,n_1}x^{m_2}$, so that

$$\sum_{i=1}^{n_2} \|x_i^{m_2}\|^p \geq 2^{2p}+\|x_{n_1}^{m_1}\|^p$$

and thus

$$\sum_{i=1}^{n_1-1} \|x_i^{m_1}\|^p+\sum_{i=1}^{n_2} \|x_i^{m_2}\|^p \geq 2^p+2^{2p}\,.$$

Continuing inductively, we construct a set $\{x_i^{m_j};\ 1\leq i\leq n_j, j=1,2,\ldots\}$ of vectors in X so that $\{x_i^{m_j};\ 1\leq i\leq n_j-1, j=1,2,\ldots\}$ are mutually disjoint,

$$\sum_{j=1}^{k}\sum_{i=1}^{n_j-1} \|x_i^{m_j}\|^p \geq 2^p+2^{2p}+\ldots+2^{kp}-\|x_{n_k}^{m_k}\|^p \geq 2^{kp}$$

and $\left\|\sum_{i=1}^{n_j-1} x_i^{m_j}\right\| \leq 1$. By our assumption, this double indexed sequence is (after normalization) equivalent to the unit vector basis of l_p. That is, for some constant A and every integer k,

$$2^{kp} \leq \sum_{j=1}^{k}\sum_{i=1}^{n_j-1} \|x_i^{m_j}\|^p \leq A \sum_{j=1}^{k}\left\|\sum_{i=1}^{n_j-1} x_i^{m_j}\right\|^p \leq kA\,,$$

which is clearly impossible.

We have thus shown that $|||\cdot|||$ is an equivalent lattice norm on X. Obviously, $|||x+y|||^p \geq |||x|||^p+|||y|||^p$, whenever $|x|\wedge|y|=0$. If $p=1$ this already proves that $(X, |||\cdot|||)$ is an abstract L_1 space. If $1<p<\infty$ we pass to the dual which, as easily verified, satisfies the assumption of 1.b.13 for q, where $1/q+1/p=1$. Moreover, $|||\cdot|||$ induces a norm on X^* (also denoted by $|||\cdot|||$) for which $|||x^*+y^*|||^q \leq |||x^*|||^q+|||y^*|||^q$, whenever $|x^*|\wedge|y^*|=0$. Starting with $|||\cdot|||$ we renorm again X^* by the procedure described above for X (replacing p by q) and arrive at a norm $|||\cdot|||_0$ for which $|||x^*|||_0^q+|||y^*|||_0^q=|||x^*+y^*|||_0^q$, whenever $|x^*|\wedge|y^*|=0$.

If $p=\infty$ we get from the preceding argument, by passing to the dual, that there is a $K<\infty$ so that $\left\|\sum_{i=1}^n x_i\right\| \leqslant K \max_{1\leqslant i\leqslant n} \|x_i\|$, whenever $\{x_i\}_{i=1}^n$ are disjoint vectors in X. We also get that X^* can be renormed so as to become an abstract L_1 space. It is not however immediately clear that the new norm on X^* is induced by a norm in X (i.e. that the unit ball of $(X^*, |||\cdot|||)$ is w^* closed) and thus it is simpler to renorm directly X. We put

$$|||x|||= \inf \max_{1\leqslant i\leqslant n} \|x_i\|,$$

where the inf is taken over all decompositions of x as a finite sum $\sum_{i=1}^n x_i$, with $|x_i| \wedge |x_j|=0$ for $i\neq j$. Clearly, $K^{-1}\|x\|\leqslant |||x|||\leqslant \|x\|$, for every $x \in X$, and $|||x+y|||=\max(|||x|||, |||y|||)$, whenever $|x|\wedge|y|=0$. In order to verify that $|||\cdot|||$ satisfies the triangle inequality it suffices to remark that the inf in the definition of $|||\cdot|||$ is actually the limit over the net of all partitions of x into a finite sum of disjoint elements (a partition $x=\sum_{i=1}^n u_i$ precedes $x=\sum_{j=1}^m v_j$ if each u_i is a sum of v_j's). The fact that X is order isomorphic to $c_0(\Gamma)$, for some Γ, follows now from 1.b.10. □

Proof of 1.b.12. It is clear that (1) implies (2) and (3) (see 1.b.8). It follows from Zippin's theorem on perfectly homogeneous bases (cf. I.2.a.9) that if X satisfies (2) then any sequence $\{x_n\}_{n=1}^\infty$ of disjoint vectors of norm one in X is equivalent to the unit vector basis of c_0 or l_p, for some $1\leqslant p<\infty$. We arrive at the same conclusion if we assume (3) and apply I.2.a.10. Thus, in order to be able to apply 1.b.13 and therefore to conclude the proof, we have just to show that the p does not depend on the particular choice of $\{x_n\}_{n=1}^\infty$. Start with one such sequence, say $\{y_n\}_{n=1}^\infty$, which is equivalent to the unit vector basis of l_{p_0} for some $1\leqslant p_0\leqslant\infty$ ($p_0=\infty$ corresponds to c_0). Let P be the band projection from X on the band spanned by $\{y_{2n}\}_{n=1}^\infty$. For every sequence of disjoint vectors $\{x_n\}_{n=1}^\infty$ of norm one in PX, the sequence $\{x_n\}_{n=1}^\infty \cup \{y_{2n+1}\}_{n=1}^\infty$ consists of disjoint vectors and thus $\{x_n\}_{n=1}^\infty$ must be equivalent to the unit vector basis of l_{p_0}. The same is true for a sequence in $(I-P)X$. Thus, if $\{u_n\}_{n=1}^\infty$ is an arbitrary sequence of disjoint vectors of norm one in X then both $\{Pu_n/\|Pu_n\|\}_{n=1}^\infty$ and $\{(I-P)u_n/\|(I-P)u_n\|\}_{n=1}^\infty$ are (we count only those indices for which the denominator is $\neq 0$) equivalent to the unit vector basis of l_{p_0}. The same is therefore true for $\{u_n\}_{n=1}^\infty$. □

The preceding theorems explain the special role of the L_p and M spaces in Banach lattice theory. It follows from these theorems that functional representation theorems like 1.b.2 and 1.b.6, which involve e.g. two sided estimates of norms of sums of disjoint elements, cannot be proved for more general classes of Banach lattices. If, however, we are satisfied with weaker estimates of the norms we can obtain representation theorems in a quite general setting. We present now a very useful general representation theorem. This was developed in the work of several authors ([4], [88], [130] and [99]).

Theorem 1.b.14. *Let X be an order continuous* (i.e. σ-complete and σ-order continuous) *Banach lattice which has a weak unit. Then there exist a probability space (Ω, Σ, μ), an (in general not closed) ideal $\tilde{X}$ of $L_1(\Omega, \Sigma, \mu)$ and a lattice norm $\|\cdot\|_{\tilde{X}}$ on $\tilde{X}$ so that*

(i) *X is order isometric to $(\tilde{X}, \|\cdot\|_{\tilde{X}})$.*
(ii) *$\tilde{X}$ is dense in $L_1(\Omega, \Sigma, \mu)$ and $L_\infty(\Omega, \Sigma, \mu)$ is dense in $\tilde{X}$.*
(iii) *$\|f\|_1 \leq \|f\|_{\tilde{X}} \leq 2\|f\|_\infty$, whenever $f \in L_\infty(\Omega, \Sigma, \mu)$.*
(iv) *The dual of the isometry given in* (i) *maps X^* onto the Banach lattice $\tilde{X}^*$ of all μ measurable functions g for which*

$$\|g\|_{\tilde{X}^*} = \sup\left\{\int_\Omega fg\, d\mu;\ \|f\|_{\tilde{X}} \leq 1\right\} < \infty .$$

The value taken by the functional corresponding to g at $f \in \tilde{X}$ is $\int_\Omega fg\, d\mu$.

The main tool in the proof of 1.b.14 is the following proposition which ensures the existence of strictly positive functionals in X^*.

Proposition 1.b.15. *For every order continuous Banach lattice X with a weak unit $e > 0$ there exists a functional $e^* > 0$ in X^* such that $e^*(|x|) = 0$ implies $x = 0$.*

Proof. Since in many applications we work with separable lattices we present first a very simple proof which is valid only under the assumption that X is separable. Let $\{x_n\}_{n=1}^\infty$ be a dense sequence in the set $\{x \in X;\ x \geq 0,\ \|x\| = 1\}$ and choose positive Hahn–Banach functionals $x_n^* \in X^*$ so that $\|x_n^*\| = 1$ and $x_n^*(x_n) = 1$ for all n (if a Hahn–Banach functional x_n^* is not positive we replace it by $|x_n^*|$). Then $e^* = \sum_{n=1}^\infty x_n^*/2^n$ is a strictly positive functional on X. Indeed, if $x > 0$ is a norm one vector in X then we can find an integer n so that $\|x - x_n\| < 1/2$ which implies that $e^*(x) \geq x_n^*(x)/2^n \geq 1/2^{n+1} > 0$.

The proof in the non-separable case is longer. We observe first that it suffices to show that there exists a sequence of mutually disjoint norm one positive functionals $\{e_n^*\}_{n=1}^\infty$ which is maximal in the sense that no more functionals can be added to this sequence without losing the disjointness. In this case, $e^* = \sum_n e_n^*/2^n$ would be a strictly positive functional on X. Indeed, otherwise $Y = \{x \in X;\ e^*(|x|) = 0\}$ is a non-trivial projection band of X and thus, there exists a positive functional $e_0^* \in X^*$ such that $\|e_0^*\| = 1$ and $e_0^* Y^\perp = 0$. This fact, however, contradicts the maximality of $\{e_n^*\}_{n=1}^\infty$ since $e^* \wedge e_0^* = 0$.

In order to complete the proof, we show that any maximal family $\{e_\alpha^*\}_{\alpha \in A}$, of disjoint norm one positive functionals in X^* is countable. Put $Y_\alpha = \{x \in X;\ e_\alpha^*(|x|) = 0\}$ and let P_α be the band projection from X onto $Y_\alpha^\perp$ i.e. $P_\alpha Y_\alpha = 0$. For each pair $\alpha, \beta \in A$ with $\alpha \neq \beta$, we get that $Y_\alpha^\perp \cap Y_\beta^\perp = \{0\}$. Indeed, for every $0 \leq u \in Y_\alpha^\perp \cap Y_\beta^\perp$, we have

$$0 = (e_\alpha^* \wedge e_\beta^*)(u) = \inf\{e_\alpha^*(v) + e_\beta^*(w);\ \ u = v + w;\ \ 0 \leq v, w \leq u\} .$$

Thus, there are sequences $\{v_n\}_{n=1}^\infty$ and $\{w_n\}_{n=1}^\infty$ so that $u=v_n+w_n$, $0\leq v_n, w_n\leq u$, $e_\alpha^*(v_n)\leq 2^{-n}$ and $e_\beta^*(w_n)\leq 2^{-n}$ for all n. Put $v_k'=\bigvee_{n=k}^\infty v_n$, $w_k'=\bigvee_{n=k}^\infty w_n$ and observe that $\{v_k'\}_{k=1}^\infty$ and $\{w_k'\}_{k=1}^\infty$ are decreasing sequences of positive elements in X. By the σ-completeness and σ-order continuity of X, these two sequences must have strong limits $v'\geq 0$, respectively $w'\geq 0$, which clearly satisfy $e_\alpha^*(v')=0$, $e_\beta^*(w')=0$ and $u\leq v'+w'$. Using the decomposition property we get that there are $0\leq v\leq v'$ and $0\leq w\leq w'$ such that $u=v+w$. Since $e_\alpha^*(v)=0$ and $e_\beta^*(w)=0$ we conclude that $v\in Y_\alpha$ and $w\in Y_\beta$. On the other hand, $Y_\alpha^\perp$ and $Y_\beta^\perp$ are ideals of X and this implies that $v\in Y_\alpha^\perp$ and $w\in Y_\beta^\perp$. Thus, $u=0$.

Put $e_\alpha=P_\alpha(e)$, $\alpha\in A$, and notice that $e_\alpha\neq 0$ for all α since e is a weak unit. The series $\sum_{\alpha\in A'} e_\alpha$ converges in X for every countable subset $A'\subset A$ since X is σ-complete and σ-order continuous. This clearly implies that A is countable. □

Remark. In the non-separable case the assumption that X is order continuous cannot be dropped. Consider e.g. $X=l_\infty(\Gamma)$ with Γ uncountable.

Proof of 1.b.14. Let X be an order continuous Banach lattice having a weak unit $e_0>0$ with $\|e_0\|=2$. By 1.b.15, X^* contains a strictly positive functional e_0^* with $\|e_0^*\|=1$. Let u^* be a positive element of norm one in X^* for which $u^*(e_0)=2$. Then $e^*=(e_0^*+u^*)/\|e_0^*+u^*\|$ is a strictly positive functional on X, $e=e_0/e^*(e_0)$ is a weak unit of X and

$$\|e^*\|=e^*(e)=1,\quad \|e\|\leq 2\,.$$

By 1.a.13, X is a cyclic space $\mathcal{M}(e)$ with respect to a σ-complete Boolean algebra of projections $\mathcal{B}$ (which, by 1.b.3, can be regarded as $\{P_\sigma\}_{\sigma\in\Sigma_0}$, where Σ_0 is the set of all simultaneously open and closed subsets of a totally disconnected space Ω). We define a probability measure μ on Ω by $\mu(\sigma)=e^*(P_\sigma e)$ (as in the proof of 1.b.2 we have that, up to sets of μ measure zero, Σ_0 is already a σ-algebra). The map which assigns to $x=\sum_{i=1}^n a_iP_{\sigma_i}e$ the function $\tilde{x}=\sum_{i=1}^n a_i\chi_{\sigma_i}$ in $L_1(\mu)$ is clearly an order isomorphism and $\|\tilde{x}\|_1=e^*(|x|)\leq\|x\|$. The map $x\to\tilde{x}$ extends thus to a positive and contractive operator from X to $L_1(\mu)$. Since e^* is strictly positive this operator is one to one and an order isomorphism. The image $\tilde{X}$ of X under this operator is clearly dense in $L_1(\mu)$ (it contains all the simple functions). For every simple function f on Ω we have $|f|\leq\|f\|_\infty\cdot 1$; if we put $f=\tilde{x}$ with $x\in X$ we get that $|x|\leq\|\tilde{x}\|_\infty e$ and thus $\|x\|\leq 2\|\tilde{x}\|_\infty$. This proves assertions (i), (ii) and (iii) of the theorem.

In order to show that $\tilde{X}$ is an ideal in $L_1(\mu)$, we observe that every $f\geq 0$ in $L_1(\mu)$ is the limit in the L_1 norm of an increasing sequence of simple functions $\{f_n\}_{n=1}^\infty$. Each f_n is of the form $\tilde{y}_n$ for some $y_n\in X$. Thus, if $f\leq\tilde{x}$ for some $x\in X$ it follows from the order continuity of X that $\{y_n\}_{n=1}^\infty$ converges in norm to some y in X. Clearly $f=\tilde{y}$.

It remains to prove (iv). For every x^* in the dual of $\tilde{X}$ we consider the (signed) measure $\nu_{x^*}(\sigma)=x^*\chi_\sigma$, $\sigma \in \Sigma$ (the σ-additivity of ν_{x^*} follows from the σ-order continuity of X). Since ν_{x^*} is absolutely continuous with respect to μ there exists a function $g \in L_1(\Omega, \Sigma, \mu)$ such that

$$\int_\Omega f\, d\nu_{x^*} = \int_\Omega fg\, d\mu\,,$$

for every f which is ν_{x^*}-integrable. By using approximation by simple functions it is easily checked that if $f \in \tilde{X}$ then f is ν_{x^*}-integrable and $\int_\Omega f\, d\nu_{x^*} = x^*(f)$. This implies that $x^*(f) = \int_\Omega fg\, d\mu$ and

$$\|x^*\| = \|g\|_{\tilde{X}^*} = \sup\left\{\int_\Omega fg\, d\mu;\ \|f\|_{\tilde{X}} \leq 1\right\}.$$

Since the converse (i.e. the fact that every g as in (iv) defines an element of $\tilde{X}^*$) is obvious this completes the proof. □

Remark. As is easily seen from the proof, the number 2 in 1.b.14(iii) can be replaced by $1+\varepsilon$ for any $\varepsilon>0$ (of course, for different ε we obtain different representations). It is also easy to see that we can obtain (iii) with 2 replaced by 1 if and only if there are a weak unit e in X and a strictly positive functional e^* in X^* with $\|e\| = \|e^*\| = e^*(e) = 1$. If, for example, $X = c_0$ then such e and e^* fail to exist.

As an application of 1.b.14 we present a proof of a result of H. Nakano [104].

Theorem 1.b.16. *A Banach lattice X is order continuous if and only if the canonical image of X into its second dual X^{**} is an ideal of X^{**}.*

Proof. Suppose that X is an order continuous Banach lattice and let i denote the canonical embedding of X into X^{**}. Let $x \in X$ and $x^{**} \in X^{**}$ be two vectors satisfying $0 \leq x^{**} \leq ix$. By the last part of 1.a.9, there exists an ideal X_0 of X with a weak unit so that $x \in X_0$ and $X = X_0 \oplus X_0^\perp$. In view of this decomposition, we can consider x^{**} as an element of X_0^{**}. By using the representation theorem 1.b.14, X_0 can be regarded as an ideal of an $L_1(\Omega, \Sigma, \mu)$-space (with $\mu(\Omega)=1$), which has all the properties described in the statement of 1.b.14. In particular, for every $\sigma \in \Sigma$, χ_σ is an element of X_0^* and therefore we can put $\nu(\sigma) = x^{**}(\chi_\sigma)$ and $\lambda(\sigma) = \int_\sigma x(\omega)\, d\mu(\omega)$, where $x(\omega)$ is the function in $L_1(\Omega, \Sigma, \mu)$ representing the element $x \in X_0$. The measure λ is clearly σ-additive and $\lambda(\sigma) \geq \nu(\sigma)$, for all $\sigma \in \Sigma$. Hence, also ν is σ-additive. Since ν is also absolutely continuous with respect to μ it follows from the Radon–Nikodym theorem that there exists a function $f \in L_1(\Omega, \Sigma, \mu)$ so that $\nu(\sigma) = \int_\sigma f(\omega)\, d\mu(\omega)$ for all $\sigma \in \Sigma$. The relation between ν and λ implies that $f(\omega) \leq x(\omega)$ for a.e. $\omega \in \Omega$, i.e. $f \in X_0$ since X_0 is an ideal of $L_1(\Omega, \Sigma, \mu)$. From this fact and assertion (iv) of 1.b.14 it follows easily that $x^{**} = if$, i.e. that iX is an ideal of X^{**}.

In order to prove the converse, assume that iX is an ideal of X^{**} and that $\{x_n\}_{n=1}^\infty$ is an increasing sequence of positive elements of X which is bounded in order by an element $x \in X$. Since $\{x_n\}_{n=1}^{\infty}$ is necessarily a weak Cauchy sequence in X there exists an $x_0^{**} \in X^{**}$ such that $ix_n \xrightarrow{w^*} x_0^{**}$. For every positive $x^* \in X^*$ we have $x_0^{**}(x^*) = \lim_{n \to \infty} x^*(x_n) \leq x^*(x)$, i.e. $0 \leq x_0^{**} \leq ix$. Since we have assumed that iX is an ideal of X^{**} we get that $x_0^{**} = ix_0$, for some $x_0 \in X$, i.e. that $x_n \xrightarrow{w} x_0$. However, for monotone sequences in a Banach lattice, weak convergence implies strong convergence (use the fact that if $x_n \xrightarrow{w} x_0$ then there exist convex combinations of the x_n's which tend strongly to x_0). This completes the proof of the order continuity of X, by 1.a.8. □

It follows from 1.b.16 that *a Banach lattice X is order continuous if and only if, for every $y, z \in X$, the order interval $[y, z] = \{x; y \leq x \leq z\}$ is weakly compact.* Indeed, if $i: X \to X^{**}$ denotes the canonical embedding and if X is order continuous then, by 1.b.16, $[iy, iz] = i[y, z]$ for every $y, z \in X$. Since any order interval in X^{**} is w^* compact it follows that $[y, z]$ is a w compact subset of X. This proves the "if" part of the above assertion. The "only if" part is obvious.

Let us also mention here another characterization of order continuous lattices (cf. [133]): *A Banach lattice $(X, \|\cdot\|)$ is order continuous if and only if there is an equivalent lattice norm $\|\ \|_1$ on X so that*

$$(*) \qquad \{x_n\}_{n=1}^{\infty} \subset X, x_n \xrightarrow{w} x \text{ and } \|x_n\|_1 \to \|x\|_1 \Rightarrow \|x_n - x\|_1 \to 0 .$$

In I.1.b.11 it was shown that every separable Banach space X admits an equivalent norm so that (*) holds. The point in the result stated here is that in case X is a Banach lattice, the new norm $\|\ \|_1$ can be chosen to be again a lattice norm if and only if X is order continuous. The proof of the "only if" part is easy. Indeed, by 1.a.5 and 1.a.7, if X is not order continuous there exists a sequence of disjoint positive vectors $\{y_n\}_{n=1}^{\infty}$ in X which is equivalent to the unit vector basis in c_0 and a vector $y \in X$ so that $y_n \leq y$ for all n. Then, for any lattice norm $\|\ \|_1$ in X, $y - y_n \xrightarrow{w} y$, $\|y - y_n\|_1 \to \|y\|_1$, but clearly $\{y - y_n\}_{n=1}^{\infty}$ does not tend strongly to y. The proof of the "if' part is somewhat long and will not be reproduced here. We just mention that this proof is based on the representation theorem 1.b.14.

The proof of 1.b.16 shows, in particular, that also the converse to 1.b.14 is true. Every lattice X, which satisfies (i)–(iv) in 1.b.14, is order continuous and has a weak unit. It is however very simple to prove this fact directly. We shall now discuss this point and some related questions in a somewhat more general context.

Definition 1.b.17. Let (Ω, Σ, μ) be a complete σ-finite measure space. A Banach space X consisting of equivalence classes, modulo equality almost everywhere, of locally integrable real valued functions on Σ is called a *Köthe function space* if the following conditions hold.

(1) If $|f(\omega)| \leq |g(\omega)|$ a.e. on Ω, with f measurable and $g \in X$, then $f \in X$ and $\|f\| \leq \|g\|$.

(2) For every $\sigma \in \Sigma$ with $\mu(\sigma) < \infty$ the characteristic function χ_σ of σ belongs to X.

Recall that a measure space is said to be *complete* if any subset of a set of measure zero is measurable. The assumption of completeness of the measure space is just a minor technical convenience; the measure space constructed in 1.b.2 (and thus in 1.b.14) can clearly be taken to be complete. A function f is called *locally integrable* if it is measurable and $\int_\sigma |f(\omega)| \, d\mu < \infty$, for every $\sigma \in \Sigma$ with $\mu(\sigma) < \infty$.

Every Köthe function space is a Banach lattice in the obvious order ($f \geqslant 0$ if $f(\omega) \geqslant 0$ a.e.). This lattice is σ-order complete. Indeed, if $\{f_n\}_{n=1}^\infty$ is an order bounded increasing sequence in X then $f(\omega) = \lim_{n\to\infty} f_n(\omega)$ is the l.u.b. of $\{f_n\}_{n=1}^\infty$.

Theorem 1.b.14 asserts, in particular, that every order continuous Banach lattice with a weak unit is order isometric to a Köthe function space. Thus, a separable Banach lattice is order isometric to a Köthe function space if and only if it is σ-order complete.

The assumption that every $f \in X$ is locally integrable implies that, for every $\sigma \in \Sigma$, the positive functional $f \to \int_\Omega f(\omega)\chi_\sigma(\omega) \, d\mu$ is well defined and thus bounded i.e. it is an element of X^*. In general, every measurable function g or Ω so that $gf \in L_1(\mu)$, for every $f \in X$, defines an element x_g^* in X^* by $x_g^*(f) = \int_\Omega f(\omega)g(\omega) \, d\mu$ (we shall often identify g with x_g^*). Any functional on X of the form x_g^* is called an *integral* and the linear space of all integrals is denoted by X'. It is an immediate consequence of the Radon–Nikodym theorem that a functional $x^* \in X^*$ is an integral if and only if, for every sequence $\{f_n\}_{n=1}^\infty$ in X with $f_n(\omega) \downarrow 0$ a.e., we have $|x^*|(f_n) \to 0$. In particular, if X is σ-order continuous then $X^* = X'$. The converse is also true. Assume that $X^* = X'$ and let $\{f_n\}_{n=1}^\infty$ be an increasing sequence of positive elements in X which converges pointwise a.e. to some f in X. Then clearly $f_n \xrightarrow{w} f$ and hence also $\|f_n - f\| \to 0$. Since every Köthe function space is σ-order complete the condition $X^* = X'$ is, by 1.a.8, also equivalent to the order continuity of X.

It is easily verified, by using the characterization of integrals given above, that, for every Köthe function space X, X' is an ideal of X^*. In the norm induced on X' by X^*, this space is also a Köthe function space on (Ω, Σ, μ). The following proposition, due to G. G. Lorentz and W. A. J. Luxemburg (cf. [86]), characterizes those Köthe spaces for which X' is a norming subspace of X^* (i.e. $\|x\| = \sup\{|x^*(x)|;\ x^* \in X',\ \|x^*\| = 1\}$, for every $x \in X$).

Proposition 1.b.18. *Let X be a Köthe function space. Then X' is a norming subspace of X^* if and only if, whenever $\{f_n\}_{n=1}^\infty$ and f are non-negative elements of X such that* $f_n(\omega) \uparrow f(\omega)$ a.e., *we have* $\|f_n\| \to \|f\|$.

Proof. Assume that X' is norming, let $f_n(\omega) \uparrow f(\omega) \in X$ a.e. and let $\varepsilon > 0$. Pick $x^* \in X'$ with $\|x^*\| = 1$ and $x^*(f) \geqslant \|f\| - \varepsilon$. Since x^* is an integral $x^*(f_n) \to x^*(f)$ and thus $\liminf_{n\to\infty} \|f_n\| \geqslant \|f\| - \varepsilon$. Consequently, $\|f\| = \lim_{n\to\infty} \|f_n\|$.

Conversely, suppose that $f_n(\omega) \uparrow f(\omega)$ a.e. in Ω implies $\|f_n\| \to \|f\|$. Consider $Y = X \cap L_1(\mu)$ as an, in general not closed, subspace of $L_1(\mu)$ with the norm $\|\cdot\|_1$

induced by $L_1(\mu)$. The set $\{f; f \in Y, \|f\| \leq 1\}$ is closed in Y. Indeed, assume that $\{f_n\}_{n=1}^\infty$ and f are in Y, with $\|f_n\| \leq 1$, for all n, and $\|f_n - f\|_1 \to 0$. By passing to a subsequence if necessary we may assume without loss of generality that $f_n(\omega) \to f(\omega)$ a.e. Hence, $g_n(\omega) \uparrow |f(\omega)|$ a.e., where $g_n(\omega) = \inf_{k \geq n} |f_k(\omega)|$, and thus $\|f\| = \lim_{n\to\infty} \|g_n\| \leq \liminf_{n\to\infty} \|f_n\| \leq 1$.

Let now $f \in X$ be an element with $\|f\| > 1$. By our assumption on X, there is a $\sigma \in \Sigma$ with $\mu(\sigma) < \infty$ so that $\|f\chi_\sigma\| > 1$. By using the separation theorem for Y, there is an $h \in L_\infty(\mu)$ so that $\int_\Omega h(\omega) f(\omega) \chi_\sigma(\omega)\, d\mu > 1$ and $|\int_\Omega h(\omega) g(\omega)\, d\mu| \leq 1$, for every $g \in Y$ with $\|g\| \leq 1$. Consequently, $h\chi_\sigma$ defines an element $x^* \in X'$ so that $\|x^*\| \leq 1$ and $x^*(f) > 1$. □

Remarks. 1. A typical example of a Köthe function space X such that $X' \neq X^*$, but X' is norming, is $L_\infty(\mu)$. An example of a Köthe function space X for which X' is not norming is the space l_∞ with the equivalent norm $\|x\|_n = \sup_k |x(k)| + n \limsup_k |x(k)|$ (here Ω is the set of integers with the discrete measure and n is a positive integer). The space $\left(\sum_{n=1}^\infty \oplus (l_\infty, \|\cdot\|_n)\right)_2$ is an example of a Köthe function space X for which $\sup \{|x^*(x)|;\, x^* \in X', \|x^*\| = 1\}$ does not even define an equivalent norm on X.

2. For every Köthe function space X, we can define also the space $X'' = (X')'$. If X' is a norming subspace of X^* then X is isometric to a subspace of X''. The space X coincides with X'' if and only if

$$(*) \qquad f_n(\omega) \uparrow f(\omega) \text{ a.e.}, \quad \{f_n\}_{n=1}^\infty \subset X, \quad f_n(\omega) \geq 0 \text{ a.e. and } \sup_n \|f_n\| < \infty$$
$$\Rightarrow f \in X \quad \text{and} \quad \|f\| = \lim_n \|f_n\|.$$

Indeed, it is easily verified directly that, for every Köthe function space Y, the space $X = Y'$ satisfies (*). Hence, if $X = X''$ then (*) holds. Conversely, if (*) holds then, by 1.b.18, X is isometric to a subspace of X''. Let $f \in X''$ be non-negative and let $\{f_n\}_{n=1}^\infty$ be a sequence of non-negative simple functions increasing to f a.e. It follows from (*), applied to this sequence $\{f_n\}_{n=1}^\infty$, that $f \in X$. Property (*) is called the *Fatou property*.

3. A very simple but useful fact is the following. If X is a Köthe function space then

$$f_n(\omega) \to f(\omega) \text{ a.e.}, \{f_n\}_{n=1}^\infty \subset X \quad \text{and} \quad \sup_n \|f_n\| < \infty \Rightarrow f \in X''.$$

Indeed, for every $g \in X'$,

$$\int_\Omega |f(\omega) g(\omega)|\, d\mu \leq \liminf_{n\to\infty} \int_\Omega |f_n(\omega) g(\omega)|\, d\mu \leq \|g\|_{X'} \cdot \sup_n \|f_n\|_X .$$

c. The Structure of Banach Lattices and their Subspaces

We begin this section by presenting some results concerning weak completeness and reflexivity of Banach lattices as well as of their subspaces. Similar theorems have already been proved for spaces with an unconditional basis in Section I.1.c but their extension to Banach lattices requires somewhat different methods of proof (for subspaces of a space with an unconditional basis we have just stated the results in I.1.c.13 without giving a proof). We also present in this section some results on complemented subspaces and basic sequences in Banach lattices. An important tool in the study of subspaces of Banach lattices is the so-called property (u), introduced in [110].

Definition 1.c.1. A Banach space X is said to have *property* (u) if, for every weak Cauchy sequence $\{x_n\}_{n=1}^\infty$ in X, there exists a sequence $\{y_n\}_{n=1}^\infty$ in X such that:

(i) the series $\sum_{n=1}^\infty y_n$ is weakly unconditionally convergent (w.u.c.), i.e. $\sum_{n=1}^\infty |y^*(y_n)|<\infty$ for every $y^*\in X^*$,

(ii) the sequence $\left\{x_n-\sum_{j=1}^n y_j\right\}_{n=1}^\infty$ converges weakly to zero.

A. Pelczynski [110] proved that every space with an unconditional basis has property (u). A similar result is valid for Banach lattices.

Proposition 1.c.2 [130]. *Any order continuous Banach lattice X has property (u).*

Proof. We first observe that it suffices to prove the assertion for separable lattices. Thus, by 1.b.14, we may assume without loss of generality that X is a Köthe function space on some probability measure space (Ω, Σ, μ) and that every element of X^* is an integral (i.e. $X^*=X'$).

Let $\{f_n\}_{n=1}^\infty$ be a weak Cauchy sequence of functions in X. This sequence is also a weak Cauchy sequence in $L_1(\Omega, \Sigma, \mu)$. Thus, since L_1 spaces are weakly sequentially complete (cf. [32] IV.8.6), there exists an $f\in L_1(\Omega, \Sigma, \mu)$ so that $\int_\Omega f_n h\,d\mu \to \int_\Omega fh\,d\mu$ as $n\to\infty$, whenever $h\in L_\infty(\Omega, \Sigma, \mu)$. Let $g\in X^*$; then $\nu_n(\sigma)=\int_\sigma f_n g\,d\mu$, $\sigma\in\Sigma$, is a sequence of σ-additive measures which converges for every $\sigma\in\Sigma$. It follows from a well-known result of Nikodym (cf. [32] III.7.4) that $\nu(\sigma)=\lim_{n\to\infty}\nu_n(\sigma)$, $\sigma\in\Sigma$, is also a σ-additive measure which is absolutely continuous with respect to μ. Since, for every set $\sigma\in\Sigma$ on which g is a bounded function, we have $\nu_n(\sigma)\to\int_\sigma fg\,d\mu$ as $n\to\infty$ it follows that $fg\in L_1(\Omega, \Sigma, \mu)$ and

$$\lim_{n\to\infty}\int_\Omega (f_n-f)g\,d\mu=0$$

(use the uniqueness of the Radon–Nikodym derivative).

Put $\eta_n=\{\omega\in\Omega; n-1\leqslant|f(\omega)|<n\}$, $\delta_n=\bigcup_{j=n+1}^{\infty}\eta_j$ and $h_n=f\chi_{\eta_n}$, $n=1, 2,\ldots$ Then, for every $g\in X^*$, we get that

$$\sum_{n=1}^{\infty}\left|\int_{\Omega} h_n g\, d\mu\right|\leqslant\int_{\Omega}|fg|\, d\mu<\infty\,.$$

and

$$\int_{\Omega}\Big(f_n-\sum_{j=1}^{n}h_j\Big)g\, d\mu=\int_{\Omega}(f_n-f)g\, d\mu+\int_{\delta_n}fg\, d\mu\to 0 \text{ as } n\to\infty\,.$$

This completes the proof. □

The fact that subspaces of an order continuous Banach lattice also have property (u) is a consequence of the following general result from [110].

Proposition 1.c.3. *Every closed subspace of a Banach space with property* (u) *has also property* (u).

Proof. Let $\{y_n\}_{n=1}^{\infty}$ be a weak Cauchy sequence in a subspace Y of a Banach space X. If X has property (u) then there exists a w.u.c. series $\sum_{i=1}^{\infty}x_i$ in X so that the sequence $u_n=y_n-\sum_{i=1}^{n}x_i$, $n=1, 2,\ldots$, converges weakly to zero in X. We would like to replace the series $\sum_{i=1}^{\infty}x_i$ by a w.u.c. series consisting of elements of Y. Since $u_n\xrightarrow{w}0$ we can find convex combinations $u'_j=\sum_{n=p_{j-1}+1}^{p_j}\lambda_n u_n$, $j=1, 2, \ldots$, with $0=p_0<p_1<\ldots<p_j<\ldots$, $\sum_{n=p_{j-1}+1}^{p_j}\lambda_n=1$, for all j, and $\sum_{j=1}^{\infty}\|u'_j\|<\infty$. Put $y'_j=\sum_{n=p_{j-1}+1}^{p_j}\lambda_n y_n$, $z_0=y'_1$ and $z_j=y'_{j+1}-y'_j$ for $j\geqslant 1$, and observe that the sequence

$$y_n-\sum_{j=0}^{n}z_j=y_n-y'_{n+1},\; n=1, 2, \ldots$$

converges weakly to 0 in Y. Since $z_j\in Y$, for all j, it remains to show that $\sum_{j=0}^{\infty}z_j$ is a w.u.c. series.

A simple computation shows that each vector z_j, $j>0$, can be written as

$$z_j=\sum_{n=p_j+1}^{p_{j+1}}\lambda_n y_n-\sum_{n=p_{j-1}+1}^{p_j}\lambda_n y_n=u'_{j+1}-u'_j+\sum_{n=p_{j-1}+1}^{p_{j+1}}\mu_n^j x_n\,,$$

where, for each j, the coefficients μ_n^j are suitable numbers between 0 and 1. Thus, for any $x^* \in X^*$, we have

$$\sum_{j=0}^{\infty} |x^*(z_j)| \leqslant 2\|x^*\| \sum_{j=1}^{\infty} \|u_j'\| + \sum_{j=0}^{\infty} \sum_{n=p_{j-1}+1}^{p_{j+1}} \mu_n^j |x^*(x_n)|$$
$$\leqslant 2\|x^*\| \sum_{j=1}^{\infty} \|u_j'\| + 2 \sum_{n=1}^{\infty} |x^*(x_n)| < \infty ,$$

since $\sum_{n=1}^{\infty} x_n$ is a w.u.c. series. □

Remark. An interesting application of the argument used in the proof of 1.c.3 was found by M. Feder [134] who showed the following. *Let X be a Banach space with an unconditional finite dimensional Schauder decomposition (F.D.D. cf. I.1.g) $\{B_n\}_{n=1}^{\infty}$ and let Y be a reflexive subspace of X. Then Y is isomorphic to a complemented subspace of a space with an unconditional F.D.D. if and only if Y has the approximation property (A.P.).*

The "only if" assertion is trivial. The "if" assertion is proved as follows. For each n let Q_n be the natural projection from X onto B_n. Let $\{T_n\}_{n=1}^{\infty}$ be a sequence of finite rank operators in $L(Y, Y)$ so that $\lim_{n\to\infty} \|T_n y - y\| = 0$, for every $y \in Y$. Such a sequence exists since Y is separable and has the M.A.P. (use the reflexivity of Y and I.1.e.15). The sequence $\left\{T_n - \sum_{i=1}^{n} Q_{i|Y}\right\}_{n=1}^{\infty}$ in $L(Y, X)$ tends to zero pointwise and hence, by the reflexivity of Y, also in the weak topology of $L(Y, X)$. (Use the fact that the map $S \to x^*(Sy)$ defines an isometry from the subspace of $L(Y, X)$ consisting of the compact operators into $C(B_Y \times B_{X^*})$, where B_Y is taken in the w topology and B_{X^*} in the w^* topology). Consider $L(Y, Y)$ as a subspace of $L(Y, X)$ and apply the argument of 1.c.3 (with $T_n = y_n$ and $Q_{i|Y} = x_i$). It follows that there exists a sequence $\{S_i\}_{i=1}^{\infty}$ of finite rank operators in $L(Y, Y)$ so that

$$\left\{T_n - \sum_{i=1}^{n} S_i\right\}_{n=1}^{\infty}$$

tends pointwise to zero and $\sup_n \sup_{\theta_i = \pm 1} \left\| \sum_{i=1}^{n} \theta_i S_i \right\| < \infty$. Thus, for every $y \in Y$, $y = \sum_{n=1}^{\infty} S_n y$ and the series converges unconditionally. An argument identical to that used in the proof of I.1.e.13 now concludes the proof. Indeed, let $W_n = S_n Y$, $n = 1, 2, \ldots$, and let W be the completion of the space of all sequences of vectors $w = (w_1, w_2, \ldots)$, which are eventually zero, so that

$$w_n \in W_n,\ n = 1, 2, \ldots \quad \text{and} \quad |||w||| = \sup_n \sup_{\theta_i = \pm 1} \left\| \sum_{i=1}^{n} \theta_i w_i \right\| < \infty .$$

Clearly, W has an unconditional *F.D.D.* The operators $U: Y \to W$ and $R: W \to Y$, defined by $Uy=(S_1 y, S_2 y, \ldots)$, $y \in Y$ and $R(w_1, w_2, \ldots)=\sum_{n=1}^{\infty} w_n$, $(w_1, w_2, \ldots) \in W$, are bounded and satisfy $RUy=y$ for every $y \in Y$. Hence, UY is a complemented subspace of W isomorphic to Y. □

We state now the main result concerning weak completeness in Banach lattices.

Theorem 1.c.4. *The following conditions are equivalent for any Banach lattice X.*

(i) *X is weakly sequentially complete.*

(ii) *No subspace of X is isomorphic to c_0.*

(iii) *Every norm-bounded increasing sequence in X has a strong limit.*

(iv) *The canonical image of X in X^{**} is a (projection) band of X^{**}.*

The equivalence (i) $\Leftrightarrow$ (ii) *remains valid in the case when X is a subspace of an order continuous Banach lattice.*

The implication (iii) $\Rightarrow$ (i) was originally proved in [106] (see also [89]) while (ii) $\Rightarrow$ (i) in [84], [99]. The result on subspaces of Banach lattices was proved in [129] and [130].

Proof. The implication (i) $\Rightarrow$ (ii) is trivial. By 1.a.5, 1.a.7 and 1.a.8, a Banach lattice satisfying (ii) is both σ-complete and σ-order continuous and thus order continuous. Hence, it suffices to prove that (ii) $\Rightarrow$ (i) only in the case when X is a subspace of an order continuous Banach lattice. By 1.c.2 and 1.c.3, such a subspace X has property (u). Thus, if there exists in X a weak Cauchy sequence which does not converge weakly to any element of X then there is in X also a w.u.c. series which does not converge. It follows from I.2.e.4 that X contains a subspace isomorphic to c_0.

We show next that (ii) $\Rightarrow$ (iii). Assume that $0 \leqslant x_1 \leqslant x_2 \leqslant \ldots$ is an increasing non convergent sequence in X with $\|x_n\| \leqslant 1$ for all n. Then there are a $\delta > 0$ and an increasing sequence $\{n_k\}_{k=1}^{\infty}$ of integers so that if $y_k = x_{n_{k+1}} - x_{n_k}$ then $\|y_k\| \geqslant \delta$, $k=1, 2, \ldots$. The sequence $\{y_k\}_{k=1}^{\infty}$ tends weakly to 0 and hence, by I.1.a.12, has a subsequence $\{y_{k_j}\}_{j=1}^{\infty}$ which is a basic sequence. Since, for every choice of scalars $\{\lambda_j\}_{j=1}^{m}$, we also have that

$$\left\|\sum_{j=1}^{m} \lambda_j y_{k_j}\right\| \leqslant \max_j |\lambda_j| \left\|\sum_{j=1}^{m} y_{k_j}\right\| \leqslant \max_j |\lambda_j| \sup_n \|x_n\| \leqslant \max_j |\lambda_j|$$

it follows that $\{y_{k_j}\}_{j=1}^{\infty}$ is equivalent to the unit vector basis of c_0, in contradiction to (ii).

Assume now that condition (iii) holds in X. Then X is a σ-complete and σ-order continuous Banach lattice and thus, by 1.b.16, the canonical image iX of X into X^{**} is an ideal of X^{**}. Let $\{x_\alpha\}_{\alpha \in A}$ be an upward directed set in X so that $\bigvee_{\alpha \in A} ix_\alpha = x^{**}$ exists in X^{**}. Then $x^{**} = \lim_\alpha ix_\alpha$ (in the strong topology of X^{**}), i.e. $x^{**} \in iX$. Indeed, otherwise there would exist an increasing subsequence $\{x_{\alpha_j}\}_{j=1}^{\infty}$ of $\{x_\alpha\}_{\alpha \in A}$ such that $\inf_j \|x_{\alpha_{j+1}} - x_{\alpha_j}\| > 0$. Since this fact contradicts (iii) we conclude that iX is a band of X^{**}. Since X^{**} is order complete it follows from 1.a.10 that this band is actually a projection band. Hence, condition (iv) holds.

It remains to show that (iv) ⇒ (i). If iX is a band of X^{**} it follows from 1.b.16 that X is order continuous. Since every separable subspace of X is contained in a band of X having weak unit we may assume without loss of generality that X has a weak unit. Hence, we can apply the functional representation theorem 1.b.14 to X. Let $\{f_n\}_{n=1}^\infty$ be a weak Cauchy sequence in X. By arguing as in the proof of 1.c.2, we can construct a function $f \in L_1(\Omega, \Sigma, \mu)$ such that, for every $g \in X^*$, $fg \in L_1(\Omega, \Sigma, \mu)$ and $\int_\Omega (f_n - f)g\, d\mu \to 0$ as $n = \infty$. This means that f is the w^*-limit in X^{**} of the sequence $\{if_n\}_{n=1}^\infty$. On the other hand, $|f|$ is the l.u.b. in X^{**} of the sequence $\{|f|\chi_{\sigma_n}\}_{n=1}^\infty$, where $\sigma_n = \{\omega \in \Omega, |f(\omega)| \leq n\}$. This implies that $f \in iX$ since $|f|\chi_{\sigma_n} \in X$ for all n and iX is assumed to be a band of X^{**}. □

Remark. The proof of 1.a.5 can be used as an alternative proof of (ii) ⇒ (iii) in 1.c.4. This proof is more complicated than the simple argument presented here. It has however the advantage that it produces a sublattice order isomorphic to c_0 (and not only a subspace isomorphic to c_0). We can deduce thus that if a Banach lattice has a subspace isomorphic to c_0 then it has also a sublattice order isomorphic to c_0.

We pass now to the characterization of reflexivity in Banach lattices and their subspaces. In this context, reflexivity is usually proved by using the well known result of Eberlein which asserts that a Banach space X is reflexive if and only if it is weakly sequentially complete and its unit ball B_X is conditionally weakly compact (the latter means that every bounded sequence contains a weak Cauchy subsequence).

Theorem 1.c.5. *The following properties are equivalent for every Banach lattice X.*

(i) *X is reflexive.*

(ii) *No subspace of X is isomorphic to l_1 or to c_0.*

(iii) *Every norm bounded increasing sequence in X has a strong limit and X^* is σ-order continuous.*

The equivalence between (i) *and* (ii) *remains valid also in the case when X is a subspace of an order continuous Banach lattice.*

The proof for (i) ⇔ (iii) was given in [106] while (i) ⇔ (ii) was proved first in [84], [99] for general lattices and in [130] for subspaces.

Proof. The implication (i) ⇒ (ii) holds trivially in every Banach space. Assume now that X is either a Banach lattice or a subspace of an order continuous Banach lattice and that X does not have any subspace isomorphic to l_1 or to c_0. Then, by 1.c.4, X is weakly sequentially complete. Furthermore, by I.2.e.5, B_X is also conditionally weakly compact. Thus, X is reflexive i.e. (ii) ⇒ (i). The fact that (i) ⇒ (iii) follows easily from 1.c.4 and 1.a.7. Therefore, it remains to prove that (iii) ⇒ (i). Using once more 1.c.4, we conclude that it suffices to prove that (iii) implies the conditional weak compactness of B_X. Let $\{x_n\}_{n=1}^\infty$ be a norm bounded sequence in X. By 1.a.9, X is an unconditional direct sum of a family of mutually disjoint ideals having a weak unit so that $[x_n]_{n=1}^\infty$ is entirely contained in one of these ideals, say X_0. Since $X=X_0 \oplus X_0^\perp$ we get that X_0^* is order isometric to a sublattice of X^* and, therefore, also σ-order continuous.

We use now the functional representation of X_0 as a Köthe function space on some probability space (Ω, Σ, μ). Put $v_n(\sigma)=\int_\sigma x_n\, d\mu$, $\sigma \in \Sigma$, $n=1, 2, \ldots$, and observe that the measures $\{v_n\}_{n=1}^\infty$ have the following properties: (1) $\{v_n\}_{n=1}^\infty$ are uniformly bounded since $|v_n(\sigma)| \leq \sup_n \|x_n\|_{X_0} < \infty$, for all n and $\sigma \in \Sigma$, (2) the σ-additivity of $\{v_n\}_{n=1}^\infty$ is uniform since $|v_n(\sigma)| \leq \|\chi_\sigma\|_{X_0^*} \cdot \sup_n \|x_n\|_{X_0}$ and X_0^* is σ-order continuous. It follows that these measures form a conditionally weakly compact set in the Banach space of all bounded measures on (Ω, Σ) (see e.g. [32] IV.9.1) and, therefore, there is a subsequence $\{v_{n_i}\}_{i=1}^\infty$ of $\{v_n\}_{n=1}^\infty$ such that $v(\sigma)=\lim_{i\to\infty} v_{n_i}(\sigma)$ exists for all $\sigma \in \Sigma$. By the σ-order continuity of X_0^*, the μ-simple functions are norm dense in X_0^*. Hence, the limit, as $i \to \infty$, of

$$g(x_{n_i}) = \int_\Omega g x_{n_i}\, d\mu = \int_\Omega g\, dv_{n_i},$$

exists for every $g \in X_0^*$. $\square$

It is clear that, in general, 1.c.4 and 1.c.5 fail to be true for subspaces of arbitrary Banach lattices. For instance, the space J of R. C. James presented in I.1.d.2, which is, as any separable space, isometric to a subspace of $C(0, 1)$ or of l_∞, is not weakly sequentially complete despite of the fact that it contains no subspace isomorphic to c_0.

It is however possible to extend 1.c.4 and 1.c.5 to complemented subspaces of a general Banach lattice. For this purpose we need first the following result from [42], [59].

Proposition 1.c.6. *Let Y be a complemented subspace of a Banach lattice X. If Y contains no subspace isomorphic to c_0 then there exists an order continuous Banach lattice X_1 which contains a complemented subspace Y_1 isomorphic to Y.*

Proof. Let $\|\cdot\|$ denote the norm in X and let P be a projection from X onto Y. Define a new semi-norm on X, by putting,

$$\|x\|_1 = \sup\{\|Pz\|;\ |z| \leq |x|\}, \quad x \in X$$

(the triangle inequality for $\|\cdot\|_1$ is proved by using the decomposition property).

Let I denote the ideal of X consisting of all $x \in X$ for which $\|x\|_1 = 0$. The quotient space X/I becomes a vector lattice when its positive cone is taken to be the image, under the quotient map $Q: X \xrightarrow{\text{onto}} X/I$, of the positive cone of X. We norm X/I by putting $\|Qx\|_1 = \|x\|_1$. The completion X_1 of X/I is a Banach lattice and the map Q, restricted to Y, is an isomorphism from Y onto a subspace Y_1 of X_1 since

$$\|y\| \leqslant \|y\|_1 = \|Qy\|_1 \leqslant \|P\| \|y\|, \quad y \in Y.$$

Observe also that if, for some $x_1, x_2 \in X$, we have $Qx_1 = Qx_2$ then $Px_1 = Px_2$. Thus, by putting $P_1(Qx) = Q(Px)$, we define a map P_1 from X/I onto Y_1 which extends uniquely to a bounded projection from X_1 onto Y_1 since

$$\|P_1(Qx)\|_1 = \|Px\|_1 = \sup\{\|Pz\|; |z| \leqslant |Px|\} \leqslant \|P\| \, \|Px\| \leqslant \|P\| \, \|x\|_1 = \|P\| \, \|Qx\|_1,$$

for all $x \in X$.

It remains to show that X_1 is order continuous. If this were not true then, by 1.a.5 and 1.a.7, X_1 would contain a sequence of positive vectors of norm one which is equivalent to the unit vector basis of c_0. Evidently, there is no loss of generality in assuming that the elements of this sequence belong to X/I, i.e. that there exist a constant $C < \infty$ and a sequence $\{x_n\}_{n=1}^\infty$ of positive elements of X so that $\|x_n\|_1 = 1$ for all n and

$$C^{-1} \max_n |a_n| \leqslant \left\| \sum_{n=1}^\infty a_n Q x_n \right\|_1 = \left\| \sum_{n=1}^\infty a_n x_n \right\|_1 \leqslant C \max_n |a_n|,$$

for every choice of $(a_1, a_2, \ldots) \in c_0$. Choose now vectors $u_n \in X$ for which $|u_n| \leqslant x_n$ and $1 \geqslant \|Pu_n\| \geqslant 1/2$ for all n. Then, for $(a_1, a_2, \ldots) \in c_0$, we have that

$$\begin{aligned} \left\| \sum_{n=1}^\infty a_n P_1 Q u_n \right\|_1 &\leqslant \|P_1\| \left\| \sum_{n=1}^\infty |a_n| \, |Qu_n| \right\|_1 \\ &\leqslant \|P_1\| \left\| \sum_{n=1}^\infty |a_n| x_n \right\|_1 \leqslant \|P_1\| C \max_n |a_n|. \end{aligned}$$

On the other hand, $Qu_n \xrightarrow{w} 0$ in X_1 (use the fact that $Qx_n \xrightarrow{w} 0$ in X_1 and that $|Qu_n| \leqslant Q(|u_n|) \leqslant Qx_n$ for all n) and therefore also $P_1 Qu_n \xrightarrow{w} 0$ in this space. Thus, by I.1.a.12, we can assume without loss of generality that $\{P_1 Qu_n\}_{n=1}^\infty$ is a basic sequence in Y_1. Since $\|P_1 Qu_n\|_1 \geqslant \|Pu_n\| \geqslant 1/2$ for all n it follows that $\{P_1 Qu_n\}_{n=1}^\infty$ is equivalent to the unit vector basis of c_0 and this contradicts our assumption on Y (which is isomorphic to Y_1). □

The following result is an immediate consequence of 1.c.4, 1.c.5 and 1.c.6.

Theorem 1.c.7. *Let Y be a complemented subspace of a Banach lattice. Then*
(i) *Y is weakly sequentially complete if and only if no subspace of Y is isomorphic to c_0.*
(ii) *Y is reflexive if and only if no subspace of Y is isomorphic to l_1 or to c_0.*

We conclude this section by presenting some results concerning the existence of unconditional basic sequences in subspaces of Banach lattices. We begin with a generalization to Banach lattices of a result of M. I. Kadec and A. Pelczynski [61] (proved originally only for L_p spaces).

Proposition 1.c.8 [42]. *Let X be an order continuous Banach lattice with a weak unit (in particular, a separable σ-complete lattice). Any closed subspace of Y of X is either isomorphic to a subspace of some L_1 space or there exist a sequence of normalized vectors $\{y_n\}_{n=1}^\infty$ in Y and a sequence of mutually disjoint elements $\{x_n\}_{n=1}^\infty$ of X such that $\{y_n\}_{n=1}^\infty$ is equivalent to the (unconditional) basic sequence $\{x_n\}_{n=1}^\infty$.*

Proof. We assume, as we may, that X is a Köthe function space over some probability measure space (Ω, Σ, μ). For $x \in X$ and $\varepsilon > 0$, put $\sigma(x, \varepsilon) = \{\omega \in \Omega; |x(\omega)| \geqslant \varepsilon \|x\|_X\}$ and consider the set $M(\varepsilon) = \{x \in X; \mu(\sigma(x, \varepsilon)) \geqslant \varepsilon\}$. If $Y \subset M(\varepsilon)$ for some $\varepsilon > 0$ then

$$\|y\|_X \geqslant \|y\|_1 = \int_\Omega |y(\omega)|\, d\mu \geqslant \int_{\sigma(y,\varepsilon)} |y(\omega)|\, d\mu \geqslant \varepsilon^2 \|y\|_X, \quad y \in Y$$

i.e. Y is isomorphic to a subspace of $L_1(\Omega, \Sigma, \mu)$. Otherwise, we can find a sequence $\{z_n\}_{n=1}^\infty \subset Y$ with $\|z_n\|_X = 1$ and $z_n \notin M(2^{-n})$ for all n. For $m > n$, put $\sigma_{n,m} = \sigma(z_n, 2^{-n}) \sim \bigcup_{k=m}^\infty \sigma(z_k, 2^{-k})$. Then, for any fixed n, $\lim_{m\to\infty} \mu(\sigma_{n,m}) = \mu(\sigma(z_n, 2^{-n}))$, which implies that $\lim_{m\to\infty} \|z_n \chi_{\sigma(z_n, 2^{-n})} - z_n \chi_{\sigma_{n,m}}\|_X = 0$ for all n (use the fact that $z_n \chi_{\sigma_{n,m}}, z_n \chi_{\sigma(z_n, 2^{-n})} \in X$, since X is an ideal in $L_1(\Omega, \Sigma, \mu)$, and the σ-order continuity of X).

We choose now, inductively, a subsequence $\{z_{n_i}\}_{i=1}^\infty$ of $\{z_n\}_{n=1}^\infty$ and a sequence of mutually disjoint sets $\{\sigma_i\}_{i=1}^\infty \subset \Sigma$ such that, for each i, $\sigma_i \subset \sigma(z_{n_i}, 2^{-n_i})$ and $\|z_{n_i} \chi_{\sigma(z_{n_i}, 2^{-n_i})} - z_{n_i} \chi_{\sigma_i}\|_X \leqslant 2^{-i}$. Put $y_i = z_{n_i}$, $x_i = z_{n_i} \chi_{\sigma_i}$, $i = 1, 2, \ldots$ and observe that

$$\|x_i - y_i\|_X \leqslant \|z_{n_i} - z_{n_i} \chi_{\sigma(z_{n_i}, 2^{-n_i})}\|_X + 2^{-i} \leqslant 2^{-n_i} + 2^{-i} \leqslant 2^{-i+1}.$$

Hence, by the perturbation result I.1.a.9(i), $\{y_i\}_{i=1}^\infty$ is equivalent to the sequence of disjoint elements $\{x_i\}_{i=1}^\infty$. □

Remark. The assumption in 1.c.8 that X has a weak unit is redundant: in the general case we get that either Y contains an unconditional basic sequence or every *separable* subspace of Y is isomorphic to a subspace of some L_1 space. As it will be shown in Vol. III, the latter condition already implies that Y itself is isomorphic to a subspace of an L_1 space.

A deep result of H. P. Rosenthal [115], to be proved in Vol. IV, states that every infinite dimensional subspace of an L_1 space contains a subspace with an unconditional basis. Combining this fact with 1.c.8 and the remark thereafter we obtain a positive solution to I.1.d.5 for subspaces of order continuous Banach lattices.

Theorem 1.c.9. *Every infinite dimensional subspace of an order continuous Banach lattice contains a subspace with an unconditional basis.*

We already mentioned above that it follows from 1.c.4 and 1.c.5 that there exist spaces (e.g. the space J) which do not embed isomorphically in an order continuous Banach lattice. We present next a result of a different nature which also allows us to deduce that certain Banach spaces do not embed in such a lattice.

Proposition 1.c.10. *Let X be an order continuous Banach lattice and let $\{x_n\}_{n=1}^\infty$ be a sequence of elements of norm one in X. Then either there exists a constant $c>0$ such that, for every choice of scalars $\{a_n\}_{n=1}^\infty$, we have*

$$2^{-n}\sum_{\varepsilon_i=\pm 1}\left\|\sum_{i=1}^n \varepsilon_i a_i x_i\right\|_X \geq c\left(\sum_{i=1}^n |a_i|^2\right)^{1/2}, \quad n=1,2,\ldots$$

or $\{x_n\}_{n=1}^\infty$ contains a subsequence $\{x_{n_j}\}_{j=1}^\infty$ which is an unconditional basic sequence equivalent to a sequence of disjoint elements of X.

Proof. We may assume without loss of generality that X has a weak unit and we can thus represent X as in 1.b.14. Using the same notation as in the proof of 1.c.8, if $\{x_n\}_{n=1}^\infty \subset M(\varepsilon)$ for some $\varepsilon>0$ then

$$1=\|x_n\|_X \geq \|x_n\|_1 \geq \varepsilon^2\|x_n\|_X=\varepsilon^2\,.$$

By Khintchine's inequality I.2.b.3 and the triangle inequality in l_2, we get, for arbitrary functions $\{f_i\}_{i=1}^\infty$ in an $L_1(\Omega,\Sigma,\mu)$-space,

$$\begin{aligned}2^{-n}\sum_{\varepsilon_i=\pm 1}\left\|\sum_{i=1}^n \varepsilon_i f_i\right\|_1 &= \int_\Omega\int_0^1\left|\sum_{i=1}^n r_i(u)f_i(\omega)\right| du\, d\mu(\omega)\\ &\geq A_1\int_\Omega\left(\sum_{i=1}^n |f_i(\omega)|^2\right)^{1/2} d\mu(\omega)\\ &\geq A_1\left(\sum_{i=1}^n\left(\int_\Omega |f_i(\omega)|\, d\mu(\omega)\right)^2\right)^{1/2}\\ &= A_1\left(\sum_{i=1}^n \|f_i\|_1^2\right)^{1/2},\end{aligned}$$

where $\{r_i\}_{i=1}^n$ denote the first n Rademacher functions. Hence,

$$2^{-n}\sum_{\varepsilon_i=\pm 1}\left\|\sum_{i=1}^n \varepsilon_i a_i x_i\right\|_X \geq 2^{-n}\sum_{\varepsilon_i=\pm 1}\left\|\sum_{i=1}^n \varepsilon_i a_i x_i\right\|_1 \geq A_1\varepsilon^2\left(\sum_{i=1}^n |a_i|^2\right)^{1/2},$$

for every choice of $\{a_i\}_{i=1}^n$. On the other hand, if there is no $\varepsilon>0$ so that $\{x_n\}_{n=1}^\infty \subset M(\varepsilon)$ then we proceed exactly as in the proof of 1.c.8 and choose a subsequence $\{x_{n_j}\}_{j=1}^\infty$ of $\{x_n\}_{n=1}^\infty$ which is equivalent to a sequence of disjoint elements in X and thus, unconditional. □

In connection with 1.c.10, consider the space E of Maurey and Rosenthal which was presented in I.1.d.6. It follows from the properties of the sequence

$\{m_i\}_{i=1}^\infty$ of integers constructed there that, for every $1>\eta>0$, it is possible to find integers h and j so that

$$\sum_{i=1}^{j-1}\left(\frac{m_i}{m_j}\right)^{1/2}\leqslant\frac{h}{m_j}\leqslant\eta^2 .$$

Therefore, for $k=\sum_{i=1}^{j-1} m_i+h$, the unit vector basis $\{e_n\}_{n=1}^\infty$ of E satisfies.

$$\max_{\varepsilon_i=\pm 1}\left\|\sum_{i=1}^k \varepsilon_i e_i\right\|=\left\|\sum_{i=1}^k e_i\right\|\leqslant\sum_{i=1}^{j-1} m_i^{1/2}+\frac{h}{m_j^{1/2}}\leqslant\frac{2h}{m_j^{1/2}}\leqslant 2\eta h^{1/2}$$

i.e.

$$k^{-1/2}\max_{\varepsilon_i=\pm 1}\left\|\sum_{i=1}^k \varepsilon_i e_i\right\|\leqslant 2\eta .$$

This fact and the property of E that no subsequence of $\{e_n\}_{n=1}^\infty$ is unconditional imply, by 1.c.10, that E is not isomorphic to a subspace of an order continuous Banach lattice.

d. *p*-Convexity in Banach Lattices

In this section we introduce and study the mutually dual notions of p-convexity and q-concavity in Banach lattices. These two notions turn out to be an important tool in the study of isomorphic properties of lattices. For example, they play a crucial role in the study of uniform convexity in Banach lattices (in Section f below) and in the study of rearrangement invariant function spaces (in Section 2.e. below). In the definition of these notions there enter expressions of the form $\left(\sum_{i=1}^n |x_i|^p\right)^{1/p}$, where $p\geqslant 1$ and the $\{x_i\}_{i=1}^n$ are elements of a lattice X. In case X is order continuous such an expression can be easily defined by using the representation theorem 1.b.14. We need however a proper definition of this expression for general Banach lattices. We start this section by presenting a method which allows us to define even more complicated expressions than $\left(\sum_{i=1}^n |x_i|^p\right)^{1/p}$ in general Banach lattices.

Let $\mathscr{H}_n$ be the family of all functions $f(t_1, \ldots, t_n)\colon R^n\to R$ which are obtained from the functions $\varphi_i(t_1, \ldots, t_n)=t_i$, $i=1, \ldots, n$, by applying finitely many operations of addition, multiplication by scalars and finite suprema and infima. It is easily seen that each $f\in\mathscr{H}_n$ is continuous on R^n and homogeneous of degree one i.e., for every $\lambda\geqslant 0$, $f(\lambda t_1, \ldots, \lambda t_n)=\lambda f(t_1, \ldots, t_n)$. Therefore, for any $f\in\mathscr{H}_n$, there exists a constant $M<\infty$ so that $|f(t_1, \ldots, t_n)|\leqslant M(|t_1|\vee\cdots\vee|t_n|)$ for all $(t_1, \ldots, t_n)\in R^n$.

Let $f \in \mathscr{H}_n$ and let $\{x_i\}_{i=1}^n$ be a finite set of elements of a Banach lattice X. By replacing formally each of the variables t_i with the corresponding vector x_i, we can give a meaning to the expression $f(x_1, \ldots, x_n) \in X$. This procedure defines the element $f(x_1, \ldots, x_n)$ in a unique manner in the sense that if $f(t_1, \ldots, t_n) = g(t_1, \ldots, t_n)$ for some $g \in \mathscr{H}_n$ and every $(t_1, \ldots, t_n) \in R^n$ then also $f(x_1, \ldots, x_n) = g(x_1, \ldots, x_n)$. The proof of this fact is trivial when X consists of functions on some set and the order in X is defined pointwise. The case of a general Banach lattice can be reduced to the function case in the following way. Put $x_0 = |x_1| \vee \cdots \vee |x_n|$ and observe that both $f(x_1, \ldots, x_n)$ and $g(x_1, \ldots, x_n)$ belong to the (in general non-closed) ideal $I(x_0)$ of all $x \in X$ for which there exists some $\lambda \geqslant 0$ so that $|x| \leqslant \lambda x_0/\|x_0\|$. In the ideal $I(x_0)$ we define the norm $\|x_0\|_\infty = \inf\{\lambda \geqslant 0; |x| \leqslant \lambda x_0/\|x_0\|\}$ and observe that the completion of $I(x_0)$, endowed with the norm $\|\cdot\|_\infty$, is an abstract M space with a strong unit. Thus, by 1.b.6, the completion of $(I(x_0), \|\cdot\|_\infty)$ is order isometric to a space of continuous functions. The observation above shows that in this M space we have $f(x_1, \ldots, x_n) = g(x_1, \ldots, x_n)$ i.e. $\|f(x_1, \ldots, x_n) - g(x_1, \ldots, x_n)\|_\infty = 0$. This implies that $f(x_1, \ldots, x_n) = g(x_1, \ldots, x_n)$ also in X since $\|x\| \leqslant \|x\|_\infty$ for every $x \in I(x_0)$.

The fact that $f(x_1, \ldots, x_n)$ is uniquely defined for every $f \in \mathscr{H}_n$ implies that the map $\tau: \mathscr{H}_n \to X$, defined by $\tau f(t_1, \ldots, t_n) = f(x_1, \ldots, x_n)$, is linear and preserves the lattice operations (e.g. if $f(t_1, \ldots, t_n) = g(t_1, \ldots, t_n) \vee h(t_1, \ldots, t_n)$ in $\mathscr{H}_n$ then one possibility, and therefore the only possibility to define $f(x_1, \ldots, x_n)$ is to put $f(x_1, \ldots, x_n) = g(x_1, \ldots, x_n) \vee h(x_1, \ldots, x_n)$).

The map $\tau: \mathscr{H}_n \to X$ can be made into a continuous map in the following way. Let B_n be the subset of R^n of all n-tuples $(t_1, \ldots, t_n)$ for which $|t_1| \vee \cdots \vee |t_n| = 1$ (i.e. the unit sphere of the real space l_∞^n) and consider $\mathscr{H}_n$ as a sublattice of $C(B_n)$, the space of all continuous functions on B_n. The map $\tau: \mathscr{H}_n \to X$ is continuous when $\mathscr{H}_n$ is endowed with the norm induced by $C(B_n)$. Indeed, if for some $f \in \mathscr{H}_n$ we have

$$\sup\{|f(t_1, \ldots, t_n)|; (t_1, \ldots, t_n) \in B_n\} \leqslant 1$$

then

$$|f(t_1, \ldots, t_n)| \leqslant |t_1| \vee \cdots \vee |t_n| = |\varphi_1(t_1, \ldots, t_n)| \vee \cdots \vee |\varphi_n(t_1, \ldots, t_n)|$$

for every $(t_1, \ldots, t_n) \in R^n$. Since τ is order preserving it follows that

$$|f(x_1, \ldots, x_n)| \leqslant |x_1| \vee \cdots \vee |x_n| .$$

Hence, for every $f \in \mathscr{H}_n$,

$$\|f(x_1, \ldots, x_n)\|_X \leqslant \|x_0\|_X \cdot \|f(t_1, \ldots, t_n)\|_{C(B_n)} ,$$

where $x_0 = |x_1| \vee \cdots \vee |x_n|$.

Observe also that $\mathscr{H}_n$, as a sublattice of $C(B_n)$, separates the points of B_n and contains the function identically equal to one (since

$$|\varphi_1(t_1, \ldots, t_n)| \vee \cdots \vee |\varphi_n(t_1, \ldots, t_n)| \equiv 1 \quad \text{for} \quad (t_1, \ldots, t_n) \in B_n).$$

Thus, by 1.b.5, $\mathcal{H}_n$ is dense in $C(B_n)$. The closure $\overline{\mathcal{H}}_n$ of $\mathcal{H}_n$, when the elements of $\mathcal{H}_n$ are considered as functions defined on all of R^n, consists of all the functions $f: R^n \to R$, which are continuous and homogeneous of degree one on R^n. This fact enables us to extend τ, in a unique manner, to a map from $\overline{\mathcal{H}}_n$ into X which is linear, continuous (when $\overline{\mathcal{H}}_n$ is identified with $C(B_n)$) and preserves the lattice operations.

We collect the observations made above in the following theorem (cf. Yudin [131] and Krivine [66]).

Theorem 1.d.1. *Let X be a Banach lattice and let $\{x_i\}_{i=1}^n$ be a finite subset of X. Then there is a unique map τ from the lattice $\overline{\mathcal{H}}_n$, of all the functions which are continuous and homogeneous of degree one on R^n, into X such that:*

(i) *$\tau\varphi_i = x_i$ for $1 \leq i \leq n$, where $\varphi_i(t_1, \ldots, t_n) = t_i$.*

(ii) *τ is linear and preserves the lattice operations.*

The map τ satisfies

$$\|\tau(f)\| \leq \| |x_1| \vee \cdots \vee |x_n| \| \sup \{|f(t_1, \ldots, t_n)|; |t_1| \vee \cdots \vee |t_n| = 1\},$$

for every $f \in \overline{\mathcal{H}}_n$.

The element $\tau(f)$ will be usually denoted by $f(x_1, \ldots, x_n)$. In many cases we shall work with functions having the form

$$f(t_1, \ldots, t_n) = \left(\sum_{i=1}^n |t_i|^p\right)^{1/p},$$

for some $p \geq 1$. It is worthwhile to remark that the vector $\left(\sum_{i=1}^n |x_i|^p\right)^{1/p}$ can be also represented by the following formula: if $1/p + 1/q = 1$ then

$$\left(\sum_{i=1}^n |x_i|^p\right)^{1/p} = \text{l.u.b.} \left\{\sum_{i=1}^n a_i x_i\right\},$$

the l.u.b. being taken (in the sense of order in a Banach lattice X containing the vectors $\{x_i\}_{i=1}^n$) over all $(a_1, \ldots, a_n) \in R^n$ for which $\sum_{i=1}^n |a_i|^q \leq 1$. Indeed, for every such $(a_1, \ldots, a_n) \in R^n$, we have that $\sum_{i=1}^n a_i t_i \leq \left(\sum_{i=1}^n |t_i|^p\right)^{1/p}$ for $(t_1, \ldots, t_n) \in R^n$ and, therefore, by 1.d.1, also that $\sum_{i=1}^n a_i x_i \leq \left(\sum_{i=1}^n |x_i|^p\right)^{1/p}$, for $\{x_i\}_{i=1}^n$ in X. Since the statement we want to prove for vectors in X is true for scalars we can find an increasing sequence of functions $f_k(t_1, \ldots, t_n) \in \overline{\mathcal{H}}_n = C(B_n)$, $k = 1, 2, \ldots$, which are finite suprema of linear combinations of the form $\sum_{i=1}^{n} a_i t_i$ with $\sum_{i=1}^{n} |a_i|^q \leq 1$, so that $\{f_k\}_{k=1}^{\infty}$ converges pointwise and thus, by Dini's theorem, in the norm of $C(B_n)$ to $\left(\sum_{i=1}^{n} |t_i|^p\right)^{1/p}$. The continuity of the map $\tau: \overline{\mathscr{H}_n} \to X$, described in 1.d.1, shows that $f_k(x_1, \ldots, x_n) \to \left(\sum_{i=1}^{n} |x_i|^p\right)^{1/p}$ as $k \to \infty$ and this completes the proof.

A special case of 1.d.1 can be used to define the notion of a complex Banach lattice. Let X be a real Banach lattice and let $\tilde{X}$ be the linear space $X \oplus X$ which is made into a complex linear space by setting $(a+ib)(x_1, x_2) = (ax_1 - bx_2, ax_2 + bx_1)$. We define the notions of absolute value and norm in $\tilde{X}$ by putting

$$|(x_1, x_2)| = (|x_1|^2 + |x_2|^2)^{1/2}, \qquad \|(x_1, x_2)\|_{\tilde{X}} = \| \, |(x_1, x_2)| \, \|_X .$$

The space $(\tilde{X}, \| \ \|_{\tilde{X}})$ is said to be a *complex Banach lattice* or, more precisely, the *complexification of the real Banach lattice X*. As expected, the complex $L_p(\mu)$ or $C(K)$ spaces are the complexifications of the real $L_p(\mu)$ or $C(K)$ with the same μ, respectively K. Since, by definition, every complex Banach lattice is the complexification of a real lattice all the notions and results of this volume can be carried over in a straightforward manner from the real case to the complex case. We shall however continue to assume in the sequel that, unless stated otherwise, the Banach lattices which we consider are real.

In connection with the functional calculus established in 1.d.1 it is often useful to apply the following set of Hölder type inequalities (cf. [66]).

Proposition 1.d.2. *Let X be a Banach lattice.*

(i) *For every* $0 < \theta < 1$ *and every* $x, y \in X$,

$$\| |x|^{\theta} |y|^{1-\theta} \| \leq \|x\|^{\theta} \|y\|^{1-\theta} .$$

(ii) *For every choice of* $1 \leq p < r < q \leq \infty$, $\{x_i\}_{i=1}^n$ *in X and positive scalars* $\{a_i\}_{i=1}^n$,

$$\left(\sum_{i=1}^{n} a_i |x_i|^r\right)^{1/r} \leq \left(\sum_{i=1}^{n} a_i |x_i|^p\right)^{\theta/p} \left(\sum_{i=1}^{n} a_i |x_i|^q\right)^{(1-\theta)/q},$$

where $0 < \theta < 1$ *is defined by* $1/r = \theta/p + (1-\theta)/q$.

(iii) *For every* $1 \leq p, q \leq \infty$ *with* $1/p + 1/q = 1$ *and every choice of* $\{x_i\}_{i=1}^n$ *in X and* $\{x_i^*\}_{i=1}^n$ *in* X^*,

$$\sum_{i=1}^{n} x_i^*(x_i) \leq \left(\left(\sum_{i=1}^{n} |x_i^*|^q\right)^{1/q}\right)\left(\left(\sum_{i=1}^{n} |x_i|^p\right)^{1/p}\right).$$

As usual, if $q=\infty$ an expression of the form $\left(\sum_{i=1}^{n}|u_i|^q\right)^{1/q}$ means $\bigvee_{i=1}^{n}|u_i|$.

Proof. (i) Since $|s|^\theta|t|^{1-\theta}$ is a homogeneous expression of degree one on R^2 and $|s|^\theta|t|^{1-\theta}\leqslant\theta|s|+(1-\theta)|t|$, for every $(s, t)\in R^2$, we get, by 1.d.1, that

$$\begin{aligned}\||x|^\theta|y|^{1-\theta}\| &= \||c^{1/\theta}x|^\theta|c^{-1/(1-\theta)}y|^{1-\theta}\|\\ &\leqslant \theta c^{1/\theta}\|x\|+(1-\theta)c^{-1/(1-\theta)}\|y\|,\end{aligned}$$

for every $c\geqslant 0$. The proof of (i) is then completed by taking $c=(\|y\|/\|x\|)^{\theta(1-\theta)}$. Assertion (ii) follows from 1.d.1 and the usual Hölder inequality. Also (iii) is an immediate consequence of Hölder's inequality when X and X^* are lattices of functions. In order to prove (iii) for general Banach lattices, put $x_0^*=\left(\sum_{i=1}^{n}|x_i^*|^q\right)^{1/q}$ and notice that

$$\|x\|_1=x_0^*(|x|)\,,\quad x\in X\,,$$

defines a seminorm on X which is additive on the positive cone of X. Hence, the completion X_1 of X endowed with $\|\cdot\|_1$ (modulo those elements $x\in X_1$ for which $\|x\|_1=0$) forms an abstract L_1 space which, by 1.b.2, is order isometric to an $L_1(\Omega, \Sigma, v)$ space. Moreover, since $\|x\|_1\leqslant\|x_0^*\|\|x\|$, for all $x\in X$, the formal identity mapping j from X into X_1 is bounded. Observe also that, for each $1\leqslant i\leqslant n$ and $x\in X$, we have

$$x_i^*(x)\leqslant\|x\|_1\,,$$

i.e. x_i^* extends to an element g_i of $L_\infty(\Omega, \Sigma, v)$ with $\|g_i\|_\infty\leqslant 1$. Let $f_i\in L_1(\Omega, \Sigma, v)$, $1\leqslant i\leqslant n$, be the functions corresponding to the elements $jx_i\in X_1$, $1\leqslant i\leqslant n$. Then, by the usual Hölder inequality, we get that

$$\sum_{i=1}^{n}x_i^*(x_i)=\sum_{i=1}^{n}\int_\Omega g_i(\omega)f_i(\omega)\,dv\leqslant\int_\Omega\left(\sum_{i=1}^{n}|g_i(\omega)|^q\right)^{1/q}\left(\sum_{i=1}^{n}|f_i(\omega)|^p\right)^{1/p}dv$$

which means that $\sum_{i=1}^{n}x_i^*(x_i)$ is $\leqslant$ than the number obtained by applying the functional $\left(\sum_{i=1}^{n}|g_i|^q\right)^{1/q}\in L_1(\Omega, \Sigma, v)^*$ to the element $\left(\sum_{i=1}^{n}|f_i|^p\right)^{1/p}$ of $L_1(\Omega, \Sigma, v)$.

However, by the uniqueness of the map τ of 1.d.1 from the lattice $\overline{\mathscr{H}}_n$ of all functions which are continuous and homogeneous of degree one into $L_1(\Omega, \Sigma, v)$, we conclude that the elements $\left(\sum_{i=1}^{n}|f_i|^p\right)^{1/p}$ and $j\left(\sum_{i=1}^{n}|x_i|^p\right)^{1/p}$ are actually the same. Similarly, it follows that also the elements $\left(\sum_{i=1}^{n}|x_i^*|^q\right)^{1/q}$ and $j^*\left(\sum_{i=1}^{n}|g_i|^q\right)^{1/q}$

coincide. Thus,

$$\int_{\Omega}\left(\sum_{i=1}^{n}|g_i(\omega)|^q\right)^{1/q}\left(\sum_{i=1}^{n}|f_i(\omega)|^p\right)^{1/p}d\nu=\left(\left(\sum_{i=1}^{n}|x_i^*|^q\right)^{1/q}\right)\left(\left(\sum_{i=1}^{n}|x_i|^p\right)^{1/p}\right)$$

and this, of course, completes the proof of (iii). □

We turn now to the main topic of this section, namely that of p-convexity and p-concavity. The starting point is the observation that, for any sequence $\{f_i\}_{i=1}^n$ of functions in $L_p(\mu)$, $1\leq p\leq\infty$, we have the equality

$$\left\|\left(\sum_{i=1}^{n}|f_i|^p\right)^{1/p}\right\|_p=\left(\sum_{i=1}^{n}\|f_i\|_p^p\right)^{1/p},\quad \text{if } 1\leq p<\infty$$

or

$$\left\|\bigvee_{i=1}^{n}|f_i|\right\|_\infty=\max_{1\leq i\leq n}\|f_i\|_\infty,\quad \text{if } p=\infty .$$

If we replace in the formulas above the equality sign by an equivalence sign (i.e. we assume the existence of a two sided estimate of $\left\|\left(\sum_{i=1}^{n}|x_i|^p\right)^{1/p}\right\|$ in terms of $\left(\sum_{i=1}^{n}\|x_i\|^p\right)^{1/p}$, for any n-tuple of vectors $\{x_i\}_{i=1}^n$ in a Banach lattice X) then we get just a characterization of spaces isomorphic to $L_p(\mu)$ spaces (use 1.b.13). The notions of p-convexity and p-concavity arise if we replace in the formulas above the equality sign by *one sided* estimates. These two notions were introduced, under different names, in [31] and [41], for spaces with an unconditional basis, and in [66], for general Banach lattices as well as for operators from and into a Banach lattice.

Definition 1.d.3. Let X be a Banach lattice, V an arbitrary Banach space and let $1\leq p\leq\infty$.

(i) A linear operator $T: V\to X$ is called *p-convex* if there exists a constant $M<\infty$ so that

$$\left\|\left(\sum_{i=1}^{n}|Tv_i|^p\right)^{1/p}\right\|\leq M\left(\sum_{i=1}^{n}\|v_i\|^p\right)^{1/p},\quad \text{if } 1\leq p<\infty$$

or

$$\left\|\bigvee_{i=1}^{n}|Tv_i|\right\|\leq M\max_{1\leq i\leq n}\|v_i\|,\quad \text{if } p=\infty ,$$

for every choice of vectors $\{v_i\}_{i=1}^n$ in V. The smallest possible value of M is denoted by $M^{(p)}(T)$.

(ii) A linear operator $T: X \to V$ is called *p-concave* if there exists a constant $M < \infty$ so that

$$\left(\sum_{i=1}^{n} \|Tx_i\|^p\right)^{1/p} \leqslant M \left\| \left(\sum_{i=1}^{n} |x_i|^p\right)^{1/p} \right\|, \quad \text{if } 1 \leqslant p < \infty$$

or

$$\max_{1 \leqslant i \leqslant n} \|Tx_i\| \leqslant M \left\| \bigvee_{i=1}^{n} |x_i| \right\|, \quad \text{if } p = \infty,$$

for every choice of vectors $\{x_i\}_{i=1}^n$ in X. The smallest possible value of M is denoted by $M_{(p)}(T)$.

(iii) We say that X is *p-convex* or *p-concave* if the identity operator I on X is p-convex, respectively, p-concave. In this case, we write $M^{(p)}(X)$ and $M_{(p)}(X)$ instead of $M^{(p)}(I)$, respectively, $M_{(p)}(I)$.

The constants $M^{(p)}(X)$ and $M_{(p)}(X)$ are called the *p-convexity*, respectively, the *p-concavity constant* of X.

Obviously, an operator T, which is p-convex or p-concave, for some $1 \leqslant p \leqslant \infty$, is necessarily bounded and $M^{(p)}(T) \geqslant \|T\|$, respectively, $M_{(p)}(T) \geqslant \|T\|$. The cases $p=1$ and $p=\infty$ are interesting only in part since every bounded operator $T: V \to X$ is 1-convex with $M^{(1)}(T) = \|T\|$ and every bounded operator $T: X \to V$ is ∞-concave with $M_{(\infty)}(T) = \|T\|$. In particular, every Banach lattice is both 1-convex and ∞-concave.

There is a simple and sometimes useful way of interpreting the notions of p-convex and p-concave operators by using some auxiliary spaces. We recall first the definition of the spaces $c_0(V)$ and $l_p(V)$, $1 \leqslant p \leqslant \infty$. These spaces consist of all the sequences $v = (v_1, v_2, \ldots)$ of elements of the Banach space V so that $a = (\|v_1\|, \|v_2\|, \ldots)$ belongs to c_0, respectively l_p, and the norm of v in $c_0(V)$ or $l_p(V)$ is, by definition, the norm of a in the respective sequence space.

For a Banach lattice X and $1 \leqslant p \leqslant \infty$, we let $\widetilde{X(l_p)}$ be the space of all sequences $x = (x_1, x_2 \ldots)$ of elements of X for which

$$\|x\|_{\widetilde{X(l_p)}} = \sup_n \left\| \left(\sum_{i=1}^{n} |x_i|^p\right)^{1/p} \right\| < \infty, \quad \text{if } 1 \leqslant p < \infty$$

or

$$\|x\|_{\widetilde{X(l_\infty)}} = \sup_n \left\| \bigvee_{i=1}^{n} |x_i| \right\| < \infty, \quad \text{if } p = \infty.$$

The closed subspace of $\widetilde{X(l_p)}$, spanned by the sequences $x = (x_1, x_2, \ldots)$ which are eventually zero, is denoted by $X(l_p)$. The space $X(l_\infty)$, which is always a proper subspace of $\widetilde{X(l_\infty)}$, is denoted also, for obvious reasons, by $X(c_0)$. For $1 \leqslant p < \infty$, the space $X(l_p)$ coincides with $\widetilde{X(l_p)}$ if and only if every norm bounded increasing sequence in X is convergent i.e. (in view of 1.c.4) if and only if X is weakly sequentially complete. In order to verify this statement we have to show that the sequence $\left\{\left(\sum_{i=1}^{n}|x_i|^p\right)^{1/p}\right\}_{n=1}^{\infty}$ converges in norm if and only if $\left\|\left(\sum_{i=m}^{n}|x_i|^p\right)^{1/p}\right\| \to 0$, as m and n tend to ∞. Both parts of this assertion are easy consequences of 1.d.1 and 1.d.2. The "only if" assertion, for example, is proved as follows. Let $f(s,t)$ be the continuous function on R^2 satisfying

$$f(s,t)||t|-|s||=||t|^p-|s|^p|,$$

for every $(s,t)\in R^2$. Then, by 1.d.1 and 1.d.2(i), we get, for every $y, z\in X$, that

$$\begin{aligned}\||\,|y|^p-|z|^p|^{1/p}\| &\le \||\,|y|-|z|\,|\|^{1/p}\|f(|y|,|z|)^{q/p}\|^{1/q}\\ &\le C_p\||\,|y|-|z|\,|\|^{1/p}\|\,|y|\vee|z|\,\|^{1/q},\end{aligned}$$

where $1/p+1/q=1$ and $C_p=\max\{f(s,t)^{1/p};\ |s|\le 1,\ |t|\le 1\}$.

A linear operator T from a Banach space V to a Banach lattice X is p-convex for some $1\le p<\infty$ if and only if the map $\hat{T}: l_p(V)\to X(l_p)$, defined by $\hat{T}(v_1, v_2, \dots)=(Tv_1, Tv_2, \dots)$, is a bounded operator. Moreover, we have $\|\hat{T}\|=M^{(p)}(T)$. The operator T is ∞-convex if and only if $\hat{T}$ is a bounded linear operator from $c_0(V)$ into $X(l_\infty)=X(c_0)$ or, alternatively, if $\hat{T}$ defines a bounded linear operator from $l_\infty(V)$ into $\widetilde{X(l_\infty)}$ (and, again, we have that $\|\hat{T}\|=M^{(\infty)}(T)$). Similarly, a linear operator $T: X\to V$ is p-concave for some $1\le p<\infty$ if and only if the map $\check{T}: X(l_p)\to l_p(V)$, defined by $\check{T}(x_1, x_2, \dots)=(Tx_1, Tx_2, \dots)$, is bounded. Moreover, $\|\check{T}\|=M_{(p)}(T)$. Note that if T is p-concave then $\check{T}$ can be actually defined as a map from $\widetilde{X(l_p)}$ into $l_p(V)$.

In order to study the behavior of the notions of p-convexity and p-concavity under duality we have to characterize the duals of the spaces introduced above. We note first the obvious fact that $l_p(V)^*$ is isometric to $l_q(V^*)$, where $1/p+1/q=1$ (for $p=\infty$ we have that $c_0(V)^*=l_1(V^*)$). We also claim that, *for every Banach lattice X and every $1\le p\le\infty$, the space $X(l_p)^*$ is order isometric to $\widetilde{X^*(l_q)}$, where again $1/p+1/q=1$.*

We shall prove this claim only for $1<p<\infty$. For $p=1$ and $p=\infty$ the proof is similar and simpler but the notation is somewhat different.

For every $(x_1^*, x_2^*, \dots)\in\widetilde{X^*(l_q)}$ and every sequence $(x_1, x_2, \dots)\in X(l_p)$ which is eventually zero put $\varphi(x_1, x_2, \dots)=\sum_{i=1}^{\infty}x_i^*(x_i)$. By 1.d.2(iii), we have

$$|\varphi(x_1, x_2, \dots)|\le\|(x_1^*, x_2^*, \dots)\|_{\widetilde{X(l_q)}}\|(x_1, x_2, \dots)\|_{X(l_p)}\,.$$

Hence, $\varphi\in X(l_p)^*$ and $\|\varphi\|\le\|(x_1^*, x_2^*, \dots)\|_{\widetilde{X(l_q)}}$. Conversely, let $\psi\in X(l_p)^*$ and, for any integer i, let $y_i^*\in X^*$ be defined by,

$$y_i^*(x)=\psi(0, \dots, 0, \underset{i}{x}, 0, \dots)\,.$$

The proof of our assertion will be completed once we show that

$$\sup_n \left\| \left(\sum_{i=1}^n |y_i^*|^q \right)^{1/q} \right\| \leq \|\psi\| .$$

By the remark following 1.d.1, $\left\| \left(\sum_{i=1}^n |y_i^*|^q \right)^{1/q} \right\|$ is the supremum of all the expressions of the form $\left\| \bigvee_{j=1}^k \sum_{i=1}^n a_{i,j} y_i^* \right\|$, where $\{a_{i,j}\}_{i=1}^n{}_{j=1}^k$ are arbitrary reals with $\sum_{i=1}^n |a_{i,j}|^p \leq 1$ for all $1 \leq j \leq k$. The definition of $\bigvee_{j=1}^k$ in X^* implies that, for every $x \geq 0$ in X,

$$\left(\bigvee_{j=1}^k \sum_{i=1}^n a_{i,j} y_i^* \right)(x) = \sup \left\{ \sum_{j=1}^k \left(\sum_{i=1}^n a_{i,j} y_i^* \right)(x_j);\ x_j \geq 0 \text{ for } 1 \leq j \leq k,\ \sum_{j=1}^k x_j = x \right\}$$
$$= \sup \left\{ \psi\left(\sum_{j=1}^k a_{1,j} x_j, \ldots, \sum_{j=1}^k a_{n,j} x_j, 0, 0, \ldots \right);\ x_j \geq 0 \text{ for } 1 \leq j \leq k,\ \sum_{j=1}^k x_j = x \right\}$$

from which we deduce that

$$\left(\bigvee_{j=1}^k \sum_{i=1}^n a_{i,j} y_i^* \right)(x) \leq \|\psi\| \sup \left\| \left(\sum_{i=1}^n \left| \sum_{j=1}^k a_{i,j} x_j \right|^p \right)^{1/p} \right\| ,$$

where the supremum is taken again over all $\{x_j\}_{j=1}^k$ as above. By the triangle inequality in l_p and 1.d.1, we get that

$$\left(\bigvee_{j=1}^k \sum_{i=1}^n a_{i,j} y_i^* \right)(x) \leq \|\psi\| \sup \left\| \sum_{j=1}^k \left(\sum_{i=1}^n |a_{i,j} x_j|^p \right)^{1/p} \right\| \leq \|\psi\| \|x\|$$

since $\sum_{i=1}^n |a_{i,j}|^p \leq 1$ for all $1 \leq j \leq k$. This completes the proof of our claim concerning $X(l_p)^*$. In particular, we get that $X(l_p)$ is reflexive if and only if X is reflexive and $1 < p < \infty$.

Suppose now that $T: V \to X$ is a linear operator and, for $1 < p < \infty$, consider the corresponding operator $\tilde{T}: l_p(V) \to X(l_p)$. It is easily checked that, by the duality relations established above, $(\tilde{T})^*$ coincides with the operator $(\widetilde{T^*}): X^*(l_q) \to l_q(V^*)$, where $1/p + 1/q = 1$. Hence, T is p-convex if and only if T^* is q-concave and $M^{(p)}(T) = M_{(q)}(T^*)$. Similarly, an operator $T: X \to V$ is p-concave if and only if T^* is q-convex. We collect these facts in the following proposition (cf. [66]).

Proposition 1.d.4. *Let X be a Banach lattice, V a Banach space and let $1 \leq p, q \leq \infty$ be so that $1/p + 1/q = 1$.*

(i) *A linear operator $T: V \to X$ is p-convex if and only if T^* is q-concave and, in this case, $M_{(q)}(T^*) = M^{(p)}(T)$.*

(ii) *A linear operator* $T: X \to V$ *is p-concave if and only if* T^* *is q-convex and, in this case,* $M^{(q)}(T^*) = M_{(p)}(T)$.

(iii) *X is p-convex (concave) if and only if* X^* *is q-concave (convex) and* $M_{(q)}(X^*) = M^{(p)}(X)$ $(M^{(q)}(X^*) = M_{(p)}(X))$.

We study next the dependence of p-convexity and p-concavity on p. For simplicity of notations, we put $M^{(p)}(T) = \infty$ or $M_{(p)}(T) = \infty$ if T is not p-convex, respectively, not p-concave.

Proposition 1.d.5. *Let X be a Banach lattice and let V be a Banach space. Let* $T: V \to X$ *and* $S: X \to V$ *be linear operators. Then the functions* $\varphi(\alpha) = \log M^{(1/\alpha)}(T)$ *and* $\psi(\beta) = \log M_{(1/\beta)}(S)$ *are convex. Consequently,* $M^{(p)}(T)$ *and* $M_{(p)}(S)$ *are non-decreasing, respectively, non-increasing continuous functions of p on any interval on which they are finite.*

Proof. Observe that, by the duality result 1.d.4, it suffices to prove that φ is convex. The definition of p-convexity shows that, whenever $M^{(p)}(T)$ is finite, we have

$$M^{(p)}(T) = \sup\left\{\left\|\left(\sum_{i=1}^{n} |Tv_i|^p\right)^{1/p}\right\|;\ v_i \in V \text{ for } 1 \leqslant i \leqslant n,\ \sum_{i=1}^{n} \|v_i\|^p = 1\right\}$$

$$= \sup\left\{\left\|\left(\sum_{i=1}^{n} a_i |Tw_i|^p\right)^{1/p}\right\|;\ w_i \in V,\ \|w_i\| = 1,\ a_i \geqslant 0 \text{ for } 1 \leqslant i \leqslant n, \text{ and } \sum_{i=1}^{n} a_i = 1\right\}.$$

But, by the Hölder type inequalities 1.d.2(i) and (ii), we get that

$$\left\|\left(\sum_{i=1}^{n} a_i |Tw_i|^r\right)^{1/r}\right\| \leqslant \left\|\left(\sum_{i=1}^{n} a_i |Tw_i|^p\right)^{1/p}\right\|^{\theta} \left\|\left(\sum_{i=1}^{n} a_i |Tw_i|^q\right)^{1/q}\right\|^{1-\theta},$$

whenever $1/r = \theta/p + (1-\theta)/q$ and $0 < \theta < 1$. It follows that $M^{(r)}(T) \leqslant M^{(p)}(T)^{\theta} \cdot M^{(q)}(T)^{1-\theta}$ and this, of course, implies that φ is convex. By taking $p=1$ and using the fact that $M^{(1)}(T) = \|T\| \leqslant M^{(q)}(T)$, we also get that $M^{(r)}(T) \leqslant M^{(q)}(T)$, i.e. that $M^{(p)}(T)$ is a non-decreasing function. That $M_{(p)}(S)$ is non-increasing follows by duality. □

Before presenting some examples, we want to illustrate the manner in which the properties introduced in 1.d.3 are generally used. A simple and nice application is the following generalization, due to B. Maurey [94], of the classical inequality of Khintchine I.2.b.3.

Theorem 1.d.6. (i) *Let X be a q-concave Banach lattice for some* $q < \infty$. *Then there exists a constant* $C < \infty$ *such that, for every sequence* $\{x_i\}_{i=1}^n$ *of elements of X, we* *have*

$$C^{-1}\left\|\left(\sum_{i=1}^{n}|x_i|^2\right)^{1/2}\right\| \leqslant \int_0^1\left\|\sum_{i=1}^{n} r_i(u)x_i\right\| du \leqslant C\left\|\left(\sum_{i=1}^{n}|x_i|^2\right)^{1/2}\right\|.$$

(ii) *Let* $\{x_i\}_{i=1}^{\infty}$ *be an unconditional basis of a Banach lattice X. Then there is a constant D so that, for every choice of scalars* $\{a_i\}_{i=1}^{n}$, *we have*

$$D^{-1}\left\|\left(\sum_{i=1}^{n}|a_ix_i|^2\right)^{1/2}\right\| \leqslant \left\|\sum_{i=1}^{n} a_ix_i\right\| \leqslant D\left\|\left(\sum_{i=1}^{n}|a_ix_i|^2\right)^{1/2}\right\|.$$

Proof. (i) Let $\{x_i\}_{i=1}^{n}$ be an arbitrary sequence of elements of X. Then, by the 1-convexity and q-concavity of X, we get that

$$\left\|\int_0^1\left|\sum_{i=1}^{n} r_i(u)x_i\right| du\right\| \leqslant \int_0^1\left\|\sum_{i=1}^{n} r_i(u)x_i\right\| du \leqslant \left(\int_0^1\left\|\sum_{i=1}^{n} r_i(u)x_i\right\|^q du\right)^{1/q}$$
$$\leqslant M_{(q)}(X)\left\|\left(\int_0^1\left|\sum_{i=1}^{n} r_i(u)x_i\right|^q du\right)^{1/q}\right\|.$$

Furthermore, by 1.d.1 and Khintchine's inequality for scalars I.2.b.3, we have

$$A_1\left(\sum_{i=1}^{n}|x_i|^2\right)^{1/2} \leqslant \int_0^1\left|\sum_{i=1}^{n} r_i(u)x_i\right| du$$

and

$$\left(\int_0^1\left|\sum_{i=1}^{n} r_i(u)x_i\right|^q du\right)^{1/q} \leqslant B_q\left(\sum_{i=1}^{n}|x_i|^2\right)^{1/2},$$

where A_1 and B_q are the constants appearing in I.2.b.3. The desired result follows from the monotonicity of the norm in X.

(ii) Let K be the unconditional constant of $\{x_i\}_{i=1}^{\infty}$. The left hand side inequality in (i) was proved in every Banach lattice with the absolute constant A_1 instead of C^{-1}. Hence, for all $\{a_i\}_{i=1}^{n}$,

$$A_1K^{-1}\left\|\left(\sum_{i=1}^{n}|a_ix_i|^2\right)^{1/2}\right\| \leqslant \left\|\sum_{i=1}^{n} a_ix_i\right\|.$$

Let $\{x_i^*\}_{i=1}^{\infty}$ be the functionals in X^* biorthogonal to the $\{x_i\}_{i=1}^{\infty}$. By the preceding remark we also get that, for all $\{b_i\}_{i=1}^{n}$,

$$A_1K^{-1}\left\|\left(\sum_{i=1}^{n}|b_ix_i^*|^2\right)^{1/2}\right\| \leqslant \left\|\sum_{i=1}^{n} b_ix_i^*\right\|.$$

We prove now the right hand side inequality of (ii). Given $\{a_i\}_{i=1}^n$, there are $\{b_i\}_{i=1}^n$ so that $\left\|\sum_{i=1}^n b_i x_i^*\right\| = 1$ and $\left|\sum_{i=1}^n a_i b_i\right| \geqslant \left\|\sum_{i=1}^n a_i x_i\right\| \Big/ K$. Hence, by 1.d.2(iii),

$$\left\|\sum_{i=1}^n a_i x_i\right\| \leqslant K \left\|\left(\sum_{i=1}^n |a_i x_i|^2\right)^{1/2}\right\| \left\|\left(\sum_{i=1}^n |b_i x_i^*|^2\right)^{1/2}\right\|$$
$$\leqslant A_1^{-1} K^2 \left\|\left(\sum_{i=1}^n |a_i x_i|^2\right)^{1/2}\right\|. \quad \square$$

Remarks. 1. The assumption that X is q-concave for some $q<\infty$ is essential in (i). Also, we cannot replace in (ii) the assumption that $\{x_i\}_{i=1}^\infty$ is an unconditional basis by the assumption that it is an unconditional basic sequence. In order to see this, consider the Rademacher system $\{r_i\}_{i=1}^\infty$ in $L_\infty(0, 1)$. We have $\left\|\sum_{i=1}^n a_i r_i\right\|_\infty = \sum_{i=1}^n |a_i|$ while $\left\|\left(\sum_{i=1}^n |a_i r_i|^2\right)^{1/2}\right\|_\infty = \left(\sum_{i=1}^n a_i^2\right)^{1/2}$. It is however clear from the proof that (ii) remains valid if $\{x_i\}_{i=1}^\infty$ is an unconditional basis of a complemented subspace of X.

2. As already remarked in the proof of (ii), the left hand side inequality of (i) is valid in every lattice X with A_1 instead of C^{-1}. The precise value of A_1 was computed by Szarek [126] (cf. also [49]) who showed that $A_1 = 1/\sqrt{2}$.

If Y is a sublattice of X then we obviously have $M^{(p)}(Y) \leqslant M^{(p)}(X)$ and $M_{(p)}(Y) \leqslant M_{(p)}(X)$ for every $1 \leqslant p \leqslant \infty$. The situation is more involved if we merely assume that Y is a subspace of X which is itself a lattice endowed with an order unrelated to that of X. It turns out that under some natural restrictions the properties of p-convexity and p-concavity are still inherited from X to Y.

Theorem 1.d.7 [58]. *Let X and Y be two Banach lattices and assume that Y is linearly isomorphic to a subspace of X.*

(i) *If X is p-convex and q-concave for some $1<p\leqslant 2$ and $q<\infty$ then Y is p-convex, too.*

(ii) *If X is p-concave for some $p\geqslant 2$ then so is Y.*

Proof. The proofs of (i) and (ii) are entirely similar so we prove only part (i). We first consider the case when Y is a space with an unconditional basis $\{y_j\}_{j=1}^n$ of finite length, whose unconditional constant is equal to one. Let T be an isomorphism from Y into X and, for $1 \leqslant j \leqslant n$, put $x_j = Ty_j$. Let $z_i = \sum_{j=1}^n a_{i,j} y_j$, $1 \leqslant i \leqslant k$, be arbitrary elements in Y and let $\{r_j\}_{j=1}^n$ denote, as usual, Rademacher functions.

Then, by 1.d.6(i) applied in X, there exists a constant $C<\infty$ so that

$$\begin{aligned}\left\|\left(\sum_{i=1}^k |z_i|^p\right)^{1/p}\right\|_Y &= \left\|\sum_{j=1}^n \left(\sum_{i=1}^k |a_{i,j}|^p\right)^{1/p} y_j\right\|_Y \\ &= \int_0^1 \left\|\sum_{j=1}^n r_j(u)\left(\sum_{i=1}^k |a_{i,j}|^p\right)^{1/p} y_j\right\|_Y du \\ &\leq \|T^{-1}\| \int_0^1 \left\|\sum_{j=1}^n r_j(u)\left(\sum_{i=1}^k |a_{i,j}|^p\right)^{1/p} x_j\right\|_X du \\ &\leq C\|T^{-1}\| \left\|\left(\sum_{j=1}^n \left(\sum_{i=1}^k |a_{i,j}|^p\right)^{2/p} |x_j|^2\right)^{1/2}\right\|_X .\end{aligned}$$

Thus, by the triangle inequality in $l_{2/p}$ and the p-convexity of X, we get that

$$\begin{aligned}\left\|\left(\sum_{i=1}^k |z_i|^p\right)^{1/p}\right\|_Y &\leq C\|T^{-1}\| \left\|\left(\sum_{i=1}^k \left(\sum_{j=1}^n |a_{i,j}x_j|^2\right)^{p/2}\right)^{1/p}\right\|_X \\ &\leq C\|T^{-1}\|M^{(p)}(X)\left(\sum_{i=1}^k \left\|\left(\sum_{j=1}^n |a_{i,j}x_j|^2\right)^{1/2}\right\|_X^p\right)^{1/p} \\ &\leq C^2\|T^{-1}\|M^{(p)}(X)\left(\sum_{i=1}^k \left(\int_0^1 \left\|\sum_{j=1}^n r_j(u)a_{i,j}x_j\right\|_X du\right)^p\right)^{1/p} \\ &\leq C^2\|T\|\,\|T^{-1}\|M^{(p)}(X)\left(\sum_{i=1}^k \left(\int_0^1 \left\|\sum_{j=1}^n r_j(u)a_{i,j}y_j\right\|_Y du\right)^p\right)^{1/p} \\ &= C^2\|T\|\,\|T^{-1}\|M^{(p)}(X)\left(\sum_{i=1}^k \|z_i\|^p\right)^{1/p} .\end{aligned}$$

We consider now the case when Y is a general Banach lattice. There is clearly no loss of generality in assuming that Y is separable. Since X is q-concave for $q<\infty$ it does not contain any subspaces isomorphic to c_0 (apply 1.c.4 and the remark following it) and, thus, the same is true for Y. Consequently, we may use the representation theorem 1.b.14 for Y. Since any finite set of disjointly supported measurable functions in Y forms an unconditional basic sequence whose unconditional constant is one, it follows from the first part of the proof that if $\{f_i\}_{i=1}^k$ is a sequence of simple functions in Y then

$$\left\|\left(\sum_{i=1}^k |f_i|^p\right)^{1/p}\right\| \leq C^2\|T\|\,\|T^{-1}\|M^{(p)}(X)\left(\sum_{i=1}^k \|f_i\|^p\right).$$

Since the simple functions are dense in Y the same inequality will hold for every choice of $\{f_i\}_{i=1}^k$ in Y. □

Remark. The restrictions imposed on p and on q in (i) and (ii) of 1.d.7 are obviously necessary since, for instance, l_2 embeds in any $L_p(0, 1)$ space, $1\leq p<\infty$, and every separable space embeds in $C(0, 1)$. However, as in 1.d.6(ii), it is easily verified that 1.d.7(i) holds without the assumption that X is q-concave for some $q<\infty$ if Y is isomorphic to a complemented subspace of X.

We present now a general procedure for constructing p-convex and p-concave lattices starting with an arbitrary Banach lattice (cf. [41] and [66]). This procedure is just an abstract description of the map $f \to |f|^s \operatorname{sign} f$ which maps $L_r(\mu)$, $1 \leqslant r < \infty$, onto $L_{rs}(\mu)$ (if $rs \geqslant 1$). In a general lattice X there is no meaning to the symbol x^s. We overcome this difficulty by introducing new algebraic operations in X and applying 1.d.1.

Let X be a Banach lattice in which the algebraic operations and the norm are, as usual, denoted by $+$, $\cdot$ and $\|\cdot\|$ and let $p>1$. For x and y in X and for a scalar α, we define

$$x \oplus y = (x^{1/p} + y^{1/p})^p, \qquad \alpha \odot x = \alpha^p \cdot x$$

where $(x^{1/p}+y^{1/p})^p$ is the element in X corresponding, by the procedure described in 1.d.1, to the function

$$f(t_1, t_2) = \big||t_1|^{1/p} \operatorname{sign} t_1 + |t_2|^{1/p} \operatorname{sign} t_2\big|^p \operatorname{sign}(|t_1|^{1/p} \operatorname{sign} t_1 + |t_2|^{1/p} \operatorname{sign} t_2)$$

and α^p is $|\alpha|^p \operatorname{sign} \alpha$. The set X, endowed with the operation $\oplus$, $\odot$ and the order $x \circledgeq 0 \Leftrightarrow x \geqslant 0$ is, as easily verified, a vector lattice denoted by $X^{(p)}$. Put $|||x||| = \|x\|^{1/p}$ for $x \in X$ and observe that $|||\cdot|||$ defines a lattice norm in $X^{(p)}$. Indeed, for $x \in X$ and a real α,

$$|||\alpha \odot x||| = \||\alpha|^p x\|^{1/p} = |\alpha| \, \|x\|^{1/p} = |\alpha| \, |||x||| \, .$$

Also, if $x, y \in X$ and α and β are positive reals with $\alpha^q + \beta^q = 1$, where $1/p + 1/q = 1$, we have by 1.d.1 and Hölder's inequality

$$(|x|^{1/p} + |y|^{1/p})^p \leqslant |x|/\alpha^p + |y|/\beta^p \, .$$

Hence, by taking $\alpha^p = \|x\|^{1/q}/\gamma$, $\beta^p = \|y\|^{1/q}/\gamma$ and $\gamma = (\|x\|^{1/p} + \|y\|^{1/p})^{p/q}$, we get

$$\begin{aligned} |||x \oplus y||| &\leqslant \|(|x|^{1/p} + |y|^{1/p})^p\|^{1/p} \leqslant (\|x\|/\alpha^p + \|y\|/\beta^p)^{1/p} \\ &= (\|x\|^{1/p} + \|y\|^{1/p})^{1/p} \gamma^{1/p} = \|x\|^{1/p} + \|y\|^{1/p} = |||x||| + |||y||| \, . \end{aligned}$$

It is also evident that $(X^{(p)}, |||\cdot|||)$ is complete (since $(X, \|\ \|)$ is complete). The lattice $(X^{(p)}, |||\cdot|||)$ will be called the *p-convexification* of X. As its name indicates it is p- convex. Indeed, if $x, y \in X^{(p)}$

$$\begin{aligned} |||(|x|^p \oplus |y|^p)^{1/p}|||^p &= |||\,|x| + |y|\,|||^p \\ &= \|\,|x| + |y|\,\| \leqslant \|x\| + \|y\| = |||x|||^p + |||y|||^p \, , \end{aligned}$$

and hence $M^{(p)}(X^{(p)}) = 1$. More generally, it is easily verified that if X is r-convex and s-concave for some $1 \leqslant r \leqslant s \leqslant \infty$ then $X^{(p)}$ is pr-convex and ps-concave with

$$M^{(pr)}(X^{(p)}) \leqslant M^{(r)}(X)^{1/p}, \qquad M_{(ps)}(X^{(p)}) \leqslant M_{(s)}(X)^{1/p}.$$

In case X is a Banach lattice of functions, $X^{(p)}$ can be obviously identified with the space of all the functions f so that $f^p = |f|^p \operatorname{sign} f \in X$ endowed with the norm $|||f||| = \| \, |f|^p \|^{1/p}$.

There is also a p-concavification procedure for a Banach lattice X which, however, can be applied only if it is known in advance that X is p-convex. Let X be a Banach lattice which is r-convex and s-concave for some $1 < p \leqslant r \leqslant s \leqslant \infty$. For vectors x and y in X and a scalar α put

$$x \oplus y = (x^p + y^p)^{1/p}, \qquad \alpha \odot x = \alpha^{1/p} \cdot x$$

and $|||x|||_0 = \|x\|^p$. Again, it can be easily verified that X, endowed with the above operations, forms a vector lattice $X_{(p)}$ provided that the order in $X_{(p)}$ is defined as in X. The "norm" $|||\cdot|||_0$ is clearly homogeneous but, instead of the triangle inequality, we can only prove that, for $\{x_i\}_{i=1}^n$ in $X_{(p)}$,

$$\begin{aligned} |||x_1 \oplus \cdots \oplus x_n|||_0 &\leqslant \|(|x_1|^p + \cdots + |x_n|^p)^{1/p}\|^p \\ &\leqslant M^{(p)}(X)^p \sum_{i=1}^n \|x_i\|^p = M^{(p)}(X)^p \sum_{i=1}^n |||x_i|||_0 . \end{aligned}$$

If $M^{(p)}(X)$ were equal to one the "norm" $|||\cdot|||_0$ would have satisfied the triangle inequality but, in general, we have to replace $|||\cdot|||_0$ by a different expression, namely

$$|||x||| = \inf \left\{ \sum_{i=1}^n |||x_i|||_0 ; \quad |x| = \sum_{i=1}^n \oplus |x_i|, \quad x_i \in X_{(p)} \text{ for } 1 \leqslant i \leqslant n \right\}.$$

It is easily seen that $|||\cdot|||$ is a lattice norm on $X_{(p)}$ such that

$$|||x|||_0 / M^{(p)}(X)^p \leqslant |||x||| \leqslant |||x|||_0 , \quad x \in X_{(p)} .$$

The Banach lattice $(X_{(p)}, |||\cdot|||)$ is called the *p-concavification* of X. A simple computation shows that $X_{(p)}$ is r/p-convex with $M^{(r/p)}(X_{(p)}) \leqslant (M^{(p)}(X) M^{(r)}(X))^p$ and s/p-concave with $M_{(s/p)}(X_{(p)}) \leqslant (M^{(p)}(X) M_{(s)}(X))^p$.

The concavification procedure described above can be used to prove the following renorming result from [41].

Proposition 1.d.8. *A Banach lattice X, which is r-convex and s-concave for some $1 \leqslant r \leqslant s \leqslant \infty$, can be renormed equivalently so that X, endowed with the new norm and the same order, is a Banach lattice whose r-convexity and s-concavity constants are both equal to one.*

Proof. We actually prove the theorem only in the case when $1<r<s<\infty$. The cases $r=1$ or $s=\infty$ are simpler since the 1-convexity and ∞-concavity constants of any lattice are always equal to one. The case $r=s$ follows from 1.b.13 (in this case X is isomorphic to $L_r(\mu)$).

Notice now that the r-concavification $Y=X_{(r)}$ of X is a Banach lattice which is $q=s/r$-concave. By 1.d.4(iii), the dual Y^* of Y is q'-convex, where $1/q'+1/q=1$. Hence, by successive q'-concavification and q'-convexification, one can find a new and equivalent lattice norm on Y^* such that, endowed with this new norm, Y^* has q'-convexity constant equal to one. By using 1.d.4(iii) again, it follows that there exist an equivalent lattice renorming on Y^{**} so that the q-concavity constant of Y^{**} becomes one and the same, of course, is true for Y since it is a sublattice of Y^{**}. In order to complete the proof, one just takes the r-convexification of Y endowed with the new norm. □

The following simple proposition will enable us to describe some classes of p-convex and p-concave operators.

Proposition 1.d.9 [66]. *Let X and Y be two Banach lattices and let $T: X \to Y$ be a positive operator. Then, for every $1 \leq p \leq \infty$ and every choice of $\{x_i\}_{i=1}^n$ in X, we have*

$$\left\|\left(\sum_{i=1}^n |Tx_i|^p\right)^{1/p}\right\| \leq \|T\| \left\|\left(\sum_{i=1}^n |x_i|^p\right)^{1/p}\right\|, \quad \text{if } p<\infty$$

and

$$\left\|\bigvee_{i=1}^n |Tx_i|\right\| \leq \|T\| \left\|\bigvee_{i=1}^n |x_i|\right\|, \quad \text{if } p=\infty .$$

Proof. Since $\left(\sum_{i=1}^n |x_i|^p\right)^{1/p} \geq \sum_{i=1}^n a_i x_i$, whenever $\sum_{i=1}^n |a_i|^q \leq 1$ and $1/p+1/q=1$, it follows that

$$T\left(\left(\sum_{i=1}^n |x_i|^p\right)^{1/p}\right) \geq \sum_{i=1}^n a_i Tx_i .$$

Hence,

$$\left(\sum_{i=1}^n |Tx_i|^p\right)^{1/p} = \text{l.u.b.}\left\{\sum_{i=1}^n a_i Tx_i\,;\ \sum_{i=1}^n |a_i|^q \leq 1\right\} \leq T\left(\left(\sum_{i=1}^n |x_i|^p\right)^{1/p}\right)$$

and the proof is readily completed by using the monotonicity of the norm. □

There are some immediate consequences of 1.d.9. Let X and Y be two Banach lattices and let $T: X \to Y$ be a positive operator.

(i) If X is p-convex then T is p-convex and $M^{(p)}(T) \leq \|T\|\, M^{(p)}(X)$.

(ii) If Y is p-concave then T is p-concave and $M_{(p)}(T) \leq \|T\|\, M_{(p)}(Y)$.

From 1.d.9 we can also derive some connections between the notions of p-concave operators and p-absolutely summing operators defined in I.2.b.1. *Every p-absolutely summing operator T from a Banach lattice X into a Banach space V is p-concave and $M_{(p)}(T) \leqslant \pi_p(T)$.* Indeed, we have just to observe that for all $\{x_i\}_{i=1}^n$ in X and $x^* \in X^*$

$$\left\|\left(\sum_{i=1}^n |x^*(x_i)|^p\right)^{1/p}\right\| \leqslant \left\|\left(\sum_{i=1}^n (|x^*|(|x_i|))^p\right)^{1/p}\right\| \leqslant \|x^*\| \left\|\left(\sum_{i=1}^n |x_i|^p\right)^{1/p}\right\|.$$

A closely related fact is the following result (cf. B. Maurey [94]).

Theorem 1.d.10. *Let V be a Banach space, X a Banach lattice, $1 \leqslant p < \infty$ and $M < \infty$. An operator $T: X \to V$ is p-concave with $M_{(p)}(T) \leqslant M$ if and only if, for every positive operator S from a $C(K)$ space into X, the composition TS is p-absolutely summing and $\pi_p(TS) \leqslant \|S\| M$. In particular, X is p-concave if and only if every positive operator S: $C(K) \to X$ is p-absolutely summing.*

Proof. We observe first that, by Hölder's inequality, for every sequence $\{f_i\}_{i=1}^n$ in $C(K)$ and every $\mu \in C(K)^*$ with $\|\mu\| = 1$, we have

$$\left(\sum_{i=1}^n |\mu(f_i)|^p\right)^{1/p} \leqslant \left(\int_K \sum_{i=1}^n |f_i|^p \, d|\mu|\right)^{1/p} \leqslant \sup\left\{\left(\sum_{i=1}^n |f_i(k)|^p\right)^{1/p}; k \in K\right\}$$

and, hence,

$$(+) \qquad \sup\left\{\left(\sum_{i=1}^n |\mu(f_i)|^p\right)^{1/p}; \mu \in C(K)^*, \|\mu\| = 1\right\} = \sup\left\{\left(\sum_{i=1}^n |f_i(k)|^p\right)^{1/p}; k \in K\right\}.$$

Suppose now that $T: X \to V$ is a p-concave operator, $S: C(K) \to X$ a positive operator and $\{f_i\}_{i=1}^n$ are arbitrary functions in $C(K)$. Then, by 1.d.9, we get that

$$\left(\sum_{i=1}^n \|TSf_i\|^p\right)^{1/p} \leqslant M_{(p)}(T) \left\|\left(\sum_{i=1}^n |Sf_i|^p\right)^{1/p}\right\| \leqslant \|S\| M_{(p)}(T) \left\|\left(\sum_{i=1}^n |f_i|^p\right)^{1/p}\right\|.$$

Thus, in view of (+), TS is p-absolutely summing and $\pi_p(TS) \leqslant \|S\| M_{(p)}(T)$.

Conversely, assume that $\pi_p(TS) \leqslant \|S\| M$ for every positive operator $S: C(K) \to X$. Let $\{x_j\}_{j=1}^m$ be vectors in X and put $x_0 = \left(\sum_{j=1}^m |x_j|^p\right)^{1/p}$. Let $I(x_0)$ be the (in general, non-closed) ideal generated by x_0 i.e. the set of all $x \in X$ for which

$|x| \leqslant \lambda x_0$, for some $\lambda \geqslant 0$. For $x \in X$, set

$$\|x\|_\infty = \inf\{\lambda \geqslant 0;\ |x| \leqslant \lambda x_0/\|x_0\|\}$$

and notice that the completion of $I(x_0)$, endowed with the norm $\|\cdot\|_\infty$, is order isometric to a $C(K)$ space. Let J denote the formal identity mapping from $I(x_0)$ into X. Then, by $(+)$ applied in $I(x_0)$, we have that

$$\left(\sum_{j=1}^m \|Tx_j\|^p\right)^{1/p} = \left(\sum_{j=1}^m \|TJx_j\|^p\right)^{1/p} \leqslant \pi_p(TJ)\|x_0\|_\infty$$
$$= \pi_p(TJ)\left\|\left(\sum_{j=1}^m |x_j|^p\right)^{1/p}\right\|,$$

i.e. T is p-concave with $M_{(p)}(T) \leqslant \pi_p(TJ) \leqslant \|J\|M = M$. □

Remarks. 1. It is easily verified that a lattice X is p-concave if every positive operator from c_0 (instead of an arbitrary $C(K)$ space) into X is p-absolutely summing.

2. The fact that a lattice is p-concave for some $p > 2$ does not necessarily imply that every bounded linear operator S from c_0 into it is p-absolutely summing (see [68] or I.2.b.8). Later on, in Section 1.f, we shall see that, for $p = 2$, we can drop the positivity assumption on S in 1.d.10.

We present now some factorization theorems for p-convex and p-concave operators, due to J. L. Krivine [66], which were inspired by results of H. P. Rosenthal [115] and B. Maurey [95].

Theorem 1.d.11. *Let X be a Banach lattice, V, W two Banach spaces and fix $1 \leqslant p < \infty$. Let T be a p-convex operator from V into X and S a p-concave operator from X into W. Then the operator ST can be factorized through an $L_p(\mu)$ space in the sense that $ST = S_1T_1$, where T_1 is an operator from V into $L_p(\mu)$ with $\|T_1\| \leqslant M^{(p)}(T)$ and S_1 is an operator from $L_p(\mu)$ into W with $\|S_1\| \leqslant M_{(p)}(S)$.*

Proof. Let I_T be the (in general non-closed) ideal of X generated by the range of T. We define new operations on I_T as in the p-concavification procedure described above. For $x, y \in I_T$ and a real α, put

$$x \oplus y = (x^p + y^p)^{1/p}, \qquad \alpha \odot x = \alpha^{1/p} \cdot x,$$

and let $\check{I}_T$ denote the vector lattice obtained when I_T is endowed with the original order and the operations defined above. Set

$$F_1 = \text{conv}\{x \in \check{I}_T;\ |x| \leqslant |Tv|, \text{ for some } v \in V \text{ with } \|v\| < 1/M^{(p)}(T)\},$$
$$F_2 = \text{conv}\{x \in \check{I}_T;\ x > 0 \text{ and } \|Sy\| \geqslant M_{(p)}(S), \text{ for some } y \text{ with } |y| \leqslant x\},$$

where both convex hulls are taken in the sense of $\check{I}_T$, i.e. by using the new operations.

If $x=\alpha_1 \odot x_1 \oplus \cdots \oplus \alpha_n \odot x_n$ is an element of $\check{I}_T$ with $\sum_{i=1}^{n} \alpha_i=1$, $\alpha_i \geq 0$, $|x_i| \leq |Tv_i|$ and $\|v_i\|<1/M^{(p)}(T)$ then

$$\|x\| \leq \|(|\alpha_1^{1/p}x_1|^p+\cdots+|\alpha_n^{1/p}x_n|^p)^{1/p}\| \leq \|(|\alpha_1^{1/p}Tv_1|^p+\cdots+|\alpha_n^{1/p}Tv_n|^p)^{1/p}\|$$
$$\leq M^{(p)}(T)\Big(\sum_{i=1}^{n} \|\alpha_i^{1/p}v_i\|^p\Big)^{1/p}<1\,.$$

On the other hand, if $x=\beta_1 \odot x_1 \oplus \cdots \oplus \beta_n \odot x_n$ is an element of $\check{I}_T$ with $\sum_{i=1}^{n} \beta_i=1$, $\beta_i \geq 0$, $x_i \geq |y_i|$ and $\|Sy_i\| \geq M_{(p)}(S)$ for all $1 \leq i \leq n$ then

$$\|x\|=\|(|\beta_1^{1/p}x_1|^p+\cdots+|\beta_n^{1/p}x_n|^p)^{1/p}\| \geq \|(|\beta_1^{1/p}y_1|^p+\cdots+|\beta_n^{1/p}y_n|^p)^{1/p}\|$$
$$\geq \Big(\sum_{i=1}^{n} \|\beta_i^{1/p}Sy_i\|^p\Big)^{1/p}\Big/M_{(p)}(S) \geq 1\,.$$

Hence, $F_1 \cap F_2=\varnothing$ and since 0 is an internal point of F_1 it follows from the separation theorem that there exists a linear functional φ on $\check{I}_T$ such that $\varphi(x) \leq 1$ for $x \in F_1$ and $\varphi(x) \geq 1$ for $x \in F_2$. Observe that, for every $\alpha>0$, $0<x \in \check{I}_T$ and every $x_0 \in F_2$, we have $\alpha\varphi(x)+\varphi(x_0) \geq 1$ since $\alpha \odot x \oplus x_0 \in F_2$. It follows that $\varphi(x) \geq 0$ whenever $x>0$ and, thus, we can define a semi-norm on I_T by putting

$$\|x\|_0=(\varphi(|x|))^{1/p}, \quad x \in I_T\,.$$

Using the linearity of φ with respect to the operations $\oplus$ and $\odot$, it is readily verified that, with respect to the original multiplication by scalars and addition, $\|\cdot\|_0$ is homogeneous and satisfies the triangle inequality (the latter fact is proved by arguments similar to those used in the p-convexification procedure).

Observe now that, for any $x, y \in I_T$, we have

$$|x|+|y| \geq (|x|^p+|y|^p)^{1/p} \geq |x| \vee |y|\,,$$

since these inequalities are valid for reals. By the fact that φ is non-negative, we get that

$$\||x|+|y|\|_0^p=\varphi(|x|+|y|) \geq \varphi((|x|^p+|y|^p)^{1/p})=\varphi(|x| \oplus |y|)=\|x\|_0^p+\|y\|_0^p$$
$$\geq \varphi(|x| \vee |y|)=\||x| \vee |y|\|_0^p\,.$$

This inequality concerning $\|\cdot\|_0$ clearly remains valid in the completion Z of I_T modulo the ideal of all $x \in I_T$ for which $\|x\|_0=0$. Therefore, if $|x| \wedge |y|=0$ for some x and y in the lattice Z then

$$\||x|+|y|\|_0^p=\|x\|_0^p+\|y\|_0^p\,,$$

i.e. Z is an abstract L_p space. It follows from 1.b.2 that Z is order isometric to an

$L_p(\mu)$ space, for a suitable measure μ. Let $T_1: V \to Z$ be defined by $T_1 v = Tv$, $v \in V$ when Tv is regarded as an element of I_T. If $\|v\| < 1/M^{(p)}(T)$ then $T_1 v \in F_1$ which implies that $\|T_1 v\|_0 \leqslant 1$ i.e. that $\|T_1\| \leqslant M^{(p)}(T)$. Let S_1 be defined by $S_1 x = Sx$, $x \in I_T$. Then, in a similar way, it can be shown that S_1 extends uniquely to an operator from Z into W such that $\|S_1\| \leqslant M_{(p)}(S)$. This completes the proof since we clearly have $S_1 T_1 = ST$. □

Corollary 1.d.12. *Let V be a Banach space and fix $1 \leqslant p < \infty$.*

(i) *Every p-convex operator T from V into a p-concave Banach lattice X can be factorized through an $L_p(\mu)$ space in the sense that $T = T_1 T_2$, where T_1 is a positive operator from $L_p(\mu)$ into X with $\|T_1\| \leqslant M_{(p)}(X)$ and T_2 is an operator from V into $L_p(\mu)$ with $\|T_2\| \leqslant M^{(p)}(T)$.*

(ii) *Every p-concave operator S from a p-convex Banach lattice X into V can be factorized through an $L_p(\mu)$ space in the sense that $S = S_1 S_2$, where S_1 is an operator from $L_p(\mu)$ into V with $\|S_1\| \leqslant M_{(p)}(S)$ and S_2 a positive operator from X into $L_p(\mu)$ with $\|S_2\| \leqslant M^{(p)}(X)$.*

Proof. Take in 1.d.11 S = identity of X, respectively, T = identity of X. □

Note that it follows from the proof of 1.d.11 that if TV is a sublattice of X then $T_1 V$ is dense $L_p(\mu)$. Hence, by taking in 1.d.11 $V = W = X$ and $S = T$ = identity of X, we recover a fact which was proved already in Section b: a Banach lattice X which is p-convex and p-concave is order isomorphic to an $L_p(\mu)$ space. Moreover, $d(X, L_p(\mu)) \leqslant M^{(p)}(X) M_{(p)}(X)$.

e. Uniform Convexity in General Banach Spaces and Related Notions

In the present section we investigate some concepts like uniform convexity, uniform smoothness, type and cotype, in the context of the theory of general Banach spaces. Those aspects which are characteristic to Banach lattices will be presented in the following section. The notions considered here have also a "local" character and, therefore, some results, which can be better understood within the framework of the local theory, will be discussed only in Vol. III.

We start with some definitions.

Definition 1.e.1. Let X be a Banach space with $\dim X \geqslant 2$.

(i) The *modulus of convexity* $\delta_X(\varepsilon)$, $0 < \varepsilon \leqslant 2$, of X is defined by

$$\delta_X(\varepsilon) = \inf\{1 - \|x+y\|/2;\ \ x, y \in X,\ \|x\| = \|y\| = 1,\ \|x-y\| = \varepsilon\}\,.$$

(ii) The *modulus of smoothness* $\rho_X(\tau)$, $\tau > 0$, of X is defined by

$$\rho_X(\tau) = \sup\{(\|x+y\| + \|x-y\|)/2 - 1;\ \ x, y \in X,\ \|x\| = 1,\ \|y\| = \tau\}.$$

(iii) X is said to be *uniformly convex* if $\delta_X(\varepsilon)>0$ for every $\varepsilon>0$, and *uniformly smooth* if $\lim_{\tau\to 0} \rho_X(\tau)/\tau=0$.

In the definition of $\delta_X(\varepsilon)$ we can as well take the infimum over all vectors $x, y \in X$ with $\|x\|, \|y\| \leqslant 1$ and $\|x-y\| \geqslant \varepsilon$. In order to verify this statement let us note first that we may clearly consider only those pairs x, y with $\|x\|=1$, $\|y\| \leqslant 1$ and $\|x-y\|=\varepsilon$. Fix such a pair x, y and let u and v be two norm one vectors in the two-dimensional space containing x and y so that $u-v=x-y$ and, in addition, y, u and v are all contained in one of the half planes determined by the line joining x with $-x$. Let $\lambda \geqslant 1$ and $\beta \geqslant 0$ be such that

$$\lambda(x+y)/2=\beta u+(1-\beta)x .$$

A quick computation shows that consequently,

$$\lambda(u+v)/2=(\beta+\lambda)u+(1-\beta-\lambda)x .$$

Since $\beta+\lambda \geqslant \max(1, \beta)$ it follows from the triangle inequality that

$$\|\beta u+(1-\beta)x\| \leqslant \|(\beta+\lambda)u+(1-\beta-\lambda)x\|$$

(consider separately the cases $\beta \leqslant 1$ and $\beta>1$). Hence,

$$\|(x+y)/2\| \leqslant \|(u+v)/2\|$$

and this proves our assertion.

Similarly, in the definition of $\rho_X(\tau)$ we may as well take the supremum over all $x, y \in X$ with $\|x\| \leqslant 1$ and $\|y\| \leqslant \tau$.

In order to motivate Definition 1.e.1, let us recall that a Banach space is called *strictly convex* if the equality $\|x\|=\|y\|=\|(x+y)/2\|=1$, for some pair of vectors x and y, implies that $x=y$. The modulus of convexity of a space measures in a certain sense its degree of strict convexity. A simple compactness argument shows that a finite dimensional Banach space is strictly convex if and only if it is uniformly convex. However, there exist many examples of infinite dimensional Banach spaces which are strictly convex but not uniformly convex (see the remark following 1.e.3).

Consider now the notion of smoothness. A Banach space X is called *smooth* if, for every $x \in X$ with $\|x\|=1$, there exists a unique $x^* \in X^*$ such that $\|x^*\|=x^*(x)=1$. Suppose that X is not smooth and let $x \in X$ and $u^*, v^* \in X^*$ be so that $\|x\|=\|u^*\|=\|v^*\|=u^*(x)=v^*(x)=1$ and $u^* \neq v^*$. Let $y \in X$ be a norm one vector for which $a=u^*(y)>0$ and $b=-v^*(y)>0$. Then, for every $t>0$, we have that $\|x+ty\| \geqslant u^*(x+ty)=1+ta$ while $\|x-ty\| \geqslant v^*(x-ty)=1+tb$. Hence, $\rho_X(\tau) \geqslant (a+b)\tau/2$ and, therefore, X is not uniformly smooth according to Definition 1.e.1. An easy compactness argument shows that, for finite dimensional spaces, smoothness is equivalent to uniform smoothness (this fact can be also deduced from 1.e.2 below since it is evident from the definitions that a reflexive space and, in particular, a finite dimensional space is smooth if and only if its dual is strictly convex).

The notions of smoothness and uniform smoothness are closely related also to the question of differentiability of the norm in a Banach space. Since we shall not use this connection in this volume we only discuss it here briefly. It is not hard to see that a Banach space X is smooth if and only if $\lim_{t\to 0}(\|x+ty\|-\|x\|)/t$ exists for every $x\neq 0$ in X and every $y\in X$. This limit, which is necessarily of the form $\varphi_x(y)$, where $\varphi_x\in X^*$, is called the Gateaux derivative of the norm in X. A Banach space is uniformly smooth if and only if the limit above exists uniformly in the set $\{(x, y);\ \|x\|=\|y\|=1\}$, i.e. if, for every $\varepsilon>0$, there exists a $\delta>0$ such that $|\|x+ty\|-\|x\|-t\varphi_x(y)|<\varepsilon|t|$, whenever $|t|<\delta$, $\|x\|=\|y\|=1$ (if the limit above exists just uniformly in $\{y;\ \|y\|=1\}$, for every x with $\|x\|=1$, the norm is said to be Fréchèt differentiable; a uniformly smooth norm is therefore called sometimes a uniformly Fréchèt differentiable norm).

For a detailed study of the notions of strict convexity, smoothness and differentiability of the norm we refer the reader to [27], [32] and [65].

We start the study of uniform convexity and uniform smoothness by proving a simple duality result.

Proposition 1.e.2 [26], [74]. *For every Banach space X we have*
(i) $\rho_{X^*}(\tau)=\sup\{\tau\varepsilon/2-\delta_X(\varepsilon),\ 0\leqslant\varepsilon\leqslant 2\},\ \tau>0.$
(ii) *X is uniformly convex if and only if X^* is uniformly smooth.*

Proof. (i) We have for $\tau>0$,

$$\begin{aligned}2\rho_{X^*}(\tau)&=\sup\{\|x^*+\tau y^*\|+\|x^*-\tau y^*\|-2;\ x^*, y^*\in B_{X^*}\}\\&=\sup\{x^*(x)+\tau y^*(x)+x^*(y)-\tau y^*(y)-2;\ x, y\in B_X,\ x^*, y^*\in B_{X^*}\}\\&=\sup\{\|x+y\|+\tau\|x-y\|-2;\ x, y\in B_X\}\\&=\sup\{\|x+y\|+\tau\varepsilon-2;\ x, y\in B_X,\ \|x-y\|=\varepsilon,\ 0\leqslant\varepsilon\leqslant 2\}\\&=\sup\{\tau\varepsilon-2\delta_X(\varepsilon);\ 0\leqslant\varepsilon\leqslant 2\}.\end{aligned}$$

The second assertion follows easily from the first. For example, if X^* is uniformly smooth then, for every $0<\varepsilon<2$, there exists a $\tau>0$ so that $\rho_{X^*}(\tau)\leqslant\tau\varepsilon/4$. Hence, by (i), we get that $\tau\varepsilon/2-\delta_X(\varepsilon)\leqslant\tau\varepsilon/4$ i.e. $\delta_X(\varepsilon)\geqslant\tau\varepsilon/4$. This proves that X is uniformly convex. The converse is proved in a similar way. □

Proposition 1.e.3 [100], [113]. *Every uniformly convex (and thus also every uniformly smooth) Banach space is reflexive.*

Proof. Assume first that X is uniformly convex and let x^{**} be a norm one element of X^{**}. Let $\{x_\alpha\}_{\alpha\in A}$ be a directed set in the canonical image iX of X in X^{**} such that $x_\alpha\xrightarrow{w^*}x^{**}$ and $\|x_\alpha\|\leqslant 1$ for all $\alpha\in A$. Since $x_\alpha+x_\beta\xrightarrow{w^*}2x^{**}$ it follows that $\lim_{\alpha,\beta}\|x_\alpha+x_\beta\|=2$. Thus, by uniform convexity, we get that $\lim_{\alpha,\beta}\|x_\alpha-x_\beta\|=0$ i.e. that $\{x_\alpha\}_{\alpha\in A}$ is norm convergent to some element of iX. This proves that X is reflexive.

If X is uniformly smooth then X^{**} is also uniformly smooth since $\rho_X(\tau) = \rho_{X^{**}}(\tau)$ (use the definition of $\rho_{X^{**}}(\tau)$ and the w* density of the unit ball of iX in that of X^{**}). Hence, by 1.e.2, X^* is uniformly convex and, thus, by the first part of the proof, X is reflexive. □

Remarks. 1. There are simple examples of strictly convex non-reflexive Banach spaces. For instance, if $\|\cdot\|$ denotes the usual norm in $C(0, 1)$ then

$$|||f||| = \|f\| + \left(\int_0^1 |f(t)|^2 \, dt \right)^{1/2}$$

defines an equivalent norm in $C(0, 1)$ which is strictly convex. Since $C(0, 1)$ is a universal space it follows that every separable space can be given an equivalent strictly convex norm.

2. The converse to 1.e.3 is false in a trivial manner since there exist even finite dimensional spaces which are not strictly convex. There are also reflexive spaces which are not uniformly convexifiable, i.e. cannot be given an equivalent norm in which they become uniformly convex. Such a space is e.g. $\left(\sum_{n=1}^{\infty} \oplus l_1^n \right)_2$ (cf. [25]). Actually, the following more general fact is true: a Banach space X which contains uniformly isomorphic copies of l_1^n for all n (i.e., for some constant $M < \infty$ and every integer n, there exists a subspace B_n of X such that $d(l_1^n, B_n) \leqslant M$) is not uniformly convexifiable. This assertion is an immediate consequence of the following lemma which is a local version of I.2.e.3.

Lemma 1.e.4 [47]. *Let B be a finite dimensional space such that $d(B, l_1^{n^2}) \leqslant M^2$, for some n and for some constant M. Then B contains an n-dimensional subspace C for which $d(C, l_1^n) \leqslant M$. Consequently, every infinite dimensional Banach space X, which contains uniformly isomorphic copies of l_1^n for all n, also contains nearly isometric copies of l_1^n for all n (and, hence, $\delta_X(\varepsilon) \leqslant \delta_{l_1^2}(\varepsilon) = 0$ for every $0 < \varepsilon < 2$).*

Proof. Let $\{e_i\}_{i=1}^{n^2}$ be the unit vector basis of $l_1^{n^2}$ and let $T: l_1^{n^2} \to B$ be an isomorphism such that

$$M^{-1}\|u\| \leqslant \|Tu\| \leqslant M\|u\|,$$

for all $u \in l_1^{n^2}$. If, for some $1 \leqslant k \leqslant n$, $d([Te_i]_{i=(k-1)n+1}^{kn}, l_1^n \leqslant M$ then the proof is already finished. Otherwise, for every $1 \leqslant k \leqslant n$, there exists a vector $u_k \in [e_i]_{i=(k-1)n+1}^{kn}$ so that $\|Tu_k\| < \|u_k\| = 1$. It follows that

$$M^{-1} \sum_{k=1}^{n} |a_k| = M^{-1} \left\| \sum_{k=1}^{n} a_k u_k \right\| \leqslant \left\| \sum_{k=1}^{n} a_k T u_k \right\| \leqslant \sum_{k=1}^{n} |a_k| \|Tu_k\| \leqslant \sum_{k=1}^{n} |a_k|,$$

for every choice of scalars $\{a_k\}_{k=1}^n$, i.e. $d([Tu_k]_{k=1}^n, l_1^n) \leqslant M$. □

The simplest example of a space, which is both uniformly convex and uniformly smooth, is the Hilbert space. The moduli of convexity and of smoothness of the Hilbert space, denoted by $\delta_2(\varepsilon)$, respectively $\rho_2(\varepsilon)$, can be computed easily by using the parallelogram identity and one obtains that

$$\delta_2(\varepsilon)=1-(1-\varepsilon^2/4)^{1/2}=\varepsilon^2/8+O(\varepsilon^4),\quad 0<\varepsilon<2,$$
$$\rho_2(\tau)=(1+\tau^2)^{1/2}-1=\tau^2/2+O(\tau^4),\quad \tau>0\,.$$

Since, by a well-known theorem of A. Dvoretzky [35] to be presented in Vol. III, every infinite-dimensional Banach space contains nearly isometric copies of l_2^n for all n it follows that the Hilbert space is the "most" uniformly convex and also the "most" uniformly smooth space in the sense that, for any Banach space X,

$$\delta_X(\varepsilon)\leqslant 1-(1-\varepsilon^2/4)^{1/2},\quad 0<\varepsilon<2 \quad\text{and}\quad \rho_X(\tau)\geqslant(1+\tau^2)^{1/2}-1,\quad \tau>0\,.$$

Dvoretzky's theorem proves these relations only if $\dim X=\infty$. Actually, these inequalities are true for every X with $\dim X\geqslant 2$. This fact is due to G. Nördlander [105] who proved it by using an elegant geometric argument (see also 1.e.5 below which implies that $\delta_X(\varepsilon)\leqslant C\varepsilon^2$, for every X and for some constant C). Other examples of spaces, being in the same time uniformly convex and uniformly smooth, are the L_p spaces, $1<p<\infty$. The exact value of the moduli of convexity $\delta_p(\varepsilon)$ and of smoothness $\rho_p(\tau)$ of L_p will be computed in Vol. IV: their asymptotical behavior is the following:

$$\delta_p(\varepsilon)=\begin{cases}(p-1)\varepsilon^2/8+o(\varepsilon^2), & 1<p<2\\ \varepsilon^p/p2^p+o(\varepsilon^p), & 2\leqslant p<\infty\end{cases}$$
$$\rho_p(\tau)=\begin{cases}\tau^p/p+o(\tau^p), & 1<p\leqslant 2\\ (p-1)\tau^2/2+o(\tau^2), & 2\leqslant p<\infty.\end{cases}$$

In applications, we do not use often the precise value of the moduli of convexity or smoothness but only a power type estimate from below or above. We say that a uniformly convex (smooth) space X has modulus of convexity (smoothness) of *power type p* if, for some $0<K<\infty$, $\delta_X(\varepsilon)\geqslant K\varepsilon^p(\rho_X(\tau)\leqslant K\tau^p)$. Using 1.e.2, it is easily seen that $\delta_X(\varepsilon)$ is of power type p if and only if $\rho_{X^*}(\tau)$ is of power type q, where $1/p+1/q=1$.

For instance, it follows from the above formulas that, in this terminology, L_p spaces have modulus of convexity of power type 2, for $1<p\leqslant 2$, and of power type p, for $p>2$. This fact will be also proved directly in the next section (cf. 1.f.1 below).

We present now a few results which describe the behavior of the functions $\delta_X(\varepsilon)$ and $\rho_X(\tau)$. These results, whose proofs are somewhat technical, are evident whenever $\delta_X(\varepsilon)$ and $\rho_X(\tau)$ behave like ε^p, respectively τ^q. Since this is the situation for L_p spaces and since, even for more general spaces, our interest will mostly be in power type estimates for $\delta_X(\varepsilon)$ and $\rho_X(\tau)$ the readers may choose to skip 1.e.5–1.e.8 when first reading this section and then consider 1.e.9 only in the case where $\delta_X(\varepsilon)$ and $\rho_X(\tau)$ behave like ε^p, respectively τ^q.

Proposition 1.e.5 [74], [40]. *The modulus of smoothness $\rho_X(\tau)$ of a Banach space X is an Orlicz function satisfying the Δ_2-condition at zero and $\rho_X(\tau)/\tau^2$ is equivalent to a decreasing function. More precisely, $\rho_X(\tau)$ is a non-decreasing convex function with $\rho_X(0)=0$, $\limsup_{\tau\to 0} \rho_X(2\tau)/\rho_X(\tau)\leq 4$ and there exists an absolute constant $C<\infty$ so that $\rho_X(\eta)/\eta^2 \leq C\rho_X(\tau)/\tau^2$, whenever $\eta>\tau>0$.*

Proof. The assertion that $\rho_X(\tau)$ is a non-decreasing convex function is a direct consequence of 1.e.2(i) and of the fact that $\rho_{X^{**}}(\tau)=\rho_X(\tau)$, $\tau>0$. In order to prove that $\rho_X(\tau)$ satisfies the Δ_2-condition at zero, we fix $\tau>0$ and let $x, y\in X$ be vectors with $\|x\|=1$ and $\|y\|=\tau$. A simple computation shows that

$$\|x+2y\|\leq 2\|x+y\|\rho_X(\tau/\|x+y\|)+2\|x+y\|-1$$

and

$$\|x-2y\|\leq 2\|x-y\|\rho_X(\tau/\|x-y\|)+2\|x-y\|-1\,.$$

Hence,

$$\begin{aligned}(\|x+2y\|&+\|x-2y\|)/2-1\\ &\leq \|x+y\|\rho_X(\tau/\|x+y\|)+\|x-y\|\rho_X(\tau/\|x-y\|)+\|x+y\|+\|x-y\|-2\\ &\leq \|x+y\|\rho_X(\tau/\|x+y\|)+\|x-y\|\rho_X(\tau/\|x-y\|)+2\rho_X(\tau)\end{aligned}$$

and, by taking the supremum over all possible choices of x and y, we get that

$$\rho_X(2\tau)\leq 2(1+\tau)\rho_X(\tau/(1-\tau))+2\rho_X(\tau)\,.$$

If $0<\tau<1/5$ then, by the convexity of $\rho_X(\tau)$, we get that

$$\rho_X(\tau/(1-\tau))\leq \rho_X(\tau(1+2\tau))\leq (1-2\tau)\rho_X(\tau)+2\tau\rho_X(2\tau)\,,$$

which further implies that

$$\rho_X(2\tau)\leq ((4-2\tau-4\tau^2)/(1-4\tau-4\tau^2))\rho_X(\tau)=(4+14\tau+O(\tau^2))\rho_X(\tau)\,,$$

i.e. there exist a $0<\tau_0<1/5$ so that

$$(*)\qquad \rho_X(2\tau)\leq (4+15\tau)\rho_X(\tau),\quad \text{whenever } 0<\tau<\tau_0\,.$$

This inequality, which already shows that $\rho_X(\tau)$ satisfies the Δ_2-condition at 0, will also be used to prove that $\rho_X(\tau)/\tau^2$ is equivalent to a decreasing function. Let $0<\tau<\eta$ and distinguish between the following cases:

Case I: $\tau\geq\tau_0$. By convexity of $\rho_X(\tau)$, we get that

$$\rho_X(\tau)/\tau\geq \rho_X(\tau_0)/\tau_0\geq \rho_2(\tau_0)/\tau_0$$

and, since $\rho_X(\eta)\leqslant\eta$, we have

$$\rho_X(\tau)/\tau^2\geqslant\rho_2(\tau_0)/\tau_0\tau\geqslant\rho_2(\tau_0)/\tau_0\eta\geqslant(\rho_2(\tau_0)/\tau_0)(\rho_X(\eta)/\eta^2)\,.$$

This completes the proof in Case I.

Case II: $0<\eta\leqslant\tau_0$. Choose an integer m such that $\eta/2^m\leqslant\tau<\eta/2^{m-1}$ and observe that, by (*),

$$\begin{aligned}\rho_X(\eta)/\eta^2&\leqslant(\rho_X(\tau)/\eta^2)\prod_{j=1}^{m}\frac{\rho_X(\eta/2^{j-1})}{\rho_X(\eta/2^j)}\leqslant(4^m\rho_X(\tau)/\eta^2)\prod_{j=1}^{m}(1+15\eta/4\cdot2^j)\\&\leqslant4(\rho_X(\tau)/\tau^2)\prod_{j=1}^{\infty}(1+15\tau_0/4\cdot2^j)\,.\end{aligned}$$

By combining the results obtained in the first two cases, we obtain, in Case III: $\tau<\tau_0<\eta$, that $\rho_X(\eta)/\eta^2\leqslant C\rho_X(\tau)/\tau^2$, where

$$C=(4\tau_0/\rho_2(\tau_0))\prod_{j=1}^{\infty}(1+15\tau_0/4\cdot2^j)<\infty\,.\quad\square$$

Proposition 1.e.6 [40]. *The modulus of convexity $\delta_X(\varepsilon)$ of a Banach space X is equivalent to the Orlicz function $\tilde\delta_X(\varepsilon)=\sup\{\varepsilon\tau/2-\rho_{X^*}(\tau);\ \tau\geqslant0\}$. The function $\tilde\delta_X(\varepsilon)$ is the maximal convex function majorated by $\delta_X(\varepsilon)$ and $\tilde\delta_X(\varepsilon)/\varepsilon^2$ is equivalent to an increasing function.*

Proof. The fact that the function $\tilde\delta_X(\varepsilon)$, as defined in the statement of 1.e.6, is nondecreasing and convex is obvious while $\tilde\delta_X(\varepsilon)\leqslant\delta_X(\varepsilon)$ follows from 1.e.2. In order to prove that $\tilde\delta_X(\varepsilon)$ is the maximal convex function majorated by $\delta_X(\varepsilon)$, it suffices to note that if a linear function $a\varepsilon+b$ with $a>0$ is majorated by $\delta_X(\varepsilon)$, for every $0<\varepsilon<2$, then it is also majorated by $\tilde\delta_X(\varepsilon)$. Indeed, if $\delta_X(\varepsilon)\geqslant a\varepsilon+b$, $0<\varepsilon<2$ then, for any $0<\eta<2$, we get that

$$\tilde\delta_X(\eta)\geqslant2a\eta/2-\rho_{X^*}(2a)=a\eta-\sup\{a\varepsilon-\delta_X(\varepsilon);\ 0<\varepsilon<2\}\geqslant a\eta+b\,.$$

We next show that $\tilde\delta_X(\varepsilon)/\varepsilon^2$ is equivalent to an increasing function. By 1.e.5, there exists a constant $C<\infty$ so that $\rho_{X^*}(\tau_1)/\tau_1^2\leqslant C\rho_{X^*}(\tau_2)/\tau_2^2$, whenever $\tau_1>\tau_2>0$. Fix $0<\varepsilon_1<\varepsilon_2<2$ and consider first the case when $\varepsilon_1C\geqslant\varepsilon_2$. Then, by the convexity of $\tilde\delta_X(\varepsilon)$, it follows that

$$\tilde\delta_X(\varepsilon_1)\leqslant\tilde\delta_X(\varepsilon_2)\varepsilon_1/\varepsilon_2\leqslant C\tilde\delta_X(\varepsilon_2)\varepsilon_1^2/\varepsilon_2^2\,.$$

In the other case, i.e., when $\varepsilon_1C<\varepsilon_2$, we put $\alpha=\varepsilon_2/C\varepsilon_1>1$. Then,

$$\begin{aligned}\tilde\delta_X(\varepsilon_1)&\leqslant\sup\{\varepsilon_1\tau/2-\rho_{X^*}(\alpha\tau)/C\alpha^2;\ \tau\geqslant0\}\\&=C^{-1}\alpha^{-2}\sup\{\varepsilon_1\tau C\alpha^2/2-\rho_{X^*}(\alpha\tau);\ \tau\geqslant0\}\\&=C^{-1}\alpha^{-2}\tilde\delta_X(\varepsilon_1C\alpha)=C^{-1}\alpha^{-2}\tilde\delta_X(\varepsilon_2)=C\tilde\delta_X(\varepsilon_2)\varepsilon_1^2/\varepsilon_2^2\,.\end{aligned}$$

It remains to show that $\delta_X(\varepsilon)$ is equivalent at zero to $\tilde{\delta}_X(\varepsilon)$. This fact is proved by the following two lemmas.

Lemma 1.e.7. *Let δ be a non-negative function on some interval $[0, T]$ such that $\delta(\varepsilon)/\varepsilon$ is equivalent to an increasing function i.e. $\delta(\varepsilon)/\varepsilon \leqslant C\,\delta(\eta)/\eta$, for some constant $C<\infty$ and for every $0<\varepsilon<\eta\leqslant T$. Let $\tilde{\delta}$ be the maximal convex function majorated by δ. Then $\tilde{\delta}(\varepsilon)\geqslant\delta(\varepsilon/2)/C$, for all $0<\varepsilon\leqslant T$.*

Proof. For each $0\leqslant\eta\leqslant T$, let $\mathscr{F}_\eta$ denote the collection of all linear functions $f(\varepsilon)$ which intersect the graph of $\delta(\varepsilon)$ in at least two distinct points $(\eta_1, \delta(\eta_1))$ and $(\eta_2, \delta(\eta_2))$ with $0\leqslant\eta_1\leqslant\eta\leqslant\eta_2\leqslant T$. By the definition of $\tilde{\delta}$, we have $\tilde{\delta}(\eta)=\inf\{f(\eta); f\in\mathscr{F}_\eta\}$. Any function $f\in\mathscr{F}_\eta$ is of the form

$$f(\varepsilon)=\delta(\eta_1)(\eta_2-\varepsilon)/(\eta_2-\eta_1)+\delta(\eta_2)(\varepsilon-\eta_1)/(\eta_2-\eta_1)\,,$$

with $\eta_1\leqslant\eta\leqslant\eta_2\leqslant T$. Suppose now that $\eta_1\leqslant\eta/2$. Then,

$$f(\eta)\geqslant\delta(\eta_2)\frac{\eta-\eta_1}{\eta_2-\eta_1}\geqslant\frac{2\eta_2}{C\eta}\,\delta(\eta/2)\,\frac{\eta}{2(\eta_2-\eta_1)}\geqslant\delta(\eta/2)/C\,.$$

In the other case, i.e. when $\eta_1>\eta/2$, we get that

$$f(\eta)\geqslant\frac{2\eta_1}{C\eta}\,\delta(\eta/2)\,\frac{\eta_2-\eta}{\eta_2-\eta_1}+\frac{2\eta_2}{C\eta}\,\delta(\eta/2)\,\frac{\eta-\eta_1}{\eta_2-\eta_1}=2\delta(\eta/2)/C\geqslant\delta(\eta/2)/C\,.$$

Consequently, also $\tilde{\delta}(\eta)$ exceeds $\delta(\eta/2)/C$. □

Lemma 1.e.8. *For every Banach space X, $\delta_X(\varepsilon)/\varepsilon$ is a non-decreasing function on $(0, 2]$.*

Proof. Fix $0<\eta<\varepsilon<2$ and vectors $x, y\in X$ such that $\|x\|=\|y\|=1$ and $\|x-y\|=\varepsilon$. In the two-dimensional subspace of X generated by x and y (see the illustration opposite) let $\overline{OA}$, $\overline{OB}$ and $\overline{OC}$ represent the vectors x, y, respectively $(x+y)/\|x+y\|$.

Since $AB=\varepsilon$ we can find points A' and B' on $\overline{AC}$, respectively $\overline{BC}$, so that $A'B'=\eta$ and $\overline{A'B'}$ is parallel to $\overline{AB}$. It is easily seen that the midpoint D' of $\overline{A'B'}$ is situated on OC.

Denote now the vectors $\overline{OA'}$ and $\overline{OB'}$ by x', respectively y', and notice that $CD'/A'B'=CD/AB$. Since $CD'=1-\|(x'+y')/2\|$ and $CD=1-\|(x+y)/2\|$ we have

$$\delta_X(\eta)/\eta\leqslant(1-\|(x'+y')/2\|)/\eta=(1-\|(x+y)/2\|)/\varepsilon$$

from which, by taking the infimum over all $x, y\in X$ as above, we get that

$$\delta_X(\eta)/\eta\leqslant\delta_X(\varepsilon)/\varepsilon\,. \quad \square$$

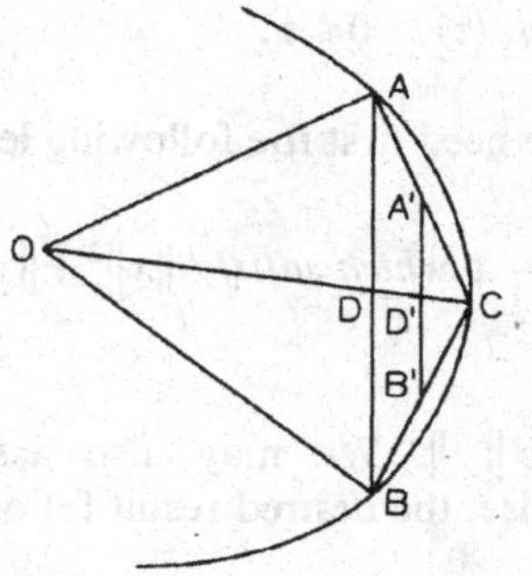

Remarks. The modulus of convexity of a Banach space need not be itself a convex function. An example of a space whose modulus of convexity is a non-convex function was given in [80].

Except for a few special cases it is in general quite difficult to compute precisely or even up to an equivalence constant the modulus of convexity of a given space. Such computations were made for Orlicz spaces in [91] (see also [40]) and for some Lorentz sequence spaces in [1]. For instance, the modulus of convexity $\delta_M(\varepsilon)$ of the Orlicz sequence space l_M with $M(t)=t^p/(1+|\log t|)$, $p\geqslant 2$, is equivalent at zero to the function $\varepsilon^p/(1+|\log \varepsilon|)$. More generally, if (i) M is a super-multiplicative function (in the sense that $M(st)\geqslant cM(s)M(t)$, for some $c>0$ and every $0\leqslant s, t\leqslant 1$) and (ii) $M(t^{1/2})$ is equivalent to a convex function, then the modulus of convexity $\delta_M(\varepsilon)$ of l_M is equivalent to $M(\varepsilon)$ itself. Results of a similar nature were obtained for Lorentz sequence spaces. Let $d(w,p)$, $p\geqslant 2$, be a Lorentz sequence space for which the sequence $w=\{w_n\}_{n=1}^{\infty}$ has the following property: there is a constant $c>0$ such that $S(kn)\geqslant cS(k)S(n)$ for all integers k and n, where $S(n)=\sum_{i=1}^{n} w_i$. The modulus of convexity $\delta_X(\varepsilon)$ of $X=d(w,p)$ with w as above is then equivalent to the function $1/S^{-1}(1/\varepsilon^p)$, where $S(t)$ is a strictly increasing function on $[0,\infty)$ coinciding with $S(n)$ for $t=n$.

We prove now a result of T. Figiel and G. Pisier [43] and T. Figiel [40] which shows that the moduli of convexity and of smoothness of a Banach space X and those of $L_2(X)$, the space of all measurable X-valued functions f on $[0,1]$ such that $\|f\|_{L_2(X)}=\left(\int_0^1 \|f(t)\|_X^2\,dt\right)^{1/2}<\infty$, are equivalent functions (a vector valued function is called measurable if it is the limit in norm of a sequence of simple measurable functions; for details see [32] III.2.10). The significance of this seemingly special result will become clear later on in this section (cf. 1.e.16).

Theorem 1.e.9. *For every Banach space X there exist constants $a, b>0$ and $C<\infty$ so that*

(i) $\delta_X(\varepsilon)\geqslant\delta_{L_2(X)}(\varepsilon)\geqslant a\delta_X(b\varepsilon),\quad 0\leqslant\varepsilon\leqslant 2,$
(ii) $\rho_X(\tau)\leqslant\rho_{L_2(X)}(\tau)\leqslant C\rho_X(\tau),\quad 0\leqslant\tau.$

For the proof of 1.e.9 we need first the following lemma.

Lemma 1.e.10. *For all $x, y\in X$ which satisfy $\|x\|^2+\|y\|^2=2$, we have $\|x+y\|^2\leqslant 4-4\delta_X(\|x-y\|/2)$.*

Proof. Suppose that $\|x\|\geqslant\|y\|$. We may also assume that $(\|x\|-\|y\|)^2<4\delta_X(\|x-y\|/2)$ since, otherwise, the desired result follows from

$$\|x+y\|^2\leqslant(\|x\|+\|y\|)^2=2(\|x\|^2+\|y\|^2)-(\|x\|-\|y\|)^2\leqslant 4-4\delta_X(\|x-y\|/2)\,.$$

Since $\delta_X(\varepsilon)\leqslant\delta_2(\varepsilon)=1-(1-\varepsilon^2/4)^{1/2}$ we get that

$$(\|x\|-\|y\|)^2<4(1-(1-\|x-y\|^2/16)^{1/2})\leqslant 4(1-(1-\|x-y\|^2/16))=\|x-y\|^2/4\,.$$

Put $z=x\|y\|/\|x\|$ and observe that, by the preceding inequality,

$$\|y-z\|\geqslant\|x-y\|-\|x-z\|=\|x-y\|-(\|x\|-\|y\|)\geqslant\|x-y\|/2\,.$$

Hence,

$$\delta_X(\|x-y\|/2\|y\|)\leqslant\delta_X(\|y-z\|/\|y\|)\leqslant 1-\|y+z\|/2\|y\|$$

from which it follows that

$$\|x+y\|\leqslant\|x-z\|+\|y+z\|\leqslant\|x\|-\|y\|+2\|y\|-2\delta_X(\|x-y\|/2\|y\|)\|y\|\,.$$

By using the fact that $\delta_X(\varepsilon)/\varepsilon$ is a non-decreasing function (cf. 1.e.8) and the fact that $\|y\|\leqslant 1$, we get

$$\|x+y\|^2\leqslant(\|x\|+\|y\|-2\delta_X(\|x-y\|/2))^2\leqslant 4(1-\delta_X(\|x-y\|/2))^2\leqslant 4-4\delta_X(\|x-y\|/2)\,.\quad\square$$

Proof of 1.e.9. The left hand inequalities in (i) and (ii) are immediate since X is isometric to a subspace of $L_2(X)$. We prove now the right hand side inequality of (i). Let f and g be two elements of $L_2(X)$ such that $\|f\|=\|g\|=1$. There is no loss of generality in assuming that $\varphi(t)=((\|f(t)\|_X^2+\|g(t)\|_X^2)/2)^{1/2}$ does not vanish (otherwise, we integrate only over $\{t;\ \varphi(t)>0\}$). By 1.e.10, we get that

$$\|f+g\|^2 \leqslant \int_0^1 \varphi^2(t)(4-4\delta_X(\|f(t)-g(t)\|_X/2\varphi(t)))\,dt\,.$$

Put $\alpha(\varepsilon)=\delta_X(\varepsilon^{1/2})$ and observe that, by the last part of 1.e.6, $\alpha(\varepsilon)/\varepsilon$ is equivalent to an increasing function. Thus, by 1.e.7, $\tilde{\alpha}(\varepsilon)$, the maximal convex function majorated by α, is equivalent to $\alpha(\varepsilon)$. This implies that there exist constants $\kappa_1, \kappa_2>0$ so that

$$\kappa_1\delta_X(\kappa_2\varepsilon)\leqslant\tilde{\alpha}(\varepsilon^2)\leqslant\delta_X(\varepsilon)\,,\quad 0\leqslant\varepsilon\leqslant 2\,.$$

Since $\int_0^1 \varphi^2(t)\,dt=1$ and $\tilde{\alpha}$ is convex it follows that

$$\begin{aligned}\int_0^1 \varphi^2(t)\delta_X(\|f(t)-g(t)\|_X/2\varphi(t))\,dt &\geqslant \int_0^1 \varphi^2(t)\tilde{\alpha}(\|f(t)-g(t)\|_X^2/4\varphi^2(t))\,dt\\ &\geqslant \tilde{\alpha}(\|f-g\|^2/4)\,.\end{aligned}$$

Thus, $\|f+g\|^2\leqslant 4-4\tilde{\alpha}(\|f-g\|^2/4)$ which implies that

$$\|f+g\|/2\leqslant 1-\tilde{\alpha}(\|f-g\|^2/4)/2$$

and hence,

$$\delta_{L_2(X)}(\varepsilon)\geqslant\kappa_1\delta_X(\varepsilon\kappa_2/2)/2\,.$$

Assertion (ii) follows from (i) by duality. Observe that, for computing $\rho_{L_2(X)}$, it is enough to consider expressions of the form $\|f+\tau g\|+\|f-\tau g\|$, with f and g being simple measurable functions, and that $(\overbrace{X\oplus X\oplus\cdots\oplus X}^{n\text{ times}})_2^*$ is isometric to $(\overbrace{X^*\oplus X^*\oplus\cdots\oplus X^*}^{n\text{ times}})_2$. □

Remark. In the last step of the proof above we did not use the duality between $L_2(X)$ and $L_2(X^*)$ since this fact is less elementary than the duality between $(X\oplus X\oplus\cdots\oplus X)_2$ and $(X^*\oplus X^*\oplus\cdots\oplus X^*)_2$. It is true that $L_2(X)^*=L_2(X^*)$ if X is reflexive (and, in particular, if X is uniformly convex) but this is no longer true for general X. It turns out that $L_2(X)^*=L_2(X^*)$ if and only if X^* has the Radon-Nikodym property (see [28] Chapter IV for a detailed discussion).

We present now some results on unconditionally convergent series in uniformly convex and uniformly smooth Banach spaces.

Theorem 1.e.11. *Let $\{x_j\}_{j=1}^{\infty}$ be a sequence of elements of a Banach space X.*

(i) *If* $\max_{\theta_j=\pm 1}\left\|\sum_{j=1}^{n}\theta_j x_j\right\|\leq 2$, *for some n, then* $\sum_{j=1}^{n}\delta_X(\|x_j\|)\leq 1$. *Consequently, if* $\sum_{j=1}^{\infty}x_j$ *is an unconditiondlly convergent series then* $\sum_{j=1}^{\infty}\delta_X(\|x_j\|)<\infty$.

(ii) *For every* $\lambda>0$ *there exists a choice of signs* $\theta_j=\pm 1$, $j=1,2,\ldots$ *so that*

$$\left\|\sum_{j=1}^{n}\theta_j x_j\right\|\leq\left(\max_{1\leq j\leq n}\|x_j\|+1/\lambda\right)\prod_{j=1}^{n}(1+\rho_X(\lambda\|x_j\|)),$$

for every integer n. Consequently, if a series $\sum_{j=1}^{\infty}\theta_j x_j$ *diverges for every choice of signs* $\theta_j=\pm 1$ *then* $\sum_{j=1}^{\infty}\rho_X(\|x_j\|)=\infty$.

Part (i) was first proved in [60] and Part (ii) in [74].

Proof. (i) Suppose that $\max_{\theta_j=\pm 1}\left\|\sum_{j=1}^{n}\theta_j x_j\right\|\leq 2$. Clearly, there is no loss of generality in assuming that $\|S_n\|=\left\|\sum_{j=1}^{n}x_j\right\|\geq\left\|\sum_{j=1}^{n}\theta_j x_j\right\|$ for every choice of $\theta_j=\pm 1$. Then, for each $1\leq j\leq n$, we have

$$\|x_j\|\leq 2\|x_j\|/\|S_n\|=\|S_n/\|S_n\|-(S_n-2x_j)/\|S_n\|\|$$

which implies that

$$\sum_{j=1}^{n}\delta_X(\|x_j\|)\leq\sum_{j=1}^{n}(1-\|S_n-x_j\|/\|S_n\|)\leq n-\left\|\sum_{j=1}^{n}(S_n-x_j)\right\|/\|S_n\|$$
$$=n-\|(n-1)S_n\|/\|S_n\|=1\,.$$

If the series $\sum_{j=1}^{\infty}x_j$ converges unconditionally then, by removing some terms if needed, we can assume that $\sup_{\theta_j=\pm 1}\left\|\sum_{j=1}^{\infty}\theta_j x_j\right\|\leq 2$ (cf. I.1.c.1) and then, by the first part of the proof, we get that $\sum_{j=1}^{\infty}\delta_X(\|x_j\|)\leq 1$.

(ii) We will choose the signs $\{\theta_j\}_{j=1}^{\infty}$ inductively. Fix $\lambda>0$ and assume that the first n signs have already been chosen so that

$$\left\|\sum_{j=1}^{k}\theta_j x_j\right\|\leq\left(\max_{1\leq j\leq k}\|x_j\|+1/\lambda\right)\prod_{j=1}^{k}(1+\rho_X(\lambda\|x_j\|)),$$

for every $1\leq k\leq n$. Put $S_n=\sum_{j=1}^{n}\theta_j x_j$ and observe that

$$(\|S_n/\|S_n\|+\lambda x_{n+1}\|+\|S_n/\|S_n\|-\lambda x_{n+1}\|)/2\leq 1+\rho_X(\lambda\|x_{n+1}\|)\,.$$

Thus, there exists a choice of $\theta_{n+1}=\pm 1$ such that $\|S_n/\|S_n\|+\lambda\theta_{n+1}x_{n+1}\|\leqslant 1+\rho_X(\lambda\|x_{n+1}\|)$ i.e.

$$\|S_n+\lambda\|S_n\|\theta_{n+1}x_{n+1}\|\leqslant\|S_n\|(1+\rho_X(\lambda\|x_{n+1}\|)) .$$

Using this inequality and the identity

$$S_{n+1}=S_n+\theta_{n+1}x_{n+1}=\frac{1}{\lambda\|S_n\|}(S_n+\lambda\|S_n\|\theta_{n+1}x_{n+1})+\left(1-\frac{1}{\lambda\|S_n\|}\right)S_n$$

we get that if $\lambda\|S_n\|>1$,

$$\begin{aligned}\|S_{n+1}\|&\leqslant\frac{1}{\lambda\|S_n\|}\|S_n\|(1+\rho_X(\lambda\|x_{n+1}\|))+\left(1-\frac{1}{\lambda\|S_n\|}\right)\|S_n\|\\&\leqslant\|S_n\|(1+\rho_X(\lambda\|x_{n+1}\|))\\&\leqslant\left(\max_{1\leqslant j\leqslant n+1}\|x_j\|+1/\lambda\right)\prod_{j=1}^{n+1}(1+\rho_X(\lambda\|x_j\|)) .\end{aligned}$$

If, on the other hand, $\lambda\|S_n\|\leqslant 1$ then every choice for θ_{n+1} is acceptable since

$$\|S_n\pm x_{n+1}\|\leqslant\|x_{n+1}\|+1/\lambda\leqslant\left(\max_{1\leqslant j\leqslant n+1}\|x_j\|+1/\lambda\right)\prod_{j=1}^{n+1}(1+\rho_X(\lambda\|x_j\|)) .$$

This proves the first part of (ii).

Suppose now that $\sum_{j=1}^{\infty}\rho_X(\|x_j\|)<\infty$. Since ρ_X satisfies the Δ_2-condition at zero, by 1.e.5, it follows that, for every integer k, $\sum_{j=1}^{\infty}\rho_X(2^k\|x_j\|)<\infty$. Observe also that $\lim_{j\to\infty}\|x_j\|=0$ for $\lim_{j\to\infty}\rho_X(\|x_j\|)=0$ and

$$\rho_X(\|x_j\|)\geqslant\rho_2(\|x_j\|)=(1+\|x_j\|^2)^{1/2}-1 .$$

These two facts imply the existence of an increasing sequence $\{n_k\}_{k=1}^{\infty}$ so that $\prod_{j=n_k}^{\infty}(1+\rho_X(2^k\|x_j\|))\leqslant 2$ and $\sup_{j\geqslant n_k}\|x_j\|\leqslant 1/2^k$ for all k. Thus, by the first part of (ii), we can find signs $\{\theta_j\}_{j=n_k}^{n_{k+1}-1}$, $k=1, 2, \ldots$ such that, for $n_k\leqslant n<n_{k+1}$, we have that $\left\|\sum_{j=n_k}^{n}\theta_jx_j\right\|\leqslant 1/2^{k-2}$. Hence, $\sum_{j=1}^{\infty}\theta_jx_j$ is a convergent series. □

The moduli of convexity and of smoothness of a Banach space are only isometric invariants and they may obviously change considerably under an equivalent renorming. Therefore, it is natural to ask whether the corresponding moduli for l_p spaces or, e.g., for the Orlicz and Lorentz sequence spaces considered in the remarks following 1.e.8, are the best ones up to equivalence, or it is possible to improve them by a suitable renorming. The key to the study of this question is usually 1.e.11 which, for instance, can be used to conclude that in a space X having a normalized unconditional basis $\{x_n\}_{n=1}^\infty$ with unconditional constant equal to K,

$$n\,\delta_X\left(1\Big/K\left\|\sum_{i=1}^n x_i\right\|\right)\leqslant 1\,,$$

for all n. In particular, for a Banach space X so that $d(X, l_p)=K$ for some $p>2$, we get that $\delta_X(\varepsilon)\leqslant K^{2p}\varepsilon^p$, whenever $\varepsilon=1/K^2n^{1/p}$, $n=1, 2, \ldots$ Hence, up to a constant, $\delta_p(\varepsilon)$ (or ε^p) is the best modulus of convexity that l_p, $p\geqslant 2$ can be given by an equivalent renorming. For $p\leqslant 2$ the modulus of convexity of l_p cannot be improved asymptotically by renorming since always $\delta_X(\varepsilon)\leqslant\delta_2(\varepsilon)\leqslant\varepsilon^2$.

It follows from a deep result of Krivine [67], to be presented in Vol. III, that in the case of an infinite dimensional $L_p(\mu)$ space, $1<p<\infty$, the modulus of convexity or smoothness cannot be improved at all (i.e. not only asymptotically) by any renorming.

By computing the value of $\left\|\sum_{i=1}^n x_i\right\|$ in the Orlicz and Lorentz sequence spaces discussed before 1.e.9 we conclude again that $M(\varepsilon)$, respectively $1/S^{-1}(\varepsilon^{-p})$, are asymptotically the largest moduli of convexity which can be achieved by an equivalent renorming.

The two cases encountered in 1.e.11 are quite extreme; the series $\sum_{j=1}^\infty \theta_j x_j$ is required to converge or to diverge for *every* choice of signs $\theta_j=\pm 1$. In many situations, it is very useful to study the behavior of the series for most choices of $\{\theta_j\}_{j=1}^\infty$ or, more precisely, to study the series $\sum_{j=1}^\infty r_j(t)x_j$ for almost all $t\in[0, 1]$, where $\{r_j\}_{j=1}^\infty$ denotes the sequence of the Rademacher functions.

In order to study this question we consider the expressions

$$\operatorname*{Average}_{\theta_j=\pm 1}\left\|\sum_{j=1}^n \theta_j x_j\right\| = 2^{-n}\sum_{\theta_j=\pm 1}\left\|\sum_{j=1}^n \theta_j x_j\right\| = \int_0^1\left\|\sum_{j=1}^n r_j(t)x_j\right\|dt$$

and introduce the following two important notions (cf. J. Hoffmann-Jørgensen [53]).

Definition 1.e.12. A Banach space X is said to be of *type p* for some $1<p\leqslant 2$, respectively, of *cotype q* for some $q\geqslant 2$, if there exists a constant $M<\infty$ so that, for every finite set of vectors $\{x_j\}_{j=1}^n$ in X, we have

$$(*)\qquad \int_0^1\left\|\sum_{j=1}^n r_j(t)x_j\right\|dt\leqslant M\left(\sum_{j=1}^n\|x_j\|^p\right)^{1/p},$$

respectively,

$$(\ast\!\!\ast)\qquad \int_0^1 \Big\|\sum_{j=1}^n r_j(t)x_j\Big\|\,dt \geqslant M^{-1}\Big(\sum_{j=1}^n \|x_j\|^q\Big)^{1/q}.$$

Any constant M satisfying (*) or (⁑) is called a type p, respectively cotype q, constant of X.

The cases $p=1$ and $q=\infty$ are not interesting since every Banach space is both of type 1 and cotype ∞ $\Big($with $\max_{1\leqslant j\leqslant n}\|x_j\|$ replacing the expression $\Big(\sum_{j=1}^n \|x_j\|^q\Big)^{1/q}$ when $q=\infty\Big)$. The conditions imposed on p, respectively q, are explained by the following remark. If all the vectors $\{x_j\}_{j=1}^n$ coincide with some vector $x\in X$ with $\|x\|=1$ then $\int_0^1 \Big\|\sum_{j=1}^n r_j(t)x_j\Big\|\,dt=\int_0^1\Big|\sum_{j=1}^n r_j(t)\Big|\,dt$ and, by Khintchine's inequality I.2.b.3 in $L_1(0,1)$, it follows that

$$A_1 n^{1/2}\leqslant \int_0^1\Big|\sum_{j=1}^n r_j(t)\Big|\,dt\leqslant n^{1/2},$$

where $A_1=2^{-1/2}$. This evidently shows that no Banach space can be of type $p>2$ or of cotype $q<2$.

The spaces $L_p(\mu)$, $1\leqslant p<\infty$ are of type min $(2, p)$ and cotype max $(2, p)$. Assume, for example, that $1\leqslant p\leqslant 2$. Then, since $L_p(\mu)$ is p-convex and p (and thus also 2)-concave, we get by Khintchine's inequality (cf. 1.d.6(i)) that for every choice of $\{f_j\}_{j=1}^n\subset L_p(\mu)$

$$\begin{aligned} A_1\Big(\sum_{j=1}^n\|f_j\|^2\Big)^{1/2} &\leqslant A_1\Big\|\Big(\sum_{j=1}^n|f_j|^2\Big)^{1/2}\Big\|\leqslant\int_0^1\Big\|\sum_{j=1}^n r_j(t)f_j\Big\|\,dt\\ &\leqslant\Big(\int_0^1\Big\|\sum_{j=1}^n r_j(t)f_j\Big\|^2dt\Big)^{1/2}\leqslant\Big\|\Big(\sum_{j=1}^n|f_j|^2\Big)^{1/2}\Big\|\\ &\leqslant\Big\|\Big(\sum_{j=1}^n|f_j|^p\Big)^{1/p}\Big\|=\Big(\sum_{j=1}^n\|f_j\|^p\Big)^{1/p}. \end{aligned}$$

Our assertion for $2<p<\infty$ is verified similarly. By considering the unit vector basis in l_p, $1\leqslant p<\infty$ it is trivially verified that an infinite dimensional $L_p(\mu)$ space is not of type r for any $r>p$ and not of cotype r for any $r<p$. It is also evident that an infinite dimensional $L_\infty(\mu)$ space (or more generally, an M space) is not of type p for any $p>1$ and not of cotype p for any $p<\infty$.

It follows from the preceding remarks that Hilbert spaces have the "best possible" type and cotype, i.e. are simultaneously of type 2 and cotype 2. The converse of this assertion is also true. We shall present in Vol. III the proof of the fact (due to Kwapien [69]) that every space, which is simultaneously of type 2 and cotype 2, is isomorphic to a Hilbert space. Since the natural context for the study of type and cotype is the local theory of Banach spaces we postpone to Vol. III also the presentation of some other important results concerning these notions. For instance, it will be shown there that a Banach space X has some type $p>1$, respectively, some cotype $q<\infty$ if and only if X contains no uniformly isomorphic copies of l_1^n, respectively l_∞^n, for all n. (For Banach lattices this result as well as Kwapien's result will be proved already in the next section).

The L_1 average $\int_0^1 \left\|\sum_{j=1}^n r_j(t)x_j\right\| dt$ can be replaced in 1.e.12 by any other L_r average, $1<r<\infty$, without affecting the definition. More precisely, *a Banach space X is of type p or cotype q if and only if there exists an $1<r<\infty$ and a constant $0<M_r<\infty$ so that, for every finite set $\{x_j\}_{j=1}^n$ in X, we have*

$$\left(\int_0^1 \left\|\sum_{j=1}^n r_j(t)x_j\right\|^r dt\right)^{1/r} \leq M_r\left(\sum_{j=1}^n \|x_j\|^p\right)^{1/p},$$

respectively,

$$\left(\int_0^1 \left\|\sum_{j=1}^n r_j(t)x_j\right\|^r dt\right)^{1/r} \geq M_r^{-1}\left(\sum_{j=1}^n \|x_j\|^q\right)^{1/q}.$$

This assertion is evident for lattices which are q-concave for some $q<\infty$. Indeed, by 1.d.6(i) and its proof, in such a lattice all the L_r averages

$$\left(\int_0^1 \left\|\sum_{j=1}^n r_j(t)x_j\right\|^r dt\right)^{1/r}, \quad 1\leq r<\infty,$$

are equivalent to $\left\|\left(\sum_{j=1}^n |x_j|^2\right)^{1/2}\right\|$ and are therefore mutually equivalent. It turns out that these L_r averages are mutually equivalent also if the lattice is not q-concave for any $q<\infty$ and, what is more interesting, this is true even in an arbitrary Banach space. This fact is due to Kahane [62].

Theorem 1.e.13. *For every $1<r<\infty$ there exists on constant $K_r<\infty$ so that, for any Banach space X and every finite subset $\{x_j\}_{j=1}^n$ of X, we have*

$$\int_0^1 \left\|\sum_{j=1}^n r_j(t)x_j\right\| dt \leq \left(\int_0^1 \left\|\sum_{j=1}^n r_j(t)x_j\right\|^r dt\right)^{1/r} \leq K_r \int_0^1 \left\|\sum_{j=1}^n r_j(t)x_j\right\| dt.$$

The left-hand side inequality is trivial. The proof of the right-hand side inequality which we present is due to C. Borell [14]. This proof is based on the following inequality from [7].

Lemma 1.e.14. *Let* $1<p\leqslant q<\infty$ *and* $\gamma=\sqrt{(p-1)/(q-1)}$. *Then, for every* $u\geqslant 0$, *we have*

$$\left(\frac{|1+\gamma u|^q+|1-\gamma u|^q}{2}\right)^{1/q}\leqslant\left(\frac{|1+u|^p+|1-u|^p}{2}\right)^{1/p}.$$

Proof. Assume first that $1<p\leqslant q\leqslant 2$. Then all the binomial coefficients $\binom{q}{2k}$ and $\binom{p}{2k}$ are positive and clearly

$$\binom{q}{2k}\gamma^2\leqslant\frac{q}{p}\binom{p}{2k},\quad k=1,2,\ldots\,.$$

Therefore, since $(1+w)^r\leqslant 1+rw$ whenever $0<r\leqslant 1$ and $w>0$, we get

$$\left(\frac{(1+\gamma u)^q+(1-\gamma u)^q}{2}\right)^{p/q}=\left(1+\sum_{k=1}^{\infty}\binom{q}{2k}\gamma^{2k}u^{2k}\right)^{p/q}$$
$$\leqslant 1+\sum_{k=1}^{\infty}\binom{p}{2k}u^{2k}=\frac{(1+u)^p+(1-u)^p}{2},$$

which completes the proof if $u\leqslant 1$.

If $u>1$ then since $|1\pm\gamma u|\leqslant|u\pm\gamma|$ and since we have already verified the desired inequality for $1/u$ we deduce that

$$\left(\frac{|1+\gamma u|^q+|1-\gamma u|^q}{2}\right)^{1/q}\leqslant\left(\frac{|u+\gamma|^q+|u-\gamma|^q}{2}\right)^{1/q}$$
$$\leqslant u\left(\frac{|1+1/u|^p+|1-1/u|^p}{2}\right)^{1/p}$$
$$=\left(\frac{|1+u|^p+|1-u|^p}{2}\right)^{1/p}.$$

This shows that 1.e.14 is true for $1<p\leqslant q\leqslant 2$. It is easily verified that 1.e.14 is equivalent to the statement that the operator T, defined by

$$Tf(s)=\int_0^1 f(t)\,dt+\gamma\int_0^1 f(t)r_1(t)\,dt\cdot r_1(s)\,,$$

is an operator of norm one from $L_p(0,1)$ into $L_q(0,1)$. Since γ is also equal to $\sqrt{(q'-1)/(p'-1)}$, where $1/p+1/p'=1$ and $1/q+1/q'=1$, we get by considering T^* that 1.e.14 is also valid for $2\leqslant p\leqslant q<\infty$. The general case $1<p\leqslant q<\infty$ is then deduced without difficulty. □

Corollary 1.e.15. *Let $1<p\leqslant q<\infty$ and $\gamma=\sqrt{(p-1)/(q-1)}$. Then, for every choice of vectors y_1 and y_2 in an arbitrary Banach space Y, we have*

$$\left(\frac{\|y_1+\gamma y_2\|^q+\|y_1-\gamma y_2\|^q}{2}\right)^{1/q}\leqslant\left(\frac{\|y_1+y_2\|^p+\|y_1-y_2\|^p}{2}\right)^{1/p}.$$

Proof. Put $z_1=y_1+y_2$, $z_2=y_1-y_2$, $u_1=(\|z_1\|+\|z_2\|)/2$ and $u_2=|\|z_1\|-\|z_2\||/2$. Then, since $0<\gamma\leqslant 1$, it follows from 1.e.14 that

$$\begin{aligned}&\left(\frac{\|y_1+\gamma y_2\|^q+\|y_1-\gamma y_2\|^q}{2}\right)^{1/q}\\&\leqslant\left(\frac{((1+\gamma)\|z_1\|/2+(1-\gamma)\|z_2\|/2)^q+((1-\gamma)\|z_1\|/2+(1+\gamma)\|z_2\|/2)^q}{2}\right)^{1/q}\\&=\left(\frac{|u_1+\gamma u_2|^q+|u_1-\gamma u_2|^q}{2}\right)^{1/q}\leqslant\left(\frac{|u_1+u_2|^p+|u_1-u_2|^p}{2}\right)^{1/p}\\&=\left(\frac{\|z_1\|^p+\|z_2\|^p}{2}\right)^{1/p}.\quad\square\end{aligned}$$

Proof of 1.e.13. Let $1<p\leqslant q<\infty$ and $\gamma=\sqrt{(p-1)/(q-1)}$, as before. The main step of the proof consists of showing that, for every choice of vectors $\{x_j\}_{j=0}^n$ in an arbitrary Banach space X, we have

$$(*)\qquad\left(\int_0^1\left\|x_0+\gamma\sum_{j=1}^n r_j(t)x_i\right\|^q dt\right)^{1/q}\leqslant\left(\int_0^1\left\|x_0+\sum_{j=1}^n r_j(t)x_i\right\|^p dt\right)^{1/p}.$$

The proof of 1.e.13 follows then from (*) by taking $x_0=0$ and by using Hölder's inequality. Indeed, for any $1<r<\infty$,

$$\left(\int_0^1\left\|\sum_{j=1}^n r_j(t)x_j\right\|^r dt\right)^{1/r}\leqslant\left(\int_0^1\left\|\sum_{j=1}^n r_j(t)x_j\right\|dt\right)^{1/(2r-1)}\left(\int_0^1\left\|\sum_{j=1}^n r_j(t)x_j\right\|^{2r}dt\right)^{(r-1)/(2r^2-r)}$$

from which, by taking $p=r$ and $q=2r$ in (*), we get that 1.e.13 is valid with $K_r=((2r-1)/(r-1))^{r-1}$.

Assertion (*) for $n=1$ is just 1.e.15. Suppose now that (*) has been already proved for n and let $\{x_j\}_{j=0}^{n+1}$ be a system of vectors in X. Then, by the induction hypothesis, it follows that

$$\begin{aligned}W_{n+1}&=\left(\int_0^1\left\|x_0+\gamma\sum_{j=1}^{n+1}r_j(t)x_j\right\|^q dt\right)^{1/q}\\&=\left(2^{-1}\int_0^1\left(\left\|x_0+\gamma x_{n+1}+\gamma\sum_{j=1}^n r_j(t)x_j\right\|^q+\left\|x_0-\gamma x_{n+1}+\gamma\sum_{j=1}^n r_j(t)x_j\right\|^q\right)dt\right)^{1/q}\\&\leqslant 2^{-1/q}(\|F\|_p^q+\|G\|_p^q)^{1/q},\end{aligned}$$

where

$$F(t)=\left\|x_0+\gamma x_{n+1}+\sum_{j=1}^{n} r_j(t)x_j\right\| \text{ and } G(t)=\left\|x_0-\gamma x_{n+1}+\sum_{j=1}^{n} r_j(t)x_j\right\|.$$

Hence, by the fact that $L_p(0,1)$ has q-concavity constant equal to one and 1.e.15 (used with $y_1=x_0+\sum_{i=1}^{n} r_j(t)x_j$ and $y_2=x_{n+1}$), we get that

$$W_{n+1}\leqslant 2^{-1/q}\|(F(t)^q+G(t)^q)^{1/q}\|_p\leqslant\left(\int_0^1\left\|x_0+\sum_{j=1}^{n+1} r_j(t)x_j\right\|^p dt\right)^{1/p}. \quad \square$$

Remark. The inequality (*), obtained in the proof of 1.e.13, yields actually a stronger result (due to Kwapien [70]), namely that if a series $\sum_{j=1}^{\infty} r_j(t)x_j$ converges in $L_1(X)$ then

$$\int_0^1 e^{C\left\|\sum_{j=1}^{\infty} r_j(t)x_j\right\|^2} dt<\infty,$$

for every $C>0$. Indeed, by using it with $q=2k$ and $p=2$, we get that

$$\begin{aligned}\int_0^1 e^{C\left\|\sum_{j=1}^{\infty} r_j(t)x_j\right\|^2} dt&=\sum_{k=0}^{\infty}\frac{C^k}{k!}\int_0^1\left\|\sum_{j=1}^{\infty} r_j(t)x_j\right\|^{2k} dt\\ &\leqslant\sum_{k=0}^{\infty}\frac{C^k(2k-1)^k}{k!}\left(\int_0^1\left\|\sum_{j=1}^{\infty} r_j(t)x_j\right\|^2 dt\right)^k.\end{aligned}$$

In view of 1.e.13, this shows that if $\sum_{j=1}^{\infty} r_j(t)x_j$ converges in $L_1(X)$ then

$$\int_0^1 e^{C\left\|\sum_{j=1}^{\infty} r_j(t)x_j\right\|^2} dt<\infty,$$

for sufficiently small values of C. To prove the convergence of the integral for arbitrary values of C one uses the fact that

$$\int_0^1 e^{C\left\|\sum_{j=1}^{\infty} r_j(t)x_j\right\|^2} dt\leqslant\left(\int_0^1 e^{2C\left\|\sum_{j=1}^{n} r_j(t)x_j\right\|^2} dt\right)\left(\int_0^1 e^{2C\left\|\sum_{j=n+1}^{\infty} r_j(t)x_j\right\|^2} dt\right)$$

and that $\int_0^1\left\|\sum_{j=n+1}^{\infty} r_j(t)x_j\right\|^2 dt$ can be made as small as we please if n is chosen large enough.

The notions of type and cotype are related to those of uniform convexity and uniform smoothness (cf. T. Figiel [40] and T. Figiel and G. Pisier [43]).

Theorem 1.e.16. (i) *A Banach space X which has modulus of convexity of power type q, for some $q \geqslant 2$, is also of cotype q.*

(ii) *A Banach space X which has modulus of smoothness of power type p, for some $1 < p \leqslant 2$, is also of type p.*

Proof. (i) The key point in the proof consists of the fact that, for any finite set $\{x_j\}_{j=1}^n$ in X, the elements $\{r_j(t)x_j\}_{j=1}^n$ form an unconditional basic sequence (with unconditional constant equal to one) in $L_2(X)$. Thus, it follows from 1.e.11 (i) applied in $L_2(X)$ that if

$$\left(\int_0^1 \left\| \sum_{j=1}^n r_j(t)x_j \right\|^2 dt\right)^{1/2} \leqslant 2 \quad \text{then} \quad \sum_{j=1}^n \delta_{L_2(X)}(\|x_j\|) \leqslant 1 .$$

By 1.e.9(i), the modulus of convexity of X and $L_2(X)$ are of the same power type. Therefore, we can find a constant $C < \infty$ such that $\delta_{L_2(X)} \geqslant \varepsilon^q/C^q$ for every $0 \leqslant \varepsilon \leqslant 2$. Consequently, $\left(\sum_{j=1}^n \|x_j\|^q\right)^{1/q} \leqslant C$, whenever $\left(\int_0^1 \left\| \sum_{j=1}^n r_j(t)x_j \right\|^2 dt\right)^{1/2} \leqslant 2$, and this completes the proof of (i) in view of the remarks following Definition 1.e.12. The proof of (ii) is quite similar. By 1.e.9 (ii), there exists a constant $D < \infty$ such that $\rho_{L_2(X)}(\tau) \leqslant D\tau^p$ for every $\tau \geqslant 0$. Thus, by applying 1.e.11 (ii) to $L_2(X)$ with $\lambda = D^{-1/p}$ and by using the identity $1 + u \leqslant e^u$, $u \geqslant 0$, we get that, for some choice of signs $\theta_j = \pm 1$,

$$\begin{aligned}
\left\| \sum_{j=1}^n \theta_j r_j x_j \right\|_{L_2(X)} &\leqslant \left(\max_{1 \leqslant j \leqslant n} \|x_j\| + D^{1/p} \right) \prod_{j=1}^n (1 + \rho_{L_2(X)}(D^{-1/p}\|x_j\|)) \\
&\leqslant \left(\max_{1 \leqslant j \leqslant n} \|x_j\| + D^{1/p} \right) \prod_{j=1}^n (1 + \|x_j\|^p) \\
&\leqslant \left(\left(\sum_{j=1}^n \|x_j\|^p \right)^{1/p} + D^{1/p} \right) e^{\sum_{j=1}^n \|x_j\|^p} .
\end{aligned}$$

Hence, if $\left(\sum_{j=1}^n \|x_j\|^p \right)^{1/p} \leqslant 1$ then

$$\left(\int_0^1 \left\| \sum_{j=1}^n r_j(t)x_j \right\|^2 dt\right)^{1/2} \leqslant (1 + D^{1/p})\, e . \quad \square$$

The converse to 1.e.16 is false. We have already verified that the non-reflexive space $L_1(0, 1)$ is of cotype 2. There is also a non-reflexive space of type 2 (cf. [55], this will be discussed in Vol. III). Also, it is not true that a Banach space X is of type p, for some $p > 1$, if its dual X^* is of cotype $q = p/(p-1)$. For instance, $X = c_0$ is of no type $p > 1$ while $X^* = l_1$ is of cotype 2. (It is an open problem whether there exist such counter-examples if we restrict ourselves to uniformly convex spaces). However, the following is true (cf. [53], [94]).

Proposition 1.e.17. *Let X be a Banach space of type p for some $p>1$. Then its dual X^* is of cotype $q=p/(p-1)$.*

Proof. For every $\varepsilon>0$ and every choice of $\{x_i^*\}_{i=1}^n$ in X^* we can find vectors $\{x_i\}_{i=1}^n$ in X so that $\|x_i^*\|<(1+\varepsilon)x_i^*(x_i)$ and $\|x_i\|=1$ for all $1\leqslant i\leqslant n$. It follows that

$$\left(\sum_{i=1}^n \|x_i^*\|^q\right)^{1/q} \leqslant (1+\varepsilon)\left(\sum_{i=1}^n x_i^*(x_i)^q\right)^{1/q}$$

$$=(1+\varepsilon)\sup\left\{\sum_{i=1}^n a_i x_i^*(x_i);\ \sum_{i=1}^n |a_i|^p\leqslant 1\right\}$$

$$=(1+\varepsilon)\sup\left\{\int_0^1\left(\sum_{i=1}^n r_i(u)x_i^*\right)\left(\sum_{j=1}^n r_j(u)a_jx_j\right)du;\ \sum_{j=1}^n |a_j|^p\leqslant 1\right\}.$$

Hence, by Hölder's inequality, we get that

$$\left(\sum_{i=1}^n \|x_i^*\|^q\right)^{1/q} \leqslant (1+\varepsilon)MK_p\left(\int_0^1\left\|\sum_{i=1}^n r_i(u)x_i^*\right\|^q du\right)^{1/q},$$

where M is a type p constant for X. □

f. Uniform Convexity in Banach Lattices and Related Notions

In this section we study the relation between the type and cotype of a space and the moduli of smoothness and convexity of renormings of the space, in the context of Banach lattices. In this framework it is very useful to compare these notions also with those of p-convexity and p-concavity, which were studied in Section d, as well as with a variant of these notions called upper, respectively, lower p-estimate, which will be defined below. The relations among these notions in a Banach lattice turn out to be very close and lead to a beautiful and useful theory. There is, for instance, a nice duality between the type and the cotype of Banach lattices (cf. 1.f.18 below, for a precise formulation) and, under some assumptions, the cotype of a Banach lattice determines the power type estimate for the modulus of convexity of a suitable renorming (cf. 1.f.10 below). We also present some examples which show that the theorems proved in this section are sharp. As happens frequently in Banach space theory, the exponent 2 plays a special role in this section.

The section ends with two diagrams which summarize the various results and examples concerning the relations among the four pairs of mutually dual notions mentioned above.

We begin with a result of T. Figiel [40] (see also [41]).

Theorem 1.f.1. *Let $1<p\leqslant 2\leqslant q<\infty$ and let X be a p-convex and q-concave Banach lattice. Then X can be renormed equivalently so that X, endowed with the new norm and the same order, becomes a Banach lattice which is uniformly convex, with modulus of convexity of power type q, and uniformly smooth, with modulus of smoothness of power type p.*

If, in addition, $M^{(p)}(X)=M_{(q)}(X)=1$ then X itself is uniformly convex and uniformly smooth with both moduli being of power type, as above.

We need first a simple lemma.

Lemma 1.f.2. *Let $q\geqslant 2$; then, for any $1<p<\infty$, there exists a constant $C=C(p,q)$ such that*

$$\left(\left|\frac{s-t}{C}\right|^q+\left|\frac{s+t}{2}\right|^q\right)^{1/q}\leqslant\left(\frac{|s|^p+|t|^p}{2}\right)^{1/p},$$

for every choice of reals s and t.

Proof. We may assume without loss of generality that $s=1>t\geqslant -1$. The function

$$\varphi(t)=\left(\frac{1+|t|^p}{2}\right)^{q/p}-\left(\frac{1+t}{2}\right)^q$$

is clearly positive on the interval $[-1, 1)$ and, since $\varphi''(1)>0$, it follows that $\varphi(t)/(1-t)^2$ (and thus also $\varphi(t)/(1-t)^q$) is bounded from below. □

Proof of 1.f.1. We consider here only the case when $M^{(p)}(X)=M_{(q)}(X)=1$. The general case can be reduced to this one by 1.d.8. We also observe that, by the duality results 1.d.4 and 1.e.2 (ii) (see, in addition, the subsequent remarks about the duality of moduli of power type), it is enough to prove that the modulus of convexity of X is of power type q.

Fix $\varepsilon>0$ and let $x, y\in X$ be such that $\|x\|=\|y\|=1$ and $\|x-y\|=\varepsilon$. By 1.f.2 and 1.d.1, it follows that there exists a constant $C=C(p,q)$ such that

$$\|((|x-y|/C)^q+(|x+y|/2)^q)^{1/q}\|\leqslant\|((|x|^p+|y|^p)/2)^{1/p}\|.$$

Thus, in view of the fact that $M^{(p)}(X)=M_{(q)}(X)=1$, we obtain

$$(\|(x-y)/C\|^q+\|(x+y)/2\|^q)^{1/q}\leqslant(\|x\|^p+\|y\|^p)^{1/p}/2^{1/p},$$

i.e.

$$\varepsilon^q/C^q\leqslant 1-\|(x+y)/2\|^q\leqslant q(1-\|(x+y)/2\|).$$

Hence, $\delta_X(\varepsilon)\geqslant\varepsilon^q/C^qq$. □

Remarks. 1. Despite its general character, 1.f.1 seems to provide one of the simplest ways to prove that the modulus of convexity of L_p spaces, $1<p<\infty$, is of power type equal to $\max\{2,p\}$.

2. T. Figiel and W. B. Johnson [41] used 1.f.1 in order to construct an example of an infinite dimensional uniformly convex Banach space with an unconditional basis which contains no isomorphic copies of l_p for $1<p<\infty$. Their approach is the following: let T be the Tsirelson type space presented in I.2.e.1 and let $1<p, r<\infty$ be arbitrary numbers. Denote by Y the dual space of the p-convexification $T^{(p)}$ of T and put $q=p/(p-1)$. Then Y is q-concave and, thus, its r-convexication $Y^{(r)}$ is r-convex and rq-concave. Hence, by 1.f.1, $Y^{(r)}$ is uniformly convex and uniformly smooth (since $M^{(r)}(Y^{(r)})=M_{(rq)}(Y^{(r)})=1$). The relation (*) of I.2.e.1, satisfied by T, easily implies that $Y^{(r)}$ contains no isomorphic copy of l_s for $s\neq rq$. The space $Y^{(r)}$ contains also no isomorphic copy of l_{rq}. Indeed, if $Y^{(r)}$ contains such a subspace there is, by I.1.a.12, a normalized block basis,

$$x_k=\sum_{j=m_k+1}^{m_{k+1}} \lambda_j e_j, \quad k=1,2,\ldots,$$

of the unit vector basis $\{e_j\}_{j=1}^\infty$ of $Y^{(r)}$, which is equivalent to the unit vector basis of l_{rq}. By abuse of language, we shall denote by $\{e_j\}_{j=1}^\infty$ also the unit vector basis of Y. By definition, the norms in $Y^{(r)}$ and Y are related by

$$\left\|\sum_{j=1}^\infty b_j e_j\right\|_{Y^{(r)}}=\left\|\sum_{j=1}^\infty |b_j|^r e_j\right\|_Y^{1/r}.$$

Hence, the normalized sequence $y_k=\sum_{j=m_k+1}^{m_{k+1}} |\lambda_j|^r e_j$, $k=1,2,\ldots$, in $Y=(T^{(p)})^*$ is equivalent to the unit vector basis in l_q. In particular, $\left\|\sum_{k=1}^\infty b_k y_k\right\|_Y \leq K\left(\sum_{k=1}^\infty |b_k|^q\right)^{1/q}$ for some constant K. Let $\{t_j\}_{j=1}^\infty$ denote the unit vector basis of T and (again, by abuse of language) also of $T^{(p)}$. Let $\{v_k\}_{k=1}^\infty$ be elements of norm one in $T^{(p)}$ so that $y_k(v_k)=1$ and $v_k=\sum_{j=m_k+1}^{m_{k+1}} \eta_j t_j$, $k=1,2,\ldots$. Then, for every choice of scalars $\{a_k\}_{k=1}^\infty$,

$$\left\|\sum_{k=1}^\infty a_k v_k\right\|_{T^{(p)}} \geq \sup\left\{\sum_{k=1}^\infty |a_k b_k|;\ \sum_{k=1}^\infty |b_k|^q\leq K^{-q}\right\}=\left(\sum_{k=1}^\infty |a_k|^p\right)^{1/p}\Big/K.$$

Consequently, $w_k=\sum_{j=m_k+1}^{m_{k+1}} |\eta_j|^p t_j$, $k=1,2,\ldots$, is a normalized block basis of $\{t_j\}_{j=1}^\infty$ (in T) which satisfies

$$\left\|\sum_{k=1}^\infty c_k w_k\right\|_T=\left\|\sum_{k=1}^\infty |c_k|^{1/p} v_k\right\|_{T^{(p)}}^p \geq \sum_{k=1}^\infty |c_k|/K^p$$

and is thus equivalent to the unit vector basis of l_1. This, however, contradicts the fact proved in I.2.e.1 that l_1 is not isomorphic to a subspace of T.

By combining 1.f.1 with 1.e.16 we deduce that a Banach lattice which is p-convex and q-concave for some $1<p\leqslant 2\leqslant q<\infty$ must be of type p and cotype q. It is however trivial to prove directly the following somewhat stronger assertion.

Proposition 1.f.3. (i) *A q-concave Banach lattice X with $q\geqslant 2$ is of cotype q.*

(ii) *A p-convex Banach lattice with $1<p\leqslant 2$, which is also q-concave for some $q<\infty$, is of type p.*

Proof. For every choice of $\{x_i\}_{i=1}^n$ in X we have, by 1.d.6 (i), that

$$\left(\sum_{i=1}^n \|x_i\|^q\right)^{1/q}\leqslant M_{(q)}(X)\left\|\left(\sum_{i=1}^n |x_i|^q\right)^{1/q}\right\|\leqslant M_{(q)}(X)\left\|\left(\sum_{i=1}^n |x_i|^2\right)^{1/2}\right\|$$

$$\leqslant M_{(q)}(X)A_1^{-1}\int_0^1\left\|\sum_{i=1}^n r_i(t)x_i\right\|dt\,.$$

This proves part (i) of the assertion. The proof of part (ii) is the same. □

Without the assumption of q-concavity for some $q<\infty$, p-convexity for $1<p\leqslant 2$ does not imply that X is of type p. Consider, for example, the space $X=L_\infty(0,1)$ which is even ∞-convex.

Notice that if in the definitions of type p or of p-convexity we consider only vectors $\{x_i\}_{i=1}^n$ with mutually disjoint supports both definitions reduce to exactly the same condition. A similar situation occurs with the definitions of cotype p and p-concavity. It turns out that the notions obtained by considering only disjoint elements in those definitions are useful in studying deeper the connection between type and convexity, respectively, cotype and concavity (e.g. in investigating to what extent the converse to 1.f.3 is true). We therefore introduce formally these notions.

Definition 1.f.4 [120], [41]. Let $1<p<\infty$. A Banach lattice X is said to satisfy an *upper*, respectively, *lower p-estimate* (for disjoint elements) if there exists a constant $M<\infty$ such that, for every choice of pairwise disjoint elements $\{x_i\}_{i=1}^n$ in X, we have

$$(*)\qquad \left\|\sum_{i=1}^n x_i\right\|\leqslant M\left(\sum_{i=1}^n \|x_i\|^p\right)^{1/p},$$

respectively,

$$(**)\qquad \left\|\sum_{i=1}^n x_i\right\|\geqslant M^{-1}\left(\sum_{i=1}^n \|x_i\|^p\right)^{1/p}.$$

The smallest constant M satisfying (*) or ($\underset{*}{*}$) is called the *upper*, respectively, *lower p-estimate constant of X*.

Proposition 1.f.5. *Let $1<p<\infty$. A Banach lattice X satisfies an upper, respectively, lower p-estimate if and only if its dual X^* satisfies a lower, respectively, upper q-estimate, where $1/p+1/q=1$.*

Proof. The proof is completely straightforward in case X is order continuous. In this case (since the proposition reduces immediately to the separable case) we may assume that X is a Köthe function space satisfying $X^*=X'$. Assume now, e.g. that X satisfies a lower p-estimate with some constant M. Let $\{g_i\}_{i=1}^n$ be positive disjoint functions in X^* and let f be a positive function in X of norm $\leqslant 1$. Then we can decompose f into a sum $\sum_{i=0}^{n} f_i$ of disjoint functions such that $\int_\Omega f_i g_j \, d\mu=0$ if $i\neq j (0\leqslant i\leqslant n,\ 1\leqslant j\leqslant n)$. Hence,

$$\int_\Omega \left(\sum_{i=1}^n g_i\right) f \, d\mu = \sum_{i=1}^n \int_\Omega f_i g_i \, d\mu \leqslant \left(\sum_{i=1}^n \|g_i\|^q\right)^{1/q} \left(\sum_{i=1}^n \|f_i\|^p\right)^{1/p}$$

$$\leqslant M\left(\sum_{i=1}^n \|g_i\|^q\right)^{1/q}$$

and thus, by taking the supremum over f, we deduce that X^* satisfies an upper q-estimate. The other assertions of 1.f.5 for function spaces are just as easy.

If a Banach lattice satisfies a lower p-estimate then, by 1.a.5 and 1.a.7, it is already σ-complete and σ-order continuous and thus, the trivial argument above shows that in the general case

$$(X \text{ satisfies a lower } p\text{-estimate}) \Rightarrow (X^* \text{ satisfies an upper } q\text{-estimate}).$$

Similarly, if X^* satisfies a lower p-estimate then X^{**}, and thus also X, satisfy an upper q-estimate. The proofs of the converse assertions in the general case do however require an additional argument which is a direct generalization of the argument used in the proof of 1.b.6.

Assume that X satisfies an upper p-estimate with constant M and let $\{x_i^*\}_{i=1}^n$ be disjoint positive elements in X^*. Fix $\varepsilon>0$ and choose positive elements $\{u_i\}_{i=1}^n$ in X so that

$$\|x_i^*\| \geqslant x_i^*(u_i) \geqslant (1-\varepsilon)\|x_i^*\| \quad \text{and} \quad \|u_i\|=1\,,$$

for all i. Since, for every $1\leqslant j\leqslant n$, we have $x_j^* \wedge \left(\sum_{i\neq j} x_i^*\right)=0$ it follows from the definition of $\wedge$ in X^* that, for all $1\leqslant j\leqslant n$, there exist $v_j,\ x_j\geqslant 0$ in X with $u_j=v_j+x_j$,

$$x_j^*(v_j)\leqslant \varepsilon\|x_j^*\| \quad \text{and} \quad \left(\sum_{i\neq j} x_i^*\right)(x_j)\leqslant \varepsilon \min_{1\leqslant i\leqslant n} \|x_i^*\|\,.$$

Clearly, $\|x_j\| \leqslant 1$ and $x_j^*(x_j) \geqslant (1-2\varepsilon)\|x_j^*\|$, for all j. Put

$$w_j = \Big(x_j - \sum_{i \neq j} x_i\Big)_+, \quad 1 \leqslant j \leqslant n .$$

Then $0 \leqslant w_j \leqslant x_j$ (and, in particular, $\|w_j\| \leqslant 1$) and, since $w_j \geqslant x_j - \sum_{i \neq j} x_i$, we get that

$$x_j^*(w_j) \geqslant (1-2\varepsilon)\|x_j^*\| - \varepsilon(n-1)\|x_j^*\| = (1-\varepsilon(n+1))\|x_j^*\| .$$

Moreover, if $j \neq k$ then

$$0 \leqslant w_j \wedge w_k \leqslant (x_j - x_k)_+ \wedge (x_k - x_j)_+ = 0 .$$

We have thus produced a sequence $\{w_j\}_{j=1}^n$ of disjoint positive elements of norm $\leqslant 1$ in X such that x_j^* almost attains its norm on w_j. Since

$$\Big\|\sum_{j=1}^n a_j w_j\Big\| \leqslant M\Big(\sum_{j=1}^n |a_j|^p\Big)^{1/p},$$

for every choice of scalars $\{a_j\}_{j=1}^n$, and since ε is arbitrary, we get, by a straightforward computation, that $\Big\|\sum_{i=1}^n x_i^*\Big\| \geqslant M^{-1}\Big(\sum_{i=1}^n \|x_i^*\|^q\Big)^{1/q}$, i.e. X^* satisfies a lower q-estimate.

Similarly, if X^* satisfies an upper p-estimate then X^{**}, and thus also X, satisfy a lower q-estimate. □

There are instances when it is preferable to express the existence of an upper or lower p-estimate without the explicit use of disjoint vectors.

Proposition 1.f.6. *Let* $1 < p < \infty$ *and* $M < \infty$.

(i) *A Banach lattice X satisfies* (*) *if and only if*

$$(*') \qquad \Big\|\bigvee_{i=1}^n |x_i|\Big\| \leqslant M\Big(\sum_{i=1}^n \|x_i\|^p\Big)^{1/p}$$

holds for every choice of $\{x_i\}_{i=1}^n$ *in* X.

(ii) *A Banach lattice X satisfies* $(\substack{*\\ *})$ *if and only if*

$$(\substack{*\\ *}') \qquad \Big\|\sum_{i=1}^n |x_i|\Big\| \geqslant M^{-1}\Big(\sum_{i=1}^n \|x_i\|^p\Big)^{1/p}$$

holds for every choice of $\{x_i\}_{i=1}^n$ *in* X.

Proof. Obviously, $(*') \Rightarrow (*)$ and $(\substack{*\\ *}') \Rightarrow (\substack{*\\ *})$. To prove that $(*) \Rightarrow (*')$ assume first that X is an order complete lattice. Let $\{x_i\}_{i=1}^n$ be vectors in X and consider the band projections $P_i = P_{\left(\bigvee_{j\neq i} |x_j| - |x_i|\right)_+}$, $1 \leq i \leq n$, on X. Let $z = \bigvee_{j=1}^n |x_j|$ and

$$y_1 = (I - P_1)z, \qquad y_2 = P_1(I - P_2)z, \qquad y_3 = P_1P_2(I - P_3)z, \ldots .$$

Then

$$\sum_{i=1}^n y_i = z, \quad 0 \leq y_i \leq |x_i|, \quad 1 \leq i \leq n \quad \text{and} \quad y_i \wedge y_k = 0 \quad \text{for } i \neq k .$$

Hence, by (*),

$$\|z\| = \left\| \sum_{j=1}^n y_j \right\| \leq M \left(\sum_{j=1}^n \|y_j\|^p \right)^{1/p} \leq M \left(\sum_{j=1}^n \|x_j\|^p \right)^{1/p},$$

i.e. (*′) holds. If X is a general lattice the implication (*) ⇒ (*′) is proved by passing to X^{**}, which is order complete and which also satisfies (*) (by 1.f.5). This proves (i). Assertion (ii) follows by duality. Indeed, a lattice X satisfies (*′) if and only if the identity map from $l_p(X)$ into $X(c_0)$ has norm $\leq M$. Thus (ii) follows from (i), 1.f.5 and the fact that $(X^*(c_0))^* = \widetilde{X^{**}(l_1)}$. □

It is evident that a Banach lattice X satisfies an upper, respectively lower p-estimate, whenever it is p-convex, respectively p-concave. The converse is false (cf. 1.f.20 below) but we have the following result of B. Maurey [94] and B. Maurey and G. Pisier [96].

Theorem 1.f.7. *If a Banach lattice X satisfies an upper, respectively, lower r-estimate for some $1 < r < \infty$ then it is p-convex, respectively q-concave, for every $1 < p < r < q < \infty$.*

The proof is based on a probabilistic lemma. Before stating the lemma let us recall some elementary notions. By a random variable we understand a measurable function defined on a probability space (Ω, Σ, μ). (Here we are going to use real-valued random variables but later on we shall consider also random variables whose range space is a Banach space X.) A set $\{f_\alpha\}_{\alpha \in A}$ of random variables is said to be *independent* if, for every finite subset $\{\alpha_i\}_{i=1}^n \subset A$ and every choice of open sets $\{D_i\}_{i=1}^n$ in X, we have

$$\mu(\{\omega \in \Omega; f_{\alpha_i}(\omega) \in D_i, 1 \leq i \leq n\}) = \prod_{i=1}^n \mu(\{\omega \in \Omega; f_{\alpha_i}(\omega) \in D_i\}) .$$

For example, the Rademacher functions $\{r_i(t)\}_{i=1}^\infty$ form a sequence of independent random variables on $[0, 1]$. In 1.f.8 we shall use a sequence of independent random variables $\{f_i\}_{i=1}^\infty$ so that $\mu(\{\omega \in \Omega; |f_i(\omega)| > \lambda\}) = 1/\lambda^p$ for every $\lambda \geq 1$. A simple way to construct such $\{f_i\}_{i=1}^\infty$ is to take $\Omega = [0, 1]^{\aleph_0}$ with the usual product measure and put $f_i(\omega) = t_i^{-1/p}$, where $\omega = (t_1, t_2, \ldots) \in [0, 1]^{\aleph_0}$.

Lemma 1.f.8. *Let $1<p<\infty$ and let $\{f_i\}_{i=1}^\infty$ be a sequence of independent random variables on some probability measure space (Ω, Σ, μ) such that*

$$\mu(\{\omega \in \Omega;\ |f_i(\omega)|>\lambda\})=1/\lambda^p\,,$$

for every i and every $\lambda \geqslant 1$. Then, for every $r>p$, there exists a constant $K=K(p,r)<\infty$ such that, for every finite sequence $\{a_i\}_{i=1}^n$ of scalars, we have

$$K^{-1}\int_\Omega \left(\sum_{i=1}^n |a_i f_i(\omega)|^r\right)^{1/r} d\mu \leqslant \left(\sum_{i=1}^n |a_i|^p\right)^{1/p} \leqslant K\int_\Omega \max_{1\leqslant i\leqslant n} |a_i f_i(\omega)|\, d\mu\,.$$

Proof. Let $\{a_i\}_{i=1}^n$ be a sequence of positive reals so that $\sum_{i=1}^n a_i^p=1$. Then, by the independence of the f_i's and the trivial fact that $1-t\leqslant e^{-t}$, whenever $t\geqslant 0$, we obtain

$$\begin{aligned}\int_\Omega \max_{1\leqslant i\leqslant n} |a_i f_i(\omega)|\, d\mu &\geqslant \mu(\{\omega\in\Omega;\ \max_{1\leqslant i\leqslant n} |a_i f_i(\omega)|>1\})\\ &=1-\mu(\{\omega\in\Omega;\ |a_i f_i(\omega)|\leqslant 1 \text{ for all } 1\leqslant i\leqslant n\})\\ &=1-\prod_{i=1}^n \mu(\{\omega\in\Omega;\ |f_i(\omega)|\leqslant 1/a_i\})=1-\prod_{i=1}^n (1-a_i^p)\geqslant 1-\prod_{i=1}^n e^{-a_i^p}\\ &=1-e^{-1}\end{aligned}$$

and this proves the right-hand side inequality of 1.f.8. Put

$$\varphi(\lambda)=\mu\left(\left\{\omega\in\Omega;\ \max_{1\leqslant i\leqslant n} |a_i f_i(\omega)|>\lambda\right\}\right).$$

Since $1-t\geqslant e^{-2t}$ for $0\leqslant t\leqslant 1/2$ we get that for $\lambda\geqslant 2^{1/p}$

$$\begin{aligned}\varphi(\lambda)&=1-\prod_{i=1}^n \mu(\{\omega\in\Omega;\ |f_i(\omega)|\leqslant \lambda/a_i\})\\ &=1-\prod_{i=1}^n (1-(a_i/\lambda)^p)\leqslant 1-\prod_{i=1}^n e^{-2a_i^p/\lambda^p}=1-e^{-2/\lambda^p}.\end{aligned}$$

Hence, by integration, it follows that

$$\int_\Omega \max_{1\leqslant i\leqslant n} |a_i f_i(\omega)|\, d\mu=\int_0^\infty \varphi(\lambda)\, d\lambda\leqslant 2^{1/p}+\int_{2^{1/p}}^\infty (1-e^{-2/\lambda^p})\, d\lambda=K_1\,,$$

where $K_1<\infty$ since $p>1$.

In order to prove the left-hand side inequality of 1.f.8, we fix $r>p$ and take a sequence $\{g_i\}_{i=1}^\infty$ of independent random variables on a probability space

(Ω', Σ', μ') so that

$$\mu(\{\omega' \in \Omega;\ |g_i(\omega')| > \lambda\}) = 1/\lambda^r\,,$$

for every i and every $\lambda \geqslant 1$. Then, by applying the right-hand side inequality to the functions $\{g_i\}_{i=1}^\infty$ instead of the $\{f_i\}_{i=1}^\infty$, we obtain

$$(1-e^{-1})\left(\sum_{i=1}^n |a_i f_i(\omega)|^r\right)^{1/r} \leqslant \int_{\Omega'} \max_{1\leqslant i\leqslant n} |a_i f_i(\omega) g_i(\omega')|\, d\mu'(\omega')\,,$$

for every $\omega \in \Omega$. Thus, by integration and the definition of K_1, it follows that

$$\begin{aligned}(1-e^{-1})\int_\Omega\left(\sum_{i=1}^n |a_i f_i(\omega)|^r\right)^{1/r} d\mu(\omega) &\leqslant \int_\Omega\int_{\Omega'} \max_{1\leqslant i\leqslant n} |a_i f_i(\omega) g_i(\omega')|\, d\mu'(\omega')\, d\mu(\omega)\\ &\leqslant K_1 \int_{\Omega'}\left(\sum_{i=1}^n |a_i g_i(\omega')|^p\right)^{1/p} d\mu'(\omega')\\ &\leqslant K_1\left(\sum_{i=1}^n |a_i|^p \int_{\Omega'} |g_i(\omega')|^p\, d\mu'(\omega')\right)^{1/p}.\end{aligned}$$

But, for every i, we have

$$\int_{\Omega'} |g_i(\omega')|^p\, d\mu'(\omega') = \int_0^1 t^{-p/r}\, dt = r/(r-p)\,.$$

Consequently,

$$(1-e^{-1})\int_\Omega\left(\sum_{i=1}^n |a_i f_i(\omega)|^r\right)^{1/r} d\mu(\omega) \leqslant K_1(r/(r-p))^{1/p}\left(\sum_{i=1}^n |a_i|^p\right)^{1/p}. \quad \square$$

Remark. Lemma 1.f.8 implies in particular that, for every $r > p$, the space l_p is isomorphic to a subspace of $L_1(l_r)$.

Proof of 1.f.7. By duality, it suffices to show that if X satisfies an upper r-estimate for some $1 < r < \infty$ then it is p-convex, for every $1 < p < r$. Fix $p < r$ and let $\{f_i\}_{i=1}^\infty$ be a sequence of independent random variables on some probability measure space (Ω, Σ, μ) so that $\mu(\{\omega \in \Omega;\ |f_i(\omega) > \lambda\}) = 1/\lambda^p$, for every i and every $\lambda \geqslant 1$. By condition (*'), there is a constant $M < \infty$ so that, for every $\omega \in \Omega$ and every choice of $\{x_i\}_{i=1}^n$ in X, we have

$$\left\|\bigvee_{i=1}^n |f_i(\omega) x_i|\right\| \leqslant M\left(\sum_{i=1}^n \|f_i(\omega) x_i\|^r\right)^{1/r}.$$

Hence, by 1.f.8, there is a constant $K=K(p,r)<\infty$ such that,

$$\left\|\left(\sum_{i=1}^n |x_i|^p\right)^{1/p}\right\| \leqslant K \left\| \int_\Omega \bigvee_{i=1}^n |f_i(\omega)x_i| \, d\mu \right\| \leqslant K \int_\Omega \left\| \bigvee_{i=1}^n |f_i(\omega)x_i| \right\| d\mu$$
$$\leqslant KM \int_\Omega \left(\sum_{i=1}^n \|f_i(\omega)x_i\|^r \right)^{1/r} d\mu \leqslant K^2 M \left(\sum_{i=1}^n \|x_i\|^p \right)^{1/p} . \ \square$$

Corollary 1.f.9. *Let $1<r<\infty$ and let X be a Banach lattice of type r, respectively, cotype r. Then X is p-convex, respectively, q-concave for every $1<p<r<q$.*

A Banach lattice satisfying an upper p-estimate for some $1<p<2$ need not be of type p (take e.g. $L_\infty(0,1)$). G. Pisier observed that, among the Lorentz spaces, there are examples of lattices which satisfy a lower 2-estimate without being of cotype 2 (cf. 1.f.19 below). However, the following assertion is true (cf. B. Maurey [94]): *A Banach lattice, which satisfies a lower q-estimate for some $q>2$, is of cotype q.* We omit the proof of this result since a slightly weaker version of it follows from the next theorem due to T. Figiel [40] and T. Figiel and W. B. Johnson [41].

Theorem 1.f.10. *Let $1<p<2<q$ and suppose that a Banach lattice X satisfies an upper p-estimate and a lower q-estimate. Then there exist two norms $\|\cdot\|_1$ and $\|\cdot\|_2$ on X which are equivalent to the original norm so that X, with the norm $\|\cdot\|_1$ and the original order, is a uniformly convex Banach lattice having modulus of convexity of power type q while X, with the norm $\|\cdot\|_2$ and the original order, is a uniformly smooth Banach lattice having modulus of smoothness of power type p. In particular, X is of type p and of cotype q.*

We need first a renorming lemma which is very similar to 1.d.8.

Lemma 1.f.11. *Let X be an r-convex Banach lattice satisfying a lower q-estimate for some $1<r<q$. Then X can be renormed equivalently so that X, endowed with the new norm and the same order, becomes a Banach lattice for which the r-convexity constant and the lower q-estimate constant are both equal to one.*

Proof. Let $X_{(r)}$ be the r-concavification of X and let $|||\cdot|||$ denote the norm in $X_{(r)}$. Since X satisfies a lower q-estimate it follows easily from the concavification procedure that $X_{(r)}$ satisfies a lower q/r-estimate. For $x \in X_{(r)}$, put

$$|||x|||_1 = \sup\left\{ \left(\sum_{i=1}^n |||x_i|||^{q/r} \right)^{r/q} \right\},$$

where the supremum is taken over all possible decompositions of x as a sum (in $X_{(r)}$) of pairwise disjoint vectors $\{x_i\}_{i=1}^n$. By using the decomposition property, it is readily verified that $|||\cdot|||_1$ is a norm in $X_{(r)}$ such that, for each $x \in X_{(r)}$, we have

$$|||x||| \leqslant |||x|||_1 \leqslant M\,|||x|||,$$

where M is the lower q/r-estimate constant of $X_{(r)}$. Furthermore, it follows from the definition of $|||\cdot|||_1$ that $Y=(X_{(r)}, |||\cdot|||_1)$ is a Banach lattice satisfying a lower q/r-estimate with constant equal to one. Thus, the r-convexification $Y^{(r)}$ of Y is a Banach lattice which is r-convex with $M^{(r)}(Y^{(r)})=1$ and satisfies a lower q-estimate with constant equal to one. The proof is now completed if we observe that $Y^{(r)}$ is order isomorphic to X. $\square$

Proof of 1.f.10. Fix $1<r<p$ and observe that, by 1.f.7, X is r-convex. By 1.f.11, we may assume that both $M^{(r)}(X)$ and the lower q-estimate constant of X are equal to one. It is clear that a Banach lattice satisfying a lower q-estimate for some $q<\infty$ does not contain a sequence of disjoint elements which is equivalent to the unit vector basis of c_0. Thus, by 1.a.5 and 1.a.7, X is σ-complete and σ-order continuous. Since, moreover, X can be assumed to be separable we get, by 1.b.14, that X is a Köthe function space on a suitable probability space (Ω, Σ, μ).

Fix $\varepsilon>0$ and let $x, y \in X$ be so that $\|x\|=\|y\|=1$ and $\|x-y\|\geqslant\varepsilon$. Put

$$u=|x+y|/2, \qquad v=(|x|^r+|y|^r)^{1/r}/2^{1/r}, \qquad w=|x-y|$$

and consider the sets

$$\sigma_0=\{\omega\in\Omega;\ u(\omega)<v(\omega)-u(\omega)\}\,,$$
$$\sigma_j=\{\omega\in\Omega;\ u(\omega)/2^j<v(\omega)-u(\omega)\leqslant u(\omega)/2^{j-1}\},\quad j=1,2,\ldots\,,$$

By 1.f.2, there exists a constant $C<\infty$ such that, for $j=0, 1, 2, \ldots$, we have

$$|w\chi_{\sigma_j}/C|^2+|u\chi_{\sigma_j}|^2\leqslant|v\chi_{\sigma_j}|^2\,,$$

from which we deduce that $\|w\chi_{\sigma_0}\|\leqslant C\|v\chi_{\sigma_0}\|$ and for $j\geqslant1$

$$|w\chi_{\sigma_j}|^2\leqslant C^2|u\chi_{\sigma_j}|^2((1+1/2^{j-1})^2-1)\leqslant 4C^2|u\chi_{\sigma_j}|^2/2^{j-1}\,,$$

i.e.

$$\|w\chi_{\sigma_j}\|\leqslant C\|u\chi_{\sigma_j}\|/2^{(j-3)/2},\quad j=1,2,\ldots.$$

Observe also that outside $\bigcup_{j=0}^{\infty}\sigma_j$, $v(\omega)=u(\omega)$ and thus, $w(\omega)=0$.

By the r-convexity of X and Hölder's inequality applied with the (conjugate) indices q/r and $q/(q-r)$, we get that

$$\begin{aligned}\varepsilon^r\leqslant\|w\|^r&=\Big\|\sum_{j=0}^{\infty}w\chi_{\sigma_j}\Big\|^r\leqslant\sum_{j=0}^{\infty}\|w\chi_{\sigma_j}\|^r\leqslant C^r\Big(\|v\chi_{\sigma_0}\|^r+\sum_{j=1}^{\infty}\|u\chi_{\sigma_j}\|^r/2^{(j-3)r/2}\Big)\\&=C^r\Big(\|v\chi_{\sigma_0}\|^r+\sum_{j=1}^{\infty}(\|u\chi_{\sigma_j}\|^r/2^{(j-3)r/q})(1/2^{(j-3)r(q-2)/2q})\Big)\\&\leqslant C^r\Big(\|v\chi_{\sigma_0}\|^q+\sum_{j=1}^{\infty}\|u\chi_{\sigma_j}\|^q/2^{j-3}\Big)^{r/q}\Big(1+\sum_{j=1}^{\infty}1/2^{(j-3)r(q-2)/2(q-r)}\Big)^{(q-r)/q}.\end{aligned}$$

It follows that there exists a constant $D<\infty$, depending only on q and r, so that

$$\varepsilon^q \leqslant D\left(\|v\chi_{\sigma_0}\|^q+\sum_{j=1}^{\infty}\|u\chi_{\sigma_j}\|^q/2^{j-3}\right).$$

Let now $x^*\in X^*$ be a positive functional such that $\|x^*\|=1$ and $x^*(u)=\|u\|$. Since X satisfies a lower q-estimate we get, for $j\geqslant 1$,

$$\begin{aligned}\|u\chi_{\sigma_j}\|^q/q&\leqslant\|u\|-(\|u\|^q-\|u\chi_{\sigma_j}\|^q)^{1/q}\leqslant\|u\|-\|u-u\chi_{\sigma_j}\|\\&\leqslant x^*(u)-x^*(u-u\chi_{\sigma_j})=x^*(u\chi_{\sigma_j})\,.\end{aligned}$$

Moreover, since $\|v\|\leqslant 1$ (by the r-convexity of X) and $0\leqslant u\leqslant v$, we have

$$\begin{aligned}\|v\chi_{\sigma_0}\|^q/q&\leqslant\|v\|-\|v-v\chi_{\sigma_0}\|\leqslant 1-\|u-u\chi_{\sigma_0}\|\\&\leqslant 1-x^*(u-u\chi_{\sigma_0})=1-\|u\|+x^*(u\chi_{\sigma_0})\,.\end{aligned}$$

In conclusion, we obtain

$$\begin{aligned}\varepsilon^q&\leqslant Dq\left(1-\|u\|+x^*(u\chi_{\sigma_0})+\sum_{j=1}^{\infty}x^*(u\chi_{\sigma_j})/2^{j-3}\right)\\&\leqslant Dq\left(1-\|u\|+\sum_{j=0}^{\infty}x^*((v-u)2^j\chi_{\sigma_j})/2^{j-3}\right)\\&\leqslant Dq(1-\|u\|+8x^*(v-u))\leqslant 9Dq(1-\|u\|)\,.\end{aligned}$$

This, of course, proves that X can be renormed equivalently as to become a uniformly convex Banach lattice with modulus of convexity of power type q. By duality, we conclude that X can be also renormed equivalently as to become a uniformly smooth lattice with modulus of smoothness of power type p. In view of 1.e.16, it is clear that X is of type p and of cotype q. □

In the assumptions required in order to be able to renorm the Banach lattice X as to have modulus of convexity of power type q, the actual value of $p>1$ does not matter (except that it affects the coefficient of ε^q in the estimate from below satisfied by the modulus of convexity). In other words, what we really need is to know that X satisfies a non-trivial upper p-estimate (i.e. with some $p>1$). The existence of non-trivial upper or lower estimates is related to the non-existence of isomorphic copies of l_1^n, respectively, l_∞^n as is shown by the following result proved by Shimogaki [120] and W. B. Johnson [56] (Shimogaki worked with a slightly different notion but used essentially the argument presented here. Johnson proved 1.f.12 for space having an unconditional basis and his proof is different.)

Theorem 1.f.12. *Let X be a Banach lattice.*

(i) *There does not exist a $p>1$ so that X satisfies an upper p-estimate if and* *only if, for every $\varepsilon>0$ and every integer n, there exists a sequence $\{x_i\}_{i=1}^n$ of pairwise disjoint elements in X such that*

$$(1-\varepsilon)\sum_{i=1}^n |a_i| \leqslant \left\| \sum_{i=1}^n a_i x_i \right\| \leqslant \sum_{i=1}^n |a_i| ,$$

for every choice of scalars $\{a_i\}_{i=1}^n$.

(ii) *There does not exist a $p<\infty$ so that X satisfies a lower p-estimate if and only if, for every $\varepsilon>0$ and every integer n, there exists a sequence $\{x_i\}_{i=1}^n$ of mutually disjoint elements in X such that*

$$\max_{1\leqslant i\leqslant n} |a_i| \leqslant \left\| \sum_{i=1}^n a_i x_i \right\| \leqslant (1+\varepsilon) \max_{1\leqslant i\leqslant n} |a_i| ,$$

for every choice of scalars $\{a_i\}_{i=1}^n$.

Proof. Observe first that (i) can be immediately deduced from (ii) by a simple duality argument. Suppose now that X satisfies no lower p-estimate for $p<\infty$. For every integer n, let α_n be the smallest constant for which every sequence $\{x_i\}_{i=1}^n$ of pairwise disjoint elements in X satisfies

$$\inf_{1\leqslant i\leqslant n} \|x_i\| \leqslant \alpha_n \left\| \sum_{i=1}^n x_i \right\| .$$

It is easily seen that $1=\alpha_1 \geqslant \alpha_2 \geqslant \ldots \geqslant 0$ and we shall show now that $\{\alpha_n\}_{n=1}^\infty$ is a sub-multiplicative sequence, i.e. that

$$\alpha_{mn} \leqslant \alpha_m \alpha_n ,$$

for all m and n. Indeed, if $\{x_{i,j}\}_{i=1,j=1}^{m\ \ n}$ is a double sequence of pairwise disjoint elements in X then

$$\inf_{1\leqslant j\leqslant n} \|x_{i,j}\| \leqslant \alpha_n \left\| \sum_{j=1}^n x_{i,j} \right\| ,$$

for every $1\leqslant i\leqslant m$. We also have

$$\inf_{1\leqslant i\leqslant m} \left\| \sum_{j=1}^n x_{i,j} \right\| \leqslant \alpha_m \left\| \sum_{i=1}^m \sum_{j=1}^n x_{i,j} \right\|$$

which, in conclusion, implies that

$$\inf_{\substack{1\leqslant i\leqslant m\\ 1\leqslant j\leqslant n}} \|x_{i,j}\| \leqslant \alpha_m \alpha_n \left\| \sum_{i=1}^m \sum_{j=1}^n x_{i,j} \right\| .$$

This proves that $\alpha_{mn} \leqslant \alpha_m \alpha_n$. Assume now that $\alpha_k < 1$ for some integer $k > 1$ and put $\gamma = -\log \alpha_k / \log k$. Let n be an arbitrary integer and choose j so that $k^j \leqslant n < k^{j+1}$. Then,

$$\alpha_n \leqslant \alpha_{k^j} \leqslant (\alpha_k)^j = 1/k^{j\gamma} \leqslant k^\gamma / n^\gamma .$$

In other words, there is a constant $K < \infty$ and a number $\gamma > 0$ so that $\alpha_n \leqslant K/n^\gamma$, for all n. Choose $p < \infty$ so that $p\gamma > 1$ and let $\{x_i\}_{i=1}^n$ be a sequence of mutually disjoint elements in X such that $\|x_1\| \geqslant \cdots \geqslant \|x_n\|$. Then, for each $1 \leqslant j \leqslant n$, we have

$$\|x_j\| = \inf_{1 \leqslant i \leqslant j} \|x_i\| \leqslant \alpha_j \left\| \sum_{i=1}^j x_i \right\| \leqslant K \left\| \sum_{i=1}^n x_i \right\| / j^\gamma$$

and consequently,

$$\left(\sum_{j=1}^n \|x_j\|^p \right)^{1/p} \leqslant K \left\| \sum_{i=1}^n x_i \right\| \left(\sum_{j=1}^\infty 1/j^{p\gamma} \right)^{1/p} .$$

This shows that X satisfies a lower p-estimate for some $p < \infty$, contrary to our assumption. Hence, we must have $\alpha_n = 1$, for all n. In other words, for every $\varepsilon > 0$ and every integer n, there exists a sequence $\{x_i\}_{i=1}^n$ of pairwise disjoint elements so that

$$1 = \inf_{1 \leqslant i \leqslant n} \|x_i\| \leqslant \left\| \sum_{i=1}^n x_i \right\| < 1 + \varepsilon .$$

It follows immediately that, for any choice of $\{a_i\}_{i=1}^n$, we have

$$\max_{1 \leqslant i \leqslant n} |a_i| \leqslant \left\| \sum_{i=1}^n a_i x_i \right\| \leqslant \max_{1 \leqslant i \leqslant n} |a_i| \cdot \left\| \sum_{i=1}^n x_i \right\| < (1+\varepsilon) \max_{1 \leqslant i \leqslant n} |a_i| .$$

This completes the proof of (ii) since the converse is trivial. □

The following corollary illustrates well the use of 1.f.12.

Corollary 1.f.13. *A Banach lattice X, which is of type p for some $p > 1$, is q-concave for some $q < \infty$.*

Proof. By 1.f.7, it suffices to show that X satisfies a lower r-estimate for some $r < \infty$. If X satisfies no such lower estimate then, by 1.f.12, it must contain nearly isometric copies of l_∞^n, for all n. However, for each k, the space $l_\infty^{2^k}$ contains an isometric copy of l_1^k and this clearly contradicts the fact that X is of type p, for some $p > 1$. (A subspace of $l_\infty^{2^k}$ which is isometric to l_1^k can be obtained by considering the span of the so-called Rademacher elements over the unit vector basis $\{e_i\}_{i=1}^{2^k}$

of $l_\infty^{2^k}$, i.e. the vectors

$$r_1 = e_1 + \cdots + e_{2^{k-1}} - e_{2^{k-1}+1} - \cdots - e_{2^k},$$
$$\cdots\cdots$$
$$r_k = e_1 - e_2 + e_3 - e_4 + \cdots + e_{2^k-1} - e_{2^k}.) \quad \square$$

We study now some questions in which properties related to the number 2, like 2-concavity, type 2 or cotype 2, etc., play a special role. We show first that, for $p=2$, Proposition 1.d.9 can be generalized so as to apply for general bounded operators (and not only positive ones).

Theorem 1.f.14 [66]. *Let X and Y be two Banach lattices and let $T: X \to Y$ be a bounded linear operator. Then, for every choice of $\{x_i\}_{i=1}^n$ in X, we have*

$$\left\| \left(\sum_{i=1}^n |Tx_i|^2 \right)^{1/2} \right\| \leq K_G \|T\| \left\| \left(\sum_{i=1}^n |x_i|^2 \right)^{1/2} \right\|,$$

where K_G is the universal Grothendieck constant (cf. I.2.b.5).

Proof. The proof will be carried out in three steps. In the first step we shall prove 1.f.14 in case $X = l_\infty^m$ and $Y = l_1^m$, for some $m < \infty$. In this case, 1.f.14 is a formally slightly stronger version of the assertion that every operator from c_0 to l_1 is 2-absolutely summing (cf. I.2.b.7). The proof of step 1 is based on Grothendieck's inequality (I.2.b.5) and is very similar to that of I.2.b.7. In the second step of the proof we apply a standard approximation argument in order to deduce from step 1 that 1.f.14 holds if X is a $C(K)$ space and Y an $L_1(\mu)$ space. In the proof of step 3 we apply the representation theorems of Kakutani in order to reduce the general case to that verified in step 2.

Step 1. Let T be an operator from l_∞^m into l_1^m, for some integer m. Let $x_i = (a_{i,1}, \ldots, a_{i,m})$, $i = 1, 2, \ldots, n$ be a sequence of elements in l_∞^m and let $(\alpha_{k,j})_{k,j=1}^m$ be the matrix representing T with respect to the unit vector bases of l_∞^m and l_1^m. Choose vectors $y_i = (b_{i,1}, \ldots, b_{i,m}) \in l_\infty^m = (l_1^m)^*$, $i = 1, 2, \ldots, n$ so that

$$\left\| \left(\sum_{i=1}^n |y_i|^2 \right)^{1/2} \right\|_\infty = 1 \quad \text{and} \quad \left\| \left(\sum_{i=1}^n |Tx_i|^2 \right)^{1/2} \right\|_1 = \sum_{i=1}^n y_i(Tx_i)$$

(use the duality between $l_1^m(l_2)$ and $l_\infty^m(l_2)$).

Consider now the vectors $u_k = (a_{1,k}, \ldots, a_{n,k})$ and $v_k = (b_{1,k}, \ldots, b_{n,k})$ as elements of l_2^n. Then,

$$\left\| \left(\sum_{i=1}^n |Tx_i|^2 \right)^{1/2} \right\|_1 = \sum_{i=1}^n \sum_{j=1}^m \sum_{k=1}^m \alpha_{k,j} a_{i,k} b_{i,j} = \sum_{j=1}^m \sum_{k=1}^m \alpha_{k,j} (u_k, v_j) .$$

Notice that

$$\max_{1\leqslant k\leqslant m} \|u_k\|_2 = \max_{1\leqslant k\leqslant m}\left(\sum_{i=1}^{n} |a_{i,k}|^2\right)^{1/2} = \left\|\left(\sum_{i=1}^{n} |x_i|^2\right)^{1/2}\right\|_\infty$$

and

$$\max_{1\leqslant k\leqslant m} \|v_k\|_2 = \left\|\left(\sum_{i=1}^{n} |y_i|^2\right)^{1/2}\right\|_\infty = 1\,.$$

It follows from Grothendieck's inequality I.2.b.5 (applied to the matrix $(\alpha_{k,j}/\|T\|)_{k,j=1}^m$) that

$$\left\|\left(\sum_{i=1}^{n} |Tx_i|^2\right)^{1/2}\right\|_1 \leqslant K_G\|T\| \max_{1\leqslant k\leqslant m}\|u_k\|_2 \max_{1\leqslant k\leqslant m}\|v_k\|_2$$
$$= K_G\|T\|\left\|\left(\sum_{i=1}^{n} |x_i|^2\right)^{1/2}\right\|_\infty .$$

Step 2. Let T be an operator from $C(K)$ into $L_1(\mu)$ and let $\{f_i\}_{i=1}^n$ be a finite sequence in $C(K)$. Then, for every $\varepsilon>0$, there exists a partition of unity $\{\varphi_j\}_{j=1}^m$ in $C(K)$ and functions $\tilde{f}_i \in [\varphi_j]_{j=1}^m$ so that $\|f_i-\tilde{f}_i\|<\varepsilon\|f_i\|$, for all i (a partition of unity in a $C(K)$ space is a set of functions $\{\varphi_j\}_{j=1}^m$ of norm one so that $\sum_{j=1}^m \varphi_j(k)=1$ and $0\leqslant\varphi_j(k)\leqslant 1$, for all $k\in K$ and $1\leqslant j\leqslant m$). Let $\{\psi_j\}_{j=1}^m$ be simple functions in $L_1(\mu)$ so that $\|T\varphi_j-\psi_j\|\leqslant\varepsilon/m$, $1\leqslant j\leqslant m$ and put $\tilde{T}\varphi_j=\psi_j$. Then $\tilde{T}$ extends to a linear operator from $[\varphi_j]_{j=1}^m$ into $[\psi_j]_{j=1}^m$ and $\|\tilde{T}-T_{|[\varphi_j]_{j=1}^m}\|\leqslant\varepsilon$. Since $[\varphi_j]_{j=1}^m$ is isometric to l_∞^m and $[\psi_j]_{j=1}^m$ is isometric to a subspace of l_1^k, for some k, it follows from step 1 that

$$\left\|\left(\sum_{i=1}^{n} |\tilde{T}\tilde{f}_i|^2\right)^{1/2}\right\| \leqslant K_G\|\tilde{T}\|\left\|\left(\sum_{i=1}^{n} |\tilde{f}_i|^2\right)^{1/2}\right\| .$$

Since this is true for every $\varepsilon>0$ we obtain, by letting $\varepsilon\to 0$, that 1.f.14 holds if $X=C(K)$ and $Y=L_1(\mu)$.

Step 3. Let T be an operator from X to Y and let $\{x_i\}_{i=1}^n$ be a finite sequence in X. Put $x_0=\left(\sum_{i=1}^{n}|x_i|^2\right)^{1/2}$ and $y_0=\left(\sum_{i=1}^{n}|Tx_i|^2\right)^{1/2}$. Let $I(x_0)$ be the (in general non-closed) ideal generated by x_0 endowed with the norm

$$\|x\|_\infty = \inf\{\lambda\geqslant 0;\, |x|\leqslant\lambda x_0/\|x_0\|\},\quad x\in I(x_0)\,,$$

whose completion is, by 1.b.6, order isometric to a $C(K)$ space. Let y_0^* be a positive functional in Y^* so that $\|y_0^*\|=1$ and $y_0^*(y_0)=\|y_0\|$. Put $\|y\|_1=y_0^*(|y|)$, for $y\in Y$, and observe that $\|\cdot\|_1$ is a semi-norm on Y such that the completion Y_0, of Y

endowed with $\|\cdot\|_1$ modulo the elements $z \in Y$ having $\|z\|_1 = 0$, is an abstract L_1 space (and, therefore, by Kakutani's Theorem 1.b.2, order isometric to a concrete $L_1(\mu)$ space). If J_1 and J_2 denote the formal identity maps from $I(x_0)$ into X, respectively, from Y into Y_0 then J_2TJ_1 is a linear operator from $I(x_0)$ into Y_0 with $\|J_2TJ_1\| \leqslant \|T\|$. Hence, by step 2,

$$\left\|\left(\sum_{i=1}^{n} |J_2TJ_1x_i|^2\right)^{1/2}\right\|_1 \leqslant K_G\|T\| \left\|\left(\sum_{i=1}^{n} |x_i|^2\right)^{1/2}\right\|_\infty .$$

This completes the proof since $\left\|\left(\sum_{i=1}^{n} |x_i|^2\right)^{1/2}\right\|_\infty = \|x_0\|_\infty = \|x_0\|$ and

$$\left\|\left(\sum_{i=1}^{n} |J_2TJ_1x_i|^2\right)^{1/2}\right\|_1 = \|y_0\|_1 = y_0^*(y_0) = \|y_0\| . \quad \square$$

Remark. It can be easily verified that K_G is the smallest constant for which 1.f.14 holds for general lattices.

Corollary 1.f.15 [66]. *Let X and Y be Banach lattices and $T: X \to Y$ a bounded linear operator.*

(i) *If X is 2-convex then T is 2-convex and*

$$M^{(2)}(T) \leqslant K_G M^{(2)}(X)\|T\| .$$

(ii) *If Y is 2-concave then T is 2-concave and*

$$M_{(2)}(T) \leqslant K_G M_{(2)}(Y)\|T\| .$$

(iii) *If X is 2-convex and Y is 2-concave then T can be factorized through an $L_2(\mu)$ space in two different ways:*

a) *There exist a probability measure μ_1, an operator $T_1: X \to L_2(\mu_1)$ with $\|T_1\| \leqslant K_G M^{(2)}(X)\|T\|$ and a positive operator $S_1: L_2(\mu_1) \to Y$ with $\|S_1\| \leqslant M_{(2)}(Y)$ such that $T = S_1T_1$.*

b) *There exist a probability measure μ_2, an operator $T_2: L_2(\mu_2) \to Y$ with $\|T_2\| \leqslant K_G M_{(2)}(Y)\|T\|$ and a positive operator $S_2: X \to L_2(\mu_2)$ with $\|S_2\| \leqslant M^{(2)}(X)$ such that $T = T_2S_2$.*

Proof. (i) For every sequence $\{x_i\}_{i=1}^n$ in X we get, by 1.f.14, that

$$\left\|\left(\sum_{i=1}^{n} |Tx_i|^2\right)^{1/2}\right\| \leqslant K_G\|T\| \left\|\left(\sum_{i=1}^{n} |x_i|^2\right)^{1/2}\right\|$$
$$\leqslant K_G\|T\|M^{(2)}(X)\left(\sum_{i=1}^{n} \|x_i\|^2\right)^{1/2} .$$

The proof of (ii) is similar while (iii) follows immediately from (i), (ii) and 1.d.12 (i) and (ii). □

We are now prepared to prove a theorem which combines a result of B. Maurey [94] ((i) ⇔ (ii)) and one of E. Dubinski, A. Pelczynski and H. P. Rosenthal [31].

Theorem 1.f.16. *The following conditions are equivalent in every Banach lattice X.*

(i) *X is of cotype 2.*

(ii) *X is 2-concave.*

(iii) *Every operator from c_0 (or from any $C(K)$ space) into X is 2-absolutely summing.*

Proof. The fact that (i) ⇔ (ii) is a direct consequence of 1.d.6 which shows that the expressions $\left\|\left(\sum_{i=1}^{n}|x_i|^2\right)^{1/2}\right\|$ and $\int_0^1\left\|\sum_{i=1}^{n}r_i(u)x_i\right\|du$ are equivalent in X. Notice that, in order to be able to use 1.d.6, we have to prove first that a lattice of cotype 2 is s-concave, for some $s<\infty$. This follows however from 1.f.7.

Suppose now that X is 2-concave and T is a bounded linear operator from a $C(K)$ space into X. Then, by 1.f.15 (ii), T is 2-concave, i.e., for every choice of $\{f_i\}_{i=1}^n$ in $C(K)$, we have

$$\left(\sum_{i=1}^{n}\|Tf_i\|^2\right)^{1/2}\leqslant M_{(2)}(T)\left\|\left(\sum_{i=1}^{n}|f_i|^2\right)^{1/2}\right\|$$

and the proof of the implication (ii) ⇒ (iii) can be completed by recalling that

$$\left\|\left(\sum_{i=1}^{n}|f_i|^2\right)^{1/2}\right\|=\sup\left\{\left(\sum_{i=1}^{n}|\mu(f_i)|^2\right)^{1/2};\ \mu\in C(K)^*,\ \|\mu\|\leqslant 1\right\}$$

(see e.g. condition (+) in the proof of 1.d.10). Finally, (iii) ⇒ (ii) follows from 1.d.10. □

There is also a partial dual version of 1.f.16 which follows easily from 1.d.6 and 1.f.13.

Proposition 1.f.17. *A Banach lattice is of type 2 if and only if it is 2-convex and q-concave, for some $q<\infty$.*

Observe that, in view of 1.b.13, 1.f.16 and 1.f.17 provide in particular a proof of Kwapien's theorem [69] for lattices: a Banach lattice, which is of type 2 and of cotype 2, is isomorphic to a Hilbert space (even order isomorphic to some $L_2(\mu)$).

We establish now the duality between type and cotype in Banach lattices which are "far" from L_∞ (cf. B. Maurey [94]).

Theorem 1.f.18. *A Banach lattice X is of type p, for some $p>1$, if and only if its dual X^* is of cotype q, where $1/p+1/q=1$, and satisfies an upper r-estimate for some $r>1$.*

Proof. Suppose that X is of type p, for some $p>1$. The fact that, in this case, X^* is of cotype $q=p/(p-1)$ is valid for every Banach space X and was proved in 1.e.17. Moreover, it follows from 1.f.13 and the duality between r-convexity and $r/(r-1)$-concavity that X^* satisfies an upper r-estimate for some $r>1$.

In order to prove the converse, we first treat the case when $q>2$. Then, by duality, X satisfies an upper p-estimate for $1<p<2$ and a lower s-estimate for $s=r/(r-1)$. Hence, by 1.f.10, X is of type p. If, on the other hand, $q=2$ then, by 1.f.16, X^* is 2-concave and thus, in view of 1.d.4 (iii), X is 2-convex. Furthermore, since X^* satisfies an upper r-estimate, it follows from 1.f.7 that X^* is r'-convex for every $r'<r$. Hence, again by 1.d.4 (iii), X is s'-concave for $s'=r'/(r'-1)$ and, therefore, of type 2, by 1.f.3. □

We conclude this section with some examples.

Example 1.f.19. (G. Pisier). *There exists a uniformly convex space with a symmetric basis which satisfies a lower 2-estimate but is not of cotype 2 (or, equivalently, 2-concave). Consequently, this space cannot be renormed equivalently as to have modulus of convexity of power type 2.*

Fix $1<p<2$ and consider the Lorentz sequence space $X=d(w,p)$, where $w=\{w_n\}_{n=1}^\infty$ is defined by $w_n=n^{p/2}-(n-1)^{p/2}$, $n=1, 2, \dots$ (Recall that the norm of $x=(a_1, a_2, \dots)$ in $d(w,p)$ is defined to be $\|x\|=\left(\sum_{n=1}^\infty a_n^{*p} w_n\right)^{1/p}$, where $a_n^*=|a_{\pi(n)}|$ with π a permutation of the integers for which $\{|a_{\pi(n)}|\}_{n=1}^\infty$ is a non-increasing sequence.) We prove first that X is not 2-concave. Let m be an integer and let $\{\tau_j\}_{j=1}^m$ be the maps of $\{1, 2, \dots, m\}$ into itself defined by $\tau_j(i)=(i+j) \bmod m$. For $1\leqslant j\leqslant m$ put

$$x_{m,j}=\sum_{i=1}^m e_i/\tau_j(i)^{1/2},$$

where $\{e_i\}_{i=1}^\infty$ denote the unit vector basis of $d(w,p)$. A simple computation shows that

$$\left\|\left(\sum_{j=1}^m |x_{m,j}|^2\right)^{1/2}\right\|=\left\|\left(\sum_{j=1}^m \sum_{i=1}^m e_i/\tau_j(i)\right)^{1/2}\right\|$$
$$=\left(\sum_{j=1}^m 1/j\right)^{1/2}\left\|\sum_{j=1}^m e_j\right\|\leqslant m^{1/2}(\log m+1)^{1/2}.$$

On the other hand,

$$\left(\sum_{j=1}^m \|x_{m,j}\|^2\right)^{1/2}\geqslant m^{1/2}\left(\sum_{i=1}^m p/2i\right)^{1/p}\geqslant (p/2)^{1/p} m^{1/2}(\log m)^{1/p}.$$

Since m is arbitrary this proves that $d(w, p)$ is not 2-concave. It follows from

1.f.16 that X is not of cotype 2 and hence, by 1.e.16, for every equivalent norm in X the modulus of convexity is not of power type 2.

In order to verify that $d(w,p)$ satisfies a lower 2-estimate it suffices to show that $d(w,1)$ (which is the p-concavification of $d(w,p)$) satisfies a lower $2/p$-estimate, i.e. (in view of 1.f.5) that $d(w,1)^*$ satisfies an upper r-estimate, where $1/r+p/2=1$. To this end, note first that if $x^*=(b_1,b_2,\dots)\in d(w,1)^*$ with $b_1\geqslant b_2\geqslant\cdots\geqslant 0$ then

$$\|x^*\|=\sup\left\{\frac{\sum_{i=1}^{\infty}a_ib_i}{\sum_{i=1}^{\infty}a_iw_i};\ \{a_i\geqslant 0\}_{i=1}^{\infty}\text{ decreasing}\right\}$$

$$=\sup\left\{\frac{(a_1-a_2)b_1+(a_2-a_3)(b_1+b_2)+\cdots+(a_i-a_{i-1})(b_1+\cdots+b_i)+\cdots}{(a_1-a_2)w_1+(a_2-a_3)(w_1+w_2)+\cdots+(a_i-a_{i-1})(w_1+\cdots+w_i)+\cdots};\right.$$
$$\left.\{a_i\geqslant 0\}_{i=1}^{\infty}\text{ decreasing}\right\}$$

$$=\sup\left\{\frac{c_1b_1+c_2(b_1+b_2)+\cdots+c_i(b_1+\cdots+b_i)+\cdots}{c_1w_1+c_2(w_1+w_2)+\cdots+c_i(w_1+\cdots+w_i)+\cdots};\quad\{c_i\geqslant 0\}_{i=1}^{\infty}\right\}$$

$$=\sup_n\left\{\sum_{j=1}^{n}b_j\Big/\sum_{j=1}^{n}w_j\right\}=\sup_n\left\{n^{-p/2}\sum_{j=1}^{n}b_j\right\}.$$

Let $u=(u_1,u_2,\dots)$ and $v=(v_1,v_2,\dots)$ be two disjointly supported elements of $d(w,1)^*$; then

$$\|u+v\|=\sup\left\{(k+l)^{-p/2}\left(\sum_{j=1}^{k}u_j^*+\sum_{j=1}^{l}v_j^*\right);\quad 0\leqslant k,l<\infty,\ k+l\geqslant 1\right\}$$
$$\leqslant\sup\{(k+l)^{-p/2}(\|u\|k^{p/2}+\|v\|l^{p/2});\quad 0\leqslant k,l<\infty,\ k+l\geqslant 1\}$$
$$\leqslant(\|u\|^r+\|v\|^r)^{1/r}$$

and this proves our assertion.

Since X is the p-convexification of $d(w,1)$ it is p-convex and $M^{(p)}(X)=1$. Therefore, by the proof of 1.f.10 (taking as q any number >2), X is uniformly convex. □

Remark. Let Y and Z be two Banach lattices and suppose that Y is linearly isomorphic to a subspace of Z. If Z satisfies a lower p-estimate, for some $p>2$, then so does Y since, by a result from [94] (stated before 1.f.10), Z, and thus also Y, are of cotype p. However, contrary to the case of p-concavity described in 1.d.7(ii), this assertion is, in general, false when $p=2$. A counterexample can be, for instance, constructed by taking $Z=(X\oplus X\oplus\dots)_2$, where X is the Lorentz sequence space $d(w,p)$ considered in 1.f.19. As is easily verified, Z, too, satisfies a lower 2-estimate. Suppose now that any Banach lattice Y, which is linearly isometric to a subspace of Z, also satisfies a lower 2-estimate. Then a simple uniformity argument proves that there exists a constant $M<\infty$ so that the lower 2-estimate constant of any Y as above is $\leqslant M$.

Let $\{x_i\}_{i=1}^n$ be vectors in X and, for $1\leqslant i\leqslant n$, put

$$u_i=2^{-n/2}(\overbrace{x_i,\ldots,x_i}^{2^{n-i}},\ \overbrace{-x_i,\ldots,-x_i}^{2^{n-i}},\ldots,\overbrace{x_i,\ldots,x_i}^{2^{n-i}},\ \overbrace{-x_i,\ldots,-x_i}^{2^{n-i}},0,0,\ldots),$$

where the number of the blocks of size 2^{n-i} is 2^i. A direct computation shows that, for every choice of scalars $\{a_i\}_{i=1}^n$,

$$\Big\|\sum_{i=1}^n a_iu_i\Big\|_Z=\Big(\int_0^1\Big\|\sum_{i=1}^n a_ir_i(t)x_i\Big\|_X^2\,dt\Big)^{1/2},$$

i.e., the unconditional constant of $\{u_i\}_{i=1}^n$ is equal to one. It follows that

$$M\Big\|\sum_{i=1}^n u_i\Big\|_Z\geqslant\Big(\sum_{i=1}^n\|u_i\|_Z^2\Big)^{1/2},\quad n=1,2,\ldots,$$

which implies that X is of cotype 2, contrary to the assertion of 1.f.19.

Example 1.f.20. *For every $q>2$ there exists a Banach space with a symmetric basis which satisfies a lower q-estimate and has a modulus of convexity of power type q, but is not q-concave.*

Proof. Let $1<p<2$, let w be the sequence defined in 1.f.19 and let $q>2$. The Lorentz sequence space $d(w,pq/2)$ is the $q/2$-convexification of $d(w,p)$. Hence, $d(w,pq/2)$ satisfies a lower q-estimate with constant one but is not q-concave. Also, $d(w,pq/2)$ is $pq/2$-convex (with $pq/2$-convexity constant equal to one) and hence, by 1.f.10, its norm has a modulus of convexity of power type q. □

We present now two diagrams. The first diagram describes the various connections between modulus of convexity, concavity, cotype and lower estimates in general Banach lattices. The second diagram describes the connections between the dual notions.

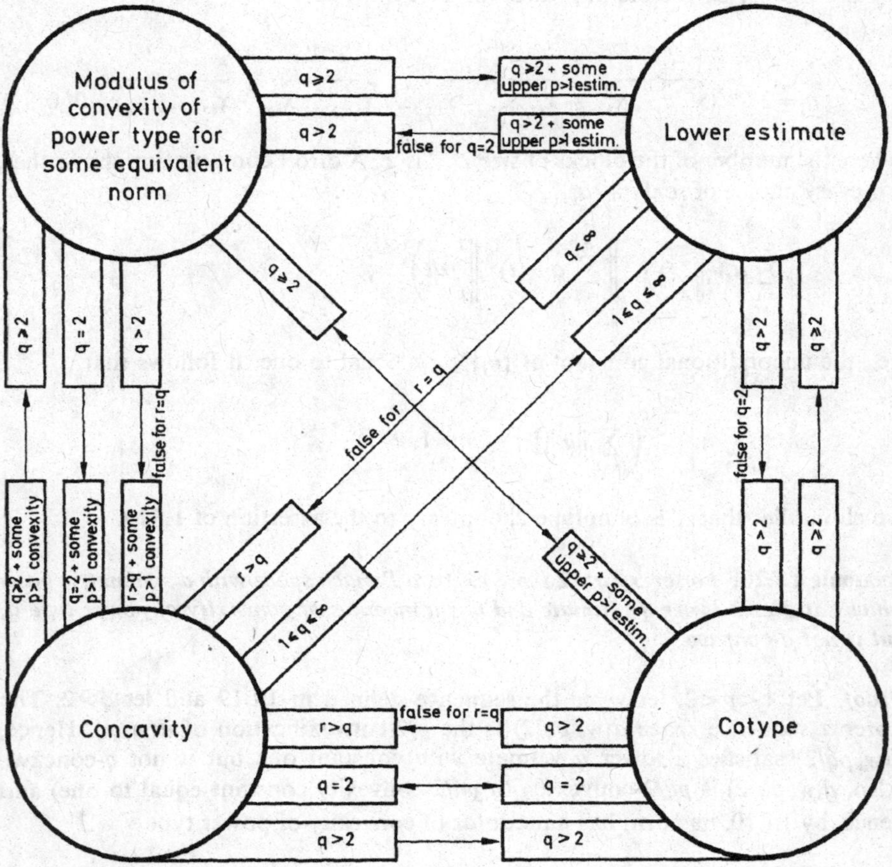

Interdependence diagram for the power type of modulus of convexity, *q*-concavity, cotype and lower *q*-estimate for general Banach lattices.

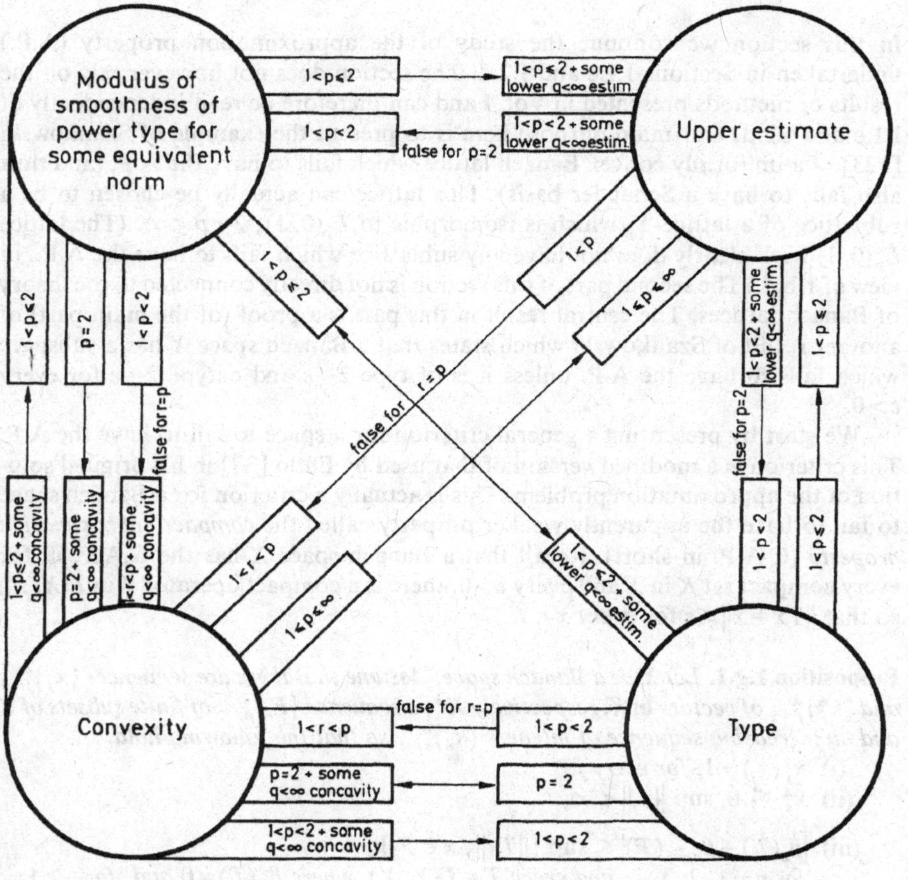

Interdependence diagram for the power type of modulus of smoothness, *p*-convexity, type and upper *p*-estimate for general Banach lattices.

g. The Approximation Property and Banach Lattices

In this section we continue the study of the approximation property (A.P.) undertaken in Sections I.1.e and I.2.d. The section does not however rely on the results or methods presented in Vol. I and can therefore be read independently of I.1.e and I.2.d. Our main purpose here is to present the example of Szankowski [123] of a uniformly convex Banach lattice which fails to have the A.P. (and thus also fails to have a Schauder basis). This lattice can actually be chosen to be a sublattice of a lattice Y_p which is isomorphic to $L_p(0, 1)$, $2<p<\infty$. (The lattice $L_p(0, 1)$ itself clearly does not have any sublattice which fails to have the A.P., in view of 1.b.4.) The second part of this section is not directly connected to the theory of Banach lattices. The central result in this part is a proof (of the main part) of another result of Szankowski which states that a Banach space X has a subspace which fails to have the A.P. unless it is of type $2-\varepsilon$ and cotype $2+\varepsilon$ for every $\varepsilon>0$.

We start by presenting a general criterion for a space to fail to have the A.P. This criterion is a modified version of that used by Enflo [37] in his original solution of the approximation problem. This is actually a criterion for a Banach space to fail to have the apparently weaker property called the *compact approximation property* (C.A.P. in short). Recall that a Banach space X has the C.A.P. if, for every compact set K in X and every $\varepsilon>0$, there is a compact operator T in $L(X, X)$ so that $\|Tx-x\|\leqslant\varepsilon$ for every $x\in K$.

Proposition 1.g.1. *Let X be a Banach space. Assume that there are sequences $\{x_j\}_{j=1}^\infty$ and $\{x_j^*\}_{j=1}^\infty$ of vectors in X, respectively X^*, a sequence $\{F_n\}_{n=1}^\infty$ of finite subsets of X and an increasing sequence of integers $\{k_n\}_{n=1}^\infty$ so that the following hold.*

(i) $x_j^*(x_j)=1$, *for every* j,

(ii) $x_j^* \xrightarrow{w^*} 0$, $\sup_j \|x_j\|<\infty$,

(iii) $|\beta_n(T)-\beta_{n-1}(T)|\leqslant \sup\{\|Tx\|;\, x\in F_n\}$,
for $n=1, 2, 3, \ldots$ *and every* $T\in L(X, X)$, *where* $\beta_0(T)=0$ *and, for* $n\geqslant 1$,

$$\beta_n(T)=k_n^{-1}\sum_{j=1}^{k_n} x_j^*(Tx_j)\,,$$

(iv) $\sum_{n=1}^{\infty}\gamma_n<\infty$, *where* $\gamma_n=\sup\{\|x\|\,;\, x\in F_n\}$.

Then X fails to have the C.A.P.

Proof. It follows from (iii) and (iv) that

$$\beta(T)=\lim_{n\to\infty}\beta_n(T)$$

exists for every $T \in L(X, X)$ and defines a linear functional on $L(X, X)$. Let $\{\eta_n\}_{n=1}^\infty$ be a sequence of positive numbers tending to ∞ so that $C = \sum_{n=1}^\infty \eta_n \gamma_n < \infty$, and put $K = \{0\} \cup \bigcup_{n=1}^\infty (\eta_n \gamma_n)^{-1} F_n$. Clearly, K is a compact set and

$$|\beta(T)| \leq C \sup\{\|Tx\|;\ x \in K\}.$$

It is also clear from (i) that if I is the identity operator on X then $\beta_n(I) = 1$ for all $n \geq 1$ and thus $\beta(I) = 1$. We shall show that $\beta(T) = 0$ whenever $T \in L(X, X)$ is compact. This will conclude the proof since it will imply that, for every compact T,

$$\sup\{\|Tx - x\|;\ x \in K\} \geq C^{-1} |\beta(I - T)| = C^{-1}.$$

Let T thus be compact and let $\delta > 0$. Pick $\{y_i\}_{i=1}^m$ in X so that, for every integer j, there is an $i(j)$ with $\|Tx_j - y_{i(j)}\| \leq \delta$. We have, for $n = 1, 2, \ldots$,

$$\beta_n(T) = k_n^{-1} \sum_{j=1}^{k_n} x_j^*(y_{i(j)}) + k_n^{-1} \sum_{j=1}^{k_n} x_j^*(Tx_j - y_{i(j)}),$$

and thus

$$|\beta_n(T)| \leq \sum_{i=1}^m k_n^{-1} \sum_{j=1}^{k_n} |x_j^*(y_i)| + \delta \sup_j \|x_j^*\|.$$

Since $x_j^* \xrightarrow{w^*} 0$ it follows that $|\beta(T)| \leq \delta \sup_j \|x_j^*\|$ and thus, since δ was arbitrary, $\beta(T) = 0$. $\square$

Remark. The proposition remains clearly valid if the $\beta_n(T)$ are defined to be of the form $k_n^{-1} \sum_{j \in \sigma_n} x_j^*(Tx_j)$, where σ_n is any set of integers of cardinality k_n, $n = 1, 2, \ldots$. For instance, in one of the applications of 1.g.1 below $\beta_n(T)$ will be defined to be $2^{-n} \sum_{j=2^n+1}^{2^{n+1}} x_j^*(Tx_j)$.

We state now the theorem of Szankowski [123].

Theorem 1.g.2. *Let $1 \leq r < p < \infty$. There is a sublattice of $l_p(L_r(0, 1)) = (L_r(0, 1) \oplus L_r(0, 1) \oplus \cdots)_p$ which fails to have the* C.A.P.

The construction of this sublattice is based on a lemma of a combinatorial nature which deals with the existence of certain partitions of $[0, 1]$. For the statement of this lemma we introduce first some notations. For every integer n, let $\mathscr{B}_n$ be the algebra of subsets of $[0, 1]$ generated by the 2^n atoms $[(i-1)/2^n, i/2^n)$, $i = 1, \ldots, 2^n$. A subset of $[0, 1]$ is $\mathscr{B}_n$-measurable if it is the union of some of these atoms. For every n, let φ_n be the permutation of $\{1, 2, \ldots, 2^n\}$ defined by $\varphi_n(2i) = 2i - 1$ and

$\varphi_n(2i-1)=2i$, $i=1, 2, \ldots, 2^{n-1}$. The map φ_n induces in an obvious way a permutation between the atoms of $\mathscr{B}_n$ and therefore also a map (denoted again by φ_n) from the set of all $\mathscr{B}_n$-measurable subsets of [0, 1] onto itself. The Lebesgue measure on [0, 1] is denoted by μ.

Lemma 1.g.3. *For every integer $n \geqslant 2^6$ there exists a partition Δ_n of [0, 1] into M_n disjoint $\mathscr{B}_n$-measurable sets of equal measure (i.e. of measure M_n^{-1}) so that*
(i) $M_n \geqslant 2^{n/16}$, $n \geqslant 2^6$,
(ii) $\mu(\varphi_n(A) \cap B) \leqslant 4\mu(A)\mu(B)$,
for every $A \in \Delta_n$, $B \in \Delta_m$, $n, m \geqslant 2^6$.

The restriction $n \geqslant 2^6$ appearing in the statement of 1.g.3 is made just for convenience and has no special significance. We postpone the proof of 1.g.3 and turn now to the construction of the sublattice in 1.g.2.

Proof of 1.g.2. Let $1 \leqslant r < p \leqslant \infty$ and let X be the space of all measurable functions f on [0,1] so that

$$\|f\| = \left(\sum_{m=2^6}^{\infty} \sum_{B \in \Delta_m} M_m^{\alpha p} \left(\int_B |f(t)|^r \, dt \right)^{p/r} \right)^{1/p} < \infty ,$$

where M_m and Δ_m are given by 1.g.3 and α is a number satisfying

$$0 < \alpha p < p/r - 1 .$$

Clearly, for every $f \in X$, we have

$$M_{2^6}^{\alpha-1} \|f\|_1 \leqslant \|f\| \leqslant M \|f\|_\infty ,$$

where $M = \left(\sum_{m=2^6}^{\infty} M_m^{1+\alpha p - p/r} \right)^{1/p} < \infty$. Hence, X is a Köthe function space on [0, 1]. It is also obvious that X is a sublattice of $l_p(L_r(0, 1))$. We shall now prove that X fails to have the C.A.P. by applying 1.g.1. We have first to define the quantities which enter into the statement of this proposition.

Let $\{w_j\}_{j=1}^\infty$ be the Walsh functions on [0, 1] defined by

$$w_1(t) = r_0(t) = 1, \quad w_2(t) = r_1(t), \quad w_3(t) = r_2(t), \quad w_4(t) = r_1(t) r_2(t), \quad w_5(t) = r_3(t)$$

and, in general,

$$w_j(t) = r_{k_1+1}(t) r_{k_2+1}(t) \ldots r_{k_l+1}(t) ,$$

where $j-1 = 2^{k_1} + 2^{k_2} + \cdots + 2^{k_l}$ with $0 \leqslant k_1 < k_2 < \cdots < k_l$ and $\{r_k\}_{k=1}^\infty$ denote the Rademacher functions. Note that $\|w_j\|_\infty = 1$ for all j and that the $\{w_j\}_{j=1}^\infty$ form an orthonormal system in $L_2(0, 1)$. Hence, $\{w_j\}_{j=1}^\infty$ is a bounded sequence if we consider it as a sequence in X or in X^* and, moreover, as a sequence in X^* we have $w_j \overset{w^*}{\to} 0$. We define for $T \in L(X, X)$ and $n=1, 2, \ldots,$

$$\beta_n(T)=2^{-n} \sum_{j=1}^{2^n} w_j(Tw_j)\,,$$

where $w_j(Tw_j)$ means of course $\int_0^1 w_j(t)\,(Tw_j)\,(t)\,dt$, and claim that (iii) and (iv) of 1.g.1 hold for a suitable choice of $\{F_n\}_{n=1}^\infty$.

In order to verify that this is the case we note first that $\{w_j\}_{j=1}^{2^n}$ is a basis of the space of all $\mathscr{B}_n$-measurable functions on $[0, 1]$. Since $\{2^{n/2}x_i^n\}_{i=1}^{2^n}$, where $x_i^n=\chi_{[(i-1)/2^n,\, i/2^n)}$, is a basis of the same space, which is also orthonormal in $L_2(0, 1)$, it follows that

$$\beta_n(T)= \sum_{i=1}^{2^n} x_i^n(Tx_i^n)$$

(recall that the trace of a linear transformation on a finite dimensional inner product space does not depend on the choice of the orthonormal basis). Using this expression for $\beta_n(T)$ we get that

$$\beta_n(T)-\beta_{n-1}(T)= \sum_{i=1}^{2^n} x_i^n(Tx_{\varphi_n(i)}^n), \quad n=1, 2, \ldots,$$

where φ_n is the permutation of $\{1, 2, \ldots, 2^n\}$ defined before the statement of 1.g.3. The partition Δ_n of $[0, 1]$ given by 1.g.3 induces in an obvious way a partition Δ_n' of $\{1, 2, \ldots, 2^n\}$. To each $A' \in \Delta_n'$ there corresponds the set $A= \bigcup_{i\in A'} [(i-1)/2^n, i/2^n)$ in Δ_n. Clearly,

$$\sum_{i\in A'} x_i^n(Tx_{\varphi_n(i)}^n)=\text{Average}\left[\left(\sum_{i\in A'} \theta_i x_i^n\right)\left(\sum_{i\in A'} T\theta_i x_{\varphi_n(i)}^n\right)\right],$$

where the average is taken over all the $2^{\bar{A}'}$ choices of the signs θ_i, $i \in A'$ ($\bar{A}'$ denotes the cardinality of A' which is $2^n/M_n$). Each of the expressions appearing in the average is in absolute value equal to at most $\int_A |Tf(t)\,dt$ with $f= \sum_{i\in A'} \theta_i x_{\varphi_n(i)}^n$ a $\mathscr{B}_n$-measurable function whose absolute value is the characteristic function of $\varphi_n(A)$. Hence, if

$$E_n=\{f; f\ \mathscr{B}_n\text{-measurable and } |f|=\chi_{\varphi_n(A)} \text{ for some } A \in \Delta_n\}$$

then

$$|\beta_n(T)-\beta_{n-1}(T)| \leq \sum_{A'\in\Delta'} \left|\sum_{i\in A'} x_i^n T x_{\varphi_n(i)}^n\right|$$

$$\leq M_n \sup\left\{\int_A |Tf(t)|\,dt;\, A \in \Delta_n, f \in E_n\right\}.$$

By the definition of the norm in X and Hölder's inequality, we have, for every $g \in X$, that

$$\|g\| \geqslant M_n^\alpha M_n^{1-1/r} \int_A |g(t)|\, dt, \quad A \in \Delta_n,$$

and, hence, for every $n \geqslant 2^6$

$$|\beta_n(T) - \beta_{n-1}(T)| \leqslant M_n^{1/r-\alpha} \sup \{\|Tf\|; f \in E_n\}.$$

Also, by (ii) of 1.g.3 we have, for every $f \in E_n$ with $|f| = \chi_{\varphi_n(A)}$

$$\|f\| = \left(\sum_{m=2^6}^{\infty} \sum_{B \in \Delta_m} M_m^{\alpha p} \mu(B \cap \varphi_n(A))^{p/r}\right)^{1/p}$$
$$\leqslant 4M_n^{-1/r}\left(\sum_{m=2^6}^{\infty} M_m M_m^{\alpha p} M_m^{-p/r}\right)^{1/p}.$$

Consequently, (iii) and (iv) of 1.g.1 hold if we put $F_n = M_n^{1/r-\alpha} E_n$ for every $n \geqslant 2^6$. □

Remark. If $r = 2 < p < \infty$ the space $l_p(L_2(0,1))$ is isomorphic to a complemented subspace of $l_p(L_p(0,1))$ which, in turn, is isometric to $L_p(0,1)$ (recall that $l_2 = L_2(0,1)$ is isomorphic to a complemented subspace of $L_p(0,1)$; take e.g. $[r_n]_{n=1}^\infty$ in $L_p(0,1)$ cf. Vol. I, p. 72). Hence, in this case, the space X of 1.g.2 is a sublattice of the lattice $Y_p = l_p(L_2(0,1)) \oplus L_p(0,1)$ and Y_p is linearly isomorphic to $L_p(0,1)$.

Proof of Lemma 1.g.3. We identify $[0,1]$ with $D = \{-1,1\}^{\aleph_0}$ endowed with the usual product measure. In this identification $\mathscr{B}_n$ corresponds to the algebra of those subsets of D which depend only on the first n coordinates. The map φ_n from the set of $\mathscr{B}_n$-measurable subsets onto itself becomes under this identification the transformation induced by the mapping

$$\varphi_n(\theta_1, \theta_2, \ldots, \theta_n, \theta_{n+1}, \ldots) = (\theta_1, \theta_2, \ldots, -\theta_n, \theta_{n+1}, \ldots)$$

of D onto itself.

We represent D as the product $\prod_{i=1}^\infty D_i$, where $D_i = \{-1,1\}^{2^i}$, and let π_i be the natural projection from D onto D_i. For each $i \geqslant 5$ we choose a system of partitions $\{\Omega_n\}_{n=2^{i+2}}^{2^{i+3}-1}$ of D_{i-1} into disjoint sets each having cardinality $(\bar{\bar{D}}_{i-1})^{1/2} = 2^{2^{i-2}}$ so that if $\sigma \in \Omega_n$ and $\eta \in \Omega_m$ with $n \neq m$ then $\overline{\overline{\sigma \cap \eta}} = 1$. To see that such partitions do indeed exist put $q = (\bar{\bar{D}}_{i-1})^{1/2}$, consider D_{i-1} as a finite field and let F be a subfield of D_{i-1} of cardinality q (this is possible since q is a power of a prime, namely of 2). We let each Ω_n be a partition of D_{i-1} into lines parallel to a fixed line of the form xF, $0 \neq x \in D_{i-1}$. The number of such lines is $(q^2-1)/(q-1) = q+1$ which is larger than the required number of partitions (namely 2^{i+2}).

We are now ready to define the partitions of D. For $2^{i+1} \leqslant n < 2^{i+2}$, $i=5, 6, \dots$ we let the elements of Δ_n be the following sets: $\{t \in D;\ \pi_{i-1}(t) \in \sigma,\ t_n = -1\}$ with $\sigma \in \Omega_{2n}$ and $\{t \in D;\ \pi_{i-1}(t) \in \sigma,\ t_n = 1\}$ with $\sigma \in \Omega_{2n+1}$. The number of the sets in Δ_n is thus $2(\bar{D}_{i-1})^{1/2} = 2^{1+2^{i-2}}$ and since $n < 2^{i+2}$ we see that (i) of 1.g.3 holds.

Let us verify that also (ii) of 1.g.3 holds. Let

$$A = \{t \in D;\ \pi_{i-1}(t) \in \sigma,\ t_n = \theta\}, \quad \sigma \in \Omega_{2n+(1+\theta)/2}, \quad 2^{i+1} \leqslant n < 2^{i+2}$$
$$B = \{t \in D;\ \pi_{j-1}(t) \in \eta,\ t_m = \theta'\}, \quad \eta \in \Omega_{2m+(1+\theta')/2} \quad 2^{j+1} \leqslant m < 2^{j+2}.$$

Clearly, $\varphi_n(A) = \{t \in D;\ \pi_{n-1}(t) \in \sigma,\ t_n = -\theta\}$ and

$$\mu(A) = 2^{-1}\mu(\pi_{i-1}^{-1}(\sigma)) = 2^{-2^{i-2}-1}, \qquad \mu(B) = 2^{-1}\mu(\pi_{j-1}^{-1}(\eta)) = 2^{-2^{j-2}-1}.$$

If $i \neq j$ then $\pi_{i-1}^{-1}(\sigma)$ and $\pi_{j-1}^{-1}(\eta)$ depend on different coordinates and hence,

$$\mu(\varphi_n(A) \cap B) \leqslant \mu(\pi_{i-1}^{-1}(\sigma) \cap \pi_{j-1}^{-1}(\eta))$$
$$= \mu(\pi_{i-1}^{-1}(\sigma)) \cdot \mu(\pi_{j-1}^{-1}(\eta)) = 4\mu(A)\mu(B).$$

If $i=j$ then either $n=m$ and $\theta=\theta'$ in which case $\varphi_n(A)$ and $B=A$ are disjoint or, by our choice of the Ω_n's, $\overline{\overline{\sigma \cap \eta}} = 1$. In this case we have

$$\mu(\varphi_n(A) \cap B) \leqslant \mu(\pi_{i-1}^{-1}(\sigma \cap \eta)) = 2^{-2^{i-1}} = 4\mu(A)\mu(B). \quad \square$$

We pass now to another result of Szankowski.

Theorem 1.g.4 [124]. *For every $1 \leqslant p < 2$ the space l_p has a subspace without the* C.A.P.

Recall that for $2 < p$ a similar result was proved in I.2.d.6. (See also remark 2 below.)

The proof of 1.g.4 is also based on a combinatorial lemma. In order to state this lemma we introduce first some notations. For $n = 1, 2, \dots$ let $\sigma_n = \{2^n, 2^n+1, \dots, 2^{n+1}-1\}$. To each integer $j \geqslant 8$ we associate nine integers $\{f_k(j)\}_{k=1}^{9}$ defined as follows:

$$f_k(4i+l) = 2i+k-1, \quad i=2, 3, 4, \dots, \quad l=0, 1, 2, 3, \quad k=1, 2$$
$$f_k(4i+l) = 4i+(l+k-2) \bmod 4,$$
$$\qquad i=2, 3, 4, \dots, \quad l=0, 1, 2, 3, \quad k=3, 4, 5$$
$$f_k(4i+l) = 8i+k-6, \quad i=2, 3, 4, \dots, \quad l=0, 1, \quad k=6, 7, 8, 9$$
$$f_k(4i+l) = 8i+k-2, \quad i=2, 3, 4, \dots, \quad l=2, 3, \quad k=6, 7, 8, 9.$$

These functions will arise from the construction of the subspace in 1.g.4. An important fact about these functions is that $f_k(j) \neq j$, for every k and j. This enables us to partition the integers into relatively large subsets so that, for every $1 \leqslant k \leqslant 9$, as j runs through one set of the partition, the corresponding integers $f_k(j)$ belong to different sets of the partition. More precisely, we have the following.

Lemma 1.g.5. *There exist partitions Δ_n and ∇_n of σ_n into disjoint sets and a sequence of integers $\{m_n\}_{n=1}^\infty$ with $m_n \geq 2^{n/8-2}$, $n=2, 3, 4, \ldots$, so that*

(i) *Each element of ∇_n has cardinality between m_n and $2m_n$.*

(ii) *Every element of ∇_n contains at most one representative from any element of Δ_n, i.e.*

$$\overline{\overline{A \cap B}} \leq 1, \quad A \in \nabla_n, \quad B \in \Delta_n, \quad n=2, 3, 4, \ldots .$$

(iii) *For every $A \in \nabla_n$, $n \geq 3$ and every $1 \leq k \leq 9$ the set $f_k(A)$ is contained entirely in an element of Δ_{n-1}, Δ_n or Δ_{n+1}.*

Note that $f_k(\sigma_n) \subset \sigma_{n-1}$ for $k=1, 2$, $f_k(\sigma_n) \subset \sigma_n$ for $k=3, 4, 5$ and $f_k(\sigma_n) \subset \sigma_{n+1}$ for $k=6, 7, 8, 9$. We postpone the proof of the lemma and pass to the

Proof of 1.g.4. Let $1 \leq p < 2$ and let X be the space of all sequences $x=(a_4, a_5, a_6, \ldots)$ so that

$$\|x\| = \left(\sum_{n=2}^{\infty} \sum_{B \in \Delta_n} \left(\sum_{j \in B} |a_j|^2 \right)^{p/2} \right)^{1/p} < \infty ,$$

where Δ_n is the partition of σ_n given by 1.g.5. The space X is a direct sum in the l_p sense of finite dimensional inner product spaces and is therefore isomorphic to a subspace of l_p. As a matter of fact, X is even isomorphic to l_p for $1<p<2$ (cf. Vol. I, p. 73). We denote by $\{e_j\}_{j=4}^\infty$ the unit vector basis of X and by $\{e_j^*\}_{j=4}^\infty$ the corresponding biorthogonal functionals in X^*. We let Z be the closed subspace of X spanned by the sequence

$$z_i = e_{2i} - e_{2i+1} + e_{4i} + e_{4i+1} + e_{4i+2} + e_{4i+3}, \quad i=2, 3, \ldots .$$

We shall prove, using 1.g.1, that Z fails to have the C.A.P. Put

$$z_i^* = \tfrac{1}{2}(e_{2i}^* - e_{2i+1}^*), \quad i=2, 3, \ldots ,$$

and, for $T \in L(Z, Z)$,

$$\beta_n(T) = 2^{-n} \sum_{i \in \sigma_n} z_i^*(Tz_i), \quad n=1, 2, 3, \ldots .$$

In order to establish that (iii) and (iv) of 1.g.1 hold (see also the remark following 1.g.1), we note first that, for every $i \geq 2$, the restriction of $(e_{4i}^* + e_{4i+1}^* + e_{4i+2}^* + e_{4i+3}^*)/4$ to Z is equal to that of z_i^* (they coincide when evaluated on z_j for every j).

Hence, for $n \geq 2$ and $T \in L(Z, Z)$,

$$\begin{aligned}
&\beta_n(T)-\beta_{n-1}(T)= \\
&2^{-n-1}\sum_{i\in\sigma_n}(e_{2i}^*-e_{2i+1}^*)T(e_{2i}-e_{2i+1}+e_{4i}+e_{4i+1}+e_{4i+2}+e_{4i+3}) \\
&-2^{-n-1}\sum_{i\in\sigma_{n-1}}(e_{4i}^*+e_{4i+1}^*+e_{4i+2}^*+e_{4i+3}^*)T(e_{2i}-e_{2i+1}+e_{4i}+e_{4i+1}+e_{4i+2}+e_{4i+3}) \\
&=2^{-n-1}\sum_{i\in\sigma_{n-1}}\left\{\begin{array}{l}
e_{4i}^*T(e_{4i}-e_{4i+1}+e_{8i}+\cdots+e_{8i+3}-e_{2i}+e_{2i+1}-e_{4i}-\cdots-e_{4i+3}) \\
+e_{4i+1}^*T(-e_{4i}+e_{4i+1}-e_{8i}-\cdots-e_{8i+3}-e_{2i}+e_{2i+1}-e_{4i}-\cdots-e_{4i+3}) \\
+e_{4i+2}^*T(e_{4i+2}-e_{4i+3}+e_{8i+4}+\cdots+e_{8i+7}-e_{2i}+e_{2i+1}-e_{4i}-\cdots-e_{4i+3}) \\
+e_{4i+3}^*T(-e_{4i+2}+e_{4i+3}-e_{8i+4}-\cdots-e_{8i+7}-e_{2i}+e_{2i+1}-e_{4i}-\cdots-e_{4i+3})
\end{array}\right. \\
&=2^{-n-1}\sum_{j\in\sigma_{n+1}}e_j^*Ty_j,
\end{aligned}$$

where

$$\sum_{k=1}^{9}\lambda_{j,k}e_{f_k(j)}=y_j\in Z,\quad j=8,9,\ldots,$$

the f_k being the functions defined before 1.g.5 and, for every j, $|\lambda_{j,k}|=1$ for eight indices k and $|\lambda_{j,k}|=2$ for the ninth k.

As in the proof of 1.g.2 we write now

$$\beta_n(T)-\beta_{n-1}(T)=2^{-n-1}\sum_{A\in\bar{V}_{n+1}}\text{Average}\left[\left(\sum_{j\in A}\theta_j e_j^*\right)T\left(\sum_{j\in A}\theta_j y_j\right)\right],$$

where the average is taken over all $2^{\bar{A}}$ choices of signs $\{\theta_j\}_{j\in A}$. By the definition of the norm in X and by (ii) of 1.g.5 we have, for every $A\in\bar{V}_{n+1}$ $(n\geq 2)$ and $\{\theta_j\}_{j\in A}$,

$$\left\|\sum_{j\in A}\theta_j e_j^*\right\|_{Z^*}\leq\left\|\sum_{j\in A}\theta_j e_j^*\right\|_{X^*}=(\bar{A})^{1/q}\leq(2m_{n+1})^{1/q},$$

where $1/p+1/q=1$. By (iii) of 1.g.5 we have, for every such A and $\{\theta_j\}_{j\in A}$ and every $1\leq k\leq 9$,

$$\left\|\sum_{j\in A}\theta_j e_{f_k(j)}\right\|=(\bar{A})^{1/2}\leq(2m_{n+1})^{1/2}$$

and, consequently,

$$\left\|\sum_{j\in A}\theta_j y_j\right\|\leq 15m_{n+1}^{1/2}.$$

Hence,

$$|\beta_n(T)-\beta_{n-1}(T)|\leqslant 2^{-n-1}(2^{n+1}m_{n+1}^{-1})(2m_{n+1})^{1/q}\sup\{\|Tz\|;\,z\in E_n\},$$

where

$$E_n=\left\{\sum_{j\in A}\theta_j y_j\,;\,A\in\mathcal{V}_{n+1},\,\theta_j=\pm 1\right\}.$$

Consequently, (iii) and (iv) of 1.g.1 hold if we put $F_n=2m_{n+1}^{-1/p}E_n$. □

Proof of 1.g.5. For $n\geqslant 2$ and $l=0,1,2,3$ we put $\sigma_n^l=\{j\in\sigma_n\,;\,j\equiv l (\mathrm{mod}\ 4)\}$ and let $\varphi_n^l\colon\sigma_n^0\to\sigma_n^l$ be the map defined by $\varphi_n^l(j)=j+l$. For $n\geqslant 2$ and $r=0,1$ we let $\psi_n^r\colon\sigma_n^0\to\sigma_{n+1}^0$ be defined by $\psi_n^r(j)=2j+4r$.

By an easy inductive procedure, we can represent σ_n^0 for $n\geqslant 2$ as a Cartesian product $C_n\times D_n$, where

$$\bar{\bar{D}}_{2m}=\bar{\bar{D}}_{2m+1}=\bar{\bar{C}}_{2m-1}=\bar{\bar{C}}_{2m}=2^{m-1},\quad m=1,2,\ldots$$

so that:

for each $c\in C_{n+1}$ there is an $r=0,1$ and a $d\in D_n$ so that $\psi_n^r(C_n\times\{d\})=\{c\}\times D_{n+1}$ and, for each $d\in D_{n+1}$, there is a $c\in C_n$ so that $\psi_n^0(\{c\}\times D_n)\cup\psi_n^1(\{c\}\times D_n)=C_{n+1}\times\{d\}$.

We represent further each D_n, $n=2,3,\ldots$ as a Cartesian product of four factors $D_n=\prod_{l=0}^{3}D_n^l$ so that

$$\bar{\bar{D}}_n^0\leqslant\bar{\bar{D}}_n^1\leqslant\bar{\bar{D}}_n^2\leqslant\bar{\bar{D}}_n^3\leqslant 2\bar{\bar{D}}_n^0\,.$$

We can now define the desired partitions.

$$\mathcal{V}_n=\left\{\varphi_n^l(\{f\}\times D_n^l);\,f\in C_n\times\prod_{i\neq l}D_n^i,\,l=0,1,2,3\right\},$$

$$\Delta_n=\left\{\varphi_n^l\left(C_n\times\prod_{i\neq l}D_n^i\times\{d\}\right);\,d\in D_n^l,\,l=0,1,2,3\right\}.$$

It is clear that (i) of 1.g.5 holds with

$$m_n=\bar{\bar{D}}_n^{\,0}\geqslant(\bar{\bar{D}}_n/8)^{1/4}\geqslant 2^{n/8-2}\,.$$

It is also evident that (ii) of 1.g.5 holds. The verification of (iii) of 1.g.5 is straightforward but somewhat long and we omit it. □

Remarks. 1. In the proof of 1.g.4 presented above it was essential that the norm in $[e_j]_{j\in\sigma_n}$, for $n=2, 3, 4, \ldots$, is given by

$$\Big\|\sum_{j\in\sigma_n} a_j e_j\Big\| = \Big(\sum_{B\in\Delta_n}\Big(\sum_{j\in B}|a_j|^2\Big)^{p/2}\Big)^{1/p},$$

i.e. that $[e_j]_{j\in\sigma_n}$ is a subspace of l_p^m for a suitable m. As for the norm in the whole space $X=[e_j]_{j=4}^\infty$ we used only the fact that X has a Schauder decomposition (cf. I.1.g) into $\{[e_j]_{j\in\sigma_n}\}_{n=1}^\infty$. Consequently, the proof of 1.g.4 shows that if, for a Banach space X, there is a $1\leqslant p<2$ and $K<\infty$ so that X has a Schauder decomposition into $\{X_n\}_{n=1}^\infty$ with $d(X_n, l_p^n)\leqslant K$ for all n, then X has a subspace which fails to have the C.A.P. The argument used in the proof of I.1.a.5 shows that if Y is a Banach space so that, for some $1\leqslant p<\infty$ and K, there is for every integer n a subspace Y_n of Y with $d(Y_n, l_p^n)\leqslant K$ then Y has a subspace X which has a Schauder decomposition into $\{X_n\}_{n=1}^\infty$ with $d(X_n, l_p^n)\leqslant K+1$. Hence, if $1\leqslant p<2$, every such Y has a subspace which fails to have the C.A.P.

2. The proof of 1.g.4 presented above can be easily modified so as to apply to the case $2<p\leqslant\infty$ (and thus to yield an independent proof of I.2.d.6). We have only to arrange the partitions ∇_n and Δ_n so that every $A\in\nabla_n$ is contained in some element of Δ_n while, for every $A\in\nabla_n$, $k=1,\ldots,9$ and every $B\in\Delta_{n-1}$, Δ_n or Δ_{n+1}, $\overline{\overline{B\cap f_k(A)}}\leqslant 1$. If this is the case we get (in the notation of the proof of 1.g.4) that, for every $A\subset\nabla_{n+1}$, $n=2, 3, \ldots$ and $\{\theta_j\}_{j\in A}$,

$$\Big\|\sum_{j\in A}\theta_j e_j^*\Big\|_{Z^*}\leqslant(2m_{n+1})^{1/2}, \qquad \Big\|\sum_{j\in A}\theta_j y_j\Big\|\leqslant 15m_{n+1}^{1/p}.$$

In Vol. III we shall present a deep result of Krivine [67] and Maurey and Pisier [96] which asserts that if, for a Banach space X,

$$p^{(X)}=\sup\{p;\ X \text{ is of type } p\}, \qquad q^{(X)}=\inf\{q;\ X \text{ is of cotype } q\}$$

then, for every n, X contains almost isometric copies of $l_{p(X)}^n$ and $l_{q(X)}^n$. An immediate consequence of this result and the preceding remarks is the following theorem.

Theorem 1.g.6. *Let X be a Banach space. If every subspace of X has the* C.A.P. *then X is of type $2-\varepsilon$ and cotype $2+\varepsilon$ for every $\varepsilon>0$.*

In connection with 1.g.6 let us recall a result of Kwapien which was already mentioned above (and actually proved in the case of lattices after 1.f.17): if a Banach space X is of type 2 and cotype 2 then X is isomorphic to a Hilbert space. Thus, 1.g.6 asserts that, unless X is "very close" to being a Hilbert space, X has a subspace which fails to have the C.A.P. There are however Banach lattices not isomorphic to Hilbert spaces in which every subspace has the C.A.P. and even the B.A.P.

Example 1.g.7 [57]. *There is a sequence of integers* $\{k_n\}_{n=1}^\infty$ *and a sequence of numbers* $\{p_n\}_{n=1}^\infty$ *with* $p_n \downarrow 2$ *so that* $X=\left(\sum_{n=1}^\infty \oplus l_{p_n}^{k_n}\right)_2$ *is not isomorphic to* l_2 *but every subspace* Y *of* X *has the bounded approximation property.*

The proof of 1.g.7 is based on the following four facts:

(1) There is a constant K (independent of $\{k_n\}_{n=1}^\infty$ and $\{p_n\}_{n=1}^\infty$ provided $p_n \leq 3$ for all n) so that, for every X of the form appearing in 1.g.7 and every finite dimensional subspace E of X with $\dim E=m$, there is a projection Q from X onto E with

$$\|Q\| \leq K\, d(E, l_2^m)\,.$$

This fact is a special case of a general result from [93] to be proved in Vol. III. The assertion is actually true for every Banach space X of type 2 and the constant K depends only on the type 2 constant of X.

(2) For every integer m there is an $\varepsilon>0$ so that if $|p-2|<\varepsilon$ and E is an m-dimensional subspace of l_p then $d(E, l_2^m)<2$. In Vol. III we shall prove a more precise version of this fact, namely that, for every $1 \leq p < \infty$ and every $E \subset l_p$ with $\dim E=m$, we have $d(E, l_2^m) \leq m^{|1/2-1/p|}$ (cf. [72]).

(3) For every $p \neq 2$, $\lim_{m\to\infty} d(l_p^m, l_2^m)=\infty$. This fact is obvious. Actually, it is not hard to verify that $d(l_2^m, l_p^m)=m^{|1/2-1/p|}$.

(4) Let $T: X \to Z$ be a quotient map and let $E \subset Z$ be a subspace of dimension m. Then there is a subspace G of X with $\dim G \leq 5^m$ so that $TG=E$ and, for every $z \in E$, there is an $x \in G$ with $Tx=z$ and $\|x\| \leq 3\|z\|$.

We prove the validity of assertion (4). Let $\{z_i\}_{i=1}^l$ be a subset of $\{z;\, z \in E,\ \|z\|=1\}$ so that $\|z_i-z_j\| \geq 1/2$ for every $i \neq j$ and which cannot be included in a larger set having this property (and thus, whenever $z \in E$ with $\|z\|=1$, there is an $1 \leq i \leq l$ with $\|z-z_i\|<1/2$). The balls $B_E(z_i, 1/4)$, $i=1, \ldots, l$ have pairwise disjoint interiors and are all contained in $B_E(0, 5/4)$. Hence, by considering the volumes of these balls, we get that $(5/4)^m \geq l(1/4)^m$ i.e. $l \leq 5^m$. For each $1 \leq i \leq l$ let x_i be an element of X so that $\|x_i\|<3/2$ and $Tx_i=z_i$ and put $G=[x_i]_{i=1}^l$. Then, clearly, $TB_G(0, 3) \supset 2\,\mathrm{conv}\,\{\pm z_i\}_{i=1}^l$ and it remains to verify that this latter set contains $B_E(0, 1)$. For every $z \in E$ with $\|z\| \leq 1$ there is a $1 \leq j_1 \leq l$ such that $\|z-\|z\|z_{j_1}\| \leq \|z\|/2$ and, by induction, we can continue and choose $\{j_k\}_{k=2}^\infty$, all integers between 1 and l, so that, for every $n \geq 1$,

$$\left\|z-\sum_{k=1}^n \lambda_k z_{j_k}\right\| \leq 2^{-n}, \qquad 0 \leq \lambda_k \leq 2^{-k+1}, \quad k=1, 2, \ldots\,.$$

Hence, $z \in 2\,\mathrm{conv}\,\{\pm z_i\}_{i=1}^l$ and this completes the proof of assertion (4).

Proof of 1.g.7. We choose inductively a sequence of numbers $\{p_n\}_{n=1}^\infty$ decreasing to 2 and a sequence of integers $\{k_n\}_{n=1}^\infty$ in the following manner. We start by taking

$p_1=3$. Then select $k_1, p_2, k_2, \ldots$ in this order so that

(i) $d(l_{p_n}^{k_n}, l_2^{k_n})>n$, $n=1, 2, \ldots$ (this is possible by (3) above)

(ii) For every $n=1, 2, \ldots, 2<p_{n+1}<p_n$ and, whenever $E\subset l_p$ with $2\leqslant p<p_{n+1}$ is a subspace of dimension $h\leqslant 2\cdot 5^{\sum_{i=1}^{n} k_i}$, then $d(E, l_2^h)\leqslant 2$. (This is possible by (2) above.)

Let now Y be any subspace of X, let F be a finite dimensional subspace of Y and put $m=\dim F$. Pick an integer n so that $m\leqslant 5^{\sum_{i=1}^{n-1} k_i}$ and let Y_n be the subspace of Y consisting of all those vectors whose components in $l_{p_i}^{k_i}$ are zero for $i<n$. Clearly, $\dim Y/Y_n\leqslant \sum_{i=1}^{n-1} k_i$. By assertion (4) above and its proof there is a subspace G of Y containing F of dimension at most

$$m+5^{\sum_{i=1}^{n-1} k_i}\leqslant 2\cdot 5^{\sum_{i=1}^{n-1} k_i}$$

so that if T denotes the quotient map from Y onto Y/Y_n then $TB_G(0, 3)\supset B_{Y/Y_n}(0, 1)$. Since

$$G\cap Y_n\subset\left(\sum_{i=n}^{\infty}\oplus \pi_i G\right)_2,$$

where π_i is the natural projection from X onto $l_{p_i}^{k_i}$, it follows that $d(G\cap Y_n, l_2^h)\leqslant 2$, where $h=\dim G\cap Y_n$. Hence, by fact (i) above, there is a projection Q from Y (even from X) onto $G\cap Y_n$ with $\|Q\|\leqslant 2K$. The restriction of T to $(I-Q)G$ is one to one (since kern $T_{|G}=G\cap Y_n$) and thus $S=(T_{|(I-Q)G})^{-1}$ is well defined. Since, for every $z\in Y/Y_n$ with $\|z\|=1$, there is a $y\in G$ with $\|y\|\leqslant 3$ and $Ty=z$, we get that $Sz=(I-Q)y$ and hence $\|S\|\leqslant 3\|I-Q\|\leqslant 9K$. The operator $P=ST+Q$ is a projection of norm $\leqslant 11K$ from Y onto G. ☐

We conclude this section by mentioning that, by an approach which is somewhat similar to the proofs of 1.g.2 and 1.g.4, it was proved in [125] that the space $L(l_2, l_2)$ with the usual operator norm fails to have the A.P. (this is the solution to a part of Problem I.1.e.10).

2. Rearrangement Invariant Function Spaces

a. Basic Definitions, Examples and Results

Many of the lattices of measurable functions which appear in analysis have an important symmetry property, namely they remain invariant if we apply a measure preserving transformation to the underlying measure space. Such lattices are called rearrangement invariant function spaces or r.i. spaces, in short. They are the natural generalization of the notion of a symmetric basis (or a symmetric sequence space) to the setting of lattices. The importance of r.i. function spaces stems mainly from two (closely related) facts: they form the natural framework for the study of some important questions concerning $L_p(\mu)$ spaces and they arise naturally in interpolation theory. In this section we present some basic facts concerning r.i. spaces and prove also a quite simple but general interpolation theorem.

Before giving the formal definition of an r.i. space we prefer to discuss in some detail the notions which enter in its definition. The main requirement imposed on an r.i. function space X will be that it is a Köthe function space on some σ-finite measure space (Ω, Σ, μ) (cf. 1.b.17) so that, for every automorphism τ of Ω into itself and every $f \in X$, the function $f(\tau^{-1}(\omega))$† also belongs to X. By an automorphism τ of a measure space Ω into itself we mean a one-to-one map from Ω onto a measurable subset $\tau(\Omega)$ of itself so that both τ and τ^{-1} are measurable and $\mu(\sigma)=\mu(\tau(\sigma))$ for every $\sigma \in \Sigma$. If $\mu(\Omega)<\infty$ then clearly $\Omega \sim \tau(\Omega)$ has measure zero and thus we can (and shall) assume that τ is onto. Observe that like any Köthe function space every r.i. space is, in particular, σ-complete. Hence, $C(0, 1)$, for example, will not be considered as an r.i. space on $[0, 1]$.

We shall restrict our attention to the case in which (Ω, Σ, μ) is a separable measure space (i.e. Σ, with the metric $d(\sigma_1, \sigma_2)=\mu(\sigma_1 \Delta \sigma_2)$, is a separable metric space, where $\sigma_1 \Delta\sigma_2=(\sigma_1 \sim \sigma_2)\cup(\sigma_2 \sim \sigma_1)$). The structure of such a measure space is simple and well known (cf. [50]): it consists of (a perhaps empty) continuous part which is isomorphic to the usual Lebesgue measure space on a finite or infinite interval on the line and of an at most countable number of atoms. (By an isomorphism of two measure spaces we mean a one-to-one correspondence between the σ-algebras which preserves the measure and the countable Boolean operations. In general (i.e. unless we have a measure theoretic pathology which is of no interest in the present context), this correspondence is induced by a point transformation between the measure spaces.) Since an automorphism τ of a measure space Ω maps the continuous part of Ω into itself and maps each atom of

† $f(\tau^{-1}(\omega))$ is defined to be zero for ω not in the range of τ.

Ω to an atom with the same mass it is clear that the study of r.i. spaces over a separable measure space reduces immediately to the study of such spaces when Ω is either a finite or infinite interval on the line or a finite or countably infinite discrete measure space in which each point has the same mass. Thus, up to some inessential normalization, we are reduced to the study of the following three cases:

(i) Ω = integers and the mass of every point is one.
(ii) $\Omega = [0, 1]$ with the usual Lebesgue measure.
(iii) $\Omega = [0, \infty)$ with the usual Lebesgue measure.

Those three cases are different as will become apparent in the sequel. Every space with a symmetric basis is, in a natural way, an r.i. space on the Ω given in (i) above. However, we will include in the definition of r.i. spaces on such Ω also some non-separable sequence spaces, e.g. l_∞ or, more generally, the Orlicz sequence spaces l_M, where M does not satisfy the Δ_2-condition at 0 (see 1.4.a). Though some of the theorems proved in the sequel have a meaning and are of interest also in case (i), our main emphasis in this chapter will be on the continuous cases, i.e. on (ii) and (iii).

It is worthwhile to make some comments on the operator $U_\tau f(\omega) = f(\tau^{-1}(\omega))$ induced by an automorphism τ of Ω on a Köthe function space X on Ω which is invariant under automorphisms. We note first that U_τ, being a positive operator, is bounded. The family of these operators is actually uniformly bounded (this fact was observed in [87]). We show this, for example, in case (ii), i.e. when $\Omega = [0, 1]$. Let k be an integer and let $P_{i,k}$, $1 \leqslant i \leqslant k$, be the projection on X defined by $P_{i,k} f = f \cdot \chi_{[(i-1)/k,\, i/k)}$. If the U_τ's are not uniformly bounded on X then, since

$$\sum_{j=1}^{k} \sum_{i=1}^{k} P_{j,k} U_\tau P_{i,k} = U_\tau ,$$

we get that for every k there are $1 \leqslant i_k, j_k \leqslant k$ so that $\sup_\tau \|P_{j_k,k} U_\tau P_{i_k,k}\| = \infty$. Let k be an integer and let σ_1 and σ_2 be two subsets of $[0, 1]$ of measure k^{-1}. Since there are automorphisms of $[0, 1]$ which map σ_1 onto $[(i_k - 1)k^{-1}, i_k k^{-1})$, respectively, σ_2 onto $[(j_k - 1)k^{-1}, j_k k^{-1})$ and since these automorphisms induce isomorphisms of X onto itself we deduce that $\sup_\tau \|P_{\sigma_2} U_\tau P_{\sigma_1}\| = \infty$, where $P_\sigma f = f\chi_\sigma$. In other words, we get that $\sup_\tau \|P_\eta U_\tau P_\sigma\| = \infty$, for every choice of subsets $\eta, \sigma \subset [0, 1]$ of positive measure. Hence, we can find a sequence $\{f_n\}_{n=1}^\infty$ of functions in X and automorphisms $\{\tau_n\}_{n=1}^\infty$ of $[0, 1]$ so that $\|f_n\| \leqslant n^{-2}$, $\|U_{\tau_n} f_n\| \geqslant n$ for every n and so that the sets $\{\sigma_n \cup \tau_n(\sigma_n)\}_{n=1}^\infty$, where $\sigma_n = \operatorname{supp} f_n$, are mutually disjoint. Thus if τ_0 is an automorphism on $[0, 1]$ so that $\tau_{0|\sigma_n} = \tau_{n|\sigma_n}$ we would get that U_{τ_0} is not bounded (i.e. cannot be defined) on X.

Since the U_τ's form a semigroup it follows from the preceding remark that

$$|||f||| = \sup\,\{\|U_\tau f\|;\ \tau \text{ an automorphism of } \Omega \text{ into } \Omega\}$$

is an equivalent lattice norm on X with respect to which each U_τ is a contraction i.e. $|||U_\tau||| \leqslant 1$. We claim that actually each U_τ is an isometry in $|||\cdot|||$. This is evident if τ is invertible, i.e. maps Ω onto Ω. To prove this fact in general, let $f \in X$, let k be an integer, write $\Omega = \bigcup_{i=1}^{k} \sigma_i$, with σ_i being mutually disjoint sets each having infinite measure, and let $f_i = f\chi_{\sigma_i}$, $1 \leqslant i \leqslant k$. Since both $f - f_1$ and $U_\tau(f - f_1)$ vanish on a set of infinite measure there is an invertible automorphism τ_1 which maps $f - f_1$ onto $U_\tau(f - f_1)$ and thus

$$|||f - f_1||| = |||U_{\tau_1}(f - f_1)||| = |||U_\tau(f - f_1)||| \leqslant |||U_\tau f||| .$$

Similarly, for every $1 \leqslant i \leqslant k$, $|||f - f_i||| \leqslant |||U_\tau f|||$. By summing over i from 1 to k we get that

$$(k-1)|||f||| \leqslant \sum_{i=1}^{k} |||f - f_i||| \leqslant k|||U_\tau f|||$$

and, since k was arbitrary, we deduce that $|||f||| = |||U_\tau f|||$.

We have thus seen that if X is a Köthe function space which is invariant under automorphisms we can renorm it so that every U_τ becomes an isometry of X. In the definition of an r.i. space given below we shall require from the outset that the U_τ's are isometries. In other words, the norm of an $f \in X$ will be assumed to depend only on the *distribution function*

$$d_f(t) = \mu(\{\omega \in \Omega; f(\omega) > t\}), \quad -\infty < t < \infty ,$$

of f or, in fact, on the distribution function of $|f|$. More precisely, if $f \in X$ and g is a measurable function such that $d_{|g|}(t) = d_{|f|}(t)$, for every $t \geqslant 0$ (i.e. $|f|$ and $|g|$ are μ-equimeasurable) then also $g \in X$ and $\|g\| = \|f\|$. This remark needs some additional explanation if $\mu(\Omega) = \infty$. In this case $d_{|f|}(t)$ may become infinite for some $t > 0$ and the identity $d_{|f|}(t) = d_{|g|}(t)$ does not necessarily imply that there is an automorphism τ of the measure space into itself which carries $|f|$ into $|g|$ or vice-versa (let e.g. $f \equiv 1$ and let g be equal to 1 on some set σ and to 1/2 on $\Omega \sim \sigma$, where $\mu(\sigma) = \mu(\Omega \sim \sigma) = \infty$). However, it is clear that, for every $f \in X$, $d_{|f|}(t)$ is finite for large enough t and that if $d_{|f|}(t) = d_{|g|}(t)$ there are, for every $\varepsilon > 0$, automorphisms τ_1 and τ_2 of Ω into itself so that $U_{\tau_2}|g| \leqslant (1+\varepsilon)U_{\tau_1}|f|$. This shows that $U_{\tau_2} g \in X$; consequently, $g \in X$ and $\|g\| = \|f\|$. (The fact that $U_{\tau_2} g \in X$ implies $g \in X$ is obvious if the complement of the support of g has infinite measure, since in this case there is an automorphism τ of Ω onto itself for which $U_\tau U_{\tau_2} g = g$. For a general g we deduce this fact by writing $g = g_1 + g_2$ with the support of each g_i having a complement of infinite measure.)

The distribution function of a non-negative function f is clearly a right continuous non-increasing function on $[0, \infty)$ (which, in case $\mu(\Omega) = \infty$, may also take the value $+\infty$; we assume, however, as will be the case for every function in an r.i. space, that it is finite for sufficiently large t). Of special importance in the investigation of r.i. spaces is the right continuous inverse f^* of d_f (for $f \geqslant 0$) which is defined by

$$f^*(s) = \inf \{t > 0;\ d_f(t) \leqslant s\}, \quad 0 \leqslant s < \mu(\Omega).$$

The function f^*, which is evidently non-increasing, right continuous and has the same distribution function as f, is called the *decreasing rearrangement* of f. If f is a general element in an r.i. space we denote by f^* the decreasing rearrangement of $|f|$. Notice that f^* is, by definition, a function on $[0, \mu(\Omega))$ even if Ω is the space of integers. In this latter case, however, f^* corresponds in a natural way to a non-decreasing function on Ω which is also denoted by f^*. We adopt here the symbol f^* for the decreasing rearrangement of a function since it is commonly used in the literature. In contrast to the notation x^*, which is used in this book to denote an element in the conjugate space X^*, the decreasing rearrangement does not have, of course, any connection to conjugate spaces. As a matter of fact, if X is an r.i. space and f a function belonging to X then f^* also belongs to X. The meaning of f^* will always be clear from the context and we trust that there will not arise any confusion between these two uses of the symbol *.

It is worthwhile to note that we do not have, in general, that $(f_1+f_2)^*(t) \leqslant f_1^*(t)+f_2^*(t)$ (take e.g. $f_1(t)=t$ and $f_2(t)=1-t$ on $[0, 1]$). The following useful identity does however hold for every choice of f_1, f_2, t_1 and t_2

$$(f_1+f_2)^*(t_1+t_2) \leqslant f_1^*(t_1)+f_2^*(t_2).$$

This identity is a consequence of the fact that

$$\{\omega \in \Omega;\ |f_1(\omega)+f_2(\omega)| > f_1^*(t_1)+f_2^*(t_2)\} \subset \{\omega \in \Omega;\ |f_1(\omega)| > f_1^*(t_1)\} \cup \{\omega \in \Omega;\ |f_2(\omega)| > f_2^*(t_2)\}.$$

In particular, we have $(f_1+f_2)^*(2t) \leqslant f_1^*(t)+f_2^*(t)$.

If a Köthe function space X on (Ω, Σ, μ) is invariant with respect to automorphisms of Ω the same is true for X', the subspace of X^* consisting of the integrals (see the end of 1.b). Indeed, if τ is an invertible automorphism of Ω and f and g are non-negative measurable functions on Ω then clearly

$$\int_\Omega f(\omega)g(\tau(\omega))\, d\mu = \int_\Omega f(\tau^{-1}(\omega))g(\omega)\, d\mu.$$

This identity and the definition of X' show that X' is invariant under invertible automorphisms of Ω. It follows however from the preceding discussion that X' is also invariant with respect to every automorphism of Ω into itself. In all interesting examples of spaces invariant with respect to automorphisms, X' is a norming subspace of X^* (or, equivalently, by 1.b.18, $0 \leqslant f_n(\omega) \uparrow f(\omega)$ a.e. with $f \in X$ implies $\|f\| = \lim_n \|f_n\|$). We shall include also this assumption in the definition of an r.i. space.

Definition 2.a.1. Let (Ω, Σ, μ) be one of the measure spaces $\{1, 2, \ldots\}$, $[0, 1]$ or

$[0, \infty)$ (with the natural measure). A Köthe function space X on (Ω, Σ, μ) is said to be a *rearrangement invariant* (r.i.) *space* if the following conditions hold.

(i) If τ is an automorphism of Ω into itself and f is a measurable function on Ω then $f \in X$ if and only if $f(\tau^{-1}(\omega)) \in X$ and if this is the case then $\|f(\omega)\| = \|f(\tau^{-1}(\omega))\|$.

(ii) X' is a norming subspace of X^* and thus X is order isometric to a subspace of X''. As a subspace of X'', X is either maximal (i.e. $X = X''$) or minimal (i.e. X is the closed linear span of the simple integrable functions of X'').

(iii) a. If $\Omega = \{1, 2, \ldots\}$ then, as sets,

$$l_1 \subset X \subset l_\infty$$

and the inclusion maps are of norm one, i.e. if $f \in l_1$ then $\|f\|_X \leq \|f\|_1$ and if $f \in X$ then $\|f\|_\infty \leq \|f\|_X$.

b. If $\Omega = [0, 1]$ then, as sets,

$$L_\infty(0, 1) \subset X \subset L_1(0, 1)$$

and the inclusion maps are of norm one i.e. if $f \in L_\infty(0, 1)$ then $\|f\|_X \leq \|f\|_\infty$ and if $f \in X$ then $\|f\|_1 \leq \|f\|_X$.

c. If $\Omega = [0, \infty)$ then, as sets,

$$L_\infty(0, \infty) \cap L_1(0, \infty) \subset X \subset L_1(0, \infty) + L_\infty(0, \infty)$$

and the inclusion maps are of norm one with respect to the natural norms in these spaces, i.e. if $f \in L_\infty \cap L_1$ then $\|f\|_X \leq \max(\|f\|_1, \|f\|_\infty)$ and if $f \in X$ then $\int_0^1 f^*(t)\, dt \leq \|f\|_X$.

We have already explained in detail the role of assumption (i) and of the restriction imposed on Ω in 2.a.1. Assumptions (ii) and (iii) require, however, additional explanation. We consider first (ii). By definition, every Köthe function space contains the simple integrable functions and thus the terminology of minimal (and clearly maximal) used in (ii) is justified. If X is separable then, since it is σ-order complete, it is also order continuous (cf. the remark preceding 1.a.8) and therefore, in this case, assumption (ii) is always satisfied with X being the minimal subspace of X''. By the remark following 1.b.18, X is a maximal subspace of X'' (i.e. $X = X''$) if and only if it has the Fatou property. The non-separable space $L_\infty(0, 1)$ is a minimal r.i. function space (it is clearly also a maximal one). Every minimal non-separable r.i. function space X on $[0, 1]$ is equal to $L_\infty(0, 1)$ (up to an equivalent norm). Indeed, if $\lim_{t \to 0} \|\chi_{[0,t]}\|_X = 0$ then X is separable (the characteristic functions of dyadic intervals span a dense set) while if $\lim_{t \to 0} \|\chi_{[0,t]}\|_X > 0$ then $X = L_\infty[0, 1]$. Similarly, every non-separable minimal r.i. function space X on $[0, \infty)$ has the property that its restriction to $[0, 1]$ is equal to $L_\infty(0, 1)$ (with an equivalent norm); however, X itself need not be isomorphic to $L_\infty(0, \infty)$.

A separable r.i. function space X is maximal if and only if X does not have a subspace isomorphic to c_0 (use 1.c.4). In particular, every reflexive r.i. space is both minimal and maximal.

The reason for assuming (ii) (besides being satisfied for separable spaces X and the common non-separable examples) is that the basic results on r.i. spaces, to be proved below, may fail to hold without it (see example 2.a.11 below).

In order to avoid possible confusion we point out that there is no connection between the terms minimal and maximal r.i. spaces defined here and the notion of a minimal Orlicz function (or of a minimal symmetric basis) introduced in I.4.b.7. Minimality in the sense of I.4.b will not be used in this volume.

Assumption (iii) in 2.a.1 is just a normalization condition. The inclusion relations in (iii) are already a consequence of (i) and the normalization is imposed just to ensure that the inclusion maps have all norm one. Condition (iii)a means that the unit vector $(1, 0, 0 \ldots)$ in X has norm one. Similarly, condition (iii)b means that $\|\chi_{[0,1]}\|_X = 1$. Indeed, for every $f \in L_\infty(0, 1)$, $|f| \leq \|f\|_\infty \chi_{[0,1]}$ and, by a simple averaging argument it follows that, for every simple function of the form $f = \sum_{i=1}^{n} a_i \chi_{[(i-1)/n,\, i/n)}$, we have $\|f\|_X \geq n^{-1} \sum_{i=1}^{n} |a_i|\, \|\chi_{[0,1]}\|_X = \|f\|_1 \|\chi_{[0,1]}\|_X$. Condition (iii)c requires a little more explanation concerning the spaces appearing in its formulation.

Proposition 2.a.2 [48]. *The space $Y = L_1(0, \infty) + L_\infty(0, \infty)$ consisting of all the functions f on $[0, \infty)$ which can be written as $g + h$ with $g \in L_1(0, \infty)$ and $h \in L_\infty(0, \infty)$, becomes a Banach space if we define the norm in it by*

$$\|f\| = \inf \{\|g\|_1 + \|h\|_\infty;\ f = g + h\}.$$

This norm on Y can also be computed by

$$\|f\| = \int_0^1 f^*(t)\, dt = \sup \left\{ \int_\sigma |f(t)|\, dt;\ \mu(\sigma) = 1 \right\}.$$

The space Y has the Fatou property and is an r.i. function space on $[0, \infty)$. The space Y' is order isometric to $L_1(0, \infty) \cap L_\infty(0, \infty)$ endowed with the norm $\max(\|f\|_1, \|f\|_\infty)$.

Proof. It is trivial to verify that Y is complete. Let us prove that the two expressions for the norm coincide, denoting for the moment $\int_0^1 f^*(t)\, dt$ by $|||f|||$. If $f = g + h$ then, for every $\sigma \subset [0, \infty)$, $\int_\sigma |f(t)|\, dt \leq \|g\|_1 + \|h\|_\infty \mu(\sigma)$, and hence $|||f||| \leq \|f\|$. Conversely, fix $f \in Y$ and put $\lambda = \|f^* - f^* \chi_{[0,1]}\|_\infty$. Then

$$\begin{aligned} \|f\| = \|f^*\| &\leq \|f^* - \min(\lambda, f^*)\|_1 + \|\min(\lambda, f^*)\|_\infty \\ &= \|(f^* - \lambda)\chi_{[0,1]}\|_1 + \lambda = \int_0^1 f^*(t)\, dt = |||f|||. \end{aligned}$$

The fact that Y has the Fatou property is evident from the form of $|||f|||$. Let now $k \in Y'$ be an element of norm one; then, in particular, $\int_0^\infty |k(t)g(t)|\,dt \leq 1$ for every $g \in L_1(0, \infty)$ with $\|g\|_1 \leq 1$ and also for every $g \in L_\infty(0, \infty)$ with $\|g\|_\infty \leq 1$. Hence, $\|k\|_\infty \leq 1$ and $\|k\|_1 \leq 1$. It is just as trivial to verify that, conversely, $\|k\|_\infty \leq 1$ and $\|k\|_1 \leq 1$ imply that $k \in Y'$ with $\|k\|_{Y'} \leq 1$. $\square$

Proposition 2.a.2 explains the notions appearing in (iii)c of 2.a.1. Condition iii(c) is equivalent to the normalization condition $\|\chi_{[0,1]}\|_X = 1$ (provided, of course, that we assume (i) of 2.a.1). Indeed, since $\|f\|_X \geq \|f^*_{|[0,1]}\|_X$ it follows by the discussion preceding 2.a.2 that $\|\chi_{[0,1]}\|_X = 1$ implies $\|f\|_X \geq \int_0^1 f^*(t)\,dt$. Consequently, we get that also $\|\chi_{[0,1]}\|_{X^*} = 1$ and hence, by duality, it follows that $\|f\|_X \leq \max(\|f\|_1, \|f\|_\infty)$.

The most commonly used r.i. function spaces on $[0, 1]$ and $[0, \infty)$, besides the L_p spaces, $1 \leq p \leq \infty$, are the Orlicz function spaces. Let M be an Orlicz function on $[0, \infty)$ (i.e. a continuous convex increasing function satisfying $M(0) = 0$ and $M(t) \to \infty$ as $t \to \infty$) and let a be either 1 or ∞. The Orlicz space $L_M(0, a)$ is the space of all (equivalence classes of) measurable functions f on $[0, a)$ so that

$$\int_0^a M(|f(t)|/\rho)\,dt < \infty\,,$$

for some $\rho > 0$. The norm in $L_M(0, a)$ is defined by

$$\|f\| = \inf\left\{\rho > 0;\ \int_0^a M(|f(t)|/\rho)\,dt \leq 1\right\}.$$

It is easily checked that $L_M(0, a)$ has the Fatou property and thus if M is normalized so that $M(1) = 1$ then $L_M(0, a)$ is a maximal r.i. function space according to 2.a.1. In the study of Orlicz function spaces, the subspace $H_M(0, a)$ of $L_M(0, a)$, which consists of all $f \in L_M(0, a)$ so that $\int_0^\infty M(|f(t)|/\rho)\,dt < \infty$ for *every* $\rho > 0$, is of particular interest. It is easily verified that $H_M(0, a)$ is the closure in $L_M(0, a)$ of the integrable simple functions and thus it is also an r.i. function space according to 2.a.1 (namely a minimal r.i. space). It is also quite easily checked (in a manner similar to the proof of I.4.a.4) that $L_M(0, 1) = H_M(0, 1)$ if and only if M satisfies the Δ_2-condition at ∞ i.e. $\limsup_{t\to\infty} M(2t)/M(t) < \infty$ and that $L_M(0, \infty) = H_M(0, \infty)$ if and only if M satisfies the Δ_2-condition both at 0 and at ∞.

Another class of r.i. function spaces which has received attention in the literature, especially in connection with interpolation theory, is that of Lorentz function spaces. Let $1 \leq p < \infty$ and let W be a positive non-increasing continuous function on $(0, \infty)$ so that $\lim_{t\to 0} W(t) = \infty$, $\lim_{t\to\infty} W(t) = 0$, $\int_0^1 W(t)\,dt = 1$ and $\int_0^\infty W(t)\,dt = \infty$.

The Lorentz function space $L_{W,p}(0, \infty)$ is the space of all measurable functions f on $[0, \infty)$ for which

$$\|f\|=\left(\int_0^\infty f^*(t)^p W(t)\,dt\right)^{1/p}<\infty .$$

If we impose on W only those conditions which involve the interval $[0, 1]$ and define the norm by integrating over $(0, 1)$ we obtain the Lorentz function space $L_{W,p}(0, 1)$. Obviously, the Lorentz function spaces have the Fatou property. The condition $\int_0^1 W(t)\,dt=1$ is imposed to ensure that the Lorentz spaces satisfy the normalization condition (iii) of 2.a.1. The other three conditions imposed on W are meant to exclude trivial cases.

If X is an r.i. function space the same is true for X'. We have already noted that (i) of 2.a.1 holds in X'. Since X' always has the Fatou property, (ii) holds too. It is evident that the normalization condition (iii) of 2.a.1 is self dual. This remark shows, in particular, that every maximal r.i. space X is of the form $X=Z'$, for some r.i. space Z (take $Z=X'$). We shall show next that, with one exception, such an X is of the form $X=Y^*$ for some r.i. space Y.

Proposition 2.a.3. *Let X be a maximal r.i. function space on $[0, 1]$ which is not order isomorphic to $L_1(0, 1)$. Then $X=Y^*$, where Y is the closed linear span of the simple functions in X'.*

Proof. Since clearly $X=Y'$ it suffices to show that $Y'=Y^*$ or, equivalently, that Y is σ-order continuous (see the discussion on general Köthe spaces preceding 1.b.18).

By its definition, Y is a minimal r.i. function space on $[0, 1]$ and, thus, if it is not σ-order continuous then, by 1.a.7, it is not separable. Hence, as remarked above, Y is, up to an equivalent renorming, equal to $L_\infty(0, 1)$ from which it is immediately deduced that X is order isomorphic to $L_1(0, 1)$. □

Remarks. 1. The unit ball of $L_1(0, 1)$ does not have any extreme point. Hence, by the Krein-Milman theorem, $L_1(0, 1)$ is not isometric to a conjugate space. This same observation concerning extreme points can be used to show that $L_1(0, 1)$ is not even isomorphic to a subspace of a separable conjugate space. (We shall discuss this matter in Vol. IV.)

2. There are maximal r.i. function spaces on $[0, \infty)$ which are not isomorphic to either $L_1(0, \infty)$ or to a conjugate space. Consider, for example, $L_1(0, \infty)+L_p(0, \infty)$ with $1<p<\infty$ (the space is separable and has a subspace isomorphic to $L_1(0, 1)$). The proof of 2.a.3 shows that if X is an r.i. function space on $[0, \infty)$ having the Fatou property then $X=Y^*$ (Y being the minimal r.i. subspace of X') provided that the restriction of X to $[0, 1]$, i.e. the subspace of all $f\in X$, which are supported on $[0, 1]$, is not isomorphic to $L_1(0, 1)$.

The special role of block bases with constant coefficients in the study of spaces with a symmetric basis is played in the theory of r.i. function spaces by σ-algebras $\mathscr{B}$ of Lebesgue measurable subsets of $[0, \infty)$ (or of $[0, 1]$). By the Radon-Nikodym Theorem, for every $f \in L_1(0, \infty)+L_\infty(0, \infty)$ and every σ-algebra $\mathscr{B}$ of measurable subsets of $[0, \infty)$ so that the Lebesgue measure restricted to $\mathscr{B}$ is σ-finite (i.e. so that $\mathscr{B}$ does not have atoms of infinite measure), there exists a unique, up to equality a.e., $\mathscr{B}$-measurable locally integrable function $E^{\mathscr{B}}f$ so that

$$\int_0^\infty gE^{\mathscr{B}}f\,dt = \int_0^\infty gf\,dt\,,$$

for every bounded, integrable and $\mathscr{B}$-measurable function g on $[0, \infty)$. (Recall that a function g is said to be $\mathscr{B}$-measurable if $g^{-1}(G) \in \mathscr{B}$, whenever G is an open subset of the line). In particular, we have that

$$\int_\sigma E^{\mathscr{B}}f\,dt = \int_\sigma f\,dt\,,$$

for every $\mathscr{B}$-measurable set σ with $\mu(\sigma)<\infty$. The function $E^{\mathscr{B}}f$, defined above, is called the *conditional expectation of f with respect to $\mathscr{B}$*. (The term conditional expectation and the notation $E^{\mathscr{B}}f$ are taken from probability theory where the term expectation, denoted by Ef, means the integral of f with respect to the underlying probability space.) The map $E^{\mathscr{B}}$, which is obviously linear, is also called sometimes an *averaging operator*. This term originates in examples of the following type: take as measure space the unit square $[0, 1]\times[0, 1]$ endowed with the usual Lebesgue measure (which is measure theoretically equivalent to $[0, 1]$) and consider the σ-algebra $\mathscr{B}$ of all subsets of $[0, 1]\times[0, 1]$ having the form $\sigma\times[0, 1]$ with σ ranging over the measurable subsets of $[0, 1]$. In this case, for every $f \in L_1([0, 1]\times[0, 1])$,

$$E^{\mathscr{B}}f(s, t) = \int_0^1 f(s, u)\,du\,.$$

The linear map $E^{\mathscr{B}}$ is clearly positive and acts as a projection of norm one in $L_\infty(0, \infty)$ and in $L_1(0, \infty)$. Therefore, $E^{\mathscr{B}}$ is a projection of norm one also in $L_1(0, \infty)+L_\infty(0, \infty)$ and in $L_1(0, \infty)\cap L_\infty(0, \infty)$. The same is true for every r.i. function space X on $[0, \infty)$ or $[0, 1]$.

Theorem 2.a.4. *Let X be an r.i. function space on the interval I, where I is either* $[0, 1]$ *or* $[0, \infty)$. *Then, for every σ-algebra $\mathscr{B}$ of measurable subsets of I so that the Lebesgue measure restricted to $\mathscr{B}$ is σ-finite, the conditional expectation $E^{\mathscr{B}}$ is a projection of norm one from X onto the subspace $X_{\mathscr{B}}$ of X consisting of all the $\mathscr{B}$-measurable functions in it.*

Proof. Assume first that X is *maximal*. We start by proving that if $f \in X$ and $\{\sigma_i\}_{i=1}^n$ are disjoint sets of finite measure in I then the conditional expectation g of f, with respect to the algebra generated by $\{\sigma_i\}_{i=1}^n$ and the points of the complement of $\bigcup_{i=1}^{n} \sigma_i$, belongs to X and satisfies $\|g\|_X \leqslant \|f\|_X$. It is clearly enough to consider the case where $n=1$ and $\sigma_1=[0, a]$. Then g is given by

$$g(t)=a^{-1}\int_0^a f(s)\,ds,\quad 0\leqslant t\leqslant a,\qquad g(t)=f(t),\quad t>a\,.$$

For every $0\leqslant s\leqslant a$ let $f_s(t)=f((t+s) \bmod a)$ if $0\leqslant t\leqslant a$ and $f_s(t)=f(t)$ for $t>a$. Then, since X is r.i., $f_s \in X$ and $\|f_s\|_X=\|f\|_X$ for every $0\leqslant s\leqslant a$. Let h be a simple integrable function on I. Then

$$\int_I g(t)h(t)\,dt=a^{-1}\int_0^a ds\int_I f_s(t)h(t)\,dt\leqslant\|f\|_X\|h\|_{X'},$$

and since $X''=X$ (as a consequence of the maximality of X which is equivalent to the Fatou property), our assertion on g is proved.

Let now $\mathscr{B}$ be a σ-algebra of subsets of I. Clearly, $X_{\mathscr{B}}$ is a Köthe function space on $(I, \mathscr{B}, \mu)$ having the Fatou property. Let $k=\sum_{i=1}^{n} b_i\chi_{\sigma_i}$ be a simple $\mathscr{B}$-measurable integrable function. Since, for every $f\in X$, there is an $\tilde{f}\in X_{\mathscr{B}}$ so that $\|\tilde{f}\|_X\leqslant\|f\|_X$ and $\int_I fk\,dt=\int_I \tilde{f}k\,dt$ (namely the restriction of the g appearing in the beginning of the proof to $\bigcup_{i=1}^{n}\sigma_i$) it follows that $\|k\|_{X'}=\|k\|_{X'_{\mathscr{B}}}$.

Hence, if $f\geqslant 0$ in X and k is as above then

$$\int_I (E^{\mathscr{B}}f)k\,dt=\int_I fk\,dt\leqslant\|f\|_X\|k\|_{X'}=\|f\|_X\|k\|_{X'_{\mathscr{B}}}\,.$$

Since $X_{\mathscr{B}}$ has the Fatou property i.e. $X''_{\mathscr{B}}=X_{\mathscr{B}}$ this proves that $E^{\mathscr{B}}f\in X_{\mathscr{B}}$ and $\|E^{\mathscr{B}}f\|_X\leqslant\|f\|_X$.

Assume now that X is a minimal subspace of X''. By what we have already shown, $E^{\mathscr{B}}$ is an operator of norm one from X into X''. Since $E^{\mathscr{B}}$ maps $L_1(I)\cap L_\infty(I)$ into itself, it maps also its closure in X'', namely X, into itself. □

Theorem 2.a.4 is actually a consequence of a general interpolation theorem (cf. 2.a.10 below). In the proof of this interpolation theorem, as well as in other investigations of r.i. spaces, a certain order-like relation in L_1+L_∞, which was introduced by Hardy, Littlewood and Polya [51], plays an important role. Before defining this relation for functions we consider briefly the simpler case of vectors in R^n which illustrates very well the general case.

Let $x=(a_1, a_2, \ldots, a_n)$ and $y=(b_1, b_2 \ldots, b_n)$ be two elements in R^n. We write $x\prec y$ if $a_1^*+a_2^*+\cdots+a_k^*\leqslant b_1^*+b_2^*+\cdots+b_k^*$ for every $k\leqslant n$, where $(a_1^*, a_2^*, \ldots, a_n^*)$, respectively $(b_1^*, b_2^*, \ldots, b_n^*)$, are the decreasing rearrangements of $(|a_1|, |a_2|, \ldots, |a_n|)$ and $(|b_1|, |b_2|, \ldots, |b_n|)$. As we shall presently see, this order-like relation is closely related to a certain set of matrices. We let $\mathscr{D}_n$ be the set of all $n\times n$ matrices

$(\alpha_{i,j})$ such that $\sum_{i=1}^{n} |\alpha_{i,j}| \leqslant 1$, for every j, and $\sum_{j=1}^{n} |\alpha_{i,j}| \leqslant 1$, for every i. It is easily seen that a matrix belongs to $\mathscr{D}_n$ if and only if the operator which it defines on R^n(with respect to the unit vector basis $\{e_k\}_{k=1}^n$) is of norm at most one in both l_1^n and l_∞^n. (We shall often identify a matrix in $\mathscr{D}_n$ with the corresponding operator T.) We denote by $\mathscr{E}_n$ the subset of $\mathscr{D}_n$ consisting of operators of the form $Te_k = \theta_k e_{\pi(k)}$, $1 \leqslant k \leqslant n$, where $|\theta_k| = 1$, for every k, and π is a permutation of $\{1, 2, \ldots, n\}$.

Proposition 2.a.5 [51]. *For vectors $x, y \in R^n$ we have $x \prec y$ if and only if $x = Ty$ for some $T \in \mathscr{D}_n$.*

Proof. To prove the "if" part assume that $a_i = \sum_{j=1}^{n} \alpha_{i,j} b_j$, $i = 1, \ldots, n$. Then, for every subset σ of $\{1, \ldots, n\}$ of cardinality k, we have

$$\sum_{i \in \sigma} |a_i| \leqslant \sum_{j=1}^{n} \left(\sum_{i \in \sigma} |\alpha_{i,j}| \right) |b_j| \leqslant \sum_{j=1}^{k} b_j^*$$

since $\sum_{i \in \sigma} |\alpha_{i,j}| \leqslant 1$ for every j and $\sum_{j=1}^{n} \left(\sum_{i \in \sigma} |\alpha_{i,j}| \right) \leqslant k$.

To prove the "only if" part, assume that $x = (a_1, a_2, \ldots, a_n) \prec y = (b_1, b_2, \ldots, b_n)$ and $x \notin \operatorname{conv}\{Ty;\ T \in \mathscr{E}_n\}$. Then, by the separation theorem, there are $\{\lambda_k\}_{k=1}^n$ so that $\sum_{k=1}^{n} \lambda_k a_k > 1$ and $\sum_{k=1}^{n} |\lambda_k b_{\pi(k)}| \leqslant 1$ for every permutation π. We may clearly assume that $|\lambda_1| \geqslant |\lambda_2| \geqslant |\lambda_3| \geqslant \cdots$. The following computation leads then to a contradiction.

$$\begin{aligned} 1 \geqslant \sum_{k=1}^{n} |\lambda_k| b_k^* &= (|\lambda_1| - |\lambda_2|) b_1^* + (|\lambda_2| - |\lambda_3|)(b_1^* + b_2^*) + \cdots \\ &\quad + |\lambda_n|(b_1^* + b_2^* + \cdots + b_n^*) \geqslant (|\lambda_1| - |\lambda_2|)|a_1| + (|\lambda_2| - |\lambda_1|)(|a_1| + |a_2|) + \\ &\quad \cdots + |\lambda_n|(|a_1| + |a_2| + \cdots + |a_n|) = \sum_{k=1}^{n} |\lambda_k||a_k| > 1 . \quad \square \end{aligned}$$

Remarks. 1. The proof shows that, for every y and every $S \in \mathscr{D}_n$, we have $Sy \in \operatorname{conv}\{Ty;\ T \in \mathscr{E}_n\}$. If $\|\cdot\|$ is any norm in R^n, with respect to which, the unit vectors have symmetric constant one then, since every $T \in \mathscr{E}_n$ is an isometry of $(R^n, \|\cdot\|)$, it follows that every $S \in \mathscr{D}_n$ is a contraction in this space. This is the essential content of the interpolation theorem 2.a.10 below (in the case of linear operators). We also get that $\|x\| \leqslant \|y\|$ whenever $x \prec y$ (again, this is the main point in 2.a.8 below).

2. From the remark above it follows also that, for every $y \in R^n$, the extreme points of the convex set $\{Sy;\ S \in \mathscr{D}_n\}$ are of the form Ty with $T \in \mathscr{E}_n$. A well known result, which essentially goes back to Birkhoff [11], states that $\mathscr{E}_n$ is precisely the set of extreme points of $\mathscr{D}_n$. Since we shall not need this somewhat stronger result we omit its proof.

3. If x and y are positive (i.e. $a_k \geq 0$, $b_k \geq 0$ for all k) and $x \prec y$ then there is a positive operator $T \in \mathscr{D}_n$ so that $x = Ty$. In order to see this, we have just to replace in the proof above the set $\mathscr{E}_n$ by that of all the operators of the form $Te_k = \theta_k e_{\pi(k)}$ with $\theta_k = 0, 1$ for all k.

We pass now to function spaces.

Definition 2.a.6. Let $f, g \in L_1(0, \infty) + L_\infty(0, \infty)$ (respectively $L_1(0, 1)$). We write $f \prec g$ if, for every $0 < s < \infty$ (respectively $0 < s \leq 1$),

$$\int_0^s f^*(t)\, dt \leq \int_0^s g^*(t)\, dt .$$

Whenever we use in the sequel the relation $\prec$ between functions we shall assume implicitly that they belong to the function spaces appearing in 2.a.6. Clearly, $f \prec g$ is equivalent to $|f| \prec |g|$, to $f^* \prec g^*$ and to $\lambda f \prec \lambda g$, for every real $\lambda \neq 0$. Also $f \prec g$ and $g \prec h$ imply $f \prec h$. The relations $f \prec g$ and $g \prec f$ hold if and only if $f^* = g^*$. For every two functions f_1 and f_2 we have $(f_1 + f_2)^* \prec f_1^* + f_2^*$. The relation $\prec$ has the following decomposition property (cf. [83]).

Proposition 2.a.7. *Assume that $g \prec f_1 + f_2$ with g, f_1 and f_2 non-negative. Then there exist non-negative g_1 and g_2 with $g_1 + g_2 = g$ and $g_i \prec f_i$, $i = 1, 2$.*

Proof. Consider first the case of vectors in R^n. If $x \prec y_1 + y_2$ with non-negative vectors then, by remark 3 following 2.a.5, there is a positive $T \in \mathscr{D}_n$ so that $x = T(y_1 + y_2)$. Put $x_i = Ty_i$, $i = 1, 2$. Then $x = x_1 + x_2$, $x_i \prec y_i$ and the x_i are non-negative, $i = 1, 2$.

We consider now the case of functions. There is no loss of generality to assume that $f_i = f_i^*$, $i = 1, 2$. Let n and $k = k(n)$ be integers and let $g^{(n)}$ be a positive function so that $g^{(n)} = \sum_{j=1}^{n} a_{j,n} \chi_{\sigma_{j,n}} \leq g$, where $\{\sigma_{j,n}\}_{j=1}^n$ are disjoint sets with $\mu(\sigma_{j,n}) = k^{-1}$, for every j, and so that $g^{(n)}(t) \uparrow g(t)$ a.e. as $n \to \infty$. Let $f_i^{(n)}$, $i = 1, 2$, be defined by $f_i^{(n)}(t) = k \int_{(j-1)/k}^{j/k} f_i(s)\, ds$ if $t \in [(j-1)/k, j/k)$, $j = 1, 2, \ldots, n$ and $f_i^{(n)}(t) = 0$ for $t \geq n/k$. Then $g^{(n)} \prec f_1^{(n)} + f_2^{(n)}$ and it follows readily from the case of vectors in R^n that $g^{(n)} = g_1^{(n)} + g_2^{(n)}$ with $g_i^{(n)} \geq 0$ and $g_i^{(n)} \prec f_i^{(n)} \prec f_i$, $i = 1, 2$. Since the functions $g_i^{(n)}$ are bounded by g there is a subsequence $\{n_j\}_{j=1}^\infty$ of the integers so that, for $i = 1, 2$, $\{g_i^{(n_j)}\}_{j=1}^\infty$ converge in the w topology of $L_1(\eta)$ for every finite interval η of $[0, \infty)$ to limits g_i, $i = 1, 2$. It is easily verified that these g_i have the desired property. □

We exhibit next the connection between the relation $\prec$ and r.i. function spaces.

Proposition 2.a.8. *Let X be an r.i. function space on I which is either $[0, 1]$ or $[0, \infty)$. Assume that $g \prec f$ and $f \in X$. Then $g \in X$ and $\|g\| \leq \|f\|$.*

Proof. Suppose first that X is maximal and let h be a simple integrable function. Then $h^* = \sum_{i=1}^{n} a_i \chi_{[0,t_i)}$ for suitable $a_i \geq 0$, $i=1, 2, \ldots, n$ and $0 < t_1 < t_2 < \cdots < t_n$. We have

$$\begin{aligned} \left|\int_I gh\, ds\right| &\leq \int_I g^*h^*\, ds = \sum_{i=1}^{n} a_i \int_0^{t_i} g^*(s)\, ds \leq \sum_{i=1}^{n} a_i \int_0^{t_i} f^*(s)\, ds \\ &= \int_I f^*h^*\, ds \leq \|f^*\|_X \|h^*\|_{X'} = \|f\|_X \|h\|_{X'}\,. \end{aligned}$$

This proves that $g \in X'' = X$ and $\|g\|_X \leq \|f\|_X$.

Assume now that X is a minimal subspace of X'' and let $\varepsilon > 0$. We may clearly assume that f and g are non-negative and since $L_1 \cap L_\infty$ is dense in X, we can write f as $f_1 + f_2$ with $f_1 \geq 0$, $f_2 \geq 0$, $f_1 \in L_1 \cap L_\infty$ and $\|f_2\|_X \leq \varepsilon$. By 2.a.7, $g = g_1 + g_2$ with $g_i \prec f_i$ for $i=1, 2$. It follows from the first part of the proof that $g_1, g_2 \in X''$, $\|g_1\|_{X''} \leq \|f_1\|_X \leq \|f\|_X$ and $\|g_2\|_{X''} \leq \varepsilon$. We also get from the first part of the proof that $g_1 \in L_1 \cap L_\infty$ and thus $g_1 \in X$ i.e. $d(g, X) \leq \varepsilon$. Since ε was arbitrary this concludes the proof. □

Before stating the interpolation theorem we need one more concept.

Definition 2.a.9. A mapping T from a Banach space X into a Banach lattice Y is said to be *quasilinear* if

(i) $|T(\alpha x)| = |\alpha| |Tx|$, $x \in X$, α scalar.
(ii) There is a constant $C < \infty$ so that

$$|T(x_1 + x_2)| \leq C(|Tx_1| + |Tx_2|), \quad x_1, x_2 \in X\,.$$

A quasilinear operator is said to be bounded if $\|T\| = \sup \{\|Tx\|; \|x\| \leq 1\} < \infty$.

It is clear that every linear operator is quasilinear. There are several important examples of non-linear quasilinear operators. The most commonly used example of this type is the so-called "square function", which is introduced in the following situation. Let $\{x_i\}$ be an unconditional basic sequence of finite or infinite length in a q-concave Banach lattice X for some $q < \infty$. Then, for every $x = \sum_i a_i x_i \in [x_i]$, the square function

$$Sx = \left(\sum_i |a_i x_i|^2\right)^{1/2}$$

is well defined (since X is q-concave it is a Köthe function space and thus $\left(\sum_i |a_i x_i|^2\right)^{1/2}$ is a well defined function even if the summation is infinite). Proposition 1.d.6 ensures that this function belongs to X and its norm is actually equivalent to $\|x\|$. It is obvious that S is quasilinear (with $C=1$) but not linear.

The following interpolation theorem is due to Calderon [22] (cf. also Mitjagin [101]).

Theorem 2.a.10. *Let X be an r.i. function space on I which is either* $[0, 1]$ *or* $[0, \infty)$. *Let T be a quasilinear operator defined on $L_\infty(I)+L_1(I)$ which is bounded on both $L_\infty(I)$ and $L_1(I)$. Then T maps X into X and*

$$\|T\|_X \leqslant C \max(\|T\|_1, \|T\|_\infty),$$

where C is the constant appearing in 2.a.9(ii).

Proof. Let $f \in X$, let $s \in$ I, put

$$g_s(t)=\begin{cases} f(t)-f^*(s) & \text{if } f(t)>f^*(s) \\ f(t)+f^*(s) & \text{if } f(t)<-f^*(s) \\ 0 & \text{if } |f(t)|\leqslant f^*(s) \end{cases}$$

and $h_s(t)=f(t)-g_s(t)$. Clearly, $\|h_s\|_\infty=f^*(s)$ and

$$\|g_s\|_1=\int_0^s f^*(t)\,dt-sf^*(s).$$

Since $|Tf|\leqslant C(|Tg_s|+|Th_s|)$ it follows that

$$\int_0^s (Tf)^*(t)\,dt\leqslant C\left(\int_0^s (Tg_s)^*(t)\,dt+\int_0^s \|Th_s\|_\infty\,dt\right)$$

$$\leqslant C(\|Tg_s\|_1+s\|Th_s\|_\infty)\leqslant C\max(\|T\|_1, \|T\|_\infty)\int_0^s f^*(t)\,dt.$$

Consequently, $Tf \prec C\max(\|T\|_1, \|T\|_\infty) f$ and the desired result follows by applying 2.a.8. □

Remarks. 1. The assumption in 2.a.10 that X is an r.i. function space is also necessary as far as the main requirement in 2.a.1 (i.e. (i) there) is concerned. Indeed, if τ is an automorphism of I into itself then $U_\tau f(t)=f(\tau^{-1}(t))$ is a linear operator of norm one in $L_1(I)$ and $L_\infty(I)$. The conclusion of 2.a.10 thus asserts, in particular, that U_τ has norm one on X.

2. Theorem 2.a.10 holds also, and with the same proof, if X is an r.i. space on the integers. For spaces X with a symmetric basis, 2.a.10 takes the following form. Let T be a quasilinear operator on c_0 which is bounded in c_0 and l_1. Let X be a Banach sequence space in which the unit vectors form a basis whose symmetric constant is M. Then T maps X into itself with $\|T\|_X\leqslant CM\max(\|T\|_{l_1}, \|T\|_{c_0})$.

3. It is instructive to note that, for a linear T and a separable r.i. space X, 2.a.10 can be easily deduced from 2.a.5. To fix ideas we assume that X is an r.i. function space on $[0, 1]$. Let T be an operator on $L_1(0, 1)$ so that $\|T\|_\infty$ and $\|T\|_1$

are $\leqslant 1$. Let h be a simple function on $[0, 1]$ of the form $\sum_{i=1}^{n} a_i \chi_{\sigma_i}$ with $\mu(\sigma_i)=n^{-1}$, for every i. Then $Th \in L_\infty(0, 1)$ and as such can be approximated in the L_∞ norm (and therefore in X) by simple functions. Since X is separable $\|\chi_{[0,t]}\|_X \to 0$ as $t \to 0$ which implies that there is, for every $\varepsilon > 0$, a finite algebra $\mathscr{B}$ of subsets of $[0, 1]$ whose atoms have all the same measure so that $h \in E^{\mathscr{B}}X$ and $\|Th\text{-}g\| \leqslant \varepsilon$ for some $g \in E^{\mathscr{B}}X$. The space $E^{\mathscr{B}}X$ has a basis with symmetric constant one and hence, by remark 1 following 2.a.5, $\|E^{\mathscr{B}}TE^{\mathscr{B}}\|_X \leqslant 1$. Therefore,

$$\|Th\| \leqslant \|E^{\mathscr{B}}Th\| + \|E^{\mathscr{B}}Th\text{-}g\| + \|g\text{-}Th\| \leqslant \|h\| + 2\varepsilon\,.$$

Since ε is arbitrary and the simple functions are dense in X it follows that $\|T\|_X \leqslant 1$.

We conclude this section by presenting an example, due to Russu [117], of a Köthe sequence space (i.e. a Köthe space on $\{1, 2, \ldots\}$) X_1 so that X_1' is a norming subspace of X_1^* and any permutation of the integers induces an isometry on X_1, but on which the conditional expectation operator fails to be defined for a suitable σ-finite algebra of sets. We present this example on $\{1, 2, \ldots\}$ since in this form it is most transparent. With only trivial notational changes this example can be presented on $[0, \infty)$ and with some more changes also on $[0, 1]$. This example shows, of course, the role played by the requirement 2.a.1(ii) that X be either a maximal or a minimal subspace of X''. Without this assumption, 2.a.4, and thus also 2.a.8 and 2.a.10, may fail to hold.

Example 2.a.11. *Let X be the Banach space of all sequences $x=(a_1, a_2, \ldots)$ so that*

$$\|x\| = \sup_k \left\{ \sum_{j=1}^{k} a_j^* \Big/ \left(\sum_{j=1}^{k} j^{-1} \right) \right\} < \infty\,.$$

There is a closed ideal X_1 in X which is invariant under permutations (i.e. $(a_1, a_2, \ldots) \in X_1$ *if and only if* $(a_{\pi(1)}, a_{\pi(2)}, \ldots) \in X_1$, *for every permutation π of the integers) but on which the conditional expectation operator cannot be defined for a suitable σ-finite algebra of sets. More precisely, there is an $x_0=(b_1, b_2, \ldots) \in X_1$ and a partition of the integers into a sequence of pairwise disjoint finite sets $\{\sigma_n\}_{n=1}^{\infty}$ so that if $c_j = \sum_{i \in \sigma_n} b_i/\bar{\bar{\sigma}}_n$, $j \in \sigma_n$, $n=1, 2, \ldots$ then $(c_1, c_2, \ldots) \notin X_1$.*

Proof. It is evident that X has the Fatou property and is an r.i. space on the integers. We take $x_0=(1, 2^{-1}, \ldots, j^{-1}, \ldots)$ and let X_1 be the smallest (closed) ideal in X which is invariant under permutations and contains x_0. It is evident that X_1 is the norm closure of the linear space of sequences $(a_1, a_2, \ldots)$ for which $\sup_j ja_j^* < \infty$. Let $n_k = 2^{k^2}$ and put

$$c_j = \sum_{i=n_k+1}^{n_{k+1}} i^{-1}/(n_{k+1}-n_k), \quad n_k < j \leqslant n_{k+1},\ k=1, 2, \ldots\,.$$

We claim that $y=(c_1, c_2, \ldots) \notin X_1$ (clearly, $y \in X$ since X is an r.i. space). We note first that, for $n_k < j \leqslant n_{k+1}$, c_j is of the order of magnitude of $(\log n_{k+1} - \log n_k)/n_{k+1}$ i.e. of k/n_{k+1}. In other words, there is an $\alpha > 0$ so that $c_j > \alpha k/n_{k+1}$, $n_k < j \leqslant n_{k+1}$, $k=1, 2, \ldots$. Let $x=(a_1, a_2, \ldots)$ be such that $ja_j^* \leqslant K$, for some K and all $j=1, 2, \ldots$. We have to show that $\|y-x\|$ is bounded from below by a constant independent of x (and thus also of K). Let π be a permutation of the integers so that $|a_j| = a^*_{\pi(j)}$, $j=1, 2, \ldots$. Let

$$\eta_k = \{j;\ n_k < j \leqslant n_{k+1},\ \pi(j) \geqslant 2Kn_{k+1}/\alpha k\}, \quad k=1, 2, \ldots.$$

It is evident that, for $k \geqslant k_0 = k_0(K)$, the cardinality $\bar{\bar{\eta}}_k$ of η_k is larger than $n_{k+1}/2$. For $j \in \eta_k$, $k=1, 2, \ldots$

$$|c_j - a_j| \geqslant \alpha k/n_{k+1} - K/\pi(j) \geqslant \alpha k/n_{k+1} - \alpha k/2n_{k+1} = \alpha k/2n_{k+1}.$$

Hence,

$$\begin{aligned}
\|y-x\| &\geqslant \sup_m \left(\sum_{j=1}^{n_{m+1}} |c_j - a_j| \Big/ \left(\sum_{j=1}^{n_{m+1}} j^{-1} \right) \right) \\
&\geqslant \sup_m \left(\sum_{k=1}^{m} (\bar{\bar{\eta}}_k \cdot \alpha k/2n_{k+1}) \Big/ \left(\sum_{j=1}^{n_{m+1}} j^{-1} \right) \right) \\
&\geqslant \sup_m \left(\sum_{k=k_0}^{m} \alpha k/4 \right) \Big/ \left(\sum_{j=1}^{n_{m+1}} j^{-1} \right) \\
&\geqslant \sup_m \left(\alpha(m(m+1) - k_0(k_0+1))/8(1+(m+1)^2 \log 2) \right) \\
&\geqslant \alpha/8 \log 2. \quad \square
\end{aligned}$$

Remarks. 1. Note that the space X of 2.a.11 is the dual of the Lorentz sequence space $d(1, w)$, where $w=(1, 1/2, \ldots, 1/j, \ldots)$.

2. Calderon [22] proved a function space analogue of 2.a.5 and showed thereby that Theorem 2.a.10 holds for a Köthe function space X on I if and only if X satisfies 2.a.1(i) and has the property that, whenever $g \prec f$ with $f \in X$, then also $g \in X$.

b. The Boyd Indices

In the previous section we proved that every operator, which is bounded on $L_1(I)$ and $L_\infty(I)$, acts also as a bounded operator on every r.i. function space on I. However, many of the interesting operators in analysis are not bounded simultaneously in both of these spaces, but only on suitable $L_p(I)$ spaces with $1 < p < \infty$. In this section we study r.i. function spaces X on I which are "between" $L_{p_1}(I)$ and $L_{p_2}(I)$ in the sense that every operator, which is defined and bounded on these two spaces, is defined and bounded also on X. This is done by assigning to each r.i. function space two indices, called the Boyd indices. The definition of these indices resembles formally the notions of upper and lower p-estimates which were studied in section 1.f. However, in spite of the formal resemblance, the Boyd indices do not coincide with the notions studied in 1.f. After investigating some simple properties of the indices and considering some examples we show that these indices really enable us to prove an interpolation theorem in the setting of r.i. function spaces for operators bounded in $L_{p_1}(I)$ and $L_{p_2}(I)$. Actually, we prove an interpolation theorem for operators which are only of weak type (p_1, p_1) and weak type (p_2, p_2) (for the definition of weak type, see 2.b.10). We thus obtain a version of the classical Marcinkiewicz interpolation theorem for r.i. function spaces. Several applications of this theorem will be presented in the following sections.

We start by defining, for every $0<s<\infty$, a linear operator D_s. If $I=[0,\infty)$ we put, for a measurable function f on I,

$$(D_s f)(t)=f(t/s), \quad 0<s<\infty, 0\leqslant t<\infty .$$

If $I=[0,1]$ we put, for a measurable f on I and $0<s<\infty$,

$$(D_s f)(t)=\begin{cases} f(t/s), & t\leqslant \min(1,s) \\ 0, & s<t\leqslant 1 \text{ (in case } s<1). \end{cases}$$

Geometrically, in the case of $[0,\infty)$ the operator D_s dilates the graph of $f(t)$ by the ratio $s:1$ in the direction of the t axis. In the case $I=[0,1]$ we have the additional effect of restricting everything to I. It is obvious that D_s acts as a linear operator of norm one on $L_\infty(I)$ and of norm s on $L_1(I)$; hence, by 2.a.10, D_s is bounded on every r.i. function space X and $\|D_s\|_X\leqslant \max(1,s)$. Clearly, $(D_s f)^*\leqslant D_s f^*$ for every f and s and hence $\|D_s\|$ on an r.i. function space X can be computed by considering only non-increasing functions f. Since, for every non-increasing $f\geqslant 0$ and every $0<r<s<\infty$, we have $D_r f\leqslant D_s f$ it is clear that $\|D_s\|$ is a non-decreasing function of s. Also note that, for every r and s, $D_r D_s=D_{rs}$, with the only exception being the case $r<1<s$ and $I=[0,1]$ in which we have $D_r D_s f=\chi_{[0,r]} D_{rs} f$. In any case we have

$$\|D_{rs}\|\leqslant\|D_r\|\,\|D_s\|$$

(if $r<1<s$ and $I=[0,1]$ simply use $D_s D_r$ instead of $D_r D_s$ in order to verify this inequality). We are now ready to define the indices (cf. [16]) of an r.i. function space.

Definition 2.b.1. Let X be an r.i. function space on an interval I which is either $[0,1]$ or $[0,\infty)$. The *Boyd indices* p_X and q_X are defined by

$$p_X=\lim_{s\to\infty}\frac{\log s}{\log\|D_s\|}=\sup_{s>1}\frac{\log s}{\log\|D_s\|}$$

$$q_X=\lim_{s\to 0^+}\frac{\log s}{\log\|D_s\|}=\inf_{0<s<1}\frac{\log s}{\log\|D_s\|}$$

The expression $\|D_s\|$ appearing above is, of course, the norm of D_s acting as an operator in X. If $\|D_s\|=1$, for some (and hence all) $s>1$, we put $p_X=\infty$. Similarly, if $\|D_s\|=1$, for all $s<1$, we put $q_X=\infty$. We have to verify that the limits in 2.b.1 exist and are equal to the respective supremum or infimum. Let $\varphi(s)=\log s/\log\|D_s\|$ and let $s, r\geqslant 1$ with $s^n\leqslant r<s^{n+1}$, for some n. Then, since $\|D_{s^{n+1}}\|\leqslant\|D_s\|^{n+1}$,

$$\varphi(r)\geqslant(\log s^n)/\log\|D_{s^{n+1}}\|\geqslant n\varphi(s)/(n+1).$$

This easily implies our assertion concerning p_X. The proof of the assertion concerning q_X is the same.

The indices p_X and q_X can be computed explicitly for many examples of concrete r.i. function spaces. We shall carry out this computation in the case of Orlicz function spaces in 2.b.5 below. Here we mention only the trivial, but important fact that if $X=L_p(I)$, $1\leqslant p\leqslant\infty$, then $p_X=q_X=p$. This fact influenced our decision to define the indices as in 2.b.1. In Boyd's paper and in several other places in the literature the indices of X are taken to be the reciprocals of the ones we use here (i.e. $\alpha_X=1/p_X$ and $\beta_X=1/q_X$).

Proposition 2.b.2. *Let X be an r.i. function space. Then*

(i) $1\leqslant p_X\leqslant q_X\leqslant\infty$

(ii) $1/p_X+1/q_{X'}=1,\qquad 1/q_X+1/p_{X'}=1.$

Proof. That $p_X\geqslant 1$ follows from $\|D_s\|\leqslant s$ while the assertion that $p_X\leqslant q_X$ can be easily deduced from $\|D_s\|\|D_{s^{-1}}\|\geqslant\|D_{ss^{-1}}\|=1$. This proves (i). To prove (ii), let $f\in X$ and $g\in X'$, pick $s<1$ and assume that $I=[0,1]$. Then

$$g(D_sf)=\int_0^s f(t/s)g(t)\,dt=s\int_0^1 f(u)g(su)\,du=s(D_{s^{-1}}g)(f).$$

By taking suprema over all f and g in the respective unit balls, we get that $\|D_s\|_X=s\|D_{s^{-1}}\|_{X'}$. This proves that $1/q_X+1/p_{X'}=1$. The proof of the other assertion in (ii) as well as the proof in case $I=[0,\infty)$ are the same. □

We considered so far only r.i. function spaces. The Boyd indices can be defined also for r.i. spaces on the integers. In this case the operators D_s are defined only if s is an integer or the reciprocal of an integer. If $f=(a_1, a_2, a_3, \ldots)$ and $n=1, 2, \ldots$ we put

$$D_nf=(\overbrace{a_1, a_1, \ldots, a_1}^{n}, \overbrace{a_2, \ldots, a_2}^{n}, a_3, \ldots),$$

$$D_{1/n}f=n^{-1}\left(\sum_{i=1}^{n} a_i, \sum_{i=n+1}^{2n} a_i, \ldots\right).$$

The indices p_X and q_X are defined as in 2.b.1 by taking the limits only over $s=n$

(respectively, $s=1/n$), $n=1, 2, \ldots$. The results proved in this section are all valid also for r.i. space on the integers. We shall however not give the proofs in this case since, while they are essentially the same as those for function spaces, they do often require a somewhat different notation.

It is worthwhile to note that if $I=[0, \infty)$ and f is any measurable function then $D_n f$ can be written as $f_1+f_2+\cdots+f_n$, where the f_i are mutually disjoint and each f_i has the same distribution function as f. The same is true if $I=[0, 1]$ and f is supported on $[0, 1/n]$. Hence, p_X is the supremum of all the numbers p which have the following property: there exists a number K so that, for every choice of an integer n and of a function f having norm one (supported on $[0, 1/n]$ if $I=[0, 1]$), we have

$$\|f_1+f_2+\cdots+f_n\| \leqslant Kn^{1/p},$$

where the $\{f_i\}_{i=1}^n$ are disjointly supported and have the same distribution function as f. Similarly, q_X is the infimum of all the numbers q for which there is a K so that, for every n and $\{f_i\}_{i=1}^n$ as above,

$$\|f_1+f_2+\cdots+f_n\| \geqslant K^{-1}n^{1/q}.$$

To justify the first assertion for $I=[0, 1]$ we have to note that, since $D_n f = D_n(\chi_{[0, 1/n]} f)$, the norm of D_n can be computed by considering only functions supported on $[0, 1/n)$.

It follows from this observation that if an r.i. function space X satisfies an upper p-estimate (cf. 1.f.4) then $p \leqslant p_X$ and if it satisfies a lower q-estimate then $q_X \leqslant q$. In general, p_X (respectively, q_X) is strictly larger (respectively, smaller) than sup $\{p;\ X$ satisfies an upper p-estimate$\}$ (respectively, inf $\{q;\ X$ satisfies a lower q-estimate$\}$). For example, consider the Lorentz function space $X=L_{W,1}(0, \infty)$, where $W(t)=1/2\sqrt{t}$. For every non-increasing f in $L_{W,1}(0, \infty)$ and every $0<s<\infty$ we clearly have

$$\int_0^\infty (f(t/s)/2\sqrt{t})\, dt = \sqrt{s} \int_0^\infty (f(t)/2\sqrt{t})\, dt$$

and thus $\|D_s\|=\sqrt{s}$, i.e. $p_X=q_X=2$. On the other hand, it is easily seen that, for any sequence $\{f_n\}_{n=1}^\infty$ of elements of norm one in X such that either sup $\{|f_n(t)|;\ 0<t<\infty\}$ or μ(support $|f_n|$) tend to zero, there is a subsequence which is equivalent to the unit vector basis of l_1. Hence, X does not satisfy an upper p-estimate for any $p>1$. Note that X^* is non-separable. Hence, there are non-reflexive and even non-separable r.i. function spaces with $1<p_X \leqslant q_X<\infty$.

Any r.i. function space X satisfies $L_1(I) \cap L_\infty(I) \subset X \subset L_1(I)+L_\infty(I)$. If we have information on the indices of X then a stronger assertion is valid.

Proposition 2.b.3. *Let X be an r.i. function space on an interval I which is either $[0, 1]$ or $[0, \infty)$. Then, for every $1 \leqslant p<p_X$ and $q_X<q \leqslant \infty$, we have*

$$L_p(I) \cap L_q(I) \subset X \subset L_p(I)+L_q(I),$$

with the inclusion maps being continuous.

The spaces $L_p(I)\cap L_q(I)$ and $L_p(I)+L_q(I)$ are defined in analogy to the case $p=1$, $q=\infty$ treated in 2.a.2. Clearly, $(L_p(I)\cap L_q(I))^*=L_{p'}(I)+L_{q'}(I)$, where $1/p+1/p'=1$, $1/q+1/q'=1$, and if $I=[0,1]$ then $L_p(I)\cap L_q(I)=L_q(I)$ and $L_p(I)+L_q(I)=L_p(I)$. We also note that in case $p_X=1$ we can take $p=1$ and, similarly, if $q_X=\infty$ we can take $q=\infty$ in 2.b.3.

Proof. It suffices to prove that $L_p(I)\cap L_q(I)\subset X$ with a continuous inclusion map; the second assertion will then follow by duality. Let $p<p_0<p_X$ and $q_X<q_0<q$. Then there is a constant K so that $\|D_s\|\leqslant Ks^{1/p_0}$ for $s\geqslant 1$ and $\|D_s\|\leqslant Ks^{1/q_0}$ for $0\leqslant s\leqslant 1$. Since $D_s\chi_{[0,1]}=\chi_{[0,s]}$ and $\|\chi_{[0,1]}\|_X=1$ we deduce that for $\sigma\subset I$

$$\|\chi_\sigma\|_X\leqslant K(\mu(\sigma))^{1/p_0} \quad \text{if } \mu(\sigma)\geqslant 1, \qquad \|\chi_\sigma\|_X\leqslant K(\mu(\sigma))^{1/q_0} \quad \text{if } \mu(\sigma)\leqslant 1 .$$

(The first of these inequalities makes sense only if $I=[0,\infty)$.)

Let now g be a non-negative simple function on I so that $\|g\|_p, \|g\|_q\leqslant 1$. Choose a simple function $\tilde{g}$ on I with $g/2\leqslant\tilde{g}\leqslant g$ so that $\tilde{g}=\sum_{k=-n}^{n} 2^k\chi_{\sigma_k}$ for some integer n and mutually disjoint sets $\{\sigma_k\}_{k=-n}^{n}$. Since

$$\mu(\sigma_k)\leqslant\min(2^{-kp},2^{-kq}), \qquad -n\leqslant k\leqslant n$$

we have that

$$\|\tilde{g}\|_X\leqslant K\left(\sum_{k=-n}^{-1} 2^k 2^{-kp/p_0}+\sum_{k=0}^{n} 2^k 2^{-kq/q_0}\right),$$

and therefore

$$\|g\|_X\leqslant 2\|\tilde{g}\|_X\leqslant 2K\sum_{k=0}^{\infty}\left(2^{-k(1-p/p_0)}+2^{-k(q/q_0-1)}\right). \ \square$$

Remarks 1. Unless $p_X=1$ (or $q_X=\infty$), we cannot take, in general, in 2.b.3 $p=p_X$ (or $q=q_X$). For example, if X is the Lorentz space $L_{W,1}(0,\infty)$ with $W(t)=1/2\sqrt{t}$ then, clearly, 2.b.3 does not hold with $p=q=2$.

2. In connection with 2.b.3 and remark 1 above we should note that the following is true. *Assume that X is an r.i. function space on* $[0,1]$ *which is p-convex and q-concave for some* $1\leqslant p\leqslant q\leqslant\infty$. *Then for every* $f\in X$

$$\|f\|_p/M^{(p)}(X)\leqslant\|f\|_X\leqslant M_{(q)}(X)\|f\|_q,$$

where $M^{(p)}(X)$, respectively, $M_{(q)}(X)$ is the p-convexity, respectively, the q-concavity constant of X. We shall prove the assertion concerning q (the assertion concerning p will follow by duality). If $q=\infty$ there is nothing to prove. If $q<\infty$ then X must be a minimal r.i. function space and thus it suffices to consider simple functions. Assume that $f=\sum_{i=1}^{n} a_i e_i$, where $e_i=\chi_{[(i-1)/n,\,i/n)}$, $1\leqslant i\leqslant n$, and let $\prod$ be the set of cyclic permutations of $\{1, 2,\ldots, n\}$. Then

$$\|f\|=\Big(\sum_{\pi\in\Pi}\Big\|\sum_{i=1}^{n}a_{\pi(i)}e_i\Big\|^q/n\Big)^{1/q}\leqslant M_{(q)}(X)\Big\|\Big(\sum_{\pi\in\Pi}\Big|\sum_{i=1}^{n}a_{\pi(i)}e_i\Big|^q/n\Big)^{1/q}\Big\|$$

$$=M_{(q)}(X)\Big\|\Big(\sum_{\pi\in\Pi}\sum_{i=1}^{n}|a_{\pi(i)}|^q e_i/n\Big)^{1/q}\Big\|=M_{(q)}(X)\Big(\sum_{i=1}^{n}|a_i|^q\Big)^{1/q}\Big\|\Big(\sum_{i=1}^{n}e_i/n\Big)^{1/q}\Big\|$$

$$=M_{(q)}(X)\Big(\sum_{i=1}^{n}|a_i|^q\Big)^{1/q}n^{-1/q}=M_{(q)}(X)\|f\|_q\,.$$

3. The converse of 2.b.3 is not true in general. If X is an r.i. function space so that

$$L_p(I)\cap L_q(I)\subset X\subset L_p(I)+L_q(I)$$

with $p<q$ then it need not be true that $p\leqslant p_X$ or that $q_X\leqslant q$. Notice that in the proof of 2.b.3 we have just used estimates for $\|\chi_{[0,a]}\|_X$ which follow from but are not equivalent to our assumptions on $\|D_s\|$ (via p_X and q_X).

Proposition 2.b.3 can be used to describe the behaviour of the Rademacher functions in an r.i. function space X on $[0, 1]$. For instance, if $q_X<\infty$ and $q>q_X$ then, by 2.b.3, there exists a constant $K<\infty$ such that $\|f\|_X\leqslant K\|f\|_q$ for all $f\in L_q(0, 1)$. Hence, by Khintchine's inequality I.2.b.3, we have that

$$A_1\Big(\sum_{i=0}^{n}a_i^2\Big)^{1/2}\leqslant\Big\|\sum_{i=0}^{n}a_ir_i\Big\|_1\leqslant\Big\|\sum_{i=0}^{n}a_ir_i\Big\|_X\leqslant K\Big\|\sum_{i=0}^{n}a_ir_i\Big\|_q\leqslant KB_q\Big(\sum_{i=0}^{n}a_i^2\Big)^{1/2},$$

for every choice of $\{a_i\}_{i=0}^n$.

This proves that in any r.i. function space X on $[0, 1]$ with $q_X<\infty$ the Rademacher functions are equivalent to the unit vector basis in l_2. If, in addition, $1<p_X$ or, equivalently, $q_{X'}<\infty$ then the same is valid in X'. This implies that the span of the Rademacher functions in such an r.i. space is also complemented.

The conditions imposed above on the Boyd indices are however far from being necessary. For example, the fact that in a given r.i. function space X on $[0, 1]$ the Rademacher functions are equivalent to the unit vector basis of l_2 does not even imply that $X\supset L_q\,(0, 1)$ for some $q<\infty$. The following is a sharp result in this direction containing the above assertions as a particular case. Part (i) of 2.b.4 was proved by V. A. Rodin and E. M. Semyonov [137].

Theorem 2.b.4. *Let X be an r.i. function space on $[0, 1]$ and let $\|\cdot\|_M$ and $\|\cdot\|_{M^*}$ denote the norms in the Orlicz function spaces $L_M(0, 1)$, respectively $L_{M^*}(0, 1)$, where $M(t)=(e^{t^2}-1)/(e-1)$ and M^* is the function complementary to M (which at ∞ is equivalent to $t(\log t)^{1/2}$).*

(i) *The Rademacher functions $\{r_i\}_{i=0}^{\infty}$ in X are equivalent to the unit vector basis in l_2 if and only if there exists a constant $K_1<\infty$ so that*

$$\|f\|_X\leqslant K_1\|f\|_M\,,$$

for all $f\in L_\infty(0, 1)$.

(ii) *The subspace* $[r_i]_{i=0}^{\infty}$ *is complemented in* X *if and only if there is a constant* $K_2<\infty$ *such that*

$$K_2^{-1}\|f\|_{M^*}\leq\|f\|_X\leq K_2\|f\|_M\,,$$

for all $f\in L_\infty(0,1)$, *in which case it is, of course, isomorphic to* l_2.

In the proof we shall use the well known central limit theorem from probability theory (cf. for example [135]). This theorem states that if $\{g_i\}_{i=1}^{\infty}$ is a sequence of independent and identically distributed random variables on a probability measure space (Ω,Σ,μ) so that $g_1\in L_2(\mu)$ and $\int_\Omega g_1(\omega)\,d\mu=0$ then the sequence

$$h_n=(g_1+g_2+\cdots+g_n)/n^{1/2}\sigma,$$

where $\sigma^2=\int_\Omega g_1^2(\omega)\,d\mu$, tends in distribution to the normal distribution i.e.

$$\lim_{n\to\infty}\mu(\{\omega\in\Omega;\,h_n(\omega)>\tau\})=\frac{1}{\sqrt{2\pi}}\int_\tau^\infty e^{-u^2/2}\,du\,,$$

for every real τ. Here we shall use only the most simple and classical special case of "coin tossing" where g_1 (and thus every g_i) takes only the values ±1, each with probability 1/2.

Proof. (i) Suppose that there exists a $K<\infty$ so that $\|f\|_X\leq K\|f\|_M$ for every $f\in L_\infty(0,1)$ and observe that a computation identical to that presented in the remark following 1.e.15 (based on the expansion of the function e^{t^2} into a power series) proves that $\sum_{i=0}^{\infty}a_ir_i\in L_M(0,1)$ whenever $\{a_i\}_{i=0}^{\infty}\in l_2$. It follows from the closed graph theorem that there is an $A<\infty$ so that

$$KA\Big(\sum_{i=0}^{\infty}a_i^2\Big)^{1/2}\geq K\Big\|\sum_{i=0}^{\infty}a_ir_i\Big\|_M\geq\Big\|\sum_{i=0}^{\infty}a_ir_i\Big\|_X\geq\Big\|\sum_{i=0}^{\infty}a_ir_i\Big\|_1\geq A_1\Big(\sum_{i=0}^{\infty}a_i^2\Big)^{1/2},$$

for every choice of $\{a_i\}_{i=0}^{\infty}$, i.e. that $\{r_i\}_{i=0}^{\infty}$ in X is equivalent to the unit vector basis of l_2.

Suppose now that there exists a constant $B<\infty$ so that

$$\Big\|\sum_{i=0}^{\infty}a_ir_i\Big\|_X\leq B\Big(\sum_{i=0}^{\infty}a_i^2\Big)^{1/2},$$

for any choice of $\{a_i\}_{i=0}^{\infty}$. In order to show that the norm in X is dominated by that in $L_M(0,1)$ it suffices to prove that the function

$$\varphi(t)=1/\|\chi_{[0,t)}\|_M=M^{-1}(1/t)=(\log(1+(e-1)/t))^{1/2}$$

belongs to X'' (the maximal r.i. function space on $[0, 1]$ containing X). Indeed, let f be a non increasing function in $L_\infty(0, 1)$ and notice that in any r.i. function space Y on $[0, 1]$ we have, by 2.a.4, that

$$t=\|\chi_{[0,t)}\|_Y \|\chi_{[0,t)}\|_{Y^*}, \quad 0\leqslant t\leqslant 1 .$$

Hence, in the particular case where $Y=L_M(0, 1)$ we find that

$$f(t)\leqslant t^{-1}\int_0^t f(u)\,du\leqslant t^{-1}\|f\|_M\|\chi_{[0,t]}\|_{M^*}\leqslant\|f\|_M\varphi(t),$$

for all $0<t\leqslant 1$. Therefore, if $\varphi\in X''$ then $\|f\|_X\leqslant K\|f\|_M$, where $K=\|\varphi\|_{X''}$ and this obviously completes the argument.

To prove that $\varphi\in X''$ or, equivalently, that the function $(\log 1/t)^{1/2}$ belongs to X'', we use the central limit theorem which, as pointed out above, asserts that if $\psi_n(t)=\left(\sum_{i=1}^n r_i(t)\right)\Big/\sqrt{n}$ then, for each $0<\tau<\infty$,

$$\lim_{n\to\infty}(\psi_n^*)^{-1}(\tau)=\lim_{n\to\infty}\mu(\{t\in[0, 1];\,|\psi_n(t)|>\tau\})=$$

$$=\frac{2}{\sqrt{2\pi}}\int_\tau^\infty e^{-u^2/2}\,du\leqslant\frac{e^{-t^2}}{\sqrt{2\pi}}\cdot$$

Thus, by passing to the inverse functions, we get that the pointwise limit ψ of $\{\psi_n^*\}_{n=1}^\infty$ satisfies

$$\psi(t)\geqslant(\log 1/t\sqrt{2\pi})^{1/2}, \quad 0<t<1/\sqrt{2\pi}\,.$$

On the other hand, since $\|\psi_n\|_X\leqslant B$ for all n (by our hypothesis) it follows that ψ, and thus also $(\log 1/t)^{1/2}$, belong to X'' (use Remark 3 following 1.b.18).

(ii) If there is a constant $K_2<\infty$ as in the statement of (ii), then in addition to having

$$\|f\|_X\leqslant K_2\|f\|_M, \quad f\in L_\infty(0, 1)\,,$$

we obtain, by duality, that also

$$\|f\|_{X'}\leqslant K_2\|f\|_M, \quad f\in L_\infty(0, 1)\,.$$

Hence, by part (i) of the theorem, the Rademacher functions in both X and X' are equivalent to the unit vector basis in l_2 i.e.

$$\left\|\sum_{i=0}^\infty a_i r_i\right\|_X\leqslant K\left(\sum_{i=0}^\infty a_i^2\right)^{1/2} \quad\text{and}\quad \left\|\sum_{i=0}^\infty a_i r_i\right\|_{X'}\leqslant K\left(\sum_{i=0}^\infty a_i^2\right)^{1/2},$$

for some constant $K<\infty$ and for every choice of $\{a_i\}_{i=0}^\infty$. This implies that the orthogonal projection P defined by

$$Pf=\sum_{i=0}^{\infty}\left(\int_0^1 f(u)r_i(u)\,du\right)r_i$$

is bounded in X. Indeed, if $a_i(f)=\int_0^1 f(u)r_i(u)\,du$ then

$$\sum_{i=0}^{\infty} a_i(f)^2=\int_0^1 f(u)\sum_{i=0}^{\infty} a_i(f)r_i(u)\,du\leqslant\|f\|_X\left\|\sum_{i=0}^{\infty} a_i(f)r_i\right\|_{X'}$$

$$\leqslant K\|f\|_X\left(\sum_{i=0}^{\infty} a_i(f)^2\right)^{1/2}$$

i.e.

$$\left(\sum_{i=0}^{\infty} a_i(f)^2\right)^{1/2}\leqslant K\|f\|_X,\quad f\in X.$$

It follows that

$$\|Pf\|_X\leqslant K\left(\sum_{i=0}^{\infty} a_i(f)^2\right)^{1/2}\leqslant K^2\|f\|_X,$$

for every $f\in X$, and this completes the proof of the assertion.

In order to prove the converse assertion it will suffice to show that if $[r_i]_{i=0}^{\infty}$ is complemented in X then the orthogonal projection P defined above is bounded. Indeed, once this is shown, the fact that

$$\left\|\sum_{i=0}^{\infty} a_ir_i\right\|_X\geqslant A_1\left(\sum_{i=0}^{\infty} a_i^2\right)^{1/2}\quad\text{and}\quad\left\|\sum_{i=0}^{\infty} a_ir_i\right\|_{X'}\geqslant A_1\left(\sum_{i=0}^{\infty} a_i^2\right)^{1/2},$$

for every choice of $\{a_i\}_{i=0}^{\infty}$, implies, by a standard duality argument, that $\{r_i\}_{i=0}^{\infty}$ is equivalent to the unit vector basis in l_2 both in X and X' (or X^*) and the desired result follows from part (i).

Actually, what we need for the duality argument in the previous paragraph is just to know that the projections $P_n=P_{|X_n}$, $n=1, 2, \ldots$, have uniformly bounded norms, where

$$X_n=[\chi_{[(k-1)2^{-n},k2^{-n})}]_{k=1}^{2^n}\subset X,\ n=1, 2, \ldots$$

Fix an integer n. Let Q be a projection from X onto $[r_i]_{i=0}^{\infty}$. Then $Q_n=R_nQ|x_n$ is a projection of norm $\leqslant\|Q\|$ from X_n onto $[r_i]_{i=0}^{n}$, where R_n is the projection of norm one on $[r_i]_{i=0}^{\infty}$ defined by $R_n\sum_{i=0}^{\infty} a_ir_i=\sum_{i=0}^{n} a_ir_i$. Let $\{w_j\}_{j=1}^{2^n}$ be the first 2^n

Walsh functions on [0, 1] introduced in the proof of 1.g.2. These functions form an (orthonormal) basis of X_n. For $1 \leq j, k \leq 2^n$ let $\theta_{j,k}$ be the value taken by w_j on the interval $[(k-1)2^{-n}, k2^{-n})\theta_{j,k} = \pm 1)$ and let T_j be the linear map on X_n defined by

$$T_j w_k = \theta_{j,k} w_k, \quad k = 1, 2, \ldots, 2^n.$$

The proof of the theorem will be concluded once we prove the following two facts:

1. $\|T_j\| = 1, \quad 1 \leq j \leq 2^n$
2. $P_n = 2^{-n} \sum_{j=1}^{2^n} T_j Q_n T_j$.

To prove statement 1, notice that, for every $1 \leq h \leq 2^n$, we have that

$$\chi_{[(h-1)2^{-n}, h2^{-n})} = 2^{-n} \sum_{k=1}^{2^n} \theta_{k,h} w_k.$$

Hence,

$$T_j \chi_{[(h-1)2^{-n}, h2^{-n})} = 2^{-n} \sum_{k=1}^{2^n} \theta_{k,h} \theta_{j,k} w_k.$$

Since, as is easily verified, $\theta_{j,k} = \theta_{k,j}$ for all k and j and since the product of two Walsh functions is again a Walsh function, we deduce that there is an index $1 \leq i \leq 2^n$ so that $\theta_{k,h} \theta_{j,k} = \theta_{k,i}$ for $1 \leq k \leq 2^n$. Consequently,

$$T_j \chi_{[(h-1)2^{-n}, h2^{-n})} = \chi_{[(i-1)2^{-n}, i2^{-n})}$$

and this means that T_j is a map induced by an automorphism of [0, 1]. Since X is an r.i. function space we deduce that T_j is an isometry on X_n thus establishing 1.

To verify 2, denote by $A_n \subset \{1, 2, \ldots, 2^n\}$ the subset of those indices j for which w_j is a Rademacher function (and not a product of two or more distinct Rademacher functions). Let $c_{i,k}$, $i \in A_n$, $1 \leq k \leq 2^n$ be such that

$$Q_n w_k = \sum_{i \in A_n} c_{i,k} w_i.$$

Then

$$2^{-n} \sum_{j=1}^{2^n} T_j Q_n T_j w_k = \sum_{i \in A_n} c_{i,k} \left(2^{-n} \sum_{j=1}^{2^n} \theta_{j,k} \theta_{i,j} \right) w_i.$$

Statement 2 follows now from the orthogonality of the matrix $(\theta_{j,k})_{j,k=1}^{2^n}$. □

Remark. The proof of part (ii) of 2.b.4 actually shows that if $[r_i]_{i=0}^\infty$ is complemented in an r.i. function space X on [0, 1] then the orthogonal projection P from X onto $[r_i]_{i=0}^\infty$ is automatically bounded.

Before continuing the study of general r.i. function spaces we evaluate the Boyd indices in an important special case.

Proposition 2.b.5 [120], [17]. *Let $X=L_M(0,1)$ be an Orlicz function space. Then*

$$p_X=\sup\{p;\ \inf_{\lambda,t\geq 1} M(\lambda t)/M(\lambda)t^p>0\}$$

$$=\sup\{p;\ X \text{ satisfies an upper } p\text{-estimate}\}\,,$$

$$q_X=\inf\{q;\ \sup_{\lambda,t\geq 1} M(\lambda t)/M(\lambda)t^q<\infty\}$$

$$=\inf\{q;\ X \text{ satisfies a lower } q\text{-estimate}\}\,.$$

Proof. Put

$$\alpha_{M,\infty}=\sup\{p;\ \inf_{\lambda,t\geq 1} M(\lambda t)/M(\lambda)t^p>0\}$$

and

$$\beta_{M,\infty}=\inf\{p;\ \sup_{\lambda,t\geq 1} M(\lambda t)/M(\lambda)t^p<\infty\}\,.$$

We shall prove now the assertion concerning p_X. The assertion on q_X will then follow by duality. Let $p<\alpha_{M,\infty}$. Then there is a $\gamma>0$ so that $M(\lambda t)\geq\gamma M(\lambda)t^p$, whenever $t,\lambda\geq 1$. By replacing γ by a possibly smaller constant we assume, as we clearly may, that this inequality holds for every $t\geq 1$ and $\lambda\geq 1/2$. Let $\{f_i\}_{i=1}^n$ be disjointly supported non-negative functions in $L_M(0,1)$ and put $\|f_i\|=a_i$, $1\leq i\leq n$, and $\left\|\sum_{i=1}^n f_i\right\|=b$. Clearly, $b\geq a_i$ for all i and thus,

$$M(f_i(u)/a_i)\geq\gamma M(f_i(u)/b)(b/a_i)^p\,,$$

for every $u\in[0,1]$ for which $f_i(u)/b\geq 1/2$. Put

$$\sigma=\left\{u\in[0,1];\ \sum_{i=1}^n f_i(u)\geq b/2\right\}.$$

Then

$$\int_\sigma M\left(\sum_{i=1}^n f_i(u)/b\right)du\leq\gamma^{-1}b^{-p}\sum_{i=1}^n a_i^p\int_0^1 M(f_i(u)/a_i)\,du=\gamma^{-1}b^{-p}\sum_{i=1}^n a_i^p\,.$$

The integral of $M\left(\sum_{i=1}^n f_i/b\right)$ over $[0,1]\sim\sigma$ is clearly less or equal to $M(1/2)$. Hence,

$$1=\int_0^1 M\left(\sum_{i=1}^n f_i(u)/b\right)du\leq M(1/2)+\int_\sigma M\left(\sum_{i=1}^n f_i(u)/b\right)du\,,$$

and therefore, if we note that $M(1/2)<M(1)=1$, we get that

$$\left\|\sum_{i=1}^{n} f_i\right\|=b\leqslant\left(\sum_{i=1}^{n}\|f_i\|^p/\gamma(1-M(1/2))\right)^{1/p},$$

which shows that X satisfies an upper p-estimate. Consequently,

$$\alpha_{M,\infty}\leqslant\sup\{p;\ X \text{ satisfies an upper } p\text{-estimate}\}\leqslant p_X .$$

In order to conclude the proof it is therefore enough to show that, whenever $p<p_X$, then $p\leqslant\alpha_{M,\infty}$. If $p<p_X$ then there is a constant K so that $\|D_s\|\leqslant Ks^{1/p}$ for $s>1$. Since $D_s\chi_{[0,u/s]}=\chi_{[0,u]}$ we deduce that

$$\|\chi_{[0,u]}\|\leqslant Ks^{1/p}\|\chi_{[0,u/s]}\|,\quad 0<u<1,\ s\geqslant 1 .$$

Observe that $1/\|\chi_{[0,u]}\|=M^{-1}(1/u)$ and thus we get that

$$M^{-1}(s/u)\leqslant Ks^{1/p}M^{-1}(1/u),\quad 0<u<1,\ s\geqslant 1 .$$

By putting $\lambda=M^{-1}(1/u)$ and $t=Ks^{1/p}$, we deduce that $p\leqslant\alpha_{M,\infty}$ since

$$M(\lambda t)\geqslant s/u=M(\lambda)t^p/K^p,\quad \lambda>1,\ t\geqslant K . \quad \square$$

Remarks. 1. An Orlicz space $L_M(0,1)$ is contained in $L_p(0,1)$, for some p, if and only if $M(t)\geqslant Kt^p$, for some constant K and every $t>1$. From the inequality $M(t)\geqslant Kt^p$ we cannot however deduce information concerning the behavior of $M(\lambda t)/M(\lambda)$. It is easy to construct Orlicz functions M so that, say $X=L_M(0,1)\subset L_2(0,1)$, but $p_X=1$. This justifies remark 2 following 2.b.3.

2. Analogous results hold for function spaces $L_M(0,\infty)$ and for sequence spaces l_M. In particular, for every Orlicz space X, we have that

$$p_X=\sup\{p;\ X \text{ satisfies an upper } p\text{-estimate}\}$$

and

$$q_X=\inf\{q;\ X \text{ satisfies a lower } q\text{-estimate}\} .$$

In the case of an Orlicz sequence space $X=l_M$, the indices p_X and q_X turn out to be the numbers α_M and β_M, respectively, which were introduced in I.4.a.9. These numbers were characterized there by the fact that l_r is isomorphic to a subspace of l_M if and only if $r\in[\alpha_M,\beta_M]$. An inspection of the proof of this fact, as given in I.4.a, yields the following additional information. If $r\in[\alpha_M,\beta_M]$ then, for every $\varepsilon>0$ and integer n, there exist disjointly supported vectors $\{x_i\}_{i=1}^n$ in l_M, all having the same distribution (i.e. they form a finite block basic sequence of the unit vector basis in l_M which is generated by one vector in the terminology of I.a.3.8), so that

$$(1-\varepsilon)\left(\sum_{i=1}^{n}|a_i|^r\right)^{1/r}\leqslant\left\|\sum_{i=1}^{n}a_ix_i\right\|\leqslant(1+\varepsilon)\left(\sum_{i=1}^{n}|a_i|^r\right)^{1/r},$$

for every choice of scalars $\{a_i\}_{i=1}^n$. Of course, if $r\notin[\alpha_M,\beta_M]=[p_X,q_X]$ then we cannot find such $\{x_i\}_{i=1}^n$ for every ε and n. Exactly the same result holds also for $X=L_M(0,1)$. For every $r\in[p_X,q_X]$, $\varepsilon>0$ and integer n, there exist n disjointly supported functions $\{x_i(t)\}_{i=1}^n$ in $L_M(0,1)$, all having the same distribution function, so that the preceding inequalities hold for every choice of scalars $\{a_i\}_{i=1}^n$. The proof is very similar to that of I.4.a.9 and thus we do not reproduce it here.

It turns out that a similar result holds for a general r.i. space X. There exist in X nice copies of l_r^n for some (but in general not all) r in $[p_X,q_X]$. More precisely, we have the following result.

Theorem 2.b.6. *Let X be an r.i. space. Then p_X, respectively q_X, is the minimum, respectively the maximum, of all the numbers p which have the following property. For every $\varepsilon>0$ and every integer n, X contains n disjointly supported functions $\{f_i\}_{i=1}^n$, having all the same distribution function, so that*

$$(1-\varepsilon)\left(\sum_{i=1}^{n}|a_i|^p\right)^{1/p}\leqslant\left\|\sum_{i=1}^{n}a_if_i\right\|\leqslant(1+\varepsilon)\left(\sum_{i=1}^{n}|a_i|^p\right)^{1/p},$$

for every choice of scalars $\{a_i\}_{i=1}^n$.

This theorem is an easy consequence of an important result of Krivine [67] (cf. also Rosenthal [116]) which will be presented (and proved) in Vol. III. We shall prove here only the following weaker version of 2.b.6 whose proof is much simpler.

Proposition 2.b.7. *Let X be an r.i. space. Then*

(i) *$q_X<\infty$ if and only if X does not contain, for all integers n, almost isometric copies of l_∞^n spanned by disjoint functions having the same distribution function.*

(ii) *$1<p_X$ if and only if X does not contain, for all integers n, almost isometric copies of l_1^n spanned by disjoint functions having the same distribution function.*

Proof. (i) If $q_X<\infty$ then, as we have already noted above, for each $q>q_X$ there is an integer K so that, for every choice of disjointly supported $\{f_i\}_{i=1}^n$, all having norm one and the same distribution function, we have $\left\|\sum_{i=1}^{n}f_i\right\|\geqslant K^{-1}n^{1/q}$. Thus, X does not contain uniformly isomorphic copies of l_∞^n spanned by disjointly supported functions with the same distribution function. The proof of the converse assertion is identical to the proof of 1.f.12(ii) if we require throughout that proof that the functions $\{x_i\}_{i=1}^n$ have the same distribution function. Assertion (ii) follows from (i) by duality. □

We turn now to the interpolation theorem which motivated the definition of the Boyd indices and which will play an important role in the sequel. We have first to define the notion of an operator of weak type (p, q). The perhaps most natural way to introduce this notion is by using the $L_{p,q}$ spaces.

Definition 2.b.8. Let (Ω, Σ, ν) be a measure space. For $1 \leq p < \infty$ and $1 \leq q < \infty$, $L_{p,q}(\Omega, \Sigma, \nu)$ is the space of all locally integrable real valued functions f on Ω for which

$$\|f\|_{p,q} = [(q/p) \int_0^\infty (t^{1/p} f^*(t))^q \, dt/t]^{1/q} < \infty .$$

For $1 \leq p \leq \infty$, $L_{p,\infty}(\Omega, \Sigma, \nu)$ is the space of all functions f as above so that

$$\|f\|_{p,\infty} = \sup_{t>0} t^{1/p} f^*(t) < \infty .$$

Note that, for $p = q$, $L_{p,q}$ coincides with L_p (with the same norm). If $1 \leq q < p$ (and $\Omega = [0, \infty)$ or $\Omega = [0, 1]$) the space $L_{p,q}$ is a Lorentz function space with weight function $W(t) = qt^{q/p-1}/p$, $0 < t < \infty$. For $q > p$, it is easily seen that $\| \ \|_{p,q}$ does not satisfy the triangle inequality and thus, it is not really a norm (the function $W(t)$ written above is increasing rather than decreasing when $q > p$). Nevertheless, $L_{p,q}$ is a linear space also for $q > p$. It can be shown that it can be made into a Banach space if $p > 1$ by introducing an actual norm $|\!|\!| \ |\!|\!|_{p,q}$ which satisfies $\|f\|_{p,q} \leq |\!|\!|f|\!|\!|_{p,q} \leq C(p, q)\|f\|_{p,q}$. We do not give these details since our interest in $L_{p,q}$ spaces lies not in their structure but only in the quantities $\| \ \|_{p,q}$ which arise naturally in interpolation theory. Note also that we have not defined $L_{\infty,q}$, for $q < \infty$, since $\int_0^\infty f^*(t)^q \, dt/t < \infty$ implies $f \equiv 0$.

We shall be interested in the sequel in inclusion relations between the spaces $L_{p,q}$.

Proposition 2.b.9 [82], [54]. *Let* $1 \leq p < \infty$ *and* $1 \leq q_1 < q_2 \leq \infty$. *Then*

$$L_{p,q_1}(\Omega, \Sigma, \nu) \subset L_{p,q_2}(\Omega, \Sigma, \nu)$$

and moreover, for every $f \in L_{p,q_1}(\Omega, \Sigma, \nu)$,

$$\|f\|_{p,q_2} \leq \|f\|_{p,q_1} .$$

Proof. If $q_2=\infty$ the result follows from

$$t^{1/p}f^*(t)=f^*(t)\left((q_1/p)\int_0^t u^{q_1/p-1}\,du\right)^{1/q_1}$$
$$\leqslant\left((q_1/p)\int_0^t (u^{1/p}f^*(u))^{q_1}\,du/u\right)^{1/q_1},\quad t>0.$$

Assume now that $q_2<\infty$. It clearly suffices to prove that $\|f\|_{p,q_2}\leqslant\|f\|_{p,q_1}$ for simple and decreasing functions. Assume that $f=\sum_{k=1}^n a_k\chi_{[t_{k-1},t_k)}$, with $a_1>a_2\ldots>a_n>0$ and $t_0=0<t_1<\cdots<t_n$, and put $\gamma=q_1/q_2(<1)$, $b_k=a_k^{q_2}$ and $s_k=t_k^{q_2/p}$, $k=1,\ldots,n$. Then the inequality, we want to establish, gets the form

$$\sum_{k=1}^n b_k(s_k-s_{k-1})\leqslant\left(\sum_{k=1}^n b_k^\gamma(s_k^\gamma-s_{k-1}^\gamma)\right)^{1/\gamma}.$$

This inequality is proved by induction on n. Indeed, the function

$$\varphi(x)=\sum_{k=1}^{n-1} b_k(s_k-s_{k-1})+x(s_n-s_{n-1})$$
$$-\left(\sum_{k=1}^{n-1} b_k^\gamma(s_k^\gamma-s_{k-1}^\gamma)+x^\gamma(s_n^\gamma-s_{n-1}^\gamma)\right)^{1/\gamma}$$

is convex (i.e. $\varphi''(x)\geqslant 0$) on $[0,\infty)$ and thus, by assuming that $\varphi(0)\leqslant 0$ and $\varphi(b_{n-1})\leqslant 0$, we get that $\varphi(b_n)\leqslant 0$. □

There are also inclusion relations between the $L_{p,q}$ spaces as p varies. These inclusion relations are proved by simply applying Hölder's inequality. If (Ω,Σ,ν) is a probability space then, for every $r<p<s$ and every q,

$$L_{s,\infty}(\Omega,\Sigma,\nu)\subset L_{p,q}(\Omega,\Sigma,\nu)\subset L_{r,1}(\Omega,\Sigma,\nu)$$

while, for a general measure space (Ω,Σ,ν), we have

$$L_{s,\infty}(\Omega,\Sigma,\nu)\cap L_{r,\infty}(\Omega,\Sigma,\nu)\subset L_{p,q}(\Omega,\Sigma,\nu)\subset L_{r,1}(\Omega,\Sigma,\nu)+L_{s,1}(\Omega,\Sigma,\nu).$$

In other words the inclusion relation between the spaces $\{L_{p_i,q_i}(\Omega,\Sigma,\nu)\}_{i=1}^3$ with $p_1<p_2<p_3$ are the same as those between $\{L_{p_i}(\Omega,\Sigma,\nu)\}_{i=1}^3$ regardless of q_i, $i=1,2,3$. The index p is thus the "main" index in $\|\cdot\|_{p,q}$ while the index q is used for a finer estimate of the size of a function once p is given.

Definition 2.b.10. Let $(\Omega_i,\Sigma_i,\nu_i)$, $i=1,2$, be two measure spaces. Let $1\leqslant p_1\leqslant\infty$ and let T be a map defined on a subset of $L_{p_1}(\Omega_1)$ which takes values in the space of all measurable functions on Ω_2.

(i) The map T is said to be of *strong type* (p_1, p_2), for some $1 \leqslant p_2 \leqslant \infty$, if there is a constant M so that

$$\|Tf\|_{p_2} \leqslant M\|f\|_{p_1},$$

for every f in the domain of definition of T.

(ii) The map T is said to be of *weak type* (p_1, p_2), for some $1 \leqslant p_2 \leqslant \infty$, if there is a constant M so that

$$\|Tf\|_{p_2, \infty} \leqslant M\|f\|_{p_1, 1},$$

for every f in the domain of definition of T with the convention that if $p_1 = \infty$ we have to replace $\|f\|_{\infty, 1}$ above, which is not defined, by $\|f\|_{\infty, \infty} = \|f\|_\infty$.

It follows immediately from the definitions involved that a map T is of weak type (p_1, p_2) if and only if there is a constant M so that

$$\sup_{t>0} t v_2(\{\omega \in \Omega_2; |Tf(\omega)| \geqslant t\})^{1/p_2} \leqslant M p_1^{-1} \int_0^\infty t^{1/p_1 - 1} f^*(t)\, dt$$

(if either p_2 or p_1 are ∞ the expression on the right-hand side, respectively left-hand side, has to be replaced by $\|Tf\|_\infty$, respectively $\|f\|_\infty$). Note that, by 2.b.9, every map of strong type (p_1, p_2) is also of weak type (p_1, p_2), with the same constant M. The definition of weak type, as presented in 2.b.10, turns out to enter naturally in the interpolation theorem proved below and is also of importance in some applications in harmonic analysis. It differs however from the classical notion of weak type which goes back to Marcinkiewicz. The original definition was: T is of Marcinkiewicz weak type (p_1, p_2) if

$$\|Tf\|_{p_2, \infty} \leqslant M\|f\|_{p_1},$$

for some constant M and every f in the domain of T or, alternatively, if

$$\sup_{t>0} t v_2(\{\omega \in \Omega_2; |Tf(\omega)| \geqslant t\})^{1/p_2} \leqslant M\|f\|_{p_1}$$

(where the left-hand term is replaced by $\|Tf\|_\infty$ in case $p_2 = \infty$). It is clear from 2.b.9 that $\|f\|_{p_1} = \|f\|_{p_1, p_1} \leqslant \|f\|_{p_1, 1}$, for every $f \in L_{p_1, 1}$. Hence, every map, which is of Marcinkiewicz weak type (p_1, p_2), is also of weak type (p_1, p_2) and thus the interpolation theorem below applies, in particular, to operators which are of the suitable Marcinkiewicz weak types. There are many important operators appearing in various parts of analysis which are of weak type but not of strong type. Here we just mention a trivial example. If $\varphi \neq 0$ is a continuous linear functional on $L_p(\Omega)$, $1 \leqslant p < \infty$ then the operator $Tf(t) = t^{-1/p}\varphi(f)$ from $L_p(\Omega)$ into the space of all measurable functions on $[0, \infty)$ is of Marcinkiewicz weak type (p, p) but not of strong type (p, p). In the proof of 2.b.13 below, we shall use simple examples of operators which are of weak type (p,p) but not of Marcinkiewicz weak type (p,p).

We are now ready to state the interpolation theorem of Boyd (cf. [16]).

Theorem 2.b.11. *Let I be either $[0,1]$ or $[1,\infty)$, let $1\leq p<q\leq\infty$ and let T be a linear operator mapping $L_{p,1}(I)+L_{q,1}(I)$ into the space of measurable functions on I. Assume that T is of weak types (p,p) and (q,q) (with respect to the Lebesgue measure on I). Then, for every r.i. function space X on I, so that $p<p_X$ and $q_X<q$, T maps X into itself and is bounded on X.*

Note that, by 2.b.3 and the observation preceding 2.b.10, $X\subset L_{p,1}(I)+L_{q,1}(I)$ i.e. T is already defined on all of X. The main step in the proof of 2.b.11 is the following lemma due to Calderon [22].

Lemma 2.b.12. *With the the same assumptions on T as in 2.b.11 there is a constant $M<\infty$ so that*

$$(Tf)^*(2t)\leq M\left(\int_0^1 f^*(tu)u^{(1-p)/p}\,du+\int_1^\infty f^*(tu)u^{(1-q)/q}\,du\right),$$

for every $0<t<\infty$ if $I=[0,\infty)$ (respectively, $0<t\leq 1/2$ if $I=[0,1]$) and every $f\in L_{p,1}(I)+L_{q,1}(I)$.

Proof. Suppose that T is of weak types (p,p) and (q,q) with the constants appearing in 2.b.10 being M_p, respectively, M_q. Let $f\in L_{p,1}(I)+L_{q,1}(I)$ and, for u, $t\in I$, set

$$g_t(u)=\begin{cases} f(u)-f^*(t) & \text{if } f(u)>f^*(t)\\ f(u)+f^*(t) & \text{if } f(u)<-f^*(t)\\ 0 & \text{if } |f(u)|\leq f^*(t)\end{cases}$$

and

$$h_t(u)=f(u)-g_t(u)\,.$$

The function h_t is the "flat" part of f (relatively to the level $f^*(t)$) while g_t is the "peaked" part of f. We apply the fact that T is of weak type (p,p) to g_t and of weak type (q,q) to h_t. Note that $g_t^*(u)$ vanishes outside the interval $[0,t]$ while, for $0<u<t$, we have $g_t^*(u)\leq f^*(u)$. Hence, for $t\in I$,

$$t^{1/p}(Tg_t)^*(t)\leq M_p p^{-1}\int_0^\infty g_t^*(s)s^{(1-p)/p}\,ds\leq M_p p^{-1}\int_0^t f^*(s)s^{(1-p)/p}\,ds$$

$$=M_p p^{-1}t^{1/p}\int_0^1 f^*(tu)u^{(1-p)/p}\,du\,.$$

Observe also that, for every $u \in I$, we have $|h_t(u)| = \min(|f(u)|, f^*(t))$. Hence, for $t \in I$,

$$\begin{aligned}
t^{1/q}(Th_t)^*(t) &\leqslant M_q q^{-1} \int_0^\infty h_t^*(s) s^{(1-q)/q}\, ds \\
&\leqslant M_q q^{-1} \left(\int_0^t f^*(t) s^{(1-q)/q}\, ds + \int_t^\infty h_t^*(s) s^{(1-q)/q}\, ds \right) \\
&= M_q q^{-1} \left(q t^{1/q} f^*(t) + t^{1/q} \int_1^\infty h_t^*(tu) u^{(1-q)/q}\, du \right) \\
&\leqslant M_q q^{-1} t^{1/q} \left(q p^{-1} \int_0^1 f^*(tu) u^{(1-p)/p}\, du + \int_1^\infty f^*(tu) u^{(1-q)/q}\, du \right).
\end{aligned}$$

Since $|Tf| \leqslant |Tg_t| + |Th_t|$ it follows that

$$\begin{aligned}
(Tf)^*(2t) &\leqslant (Tg_t)^*(t) + (Th_t)^*(t) \\
&\leqslant (M_p p^{-1} + M_q p^{-1}) \int_0^1 f^*(tu) u^{(1-p)/p}\, du \\
&\quad + M_q q^{-1} \int_1^\infty f^*(tu) u^{(1-q)q}\, du\,.
\end{aligned}$$

This proves our assertion with $M = p^{-1}(M_p + M_q)$. $\square$

Proof of 2.b.11. Choose p_0 and q_0 so that $p < p_0 < p_X$ and $q_X < q_0 < q$. Then there is a constant K so that

$$\|D_s\| \leqslant K s^{1/p_0}, \qquad 2 \leqslant s < \infty; \qquad \|D_s\| \leqslant K s^{1/q_0}, \quad 0 \leqslant s \leqslant 2\,.$$

Let $g \in X'$ with $\|g\|_{X'} = 1$. We have

$$\begin{aligned}
\int_0^\infty \int_0^1 f^*(tu/2) g(t) u^{(1-p)/p}\, du\, dt &= \int_0^1 u^{(1-p)/p} \left(\int_0^\infty (D_{2/u} f^*)(t) g(t)\, dt \right) du \\
&\leqslant \|f\|_X 2^{1/p_0} K \int_0^1 u^{1/p - 1/p_0 - 1}\, du \\
&= \|f\|_X 2^{1/p_0} K (1/p - 1/p_0)^{-1},
\end{aligned}$$

and similarly,

$$\int_0^\infty \int_1^\infty f^*(tu/2) g(t) u^{(1-q)/q}\, du\, dt \leqslant \|f\|_X 2^{1/q_0} K (1/q_0 - 1/q)^{-1}\,.$$

(If $I = [0, 1]$ we take in the formula above $g(t) = 0$ for $t > 1$.)

Hence, by 2.b.12, we get that

$$\int_0^\infty (Tf)^*(t)g(t)\,dt \leq M_0\|f\|_X ,$$

for every $g \in X'$ with $\|g\|_{X'}=1$, where $M_0=KM(2^{1/p_0}(1/p-1/p_0)^{-1}+2^{1/q_0}(1/q_0-1/q)^{-1})$. If X is a maximal r.i. function space this already proves that $Tf \in X$ and that $\|Tf\|_X \leq M_0\|f\|_X$, as desired.

If X is a minimal r.i. function space we get, by what has already been proved, that T maps X'' into itself and is bounded there and also that T maps $L_{p_0}(I)\cap L_{q_0}(I)$ into itself. Since X is the closure of $L_{p_0}(I)\cap L_{q_0}(I)$ in X'' it follows that T also maps X into itself. □

Remarks. 1. If X is a maximal r.i. space then the proof above works also when T is only a quasilinear operator. The only change needed in the proof is to replace the M of 2.b.12 by the constant CM, where C is the constant appearing in the definition of quasilinearity. In the case of a minimal r.i. space the proof given above does not work for a quasilinear operator. A bounded quasilinear operator need not be continuous. Thus, the fact that it maps $L_{p_0}(I)\cap L_{q_0}(I)$ into itself does not ensure immediately that the same is true for its closure in X''.

2. Suppose that T is a linear operator defined on (a non-closed) linear subspace Y of $L_p(I)+L_q(I)$. Assume that T is of weak types (p, p) and (q, q) and that $f \in Y$ implies $\max(1, f) \in Y$. Then the proof of 2.b.11 shows that, whenever X is a maximal r.i. function space on I, T maps $X\cap Y$ into X and is bounded on $X\cap Y$. If, in addition, $L_p(I)\cap L_q(I)\cap Y$ is dense in $X\cap Y$ (in the norm induced by X) the same is true when X is a minimal r.i. function space on I. The operator T can then, of course, be extended in a unique way to a bounded linear operator from the closure of $X\cap Y$ in X into X. A typical example where this remark is used (and in which the two assumptions made on Y are satisfied) is the case where Y is the space of all simple integrable functions or, alternatively, the space of all finite linear combinations of characteristic functions of dyadic intervals.

We shall prove now a converse to 2.b.11.

Proposition 2.b.13 [15], [16]. *Let X be an r.i. function space on I (where $I=[0, 1]$ or $[0, \infty)$). Assume that, for some $1\leq p<q<\infty$, every linear operator defined on $L_{p,1}(I)+L_{q,1}(I)$ which is of weak types (p, p) and (q, q) maps X into itself. Then $p<p_X$ and $q_X<q$.*

The main point in 2.b.13 is the assertion that we have strict inequalities (not only $p\leq p_X$ and $q_X\leq q$).

Proof. We shall prove only the assertion concerning p_X. The assertion concerning q_X can be proved in an entirely similar way, by using the formal adjoint of the operators T_γ appearing below.

Consider the operators

$$(T_\gamma f)(t)=t^{-\gamma}\int_0^t s^{\gamma-1}f(s)\,ds=\int_0^1 s^{\gamma-1}f(st)\,ds, \quad 0<\gamma\leq 1,\ t\in I .$$

We claim first that T_γ is of weak type (r, r) for every $r \geqslant \gamma^{-1}$. Indeed, for $f \in L_{r,1}(I)$ and $r \geqslant \gamma^{-1}$, we have

$$t^{1/r}(T_\gamma f)^*(t) \leqslant t^{1/r-\gamma} \int_0^t s^{\gamma-1} f^*(s)\, ds \leqslant \int_0^t s^{1/r-1} f^*(s)\, ds .$$

We used here the fact (proved e.g. by differentiation) that $(T_\gamma f^*)^* = T_\gamma f^*$ and $T_\gamma |f| \leqslant T_\gamma f^*$. Note further that, for $\gamma > \varepsilon > 0$ and for every simple integrable f on I,

$$\begin{aligned} T_\gamma T_{\gamma-\varepsilon} f(u) &= \int_0^1 t^{\gamma-1} \int_0^1 s^{\gamma-\varepsilon-1} f(stu)\, ds\, dt \\ &= \int_0^1 v^{\gamma-1} \int_v^1 s^{-\varepsilon-1} f(vu)\, ds\, dv \\ &= \varepsilon^{-1}\left(\int_0^1 v^{\gamma-\varepsilon-1} f(vu)\, dv - \int_0^1 v^{\gamma-1} f(vu)\, dv \right) \\ &= \varepsilon^{-1}((T_{\gamma-\varepsilon} f)(u) - (T_\gamma f)(u)) \end{aligned}$$

and hence, $T_\gamma f = T_{\gamma-\varepsilon} f - \varepsilon T_\gamma T_{\gamma-\varepsilon} f$.

It follows from our assumption that $T_{1/p}$ is a bounded operator on X. If $\varepsilon < \|T_{1/p}\|^{-1}$ then $(I_X - \varepsilon T_{1/p})^{-1}$ is also a bounded operator on X. Fix now an ε with $0 < \varepsilon < \min(p^{-1}, \|T_{1/p}\|^{-1})$ and put $\eta = 1/p - \varepsilon$. If f is a simple integrable function such that $T_\eta f \in X$ then we get from what we have shown above that $T_{1/p} f = (I_X - \varepsilon T_{1/p}) T_\eta f$ and thus

$$T_\eta f = (I_X - \varepsilon T_{1/p})^{-1} T_{1/p} f .$$

In particular,

$$\|T_\eta f\|_X \leqslant C \|f\|_X, \quad \text{where } C = \|T_{1/p}\| \, \|(I_X - \varepsilon T_{1/p})^{-1}\| .$$

It should be noted that in the preceding step it is essential to know *a-priori* that $T_\eta f \in X$. In case $I = [0, 1]$ it is clear that if f is a simple function then $T_\eta f$ is a bounded function and hence $T_\eta f \in X$. Thus the estimate above for $\|T_\eta f\|_X$ applies for every simple f. Since, for a non-increasing simple positive f and $0 \leqslant s, t \leqslant 1$,

$$s^\eta (D_{1/s} f)(t)/\eta = f(st) \int_0^s u^{\eta-1}\, du \leqslant \int_0^1 u^{\eta-1} f(ut)\, du = (T_\eta f)(t)$$

it follows that $\|D_{1/s}\| \leqslant \eta C (1/s)^\eta$ and hence $1/p_X \leqslant \eta = 1/p - \varepsilon$ i.e. $p < p_X$.

In case $I = [0, \infty)$ it is not *a-priori* clear that $T_\eta f \in X$. We overcome this difficulty by considering the operators $(T_\gamma^a f)(t) = \chi_{[0,a]}(T_\gamma f)(t)$ for $0 < a < \infty$, $0 < \gamma \leqslant 1$. Note that for simple integrable f, $T_{\gamma_1}^a T_{\gamma_2}^a f = T_{\gamma_1}^a T_{\gamma_2} f$ and that $T_\gamma^a f \in X$ for every a and γ. Hence, with the same ε and η as above,

$$T_\eta^a f = (I_X - \varepsilon T_{1/p}^a)^{-1} T_{1/p}^a f$$

and thus, for a constant C independent of a and f, $\|T_\eta^a f\|_X \leq C\|f\|_X$. By letting $a \to \infty$ we deduce that $T_\eta f \in X''$ for every simple integrable f and $\|T_\eta f\|_{X''} \leq C\|f\|_X$. The proof is now concluded as in the case $I=[0, 1]$. □

Remark. The assumption made in 2.b.13 that X is an r.i. function space on I is necessary to the extent that the main assumption in 2.b.13 already implies that, for every automorphism τ or I, the operator $U_\tau f(t)=f(\tau^{-1}(t))$ maps X into itself. Moreover, it follows from the main assumption on X in 2.b.13 that $f \in X$ and $g \prec f$ imply that $g \in X$ (see remark 2 following 2.a.11).

Theorem 2.b.11 reduces for $X=L_r$ to a special case of the two classical interpolations theorems of Riesz-Thorin and Marcinkiewicz. These two fundamental theorems inspired all the subsequent development of interpolation theory. The Riesz–Thorin theorem, and often also the Marcinkiewicz theorem, are proved in most textbooks on functional analysis and harmonic analysis. We shall apply them in the following sections only in the special case contained in 2.b.11 and therefore we shall not reproduce here their proofs. We find it however appropriate to state these theorems here in their general form. A very detailed discussion of these theorems and other material on interpolation in L_p spaces and related spaces can be found in the book [10]. We refer to this book also for a discussion of the history of the various variants of these theorems. (In Section g below we outline the general Lions–Peetre interpolation theory from which, in particular, 2.b.15 follows.)

Theorem 2.b.14 (The Riesz–Thorin interpolation theorem). *Let* $(\Omega_i, \Sigma_i, \mu_i)$, $i=1, 2$, *be two measure spaces, let* $p_1 \neq q_1$ *and* $p_2 \neq q_2$ *be numbers in* $[1, \infty]$ *(*∞ *is included). Let* T *be a linear operator mapping* $L_{p_1}(\mu_1)+L_{q_1}(\mu_1)$ *into* $L_{p_2}(\mu_2)+L_{q_2}(\mu_2)$, *which is of strong types* (p_1, p_2) *and* (q_1, q_2). *Then, for every* $0<\theta<1$, T *is of strong type* (r_1, r_2), *where*

$$\frac{1}{r_i}=\frac{\theta}{p_i}+\frac{(1-\theta)}{q_i}, \quad i=1, 2 .$$

Moreover,

$$\|T\|_{r_1, r_2} \leq \|T\|_{p_1, p_2}^{\theta} \|T\|_{q_1, q_2}^{1-\theta} .$$

The symbol $\|T\|_{r_1, r_2}$ means of course $\sup\{\|Tf\|_{r_2};\ \|f\|_{r_1} \leq 1\}$. The preceding inequality concerning the norms holds if we use complex scalars. For real scalars the factor 2 has to be added to the right-hand side of the inequality.

Theorem 2.b.15 (The Marcinkiewicz interpolation theorem). *Let* $(\Omega_i, \Sigma_i, \mu_i)$, $i=1, 2$, *be two measure spaces and let* $p_1 \neq q_1$ *and* $p_2 \neq q_2$ *be numbers in* $[1, \infty]$ *(*∞ *is included). Let* T *be a quasilinear operator mapping* $L_{p_1}(\mu_1)+L_{q_1}(\mu_1)$ *into the space of measurable functions on* Ω_2, *which is of weak types* (p_1, p_2) *and* (q_1, q_2). *Then,* *for every* $0<\theta<1$, *T is of strong type* (r_1, r_2) *where*

$$\frac{1}{r_i}=\frac{\theta}{p_i}+\frac{(1-\theta)}{p_i}, \quad i=1,2,$$

provided that $r_1 \leqslant r_2$.

c. The Haar and the Trigonometric Systems

The present section is devoted mostly to the study of the unconditionality of the Haar system in r.i. function spaces on [0, 1]. We treat this matter in the more general setting of martingales and present a complete characterization of those r.i. function spaces on [0, 1] for which the Haar basis is unconditional. Next, we introduce the notion of reproducibility for bases and show that the Haar basis has this property in every separable r.i. function space. We conclude by presenting some applications of reproducibility and also by discussing briefly the trigonometric system.

The Haar system $\{\chi_n\}_{n=1}^\infty$ was introduced in I.1.a.4. For convenience, we recall here that $\chi_1(t)\equiv 1$ and, for $l=1, 2, \ldots, 2^k$ and $k=0, 1, \ldots,$

$$\chi_{2^k+l}(t)=\begin{cases} 1 & \text{if } t\in[(2l-2)2^{-k-1}, (2l-1)2^{-k-1}) \\ -1 & \text{if } t\in[(2l-1)2^{-k-1}, 2l\cdot 2^{-k-1}) \\ 0 & \text{otherwise}. \end{cases}$$

The vectors $\{\chi_n\}_{n=1}^\infty$, as defined above, are normalized in $L_\infty(0, 1)$ and, unless stated otherwise, we shall always use this normalization.

The Haar system can be associated in a natural way with an increasing sequence of σ-algebras $\{\mathscr{A}_n\}_{n=1}^\infty$ of measurable subsets of [0, 1]. The σ-algebra $\mathscr{A}_1$ consists only of the sets $\emptyset$ and [0, 1] and if $n=2^k+l$, for some $1\leqslant l\leqslant 2^k$ and $k\geqslant 0$, then $\mathscr{A}_n$ is defined to be the σ-algebra generated by $\mathscr{A}_{n-1}$ and the two intervals $[(2l-2)2^{-k-1}, (2l-1)2^{-k-1})$ and $[(2l-1)2^{-k-1}, 2l\cdot 2^{-k-1})$. It is evident that $\mathscr{A}_n$ is, for every n, the smallest σ-algebra $\mathscr{A}$ for which the functions $\{\chi_1, \ldots, \chi_n\}$ are $\mathscr{A}$-measurable.

We have seen in I.1.a that the Haar system forms a monotone basis of every $L_p(0, 1)$ space, $1<p<\infty$. This fact is true in every separable r.i. function space on [0, 1].

Proposition 2.c.1 [36]. *The Haar system is a monotone basis of every separable r.i. function space on* [0, 1].

Proof. Let X be a separable r.i. function space on [0, 1]. Since X is not isomorphic to $L_\infty(0, 1)$, we have that $\lim_{t\to 0} \|\chi_{[0,t]}\|=0$ and thus, every simple function on [0, 1] can be approximated, in the norm of X, by step functions over the dyadic intervals

$[l\cdot 2^{-k}, (l+1)2^{-k})$, $0\leqslant l\leqslant 2^k-1$, $k=0, 1, \ldots$. It follows that the linear span of the step functions over the dyadic intervals or, equivalently, of the Haar functions is dense in X and, therefore, it remains to show that $\{\chi_n\}_{n=1}^\infty$ forms a monotone basic sequence in X. This fact can be proved e.g. by using the observation that, for every $n<m$ and every choice of scalars $\{a_i\}_{i=1}^m$, we have

$$E^{\mathscr{A}_n}\left(\sum_{i=1}^m a_i\chi_i\right)=\sum_{i=1}^n a_i\chi_i$$

and by the assertion of 2.a.4 that the conditional expectations $E^{\mathscr{A}_n}$ act as norm one operators in X. Actually, it is easier to check directly that $\{\chi_n\}_{n=1}^\infty$ is a basic sequence. Put $f=\sum_{i=1}^n a_i\chi_i$ and $g=\sum_{i=1}^{n+1} a_i\chi_i$ and notice that f and g coincide in $[0, 1]$ with the exception of some dyadic interval η on which f is a constant, say it takes the value b there, and g is equal to $b+a_{n+1}$ on the first half of η and to $b-a_{n+1}$ on the second half of η. Let τ be an automorphism of $[0, 1]$ which permutes the first half of η with its second half and leaves invariant every point outside η. Then it is easily seen that

$$f(t)=(g(t)+g(\tau(t)))/2,$$

for every $t\in[0, 1]$, and thus $\|f\|\leqslant\|g\|$. □

In order to characterize those r.i. function spaces on $[0, 1]$ for which the Haar basis is unconditional we work in the framework of martingale theory. We first recall the definition of a martingale. Let (Ω, Σ, ν) be a probability space and let

$$\mathscr{B}_1\subset\mathscr{B}_2\subset\cdots\subset\mathscr{B}_n\subset\cdots$$

be an increasing sequence of σ-subalgebras of Σ. A sequence $\{f_n\}_{n=1}^\infty$ of integrable random variables over (Ω, Σ, ν) is said to be a *martingale with respect to* $\{\mathscr{B}_n\}_{n=1}^\infty$ if

$$E^{\mathscr{B}_n}f_{n+1}=f_n,$$

for all n. Also a finite sequence $\{f_n\}_{n=1}^k$ satisfying $E^{\mathscr{B}_n}f_{n+1}=f_n$ for $1\leqslant n<k$ will be called a martingale. (This finite sequence can be identified with an infinite martingale which is constant for $n\geqslant k$.)

It follows immediately from the definition of a martingale $\{f_n\}_{n=1}^\infty$ that f_n is $\mathscr{B}_n$-measurable, for every n, and that

$$E^{\mathscr{B}_j}f_n=f_j,$$

for $1\leqslant j\leqslant n$. In particular, we get that any subsequence $\{f_{n_i}\}_{i=1}^\infty$ of $\{f_n\}_{n=1}^\infty$ is a martingale with respect to $\{\mathscr{B}_{n_i}\}_{i=1}^\infty$.

The connection between the study of the Haar system and the theory of martingales is quite obvious. If X is a separable r.i. function space on $[0, 1]$ and

$\{\mathscr{A}_n\}_{n=1}^{\infty}$ is, as above, the (increasing) sequence of σ-algebras associated to the Haar basis $\{\chi_n\}_{n=1}^{\infty}$ of X then, for every $f=\sum_{i=1}^{\infty} a_i\chi_i \in X$, the sequence

$$f_n = E^{\mathscr{A}_n} f = \sum_{i=1}^{n} a_i\chi_i, \quad n=1, 2, \ldots$$

defines clearly a martingale with respect to $\{\mathscr{A}_n\}_{n=1}^{\infty}$.

We begin the study of martingales by proving the following inequality (cf. [29]):

Proposition 2.c.2. *Let $\{f_n\}_{n=1}^{k}$ be a (finite) martingale with respect to a sequence $\{\mathscr{B}_n\}_{n=1}^{k}$ of σ-subalgebras of some probability space (Ω, Σ, v). Then, for every $t>0$, we have*

$$t v(\sigma_t) \leqslant \int_{\sigma_t} |f_k(\omega)|\, dv\,,$$

where $\sigma_t = \{\omega \in \Omega;\ \max_{1\leqslant n\leqslant k} |f_n(\omega)| > t\}$.

Proof. Fix $t>0$ and define a function T on Ω with values in the set of integers $\{1, 2, \ldots, k, k+1\}$ by putting, for $\omega \in \Omega$,

$$T(\omega) = \begin{cases} \min\{j;\ |f_j(\omega)|>t\} & \text{if } \omega \in \sigma_t \\ k+1 & \text{if } \omega \notin \sigma_t\,. \end{cases}$$

Let $1\leqslant j\leqslant k$ and observe that

$$\Omega_j = \{\omega \in \Omega;\ T(\omega)=j\}$$

is precisely the set where $|f_j(\omega)|>t$ and $\max_{1\leqslant n<j} |f_n(\omega)| \leqslant t$. This implies that $\Omega_j \in \mathscr{B}_j$ for $1\leqslant j\leqslant k$. Furthermore, since $f_j = E^{\mathscr{B}_j} f_k$ and since the conditional expectation is a positive operator, it follows that $|f_j| \leqslant E^{\mathscr{B}_j}|f_k|$, $1\leqslant j\leqslant k$. Using these facts we get that

$$\begin{aligned} t v(\sigma_t) &= t v(\{\omega\in\Omega;\ T(\omega)\leqslant k\}) = t\sum_{j=1}^{k} v(\Omega_j) \\ &\leqslant \sum_{j=1}^{k} \int_{\Omega_j} |f_j(\omega)|\, dv \leqslant \sum_{j=1}^{k} \int_{\Omega_j} (E^{\mathscr{B}_j}|f_k|)(\omega)\, dv \\ &= \sum_{j=1}^{k} \int_{\Omega_j} |f_k(\omega)|\, dv = \int_{\sigma_t} |f_k(\omega)|\, dv\,. \quad \square \end{aligned}$$

Remarks. 1. The function T introduced in the proof above is a special instance of a fundamental notion used in probability theory, namely that of a *stopping time*.

Specifically, if $\{\mathscr{B}_j\}_{j=1}^k$ (with k finite or ∞) is an increasing sequence of σ-subalgebras of Σ then a function $T: \Omega \to \{1, 2, \ldots, k\}$ is called a stopping time with respect to $\{\mathscr{B}_j\}_{j=1}^k$ if $\{\omega;\, T(\omega)=j\} \in \mathscr{B}_j$, for every j.

2. The proof of 2.c.2 shows also that if $\{\mathscr{B}_n\}_{n=1}^\infty$ is an increasing sequence of σ-subalgebras of Σ then, for every $f \in L_1(\Omega, \Sigma, \nu)$,

$$t\nu\left(\left\{\omega \in \Omega;\, \sup_n |E^{\mathscr{B}_n} f|(\omega) > t\right\}\right) \leqslant \int_\Omega |f(\omega)|\, d\nu$$

i.e. the quasilinear operator $Uf = \sup_n |E^{\mathscr{B}_n} f|$ is of weak type (1, 1). It is evident that U is of strong type (∞, ∞). Hence, by the Marcinkiewicz interpolation theorem, U is of strong type (p, p) for every $p > 1$. This is essentially the assertion of 2.c.3 below which is called the maximal inequality of Doob (cf. [29]). We prefer to prove it directly since the proof is simple and gives also information on $\|U\|_p$.

We recall that, for every $1 \leqslant p < \infty$ and every $g \in L_p(\Omega, \Sigma, \nu)$, we have

$$(*) \qquad \int_\Omega |g(\omega)|^p\, d\nu = \int_0^\infty p t^{p-1} \nu(\{\omega \in \Omega;\, |g(t)| > t\})\, dt\,.$$

This identity can be easily checked for simple functions while for a general g is deduced by approximation with suitable simple functions.

Proposition 2.c.3. *Let $\{f_n\}_{n=1}^\infty$ be a martingale with respect to a sequence $\{\mathscr{B}_n\}_{n=1}^\infty$ of σ-subalgebras of some probability space (Ω, Σ, ν). Then, for every $1 < p < \infty$,*

$$\left\|\sup_n |f_n|\right\|_p \leqslant q \sup_n \|f_n\|_p\,,$$

where $\|\cdot\|_p$ denotes the norm in $L_p(\Omega, \Sigma, \nu)$ and q is the conjugate index of p i.e. $1/p + 1/q = 1$.

Proof. Observe first that it suffices to prove the maximal inequality of Doob only for finite martingales since the general case follows from it by Fatou's lemma. Fix k and put

$$f = \max_{1 \leqslant n \leqslant k} |f_n| \quad \text{and} \quad \sigma_t = \{\omega \in \Omega;\, f(\omega) > t\}, \quad t > 0\,.$$

By (*) and 2.c.2, we get that

$$\begin{aligned}\|f\|_p^p &= \int_0^\infty p t^{p-1} \nu(\sigma_t)\, dt \leqslant \int_0^\infty p t^{p-2} \int_{\sigma_t} |f_k(\omega)|\, d\nu(\omega)\, dt \\ &= \int_\Omega |f_k(\omega)| \int_0^\infty p t^{p-2} \chi_{\sigma_t}(\omega)\, dt\, d\nu(\omega) = \int_\Omega |f_k(\omega)| \int_0^{f(\omega)} p t^{p-2}\, dt\, d\nu(\omega) \\ &= q \int_\Omega |f_k(\omega)| f(\omega)^{p-1}\, d\nu(\omega)\,.\end{aligned}$$

Hence, by Hölder's inequality, it follows that

$$\|f\|_p^p \leqslant q\|f_k\|_p \cdot \|f\|_p^{p/q} \quad \text{i.e. that} \quad \|f\|_p \leqslant q\|f_k\|_p . \quad \square$$

We are prepared now to state a lemma which is crucial in the proof of the unconditionality of the Haar basis in $L_p(0, 1)$, $1<p<\infty$.

Lemma 2.c.4. *For every p of the form* 2^m, $m=1, 2, \ldots$ *there exists a constant* $C_p<\infty$ *such that if* $\{f_n\}_{n=1}^k$ *is a finite martingale with respect to a sequence* $\{\mathscr{B}_n\}_{n=1}^k$ *of σ-subalgebras of some probability space* (Ω, Σ, ν) *then*

$$C_p^{-1}\|f_k\|_p \leqslant \|S\|_p \leqslant C_p\|f_k\|_p ,$$

where S is the square function of the differences of $\{f_n\}_{n=1}^k$ *i.e.*

$$S=\left(|f_1|^2+\sum_{j=2}^k |f_j-f_{j-1}|^2\right)^{1/2} .$$

Proof. Put $f_0 \equiv 0$ and $\Delta f_j = f_j - f_{j-1}$, $1 \leqslant j \leqslant k$, so that

$$S=\left(\sum_{j=1}^k |\Delta f_j|^2\right)^{1/2} .$$

For $p=2$ (i.e. $m=1$) the assertion is true with $C_2=1$. Indeed, if $\{f_n\}_{n=1}^k$ is a martingale we get, for any $1 \leqslant j<n \leqslant k$, that

$$\int_\Omega \Delta f_j\, \Delta f_n\, d\nu = \int_\Omega E^{\mathscr{B}_j}(\Delta f_j\ \Delta f_n)\, d\nu = \int_\Omega \Delta f_j\ E^{\mathscr{B}_j}(\Delta f_n)\, d\nu = 0 ,$$

and the orthogonality in $L_2(\Omega, \Sigma, \nu)$ of $\{\Delta f_n\}_{n=1}^k$ clearly implies that

$$\|S\|_2^2=\sum_{j=1}^k \|\Delta f_j\|_2^2=\left\|\sum_{j=1}^k \Delta f_j\right\|_2^2=\|f_k\|_2^2 .$$

For a general p of the form 2^m, $m=1, 2, \ldots$ the proof is done by induction on m. To this end, suppose that the assertion of 2.c.4 is valid for some p and for *every* finite martingale. Let $\{f_n\}_{n=1}^k$ be a finite martingale and observe that

$$\begin{aligned}|f_k|^2 &= \left|\sum_{j=1}^k \Delta f_j\right|^2 = S^2+2\sum_{\substack{m,n=1\\ m<n}}^k \Delta f_m\ \Delta f_n\\ &= S^2+2\sum_{n=2}^k \Delta f_n \sum_{m=1}^{n-1} \Delta f_m = S^2+2\sum_{n=2}^k f_{n-1}\ \Delta f_n\end{aligned}$$

from which it follows that

$$\|S^2\|_p - 2\left\|\sum_{n=2}^{k} f_{n-1}\,\Delta f_n\right\|_p \leqslant \|f_k^2\|_p \leqslant \|S^2\|_p + 2\left\|\sum_{n=2}^{k} f_{n-1}\,\Delta f_n\right\|_p .$$

Notice now that the sequence $\{g_j\}_{j=1}^k$ defined by $g_1 \equiv 0$ and

$$g_j = \sum_{n=2}^{j} f_{n-1}\,\Delta f_n, \quad j=2, \ldots, k,$$

is also a martingale with respect to the same sequence of σ-subalgebras as that corresponding to $\{f_n\}_{n=1}^k$. Thus, by applying the induction hypothesis to this martingale, we get that

$$\left\|\sum_{n=2}^{k} f_{n-1}\,\Delta f_n\right\|_p \leqslant C_p\left\|\left(\sum_{n=2}^{k} |f_{n-1}\,\Delta f_n|^2\right)^{1/2}\right\|_p .$$

Put $f = \max_{1 \leqslant n \leqslant k} |f_n|$. Then, by using the Cauchy-Schwarz inequality and Doob's maximal inequality 2.c.3, it follows that

$$\begin{aligned}\left\|\sum_{n=2}^{k} f_{n-1}\,\Delta f_n\right\|_p &\leqslant C_p\|fS\|_p \leqslant C_p\|S\|_{2p}\|f\|_{2p}\\ &\leqslant C_p(2p)(2p-1)^{-1}\|S\|_{2p}\|f_k\|_{2p} .\end{aligned}$$

By combining this inequality with a preceding one, we obtain

$$\begin{aligned}\|S\|_{2p}^2 - 2C_p(2p)(2p-1)^{-1}\|S\|_{2p}\|f_k\|_{2p} &\leqslant \|f_k\|_{2p}^2 \leqslant\\ &\|S\|_{2p}^2 + 2C_p(2p)(2p-1)^{-1}\|S\|_{2p}\|f_k\|_{2p} .\end{aligned}$$

Set $\alpha = 2C_p(2p)(2p-1)^{-1}$ and $\Lambda = \|S\|_{2p}/\|f_k\|_{2p}$. Then the above inequalities become

$$\Lambda^2 - \alpha\Lambda \leqslant 1 \leqslant \Lambda^2 + \alpha\Lambda$$

which easily yields that

$$(\alpha+1)^{-1} \leqslant \frac{-\alpha+\sqrt{\alpha^2+4}}{2} \leqslant \Lambda \leqslant \frac{\alpha+\sqrt{\alpha^2+4}}{2} \leqslant \alpha+1 .$$

This proves that the assertion of 2.c.4 is true for $2p$ with

$$C_{2p} \leqslant 1 + 2C_p(2p)\cdot(2p-1)^{-1} . \quad \square$$

Theorem 2.c.5. *The Haar system* $\{\chi_n\}_{n=1}^\infty$ *is an unconditional basis of* $L_p(0, 1)$, *for every* $1 < p < \infty$.

Proof. First, we assume that p is of the form 2^m, $m=1, 2, \ldots$. Then, for every choice of scalars $\{a_n\}_{n=1}^k$ and every choice of signs $\theta_n=\pm 1$, $1 \leqslant n \leqslant k$, the functions $f_j=\sum_{n=1}^{j} a_n\chi_n$, $1\leqslant j\leqslant k$ and $g_j=\sum_{n=1}^{j} a_n\theta_n\chi_n$, $1\leqslant j\leqslant k$ form martingales with respect to $\{\mathscr{A}_n\}_{n=1}^k$. Since both these martingales have clearly the same square function $\left(\sum_{n=1}^{k} |a_n\chi_n|^2\right)^{1/2}$ it follows from 2.c.4 that

$$C_p^{-2}\|f_k\|_p \leqslant \|g_k\|_p \leqslant C_p^2\|f_k\|_p .$$

This, of course, proves the unconditionality of $\{\chi_n\}_{n=1}^\infty$. Since the unconditionality of $\{\chi_n\}_{n=1}^\infty$ is equivalent to the uniform boundedness of the natural projections associated to $\{\chi_n\}_{n=1}^\infty$ it follows from the Riesz-Thorin interpolation theorem 2.b.14 that the Haar system is an unconditional basis of $L_p(0, 1)$, for every $p \geqslant 2$. For $1<p<2$ the proof is achieved by duality since the Haar system, normalized in $L_p(0, 1)$, $1<p<2$, is clearly biorthogonal to the Haar system normalized in $L_q(0, 1)$, where $1/p+1/q=1$. □

The remarkable result 2.c.5 was originally proved by Paley [108]. Later on, proofs in the more general setting of martingales were given by Burkholder and Gundy [20], Burkholder [19], Garsia [46] and others. It follows from these proofs that the unconditional constant of the Haar basis in $L_p(0, 1)$, $1<p<\infty$ is $\leqslant K \max(p, 1/(p-1))$, for some universal constant K. It can be shown that this is the right order of magnitude. The idea of the proof presented here is taken from Cotlar [23]. It is easily seen that this proof gives (for $p \geqslant 2$) $C_p \leqslant Kp$ but for the unconditional constant of the Haar basis we get only the estimate $\leqslant K^2p^2$.

In some of the proofs of 2.c.5 it is shown first that, for every choice of signs $\{\theta_i\}_{i=1}^\infty$, the operator $T_\theta\left(\sum_{i=1}^{\infty} a_i\chi_i\right)=\sum_{i=1}^{\infty} a_i\theta_i\chi_i$ is of weak type (1, 1). The proof is then completed by using the Marcinkiewicz interpolation theorem and duality since obviously T_θ is of strong type (2, 2). The operator T_θ is not of strong type (1, 1) for every choice of signs $\{\theta_i\}_{i=1}^\infty$, i.e. the Haar basis is not unconditional in $L_1(0, 1)$. This follows from I.1.d.1 but can also be easily verified directly. Indeed, for every integer k,

$$\|\chi_1+\chi_2+2\chi_3+4\chi_5+8\chi_9+\cdots+2^k\chi_{2^k+1}\|_1=1$$

while

$$\lim_{k\to\infty} \|\chi_1+2\chi_3+8\chi_9+\cdots+2^{2k+1}\chi_{2^{2k+1}+1}\|_1=\infty .$$

Theorem 2.c.5 can be used in conjunction with the interpolation theorem 2.b.11 in order to give a characterization of the separable r.i. function spaces on [0, 1] for which the Haar system is unconditional.

Theorem 2.c.6. *The Haar system* $\{\chi_n\}_{n=1}^\infty$ *is an unconditional basis in a separable r.i. function space* X *on* [0, 1] *if and only if* X *does not contain uniformly isomorphic* *copies of l_1^n or of l_∞^n, for all n, on disjoint vectors with the same distribution function or, equivalently, if and only if* $1<p_X$ *and* $q_X<\infty$.

Proof. Suppose that X does not contain l_1^n's and l_∞^n's as in the statement. Then, by 2.b.7, its Boyd indices p_X and q_X satisfy $1<p_X$ and $q_X<\infty$. Hence, by using 2.b.11, for any $1<p<p_X$ and $q_X<q<\infty$, and 2.c.5 (in $L_p(0, 1)$ and in $L_q(0, 1)$), it follows that the Haar basis is unconditional in X.

Conversely, suppose e.g. that X contains uniformly isomorphic copies of l_1^n, for all n, on mutually disjoint vectors having the same distribution function. In other words, there exists a constant $M<\infty$ with the property that, for every n, there exist 2^n mutually disjoint positive functions $\{u_i\}_{i=1}^{2^n}$ all having the same distribution function such that $\|u_i\|=1$ for $1\leq i\leq 2^n$ and

$$M\left\|\sum_{i=1}^{2^n} u_i\right\|\geq 2^n .$$

Consider now the "Haar system $\{h_k\}_{k=1}^{2^n}$ over the vectors $\{u_i\}_{i=1}^{2^n}$", which is defined by

$$\begin{aligned}
&h_1=(u_1+\cdots+u_{2^n})/2^n\\
&h_2=(u_1+\cdots+u_{2^{n-1}}-u_{2^{n-1}+1}-\cdots-u_{2^n})/2^n\\
&\vdots\\
&h_{2^{n-1}+1}=(u_1-u_2)/2^n\\
&\vdots\\
&h_{2^n}=(u_{2^n-1}-u_{2^n})/2^n .
\end{aligned}$$

We may assume without loss of generality that each u_i is a finite linear combination of characteristic functions of intervals of the form $[(l_j-1)2^{-k}, l_j2^{-k})$ for some fixed k independent of i. Hence, by applying a suitable automorphism to $[0, 1]$, we may assume that on the first 2^n dyadic intervals of length 2^{-k} each u_i is non-zero on exactly one of these intervals and takes there a value independent of i (say β_1), that the same is true on the next 2^n dyadic intervals of length 2^{-k} (with β_1 replaced by β_2) and so on. In other words, we may assume that, for some integer m and scalars $\{\beta_j\}_{j=1}^m$, we have

$$u_i=\sum_{j=1}^m \beta_j\chi_{[(i-1+(j-1)2^n)2^{-k},\,(i+(j-1)2^n)2^{-k})},\quad 1\leq i\leq 2^n .$$

Note that

$$2^nh_2=u_1+u_2+\cdots u_{2^{n-1}}-u_{2^{n-1}+1}-\cdots-u_{2^n}=\sum_{j=1}^m \beta_j\chi_{2^{k-n+1}+j}$$

$$2^nh_3=u_1+u_2+\cdots+u_{2^{n-2}}-u_{2^{n-2}+1}-\cdots-u_{2^{n-1}}=\sum_{j=1}^m \beta_j\chi_{2^{k-n+2}+2j-1}$$

$$\begin{aligned}
2^nh_4&=u_{2^{n-1}+1}+\cdots+u_{2^{n-1}+2^{n-2}}-u_{2^{n-1}+2^{n-2}+1}-\cdots-u_{2^n}\\
&=\sum_{j=1}^m \beta_j\chi_{2^{k-n+2}+2j}
\end{aligned}$$

and so on. In other words, $\{h_i\}_{i=2}^{2^n}$ forms a block basis of a permutation of the Haar basis $\{\chi_s\}_{s=1}^{\infty}$ of X (observe that necessarily $m \leqslant 2^{k-n}$). Hence, the unconditional constant K_n of $\{h_i\}_{i=2}^{2^n}$ does not exceed that of the Haar basis of X. On the other hand, the definition of the $\{h_i\}_{i=1}^{2^n}$ in terms of the $\{u_i\}_{i=1}^{2^n}$ and the fact that the $\{u_i\}_{i=1}^{2^n}$ are M-equivalent to the unit vector basis of $l_1^{2^n}$ shows that, up to a factor M, the unconditionality constant of $\{h_i\}_{i=1}^{2^n}$ is the same as that of the first 2^n elements of the Haar basis in $L_1(0, 1)$. Consequently, $K_n \to \infty$ as $n \to \infty$ and we conclude that the Haar basis of X is not unconditional.

The case when X contains uniformly isomorphic copies of l_∞^n, for all n, on disjoint vectors having the same distribution function is treated similarly. □

We introduce next a notion of reproducibility for general bases and study it mainly in connection with the Haar basis. It is easily checked (see e.g. I.1.a.12) that when a Banach space X, with a normalized basis $\{x_n\}_{n=1}^{\infty}$ which tends weakly to 0, is isomorphic to a subspace of some space Y, with a basis $\{y_k\}_{k=1}^{\infty}$, then there exists a subsequence $\{x_{n_j}\}_{j=1}^{\infty}$ of $\{x_n\}_{n=1}^{\infty}$ which is equivalent to a block basis of $\{y_k\}_{k=1}^{\infty}$. We describe this situation by saying that a subsequence of $\{x_n\}_{n=1}^{\infty}$ can be "reproduced" as a block basis of $\{y_k\}_{k=1}^{\infty}$. Of particular interest is the case when $\{x_n\}_{n=1}^{\infty}$ itself can be reproduced as a block basis of any basis $\{y_k\}_{k=1}^{\infty}$ of an arbitrary space Y containing an isomorphic copy of X.

More precisely, we have the following definition from [76].

Definition 2.c.7. A Schauder basis $\{x_n\}_{n=1}^{\infty}$ of a Banach space X is said to be *K-reproducible*, for some $K \geqslant 1$, if, for every isometric embedding of X into a space Y with a basis $\{y_k\}_{k=1}^{\infty}$ and every $\varepsilon > 0$, there exists a block basis $\{z_n\}_{n=1}^{\infty}$ of $\{y_k\}_{k=1}^{\infty}$ which is $K+\varepsilon$-equivalent to $\{x_n\}_{n=1}^{\infty}$. A basis is said to be *reproducible* if it is K-reproducible for some $K < \infty$. When $K = 1$ the basis is said to be *precisely reproducible*.

We recall that two bases $\{x_n\}_{n=1}^{\infty}$ and $\{z_n\}_{n=1}^{\infty}$ are said to be K-equivalent for some constant $K < \infty$, whenever there exist constants K_1 and K_2 whose product is equal to K, such that

$$K_1^{-1}\left\|\sum_{n=1}^{\infty} a_n z_n\right\| \leqslant \left\|\sum_{n=1}^{\infty} a_n x_n\right\| \leqslant K_2\left\|\sum_{n=1}^{\infty} a_n z_n\right\|,$$

for every choice of scalars $\{a_n\}_{n=1}^{\infty}$.

It is clear that all subsymmetric bases with subsymmetric constant equal to one and, in particular, the unit vector bases in c_0 and in l_p, $1 \leqslant p \leqslant \infty$, are precisely reproducible. While this fact is entirely trivial, the reproducibility of the Haar basis is less obvious (cf. [107] and [76]).

Theorem 2.c.8. *The Haar basis of every separable r.i. function space X on $[0, 1]$ is precisely reproducible.*

The proof of 2.c.8 will be based on a theorem of Liapounoff [73] which has many other applications in functional analysis.

Theorem 2.c.9. *Let $\{\mu_i\}_{i=1}^n$ be a sequence of finite (not necessarily positive) non-atomic measures on a measure space (Ω, Σ). Then the set*

$$r(\{\mu_i\}_{i=1}^n) = \{(\mu_1(\sigma), \mu_2(\sigma), \ldots, \mu_n(\sigma)); \sigma \in \Sigma\}$$

is a compact convex subset of R^n.

Proof [75]. By decomposing each μ_i into its positive and negative part we reduce the case of general measures to that of positive measures (with n replaced by $2n$). It suffices therefore to prove 2.c.9 for positive measures. The proof will be done by induction on n. The proof for $n=1$ is the same as that of the induction step so we present only the induction step.

Let $\mu = \sum_{i=1}^n \mu_i$, W be the subset $\{g; 0 \leqslant g \leqslant 1\}$ of $L_\infty(\Omega, \Sigma, \mu)$ and let T be the map from $L_\infty(\mu)$ to R^n defined by

$$Tg = \left(\int_\Omega g\, d\mu_1, \int_\Omega g\, d\mu_2, \ldots, \int_\Omega g\, d\mu_n \right).$$

The set W is w^* compact and convex and T is linear and w^* continuous (since the $\{\mu_i\}_{i=1}^n$ are absolutely continuous with respect to μ). Hence, $T(W)$ is a convex and compact subset of R^n.

It is clear that $T(W) \supset r(\{\mu_i\}_{i=1}^n)$. In order to complete the proof, it suffices to show that these two sets coincide, i.e. that, for every $(a_1, a_2, \ldots, a_n) \in T(W)$ $W_0 = T^{-1}(a_1, a_2, \ldots, a_n) \cap W$ contains a characteristic function. The set W_0 is w^* compact and convex; thus it has extreme points, by the Krein-Milman theorem, and it is enough to prove that any extreme point g of W_0 must be a characteristic function.

Assume that $g \in \operatorname{ext} W_0$ is not a characteristic function. Then there is an $\varepsilon > 0$ and a subset σ_0 in Σ so that $\mu(\sigma_0) > 0$ and $\varepsilon \leqslant g(\omega) \leqslant 1 - \varepsilon$ for $\omega \in \sigma_0$. Since μ is non-atomic there exists a $\sigma_1 \subset \sigma_0$ so that $\mu(\sigma_1) > 0$ and $\mu(\sigma_2) > 0$, where $\sigma_2 = \sigma_0 \sim \sigma_1$. By the induction hypothesis, there are $\eta_1 \subset \sigma_1$ and $\eta_2 \subset \sigma_2$ so that

$$\mu_i(\eta_1) = \mu_i(\sigma_1)/2, \qquad \mu_i(\eta_2) = \mu_i(\sigma_2)/2, \quad i = 2, 3, \ldots, n.$$

Pick real s, t so that $|s|, |t| < \varepsilon$, $s^2 + t^2 > 0$ and

$$s(\mu_1(\sigma_1) - 2\mu_1(\eta_1)) + t(\mu_1(\sigma_2) - 2\mu_1(\eta_2)) = 0.$$

Let $h = s(\chi_{\sigma_1} - 2\chi_{\eta_1}) + t(\chi_{\sigma_2} - 2\chi_{\eta_2})$. Then $\int_\Omega h\, d\mu_i = 0$, for $1 \leqslant i \leqslant n$, and $|h| \leqslant g \leqslant 1 - |h|$ on Ω. Hence, $g \pm h \in W_0$ and, since $h \neq 0$, this contradicts the assumption that $g \in \operatorname{ext} W_0$. $\square$

Proof of 2.c.8. Assume that X is a subspace of a Banach space Y having a basis $\{y_k\}_{k=1}^\infty$ whose basis constant is K. Fix $\varepsilon > 0$ and let $z_1 = \sum_{k=1}^{p_1} a_k y_k$ be a vector in Y so that $\|z_1-h_1\|<\varepsilon\|h_1\|/2^3K$, where h_1 is the function identically equal to one on $[0, 1]$ (h_1 is an element of X and thus also of Y). Observe next that, for every $y^*\in Y^*$, the function $v(\sigma)=y^*(\chi_\sigma)$ defines a finite non-atomic measure on $[0, 1]$ (since $\|\chi_\sigma\|\to 0$ as $\mu(\sigma)\to 0$, where μ denotes the Lebesgue measure). Hence, by 2.c.9, there exists a $\sigma_1\subset[0, 1]$ so that if we put $\sigma_2=[0, 1]\sim\sigma_1$ we have

$$\mu(\sigma_1)=\mu(\sigma_2)=1/2, \qquad y_k^*(\chi_{\sigma_1})=y_k^*(\chi_{\sigma_2}), \quad 1\leq k\leq p_1\,,$$

where $\{y_k^*\}_{k=1}^\infty$ denote the functionals biorthogonal to $\{y_k\}_{k=1}^\infty$. Hence, if we put $h_2=\chi_{\sigma_1}-\chi_{\sigma_2}$ then $y_k^*(h_2)=0$, $1\leq k\leq p_1$. It follows that there exists a $p_2>p_1$ and a vector $z_2\in Y$ so that $z_2=\sum_{k=p_1+1}^{p_2} a_k y_k$ and $\|z_2-h_2\|<\varepsilon\|h_2\|/2^4K$. By applying 2.c.9 again, we find disjoint sets $\sigma_{1,1}$ and $\sigma_{1,2}$ with $\sigma_{1,1}\cup\sigma_{1,2}=\sigma_1$ and

$$\mu(\sigma_{1,1})=\mu(\sigma_{1,2})=1/4, \qquad y_k^*(\chi_{\sigma_{1,1}})=y_k^*(\chi_{\sigma_{1,2}}), \quad 1\leq k\leq p_2\,.$$

Hence, $y_k^*(h_3)=0$, $1\leq k\leq p_2$, where $h_3=\chi_{\sigma_{1,1}}-\chi_{\sigma_{1,2}}$. We continue in an obvious inductive procedure and find a sequence $\{h_n\}_{n=1}^\infty$ in X and a block basis $\{z_n\}_{n=1}^\infty$ of $\{y_k\}_{k=1}^\infty$ so that $\|z_n-h_n\|<\varepsilon\|h_n\|/2^{n+2}K$. Thus, $\{z_n\}_{n=1}^\infty$ is $(1+\varepsilon)(1-\varepsilon)^{-1}$ equivalent to $\{h_n\}_{n=1}^\infty$. The proof is completed by observing that the $\{h_n\}_{n=1}^\infty$ are, by their construction, isometrically equivalent to the Haar basis of X. □

Note that, in the proof above, it was not essential that e.g. $y_k^*(h_1)=0$ for $1\leq k\leq p_1$. It would have been enough to construct σ_1, and thus h_1, so that $|y^*(h_1)|$ becomes as small as we wish (say $<\varepsilon\|h_1\|/2^3Kp_1$). This can be done by using, instead of 2.c.9, the following proposition which is also of some independent interest.

Proposition 2.c.10. [137] *Let X be an r.i. function space on* $[0, 1]$. *The Rademacher functions form a weakly null sequence in X if and only if X is not equal to $L_\infty(0, 1)$, up to an equivalent norm.*

Proof. We have already studied in 2.b.4 the case when the Rademacher functions form a sequence which is equivalent to the unit vector basis of l_2. For a general r.i. function space X, the Rademacher functions $\{r_n\}_{n=1}^\infty$ form a symmetric basic sequence with symmetric constant equal to one since they clearly are symmetric and independent random variables with the same distribution.

Suppose now that $\{r_n\}_{n=1}^\infty$ does not converge weakly to zero in X. Then, by the unconditionality and symmetricity of $\{r_n\}_{n=1}^\infty$, we easily conclude that the Rademacher functions in X form a sequence which is equivalent to the unit vector basis of l_1. On the other hand, since $\left\|\sum_{n=1}^k r_n\right\|_2=k^{1/2}$, $k=1, 2, \ldots$ we get that, for every $\varepsilon>0$, $\lim_{k\to\infty}\mu(\sigma(k, \varepsilon))=0$, where

$$\sigma(k, \varepsilon)=\left\{t\in[0, 1];\ \left|\sum_{n=1}^k r_n(t)/k\right|>\varepsilon\right\}.$$

But

$$\left\|\left(\sum_{n=1}^{k} r_n\right)\Big/ k\right\|_X \leqslant \varepsilon + \|\chi_{\sigma(k,\varepsilon)}\|_X$$

from which it follows that if X is not equal to $L_\infty(0, 1)$, up to an equivalent norm, then $\|\chi_{\sigma(k,\varepsilon)}\|_X$, and thus also $\left\|\left(\sum_{n=1}^{k} r_n\right)\Big/ k\right\|_X$, tend to zero as $k \to \infty$. This contradicts the fact that $\{r_n\}_{n=1}^\infty$ is equivalent to the unit vector basis of l_1.

The converse assertion is trivial since in $L_\infty(0, 1)$ the sequence $\{r_n\}_{n=1}^\infty$ is clearly isometrically equivalent to the unit vector basis of l_1. □

A simple consequence of 2.c.8 is the following.

Corollary 2.c.11. *Let Y be a Banach space which has a basis whose unconditional constant K is finite. Then, for every r.i. function space X which is isomorphic to a subspace of Y, the Haar system forms an unconditional basis. Moreover, $K_X \leqslant Kd$, where K_X is the unconditional constant of the Haar basis of X and $d = \inf\{d(X, Z);\ Z$ is a subspace of $Y\}$.*

In particular, we get that an r.i. function space X on $[0, 1]$ has an unconditional basis if and only if $p_X > 1$ and $q_X < \infty$. Note also that, since the Haar basis is not unconditional in $L_1(0, 1)$ or in $L_\infty(0, 1)$, the unconditionality constant of the Haar basis in $L_p(0, 1)$ tends to ∞ if p tends to either 1 or ∞. Hence, if either $\inf_n \{p_n\}_{n=1}^\infty = 1$ or $\sup_n \{p_n\}_{n=1}^\infty = \infty$ the space $\left(\sum_{n=1}^\infty \oplus L_{p_n}(0, 1)\right)_2$ (which is reflexive if $1 < p_n < \infty$, for every n) is not isomorphic to a subspace of a space with an unconditional basis.

It turns out that not only the Haar basis of an r.i. function space on $[0, 1]$ is reproducible. In fact, the same is true for every unconditional basis of such a space. This result is due to G. Schechtman (cf. [119] for L_p spaces).

Theorem 2.c.12. *Let $\{x_n\}_{n=1}^\infty$ be an unconditional basis of an r.i. function space X on $[0, 1]$. Then $\{x_n\}_{n=1}^\infty$ is reproducible. In particular, $\{x_n\}_{n=1}^\infty$ is equivalent to a block basis of the Haar basis in X.*

Proof. We first observe that, since the interval $I = [0, 1]$ and the square $I \times I$ are measure theoretically equivalent, X is order isometric to the r.i. function space $X(I \times I)$ on the unit square in which the norm $\|f\|_{X(I \times I)}$ of a measurable function $f(s, t)$ on $I \times I$ is taken to be equal to $\|f^*\|_X$.

Suppose now that $X(I \times I)$ is a subspace of a Banach space Y which has a basis $\{y_i\}_{i=1}^\infty$ with basis constant K. The basis $\{x_n\}_{n=1}^\infty$ of X is, of course, isometrically equivalent to the sequence $\{x_n(s)\}_{n=1}^\infty$ in $X(I \times I)$. Fix $\varepsilon > 0$ and let $v_1 = \sum_{i=1}^{q_1} b_i y_i$ be a vector in Y so that $\|v_1 - x_1(s)\| < \varepsilon \|x_1\|/2^3 K$. Then notice that, by 2.c.10, the sequence $\{x_2(s)r_k(t)\}_{k=1}^{\infty}$, where $\{r_k(t)\}_{k=1}^{\infty}$ are the Rademacher functions on I, tends weakly to zero in $X(I\times I)$. Hence, there is an integer k_2 and a vector $v_2=\sum_{i=q_1+1}^{q_2} b_iy_i\in Y$ such that $\|v_2-x_2(s)r_{k_2}(t)\|<\varepsilon\|x_2\|/2^4K$. By repeating the argument for the sequence $\{x_3(s)r_{k_2+k}(t)\}_{k=1}^{\infty}$ and by continuing in this manner, we are able to construct a block basis $\{v_n\}_{n=1}^{\infty}$ of $\{y_i\}_{i=1}^{\infty}$ and an increasing sequence $\{k_n\}_{n=1}^{\infty}$ of integers so that $k_1=0$ and $\|v_n-x_n(s)r_{k_n}(t)\|<\varepsilon\|x_n\|/2^{n+2}K$, for all n (where $r_0(t)=1$). This implies that the sequence $\{x_n(s)r_{k_n}(t)\}_{n=1}^{\infty}$ is $(1+\varepsilon)(1+\varepsilon)^{-1}$-equivalent to the block basis $\{v_n\}_{n=1}^{\infty}$. Since $\{x_n\}_{n=1}^{\infty}$ is unconditional it is equivalent to $\{x_n(s)r_{k_n}(t)\}_{n=1}^{\infty}$. This fact is evident if $X=L_p(0, 1)$. For a general X this is a consequence of 1.d.6(ii) and of 2.d.1 below. □

It should be mentioned that if 2.c.12 is applied to the Haar basis $\{x_n\}_{n=1}^{\infty}$ of an r.i. function space in which this basis is unconditional we still do not obtain the precise reproducibility of $\{x_n\}_{n=1}^{\infty}$, as given by 2.c.8. Note also that the proof of 2.c.12 shows that the last assertion in its statement is valid also if $\{x_n\}_{n=1}^{\infty}$ is an unconditional basic sequence, provided that X is q-concave for some $q<\infty$. However, an unconditional basic sequence in X need not be reproducible even when X is q-concave for some $q<\infty$. For instance, it follows from the discussion preceding I.2.b.10 that, for each $p\geqslant 1$, the space $L_p(0, 1)$ contains an unconditional basic sequence which is equivalent to the unit vector basis of $\left(\sum_{n=1}^{\infty}\oplus l_2^n\right)_p$. Obviously, this basic sequence is not reproducible.

While unconditional bases in an r.i. function space as above are reproducible this is not the case for conditional bases. We conclude the study of the reproducibility with the following proposition, whose proof is evident.

Proposition 2.c.13. *Any conditional basic sequence in a Banach space with an unconditional basis is not reproducible.*

A result of Pelczynski and Singer [112] (which we partially proved at the end of section I.2.b) states that every infinite dimensional Banach space with a basis has a conditional basis and, actually, even uncountably many mutually non-equivalent conditional bases. Hence, by the observation 2.c.13, every space with an unconditional basis has uncountably many mutually non-equivalent non-reproducible bases. In Vol. IV we shall present a result from [76] which asserts that $C(0, 1)$ has the remarkable property that all its Schauder bases are reproducible.

If X is a separable r.i. function space on [0, 1] with $1<p_X$ and $q_X<\infty$ it follows from 2.c.6 and I.3.b.1 that X is isomorphic to a complemented subspace of a space Y with a symmetric basis. Moreover, if X is uniformly convex then it follows from I.3.b.2 that Y may be chosen to be, in addition, also uniformly convex. It will be of interest for us in the sequel to know that Y cannot be taken to be a separable Orlicz sequence space, unless $X=L_2(0, 1)$.

Theorem 2.c.14 [78]. *A separable r.i. function space X on* [0, 1] *is isomorphic to a* *subspace of some separable Orlicz sequence space h_M if and only if $X=L_2(0,1)$ (up to an equivalent norm).*

Proof. The "if" part is trivial. To prove the "only if" assertion, we notice first that, by 2.c.11, we have $1<p_X$ and $q_X<\infty$. Let m be an integer and let X_m be the r.i. function space on $[0,1]$, whose norm is defined by

$$\|f\|_{X_m}=\|D_{1/m}f\|_X/\|D_{1/m}1\|_X\ ,$$

where $D_{1/m}$ is the dilation operator introduced in the previous section. Since $\|D_s\|_{X_m}\leqslant\|D_s\|_X$ for every $0<s<1$ it follows, by 2.b.3 and the discussion following it, that there is a constant K, independent of m, so that, for every choice of scalars $\{a_n\}_{n=1}^l$,

$$K^{-1}\left(\sum_{n=1}^{l} a_n^2\right)^{1/2}\leqslant\left\|\sum_{n=1}^{l} a_n r_n\right\|_{X_m}\leqslant K\left(\sum_{n=1}^{l} a_n^2\right)^{1/2}.$$

In view of the definition of $\|\cdot\|_{X_m}$, it follows that if we put

$$r_n^m(t)=\begin{cases}\text{sign}\sin 2^n m\pi t, & 1\leqslant t\leqslant 1/m\\ 0, & \text{otherwise}\end{cases},\quad n=1,2,\ldots$$

and

$$\lambda_m=\|D_{1/m}1\|_X=\|\chi_{[0,1/m]}\|_X=\|r_n^m\|_X$$

then

$$K^{-1}\left(\sum_{n=1}^{l} a_n^2\right)^{1/2}\leqslant\left\|\sum_{n=1}^{l} a_n r_n^m/\lambda_m\right\|_X\leqslant K\left(\sum_{n=1}^{l} a_n^2\right)^{1/2}.$$

The same inequalities remain clearly valid if we replace the $\{r_n^m\}_{n=1}^\infty$ by their "translations" $\{r_{i,n}^m\}_{n=1}^\infty$ to the interval $[(i-1)/m, i/m]$, for $1\leqslant i\leqslant m$. More precisely,

$$r_{i,n}^m(t)=\begin{cases}\text{sign}\sin 2^n m\pi(t-(i-1)/m) & \text{if } t\in[(i-1)/m, i/m]\\ 0 & \text{otherwise}\ .\end{cases}$$

Assume now that T is an isomorphism from X into a separable Orlicz sequence space h_M. Since, for every $1\leqslant i\leqslant m$, $w\lim_{n\to\infty} r_{i,n}^m=0$ it follows that, for every $\varepsilon>0$, there are, by I.1.a.12, increasing sequences of integers $\{n_{k,i}\}_{k=1}^\infty$ and vectors $x_{i,k}\in h_M$ so that

$$(*)\qquad \|x_{i,k}-Tr_{i,n_{k,i}}^m/\lambda_m\|\leqslant\varepsilon 2^{-k},\ x_{i,k}=\sum_{j\in\sigma_{i,k}} b_j e_j,\quad 1\leqslant i\leqslant m,\ k=1,2,\ldots,$$

where $\{e_j\}_{j=1}^{\infty}$ denotes the unit vector basis of h_M and $\{\sigma_{i,k}\}_{i=1}^{m},{}_{k=1}^{\infty}$ are mutually disjoint subsets of the integers (i.e. $\sigma_{i_1,k_1}\cap\sigma_{i_2,k_2}=\varnothing$, unless $i_1=i_2$ and $k_1=k_2$). We shall use this fact for a fixed but sufficiently small ε (depending only on m). The required order of magnitude of ε will be pointed out as the proof proceeds.

Let $F_{i,k}$ be the Orlicz functions defined by

$$F_{i,k}(t)=\sum_{j\in\sigma_{i,k}} M(|b_j|t), \quad 1\leqslant i\leqslant m,\ k=1,2,\dots .$$

Recall that, for a finite set η of pairs (i, k) and scalars $\{a_{(i,k)}\}_{(i,k)\in\eta}$, we have

$$\Big\|\sum_{(i,k)\in\eta} a_{i,k}x_{i,k}\Big\|=1\Leftrightarrow \sum_{(i,k)\in\eta} F_{i,k}(|a_{i,k}|)=1 .$$

It follows from (*) that if ε is small enough then, for every $1\leqslant i\leqslant m$ and every choice of $\{a_k\}_{k=1}^{n}$, we have

$$C^{-1}\Big(\sum_{k=1}^{n} a_k^2\Big)^{1/2}\leqslant\Big\|\sum_{k=1}^{n} a_k x_{i,k}\Big\|\leqslant C\Big(\sum_{k=1}^{n} a_k^2\Big)^{1/2},$$

where $C=2K\|T\|\|T^{-1}\|$. For every $0\leqslant t<C^{-1}$ and every $1\leqslant i\leqslant m$ there are integers $k_1(i, t)$ and $k_2(i, t)$ so that

$$F_{i,k_1(i,t)}(t)\leqslant 2C^2t^2, \qquad F_{i,k_2(i,t)}(t)\geqslant t^2/2C^2 .$$

Let us verify e.g. the existence of $k_1(i, t)$. Given $t<C^{-1}$ we pick an integer n so that $1/2n\leqslant t^2C^2<1/n$. Then $\big\|\sum_{k=1}^{n} tx_{i,k}\big\|\leqslant Ctn^{1/2}<1$ and thus $\sum_{k=1}^{n} F_{i,k}(t)<1$; in particular, $F_{i,k}(t)<1/n\leqslant 2C^2t^2$, for some $1\leqslant k\leqslant n$.

Let now $\{t_i\}_{i=1}^{m}$ be reals so that $\sum_{i=1}^{m} t_i^2\leqslant 1/2C^2$. Then $\sum_{i=1}^{m} F_{i,k_1(i,t_i)}(t_i)\leqslant 1$ and thus also $\big\|\sum_{i=1}^{m} t_i x_{i,k_1(i,t_i)}\big\|\leqslant 1$. Similarly, if $\sum_{i=1}^{m} t_i^2\geqslant 2C^2$ then $\big\|\sum_{i=1}^{m} t_i x_{i,k_2(i,t_i)}\big\|\geqslant 1$. Note that since X is, in particular, a Köthe function space we have, for every choice of $\{t_i\}_{i=1}^{m}$ and $\{k_i\}_{i=1}^{m}$, that

$$\Big\|\sum_{i=1}^{m} t_i r^m_{i,k_i}\Big\|=\Big\|\sum_{i=1}^{m} t_i\chi_{[(i-1)/m,\,i/m]}\Big\| .$$

Hence, by combining the preceding observations with (*), we get that if ε is small enough ($\varepsilon<1/4Cm^{1/2}$, to be precise) then

$$\sum_{i=1}^{m} t_i^2=1/2C^2 \Rightarrow \Big\|\sum_{i=1}^{m} t_i\chi_{[(i-1)/m,\,i/m]}\Big\|\leqslant\lambda_m\|T^{-1}\|\Big(1+\varepsilon\sum_{i=1}^{m} t_i\Big)$$
$$\leqslant 2\lambda_m\|T^{-1}\|$$

and

$$\sum_{i=1}^{m} t_i^2 = 2C^2 \Rightarrow \left\| \sum_{i=1}^{m} t_i \chi_{[(i-1)/n,\, i/m]} \right\| \geqslant \lambda_m \|T\|^{-1} \left(1 - \varepsilon \sum_{i=1}^{m} t_i \right) \geqslant \lambda_m / 2\|T\| .$$

In other words, there exists an absolute constant K_0 (which, in particular, is independent of m) so that, for every choice of $\{t_i\}_{i=1}^m$, we have

$$K_0^{-1} \lambda_m \left(\sum_{i=1}^{m} t_i^2 \right)^{1/2} \leqslant \left\| \sum_{i=1}^{m} t_i \chi_{[(i-1)/m,\, i/m]} \right\| \leqslant K_0 \lambda_m \left(\sum_{i=1}^{m} t_i^2 \right)^{1/2} .$$

By taking, in particular, $t_i = 1$, $1 \leqslant i \leqslant m$, we deduce that

$$K_0^{-1} m^{-1/2} \leqslant \lambda_m \leqslant K_0 m^{-1/2},$$

for every m. Hence, for every simple dyadic function f on $[0, 1]$, we have

$$K_0^{-2} \|f\|_2 \leqslant \|f\|_X \leqslant K_0^2 \|f\|_2$$

and this proves that $X = L_2(0, 1)$, up to an equivalent norm. $\square$

We shall see later on in this chapter that, in the general theory of r.i. function spaces on $[0, 1]$, a certain exceptional role is played by those spaces X for which the Haar basis $\{\chi_n\}_{n=1}^\infty$ is equivalent to a sequence of disjointly supported elements $\{f_n\}_{n=1}^\infty$ in X. From 2.c.14 we can deduce, in particular, that this cannot happen in an Orlicz function space $L_M(0, 1)$ (unless, it is $L_2(0, 1)$, up to an equivalent norm). Indeed, if $\{f_n\}_{n=1}^\infty$ are disjointly supported elements in $L_M(0, 1)$ with $\sigma_n = \operatorname{supp} f_n$ then $\left\| \sum_{n=1}^{k} a_n f_n \right\| = 1$ if and only if $\sum_{n=1}^{k} F_n(|a_n|) = 1$, where $F_n(t) = \int_{\sigma_n} M(t|f_n(s)|)\, ds$ $n = 1, 2, \ldots$. In other words, the span of $\{f_n\}_{n=1}^\infty$ in $L_M(0, 1)$ is a modular sequence space (cf. I.4.d.1). The proof of 2.c.14 actually shows that an r.i. function space on $[0, 1]$ which embeds in a modular sequence space is $L_2(0, 1)$. (This is not really a stronger statement than that appearing in 2.c.14 in view of I.4.d.5.)

The Haar system is by far the most convenient basis for studying the structure of r.i. function spaces on $[0, 1]$. Therefore, this system has a central role in Banach space theory. However, from the point of view of analysis in general and, of course, from that of harmonic analysis, the most important basis or system of functions is the trigonometric system. We shall discuss here very briefly this system. For a more detailed study of the trigonometric system we refer to books on harmonic analysis (especially, Zygmund [132]).

It is somewhat more convenient to treat the trigonometric system in the setting of complex scalars. Thus, in the rest of this section we shall consider complex r.i. function spaces on $[0, 1]$. The trigonometric system on $[0, 1]$ consists of the functions $1, e^{2\pi i t}, e^{-2\pi i t}, e^{4\pi i t}, \ldots$. The order in which the trigonometric functions are enumerated is important. They form an unconditional basis of an r.i. function space X on [0, 1] only if $X=L_2(0, 1)$ (up to an equivalent norm). In order to verify this, we have just to remark that the square function corresponding to the expansion $\sum_k a_k e^{2\pi kit}$ is simply $\left(\sum_k |a_k|^2\right)^{1/2}$ and to apply 1.d.6(ii).

A major tool in the study of the trigonometric system is the following classical result of M. Riesz.

Theorem 2.c.15. *Let R be the linear projection defined by*

$$R\left(\sum_{k=-n}^{n} a_k e^{2\pi kit}\right)=\sum_{k=0}^{n} a_k e^{2\pi kit}$$

on the linear space of the trigonometric polynomials. Then, for any $1<p<\infty$, *R extends uniquely to a bounded projection in* $L_p(0, 1)$.

Proof. We shall present a proof which is due essentially to Bochner. It is very similar to, but simpler than, the proof given for 2.c.4 (obviously, it motivated Cotlar's proof of 2.c.4).

We note first that, by the Riesz–Thorin theorem and duality, it suffices to prove the theorem for p being a power of 2 and that for $p=2$ the assertion is obvious. We observe next that it suffices to prove that there exists a constant C_p so that, whenever f is of the form $\sum_{k=1}^{n} \alpha_k e^{2\pi kit}$, then

$$C_p^{-1}\|v\|_p \leqslant \|u\|_p \leqslant C_p\|v\|_p ,$$

where u and v are the real, respectively, the purely imaginary parts of f (i.e. $f=u+iv$). Indeed, once this is shown we put, given any function h of the form $\sum_{k=-n}^{n} a_k e^{2\pi kit}$ with $a_0=0$,

$$\sum_{k=1}^{n} (a_k+\overline{a_{-k}})\, e^{2\pi kit}=f=u+iv, \qquad \sum_{k=1}^{n} (a_k-\overline{a_{-k}})\, e^{2\pi kit}=g=r+is .$$

Then

$$u+is=h, \qquad r+iv=\sum_{k=1}^{n} a_k e^{2\pi kit}-\sum_{k=-n}^{-1} a_k e^{2\pi kit}$$

and

$$\|r+iv\|_p \leqslant C_p(\|u\|_p+\|s\|_p) \leqslant 2C_p\|h\|_p .$$

This shows that $\|R\|_p \leqslant 2C_p$ when restricted to those functions with $a_0=0$ and therefore R is bounded also without this restriction.

We prove the existence of C_p for $p=2^m$ by induction on m. Assume that a suitable C_p exists and let $f=u+iv$. Then $f^2=u^2-v^2+2iuv$. Since

$$\|v^2\|_p-\|u^2-v^2\|_p\leq\|u^2\|_p\leq\|v^2\|_p+\|u^2-v^2\|_p,$$

it follows from the induction hypothesis, applied to f^2, that

$$\|v\|_{2p}^2-2C_p\|uv\|_p\leq\|u\|_{2p}^2\leq\|v\|_{2p}^2+2C_p\|uv\|_p\,.$$

Hence, by the Cauchy-Schwarz inequality,

$$\|v\|_{2p}^2-2C_p\|u\|_{2p}\|v\|_{2p}\leq\|u\|_{2p}^2\leq\|v\|_{2p}^2+2C_p\|u\|_{2p}\|v\|_{2p}\,,$$

which clearly implies the existence of C_{2p}. □

An immediate consequence of 2.c.15 and 2.b.11 is

Theorem 2.c.16. *Let X be a separable r.i. function space on $[0, 1]$ with $1<p_X$ and $q_X<\infty$. Then the trigonometric system is a Schauder basis of X.*

Proof. By 2.b.11 and 2.c.15, R extends uniquely to a bounded operator on X. Since multiplication by $e^{2\pi mit}$ acts as an isometry on X for every integer m we get that, for every choice of scalars $\{a_k\}_{k=0}^n$ and $1<m<n$, we have

$$\left\|\sum_{k=m}^{n} a_k\, e^{2\pi kit}\right\|=\left\|\sum_{k=0}^{n-m} a_{k+m}\, e^{2\pi kit}\right\|\leq\|R\|\left\|\sum_{k=-m}^{n-m} a_{k+m}\, e^{2\pi kit}\right\|$$
$$=\|R\|\left\|\sum_{k=0}^{n} a_k\, e^{2\pi kit}\right\|.$$

Consequently, the trigonometric system is a basic sequence in X. Since X is separable $C(0, 1)$ is dense in X and this proves that the trigonometric system spans X i.e. is a basis of X. □

The operator R defined in 2.c.15 is not bounded in $L_1(0, 1)$ and thus, the trigonometric system is not a basis of $L_1(0, 1)$. It can e.g. be verified that

$$f(t)=\sum_{k=2}^{\infty}\cos 2\pi kt/\log k=\sum_{k=2}^{\infty}(e^{2\pi kit}+e^{-2\pi kit})/2\log k$$

belongs to $L_1(0, 1)$ while $\sum_{k=2}^{\infty} e^{2\pi kit}/\log k$ does not belong to $L_1(0, 1)$. The operator R is, however, of weak type (1, 1). Let us also mention that, actually, the converse of 2.c.16 is valid. It was proved in [39] *that the operator R is bounded in an r.i. function space X on $[0, 1]$ (or, equivalently, that the trigonometric system is a basic sequence in X) only if $1<p_X$ and $q_X<\infty$.*

We conclude this section by mentioning another simple but interesting consequence of 2.c.15. Let X be an r.i. function space on [0, 1]. We denote by X_a the closed linear span of $\{e^{2\pi kit}\}_{k=0}^{\infty}$ in X. The space X_a can be viewed as the analytic part of X. It can be identified with the space of all functions $f(z)$, which are analytic in the open unit disc $|z|<1$ and for which,

$$\|f\|_{X_a}=\sup_{r<1}\|f_r\|_X<\infty$$

where $f_r(t)=f(re^{2\pi it})$, $0\leqslant t\leqslant 1$. In case $X=L_p(0,1)$, $1\leqslant p\leqslant\infty$, the space X_a is commonly denoted by H_p, $1\leqslant p\leqslant\infty$.

Proposition 2.c.17 [12]. *Let X be a separable r.i. function space on* [0, 1] *with* $1<p_X$ *and* $q_X<\infty$. *Then X is isomorphic to X_a.*

Proof. Let R be the projection on X defined in 2.c.15. The map $(Sg)(t)=e^{-2\pi it}g(1-t)$ is an isometry from X_a onto $(I_X-R)X=[e^{-2\pi kit}]_{k=1}^{\infty}$. Hence $X\approx X_a\oplus X_a$. We have also that $X_a\approx X_a\oplus X_a$. Indeed, the map $f(t)\rightarrow f(2t(\mathrm{mod}\ 1))$ defines an isomorphism between X_a and its subspace $[e^{4\pi kit}]_{k=0}^{\infty}$. This subspace is complemented by the projection Q defined by

$$Qf(t)=(f(t)+f((t+1/2)(\mathrm{mod}\ 1)))/2$$

and $\ker Q=[e^{2\pi(2k+1)it}]_{k=0}^{\infty}$ is also isomorphic to X_a. □

It follows, in particular, that, for $1<p<\infty$, the spaces $L_p(0,1)$ and H_p are isomorphic. It is known that this is not the case if $p=1$ or $p=\infty$. For a detailed study of the isomorphic properties of H_1 and H_∞ we refer to [111]. For the classical theory of H_p spaces the reader is referred to [34] and [52].

d. Some Results on Complemented Subspaces

In this section we present several results on the structure of unconditional basic sequences and complemented subspaces of r.i. function spaces X on [0, 1]. Nearly all the proofs rely on the unconditionality of the Haar basis and therefore we shall assume in most places that X is separable and its Boyd indices satisfy $1<p_X$ and $q_X<\infty$. For instance, we prove that under those assumptions X is a primary space. The results presented here were originally proved for L_p spaces. In many cases the extension to more general r.i. function spaces is easily achieved by a suitable application of the interpolation theorem 2.b.11.

We begin with some preliminary material involving the space $X(l_2)$ associated to a Banach lattice X. Recall that $X(l_2)$ is the completion of the space of all sequences

$(x_1, x_2, \ldots)$ of elements of X which are eventually zero, with respect to the norm,

$$\|(x_1, x_2, \ldots)\|_{X(l_2)} = \left\| \left(\sum_{i=1}^{\infty} |x_i|^2 \right)^{1/2} \right\|_X .$$

In other words, $X(l_2)$ consists of all the sequences $(x_1, x_2, \ldots)$ for which

$$\lim_{m \to \infty} \sup_{n > m} \left\| \left(\sum_{i=m}^{n} |x_i|^2 \right)^{1/2} \right\|_X = 0 .$$

In the case when X is an r.i. function space on the interval $I = [0, 1]$ and $q_X < \infty$, the space $X(l_2)$ can be identified (up to an equivalent norm) with the subspace of the r.i. space $X(I \times I)$ spanned by the functions of the form $x(s)r_n(t)$, where $x(s) \in X$ and $n = 1, 2, \ldots$ ($r_n(t)$ denotes as usual the nth Rademacher function). This subspace of $X(I \times I)$ is denoted by Rad X. More precisely, we have the following result.

Proposition 2.d.1. *Let X be an r.i. function space on $I = [0, 1]$ with $q_X < \infty$. Then there exists a constant $M < \infty$ so that, for every $x = (x_1, x_2, \ldots) \in X(l_2)$, we have*

$$M^{-1} \|x\|_{X(l_2)} \leq \left\| \sum_{i=1}^{\infty} x_i(s) r_i(t) \right\|_{X(I \times I)} \leq M \|x\|_{X(l_2)} .$$

Proof. Let $\{x_i\}_{i=1}^n$ be a finite sequence of elements of X. Since $\{x_i(s)r_i(t)\}_{i=1}^n$ has unconditional constant equal to one in $X(I \times I)$ it follows from Khintchine's inequality that

$$\begin{aligned} \left\| \sum_{i=1}^{n} x_i(s) r_i(t) \right\|_{X(I \times I)} &= \int_0^1 \left\| \sum_{i=1}^{n} x_i(s) r_i(t) r_i(u) \right\|_{X(I \times I)} du \\ &\geq \left\| \int_0^1 \left| \sum_{i=1}^{n} x_i(s) r_i(t) r_i(u) \right| du \right\|_{X(I \times I)} \\ &\geq A_1 \left\| \left(\sum_{i=1}^{n} |x_i(s)|^2 \right)^{1/2} \right\|_{X(I \times I)} \\ &= A_1 \left\| \left(\sum_{i=1}^{n} |x_i|^2 \right)^{1/2} \right\|_X . \end{aligned}$$

The other estimate can be obtained in a similar manner if X is q-concave, for some $q < \infty$. Our assumption that $q_X < \infty$ is, however, weaker and thus the proof requires additional work.

Let $\{x_i\}_{i=1}^n \subset X$ be chosen so that the function $f = \left(\sum_{i=1}^{n} |x_i|^2 \right)^{1/2}$ satisfies $\|f\|_X \leq 1$. It follows easily from Khintchine's inequality in $L_q(0, 1)$ that

$$\mu\left(\left\{ t \in I;\ \left| \sum_{i=1}^{n} x_i(s) r_i(t) \right| > v \right\} \right) \leq B_q^q f(s)^q v^{-q} ,$$

for every $0<v<\infty$, $s \in I$ and for every $1 \leq q \leq \infty$. On the other hand, since $t^{-1/q}$ is a decreasing function, we have that

$$\mu(\{t \in I;\ f(s)t^{-1/q}>v\})=f(s)^q v^{-q}.$$

Hence, for every $v>0$ and $q \geq 1$,

$$d_{\left|\sum_{i=1}^{n} x_i(s)r_i(t)\right|}(v) \leq B_q^q\, d_{f(s)t^{-1/q}}(v)$$

from which we deduce (since the dilation operator $D_{B_q^q}$ has a norm $\leq B_q^q$) that

$$\begin{aligned}\left\|\sum_{i=1}^{n} x_i(s)r_i(t)\right\|_{X(I\times I)} &\leq B_q^q \|f(s)t^{-1/q}\|_{X(I\times I)}\\ &\leq B_q^q \left\| f(s) \sum_{n=0}^{\infty} 2^{(n+1)/q} \chi_{(2^{-n-1},2^{-n}]}(t)\right\|_{X(I\times I)}\\ &\leq B_q^q \sum_{n=0}^{\infty} 2^{(n+1)/q} \|f(s)\chi_{(2^{-n-1},2^{-n}]}(t)\|_{X(I\times I)}.\end{aligned}$$

Fix now $q>q_0>q_X$ and recall that the definition of q_X implies the existence of a constant $C_{q_0}<\infty$ so that

$$\|D_u\|_X \leq C_{q_0} u^{1/q_0},$$

for all $0 \leq u \leq 1$. Therefore, by using the fact that $f(s)\chi_{(2^{-n-1},2^{-n}]}(t)$ in $X(I\times I)$ has the same distribution function as $D_{2^{-n-1}}f$ in X, we get that

$$\begin{aligned}\left\|\sum_{i=1}^{n} x_i(s)r_i(t)\right\|_{X(I\times I)} &\leq B_q^q \sum_{n=0}^{\infty} 2^{(n+1)/q} \|D_{2^{-n-1}}f\|_X\\ &\leq B_q^q C_{q_0} \sum_{n=0}^{\infty} 2^{(n+1)/q} 2^{-(n+1)/q_0}.\end{aligned}$$

Since $q>q_0$ the series above converges. □

Remark. In case X is q-concave for some $q<\infty$ then $X(l_2)$ admits another representation which we encountered already in Section 1.e (but which will not be used in this section). The space $X(l_2)$ is, for every $1 \leq p<\infty$, isomorphic to the subspace of $L_p(X)$ spanned by the functions of the form $xr_n(t)$ with $x \in X$ and $n=1, 2, \ldots$ (use 1.d.6 and 1.e.13).

An important fact concerning Rad X is the following.

Proposition 2.d.2. *Let X be a separable r.i. function space on $I=[0, 1]$ such that $1<p_X$ and $q_X<\infty$. Then* Rad X *is a complemented subspace of* $X(I\times I)$.

Proof. For $f(s, t) \in \bigcup_{p>1} L_p(I \times I)$ put

$$Pf(s, t) = \sum_{n=1}^{\infty} \left(\int_0^1 f(s, u) r_n(u)\, du \right) r_n(t).$$

We have already observed in the proof of 2.b.4(ii) that, for $1 < p < \infty$, there is a constant C_p so that for $s \in I$

$$\int_0^1 |Pf(s, t)|^p\, dt \leq C_p \int_0^1 |f(s, t)|^p\, dt.$$

By integrating this inequality with respect to s over $[0, 1]$, we deduce that P is bounded on $L_p(I \times I)$ for every $1 < p < \infty$. Hence, by 2.b.11, P is a bounded operator on $X(I \times I)$. It is clear that the range of P contains Rad X and that every function in the range of P is of the form $g(s, t) = \sum_{n=1}^{\infty} x_n(s) r_n(t)$. It remains to show that if this infinite sum (taken in the sense of convergence in measure) belongs to $X(I \times I)$ then the series converges in the norm of $X(I \times I)$ and thus its sum $g(s, t)$ must belong to Rad X. Indeed, assume that there is an $\varepsilon > 0$ and a sequence of integers $0 = m_1 < m_2 < \cdots$ so that $g_i(s, t) = \sum_{n=m_i+1}^{m_{i+1}} x_n(s) r_n(t)$ satisfies $\|g_i\|_{X(I \times I)} \geq \varepsilon$, for every i. Since the sum $\sum_{i=1}^{\infty} g_i(s, t) = g(s, t)$ belongs to $X(I \times I)$ and since, for every choice of signs $\theta = \{\theta_i\}_{i=1}^{\infty}$, $g_\theta(s, t) = \sum_{i=1}^{\infty} \theta_i g_i(s, t)$ has the same distribution function as $g(s, t)$ we would deduce that $g_\theta(s, t) \in X(I \times I)$, too. But, for any two distinct sequences of signs θ and θ', we have that $\|g_\theta - g_{\theta'}\|_{X(I \times I)} \geq 2\varepsilon$ and this contradicts the separability of X. $\square$

Remark. If X is a non-separable r.i. function space on $[0, 1]$ with $1 < p_X$ and $q_X < \infty$ then the projection P above is still bounded. However, its range is not Rad X but the strictly larger space consisting of all the functions of the form $\sum_{n=1}^{\infty} x_n(s) r_n(t)$ with $\left(\sum_{n=1}^{\infty} |x_n(s)|^2 \right)^{1/2} \in X$.

From 2.d.1 and 2.d.2 and the interpolation theorem 2.b.11 it is easy to deduce an interpolation theorem for operators defined on $X(l_2)$. For simplicity of notation we shall write $L_p(l_2)$ instead of $L_p(0, 1)(l_2)$.

Proposition 2.d.3. *Let X be a separable r.i. function space on* $[0, 1]$ *with* $1 < p_X$ *and* $q_X < \infty$. *Let* $1 < p < p_X$, $q_X < q < \infty$ *and let T be a bounded linear operator from $L_p(l_2)$ into itself which also acts as a bounded operator from $L_q(l_2)$ into itself. Then T maps $X(l_2)$ into itself and is bounded on this space.*

Observe that, since $L_q(0, 1) \subset X \subset L_p(0, 1)$ (cf. 2.b.3), we have also, in a natural way, the inclusions

$$L_q(l_2) \subset X(l_2) \subset L_p(l_2)$$

and thus it makes sense to talk of the restriction to $L_q(l_2)$ or to $X(l_2)$ of an operator defined on $L_p(l_2)$.

Proof. We first identify the three spaces appearing in the statement of 2.d.3 with the respective spaces Rad $L_p(I)$, Rad $L_q(I)$ and Rad X, of functions on $I \times I$, via the correspondence established in 2.d.1. Let P be the projection defined in 2.d.2. Since TP is a bounded operator from $L_p(I \times I)$ into itself and from $L_q(I \times I)$ into itself it follows, by 2.b.11, that TP is also a bounded operator from $X(I \times I)$ into itself. Consequently, T is a bounded operator from Rad X into itself. □

Remark. A similar result holds, with the same proof, for operators from X into $X(l_2)$ or from $X(l_2)$ into X. Thus, e.g. (with X, p and q as in 2.d.3) if T is a bounded linear operator from $L_p(0, 1)$ into $L_p(l_2)$ which also acts as a bounded operator from $L_q(0, 1)$ to $L_q(l_2)$ then T is a bounded linear operator from X into $X(l_2)$.

Another consequence of 2.d.1 and 2.d.2 is the following result (proved by B. S. Mitjagin [102] under somewhat different assumptions).

Proposition 2.d.4. *Let X be a separable r.i. function space on* $[0, 1]$ *with* $1 < p_X$ *and* $q_X < \infty$. *Then X is isomorphic to* $X(l_2)$.

Proof. By 2.d.1 and 2.d.2, $X(l_2)$ is isomorphic to a complemented subspace of X. On the other hand, X is clearly isometric to a complemented subspace of $X(l_2)$. Since X and $X(l_2)$ are obviously isomorphic to their respective squares the desired result follows by using Pelczynski's decomposition method. □

Remark. The isomorphism between X and $X(l_2)$, established in 2.d.4, is 'canonical' in the terminology of category theory. In other words, the isomorphism from X onto $X(l_2)$ is induced by an operator whose formal form is independent of X. Proposition 2.d.3 is a direct consequence of this fact and 2.b.11. The proof of 2.d.3, presented above, is, however, conceptually simpler.

From 2.d.4 we can get some information on general complemented subspaces of r.i. function spaces.

Proposition 2.d.5. *Let X be an r.i. function space on* $[0, 1]$ *with* $1 < p_X$ *and* $q_X < \infty$. *Let Y be a complemented subspace of X which, in turn, contains a complemented subspace isomorphic to X. Then Y itself is isomorphic to X.*

Proof. We assume first that X is separable. We let $Y(l_2)$ denote the subspace of $X(l_2)$ consisting of all $x = (x_1, x_2, \ldots, x_n, \ldots) \in X(l_2)$, with $x_n \in Y$ for all n. (Note that the vectors $\left(\sum_{n=1}^{k} |x_n|^2\right)^{1/2}$ need not belong to Y and that $Y(l_2)$ is not a space intrinsically associated to Y in the sense that it depends on the embedding of Y in

X.) Let Q be a projection from X onto Y. By 1.f.14, the operator

$$\hat{Q}(x_1, x_2, \ldots, x_n, \ldots) = (Qx_1, Qx_2, \ldots, Qx_n, \ldots)$$

defines a projection from $X(l_2)$ onto $Y(l_2)$ whose norm is $\leqslant K_G\|Q\|$ (as we have already mentioned, 1.f.14 was proved by J. L. Krivine [66]; prior to this, B. S. Mitjagin [102] had proved this fact for X being an r.i. function space on $[0, 1]$). Hence, there exist Banach spaces Z and W so that

$$X(l_2) \approx Y(l_2) \oplus W, \quad Y \approx X \oplus Z\,.$$

It follows that

$$Y \approx X \oplus Z \approx X \oplus X \oplus Z \approx X \oplus Y$$

and, by 2.d.4,

$$X \approx X(l_2) \approx Y(l_2) \oplus W \approx Y \oplus Y(l_2) \oplus W \approx X \oplus Y\,.$$

This concludes the proof if X is separable. If X is non-separable we can apply the same argument provided we replace $X(l_2)$ by the space $\widetilde{X(l_2)}$ of all sequences $x = (x_1, x_2, \ldots, x_n, \ldots)$ such that $\left(\sum_{n=1}^{\infty} |x_n|^2\right)^{1/2} \in X$ and $Y(l_2)$ by $\widetilde{Y(l_2)}$. For non-separable X with $1 < p_X$ and $q_X < \infty$ we have, in analogy to 2.d.4, that $X \approx \widetilde{X(l_2)}$. The proof is similar to 2.d.4; this can be also deduced from 2.d.4 by duality (apply 2.a.3 and the discussion preceding 1.d.4). □

We pass now to a discussion of some facts concerning unconditional bases in X and $X(l_2)$. To this end, we need the two dimensional version of the classical Khintchine inequality.

Proposition 2.d.6. *For every $1 \leqslant p < \infty$ there exist constants $A_p^{(2)}$ and $B_p^{(2)}$ so that, for every matrix of scalars $(a_{m,n})_{m,n=1}^{\infty}$ with only finitely many non-zero entries,*

$$A_p^{(2)}\left(\sum_{m=1}^{\infty}\sum_{n=1}^{\infty} |a_{m,n}|^2\right)^{1/2} \leqslant \left(\int_0^1\int_0^1 \left|\sum_{m=1}^{\infty}\sum_{n=1}^{\infty} a_{m,n} r_m(t) r_n(s)\right|^p dt\, ds\right)^{1/p}$$
$$\leqslant B_p^{(2)}\left(\sum_{m=1}^{\infty}\sum_{n=1}^{\infty} |a_{m,n}|^2\right)^{1/2}.$$

Proof. Since the functions $\{r_m(t) r_n(s)\}_{m=1,n=1}^{\infty\ \ \infty}$ form an orthogonal system in $L_2([0, 1] \times [0, 1])$ we have that

$$\left(\int_0^1\int_0^1 \left|\sum_{m=1}^{\infty}\sum_{n=1}^{\infty} a_{m,n} r_m(t) r_n(s)\right|^2 dt\, ds\right)^{1/2} = \left(\sum_{m=1}^{\infty}\sum_{n=1}^{\infty} |a_{m,n}|^2\right)^{1/2}.$$

Therefore, for $p>2$, it suffices to prove the right-hand side inequality of 2.d.6. Using the one-dimensional inequality of Khintchine and the triangle inequality in $L_{p/2}$, we get

$$\begin{aligned}
&\left(\int_0^1\int_0^1\left|\sum_{m=1}^\infty\sum_{n=1}^\infty a_{m,n}r_m(t)r_n(s)\right|^p dt\,ds\right)^{1/p}\\
&\leqslant B_p\left(\int_0^1\left(\sum_{m=1}^\infty\left|\sum_{n=1}^\infty a_{m,n}r_n(s)\right|^2\right)^{p/2} ds\right)^{1/p}\\
&\leqslant B_p\left(\sum_{m=1}^\infty\left(\int_0^1\left|\sum_{n=1}^\infty a_{m,n}r_n(s)\right|^p ds\right)^{2/p}\right)^{1/2}\\
&\leqslant B_p^2\left(\sum_{m=1}^\infty\sum_{n=1}^\infty |a_{m,n}|^2\right)^{1/2}.
\end{aligned}$$

It remains to prove the left-hand side inequality for $1\leqslant p<2$. It suffices to consider $p=1$. The desired inequality follows from what we have already proved for $p=4$ and the fact that $\|f\|_2\leqslant\|f\|_1^{1/3}\|f\|_4^{2/3}$ for every function f (by Hölder's inequality). □

The two dimensional Khintchine inequality 2.d.6 and its proof extend in an obvious way to any finite dimension.

Our first result on bases in $X(l_2)$ is valid for general lattices.

Proposition 2.d.7. *Let X be a q-concave Banach lattice for some $q<\infty$. Let $\{y_m\}_{m=1}^\infty$ be an unconditional basic sequence in X. Then the double sequence*

$$y_{m,n}=(0,\ldots,0,\overset{n}{y_m},0,\ldots),\quad m,n=1,2,\ldots$$

forms an unconditional basic sequence in $X(l_2)$. In particular, if $\{y_m\}_{m=1}^\infty$ is an unconditional basis of X then $\{y_{m,n}\}_{m,n=1}^\infty$ is an unconditional basis of $X(l_2)$.

Proof. Let K be the unconditional constant of $\{y_m\}_{m=1}^\infty$ and let M denote the q-concavity constant of X. Then, by Khintchine's inequality in $L_1(I)$ and $L_q(I\times I)$, we get, for every choice of scalars $\{a_{m,n}\}_{m,n=1}^\infty$ with only finitely many different from zero, that

$$\begin{aligned}
\left\|\sum_{m=1}^\infty\sum_{n=1}^\infty a_{m,n}y_{m,n}\right\|_{X(l_2)}&=\left\|\left(\sum_{n=1}^\infty\left|\sum_{m=1}^\infty a_{m,n}y_m\right|^2\right)^{1/2}\right\|_X\\
&\leqslant A_1^{-1}\left\|\int_0^1\left|\sum_{n=1}^\infty r_n(u)\sum_{m=1}^\infty a_{m,n}y_m\right|du\right\|_X\\
&\leqslant A_1^{-1}K\int_0^1\int_0^1\left\|\sum_{m=1}^\infty r_m(v)\sum_{n=1}^\infty a_{m,n}r_n(u)y_m\right\|_X du\,dv\\
&\leqslant A_1^{-1}K\left(\int_0^1\int_0^1\left\|\sum_{m=1}^\infty r_m(v)\sum_{n=1}^\infty a_{m,n}r_n(u)y_m\right\|_X^q du\,dv\right)^{1/q}
\end{aligned}$$

$$\leqslant A_1^{-1}KM\left\|\left(\int_0^1\int_0^1\left|\sum_{m=1}^{\infty}\sum_{n=1}^{\infty} r_m(v)r_n(u)a_{m,n}y_m\right|^q du\,dv\right)^{1/q}\right\|_X$$
$$\leqslant A_1^{-1}B_q^{(2)}KM\left\|\left(\sum_{m=1}^{\infty}\sum_{n=1}^{\infty}|a_{m,n}|^2|y_m|^2\right)^{1/2}\right\|_X.$$

In a similar manner, it can be shown that

$$\left\|\sum_{m=1}^{\infty}\sum_{n=1}^{\infty} a_{m,n}y_{m,n}\right\|_{X(l_2)} \geqslant A_1^{(2)}B_q^{-1}K^{-1}M^{-1}\left\|\left(\sum_{m=1}^{\infty}\sum_{n=1}^{\infty}|a_{m,n}|^2|y_m|^2\right)^{1/2}\right\|_X.$$

These two estimates for the norm of $\sum_{m=1}^{\infty}\sum_{n=1}^{\infty} a_{m,n}y_{m,n}$ in $X(l_2)$ prove our assertion. □

Note that if we put $Y=[y_m]_{m=1}^{\infty}$ then, with the notation used in the proof of 2.d.5, $Y(l_2)=[y_{m,n}]_{m,n=1}^{\infty}$.

In the case of the Haar basis in an r.i. function space we can prove the same result with the assumption of q-concavity replaced by the weaker one that $q_X<\infty$.

Proposition 2.d.8. *Let X be a separable r.i. function space on $I=[0, 1]$ with $1<p_X$ and $q_X<\infty$. Let $\{\chi_m\}_{m=1}^{\infty}$ be the Haar basis of X. Then the vectors*

$$\chi_{m,n}=(0, \ldots, 0, \overset{n}{\chi_m}, 0, \ldots), \quad m, n=1, 2, \ldots$$

form an unconditional basis of $X(l_2)$.

Proof. Let $1<p<p_X$ and $q_X<q<\infty$. By 2.d.7, $\{\chi_{m,n}\}_{m,n=1}^{\infty}$ is an unconditional basis of $L_p(l_2)$ and $L_q(l_2)$. Hence, by 2.d.3, this double sequence is also an unconditional basis of $X(l_2)$. □

Before we proceed, let us recall that, by 1.d.6 and the remark thereafter, for every unconditional basis $\{u_n\}_{n=1}^{\infty}$ of a complemented subspace of a Banach lattice X (and thus, in particular, for the Haar basis of a separable r.i. function space on $[0, 1]$ with $1<p_X\leqslant q_X<\infty$), the expression $\left\|\sum_{n=1}^{k} a_nu_n\right\|_X$ is equivalent to the norm of the square function $\left\|\left(\sum_{n=1}^{k}|a_nu_n|^2\right)^{1/2}\right\|_X$.

In the previous section we have seen that every unconditional basic sequence $\{y_m\}_{m=1}^{\infty}$ in an r.i. function space X on $[0, 1]$ which is q-concave for some $q<\infty$ is equivalent to a block basis $\{z_m\}_{m=1}^{\infty}$ of the Haar system of X. We shall prove now, under the usual assumptions made on X in this section, that if $[y_m]_{m=1}^{\infty}$ is complemented in X then $\{z_m\}_{m=1}^{\infty}$ can be chosen so that, in addition, $[z_m]_{m=1}^{\infty}$ is complemented in X. For the case of $X=L_p(0, 1)$, $1<p<\infty$, this was proved by G. Schechtman [119].

Theorem 2.d.9. *Let X be a separable r.i. function space on $[0, 1]$ with $1<p_X$ and $q_X<\infty$. Let Y be a complemented subspace of X having an unconditional basis $\{y_m\}_{m=1}^\infty$. Then $\{y_m\}_{m=1}^\infty$ is equivalent to a block basis $\{z_m\}_{m=1}^\infty$ of the Haar basis in X with $[z_m]_{m=1}^\infty$ complemented in X.*

Proof. Since X is separable the step functions on the dyadic intervals are dense in X. Hence, by a standard perturbation argument (cf. I.1.a.9(ii)), there is no loss of generality in assuming that

$$y_m=\sum_{l=1}^{2^{k_m}} c_{2^{k_m}+l}|\chi_{2^{k_m}+l}|,\quad m=1, 2, \dots,$$

for a suitable sequence of integers $k_1<k_2<\cdots$ and for suitable scalars $\{c_i\}$. Put

$$z_m=\sum_{l=1}^{2^{k_m}} c_{2^{k_m}+l}\chi_{2^{k_m}+l},\quad m=1, 2, \dots$$

and notice that $\{z_m\}_{m=1}^\infty$ is a block basis of the Haar basis of X so that

$$|z_m|=\sum_{l=1}^{2^{k_m}} |c_{2^{k_m}+l}||\chi_{2^{k_m}+l}|=|y_m|,$$

for every m. Hence, for every choice of scalars $\{a_m\}_{m=1}^\infty$,

$$\Big(\sum_{m=1}^\infty |a_m y_m|^2\Big)^{1/2}=\Big(\sum_{m=1}^\infty |a_m z_m|^2\Big)^{1/2}.$$

By applying the remark following 2.d.8 to the unconditional basis $\{y_m\}_{m=1}^\infty$, of a complemented subspace of X, and to the Haar basis of X, we deduce that $\{y_m\}_{m=1}^\infty$ is equivalent to $\{z_m\}_{m=1}^\infty$. It remains to prove that $[z_m]_{m=1}^\infty$ is complemented in X.

For every $m, n=1, 2, \dots$, put

$$y_{m,n}=(0, \dots, 0, \overset{n}{y_m}, 0, \dots).$$

These vectors belong to $Y(l_2)$. As noted in the proof of 2.d.5, $Y(l_2)$ is a complemented subspace of $X(l_2)$. We shall show next that $[y_{m,m}]_{m=1}^\infty$ is a complemented subspace of $Y(l_2)$ and thus of $X(l_2)$. This would be clear if $\{y_{m,n}\}_{m,n=1}^\infty$ were an unconditional basis. However, this fact is ensured by 2.d.7 only under the assumption that X is q-concave for some $q<\infty$ which is stronger than our assumption that $q_X<\infty$. We shall show that there is a constant M so that, for every choice of scalars $\{a_{m,n}\}_{m,n=1}^\infty$ with only finitely many non-zero entries and of signs $\{\theta_m\}_{m=1}^\infty$ and $\{\eta_n\}_{n=1}^\infty$, we have

$$(*)\qquad \Big\|\sum_{m=1}^\infty\sum_{n=1}^\infty \theta_m\eta_n a_{m,n}y_{m,n}\Big\|_{X(l_2)}\leq M\Big\|\sum_{m=1}^\infty\sum_{n=1}^\infty a_{m,n}y_{m,n}\Big\|_{X(l_2)}.$$

From (*) it follows easily that $\{y_{m,n}\}_{m,n=1}^{\infty}$ is a Schauder basis of $Y(l_2)$ if ordered as

$$y_{1,1}, y_{1,2}, y_{2,2}, y_{2,1}, y_{1,3}, y_{2,3}, y_{3,3}, y_{3,2}, y_{3,1}, y_{1,4}, \ldots$$

and that the canonical projection from $Y(l_2)$ onto $[y_{m,m}]_{m=1}^{\infty}$ is bounded (the argument is the same as the one used in I.1.c.8 to prove the boundedness of the diagonal operator). In order to prove (*), note first that the operator T_θ on Y defined by

$$T_\theta \sum_{m=1}^{\infty} a_m y_m = \sum_{m=1}^{\infty} \theta_m a_m y_m$$

satisfies $\|T_\theta\| \leq K$, where K is the unconditional constant of $\{y_m\}_{m=1}^{\infty}$. Hence, by 1.f.14,

$$\begin{aligned}\left\| \sum_{m=1}^{\infty} \sum_{n=1}^{\infty} \theta_m \eta_n a_{m,n} y_{m,n} \right\|_{X(l_2)} &= \left\| \left(\sum_{n=1}^{\infty} \left| \sum_{m=1}^{\infty} \theta_m a_{m,n} y_m \right|^2 \right)^{1/2} \right\|_X \\ &\leq K K_G \left\| \left(\sum_{n=1}^{\infty} \left| \sum_{m=1}^{\infty} a_{m,n} y_m \right|^2 \right)^{1/2} \right\|_X \\ &= K K_G \left\| \sum_{m=1}^{\infty} \sum_{n=1}^{\infty} a_{m,n} y_{m,n} \right\|_{X(l_2)}.\end{aligned}$$

This proves that $[y_{m,m}]_{m=1}^{\infty}$ is complemented in $X(l_2)$.

Consider now the vectors

$$h_{2^{k_m}+l} = (0, \ldots, 0, \overset{m}{|\chi_{2^{k_m}+l}|}, 0, \ldots), \quad l = 1, 2, \ldots, 2^{k_m}, \quad m = 1, 2, \ldots$$

and notice that $\{h_{2^{k_m}+l}\}_{l=1, m=1}^{2^{k_m}, \infty}$ is equivalent to $\{\chi_{2^{k_m}+l}\}_{l=1, m=1}^{2^{k_m}, \infty}$. Indeed, this follows from

$$\left\| \sum_{m=1}^{\infty} \sum_{l=1}^{2^{k_m}} a_{2^{k_m}+l} h_{2^{k_m}+l} \right\|_{X(l_2)} = \left\| \left(\sum_{m=1}^{\infty} \sum_{l=1}^{2^{k_m}} |a_{2^{k_m}+l} \chi_{2^{k_m}+l}|^2 \right)^{1/2} \right\|_X$$

and the remark following 2.d.8. The canonical isomorphism from $[\chi_{2^{k_m}+l}]_{l=1, m=1}^{2^{k_m}, \infty}$ onto $[h_{2^{k_m}+l}]_{l=1, m=1}^{2^{k_m}, \infty}$ maps z_m to $y_{m,m}$, for every m. Since $[y_{m,m}]_{m=1}^{\infty}$ is complemented in $X(l_2)$ and thus, in particular, also in $[h_{2^{k_m}+l}]_{l=1, m=1}^{2^{k_m}, \infty}$ the space $[z_m]_{m=1}^{\infty}$ is complemented in $[\chi_{2^{k_m}+l}]_{l=1, m=1}^{2^{k_m}, \infty}$ and thus in X, by the unconditionality of the Haar basis in X. □

Theorem 2.d.9 should, in principle, be of help in classifying the complemented subspaces of an r.i. function space having an unconditional basis. However, in practice it is usually hard to determine which block bases of the Haar basis span a complemented subspace and to classify the subspaces they span. A subsequence of the Haar basis obviously spans a complemented subspace if the Haar basis is unconditional. The spaces spanned by subsequences of the Haar basis have been characterized by J. L. B. Gamlen and R. J. Gaudet [45] in the case of $L_p(0, 1)$, $1<p<\infty$. The extension of their result to r.i. function spaces is straightforward.

Theorem 2.d.10. *Let* $\{\varphi_n\}_{n=1}^\infty$ *be a subsequence of the Haar system and let*

$$\sigma=\{t\in[0,1];\ t\in \operatorname{supp}\varphi_n \textit{ for infinitely many } n\}\ .$$

Let X be a separable r.i. function space on $[0, 1]$ *so that* $1<p_X$ *and* $q_X<\infty$. *If* $\mu(\sigma)=0$ *then there exists a sequence of pairwise disjoint characteristic functions in X whose closed linear span Y contains* $[\varphi_n]_{n=1}^\infty$ *(obviously, both* $[\varphi_n]_{n=1}^\infty$ *and Y are complemented in X) while if* $\mu(\sigma)>0$ *then* $[\varphi_n]_{n=1}^\infty$ *is isomorphic to X. In particular, every subsequence of the Haar basis in* $L_p(0, 1)$, $1<p<\infty$, *spans a subspace which is isomorphic either to* l_p *(if* $\mu(\sigma)=0$*) or to* $L_p(0, 1)$ *itself (if* $\mu(\sigma)>0$*).*

Proof. *Case* $\mu(\sigma)=0$. For every positive integer m, put

$$\sigma_m=\{t\in[0,1];\ t\in \operatorname{supp}\varphi_n \text{ for exactly } m \text{ distinct values of } n\}\ .$$

For $t\in\sigma_m$, $m=1, 2, \ldots$, let $n(t, m)$ be the largest integer n for which $t\in\operatorname{supp}\varphi_n$. Since two distinct Haar functions have either disjoint supports or the support of one of them is entirely contained in that of the other it follows that, for any $s\in\sigma_m\cap\operatorname{supp}\varphi_{n(t,m)}$, we have $\varphi_{n(s,m)}=\varphi_{n(t,m)}$. Hence, for every m, the distinct functions among $\{\varphi_{n(t,m)}\}_{t\in\sigma_m}$ have pairwise disjoint supports. Consequently,

$$\{(\varphi_{n(t,m)})_+\chi_{\sigma_m},\ (\varphi_{n(t,m)})_-\chi_{\sigma_m};\ t\in\sigma_m,\ m=1, 2, \ldots\}$$

forms a sequence of pairwise disjoint characteristic functions whose closed linear span Y is, by 2.a.4, complemented in X. Moreover, if $t\in\sigma_m\cap\operatorname{supp}\varphi_k$, for some m and k, then $\operatorname{supp}\varphi_{n(t,m)}\subset\operatorname{supp}\varphi_k$ which implies that $\varphi_k\chi_{\sigma_m}|\varphi_{n(t,m)}|\in Y$. Since $\bigcup_{m=1}^\infty\sigma_m$ coincides with $[0, 1]$, except the set σ which is assumed to have measure zero and except the set σ_0 on which all the $\{\varphi_n\}_{n=1}^\infty$ vanish, it follows that $[\varphi_n]_{n=1}^\infty\subset Y$. This completes the proof in case $\mu(\sigma)=0$ for a general X. When $X=L_p(0, 1)$, for some $1<p<\infty$, then Y is clearly isometric to l_p and, therefore, $[\varphi_n]_{n=1}^\infty$ is isometric to a complemented subspace of l_p. Hence, by I.2.a.3, $[\varphi_n]_{n=1}^\infty\approx l_p$.

Case $\mu(\sigma)>0$. We begin by defining the notion of a tree

$$\{\eta_{i_1,i_2,\ldots,i_n};\ i_j=0, 1,\ 1\leqslant j\leqslant n,\ n=1, 2, \ldots\}$$

of subsets of σ. By this we mean a collection of measurable sets so that $\sigma=\eta_0\cup\eta_1$, $\mu(\eta_{i_1,i_2,\ldots,i_n})=2^{-n}\mu(\sigma)$ and

$$\eta_{i_1,i_2,\ldots,i_n}=\eta_{i_1,i_2,\ldots,i_n,0}\cup\eta_{i_1,i_2,\ldots,i_n,1}\,,$$

for every n and $\{i_j\}_{j=1}^n$. For every such tree the functions $\{h_m\}_{m=1}^\infty$, defined by $h_1=\chi_\sigma$, $h_2=\chi_{\eta_0}-\chi_{\eta_1}$, $h_3=\chi_{\eta_{0,0}}-\chi_{\eta_{0,1}}$, $h_4=\chi_{\eta_{1,0}}-\chi_{\eta_{1,1}}$ and so on, form clearly a monotone basis equivalent to the Haar basis of X. Moreover, $[h_m]_{m=1}^\infty$ is complemented in X since it is the range of the projection $Pf=\chi_\sigma E^{\mathscr{B}}f$ of norm one, where $E^{\mathscr{B}}$ denotes the conditional expectation operator with respect to the σ-algebra $\mathscr{B}$ generated by the tree. Hence, by the perturbation result I.1.a.9(ii) and 2.d.5, it suffices to find a tree of subsets of σ and a sequence $\{y_m\}_{m=2}^\infty$ of vectors in $[\varphi_n]_{n=1}^\infty$ so that $\|y_m-h_m\|<\varepsilon_m$ for $m=2, 3, \ldots$, where $\{h_m\}_{m=2}^\infty$ are the functions associated to the tree and $\{\varepsilon_m\}_{m=2}^\infty$ are small enough positive numbers given in advance (to be specific, $\varepsilon_m=\mu(\sigma)4^{-m-2}$, $m=2, 3, \ldots$).

To construct η_0 (and thus η_1) we proceed as follows. Given $\delta>0$ we find a relatively open subset G of $[0, 1]$ so that $G\supset\sigma$ and $\mu(G\sim\sigma)<\delta$. For every $t\in\sigma$, let $n(t)$ be the smallest integer n for which $t\in \operatorname{supp}\varphi_n\subset G$. Clearly, the distinct functions among $\{\varphi_{n(t)}\}_{t\in\sigma}$ have mutually disjoint supports and thus their sum y_2 belongs to $[\varphi_n]_{n=1}^\infty$ and satisfies $\chi_\sigma\leqslant|y_2|\leqslant\chi_G$. The function y_2 takes the value 1 on half of its support and -1 on the second half. It is clear, therefore, that if δ is small enough we can find η_0 so that the corresponding function h_2 satisfies $\|y_2-h_2\|<\varepsilon_2$ (we used here the fact that X is separable and thus $\|\chi_{G\sim\sigma}\|=o(1)$). By replacing in the above construction σ with η_0 we can find in exactly the same manner a subset $\eta_{0,0}$ of η_0 and a vector $y_3\in[\varphi_n]_{n=1}^\infty$ so that $\|y_3-h_3\|<\varepsilon_3$. The construction of the entire tree and of the vectors $\{y_n\}_{n=4}^\infty$ is continued by an obvious inductive argument. □

Remark. The assumption that $1<p_X$ and $q_X<\infty$ was used only in the proof of the case $\mu(\sigma)>0$.

We conclude this section by proving that separable r.i. function spaces on $[0, 1]$ with non-trivial Boyd indices are primary. Recall that a Banach space X is said to be primary if, for every decomposition of X into a direct sum of two subspaces, at least one of the factors is isomorphic to X itself.

Theorem 2.d.11. *Every separable r.i. function space X on $[0, 1]$ with $1<p_X$ and $q_X<\infty$ is primary.*

Originally, 2.d.11 was proved by P. Enflo for L_p spaces, $1\leqslant p<\infty$. A proof in the case $p=1$, which is similar in spirit to Enflo's original argument, is given in P. Enflo and T. Starbird [38] (cf. also B. Maurey [92]). The present proof as well as the extension to r.i. function spaces (under slightly more restrictive conditions) is due to D. Alspach, P. Enflo and E. Odell [2].

Proof. By 2.d.4, X is isomorphic to $X(l_2)$ and, therefore, it is equivalent to prove the primariness of $X(l_2)$. Suppose that

$$X(l_2)=Y\oplus Z\,,$$

for some subspaces Y and Z, and let P be the projection from $X(l_2)$ onto Y which vanishes on Z. By 2.d.8, the vectors

$$\chi_{m,n}=(0,\ldots,0,\overset{n}{\chi_m},0,\ldots),\quad m,n=1,2,\ldots$$

form an unconditional basis of $X(l_2)$. Let $\{\chi_{m,n}^*\}_{m,m=1}^\infty$ be the sequence of the biorthogonal functionals associated to $\{\chi_{m,n}\}_{m,n=1}^\infty$. Then, for each m and n, at least one of the numbers $\chi_{m,n}^* P\chi_{m,n}$ and $\chi_{m,n}^*(I-P)\chi_{m,n}$ is $\geq 1/2$. Put

$$\sigma_Y=\{m\in N;\ \chi_{m,n}^* P\chi_{m,n}\geq 1/2 \text{ for infinitely many values of } n\}$$

and

$$\sigma_Z=N\sim\sigma_Y.$$

Notice that, by 2.d.10, either $[\chi_m]_{m\in\sigma_Y}$ or $[\chi_m]_{m\in\sigma_Z}$ is isomorphic to X. We may assume without loss of generality that $\sigma_Y=\{m_1<m_2<\cdots\}$ and $[\chi_{m_j}]_{j=1}^\infty\approx X$. Since, for each j, $\{\chi_{m_j,n}/\|\chi_{m_j,n}\|\}_{n=1}^\infty$ is equivalent to the unit vector basis of l_2 and thus weakly null, it follows that there exists an increasing sequence $\{n_j\}_{j=1}^\infty$ of integers so that $\{P\chi_{m_j,n_j}\}_{j=1}^\infty$ is equivalent to a block basis $\{z_j\}_{j=1}^\infty$ of $\{\chi_{m,n}\}_{m,n=1}^\infty$ and, in addition,

$$\chi_{m_j,n_j}^* z_j\geq 1/3\,,$$

for every j. For any sequence $\{a_j\}_{j=1}^\infty$ of scalars we have

$$\sum_{j=1}^\infty a_j\chi_{m_j,n_j} \text{ converges} \Rightarrow \sum_{j=1}^\infty a_j P\chi_{m_j,n_j} \text{ converges}$$
$$\Rightarrow \sum_{j=1}^\infty a_j z_j \text{ converges} \Rightarrow \sum_{j=1}^\infty a_j\chi_{m_j,n_j} \text{ converges.}$$

Hence, $\{z_j\}_{j=1}^\infty$ is equivalent to $\{\chi_{m_j,n_j}\}_{j=1}^\infty$ and, moreover,

$$Qx=\sum_{j=1}^\infty \chi_{m_j,n_j}^*(x)z_j/\chi_{m_j,n_j}^*(z_j)$$

defines a bounded projection from $X(l_2)$ onto $[z_j]_{j=1}^\infty$. Thus, if the $\{z_j\}_{j=1}^\infty$ were chosen sufficiently close to $\{P\chi_{m_j,n_j}\}_{j=1}^\infty$ we could deduce, by I.1.a.9(ii), that $[P\chi_{m_j,n_j}]_{j=1}^\infty$ is complemented in $X(l_2)$. Therefore, in order to prove that $X\approx Y$, it suffices, in view of 2.d.5, to show that $\{\chi_{m_j,n_j}\}_{j=1}^\infty$ is equivalent to $\{\chi_{m_j}\}_{j=1}^\infty$. Note however that

$$\left\|\sum_{j=1}^\infty b_j\chi_{m_j,n_j}\right\|_{X(l_2)}=\left\|\left(\sum_{j=1}^\infty |b_j\chi_{m_j}|^2\right)^{1/2}\right\|_X$$

and thus the desired result is an immediate consequence of the remark following 2.d.8. □

e. Isomorphisms Between r.i. Function Spaces; Uniqueness of the r.i. Structure

The first topic considered in this section is that of isomorphic embeddings of an r.i. function space X on $[0, 1]$ into another r.i. function space Y. The simplest example of such an embedding is that of $X=L_2(0,1)$ into an arbitrary r.i. function space Y with $q_Y<\infty$. Another well-known example of an isomorphic embedding, which will be discussed in detail in Section f below and especially in Vol. IV, is that of $X=L_p(0, 1)$ into $Y=L_r(0, 1)$, for $1\leqslant r<p<2$. A characteristic of this example is that the identity mapping is a continuous operator (but in general not an isomorphism) from X into Y i.e. there exists a constant $c>0$ such that $\|f\|_X\geqslant c\|f\|_Y$, for every $f\in X$. There is also a third possibility which was already mentioned briefly in 2.c. This is the case where the Haar basis of X is equivalent to a sequence of pairwise disjoint functions in Y. Concrete examples of this type will be constructed in Section g below.

The three cases described above essentially exhaust all the possibilities of embedding isomorphically X into Y provided some natural conditions are imposed on X and Y (e.g. one clearly has to exclude the case $Y=L_\infty(0, 1)$ when no information on X can be obtained). This result from [58] Section 6 will be stated below in a precise way but without giving a proof since its proof is too complicated to be reproduced here. Instead, we shall prove a version of it which, though weaker, still has several interesting applications. This theorem as well as its applications will deal only with r.i. function spaces on $[0, 1]$ which are, in a sense, on one side of the Hilbert space $L_2(0, 1)$ (for example, r.i. function spaces of type 2 or of cotype 2). In many cases, we have actually to ensure that X is even "far" from $L_2(0,1)$ (for instance, in the sense that it is q-concave for some $q<2$ or r-convex for some $r>2$).

The results proved on isomorphisms between r.i. function spaces can be used to study the uniqueness of the r.i. structure of a given r.i. function space X on $[0, 1]$, i.e. the question when any two representations of X as an r.i. function space on $[0, 1]$ must coincide (except, perhaps, for an equivalent renorming). Generally speaking, the uniqueness of the r.i. structure on $[0, 1]$ can be proved for those spaces X for which the Haar system of any representation of X as an r.i. function space on $[0, 1]$ is not equivalent to a sequence of pairwise disjoint (relative to the representation under consideration) functions in X. For example, by 2.c.14 and the remark thereafter, this is the case for reflexive Orlicz function spaces $L_M(0, 1)$ and, indeed, these spaces have a unique r.i. structure.

Before stating the first result we discuss some facts concerning the notion of a p-stable random variable which will be treated in detail only in Vol. IV. Let (Ω, Σ, ν) be a probability space having no atoms. A real valued random variable g on Ω is called p-stable, for some $0<p\leqslant 2$, if

$$\int_\Omega e^{itg(\omega)}\, d\nu(\omega)=e^{-c|t|^p},$$

for some constant $c>0$ and for every $-\infty<t<\infty$. That such random variables do indeed exist and that any p-stable variable, $1<p\leq 2$ belongs to the space $L_r(\Omega, \Sigma, \nu)$, for any $1\leq r<p$, will be proved in Vol. IV. (See also [135] and the remark following 2.f.5.) The value of the constant c determines the distribution function of g and thus also the norm of g in $L_r(\Omega, \Sigma, \nu)$ i.e. two p-stable random variables with the same c have the same norm in $L_r(\Omega, \Sigma, \nu)$. This statement is a consequence of the fact that the Fourier transform on $L_1(-\infty, \infty)$ is one to one. Unless $p=2$, a p-stable random variable does not belong to $L_p(\Omega, \Sigma, \nu)$ itself.

The importance of p-stable random variables in Banach space theory stems from the fact that they can be used to embed isometrically l_p into $L_r(0, 1)$, for $1\leq r<p\leq 2$. If $\{g_n\}_{n=1}^\infty$ is a sequence of identically distributed independent p-stable random variables with $\|g_1\|_r=1$ then, for every choice of scalars $\{a_n\}_{n=1}^\infty$,

$$\left\|\sum_{n=1}^\infty a_n g_n\right\|_r=\left(\sum_{n=1}^\infty |a_n|^p\right)^{1/p}.$$

Indeed, if we suppose that $\sum_{n=1}^\infty |a_n|^p=1$ then, by the independence of the sequence $\{g_n\}_{n=1}^\infty$, we get that

$$\int_\Omega e^{it\sum_{n=1}^\infty a_n g_n(\omega)}\,d\nu(\omega)=\prod_{n=1}^\infty \int_\Omega e^{ita_n g_n(\omega)}\,d\nu(\omega)=\prod_{n=1}^\infty e^{-c|a_n t|^p}$$

$$=e^{-c|t|^p\sum_{n=1}^\infty |a_n|^p}=e^{-c|t|^p},\qquad -\infty<t<\infty,$$

i.e. $\sum_{n=1}^\infty a_n g_n$ is again a p-stable random variable with the same constant c as all of the g_n's. Hence, $\left\|\sum_{n=1}^\infty a_n g_n\right\|_r=1$.

The above property of the p-stable random variables will be used in the proof of the following result (cf. [58] Section 5).

Theorem 2.e.1. *Let X be a separable r.i. function space on $[0, 1]$ having the property that, for some $1<q<2$, $C<\infty$ and for every $f\in L_q(0, 1)$,*

$$\|f\|_X\leq C\|f\|_q\,.$$

Let Y be an r.i. function space on $[0, 1]$ or on $[0, \infty)$ which does not contain uniformly isomorphic copies of l_∞^n for all n. If X is isomorphic to a subspace of Y then either

(i) *there exists a constant $D<\infty$ so that*

$$\|f\|_Y\leq D\|f\|_X,$$

for every $f\in X$, or

(ii) *the Haar basis of X is equivalent to a sequence of pairwise disjoint functions in Y.*

The proof of 2.e.1 is based mainly on the following lemma.

Lemma 2.e.2. *Let T be a bounded linear operator from a separable r.i. function space X on $[0, 1]$ into an r.i. function space Y on $[0, 1]$ or on $[0, \infty)$ so that $\lim_{t \to 0} \|\chi_{[0,t)}\|_Y = 0$ (i.e. the restriction of Y to $[0, 1]$ is not equal to $L_\infty(0, 1)$, even up to an equivalent norm). Suppose that there exist numbers $s \geq 1$ and $R < \infty$ such that if, for $n = 0, 1, 2, \ldots$, we put*

$$y_n = \bigvee_{i=1}^{2^n} |T\chi_{[(i-1)2^{-n}, i2^{-n})}| \quad \text{and} \quad \delta_n = \{t \in [0, s];\ y_n(t) \leq R\}$$

then

$$\alpha = \inf_n \|y_n \chi_{\delta_n}\|_Y > 0 .$$

Then there exists a constant $D < \infty$ so that

$$\|f\|_Y \leq D\|f\|_X ,$$

for every $f \in X$.

Proof. Fix n and put

$$\eta_n = \{t \in [0, s];\ \alpha/2\|\chi_{[0,s]}\|_Y \leq y_n(t) \leq R\} .$$

Observe that $\|y_n \chi_{\eta_n}\|_Y \geq \alpha/2$. Let $\{\eta_{n,i}\}_{i=1}^{2^n}$ be a partition of η_n into mutually disjoint measurable subsets such that

$$|T\chi_{[(i-1)2^{-n}, i2^{-n})}|(t) = y_n(t) ,$$

for $t \in \eta_{n,i}$, $i = 1, 2, \ldots, 2^n$. We assume, as we clearly may without loss of generality, that $\{\mu(\eta_{n,i})\}_{i=1}^{2^n}$ is a non-increasing sequence. For any $1 < m \leq 2^n$, we have

$$\begin{aligned}
\left\| y_n \sum_{i=1}^{m-1} \chi_{\eta_{n,i}} \right\|_Y &\leq \left\| \bigvee_{i=1}^{m} |T\chi_{[(i-1)2^{-n}, i2^{-n})}| \right\|_Y \\
&\leq \left\| \int_0^1 \left| \sum_{i=1}^{m-1} r_i(u) T\chi_{[(i-1)2^{-n}, i2^{-n})} \right| du \right\|_Y \\
&\leq \|T\| \int_0^1 \left\| \sum_{i=1}^{m-1} r_i(u) \chi_{[(i-1)2^{-n}, i2^{-n})} \right\|_X du \\
&= \|T\| \, \|\chi_{[0,(m-1)2^{-n})}\|_X
\end{aligned}$$

from which it follows that

$$\begin{aligned}
\alpha/2 \leq \|y_n \chi_{\eta_n}\|_Y &\leq \|T\| \, \|\chi_{[0,(m-1)2^{-n})}\|_X + \left\| y_n \sum_{i=m}^{2^n} \chi_{\eta_{n,i}} \right\|_Y \\
&\leq \|T\| \, \|\chi_{[0,(m-1)2^{-m})}\|_X + R\|\chi_{[0, 2^n \mu(\eta_{n,m}))}\|_Y .
\end{aligned}$$

Since X is assumed to be separable there exists a $\beta>0$ so that

$$\|\chi_{[0,\beta)}\|_X<\alpha/4\|T\|.$$

Hence, by letting $m>1$ be the smallest integer for which $m2^{-n}>\beta$, we get that

$$\alpha/4\leqslant R\|\chi_{[0,2^n\mu(\eta_{n,m}))}\|_Y.$$

Since $\|\chi_{[0,t)}\|_Y\to 0$ as $t\to 0$ it follows that there exists a $\gamma>0$, independent of n and m, so that

$$2^n\mu(\eta_{n,m})\geqslant\gamma.$$

This means that we have found positive reals β and γ, independent of n, such that a fixed proportion $\beta<m2^{-n}$ of the 2^n sets $\{\eta_{n,i}\}_{i=1}^{2^n}$ have Lebesgue measure $\geqslant\gamma\cdot 2^{-n}$.

Let now N be an integer for which $N\gamma\geqslant 1$. Then, by taking into account the definition of the sets $\{\eta_{n,i}\}_{i=1}^{2^n}$, it follows that, for any choice of scalars $\{a_i\}_{i=1}^m$, the function

$$g=\sum_{i=1}^m a_i\chi_{[(i-1)N^{-1}2^{-n},\,iN^{-1}2^{-n})}$$

satisfies

$$\begin{aligned}\|g\|_Y&\leqslant\Big\|\sum_{i=1}^m a_i\chi_{\eta_{n,i}}\Big\|_Y\leqslant 2\alpha^{-1}\|\chi_{[0,s]}\|_Y\Big\|\sum_{i=1}^m a_i\chi_{\eta_{n,i}}T\chi_{[(i-1)2^{-n},\,i2^{-n})}\Big\|_Y\\&\leqslant 2\alpha^{-1}\|\chi_{[0,s]}\|_Y\Big\|\bigvee_{i=1}^m|T(a_i\chi_{[(i-1)2^{-n},\,i2^{-n})})|\Big\|_Y\\&\leqslant 2\alpha^{-1}\|\chi_{[0,s]}\|_Y\int_0^1\Big\|\sum_{i=1}^m r_i(u)a_iT\chi_{[(i-1)2^{-n},\,i2^{-n})}\Big\|_Y\,du\\&\leqslant 2\|T\|\alpha^{-1}\|\chi_{[0,s]}\|_Y\Big\|\sum_{i=1}^n a_i\chi_{[(i-1)2^{-n},\,i2^{-n})}\Big\|_X\\&\leqslant 2N\|T\|\alpha^{-1}\|\chi_{[0,s]}\|_Y\|g\|_X.\end{aligned}$$

This proves our assertion for functions g as above. For a general step function of the form

$$f=\sum_{i=1}^{N2^n} a_i\chi_{[(i-1)N^{-1}2^{-n},\,iN^{-1}2^{-n})}$$

we split the interval $[0, 1]$ into $[N2^nm^{-1}]+1(\leqslant N\beta^{-1}+1)$ intervals each having measure $\leqslant mN^{-1}2^{-n}$ and then use the estimate established above in each of these intervals. It follows that $\|f\|_Y\leqslant D\|f\|_X$, where $D=2N\|T\|\alpha^{-1}\|\chi_{[0,s]}\|_Y(N\beta^{-1}+1)$ is independent of n. This completes the proof since the step functions f as above, with $\{a_i\}_{i=1}^{N2^n}$ and n arbitrary, are dense in X. □

Proof of 2.e.1. Let T_0 be an isomorphism from X into Y and, for every measurable subset E of $[0, 1]$ with $\mu(E)>0$, let X_E denote the subspace of X of all functions supported by E. Let τ_E be an invertible transformation of $[0, 1]$ onto E so that $\mu(\tau_E^{-1}\delta)=\mu(\delta)/\mu(E)$, for every measurable subset δ of E. Then the mapping $f\to f(\tau_E^{-1})$ defines an order isomorphism S_E from X onto X_E.

We distinguish now between two cases.

Case I. There exist a measurable subset E of $[0, 1]$ with $\mu(E)>0$, a transformation τ_E as above and numbers $s>1$ and $R<\infty$ so that if, for $n=0, 1, 2, \ldots$, we put

$$z_n^E=\left(\sum_{i=1}^{2^n}|T_0S_E\chi_{[(i-1)2^{-n},\, i2^{-n})}|^2\right)^{1/2} \quad \text{and} \quad \sigma_n^E=\{t\in[0, s];\ z_n^E(t)\leqslant R\}$$

then

$$\inf_n \|z_n^E\chi_{\sigma_n^E}\|_Y>0.$$

In this case assertion (i) of 2.e.1 holds. This is proved by applying 2.e.2 to the operator $T=T_0S_E$. In order to verify that T satisfies the hypotheses of 2.e.2, fix $q<p<2$ and let $\{g_n\}_{n=1}^\infty$ be a sequence of identically distributed independent p-stable random variables so that $\|g_1\|_q=1$. The properties of the p-stable random variables described above imply that the functions

$$w_n=\left(\sum_{i=1}^{2^n}|T\chi_{[(i-1)2^{-n},\, i2^{-n})}|^p\right)^{1/p}, \quad n=1, 2, \ldots$$

satisfy

$$\begin{aligned}
\|w_n\|_Y&=\|g_1\|_1^{-1}\left\|\int_\Omega\left|\sum_{i=1}^{2^n}g_i(\omega)T\chi_{[(i-1)2^{-n},\, i2^{-n})}\right|dv(\omega)\right\|_Y\\
&\leqslant\|g_1\|_1^{-1}\|T\|\int_\Omega\left\|\sum_{i=1}^{2^n}g_i(\omega)\chi_{[(i-1)2^{-n},\, i2^{-n})}\right\|_X dv(\omega)\\
&\leqslant C\|g_1\|_1^{-1}\|T\|\left(\int_\Omega\left\|\sum_{i=1}^{2^n}g_i(\omega)\chi_{[(i-1)2^{-n},\, i2^{-n})}\right\|_q^q dv(\omega)\right)^{1/q}\\
&=C\|g_1\|_1^{-1}\|T\|\left(\int_0^1\left(\sum_{i=1}^{2^n}|\chi_{[(i-1)2^{-n},\, i2^{-n})}|^p\right)^{q/p}dt\right)^{1/q}\\
&=C\|g_1\|_1^{-1}\|T\|.
\end{aligned}$$

Let y_n and δ_n have the same meaning as in the statement of 2.e.2. Then, by 1.d.2 and the fact that $\sigma_n^E\subset\delta_n$ for all n, we get that

$$\|z_n^E\chi_{\sigma_n^E}\|_Y\leqslant\|w_n\chi_{\sigma_n^E}\|_Y^{p/2}\|y_n\chi_{\sigma_n^E}\|_Y^{1-p/2}\leqslant(C\|g_1\|_1^{-1}\|T\|)^{p/2}\|y_n\chi_{\delta_n}\|_Y^{1-p/2}$$

i.e.

$$\inf_n \|y_n \chi_{\delta_n}\|_Y > 0 .$$

Case II. This is the case when the assumptions of Case I are not satisfied. In this case we construct inductively a system of vectors $\{\chi'_n\}_{n=1}^\infty$ in X, which is obtained from the Haar system $\{\chi_n\}_{n=1}^\infty$ by a suitable automorphism of [0, 1], and a family $\{f_{i,n}\}_{i=1,\,n=1}^{n\ \ \ \infty}$ of elements of Y such that, for each n,

(a) $\|f_{i,n} - T_0\chi'_i/\|\chi'_i\|_X\|_Y < 1/2^{i+2}\|T_0^{-1}\|$, $i = 1, 2, \ldots, n$.
(b) $\{f_{i,n}\}_{i=1}^n$ are pairwise disjoint.
(c) $f_{i,n+1} = f_{i,n\,|\,\mathrm{supp}\, f_{i,n+1}}$, $i = 1, 2, \ldots, n$.

Once this construction is completed, we put $f_i = \lim_{n\to\infty} f_{i,n}$ (the limit exists since Y is order continuous by 1.a.5, 1.a.7 and the conditions imposed on Y) and observe that $\{f_i\}_{i=1}^\infty$ is a sequence of mutually disjoint functions in Y which is equivalent to $\{\chi'_i/\|\chi'_i\|_X\}_{i=1}^\infty$. This, of course, shows that in Case II assertion (ii) of 2.e.1 holds.

The possibility of constructing inductively the $\{\chi'_n\}_{n=1}^\infty$ and $\{f_{i,n}\}_{i=1,\,n=1}^{n\ \ \ \infty}$ as above follows immediately from the next lemma.

Lemma 2.e.3. *Under the hypotheses of Case II above, if $\{f_i\}_{i=1}^n$ is a sequence of pairwise disjoint functions in Y then, for every $\varepsilon > 0$ and every measurable subset E of [0, 1] with $\mu(E) > 0$, there exist a sequence $\{g_i\}_{i=1}^{n+1}$ of pairwise disjoint elements of Y satisfying $\|f_i - g_i\|_Y < \varepsilon$ and $g_i = f_{i\,|\,\mathrm{supp}\, g_i}$, for $i = 1, 2, \ldots, n$, and a partition of E into two disjoint measurable subsets E_1 and E_2, each of measure $\mu(E)/2$, so that*

$$\|T_0(\chi_{E_1} - \chi_{E_2}) - g_{n+1}\|_Y < \varepsilon .$$

Proof. We have already pointed out that the conditions imposed on Y ensure that it is an order continuous lattice. Actually, by 1.f.7 and 1.f.12, Y is even r-concave for some $r < \infty$. It follows that, for every $\varepsilon > 0$, there exist reals $s \geq 1$ and $\rho > 0$ such that

$$\|f_i \chi_\sigma\|_Y < \varepsilon/2, \quad i = 1, 2, \ldots, n ,$$

whenever $\sigma \subset [s, \infty)$ or $\mu(\sigma) < \rho$. Furthermore, by the assumptions characterizing Case II, for every subset E of [0, 1] with $\mu(E) > 0$, there exist an integer m and a subset $\delta \subset [0, s]$ with $\mu(\delta) > s - \rho$ so that

$$\|z_m^E \chi_\delta\|_Y < \varepsilon/4MB_r ,$$

where z_m^E is the expression appearing in the definition of Case 1, M the r-concavity constant of Y and B_r the Khintchine constant in $L_r(0, 1)$. (Use the fact that, e.g. by 1.f.14, $\sup_m \|z_m^E\|_Y < \infty$ and thus if R is large enough $\mu(\sigma_m^E) > s - \rho$ for every m.)

Let S_E have the same meaning as in the proof of Case I and set

$$h_1(u,t)=\sum_{i=1}^{2^{m-1}} r_i(u)S_E\chi_{[(i-1)2^{-m},\, i2^{-m})}(t)$$

$$h_2(u,t)=\sum_{i=1}^{2^{m-1}} r_i(u)S_E\chi_{[(2^{m-1}+i-1)2^{-m},\,(2^{m-1}+i)2^{-m})}(t)\,.$$

Then, by the r-concavity of Y and Khintchine's inequality, we get, for $j=1, 2$, that

$$\left(\int_0^1 \|\chi_\delta T_0 h_j\|_Y^r\,du\right)^{1/r} \leqslant M\left\|\left(\int_0^1 |\chi_\delta T_0 h_j|^r\,du\right)^{1/r}\right\|_Y$$
$$\leqslant MB_r\left\|\chi_\delta\left(\sum_{i=1}^{2^m}|T_0S_E\chi_{[(i-1)2^{-m},\, i2^{-m})}|^2\right)^{1/2}\right\|_Y$$
$$=MB_r\|\chi_\delta z_m^E\|_Y<\varepsilon/4\,.$$

Thus, it is possible to find an $u_0\in[0,1]$ so that

$$\|\chi_\delta T_0 h_j(u_0,t)\|_Y<\varepsilon/2\,,$$

simultaneously for $j=1$ and $j=2$. We are now able to define the functions $\{g_i\}_{i=1}^{n+1}$. Put $g_i=f_i\chi_\delta$, for $i=1,2,\dots,n$, and $g_{n+1}=(1-\chi_\delta)T_0h$, where $h(t)=h_1(u_0,t)-h_2(u_0,t)$. It is easily seen that $\{g_i\}_{i=1}^{n+1}$ are mutually disjoint functions in Y with $g_i=f_{i|_{\operatorname{supp} g_i}}$ and

$$\|f_i-g_i\|_Y=\|f_i(1-\chi_\delta)\|_Y<\varepsilon\,,$$

for $i=1, 2, \dots, n$, since $\mu([0,s]\sim\delta)<\rho$. Moreover, by its definition $h(t)$ takes only two values, namely $+1$ on a subset E_1 and -1 on a subset E_2, each having measure $\mu(E)/2$. This completes the proof since

$$\|T_0(\chi_{E_1}-\chi_{E_2})-g_{n+1}\|_Y=\|T_0h-g_{n+1}\|_Y=\|\chi_\delta T_0h\|_Y<\varepsilon\,.\ \square$$

Corollary 2.e.4. *Let Y be an r.i. function space on $[0,1]$ which does not contain uniformly isomorphic copies of l_∞^n for all n. If Y contains a subspace isomorphic to $L_1(0,1)$ then Y itself coincides, up to an equivalent norm, with $L_1[0,1]$.*

Proof. Since the Haar basis of $L_1(0,1)$ is not unconditional, assertion (ii) of 2.e.1 cannot hold in the present case. Hence, there is a $D<\infty$ so that $\|f\|_Y\leqslant D\|f\|_1$, for every $f\in L_1(0,1)$. Since $\|f\|_1\leqslant\|f\|_Y$, for every f, (by 2.a.1) this concludes the proof. $\square$

N. J. Kalton [136] has independently proved 2.e.4 under the less restrictive assumption that Y does not contain an isomorphic copy of c_0.

A weaker version of 2.e.1 was proved in [78] for the case when both X and Y are Orlicz function spaces on $[0,1]$. More precisely, it was shown there that if

$L_M(0, 1)$ is isomorphic to a subspace of $L_N(0, 1)$ and $\beta_{N,\infty}<2$ (the indices $\alpha_{N,\infty}$ and $\beta_{N,\infty}$ have been defined in 2.b.5 and, as proved there, they coincide with the Boyd indices p_{L_N}, respectively q_{L_N}; moreover, the interval $[\alpha_{N,\infty}, \beta_{N,\infty}]$ is the set of all numbers r for which the unit vector basis of l_r is equivalent to a sequence of mutually disjoint functions in $L_N(0, 1)$) then $N(t)\leqslant DM(t)$, for some constant $D<\infty$ and every $t\geqslant 1$. This result was used in [78] to show that if $L_M(0, 1)\approx L_N(0, 1)$ and $1<\alpha_{N,\infty}\leqslant\beta_{N,\infty}<2$ then M and N are equivalent at ∞ i.e. $L_M(0, 1)= L_N(0, 1)$, up to an equivalent norm. The proof of this fact is obvious once we show that also $\beta_{M,\infty}<2$. In view of the reflexivity of $L_N(0, 1)$, which follows from the fact that $1<\alpha_{N,\infty}$ and $\beta_{N,\infty}<\infty$, and the duality relation between $\alpha_{N^*,\infty}$ and $\beta_{N,\infty}$ this is equivalent to showing that $2<\alpha_{N^*,\infty}$ implies that also $2<\alpha_{M^*,\infty}$. This later assertion is a consequence of 1.c.10 and the characterizations of the intervals $[\alpha_{N^*,\infty}, \beta_{N^*,\infty}]$ (respectively $[\alpha_{M^*,\infty}, \beta_{M^*,\infty}]$). Indeed, by 1.c.10, *every symmetric basic sequence in a Köthe space of type 2 is equivalent either to a sequence of disjointly supported functions or to the unit vector basis in* l_2. Hence, if $\alpha_{N^*,\infty}>2$ then any symmetric basic sequence in $L_{N^*}(0, 1)$ is equivalent to the unit vector basis of an Orlicz sequence space and l_p is isomorphic to a subspace of $L_{N^*}(0, 1)$ if and only if $p=2$ or $p\in[\alpha_{N^*,\infty}, \beta_{N^*,\infty}]$.

We have described in some detail the preceding argument since it indicates how to proceed in order to prove the uniqueness of the r.i. structure on $[0, 1]$ of an r.i. function space X on $[0, 1]$ which is reflexive and q-concave some $q<2$: passing to the dual X^* of X which, by 1.d.4, is r-convex with $1/r+1/q=1$, we would have to check whether any r.i. function space Z on $[0, 1]$, which embeds isomorphically into X^*, is also r-convex or it is isomorphic to $L_2(0, 1)$ (recall that, by remark 2 following 2.b.3, the q-concavity of X implies that $\|f\|_X\leqslant C\|f\|_q$, for some constant $C<\infty$ and every $f\in L_q(0, 1)$).

The preceding remarks lead us naturally to the study of r-convex r.i. function spaces for $r\geqslant 2$. The study of such spaces in facilitated by the fact that symmetric basic sequences in such spaces can be fully described. We have already seen above that 1.c.10 gives much information on infinite symmetric basic sequences in such spaces and that such sequences have a relatively simple form (e.g. in $L_p(0, 1)$, $2<p<\infty$, they are equivalent to the unit vector basis in either l_2 or l_p). Let us point out that, for spaces which are not r-convex for $r\geqslant 2$, the situation is considerably more complicated. As shown in the beginning of this section, if $1\leqslant r<p\leqslant 2$ then l_p is isometric to a subspace of $L_r(0, 1)$. There are even symmetric basic sequences in $L_r(0, 1)$ with $1<r<2$ which are not equivalent to the unit vector basis of an Orlicz sequence space. We shall treat this question in detail in Vol. IV. For our purposes here we need a quantitative description of all the finite symmetric basic sequences in a Banach lattice of type 2 (i.e., by 1.f.17, a Köthe space which is 2-convex and q-concave for some $q<\infty$). Such a description cannot be derived from 1.c.10. The following result (cf. [58] Section 2) has, despite its somewhat complicated statement, several interesting consequences.

Theorem 2.e.5. *For every $K\geqslant 1$, $M\geqslant 1$ and every integer m there exists a constant $D=D(K, M, m)$ so that if $\{x_i\}_{i=1}^n$ is a K-symmetric normalized basic sequence of finite length in a Banach lattice X, which is 2-convex and $2m$-concave with both*

2-convexity and 2m-concavity constants $\leqslant M$, *then, for every choice of scalars* $\{a_i\}_{i=1}^n$,

$$D^{-1}\left\|\sum_{i=1}^n a_i x_i\right\| \leqslant \max\left\{\left(\sum_\pi \left\|\bigvee_{i=1}^n |a_{\pi(i)} x_i|\right\|^{2m} \Big/ n!\right)^{1/2m}, w_n\left(\sum_{i=1}^n |a_i|^2\right)^{1/2}\right\}$$
$$\leqslant D\left\|\sum_{i=1}^n a_i x_i\right\|,$$

where $w_n = \left\|\sum_{i=1}^n x_i\right\| \Big/ n^{1/2}$ *and* $\sum_\pi$ *refers to summation over all the permutations* π *of the integers* $\{1, 2, \ldots, n\}$.

Before proving 2.e.5, we present some of its consequences. For instance, if $X = L_p(0, 1)$, $p > 2$ then, for every choice of scalars $\{a_i\}_{i=1}^n$, we clearly have

$$\left\|\bigvee_{i=1}^n |a_{\pi(i)} x_i|\right\|_p \leqslant \left\|\left(\sum_{i=1}^n |a_{\pi(i)} x_i|^p\right)^{1/p}\right\|_p = \left(\sum_{i=1}^n |a_i|^p\right)^{1/p}.$$

On the other hand, since $\{x_i\}_{i=1}^n$ is also K-unconditional and $L_p(0, 1)$ is of cotype p, we get that

$$K\left\|\sum_{i=1}^n a_i x_i\right\|_p \geqslant \left(\sum_{i=1}^n |a_i|^p\right)^{1/p}.$$

Combining these two estimates with the statement of 2.e.5 we obtain the following result (for which a simple direct proof can be found in [58] Section 1).

Theorem 2.e.6. *For every* $K > 1$ *and* $p > 2$ *there exists a constant* $D = D(K, p)$ *so that, for every K-symmetric normalized sequence* $\{x_i\}_{i=1}^n$ *in* $L_p(0, 1)$ *and every choice of scalars* $\{a_i\}_{i=1}^n$, *we have*

$$D^{-1}\left\|\sum_{i=1}^n a_i x_i\right\|_p \leqslant \max\left\{\left(\sum_{i=1}^n |a_i|^p\right)^{1/p}, w_n\left(\sum_{i=1}^n |a_i|^2\right)^{1/2}\right\} \leqslant D\left\|\sum_{i=1}^n a_i x_i\right\|_p,$$

where $w_n = \left\|\sum_{i=1}^n x_i\right\|_p \Big/ n^{1/2}$.

This theorem can be used to characterize those r.i. function spaces which embed isomorphically into $L_p(0, 1)$, $p > 2$.

Theorem 2.e.7. *Let X be an r.i. function space on an interval I, where I is either* $[0, 1]$ *or* $[0, \infty)$, *and suppose that X is isomorphic to a subspace of* $L_p(0, 1)$, $p > 2$. *Then, up to an equivalent norm, X is equal to either* $L_p(I)$, $L_2(I)$ *or* $L_p(I) \cap L_2(I)$.

Clearly, when $I = [0, 1]$ only the first two possibilities are of interest since $L_p(0, 1) \cap L_2(0, 1) = L_p(0, 1)$.

Proof. Let T be an isomorphism from X into $L_p(0,1)$, $p>2$. We treat first the case $I=[0,1]$. Since, for every n, the sequence $\{T\chi_{[(i-1)2^{-n},i2^{-n})}\}_{i=1}^{2^n}$ is K-symmetric with $K\leq\|T\|\|T^{-1}\|$ it follows from 2.e.6 that there exists a constant $D<\infty$ so that

$$\begin{aligned}D^{-1}\left\|\sum_{i=1}^{2^n}a_i\chi_{[(i-1)2^{-n},i2^{-n})}\right\|_X&\\ \leq\max\left\{\left\|\chi_{[0,2^{-n})}\right\|_X\left(\sum_{i=1}^{2^n}|a_i|^p\right)^{1/p},\left(\sum_{i=1}^{2^n}|a_i|^2/2^n\right)^{1/2}\right\}&\\ \leq D\left\|\sum_{i=1}^{2^n}a_i\chi_{[(i-1)2^{-n},i2^{-n})}\right\|_X,&\end{aligned}$$

for every n and every choice of scalars $\{a_i\}_{i=1}^{2^n}$. Hence, for any simple function f which is measurable with respect to the algebra generated by the intervals $[(i-1)2^{-n},i2^{-n})$, $i=1,2,\ldots,2^n$, we get that

$$D^{-1}\|f\|_X\leq\max\{\alpha_n\|f\|_p,\|f\|_2\}\leq D\|f\|_X,$$

where $\alpha_n=2^{n/p}\|\chi_{[0,2^{-n})}\|_X$. Since $0\leq\alpha_n\leq D$ for all n (set $f\equiv1$ in the above inequality) we deduce that, for any simple function f over the dyadic intervals,

$$D^{-1}\|f\|_X\leq\max\{\alpha\|f\|_p,\|f\|_2\}\leq D\|f\|_X,$$

where $\alpha=\liminf_{n\to\infty}\alpha_n$. This completes the proof for the case $I=[0,1]$ since, up to an equivalent renorming, X is $L_2(0,1)$ if $\alpha=0$ or $L_p(0,1)$, otherwise.

The case $I=[0,\infty)$ can be deduced from the previous one. Fix $s\geq1$ and observe that the map $f(t)\to f(t/s)\|\chi_{[0,s]}\|_X^{-1}$ induces an order isometry U between an r.i. function space on $[0,1]$ and the restriction X_s of X to the interval $[0,s)$. This isometry has the additional property that there exist numbers β_s and γ_s, depending only on s, so that, for every simple function f on $[0,1]$, we have

$$\|f\|_p=\beta_s\|Uf\|_p\quad\text{and}\quad\|f\|_2=\gamma_s\|Uf\|_2.$$

Thus, it follows from the case $I=[0,1]$ that

$$D^{-1}\|g\|_X\leq\max\{\alpha_s\beta_s\|g\|_p,\gamma_s\|g\|_2\}\leq D\|g\|_X,$$

for every simple function g which is supported by $[0,s]$ and for some $0\leq\alpha_s\leq D$. Passing to lim inf as $s\to\infty$, we get that there are non-negative constants β and γ such that

$$D^{-1}\|g\|_X\leq\max\{\beta\|g\|_p,\gamma\|g\|_2\}\leq D\|g\|_X,$$

for every simple function g on $[0,\infty)$ which has a bounded support. The alternatives $\gamma=0$ and $\beta=0$ yield that, up to an equivalent norm, X is equal to $L_p(0,\infty)$, respectively $L_2(0,\infty)$. When both β and γ are strictly positive we get that X is $L_p(0,\infty)\cap L_2(0,\infty)$, up to an equivalent norm. □

Remark. The space $L_p(0,\infty)\cap L_2(0,\infty)$ can be realized as the subspace of all the pairs $(f,f)\in L_p(0,\infty)\oplus L_2(0,\infty)$. This proves that $L_p(0,\infty)\cap L_2(0,\infty)$ is actually isomorphic to a subspace of $L_p(0,1)$ for every $1\leqslant p\leqslant\infty$ since $L_p(0,\infty)\oplus L_2(0,\infty)$ is isomorphic to a subspace of $L_p(0,\infty)$. In Section 2.f it will be shown that $L_p(0,\infty)\cap L_2(0,\infty)$ is even isomorphic to $L_p(0,\infty)$ if $2\leqslant p<\infty$.

Corollary 2.e.8. (i) *For every* $1\leqslant p\leqslant\infty$, *the space* $L_p(0,1)$ *has a unique representation as an r.i. function space on* $[0,1]$ *i.e. every r.i. function space* Y *on* $[0,1]$, *which is isomorphic to* $L_p(0,1)$, *is already equal to* $L_p(0,1)$, *up to an equivalent renorming.*

(ii) *For every* $1<p<\infty, p\neq 2$, *the space* $L_p(0,\infty)$ *has exactly two representations as an r.i. function space on* $[0,\infty)$, *namely* $L_p(0,\infty)$ *and* $L_p(0,\infty)\cap L_2(0,\infty)$ *if* $p>2$ *or* $L_p(0,\infty)$ *and* $L_p(0,\infty)+L_2(0,\infty)$ *if* $1<p<2$. *The spaces* $L_1(0,\infty)$, $L_2(0,\infty)$ *and* $L_\infty(0,\infty)$ *have a unique representation as r.i. function spaces on* $[0,\infty)$.

Proof. Both assertions follow in the case $2<p<\infty$ from 2.e.7 and the preceding remark. The case $1<p<2$ is obtained by duality. That $L_1(0,1)$ has a unique r.i. structure on $[0,1]$ follows immediately from 2.e.4.

Suppose now that Y is an r.i. function space on $[0,1]$ which is isomorphic to $L_\infty(0,1)$. Since the images in $L_\infty(0,1)$, under isomorphism, of the subspaces $[\chi_{[(i-1)n^{-1},\,in^{-1})}]_{i=1}^n$, $n=1,2,\ldots$ of Y are uniformly complemented in $L_\infty(0,1)$ it follows from the argument used to prove the uniqueness of the unconditional basis of c_0 (cf. I.2.b.9) that there exists a constant $K<\infty$, depending only on the distance between Y and $L_\infty(0,1)$, so that, for every n and every choice of scalars $\{a_i\}_{i=1}^n$,

$$\Big\|\sum_{i=1}^n a_i\chi_{[(i-1)n^{-1},\,in^{-1})}\Big\|_Y \Big/ \|\chi_{[0,n^{-1})}\|_Y\leqslant K\max_{1\leqslant i\leqslant n}|a_i|\,.$$

Hence, $K^{-1}\leqslant\|\chi_{[0,n^{-1})}\|_Y$, for all n, and this shows that Y is equal to $L_\infty(0,1)$, up to an equivalent norm. Similar arguments can be used to show that $L_1(0,\infty)$, $L_2(0,\infty)$ and $L_\infty(0,\infty)$ have a unique r.i. structure on $[0,\infty)$. □

We now present the *proof of 2.e.5*. Since $\{x_i\}_{i=1}^n$ is a K-unconditional basis there exists a new norm $|||\cdot|||$ on $[x_i]_{i=1}^n$ so that $[x_i]_{i=1}^n$ endowed with this norm becomes a Banach lattice and

$$K^{-1}\|x\|\leqslant|||x|||\leqslant K\|x\|,\quad x\in[x_i]_{i=1}^n\,.$$

By 1.f.17, this lattice is 2-convex with 2-convexity constant M_0 depending only on K, M and m. Hence, by the argument used in the proof of the assertion of remark 2 following 2.b.3, we obtain

$$\left\|\sum_{i=1}^{n} a_i x_i\right\| \geq K^{-1}\left\|\sum_{i=1}^{n} a_i x_i\right\| \geq K^{-1} M_0^{-1}\left(\sum_{i=1}^{n} |a_i|^2\right)^{1/2}\left\|\sum_{i=1}^{n} x_i\right\|/n^{1/2}$$

$$\geq K^{-2} M_0^{-1} w_n\left(\sum_{i=1}^{n} |a_i|^2\right)^{1/2},$$

for every choice of scalars $\{a_i\}_{i=1}^n$. On the other hand, it is easily seen that, for any permutation π of the integers $\{1, 2, \ldots, n\}$ and any $\{a_i\}_{i=1}^n$ as above,

$$\left\|\sum_{i=1}^{n} a_i x_i\right\| \geq K^{-1}\int_0^1\left\|\sum_{i=1}^{n} a_{\pi(i)} r_i(u) x_i\right\| du \geq K^{-1}\left\|\int_0^1\left|\sum_{i=1}^{n} a_{\pi(i)} r_i(u) x_i\right| du\right\|$$

$$\geq K^{-1}\left\|\bigvee_{i=1}^{n} |a_{\pi(i)} x_i|\right\|.$$

This proves the right-hand side inequality of 2.e.5. The proof of the opposite inequality is more difficult. Fix the scalars $\{a_i\}_{i=1}^n$. By 1.d.6, there is a constant C, depending only on m and M, so that, for any permutation π of $\{1, 2, \ldots, n\}$, we have

$$\left\|\sum_{i=1}^{n} a_i x_i\right\| \leq K\left\|\sum_{i=1}^{n} a_{\pi(i)} x_i\right\| \leq KC\left\|\left(\sum_{i=1}^{n} |a_{\pi(i)} x_i|^2\right)^{1/2}\right\|$$

$$= KC\left\|\left(\left(\sum_{i=1}^{n} |a_{\pi(i)} x_i|^2\right)^m\right)^{1/2m}\right\|$$

$$= KC\left\|\left(\sum_{l=1}^{m} \sum_{\substack{m_1+\cdots+m_l=m \\ m_1,\ldots,m_l \geq 1}} \sum_{\substack{i_1,\ldots,i_l \\ \text{distinct}}} |a_{\pi(i_1)} x_{i_1}|^{2m_1} \ldots |a_{\pi(i_l)} x_{i_l}|^{2m_l}\right)^{1/2m}\right\|.$$

Hence, by averaging in the sense of $l_{2m}^{n!}$ over all possible permutations π of the integers $\{1, 2, \ldots, n\}$ and by separating the term corresponding to $l=m$ from the other terms, we get that

$$\left\|\sum_{i=1}^{n} a_i x_i\right\| \leq KC(n!)^{-1/2m}(S_1+S_2),$$

where

$$S_1=\left(\sum_{\pi}\left\|\left(\sum_{\substack{i_1,\ldots,i_m \\ \text{distinct}}} |a_{\pi(i_1)} x_{i_1}|^2 \ldots |a_{\pi(i_m)} x_{i_m}|^2\right)^{1/2m}\right\|^{2m}\right)^{1/2m}$$

and

$$S_2=\left(\sum_{\pi}\left\|\left(\sum_{l=1}^{m-1}\ \sum_{\substack{m_1+\cdots+m_l=m\cdot\\ m_1,\ldots,m_l\geq 1}}\ \sum_{\substack{i_1,\ldots,i_l\\ \text{distinct}}}|a_{\pi(i_1)}x_{i_1}|^{2m_1}\ldots|a_{\pi(i_l)}x_{i_l}|^{2m_l}\right)^{1/2m}\right\|^{2m}\right)^{1/2m}.$$

Suppose now for simplicity of notation that n is an even integer such that $n/2>m$. We first evaluate the expression S_1 by using the $2m$-concavity of X.

$$\begin{aligned}S_1&\leq M\left\|\left(\sum_{\pi}\sum_{\substack{i_1,\ldots,i_m\\ \text{distinct}}}|a_{\pi(i_1)}x_{i_1}|^2\ldots|a_{\pi(i_m)}x_{i_m}|^2\right)^{1/2m}\right\|\\&=M\left\|\left(\sum_{\substack{i_1,\ldots,i_m\\ \text{distinct}}}\left(\sum_{\pi}|a_{\pi(i_1)}|^2\ldots|a_{\pi(i_m)}|^2\right)|x_{i_1}|^2\ldots|x_{i_m}|^2\right)^{1/2m}\right\|\\&=M\left((n-m)!\sum_{\substack{j_1,\ldots,j_m\\ \text{distinct}}}|a_{j_1}|^2\ldots|a_{j_m}|^2\right)^{1/2m}\left\|\left(\sum_{\substack{i_1,\ldots,i_m\\ \text{distinct}}}|x_{i_1}|^2\ldots|x_{i_m}|^2\right)^{1/2m}\right\|\\&\leq M((n-m)!)^{1/2m}\left(\sum_{i=1}^n|a_i|^2\right)^{1/2}\left\|\left(\sum_{i=1}^n|x_i|^2\right)^{1/2}\right\|.\end{aligned}$$

Since $n-m+1>n-n/2+1>n/2$ it follows that

$$\begin{aligned}S_1&\leq 2^{1/2}M(n!)^{1/2m}\ n^{-1/2}\left(\sum_{i=1}^n|a_i|^2\right)^{1/2}A_1^{-1}\int_0^1\left\|\sum_{i=1}^n r_i(u)x_i\right\|du\\&\leq 2MK(n!)^{1/2m}w_n\left(\sum_{i=1}^n|a_i|^2\right)^{1/2}.\end{aligned}$$

In order to evaluate the expression S_2, we fix $1\leq l<m$ and $m_1,\ldots,m_l$ and put

$$S(l;m_1,\ldots,m_l)=\left(\sum_{\pi}\left\|\left(\sum_{\substack{i_1,\ldots,i_l\\ \text{distinct}}}|a_{\pi(i_1)}x_{i_1}|^{2m_1}\ldots|a_{\pi(i_l)}x_{i_l}|^{2m_l}\right)^{1/2m}\right\|^{2m}\right)^{1/2m}.$$

Let $\theta_j=m_j/m$ and $z_j(\pi)=\left(\sum_{i=1}^n|a_{\pi(i)}x_i|^{2m_j}\right)^{1/2m_j}$, $j=1,2,\ldots,l$. Then, by 1.d.2(i) applied in the Banach lattice $l_{2m}^{n!}(X)$ for the vectors $z_j=(z_j(\pi))_\pi$ and θ_j as above, we get that

$$S(l;m_1,\ldots,m_l)\leq\|z_1^{\theta_1}\ldots z_l^{\theta_l}\|_{l_{2m}^{n!}(X)}\leq\|z_1\|_{l_{2m}^{n!}(X)}^{\theta_1}\ldots\|z_l\|_{l_{2m}^{n!}(X)}^{\theta_l}.$$

Since $l<m$ at least one of the integers $m_1,\ldots,m_l$, say m_1, is ≥ 2. Therefore, by

estimating from above the norm in $l_{2m_1}^n$ by that in l_4^n and the norm in $l_{2m_j}^n$, $1<j\leqslant l$ by that in l_2^n, it follows that

$$S(l;m_1,\ldots,m_l)\leqslant\left(\sum_\pi\left\|\left(\sum_{i=1}^n|a_{\pi(i)}x_i|^4\right)^{1/4}\right\|^{2m}\right)^{\theta_1/2m}\left(\sum_\pi\left\|\left(\sum_{i=1}^n|a_{\pi(i)}x_i|^2\right)^{1/2}\right\|^{2m}\right)^{(1-\theta_1)/2m}$$

Note that, by 1.d.2, we have that

$$\left\|\left(\sum_{i=1}^n|a_{\pi(i)}x_i|^4\right)^{1/4}\right\|\leqslant\left\|\left(\sum_{i=1}^n|a_{\pi(i)}x_i|^2\right)^{1/2}\right\|^{1/2}\left\|\bigvee_{i=1}^n|a_{\pi(i)}x_i|\right\|^{1/2}$$

and consequently,

$$S(l;m_1,\ldots,m_l)\leqslant(n!)^{1/2m}(A_1^{-1}K)^{1-\theta_1/2}\left\|\sum_{i=1}^n a_ix_i\right\|^{1-\theta_1/2}R^{\theta_1/2},$$

where $R=\left(\sum_\pi\left\|\bigvee_{i=1}^n|a_{\pi(i)}x_i|\right\|^{2m}/n!\right)^{1/2m}$.

Combining the preceding estimates we get that there exist finite constants $C_1=C_1(K,M,m)$ and $C_2=C_2(K,M,m)$ so that

$$\begin{aligned}\left\|\sum_{i=1}^n a_ix_i\right\|&\leqslant C_1\max\left\{(n!)^{-1/2m}\max_{\substack{1\leqslant l<m\\ m_1,\ldots,m_l}}S(l;m_1,\ldots,m_l),\ w_n\left(\sum_{i=1}^n|a_i|^2\right)^{1/2}\right\}\\&\leqslant C_2\max\left\{\left\|\sum_{i=1}^n a_ix_i\right\|^{1-1/m}R^{1/m},\ w_n\left(\sum_{i=1}^n|a_i|^2\right)^{1/2}\right\},\end{aligned}$$

and therefore

$$\left\|\sum_{i=1}^n a_ix_i\right\|\leqslant C_2^m\max\left\{R,\ w_n\left(\sum_{i=1}^n|a_i|^2\right)^{1/2}\right\}.\ \square$$

In order to present further applications of 2.e.5, we have to study the $l_{2m}^{n!}$ average appearing in its statement.

Lemma 2.e.9. *Let X be an r-convex Banach lattice with r-convexity constant $\leqslant M$, for some $r>1$ and some $M<\infty$. Let $q\geqslant r$ and let $\{x_i\}_{i=1}^n$ be a fixed sequence of vectors in X. Then, for every $1<k\leqslant n$, the formula*

$$|||a|||=\left(\sum_\pi\left\|\bigvee_{i=1}^k|a_ix_{\pi(i)}|\right\|^q\right)^{1/q},\quad a=(a_1,\ldots,a_k)\in R^k,$$

with $\sum_\pi$ meaning summation over all permutation π of $\{1,2,\ldots,n\}$, defines a norm on R^k such that R^k, endowed with this norm and the pointwise order, is an r-convex *lattice whose r-convexity constant is $\leqslant M$. The same assertion holds if r-convexity is replaced by the existence of an upper r-estimate.*

Proof. Since a direct sum in the sense of l_q, $q \geqslant r$ of a sequence of r-convex Banach lattices, each with r-convexity constant $\leqslant M$, is clearly r-convex with r-convexity constant $\leqslant M$ it suffices to prove the assertion for a norm on R^k having the form

$$|||a|||_0 = \left\| \bigvee_{i=1}^{k} |a_i x_i| \right\|, \quad a = (a_1, \ldots, a_k) \in R^k .$$

Let $a^j = (a_1^j, \ldots, a_k^j)$, $j = 1, 2, \ldots, m$ be a sequence of vectors in R^k. Then

$$\left|\left|\left| \left(\sum_{j=1}^{m} |a^j|^r \right)^{1/r} \right|\right|\right|_0 = \left\| \bigvee_{i=1}^{k} \left(\sum_{j=1}^{m} |a_i^j x_i|^r \right)^{1/r} \right\| \leqslant \left\| \left(\sum_{j=1}^{m} \bigvee_{i=1}^{k} |a_i^j x_i|^r \right)^{1/r} \right\|$$
$$\leqslant M \left(\sum_{j=1}^{m} \left\| \bigvee_{i=1}^{k} |a_i^j x_i| \right\|^r \right)^{1/r} = M \left(\sum_{j=1}^{m} |||a^j|||_0^r \right)^{1/r} .$$

The case when X satisfies an upper r-estimate is treated in the same manner. □

We are prepared now to present the result needed in addition to 2.e.1 for proving the uniqueness of the r.i. structure.

Proposition 2.e.10 ([58] Section 2). *Let Y be a Banach lattice of type 2 which is r-convex for some $r > 2$. An r.i. function space X on $[0, 1]$, which is isomorphic to a subspace of Y, must also be r-convex unless it is equal to $L_2(0, 1)$, up to an equivalent renorming.*

Proof. Since a Banach lattice of type 2 is, by 1.f.13, also q-concave for some $q < \infty$ there is no loss of generality in assuming that Y is $2m$-concave for some integer m. Let M be a number exceeding both the r-convexity and $2m$-concavity constants of Y as well as its type 2 constant. Let T be an isomorphism from X into Y and put $K = \max\{\|T\|, \|T^{-1}\|\}$.

Since, for every integer n, $\{T\chi_{[(i-1)2^{-n}, i2^{-n})}\}_{i=1}^{2^n}$ is a K^2-symmetric sequence in Y it follows from 2.e.5 that there exists a constant D, depending only on K, M and m, so that

$$D^{-1} \left\| \sum_{i=1}^{2^n} a_i T\chi_{[(i-1)2^{-n}, i2^{-n})} \right\|_Y$$
$$\leqslant \max \left\{ \left(\sum_{\pi} \left\| \bigvee_{i=1}^{2^n} |a_{\pi(i)} T\chi_{[(i-1)2^{-n}, i2^{-n})}| \right\|_Y^{2m} \Big/ (2^n)! \right)^{1/2m}, \left(\sum_{i=1}^{2^n} |a_i|^2 / 2^n \right)^{1/2} \right\}$$
$$\leqslant D \left\| \sum_{i=1}^{2^n} a_i T\chi_{[(i-1)2^{-n}, i2^{-n})} \right\|_Y ,$$

for every choice of scalars $\{a_i\}_{i=1}^{2^n}$. In other words, for any simple function of the form $f=\sum_{i=1}^{2^n} a_i\chi_{[(i-1)2^{-n},\, i2^{-n})}$, we have

$$(*)\qquad (DK)^{-1}\|f\|_X \leqslant \max\left\{\left(\sum_{\pi}\left\|\bigvee_{i=1}^{2^n}|a_{\pi(i)}T\chi_{[(i-1)2^{-n},\, i2^{-n})}|\right\|_Y^{2m}\Big/(2^n)!\right)^{1/2m}, \|f\|_2\right\} \leqslant DK\|f\|_X\,.$$

The idea of the proof is to show that if X is not $L_2(0, 1)$ then the norm of any simple function f as above, which is supported by a sufficiently small interval, is given by the first term of the inner expression in (*) which, by 2.e.9, defines an r-convex norm. We distinguish between two possible alternatives, as follows.

Case I. Suppose that, for every n, we have

$$\|\chi_{[0,\, 2^{-n})}\|_X \leqslant 2DKM\,\|\chi_{[0,\, 2^{-n})}\|_2\,.$$

Then, for every simple function f as above,

$$\|f\|_X \leqslant M\left(\sum_{i=1}^{2^n}\|a_i\chi_{[(i-1)2^{-n},\, i2^{-n})}\|_X^2\right)^{1/2} \leqslant 2DKM^2\left(\sum_{i=1}^{2^n}\|a_i\chi_{[(i-1)2^{-n},\, i2^{-n})}\|_2^2\right)^{1/2} = 2DKM^2\|f\|_2\,.$$

On the other hand, it follows from (*) that

$$\|f\|_X \geqslant (DK)^{-1}\|f\|_2$$

which shows that X is, up to an equivalent renorming, equal to $L_2(0, 1)$.

Case II. There exists an integer k such that

$$\|\chi_{[0,\, 2^{-k})}\|_X > 2DKM\,\|\chi_{[0,\, 2^{-k})}\|_2\,.$$

In this case, it is clear that when we evaluate the norm of

$$\chi_{[0,\, 2^{-k})} = \sum_{i=1}^{2^{n-k}} \chi_{[(i-1)2^{-n},\, i2^{-n})}, \quad n \geqslant k\,,$$

by using formula (*) then the maximum in the inner expression is necessarily attained in the first term, i.e.

$$(DK)^{-1}\|\chi_{[0,\, 2^{-k})}\|_X \leqslant \left(\sum_{\pi}\left\|\bigvee_{i=1}^{2^{n-k}}|T\chi_{[(\pi(i)-1)2^{-n},\, \pi(i)2^{-n})}|\right\|_Y^{2m}\Big/(2^n)!\right)^{1/2m} \leqslant DK\|\chi_{[0,\, 2^{-k})}\|_X\,.$$

By 2.e.9, for every $n \geq k$, the expression

$$|||a||| = \left(\sum_{\pi} \left\| \bigvee_{i=1}^{2^{n-k}} |a_i T\chi_{[(\pi(i)-1)2^{-n}, \pi(i)2^{-n})}| \right\|_Y^{2m} \Big/ (2^n)! \right)^{1/2m},$$

$$a = (a_1, \ldots, a_{2^{n-k}}) \in R^{2^{n-k}},$$

where $\sum_{\pi}$ means summation over all permutations π of $\{1, 2, \ldots, 2^n\}$, defines an r-convex symmetric norm on $R^{2^{n-k}}$ with r-convexity constant $\leq M$. Thus, by remark 2 following 2.b.3, we get that

$$M|||a|||/|||(\overbrace{1, \ldots, 1}^{2^{n-k} \text{ times}})||| \geq \left(\sum_{i=1}^{2^{n-k}} |a_i|^r / 2^{n-k} \right)^{1/r},$$

for every $a \in (R^{2^{n-k}}, |||\cdot|||)$. It follows that, for any simple function of the form $\psi = \sum_{i=1}^{2^{n-k}} b_i \chi_{[(i-1)2^{-n}, i2^{-n})}$, i.e. supported entirely by the interval $[0, 2^{-k})$, we have

$$\left(\sum_{\pi} \left\| \bigvee_{i=1}^{2^{n-k}} |b_i T\chi_{[(\pi(i)-1)2^{-n}, \pi(i)2^{-n})}| \right\|_Y^{2m} \Big/ (2^n)! \right)^{1/2m} = |||(b_1, \ldots, b_{2^{n-k}})|||$$

$$\geq M^{-1} \left(\sum_{\pi} \left\| \bigvee_{i=1}^{2^{n-k}} |T\chi_{[(\pi(i)-1)2^{-n}, \pi(i)2^{-n})}| \right\|_Y^{2m} \Big/ (2^n)! \right)^{1/2m} \left(\sum_{i=1}^{2^{n-k}} |b_i|^r / 2^{n-k} \right)^{1/r}$$

$$\geq (DKM)^{-1} \|\chi_{[0, 2^{-k})}\|_X \left(\sum_{i=1}^{2^{n-k}} |b_i|^2 / 2^{n-k} \right)^{1/2} \geq 2\|\psi\|_2 .$$

This inequality shows that if we restrict ourselves to simple functions ψ as above the maximum in the inner expression of (*) is always attained by the first term. In view of 2.e.9, it follows that the restriction of X to the interval $[0, 2^{-k})$ is r-convex. This, of course, implies that X is r-convex too. □

We state now the result on the uniqueness of the r.i. structure (cf. [58] Section 5). As in 2.e.8, we shall say that an r.i. function space X on an interval I has a *unique representation* as an r.i. function space on I if any r.i. function space Y on I, which is isomorphic to X, is already equal to X, up to an equivalent norm.

Theorem 2.e.11. *Every r.i. function space X on* $[0, 1]$, *which is q-concave for some $q < 2$, has a unique representation as an r.i. function space on* $[0, 1]$.

Proof. We present here a simplified proof which requires however the additional assumption that X does not contain uniformly isomorphic copies of l_1^n for all n.

Suppose that an r.i. function space Y on $[0, 1]$ is isomorphic to X. By 1.f.12, both X and Y satisfy an upper p-estimate for some $p > 1$ and, by 1.d.7, Y is 2-concave. Thus, by 1.f.18, their duals X^* and Y^* are r.i. function spaces on $[0, 1]$ of type 2 and, by 1.d.4, X^* is r-convex for r satisfying $1/r + 1/q = 1$. It follows from 2.e.10

that also Y^* is r-convex (Y^* cannot coincide with $L_2(0,1)$ since it is isomorphic to an r-convex lattice) and, therefore, Y is q-concave, too. By remark 2 following 2.b.3, there is a constant M such that

$$\|f\|_X \leq M\|f\|_q \quad \text{and} \quad \|f\|_Y \leq M\|f\|_q\,,$$

for every $f \in L_q(0,1)$. The Haar basis of X or of Y cannot be equivalent to a sequence of mutually disjoint functions in Y, respectively X, since it contains the Rademacher functions as a block basis and this would contradict the q-concavity of X and Y. This means that assertion (ii) of 2.e.1 does not hold for any isomorphism from X into Y or, vice-versa, from Y into X. Consequently, assertion (i) of 2.e.1 is satisfied by any such isomorphisms and this, obviously, completes the proof. □

Remarks. 1. It is interesting to compare the behavior of r.i. function spaces on $[0,1]$ with that of spaces having a symmetric basis from the point of view of the uniqueness of the symmetric structure: in I.4.c, we presented several spaces with a non-unique symmetric basis and it can be easily checked that those examples can be constructed as to be, for instance, q-concave for any $1<q<2$ given in advance.

2. As we have seen in 2.e.8, even the spaces $L_p(0,\infty)$, $1<p\neq 2<\infty$ do not have a unique representation as an r.i. function space on $[0,\infty)$. Additional information on this matter as well as on the class of spaces which have simultaneous representations as r.i. function spaces on both $[0,1]$ and $[0,\infty)$ will be given in 2.f.

The proof of 2.e.11 in the general case is more complicated. Actually, it is a particular case of a more general result from [58] Section 5 which is reproduced here without a proof.

Theorem 2.e.12. *Let X be an r.i. function space on $[0,1]$ which is q-concave for some $q<2$. Then every r.i. function space Y on $[0,1]$, which is isomorphic to a complemented subspace of X, is equal, up to an equivalent norm, to either X or $L_2(0,1)$.*

The results presented above apply mostly to r.i. function spaces X sitting on "one side" of 2. These theorems can be extended to arbitrary r.i. function spaces but, since their proofs are too complicated to be presented in the book, we shall limit ourselves to the presentation of the statements. The following theorem is (almost) a generalization of 2.e.1. (The word almost is used since in 2.e.1 we did not assume $p_X>1$.)

Theorem 2.e.13 ([58] Section 6). *Let X be a separable r.i. function space on $[0,1]$ such that $1<p_X$ and $q_X<\infty$. Let Y be an r.i. function space on $[0,1]$ or on $[0,\infty)$ which does not contain uniformly isomorphic copies of l_∞^n for all n. If X is isomorphic to a subspace of Y then either*

(i) *there exists a constant* $D<\infty$ *so that*

$$\|f\|_Y \leq D\|f\|_X,$$

for every $f \in X$, *or*

(ii) *the Haar basis of* X *is equivalent to a sequence of pairwise disjoint functions in* Y, *or*

(iii) X *is equal to* $L_2(0, 1)$, *up to an equivalent renorming.*

In view of 2.c.14 and the observation made thereafter it follows that the alternative (ii) of 2.e.13 cannot hold if Y is an Orlicz function space (unless, of course, X is $L_2(0, 1)$). We also recall that, for an Orlicz function space X, the conditions $1<p_X$ and $q_X<\infty$ are equivalent to the reflexivity of X.

Corollary 2.e.14 ([58] Section 7). *Let* X *be a separable r.i. function space on* $[0, 1]$ *which is different from* $L_2(0, 1)$, *even up to an equivalent renorming. Let* M *be an Orlicz function satisfying the* Δ_2*-condition both at* 0 *and at* ∞.

(i) *If* $1<p_X$ *and* $q_X<\infty$ *and* X *is isomorphic to a subspace of* $L_M(0, \infty)$ *then*

$$\|f\|_{L_M(0,1)} \leq D\|f\|_X,$$

for some $D<\infty$ *and every* $f \in X$.

(ii) *If* X *is isomorphic to a complemented subspace of* $L_M(0, \infty)$ *and* $L_M(0, \infty)$ *is reflexive then, up to an equivalent norm* $X=L_M(0, 1)$. *In particular, any reflexive Orlicz function space on* $[0, 1]$ *has a unique representation as an r.i. function space on* $[0, 1]$.

Let X be a uniformly convexifiable r.i. function space on $[0, 1]$ and suppose that another r.i. function space Y on $[0, 1]$ is isomorphic to X. By the discussion preceding 1.e.4, 2.b.7 and 2.b.2, both X and X^* have non-trivial Boyd indices (i.e. $1<p_X, p_{X^*}$ and $q_X, q_{X^*}<\infty$). Thus, it follows from 2.e.13 that if X is not equal to Y, up to an equivalent norm, then the Haar basis of X or of X^* must be equivalent to a sequence of pairwise disjoint functions in Y, respectively, Y^*. However, it can be shown with some additional effort that the following generalization of 2.e.11 is true (cf. [58] Section 6).

Theorem 2.e.15. *Let* X *be a uniformly convexifiable r.i. function space on* $[0, 1]$. *Then either* X *has a unique representation as an r.i. function space on* $[0, 1]$ *or the Haar basis of* X *is equivalent to a sequence of mutually disjoint functions in* X.

Remark. In Volume IV we shall present, for every $1<p<2$, an example of an r.i. function space X on $[0, 1]$ which embeds isomorphically into $L_p(0, 1)$, does not have a unique representation as an r.i. function space on $[0, 1]$ (actually, it has uncountably many mutually non-equivalent such representations) and the Haar basis of X is equivalent to a sequence of pairwise disjoint elements in X.

We have seen in 2.d.5 that a complemented subspace Y of a separable r.i. function space X on $[0, 1]$ whose Boyd indices are non-trivial is isomorphic to X provided that Y contains a *complemented* copy of X. The complementation of X in Y is actually redundant if we exclude the case when the Haar basis in X is equivalent to a sequence of pairwise disjoint vectors in X (the example mentioned in the remark above will also show that we have to exclude this case). This is a consequence of the following theorem from [58] Section 9 which, again, is stated without a proof.

Theorem 2.e.16. *Let X be an r.i. function space on* $[0, 1]$ *such that*

(a) $1<p_X$,

(b) *X is q-concave for some $q<\infty$, and*

(c) *the Haar basis of X is not equivalent to any sequence of disjointly supported functions in X.*

Then every subspace of X, which is isomorphic to X, contains a further subspace which is complemented in X and still isomorphic to X. In particular, the theorem is valid for $L_p(0, 1)$ spaces, $1<p<\infty$, or, more generally, (by 2.c.14) for any reflexive Orlicz function space on $[0, 1]$.

The conclusion of 2.e.16 is valid also for $X=L_1(0, 1)$ but this fact, which is due to Enflo and Starbird [38], does not follow from the general assertion of 2.e.16.

The following corollary is an immediate consequence of 2.d.5 and 2.e.1.

Corollary 2.e.17. *Let X be an r.i. function space on* $[0, 1]$ *which satisfies the conditions* (a), (b) *and* (c) *of 2.e.16. Then every complemented subspace Y of X, which contains an isomorphic copy of X, is already isomorphic to X.*

We conclude this section by presenting a result proved independently in [30] for subspaces of $L_p(0, 1)$, $2<p<\infty$, with an unconditional basis and, later on, in [58] Section 2 for general Banach lattices. An interesting thing about this theorem, which actually generalizes 2.e.10 to the case when X is an arbitrary lattice, is the fact that, though its statement has nothing to do with r.i. spaces or other symmetric structures, the proof from [58], which is reproduced here, is based on the characterization of symmetric basic sequences in a Banach lattice of type 2, presented in 2.e.5.

Theorem 2.e.18. *Let Y be a Banach lattice of type* 2 *which satisfies an upper r-estimate for some $r>2$. Then every lattice X, which is isomorphic to a subspace of Y, satisfies itself an upper r-estimate or it contains uniformly isomorphic copies of l_2^n on disjointly supported vectors for all n.*

Proof. Let $\{x_i\}_{i=1}^n$ be a sequence of mutually disjoint vectors in X such that $\sum_{i=1}^n \|x_i\|^r=n$. For every $1\leq i\leq n$, consider the sequence $\hat{x}_i=(x_{\pi(i)}/(n!)^{1/r})_\pi$, where π ranges over all permutations of $\{1, 2, \dots, n\}$, as an element of $l_r^{n!}(X)$. Since, for arbitrary values of i and j, the vectors $\hat{x}_i$ and $\hat{x}_j$ consist actually of the same sequence of vectors in X arranged in a different manner it follows that $\{\hat{x}_i\}_{i=1}^n$ forms a symmetric basic sequence in $l_r^{n!}(X)$ (with symmetric constant equal to one). Observe also that the factor $(n!)^{1/r}$ appearing in the definition of the $\hat{x}_i$'s has been chosen to ensure that $\|\hat{x}_i\|=1$ for all $1\leqslant i\leqslant n$.

Since $l_r^{n!}(Y)$ is also of type 2 and, therefore, $2m$-concave for some integer m and since there clearly exists an isomorphism $\hat{T}$ from $l_r^{n!}(X)$ into $l_r^{n!}(Y)$ we can apply 2.e.5 and conclude the existence of a constant D, independent of the x_i's, so that, for every choice of scalars $\{a_i\}_{i=1}^n$, we have

$$(*)\qquad D^{-1}\left\|\sum_{i=1}^n a_i\hat{x}_i\right\|\leqslant \max\left\{\left(\sum_\pi\left\|\bigvee_{i=1}^n |a_{\pi(i)}\hat{T}\hat{x}_i|\right\|^{2m}/n!\right)^{1/2m}, w_n\left(\sum_{i=1}^n |a_i|^2\right)^{1/2}\right\}$$
$$\leqslant D\left\|\sum_{i=1}^n a_i\hat{x}_i\right\|,$$

where $w_n=\left\|\sum_{i=1}^n \hat{x}_i\right\|/n^{1/2}$.

Suppose now that X does not contain uniformly isomorphic copies of l_2^n on disjoint vectors, for all n. Then the 2-concavification $X_{(2)}$ of X does not contain uniformly isomorphic copies of l_1^n on disjoint vectors, for all n, and thus, by 1.f.12, $X_{(2)}$ satisfies a non-trivial upper estimate. It follows that X, which is order isomorphic to the 2-convexification of $X_{(2)}$, satisfies an upper p-estimate for some $p>2$. (We may assume that $p<r$; if $p\geqslant r$ there is nothing more to prove.) Thus also $l_r^{n!}(X)$ satisfies an upper p-estimate and, therefore, for sufficiently large n we have $w_n<1$. For such n it is possible to choose integers h and k so that $1\leqslant hw_n^{2r/(r-2)}<2$ and $kh\leqslant n<(k+1)h$.

Let M be the (joint) upper r-estimate constant of Y and $l_r^{n!}(Y)$ and, for $j=1,2,\ldots,k$, put

$$\hat{u}_j=\sum_{i=(j-1)h+1}^{jh}\hat{x}_i\,.$$

By (*) and 2.e.9, we get, for every $1\leqslant j\leqslant k$, that (assuming, as we clearly may, that $M\|\hat{T}\|\geqslant 1$)

$$\|\hat{u}_j\|\leqslant D\max\left\{\left(\sum_\pi\left\|\bigvee_{i=(j-1)h+1}^{jh}|\hat{T}\hat{x}_{\pi(i)}|\right\|^{2m}/n!\right)^{1/2m}, w_nh^{1/2}\right\}$$
$$\leqslant DM\|\hat{T}\|\max\left\{\left(\sum_{i=(j-1)h+1}^{jh}\|\hat{x}_i\|^r\right)^{1/r}, w_nh^{1/2}\right\}$$
$$\leqslant DM\|\hat{T}\|\max\{h^{1/r}, w_nh^{1/2}\}$$

and, in view of the conditions imposed on h, it follows that $h^{1/r}\leqslant w_nh^{1/2}$, i.e. $\|\hat{u}_j\|\leqslant DM\|\hat{T}\|w_nh^{1/2}$ for all $1\leqslant j\leqslant k$. Since the $\{\hat{x}_i\}_{i=1}^n$ and therefore also the

$\{\hat{u}_j\}_{j=1}^n$ are mutually disjoint vectors in $l_r^{n!}(X)$ we get that

$$\left\|\sum_{j=1}^{k} \hat{u}_j\right\| \leqslant M_1\left(\sum_{j=1}^{k} \|\hat{u}_j\|^p\right)^{1/p} \leqslant DMM_1\|\hat{T}\|w_n h^{1/2}k^{1/p},$$

where M_1 is the upper p-estimate constant of $l_r^{n!}(X)$. On the other hand, again by (*), we have that

$$\left\|\sum_{j=1}^{k} \hat{u}_j\right\| = \left\|\sum_{i=1}^{kh} \hat{x}_i\right\| \geqslant D^{-1}w_n k^{1/2}h^{1/2}.$$

Hence, by the choice of h and k above,

$$nw_n^{2r/(r-2)} < 2(k+1) \leqslant 2((D^2MM_1\|\hat{T}\|)^{2p/(p-2)}+1) = C,$$

where C is a constant independent of the choice of the sequence $\{x_i\}_{i=1}^n$. This completes the proof since

$$\left\|\sum_{i=1}^{n} x_i\right\| = \left\|\sum_{i=1}^{n} \hat{x}_i\right\| = n^{1/2}w_n \leqslant C^{(r-2)/2r}n^{1/r} = C^{(r-2)/2r}\left(\sum_{i=1}^{n} \|x_i\|^r\right)^{1/r}. \quad \square$$

In the special case when Y is an $L_r(0, 1)$ space, $r>2$, Theorem 2.e.18 can be restated in a stronger form.

Corollary 2.e.19. *A Banach lattice X, which is linearly isomorphic to a subspace of $L_r(0, 1)$ for some $2<r<\infty$, is itself order isomorphic to an $L_r(\nu)$ space for a suitable measure ν, unless it contains uniformly isomorphic copies of l_2^n on disjointly supported vectors for all n.*

Assertion 2.e.19 follows immediately from 2.e.18, the fact that a lattice X as above is of cotype r and 1.b.13.

f. Applications of the Poisson Process to r.i. Function Spaces

The principal motivation for this section is to provide a proof for the claim made in 2.e.8(ii) that $L_p(0, 1)$ is isomorphic to $L_p(0, \infty)+L_2(0, \infty)$ or to $L_p(0, \infty)\cap L_2(0, \infty)$ when $1<p<2$, respectively, $p>2$ and, more generally, to study those r.i. function spaces on [0, 1] which admit also a representation as an r.i. function space on [0, ∞). As we shall see below, this class contains in particular all the r.i. function spaces on [0, 1] whose Boyd indices are non-trivial. This representation of r.i. function spaces on [0, 1] as suitable r.i. function spaces on [0, ∞) is used later on in this section for proving that if $t^{-1/r}$ belongs to some r.i. function space X on $[0, 1]$ for some $1<r<2$ then X contains an isometric copy of $L_r(0, 1)$. In particular, $L_r(0, 1)$ is isometric to a subspace of $L_p(0, 1)$ if $p<r<2$. The section ends with a discussion of Rosenthal's spaces $X_{p,2}$ (which were introduced in I.4.d) and proper generalizations of them in the context of arbitrary r.i. function spaces on $[0, \infty)$.

Considerations based on 2.e.13 show that if we want to find an r.i. function space $\mathscr{X}$ on $[0, \infty)$ which is isomorphic to a given r.i. function space X on $[0, 1]$ then it is very natural to look for an $\mathscr{X}$ whose restriction to $[0, 1]$ coincides with X. The problem is, of course, to determine a way to define $\mathscr{X}$ on the entire half line $[0, \infty)$ so that it is isomorphic to X. We define $\mathscr{X}$ so that it behaves like $L_2(0, \infty)$ in the neighborhood of ∞. More precisely, the norm of a function $f \in \mathscr{X}$ will be equivalent to the expression

$$[\![f]\!]_{\mathscr{X}} = \|f^* \chi_{[0,1]}\|_X + \|f^* \chi_{(1,\infty)}\|_2 ,$$

which, in general, is not a norm.

Theorem 2.f.1 ([58] Section 8). *Let X be an r.i. function space on $[0, 1]$ and let $\mathscr{X}$ be the r.i. function space on $[0, \infty)$ of all the measurable f for which $f^*\chi_{[0,1]} \in X$ and $f^*\chi_{(1,\infty)} \in L_2(0, \infty)$, endowed with the norm*

$$\|f\|_{\mathscr{X}} = \max\left\{\|f^*\chi_{[0,1]}\|_X, \left(\sum_{n=0}^{\infty}\left(\int_n^{n+1} f^*(u)\,du\right)^2\right)^{1/2}\right\}.$$

(i) *If $q_X<\infty$ then X contains a subspace isometric to an r.i. function space $\mathscr{X}_0$ on $[0, \infty)$ which, up to an equivalent norm, is equal to $\mathscr{X}$.*
(ii) *If $1<p_X$ and $q_X<\infty$ then $\mathscr{X}$ is isomorphic to X.*

Before proving 2.f.1 we would like to make some comments. First, we point out that $\|\cdot\|_{\mathscr{X}}$ is indeed a norm since both expressions inside the maximum defining $\|\cdot\|_{\mathscr{X}}$ can be written as suprema in terms of f. For instance, it is easily verified that the second expression is equal to

$$\sup\left\{\left(\sum_{n=0}^{\infty}\left(\int_{\eta_n} |f(u)|\,du\right)^2\right)^{1/2}\right\},$$

where the supremum is taken over all partitions of $[0, \infty)$ into pairwise disjoint subsets $\{\eta_n\}_{n=0}^{\infty}$ each having measure equal to one. It is also easily checked that

$$[\![f]\!]_{\mathscr{X}}/2 \le \|f\|_{\mathscr{X}} \le [\![f]\!]_{\mathscr{X}} ,$$

for every $f \in \mathscr{X}$. For example, the left-hand side inequality follows from the fact that, for any $f \in \mathscr{X}$, we have

$$\|f^*\chi_{(1,\infty)}\|_2 = \left(\sum_{n=1}^{\infty}\int_n^{n+1} f^*(u)^2\,du\right)^{1/2} \leqslant \left(\sum_{n=1}^{\infty} f^*(n)^2\right)^{1/2}$$
$$\leqslant \left(\sum_{n=0}^{\infty}\left(\int_n^{n+1} f^*(u)\,du\right)^2\right)^{1/2}.$$

The actual form of the norm in $\mathscr{X}$ is not really important and we may use the equivalent expression $[\![\cdot]\!]_{\mathscr{X}}$ for all practical purposes. The fact that $\|\cdot\|_{\mathscr{X}}$ is equivalent to $[\![\cdot]\!]_{\mathscr{X}}$ yields immediately that $\mathscr{X}$ is a minimal or a maximal r.i. function space on $[0, \infty)$ according to whether X is minimal, respectively, maximal.

Our approach to prove 2.f.1 is to embed each of the subspaces X_n of $\mathscr{X}$ obtained by restricting $\mathscr{X}$ to $[n-1, n)$ into "independent" copies of X in X. A simple example of such an embedding is given by $Sf(u, v) = \sum_{n=1}^{\infty} r_n(u) f(n-1+v)$, $0 \leqslant u, v \leqslant 1, f \in \mathscr{X}$, where $\{r_n\}_{n=1}^{\infty}$ denote as usual the Rademacher functions and Sf is considered as an element in $X([0,1]\times[0,1])$ (recall that $X([0,1]\times[0,1])$ is isometric to X and $\|g\|_{X([0,1]\times[0,1])} = \|g^*\|_X$). It is easily verified, by using Khintchine's inequality, that if f is a decreasing non-negative element in $\mathscr{X}$ then $C^{-1}\|f\| \leqslant \|Sf\| \leqslant C\|f\|$ for some constant C independent of f. The difficulty arises if we consider general (i.e. not necessarily decreasing) functions $f \in \mathscr{X}$. In order to deal with such functions we need that not only $\{\chi_{[n-1,n)}\}_{n=1}^{\infty}$ are mapped into independent random variables (as in the case with the S above) but that $\{\chi_{[s_j,t_j)}\}_{j=1}^{k}$ for any choice of disjoint intervals $\{[s_j, t_j)\}_{j=1}^{k}$ are mapped into independent random variables and that the distribution function of the image of $\chi_{[s,t)}$ depends only on $t-s$. These considerations lead us to replace S by a more sophisticated operator which is built by using a stationary random process $\{N_t\}_{0\leqslant t<\infty}$ with independent increments. Let us first explain this probabilistic terminology. A *random process* $\{N_t\}_{0\leqslant t<\infty}$ is just a family of measurable functions on some probability space. The process is called *stationary* if the distribution function of $N_t - N_s$ depends only on $t-s$. The process is said to have *independent increments* if $\{N_{t_j} - N_{s_j}\}_{j=1}^{k}$ are independent random variables whenever $\{[s_j, t_j)\}_{j=1}^{k}$ are mutually disjoint intervals.

We shall employ here the *Poisson process* which is one of the simplest and most important examples of a stationary process with independent increments. Let us recall that an integer valued random variable N on (Ω, Σ, ν) is said to have the *Poisson distribution with parameter* $\alpha > 0$ if

$$\nu(\{\omega \in \Omega;\, N(\omega) = k\}) = e^{-\alpha}\alpha^k/k!, \quad k = 0, 1, 2, \ldots .$$

The characteristic function of such a random variable (i.e. the Fourier transform of its density N) is given by

$$F_\alpha(x) = \sum_{k=0}^{\infty} e^{ikx-\alpha}\alpha^k/k! = e^{\alpha(e^{ix}-1)}, \quad -\infty < x < \infty .$$

Since $F_\alpha(x)F_\beta(x)=F_{\alpha+\beta}(x)$ it follows that if N_1 and N_2 are independent random variables with the Poisson distribution with parameter α_1, respectively α_2, then N_1+N_2 has the Poisson distribution with parameter $\alpha_1+\alpha_2$. Consequently, if we consider the requirements that $\{N_t\}_{0\le t<\infty}$ be a random process with independent increments and that N_t-N_s have a Poisson distribution with parameter $\alpha(t-s)$ for some fixed $\alpha>0$ and every $0\le s<t<\infty$ then, for every choice of a finite number of reals $\{t_j\}_{j=1}^k$, these requirements (concerning only $\{N_{t_j}\}_{j=1}^k$) are mutually consistent. It follows therefore from a general consistency theorem of Kolmogorov (see Doob [29] for a detailed discussion) that such a process (called the Poisson process with parameter α) really exists on a suitable probability space (Ω, Σ, ν). In our case it is however simpler (and of advantage to the proof of 2.f.1 presented below) to replace the general existence theorem by a concrete representation of the Poisson process on $\Omega=[0, 1]\times[0, 1]$ with the usual Lebesgue measure.

Let $\{\sigma_k\}_{k=0}^\infty$ be a partition of $[0, 1]$ into disjoint sets so that

$$\mu(\sigma_k)=e^{-\alpha}\alpha^k/k!,\quad k=0, 1, 2, \ldots$$

and let $\{\varphi_j\}_{j=1}^\infty$ be a sequence of independent and uniformly distributed random variables on $[0, 1]$ (a variable φ is said to be *uniformly distributed* on $[0, 1]$ if its distribution function is the same as that of $f(v)=v$). Consider the function

$$N_t(u, v)=\sum_{k=1}^\infty \chi_{\sigma_k}(u)\sum_{j=1}^k \chi_{[0, t)}(\varphi_j(v)),\quad 0\le u, v, t\le 1 .$$

The family $\{N_t\}_{0\le t\le 1}$ forms a Poisson process with parameter α (restricted to $0\le t\le 1$) on $[0, 1]\times[0, 1]$. In order to verify this we have just to note (since the characteristic function of a random variable determines its distribution) that for every choice of $0=t_0<t_1<\cdots<t_h=1$ and reals $\{x_l\}_{l=1}^h$ we have that

$$\int_0^1\int_0^1 e^{i\sum_{l=1}^h x_l(N_{t_l}(u, v)-N_{t_{l-1}}(u, v))}\,du\,dv$$

$$=\mu(\sigma_0)+\sum_{k=1}^\infty \mu(\sigma_k)\int_0^1 e^{i\sum_{l=1}^h x_l\sum_{j=1}^k \chi_{[t_{l-1}, t_l)}(\varphi_j(v))}\,dv$$

$$=e^{-\alpha}+\sum_{k=1}^\infty e^{-\alpha}\frac{\alpha^k}{k!}\left(\sum_{l=1}^h e^{ix_l}(t_l-t_{l-1})\right)^k$$

$$=e^{-\alpha\left(1-\sum_{l=1}^h e^{ix_l}(t_l-t_{l-1})\right)}=\prod_{l=1}^h F_{\alpha(t_l-t_{l-1})}(x_l) .$$

In order to define N_t for every $t\ge 0$ we construct a sequence $\{N_t^{(n)}\}_{0\le t<1}$, $n=0, 1, 2, \ldots$ of independent copies of the family $\{N_t\}_{0\le t\le 1}$ defined above and put, for every integer m and $0\le t<1$,

$$N_{m+t}=\sum_{n=0}^{m-1} N_1^{(n)}+N_t^{(m)} .$$

(In order to avoid conflicting notation we let $N_t^{(0)}$ be equal to the N_t defined above for $0 \leqslant t \leqslant 1$.)

We are now prepared to give the *proof of 2.f.1.* We start with assertion (i). Let $\{N_t'\}_{0 \leqslant t < \infty}$ and $\{N_t''\}_{0 \leqslant t < \infty}$ be two independent copies of the Poisson process with parameter 1. For $0 \leqslant s < t < \infty$, put

$$T'\chi_{[s,t)} = N_t' - N_s', \qquad T''\chi_{[s,t)} = N_t'' - N_s''.$$

The mappings T' and T'' can be extended by linearity to linear operators defined on all integrable step functions on $[0, \infty)$. We shall show that if $q_X < \infty$ then $T = T' - T''$ extends to an isomorphism from $\mathscr{X}$ onto a subspace of $X([0,1] \times [0,1])$. In the language of probability theory (which we shall however not use below) Tf is the so-called stochastic integral $\int_0^\infty f(t)\, d(N_t' - N_t'')$ with respect to the symmetric stationary process with independent increments $\{N_t' - N_t''\}_{0 \leqslant t \leqslant \infty}$. (Recall that a random variable G is said to be *symmetric* if G and $-G$ have the same distribution.)

Step I. We shall first prove that the restriction of T to the step functions on $[0, 1]$ is an isomorphism. By the concrete representation of the Poisson process given above, we have that

$$N_t'(u, v) = \sum_{k=1}^{\infty} \chi_{\sigma_k'}(u) \sum_{j=1}^{k} \chi_{[0,t)}(\varphi_j'(v)),$$

for every $t, u, v \in [0, 1]$, where $\{\sigma_k'\}_{k=0}^\infty$ is a partition of $[0, 1]$ into mutually disjoint subsets such that $\mu(\sigma_k') = 1/ek!$, $k = 0, 1, 2, \ldots$ and $\{\varphi_j'\}_{j=1}^\infty$ a sequence of uniformly distributed independent random variables on $[0, 1]$. We shall use an analogous notation for $N_t''(u, v)$. It is readily verified that, for every step function g on $[0, 1]$, we have

$$(T'g)(u, v) = \sum_{k=1}^{\infty} \chi_{\sigma_k'}(u) \sum_{j=1}^{k} g(\varphi_j'(v)), \quad u, v \in [0, 1].$$

Observe now that if D_s denotes, as in 2.b, the dilation operator (acting on functions defined on the unit square, say via a fixed measure preserving isomorphism between the unit interval and the unit square), then, for each k, $\chi_{\sigma_k'}(u) \sum_{j=1}^{k} g(\varphi_j'(v))$ has the same distribution function as $D_{\mu(\sigma_k)} \sum_{j=1}^{k} g(\varphi_j'(v))$. Hence,

$$\|T'g\|_X \leqslant \sum_{k=1}^{\infty} \|D_{\mu(\sigma_k)}\|_X \left\| \sum_{j=1}^{k} g(\varphi_j'(v)) \right\|_X \leqslant \sum_{k=1}^{\infty} k \|D_{\mu(\sigma_k)}\|_X \|g\|_X.$$

By using now the assumption that $q_X < \infty$, we can estimate the norm of $T'g$. Fix $q_X < q < \infty$ and recall that there exists a constant K_q so that $\|D_s\|_X \leqslant K_q s^{1/q}$, for every $0 \leqslant s \leqslant 1$. It follows that

$$\|T'g\|_X \leqslant K_q \|g\|_X \sum_{k=1}^{\infty} k/(ek!)^{1/q}$$

from which we get

$$\|T'g\|_X \leqslant K\|g\|_X ,$$

where $K = K_q \sum_{k=1}^{\infty} k/(ek!)^{1/q}$. On the other hand, we also have

$$\|T'g\|_X \geqslant \|\chi_{\sigma_i'}(u) g(\varphi_1'(v))\|_X \geqslant \|g\|_X / \|D_{1/\mu(\sigma_i')}\|_X \geqslant \|g\|_X / e .$$

Notice that, since $T'g$ and $T''g$ are independent, we have

$$\begin{aligned}\mu(\{(u, v); |Tg(u, v)| > \lambda\}) &\geqslant \mu(\{(u, v); |T'g(u, v)| > \lambda \quad \text{and} \quad u \in \sigma_0''\}) \\ &= e^{-1} \mu(\{(u, v); |T'g(u, v)| > \lambda\})\end{aligned}$$

and thus

$$\|T'g\|_X \leqslant \|D_e\|_X \|Tg\|_X \leqslant e\|Tg\|_X .$$

Consequently,

$$\|g\|_X / e^2 \leqslant \|Tg\|_X \leqslant 2K\|g\|_X .$$

Step II. We shall now consider the action of T on an arbitrary integrable step function f on $[0, \infty)$. We note first that, by the properties of the Poisson process and the definition of T, the functions Tf and Tf^* have the same distribution function. Thus, in order to get estimates on $\|Tf\|$, we may assume without loss of generality that $f = f^*$.

The sequence $\{Tf_n\}_{n=1}^{\infty}$, where $f_n = f\chi_{[n-1, n)}$, $n = 1, 2, \ldots$ is a sequence of symmetric independent random variables and therefore forms an unconditional basic sequence (with unconditional constant one) in every r.i. function space. Note also that, by 2.b.3, for every $h \in X$

$$\|h\|_1 \leqslant \|h\|_X \leqslant C\|h\|_q ,$$

for some C, and that there is no loss of generality to assume that q was chosen to be $\geqslant 2$. Since L_q is of type 2 we get that

$$\|T(f\chi_{[1, \infty)})\|_X \leqslant C \left\| \sum_{n=2}^{\infty} Tf_n \right\|_q \leqslant CB_q \left(\sum_{n=2}^{\infty} \|Tf_n\|_q^2 \right)^{1/2} .$$

By Step I (applied in L_q) and the fact that f is non-increasing, we deduce that

$$\left(\sum_{n=2}^{\infty}\|Tf_n\|_q^2\right)^{1/2}\leqslant 2K\left(\sum_{n=2}^{\infty}\|f_n\|_q^2\right)^{1/2}\leqslant 2K\left(\sum_{n=2}^{\infty}f(n-1)^2\right)^{1/2}.$$

Consequently,

$$\begin{aligned}\|Tf\|_X&\leqslant\|Tf\chi_{[0,1)}\|_X+\|Tf\chi_{[1,\infty)}\|_X\\&\leqslant 2K\|f\chi_{[0,1)}\|_X+2KCB_q\left(\sum_{n=0}^{\infty}\left(\int_n^{n+1}f(u)\,du\right)^2\right)^{1/2}\\&\leqslant 2K(1+CB_q)\|f\|_{\mathscr{X}}.\end{aligned}$$

Similarly, since L_1 is of cotype 2 we get by Step I (applied to L_1) that

$$\|Tf\|_X\geqslant\left\|\sum_{n=1}^{\infty}Tf_n\right\|_1\geqslant A_1\left(\sum_{n=1}^{\infty}\|Tf_n\|_1^2\right)^{1/2}\geqslant A_1\,e^{-2}\left(\sum_{n=1}^{\infty}\|f_n\|_1^2\right)^{1/2}.$$

Also, by Step I (applied to X),

$$\|Tf\|_X\geqslant\|Tf\chi_{[0,1)}\|_X\geqslant e^{-2}\|f\chi_{[0,1)}\|_X$$

and hence

$$\|Tf\|_X\geqslant 2^{-1}A_1\,e^{-2}\|f\|_{\mathscr{X}}.$$

It is clear from the preceding inequalities that if X (and thus $\mathscr{X}$) is minimal then T extends to an isomorphism from $\mathscr{X}$ into X. If X is a maximal r.i. function space the extension of T from the closure of the simple integrable functions to all of $\mathscr{X}$ is obtained as follows: Let $f\geqslant 0$ belong to $\mathscr{X}$ and $\{h_n\}_{n=1}^{\infty}$ be an increasing sequence of simple integrable functions which converges to f a.e. Since T' and T'' are positive, the sequences $\{T'h_n\}_{n=1}^{\infty}$ and $\{T''h_n\}_{n=1}^{\infty}$ are increasing and norm bounded. Hence, since X has the Fatou property, they converge a.e. to some elements in X denoted by $T'f$ and $T''f$, respectively. It is easily verified that these elements do not depend on the choice of $\{h_n\}_{n=1}^{\infty}$ and that $T=T'-T''$ extends to a bounded operator on $\mathscr{X}$. Once we know this, the computation done in Steps I and II above is valid if f is an arbitrary element of $\mathscr{X}$ and hence T is an isomorphism on $\mathscr{X}$.

In order to complete the proof of 2.f.1(i), put

$$\|f\|_{\mathscr{X}_0}=\|Tf\|_X/\|T\chi_{[0,1]}\|_X,\quad f\in\mathscr{X}.$$

Then $\mathscr{X}_0=(\mathscr{X},\|\cdot\|_{\mathscr{X}_0})$ is isomorphic to $\mathscr{X}$ and isometric to a subspace of X. It remains to verify that $\mathscr{X}_0$ is an r.i. function space on $[0,\infty)$. As already mentioned in this proof, if f is a step function on $[0,\infty)$ with bounded support then Tf

and Tf^* have the same distribution function and, thus, f and f^* have the same $\|\cdot\|_{\mathscr{X}_0}$-norm. It is easily deduced from this, by an approximation argument, that Tf and Tf^* have the same norms in X, for every $f \in \mathscr{X}$. This proves 2.f.1(i).

The *proof of 2.f.1(ii)* uses the decomposition method. It is evident that each of the spaces X and $\mathscr{X}$ is isomorphic to its square and that $\mathscr{X}$ contains a complemented subspace isomorphic to X. Therefore, in order to prove that X and $\mathscr{X}$ are isomorphic, it suffices to show that if, in addition, $1 < p_X$ then the image Z_X of $\mathscr{X}$ under T is a complemented subspace of $X([0, 1] \times [0, 1])$. This is proved by interpolation as follows.

Let Z_r denote the space Z_X when $X = L_r(0, 1)$ and let P be the orthogonal projection from $L_2([0, 1] \times [0, 1])$ onto Z_2. This projection has evidently norm one. If P were also bounded as an operator in $L_r([0, 1] \times [0, 1])$ for every $r > 2$ then, since P coincides with its adjoint, we would get that P extends to a bounded operator in every $L_r([0, 1] \times [0, 1])$ space, $1 < r < \infty$. It would then easily follow from 2.b.11 that P is a bounded projection in X whose range is precisely Z_X.

Fix $r \geq 2$ and, for $n = 1, 2, \ldots$, let $Z_r^{(n)}$ be the closed linear span of the sequence $w_k = T\chi_{[(k-1)2^{-n}, k2^{-n})}$, $k = 1, 2, \ldots$ in $L_r([0, 1] \times [0, 1])$. Let P_n be the orthogonal projection from $L_2([0, 1] \times [0, 1])$ onto $Z_2^{(n)}$ which is given by

$$P_n f = \sum_{k=1}^{\infty} \left(\int_0^1 \int_0^1 f w_k \, du \, dv \right) w_k / \|w_k\|_2^2, \quad f \in L_2([0, 1] \times [0, 1]).$$

Since $P_n f = Th$, where

$$h = \sum_{k=1}^{\infty} \left(\int_0^1 \int_0^1 f w_k \, du \, dv \right) \chi_{[(k-1)2^{-n}, k2^{-n})} / \|w_k\|_2^2$$

we get that

$$\|P_n f\|_r \leq \|T\|_r \max\{\|h\|_r, \|h\|_2\}$$

(use the fact that the space $\mathscr{X}$ corresponding to $X = L_r(0, 1)$ is isomorphic to $L_r(0, \infty) \cap L_2(0, \infty)$). Clearly,

$$\|h\|_2 = \|T^{-1} P_n f\|_2 \leq \|T^{-1}\|_2 \|f\|_2 \leq \|T^{-1}\|_2 \|f\|_r .$$

We also have that

$$\begin{aligned} \|h\|_r &= 2^{-n/r} \|w_1\|_2^{-2} \left(\sum_{k=1}^{\infty} \left| \int_0^1 \int_0^1 f w_k \, du \, dv \right|^r \right)^{1/r} \\ &\leq 2^{n/r'} \|T^{-1}\|_2^2 \sup \int_0^1 \int_0^1 f \left(\sum_{k=1}^{\infty} c_k w_k \right) du \, dv \\ &\leq 2^{n/r'} \|T^{-1}\|_2^2 \|f\|_r \sup \left\| \sum_{k=1}^{\infty} c_k w_k \right\|_{r'} , \end{aligned}$$

where both suprema are taken over all choices of $\{c_k\}_{k=1}^\infty$ satisfying $\sum_{k=1}^\infty |c_k|^{r'} \leq 1$ and $1/r+1/r'=1$. Since $L_{r'}([0,1]\times[0,1])$ is of type r' we obtain

$$\left\|\sum_{k=1}^\infty c_k w_k\right\|_{r'} \leq \left(\sum_{k=1}^\infty |c_k|^{r'}\|w_k\|_{r'}^{r'}\right)^{1/r'} \leq \|T\|_{r'}2^{-n/r'}\left(\sum_{k=1}^\infty |c_k|^{r'}\right)^{1/r'}$$

from which it follows that

$$\|h\|_r \leq \|T\|_{r'}\|T^{-1}\|_2^2\|f\|_r \,.$$

This means that we have just proved that

$$M_r = \sup_n \|P_n\|_r < \infty \,.$$

Since this holds for every $r>2$ it follows from Hölder's inequality that, for each simple function g and all integers m and n,

$$\begin{aligned}\|P_m g - P_n g\|_r &\leq \|P_m g - P_n g\|_2^{1/(r-1)}\|P_m g - P_n g\|_{2r}^{(r-2)/(r-1)}\\ &\leq (2M_{2r}\|g\|_{2r})^{(r-2)/(r-1)}\|P_m g - P_n g\|_2^{1/(r-1)}\end{aligned}$$

and this implies that $\{P_n\}_{n=1}^\infty$ converges in the strong operator topology of $L_r([0,1]\times[0,1])$ since it obviously tends to P in the strong operator topology of $L_2([0,1]\times[0,1])$. This proves that P acts as a bounded operator in $L_r([0,1]\times[0,1])$. □

Remarks. 1. It is clear that 2.f.1(ii) does not hold for every r.i. function space X on $[0,1]$ with $q_X<\infty$ (take, e.g. $X=L_1(0,1)$). However, the conditions imposed in 2.f.1(ii) that $1<p_X$ and $q_X<\infty$ are not always needed in order to prove that X is isomorphic to $\mathscr{X}$. It is proved in [58] Section 8 that X is isomorphic to $\mathscr{X}$ if, for example, X is the Orlicz function space $L_F(0,1)$ with $F(t)=e^t-1$, for which $p_X=q_X=\infty$.

2. In general, the space $\mathscr{X}$ associated to a given r.i. function space X on $[0,1]$ by 2.f.1 is not the only r.i. function space on $[0,\infty)$ which is isomorphic to X. For some spaces, however, as e.g. $X=L_F(0,1)$ with $F(t)=t^p(1+|\log t|)$, $1<p<\infty$, the corresponding space $\mathscr{X}$ is, up to an equivalent norm, the unique r.i. function space on $[0,\infty)$ isomorphic to X (cf. [58] Section 8).

3. There are also Orlicz function spaces on $[0,1]$, as e.g. $L_G(0,1)$ with $G(t)=t(1+|\log t|)^{1/4}$, which are isomorphic to no r.i. function space on $[0,\infty)$ (cf. [58], Section 8).

4. The space $Y=L_p(0,\infty)+L_q(0,\infty)$ with $1<p<q<2$ is not isomorphic to an r.i. function space on $[0,1]$. Indeed, assume that $Y\approx X$ with X being an r.i. function space on $[0,1]$. Since Y is p-convex and q-concave it follows, by applying 2.e.10 to X^* and Y^* and by 1.d.4(iii), that X is also q-concave. Hence, by the

second remark following 2.b.3, $\|f\|_X \leqslant C\|f\|_q$ for some $C<\infty$ and all $f \in L_q(0, 1)$ and thus, by 2.e.1, $\|f\|_q = \|f\|_Y \leqslant D\|f\|_X$ for some $D<\infty$ and all $f \in X$. Consequently, X is equal to $L_q(0, 1)$ up to an equivalent norm. This however is impossible since Y contains a subspace isomorphic to l_p which does not embed into $L_q(0, 1)$ (since e.g. l_p is not of type q).

As a first application of 2.f.1, we present some results on the possibility to embed isometrically the space $L_r(0, 1)$ into an r.i. function space on $[0, 1]$. We begin with the case of r.i. function spaces on $[0, \infty)$ (cf. [58] Section 8).

Theorem 2.f.2. *Let Y be an r.i. function space on $[0, \infty)$. If, for some $1 \leqslant r < \infty$, the function $t^{-1/r}$ belongs to Y then $L_r(0, \infty)$ is order isometric to a sublattice of Y.*

The proof of 2.f.2 is based on a general method for the embedding of r.i. function spaces which is described in the following simple proposition.

Proposition 2.f.3. *Let Y be an r.i. function space on some interval I, where I is either $[0, 1]$ or $[0, \infty)$. Let ψ be a positive element of norm one in Y and let Y_ψ be the space of all measurable functions f on I such that $f(s)\psi(t) \in Y(I \times I)$. Then Y_ψ, endowed with the norm*

$$\|f\|_{Y_\psi} = \|f(s)\psi(t)\|_{Y(I\times I)},$$

is an r.i. function space on I which is order isometric to a sublattice of Y. If Y is minimal, respectively, maximal then so is Y_ψ.

The proof of 2.f.3 is straightforward.

Proof of 2.f.2. In view of 2.f.3, it suffices to show that the space Y_{ψ_r}, where $\psi_r(t) = t^{-1/r}/\|t^{-1/r}\|_Y$, is equal to $L_r(0, \infty)$.

We evaluate the distribution function of $f(s)\psi_r(t)$ on $[0, \infty) \times [0, \infty)$, where f is an arbitrary function in $L_r(0, \infty)$. We have, for every $\lambda > 0$,

$$\begin{aligned}
&\mu(\{(s, t) \in [0, \infty) \times [0, \infty);\ |f(s)|\psi_r(t) > \lambda\})\\
&= \int_0^\infty \mu(\{s \in [0, \infty);\ |f(s)|\psi_r(t) > \lambda\})\, dt\\
&= \int_0^\infty \mu(\{s \in [0, \infty);\ |f(s)| > u\}) r u^{r-1} \lambda^{-r} \|t^{-1/r}\|_Y^{-r}\, du = \lambda^{-r}\|t^{-1/r}\|_Y^{-r}\|f\|_r^r\\
&= \mu(\{t \in [0, \infty);\ \psi_r(t) > \lambda\}) \cdot \|f\|_r^r .
\end{aligned}$$

Hence, if $\|f\|_r = 1$ then $|f(s)|\psi_r(t)$ and $\psi_r(t)$ have the same distribution function which implies that $\|f\|_{Y_{\psi_r}} = 1$, thus completing the proof. □

We pass now to the case of r.i. function spaces on $[0, 1]$.

Theorem 2.f.4 ([58] Section 8). *Let X be an r.i. function space on $[0, 1]$. If, for some $1<r<2$, the function $t^{-1/r}$, $0<t\leq 1$ belongs to X then $L_r(0, 1)$ embeds isometrically into X.*

Proof. Suppose first that $q_X<\infty$. Then, by 2.f.1(i), X contains a subspace isometric to the r.i. function space $\mathscr{X}_0$ on $[0, \infty)$, defined there. Since $1<r<2$ we have that $\|t^{-1/r}\chi_{(1,\infty)}\|_2<\infty$ i.e. the function $t^{-1/r}$, $0<t<\infty$ belongs to $\mathscr{X}_0$ and therefore, by 2.f.2, $L_r(0, \infty)$ embeds isometrically into $\mathscr{X}_0$. This proves that X contains an isometric copy of $L_r(0, 1)$.

In the case of a general r.i. function space X on $[0, 1]$ we could have defined as well the r.i. function space Y_0 on $[0, \infty)$ of all measurable f on $[0, \infty)$ for which

$$\|f\|_{Y_0}=\|Tf\|_X/\|T\chi_{[0,1]}\|_X<\infty\ ,$$

where T is the operator introduced in the proof of 2.f.1(i) with the aid of the symmetrized Poisson process, provided we know that $T\chi_{[0,1]}\in X$. In this case, Y_0 would be isometric to a subspace of X and the present proof would be completed, by 2.f.2, once it is shown that $t^{-1/r}\in Y_0$.

That $T\chi_{[0,1]}\in X$ is easily verified directly. Indeed, by its definition, $T'\chi_{[0,1]}$ takes the value k on a subset of measure $1/ek!$, $k=0, 1, \ldots$, and thus $(T'\chi_{[0,1]})^*(t)\leq At^{-1/r}$ for some constant A. Consequently, $T'\chi_{[0,1]}$ and hence also $T''\chi_{[0,1]}$ and $T\chi_{[0,1]}$ belong to X. In order to show that $Tg\in X$, where $g(t)=t^{-1/r}$, it suffices again to prove that $(Tg)^*(t)\leq A_0t^{-1/r}$ for some constant A_0, i.e. that Tg belongs to the space $L_{r,\infty}$ defined in 2.b.8 (recall that $\|f\|_{r,\infty}=\sup_{0\leq t\leq 1} t^{1/r}f^*(t)$). Since obviously $q_{L_{r,\infty}(0,1)}=r$ and $g\chi_{(0,1]}\in L_{r,\infty}(0, 1)$ we get from the first part of the proof that indeed $Tg\in L_{r,\infty}(0, 1)$. We have only to clarify one point which was mentioned already in 2.b.8. The expression $\|\cdot\|_{r,\infty}$ is not a norm since it does not satisfy the triangle inequality. However,

$$|||f|||_{r,\infty}=\sup_{0<t\leq 1} t^{1/r-1}\int_0^t f^*(s)\,ds$$

is a norm and satisfies

$$\|f\|_{r,\infty}\leq|||f|||_{r,\infty}\leq r\|f\|_{r,\infty}/(r-1)\,,$$

for every $f\in L_{r,\infty}(0, 1)$. The Boyd indices are clearly not affected by the passage from $\|\ \|_{r,\infty}$ to $|||\ |||_{r,\infty}$. $\square$.

The most interesting class of r.i. function spaces on $[0, 1]$ which contain the function $t^{-1/r}$ for some $1<r<2$ is, of course, that of $L_p(0, 1)$ spaces with $1\leq p<r$.

Corollary 2.f.5. *For $1\leq p<r<2$, the space $L_r(0, 1)$ embeds isometrically into $L_p(0, 1)$.*

This result was first stated explicitly by J. Bretagnolle, D. Dacunha-Castelle and J. L. Krivine [18] who obtained it by using sequences of independent r-stable random variables and ultrapowers of Banach spaces. We shall discuss in detail this method of embedding $L_r(0, 1)$ into $L_p(0, 1)$ in Volume IV. Essentially speaking, there is no difference between the present approach and that which uses r-stable random variables: the embedding of $L_r(0, \infty)$ into an r.i. function space X on $[0, 1]$, given by 2.f.4, consists of a composition between the map generated by the symmetrized Poisson process in 2.f.1(i) and the map $f(s) \to f(s)t^{-1/r}/\|t^{-1/r}\|_{\mathscr{X}}$. If R_w denotes the image of the characteristic function $\chi_{[0,w)}$ under this composition then $\{R_w\}_{0 \leq w < \infty}$ is an r-stable random process i.e. a stationary process with independent increments so that

$$\int_0^1 \int_0^1 e^{i\lambda R_w(s,t)}\, ds\, dt = e^{-cw|\lambda|^r},$$

for some $c > 0$. The verification of this fact is direct.

Corollary 2.f.5 holds also for $r = 2$ and $1 \leq p < 2$ but this fact does not obviously follow from 2.f.4. Of course, we have already used several times the fact that $L_p(0, 1)$, $1 \leq p < 2$ contains an isomorphic copy of $L_2(0, 1)$ (e.g. the span of the Rademacher functions). The existence of an isometric embedding of $L_2(0, 1) = l_2$ in $L_p(0, 1)$ is obtained by considering the span in $L_p(0, 1)$ of independent 2-stable random variables (i.e. variables having the normal distribution).

We turn now our attention to another application of 2.f.1 which concerns the spaces $X_{p,2,w}$, $p > 2$ of H. P. Rosenthal [114]. These spaces were introduced in I.4.d as the closed linear span in $(l_p \oplus l_2)_\infty$ of the sequence $\{f_n + w_n g_n\}_{n=1}^\infty$, where $\{f_n\}_{n=1}^\infty$ and $\{g_n\}_{n=1}^\infty$ denote the unit vector bases of l_p, respectively, l_2 and $w = \{w_n\}_{n=1}^\infty$ is a sequence of positive reals satisfying the condition

$$(*) \qquad \sum_{n=1}^\infty w_n^{2p/(p-2)} = \infty, \qquad \lim_{n\to\infty} w_n = 0 \quad \text{and} \quad w_n < 1 \text{ for all } n\,.$$

We recall that, by I.4.d.6, the isomorphism type of the space $X_{p,2,w}$ does not depend on the particular sequence $w = \{w_n\}_{n=1}^\infty$ satisfying (*), which appears in the definition of $X_{p,2,w}$, and, thus, we use the notation $X_{p,2}$ instead of $X_{p,2,w}$ (sometimes, this space is even denoted by X_p). It is easy to check directly that $X_{p,2,w}$ is not complemented in $(l_p \oplus l_2)_\infty$. It follows from I.2.c.14 that $X_{p,2}$ is not even isomorphic to a complemented subspace of $(l_p \oplus l_2)_\infty$. On the other hand, it was proved by H. P. Rosenthal [114] that $L_p(0, 1)$, $p > 2$ has a complemented subspace isomorphic to $X_{p,2}$. This assertion can also be deduced from the fact, which follows from 2.f.1(ii), that $L_p(0, 1)$ is isomorphic to the space $L_p(0, \infty) \cap L_2(0, \infty)$ for every $p \geq 2$. Indeed, fix $p > 2$ and let $\{\eta_n\}_{n=1}^\infty$ be a sequence of pairwise disjoint subsets of $[0, \infty)$ so that

$$\sum_{n=1}^\infty \mu(\eta_n) = \infty, \ \lim_{n\to\infty} \mu(\eta_n) = 0 \quad \text{and} \quad 0 < \mu(\eta_n) < 1 \text{ for all } n\,.$$

Put $w_n=\mu(\eta_n)^{(p-2)/2p}$ and observe that the sequence $w=\{w_n\}_{n=1}^\infty$, defined in this way, satisfies the condition (*), stated above, and, for every choice of scalars $\{a_n\}_{n=1}^\infty$, we have

$$\begin{aligned}\Big\|\sum_{n=1}^\infty a_n\mu(\eta_n)^{-1/p}\chi_{\eta_n}\Big\|_{L_p(0,\infty)\cap L_2(0,\infty)} &= \max\left\{\left(\sum_{n=1}^\infty |a_n|^p\right)^{1/p}, \left(\sum_{n=1}^\infty |a_n|^2\mu(\eta_n)^{1-2/p}\right)^{1/2}\right\}\\ &= \max\left\{\left(\sum_{n=1}^\infty |a_n|^p\right)^{1/p}, \left(\sum_{n=1}^\infty |a_n w_n|^2\right)^{1/2}\right\}\end{aligned}$$

i.e. the sequence $\{\mu(\eta_n)^{-1/p}\chi_{\eta_n}\}_{n=1}^\infty$ in $L_p(0,\infty)\cap L_2(0,\infty)$ is isometrically equivalent to the unit vector basis of $X_{p,2,w}$. Moreover, by 2.a.4, $[\chi_{\eta_n}]_{n=1}^\infty$ is a complemented subspace of $L_p(0,\infty)\cap L_2(0,\infty)$.

Let us state this result explicitly.

Theorem 2.f.6 [114]. *The space $X_{p,2}$, $p>2$ is isomorphic to a complemented subspace of $L_p(0,1)$.*

The analysis of the way in which $X_{p,2}$ embeds isomorphically in $L_p(0,1)$ via the operator T, defined in the proof of 2.f.1(i) with the help of the Poisson process, leads to the conclusion that the unit vector basis of $X_{p,2,w}$ is realized in $L_p(0,1)$ as a sequence $\{\psi_n\}_{n=1}^\infty$ of symmetric and independent random variables with Poisson distribution such that $\lim_{n\to\infty}\|\psi_n\|_2=0$ and $\sum_{n=1}^\infty\|\psi_n\|_2^2=\infty$. Rosenthal's embedding, however, is based on a sequence of symmetric and independent random variables satisfying the same conditions but taking only three values: $+1$, 0 and -1.

The arguments used here to prove 2.f.6 have actually a more general character. Let Y be an r.i. function space on $[0,\infty)$ and, for every sequence $\sigma=\{\sigma_n\}_{n=1}^\infty$ of pairwise disjoint subsets of $[0,\infty)$ of finite measure so that

$$(\ast)\qquad \sum_n\{\mu(\sigma_n);\ \mu(\sigma_n)<\varepsilon\}=\infty,\quad \text{for every } \varepsilon>0\,,$$

let $X_{Y,\sigma}$ be the subspace of Y consisting of all $f\in Y$ which are constant on each of the sets σ_n, $n=1,2,\dots$ and zero elsewhere. Observe that $\{\chi_{\sigma_n}\}_{n=1}^\infty$ is an unconditional basis for $X_{Y,\sigma}$ whenever Y is separable. Though, *a-priori*, $X_{Y,\sigma}$ depends on the particular choice of σ, the spaces $X_{Y,\sigma}$ generate the same isomorphism type X_Y provided we consider only sequences σ satisfying the condition $(\ast)$.

Proposition 2.f.7 ([58], Section 8). *For every r.i. function space Y on $[0,\infty)$, the space $X_{Y,\sigma}$ is unique, up to isomorphism, i.e. if σ' and σ'' are two sequences of disjoint sets of finite measure which satisfy the condition $(\ast)$ above then $X_{Y,\sigma'}$ and $X_{Y,\sigma''}$ are isomorphic. Moreover, each of the spaces $X_{Y,\sigma}$ is complemented in Y.*

Proof. The proof of 2.f.7 resembles that of I.4.d.6. By 2.a.4, each space $X_{Y,\sigma}$ is the range of a conditional expectation in Y. This fact also implies that they have the same isomorphism type. Indeed, if $\sigma'=\{\sigma'_n\}_{n=1}^\infty$ and $\sigma''=\{\sigma''_n\}_{n=1}^\infty$ are two sequences of subsets of $[0,\infty)$ as above then, by (⁑), one can find, for each m, a subsequence $\{\sigma''_{n(m,i)}\}_{i=1}^\infty$ of $\{\sigma''_n\}_{n=1}^\infty$ so that $\mu(\sigma'_m)=\sum_{i=1}^\infty \mu(\sigma''_{n(m,i)})$ and $n(m_1,i_1)\neq n(m_2,i_2)$ unless $m_1=m_2$ and $i_1=i_2$. Put $\eta=\left\{\bigcup_{i=1}^\infty \sigma''_{n(m,i)}\right\}_{m=1}^\infty$ and notice that the space $X_{Y,\eta}$, which is clearly isometric to $X_{Y,\sigma'}$, is, by the remark made in the beginning of the proof, a complemented subspace of $X_{Y,\sigma''}$. Hence, $X_{Y,\sigma''}\approx X_{Y,\sigma'}\oplus V$ for a suitable space V and, of course, also $X_{Y,\sigma'}\approx X_{Y,\sigma''}\oplus W$ for a suitable W. It remains to show that each of the spaces $X_{Y,\sigma}$ is isomorphic to its square.

Let $\sigma=[\sigma_n]_{n=1}^\infty$ be a sequence satisfying (⁑) and let $\delta=\{\delta_{n,j}\}_{j=1,n=1}^{\infty\ \ \infty}$ be a double sequence of mutually disjoint subsets of $[0,\infty)$ such that $\mu(\delta_{n,j})=\mu(\sigma_n)$ for all j and n. It follows directly from the definition of δ that

$$X_{Y,\delta}\approx X_{Y,\delta}\oplus X_{Y,\delta}\approx X_{Y,\sigma}\oplus X_{Y,\delta},$$

while from the first part of the proof we get that

$$X_{Y,\sigma}\approx X_{Y,\delta}\oplus Z,$$

for a suitable Z, since also δ satisfies the condition (⁑). Hence,

$$X_{Y,\sigma}\approx X_{Y,\delta}\oplus X_{Y,\delta}\oplus Z\approx X_{Y,\delta}\oplus X_{Y,\sigma}\approx X_{Y,\delta}$$

i.e. $X_{Y,\sigma}$ is isomorphic to its square. □

g. Interpolation Spaces and their Applications

The main purpose of interpolation theory is to prove results of the following general nature (stated here in an imprecise way).

(+) For suitable triples X_1, X, X_2 and Y_1, Y, Y_2 of Banach spaces every linear operator T, which is bounded as a map from X_i to Y_i, $i=1,2$, is also a bounded operator from X to Y.

By 2.a.10 this is, for example, the case if $X_1=Y_1=L_1(0,1)$, $X_2=Y_2=L_\infty(0,1)$ and $X=Y$ is any r.i. function space on $[0,1]$. Several other such cases were exhibited in Section 2.b. The importance of statements of the form (+) stems from the fact that it is often much simpler to verify the boundedness of T from X_i to Y_i, $i=1,2$, than to verify directly the boundedness of T from X to Y. In the preceding sections we encountered very many examples of this nature.

In general (and again, imprecise) terms we have that if (+) holds then X is in "between" X_1 and X_2 and Y is in "between" Y_1 and Y_2. Interpolation theory contains many "interpolation methods" of constructing spaces in between given spaces X_1 and X_2 (or, more precisely, a given interpolation pair (X_1, X_2) cf. Definition 2.g.1 below). If X is obtained by such a construction from (X_1, X_2) and if this same construction applied to the pair (Y_1, Y_2) yields the space Y then it is usually easy to verify that (+) holds. In order to get in this way a useful result for applications to hard analysis it is crucial to identify concretely the space obtained in this procedure from (X_1, X_2). In Banach space theory there is another aspect of this construction of interpolation spaces which is very useful. It is usually quite simple (and often even straightforward) to establish some geometrical properties of the spaces obtained by various interpolation methods even if these spaces cannot be described or represented explicitly. The interpolation methods often enable us therefore to produce interesting examples of Banach spaces and also lead to useful theorems of a general character. This section is devoted mainly to an exposition of this latter aspect of interpolation theory.

After establishing some preliminary notations of interpolation theory we present a quite general method of constructing interpolation spaces. This construction is a particular case of the so-called real method in interpolation theory. We apply this construction for giving examples of r.i. function spaces which illustrate some of the results of Section 2.e and clarify the role of the conditions appearing in the statement of these results. As other illustrations of the use of the general construction of interpolation spaces we prove a result concerning the notion of equi-integrable sets in Köthe function spaces and a theorem concerning factorization of weakly compact operators. In the discussion of the general construction all that we use from interpolation theory is just the definition of the spaces. The proofs that these spaces have the desired properties are easy direct verifications.

After treating the general construction we pass to a special case which was introduced and studied in depth by Lions and Peetre [81]. In this special case much more can be said than in the general case and, actually, Lions and Peetre were able to produce a well-rounded theory of the interpolation spaces they introduced. We outline briefly their theory and then discuss in some more detail the Banach space properties of the Lions–Peetre interpolation spaces. At the end of this section a typical application of the Lions–Peetre interpolation spaces to Banach space theory is presented, namely a "uniform convexification" of the example of Maurey and Rosenthal introduced in I.1.d.6.

Definition 2.g.1. A pair (X_1, X_2) of Banach spaces is called an *interpolation pair* if these Banach spaces are given as subspaces of a common Hausdorff topological vector space Z so that the embeddings of X_1 and of X_2 into Z are continuous.

Note that, by definition, the interpolation pair depends not only on X_1 and X_2 themselves but also on the space Z and the specific embeddings of X_1 and X_2 into Z. It is however convenient not to include Z explicitly in the notation.

Typical examples of interpolation pairs are the following.

1. X_1 and X_2 are Köthe function spaces on the same σ-finite measure space (Ω, Σ, μ). In this case we may take as Z the space of all locally integrable functions

on Ω and define the topology on Z by requiring that $f_\alpha \to 0$ in Z if and only if $\int_\sigma f_\alpha(\omega)\, d\mu \to 0$ for every $\sigma \in \Sigma$ with $\mu(\sigma) < \infty$. In particular, if X_1 and X_2 are Köthe sequence spaces (i.e. $\Omega =$ the integers with the discrete measure) Z will be the space of all sequences of scalars endowed with the topology of pointwise convergence. In case where X_1 and X_2 are r.i. function spaces on an interval I we may take as Z also the space $L_\infty(I) + L_1(I)$.

2. Given a continuous one-to-one linear map j from a Banach space X_1 into a Banach space X_2 we can take the space X_2 itself as the Z corresponding to the pair (X_1, X_2).

Another possibility, which however in the present context is trivial and uninteresting, is to consider an arbitrary pair of Banach spaces (X_1, X_2) taking as Z the direct sum $X_1 \oplus X_2$. This is uninteresting since interpolation theorems derive their importance from the existence of a big overlap between X_1 and X_2. In a typical situation the set theoretical intersection $X_1 \cap X_2$ is dense in either X_1 or X_2 (usually, in both).

Definition 2.g.2. (i) Let (X_1, X_2) be an interpolation pair. *The space* $X_1 + X_2$ is defined to be the linear span of X_1 and X_2 in Z, i.e. $\{z \in Z;\ z = x_1 + x_2,\ x_1 \in X_1,\ x_2 \in X_2\}$, normed by

$$\|z\|_{X_1+X_2} = \inf\{\|x_1\|_{X_1} + \|x_2\|_{X_2};\ z = x_1 + x_2\}.$$

The space $X_1 \cap X_2$ is the set theoretical intersection of X_1 and X_2 normed by

$$\|z\|_{X_1 \cap X_2} = \max\{\|z\|_{X_1}, \|z\|_{X_2}\}.$$

(ii) A linear operator T is said to be a *bounded operator from the interpolation pair* (X_1, X_2) *into the interpolation pair* (Y_1, Y_2) if T is defined on $X_1 + X_2$ and acts as a bounded operator from X_1 into Y_1 and from X_2 into Y_2. The norm $\|T\|_{1,2}$ of T as an operator between pairs is defined to be

$$\|T\|_{1,2} = \max\{\|T\|_1, \|T\|_2\},$$

where $\|T\|_i$, $i = 1, 2$, is the norm of $T\colon X_i \to Y_i$.

It is easily verified that both $X_1 + X_2$ and $X_1 \cap X_2$ are Banach spaces. Let us check e.g. that $\|\cdot\|_{X_1+X_2}$ is an actual norm (and not only a semi-norm) on $X_1 + X_2$. Assume that $\|x_1 + x_2\|_{X_1+X_2} = 0$. Then there exist $\{x_{1,n}\}_{n=1}^\infty$ in X_1 and $\{x_{2,n}\}_{n=1}^\infty$ in X_2 so that $x_1 + x_2 = x_{1,n} + x_{2,n}$ for every n and $\|x_{1,n}\|_{X_1} \to 0$, $\|x_{2,n}\|_{X_2} \to 0$. Hence, $x_1 - x_{1,n} = -x_2 + x_{2,n}$ tends in X_1 (and thus in Z) to x_1 and tends in X_2 (and thus in Z) to $-x_2$. Since Z is Hausdorff we deduce that $x_1 + x_2 = 0$.

Clearly,

$$X_1 \cap X_2 \subset X_1,\ X_2 \subset X_1 + X_2$$

and all the inclusion mappings are operators of norm one. Note that every operator

T between the pairs (X_1, X_2) and (Y_1, Y_2) defines an operator (also denoted by T) from X_1+X_2 and $X_1 \cap X_2$ into Y_1+Y_2, respectively $Y_1 \cap Y_2$. The norms $\|T\|_{X_1+X_2}$ and $\|T\|_{X_1 \cap X_2}$ are both dominated by the norm $\|T\|_{1,2}$ of T as an operator between pairs. The notation in 2.g.2 is consistent with the notation $L_p(I)+L_q(I)$ and $L_p(I) \cap L_q(I)$ used in previous sections. The only role played by the space Z of Definition 2.g.1 in interpolation theory is in the definition of X_1+X_2 and $X_1 \cap X_2$. If X_1 and X_2 are r.i. function spaces on some interval I then the two possible choices of Z mentioned in example 1 above lead to exactly the same spaces X_1+X_2 and $X_1 \cap X_2$ (with obviously the same norm). Unless stated otherwise, we shall assume in the sequel that if X_1 and X_2 are Köthe function spaces on (Ω, Σ, μ) and (X_1, X_2) is considered as an interpolation pair then Z is chosen so that it yields the same X_1+X_2 and $X_1 \cap X_2$ as the choice indicated in example 1 above. Also, if X_1 is given as subspace of X_2 with a continuous embedding we shall assume (unless stated otherwise) that $Z=X_2$. In this case $X_1+X_2=X_2$ and $X_1 \cap X_2=X_1$, up to equivalent norms. The norms will coincide in case $\|x\|_{X_2} \leqslant \|x\|_{X_1}$ for every $x \in X_1$ i.e. if the embedding of X_1 into X_2 is of norm at most one. (If X_1 and X_2 are Köthe function spaces on (Ω, Σ, μ) and $X_1 \subset X_2$ as sets both conventions made above are in agreement.) Another convention we make is the following: If X_1 and X_2 are Banach lattices we say that they form *an interpolation pair* (X_1, X_2) *of Banach lattices* if the space Z is a vector lattice and X_1 and X_2 are both ideals in Z. In this case, of course, also X_1+X_2 and $X_1 \cap X_2$ are Banach lattices.

We present now a general definition of interpolation spaces which is useful in Banach space theory. It is a particular case of a more general definition introduced by Peetre [109].

Definition 2.g.3. Let (X_1, X_2) be an interpolation pair of Banach spaces. For every choice of positive scalars a, b let $k(\cdot, a, b)$ denote the equivalent norm on X_1+X_2 defined by

$$k(z, a, b)=\inf \{a\|x_1\|_{X_1}+b\|x_2\|_{X_2};\ z=x_1+x_2\}.$$

Let, in addition, Y be a space with a normalized unconditional basis $\{y_n\}_{n=1}^\infty$ whose unconditional constant is one and let $\{a_n\}_{n=1}^\infty$ and $\{b_n\}_{n=1}^\infty$ be sequences of positive numbers so that $\sum_{n=1}^\infty \min(a_n, b_n)<\infty$. *The space* $K(X_1, X_2, Y, \{a_n\}, \{b_n\})$, respectively $\tilde{K}(X_1, X_2, Y, \{a_n\}, \{b_n\})$, is defined to be the space of all elements $z \in X_1+X_2$ such that $\sum_{n=1}^\infty k(z, a_n, b_n) y_n$ converges, respectively $\left\{\sum_{n=1}^m k(z, a_n, b_n) y_n\right\}_{m=1}^\infty$ is bounded, normed by

$$\|z\|_{K(X_1, X_2)}=\|z\|_{\tilde{K}(X_1, X_2)}=\sup_m \left\|\sum_{n=1}^m k(z, a_n, b_n) y_n\right\|_Y.$$

Obviously, if $\{y_n\}_{n=1}^\infty$ is boundedly complete then $K(X_1, X_2, Y, \{a_n\}, \{b_n\})=\tilde{K}(X_1, X_2, Y, \{a_n\}, \{b_n\})$. It is also clear that the spaces defined in 2.g.3 are Banach spaces satisfying

$$X_1 \cap X_2 \subset K(X_1, X_2, Y, \{a_n\}, \{b_n\}) \subset \tilde{K}(X_1, X_2, Y, \{a_n\}, \{b_n\}) \subset X_1 + X_2,$$

and

$$\min(a_1, b_1)\|z\|_{X_1+X_2} \leq \|z\|_{K(X_1, X_2)} \leq \sum_{n=1}^{\infty} \min(a_n, b_n)\|z\|_{X_1 \cap X_2}.$$

Notice that if (X_1, X_2) is an interpolation pair of Banach lattices then both $K(X_1, X_2, Y, \{a_n\}, \{b_n\})$ and $\tilde{K}(X_1, X_2, Y, \{a_n\}, \{b_n\})$ are also Banach lattices.

It is a trivial, but important, fact that the spaces defined in 2.g.3 are interpolation spaces between X_1 and X_2 in the following sense.

Proposition 2.g.4. *Let (X_1, X_2) and (W_1, W_2) be two pairs of interpolation spaces and let Y, $\{y_n\}_{n=1}^{\infty}$, $\{a_n\}_{n=1}^{\infty}$ and $\{b_n\}_{n=1}^{\infty}$ be as in 2.g.3. Then every bounded operator T from the pair (X_1, X_2) into (W_1, W_2) maps $K(X_1, X_2, Y, \{a_n\}, \{b_n\})$, respectively $\tilde{K}(X_1, X_2, Y, \{a_n\}, \{b_n\})$, into $K(W_1, W_2, Y, \{a_n\}, \{b_n\})$, respectively $\tilde{K}(W_1, W_2, Y, \{a_n\}, \{b_n\})$, and the norm of T on these spaces does not exceed $\|T\|_{1,2}$.*

Proof. It is enough to note that for every $z \in X_1 + X_2$ and every choice of scalars a, b

$$k(Tz, a, b) \leq \|T\|_{1,2} k(z, a, b). \quad \square$$

It follows in particular from 2.g.4 that if X_1 and X_2 are minimal (respectively maximal) r.i. function spaces on an interval I the same is true for $K(X_1, X_2, Y, \{a_n\}, \{b_n\})$ (respectively $\tilde{K}(X_1, X_2, Y, \{a_n\}, \{b_n\})$) provided we normalize the norm in the space by putting

$$\|f\|_0 = \|f\|_{K(X_1, X_2)} / \|\chi_{[0,1]}\|_{K(X_1, X_2)}.$$

By applying 2.g.4 to the dilation operators D_s of Section 2.b, we obtain in this situation the following estimates for the Boyd indices

$$\min(p_{X_1}, p_{X_2}) \leq p_{K(X_1, X_2)} \leq q_{K(X_1, X_2)} \leq \max(q_{X_1}, q_{X_2}).$$

The spaces introduced in 2.g.4 were already defined in a slightly modified form (and with a somewhat different notation) on page 126 of Vol. I. They were used there in particular to show that the universal space U_1 for all unconditional basic sequences (defined in I.2.d.10) has uncountably many mutually non-equivalent symmetric bases. We shall show now that exactly the same argument as that used in Vol. I proves that U_1 has also uncountably many mutually non-equivalent representations as an r.i. function space. Recall that U_1, by its definition, has a normalized basis $\{e_n\}_{n=1}^{\infty}$ whose unconditional constant is one and every normalized unconditional basis in an arbitrary Banach space is equivalent to a subsequence of $\{e_n\}_{n=1}^{\infty}$. This property characterizes U_1 in the following strong sense.

Every Banach space with an unconditional basis which has a complemented subspace isomorphic to U_1 is already isomorphic to U_1.

Theorem 2.g.5. *The universal space U_1 for all unconditional basic sequences has uncountably many mutually non-equivalent representations as an r.i. function space on* $[0, 1]$ *and on* $[0, \infty)$.

Proof. Let X_1 and X_2 be two minimal r.i. function spaces on $[0, 1]$ with non-trivial Boyd indices (i.e. $\min(p_{X_1}, p_{X_2}) > 1$ and $\max(q_{X_1}, q_{X_2}) < \infty$) so that

(i) $X_1 \subset X_2$ and $\|f\|_{X_2} \leq \|f\|_{X_1}$ for $f \in X_1$,

(ii) $\lim_{t \to 0} \|\chi_{[0,t]}\|_{X_2} / \|\chi_{[0,t]}\|_{X_1} = 0$

and let $\{m_n\}_{n=1}^{\infty}$ be an increasing sequence of numbers satisfying the following lacunarity condition

$$m_n^{-1} \sum_{i=1}^{n-1} m_i + m_n \sum_{i=n+1}^{\infty} m_i^{-1} < 2^{-n-1}, \quad n = 1, 2, \ldots$$

Consider the space $W = K(X_1, X_2, U_1, \{m_n^{-1}\}, \{m_n\})$ with the norm normalized so that $\|\chi_{[0,1]}\|_W = 1$. By the preceding remarks W is a minimal r.i. function space on $[0, 1]$ with non-trivial Boyd indices and thus, by 2.c.6, the Haar basis is an unconditional basis of W.

Exactly as in the proof of I.3.b.4 it follows from the lacunarity condition that there exists a sequence $\{\sigma_n\}_{n=1}^{\infty}$ of pairwise disjoint measurable subsets of $[0, 1]$ so that $\{e_n\}_{n=1}^{\infty}$ is equivalent to $\{\chi_{\sigma_n} / \|\chi_{\sigma_n}\|_W\}_{n=1}^{\infty}$. (The sets σ_n are chosen so that $k(\chi_{\sigma_n}, m^{-1}, m)$ attains its maximum at $m = m_n$. The lacunarity condition implies then that $k(\chi_{\sigma_n}, m_i^{-1}, m_i)$ for $i \neq n$ is negligible i.e. that $\|\chi_{\sigma_n}\|_W$ can be estimated very well by using only $k(\chi_{\sigma_n}, m_n^{-1}, m_n)$.) By 2.a.4 there is a projection of norm one from W onto $[\chi_{\sigma_n}]_{n=1}^{\infty}$. Hence, W has a complemented subspace isomorphic to U_1 and therefore W itself is isomorphic to U_1.

This proves that U_1 can be represented as an r.i. function space on $[0, 1]$. That U_1 has uncountably many different such representations follows from the fact that, for every $1 < p < \infty$ given in advance, the W above can be chosen so that $p_W = q_W = p$ (take e.g. $X_1 = L_M(0, 1)$ with M an Orlicz function equivalent at infinity to $t_p \log t$ and $X_2 = L_p(0, 1)$). By 2.f.1, it follows that U_1 has also uncountably many non-equivalent representations as an r.i. function space on $[0, \infty)$. □

It is often easy to prove that if X_1, X_2 and Y have a nice property then the same is true for $K(X_1, X_2, Y, \{a_n\}, \{b_n\})$. For example, in the proof of I.3.b.2 we verified that if X_1, X_2 and Y are uniformly convex then $K(X_1, X_2, Y, \{a_n\}, \{b_n\})$ has an equivalent uniformly convex norm. (In I.3.b.2 we used instead of $k(z, a, b)$ of 2.g.3 the equivalent expression $\inf\{(a^2\|x_1\|_{X_1}^2 + b^2\|x_2\|_{X_2}^2)^{1/2}, z = x_1 + x_2\}$ and showed that if this expression is used in 2.g.3(ii) the equivalent norm obtained in this manner on $K(X_1, X_2, Y, \{a_n\}, \{b_n\})$ is uniformly convex.) We prove now a similar result concerning p-convexity and concavity in lattices.

Proposition 2.g.6. *Let $1 \leqslant p \leqslant \infty$, let (X_1, X_2) be an interpolation pair of p-convex (respectively p-concave) lattices. Assume also that the lattice structure on Y induced by $\{y_n\}_{n=1}^{\infty}$ is p-convex (respectively p-concave). Then, for every choice of positive $\{a_n\}_{n=1}^{\infty}$ and $\{b_n\}_{n=1}^{\infty}$ with $\sum_{n=1}^{\infty} \min(a_n, b_n) < \infty$, the spaces $K(X_1, X_2, Y, \{a_n\}, \{b_n\})$ and $\tilde{K}(X_1, X_2, Y, \{a_n\}, \{b_n\})$ are p-convex (respectively p-concave).*

Proof. The verification of 2.g.6 is essentially a straightforward computation. We carry it out in detail in the case of p-convexity, $1 < p < \infty$, and $K(X_1, X_2, Y, \{a_n\}, \{b_n\})$. Let $\{z_i\}_{i=1}^m$ be vectors in $X_1 + X_2$ and let a, b and ε be positive numbers. Let $x_{1,i} \in X_1$, $x_{2,i} \in X_2$ be so that, for $1 \leqslant i \leqslant m$,

$$z_i = x_{1,i} + x_{2,i}, \qquad a\|x_{1,i}\|_{X_1} + b\|x_{2,i}\|_{X_2} \leqslant k(z_i, a, b) + \varepsilon .$$

Since

$$\left(\sum_{i=1}^m |z_i|^p\right)^{1/p} \leqslant \left(\sum_{i=1}^m |x_{1,i}|^p\right)^{1/p} + \left(\sum_{i=1}^m |x_{2,i}|^p\right)^{1/p}$$

we get that

$$\begin{aligned} k\left(\left(\sum_{i=1}^m |z_i|^p\right)^{1/p}, a, b\right) &\leqslant a\left\|\left(\sum_{i=1}^m |x_{1,i}|^p\right)^{1/p}\right\|_{X_1} + b\left\|\left(\sum_{i=1}^m |x_{2,i}|^p\right)^{1/p}\right\|_{X_2} \\ &\leqslant aM^{(p)}(X_1)\left(\sum_{i=1}^m \|x_{1,i}\|_{X_1}^p\right)^{1/p} + bM^{(p)}(X_2)\left(\sum_{i=1}^m \|x_{2,i}\|_{X_2}^p\right)^{1/p} \\ &\leqslant (M^{(p)}(X_1) + M^{(p)}(X_2))\left(\left(\sum_{i=1}^m k(z_i, a, b)^p\right)^{1/p} + m^{1/p}\varepsilon\right). \end{aligned}$$

Hence, since $\varepsilon > 0$ was arbitrary, we deduce that

$$\begin{aligned} \left\|\left(\sum_{i=1}^m |z_i|^p\right)^{1/p}\right\|_{K(X_1, X_1)} &\leqslant (M^{(p)}(X_1) + M^{(p)}(X_2))\left\|\sum_{n=1}^{\infty}\left(\sum_{i=1}^m k(z_i, a_n, b_n)^p\right)^{1/p} y_n\right\|_Y \\ &\leqslant M^{(p)}(Y)(M^{(p)}(X_1) + M^{(p)}(X_2))\left(\sum_{i=1}^m \|z_i\|_{K(X_1, X_2)}^p\right)^{1/p}. \end{aligned}$$

The proof of the p-concavity assertion in 2.g.6 is similar. We just have to note that by applying the decomposition property in the p-concavification $(X_1 + X_2)_{(p)}$ of $X_1 + X_2$ (considered just as a vector lattice) we deduce that, whenever $\{z_i\}_{i=1}^m \in X_1 + X_2$, $x_1 \in X_1$ and $x_2 \in X_2$ satisfy $\left(\sum_{i=1}^m |z_i|^p\right)^{1/p} = x_1 + x_2$, then there exist $\{x_{1,i}\}_{i=1}^m \in X_1$ and $\{x_{2,i}\}_{i=1}^m \in X_2$ so that $|z_i| \leqslant x_{1,i} + x_{2,i}$ for every i, and

$$\left(\sum_{i=1}^m |x_{i,1}|^p\right)^{1/p} \leqslant 2^{(p-1)/p}|x_1|, \qquad \left(\sum_{i=1}^m |x_{2,i}|^p\right)^{1/p} \leqslant 2^{(p-1)/p}|x_2|$$

and consequently,

$$\left(\sum_{i=1}^{m} k(z_i, a, b)^p\right)^{1/p} \leq 2^{(p-1)/p} \max\,(M_{(p)}(X_1), M_{(p)}(X_2))\, k\!\left(\left(\sum_{i=1}^{m} |z_i|^p\right)^{1/p}, a, b\right). \square$$

We shall apply 2.g.6 in proving a variant of 2.g.5 which is of interest in connection with the results of Section 2.e. We have seen in 2.e.11 that every r.i. function space on $[0, 1]$, which is q-concave for some $q<2$, has a unique representation as an r.i. function space on $[0, 1]$. That this fact is not true for $q=2$ is shown by the following example from [58] Section 10.

Example 2.g.7. *For every $1<p<2$ there exists an r.i. function space X on $[0, 1]$ which is p-convex and 2-concave but has uncountably many mutually non-equivalent representations as an r.i. function space on $[0, 1]$.*

The construction of this example is similar to that presented in the proof of 2.g.5 except that U_1 must be replaced by a suitable space whose existence is asserted in the following proposition.

Proposition 2.g.8. *For every $1\leq p\leq q\leq\infty$ there exists a space $U_{p,q}$ (universal for all unconditional bases which are p-convex and q-concave), with a normalized basis $\{e_n^{p,q}\}_{n=1}^{\infty}$ having an unconditional constant equal to one, such that*

(i) *the lattice structure on $U_{p,q}$ generated by $\{e_n^{p,q}\}_{n=1}^{\infty}$ is p-convex and q-concave,*

(ii) *every normalized unconditional basis $\{y_i\}_{i=1}^{\infty}$ with unconditional constant equal to one which generates a p-convex and q-concave lattice structure is equivalent to a subsequence $\{e_{n_i}^{p,q}\}_{i=1}^{\infty}$ of $\{e_n^{p,q}\}_{n=1}^{\infty}$.*

Proof. If $p=q$ then l_p (respectively c_0, if $p=\infty$) has the desired property. We can assume therefore that $p<q$. If $p=1$ and $q=\infty$ we of course take the universal space U_1 as $U_{1,\infty}$. In order to treat the general case, fix $r>1$ and consider the r-convexification $U_1^{(r)}$ of U_1. The vectors $\{e_n\}_{n=1}^{\infty}$ form an unconditional basis in $U_1^{(r)}$ and if Y is an arbitrary Banach space with a normalized r-convex unconditional basis $\{y_i\}_{i=1}^{\infty}$ then $\{y_i\}_{i=1}^{\infty}$, considered as an unconditional basis in the r-concavification $Y_{(r)}$ of Y, is equivalent to a subsequence $\{e_{n_i}\}_{i=1}^{\infty}$ of $\{e_n\}_{n=1}^{\infty}$ (considered in U_1). Hence, as readily verified, $\{y_i\}_{i=1}^{\infty}$ in Y is equivalent to $\{e_{n_i}\}_{i=1}^{\infty}$ in $U_1^{(r)}$. This proves that $U_1^{(r)}$ is universal (in the sense of 2.g.8) for all the unconditional bases which are r-convex, thus completing the proof if $q=\infty$. If $q<\infty$ we choose r so that $pr/(r-1)=q$. Then, $V=[e_n^*]_{n=1}^{\infty}\subset(U_1^{(r)})^*$, where $\{e_n^*\}_{n=1}^{\infty}$ are the biorthogonal functionals corresponding to the basis $\{e_n\}_{n=1}^{\infty}$ of $U_1^{(r)}$ is universal for all the unconditional bases which are $r/(r-1)$-concave. Hence, we get, in the same way as above, that $V^{(p)}$ is universal for all the unconditional bases which are p-convex and $q=pr/(r-1)$-concave. $\square$

Remark. By using the decomposition method we get easily the following strong uniqueness assertion. If W is a Banach space with an unconditional basis which generates a p-convex and q-concave lattice structure and if W has a complemented subspace isomorphic to $U_{p,q}$ then W itself is isomorphic to $U_{p,q}$.

Proof of 2.g.7. Fix $1<p<2$ and let X_1 and X_2 be two minimal r.i. function spaces on $[0, 1]$ which are p-convex and 2-concave and satisfy in addition the requirements (i) and (ii) appearing in the proof of 2.g.5. Let $\{m_n\}_{n=1}^{\infty}$ be a sequence of numbers satisfying the lacunarity condition stated in the proof of 2.g.5. Consider the space $W=K(X_1, X_2, U_{p,2}, \{m_n^{-1}\}, \{m_n\})$, where $U_{p,2}$ is the space constructed in 2.g.8. By 2.g.6, W is p-convex and 2-concave. It follows from 1.d.7 that the lattice structure generated by the Haar basis of W is also p-convex and 2-concave (to be precise we first have to renorm W so that in the new norm the unconditional constant of the Haar basis becomes one). As in the proof of 2.g.5, we have that W contains a complemented subspace isomorphic to $U_{p,2}$ and hence, by the remark following 2.g.8, W is isomorphic to $U_{p,2}$. This fact is true for every choice of X_1 and X_2 as above. By letting r vary between p and 2 and taking $X_1=L_M(0, 1)$ with M equivalent at infinity to $t^r \log t$ and $X_2=L_r(0, 1)$, we obtain that $U_{p,2}$ has uncountably many mutually non-equivalent representations as an r.i. function space on $[0, 1]$. □

Remark. If in the proof above we would have used $U_{p,q}$ with $1<p<q<2$ instead of $U_{p,2}$ then we would have obtained an r.i. function space W on $[0, 1]$ which is p-convex and q-concave (provided, of course, that both X_1 and X_2 are p-convex and q-concave). This space W still contains $U_{p,q}$ as a complemented subspace but W is not isomorphic to $U_{p,q}$ (which would contradict 2.e.11), since the Haar basis of W is not q-concave but only 2-concave.

We pass now to another situation where the spaces defined in 2.g.3 are used in Banach space theory. This application involves the following notion.

Definition 2.g.9. A set F in a Köthe function space X on a probability space (Ω, Σ, μ) is said to be *equi-integrable* if for every $\varepsilon>0$ there is a $\lambda<\infty$ so that

$$\|f\chi_{\sigma(f,\lambda)}\|<\varepsilon, \quad f\in F,$$

where

$$\sigma(f, \lambda)=\{\omega\in\Omega; |f(\omega)|\geq\lambda\}.$$

Note that every equi-integrable set is bounded. Equi-integrable sets enter in several arguments in functional analysis, usually in compactness arguments. Recall e.g. that a set F in $L_1(\Omega, \Sigma, \mu)$ is equi-integrable if and only if its w closure is w compact (cf. [32] IV.8.11). The next proposition enables sometimes to reduce the study of equi-integrable sets to that of bounded sets.

Proposition 2.g.10 ([58] Section 6). *Let X be a Köthe function space on a probability space (Ω, Σ, μ) and let F be an equi-integrable subset of X. Then there is a Köthe function space W on (Ω, Σ, μ) so that*

(i) *The unit ball of W is equi-integrable in X (and thus in particular $W\subset X$ with a continuous embedding).*

(ii) *F is bounded (even equi-integrable) in W.*

Proof. Let $\{\lambda_n\}_{n=1}^\infty$ be such that

$$\|f\chi_{\sigma(f,\lambda_n)}\|_X < 4^{-n}, \quad f \in F, n=1,2,\ldots .$$

We claim that $W = K(L_\infty(\Omega, \Sigma, \mu), X, l_2, \{2^{-n}\lambda_n^{-1}\}, \{2^n\})$ has the desired properties. Note first that since $L_\infty(\Omega, \Sigma, \mu)$ is contained in X with a continuous embedding the same is true for W. For $f \in F$ and $n=1,2,\ldots$ we have

$$k(f, 2^{-n}\lambda_n^{-1}, 2^n) \leq 2^{-n}\lambda_n^{-1}\|f - f\chi_{\sigma(f,\lambda_n)}\|_\infty + 2^n\|f\chi_{\sigma(f,\lambda_n)}\|_X \leq 2^{-n+1}$$

and therefore F is a bounded subset of W. Moreover, if $m \geq n$

$$k(f\chi_{\sigma(f,\lambda_m)}, 2^{-n}\lambda_n^{-1}, 2^n) \leq 2^n\|f\chi_{\sigma(f,\lambda_m)}\|_X \leq 2^n 4^{-m}$$

and consequently, for $m=1,2,\ldots$

$$\|f\chi_{\sigma(f,\lambda_m)}\|_W^2 \leq \sum_{n=1}^{m} 2^{2n}4^{-2m} + \sum_{n=m+1}^{\infty} 2^{-2(n-1)} \leq 2^{-2m+3}$$

and this proves (ii). To prove (i) note that if $\|f\|_W < 1$ then $k(f, 2^{-n}\lambda_n^{-1}, 2^n) < 1$ for every n and hence $f = g_n + h_n$ with $2^{-n}\lambda_n^{-1}\|h_n\|_\infty + 2^n\|g_n\|_X < 1$. In particular, $\|h_n\|_\infty \leq \lambda_n 2^n$ and thus whenever $|f(\omega)| \geq \lambda_n 2^{n+1}$ we have $|f(\omega)| \leq 2|g_n(\omega)|$ and hence

$$\|f\chi_{\sigma(f,\lambda_n 2^{n+1})}\|_X \leq 2^{-n+1}, \quad n=1,2,\ldots . \quad \square$$

Remarks. 1. It follows immediately from 2.g.10 that the convex hull of an equi-integrable set is again equi-integrable.

2. If X is p-convex then, by 2.g.6, the space W constructed above is also p-convex provided $p \leq 2$. If $2 < p < \infty$ the same is true provided we replace l_2 in the definition of W by l_p.

As a further application of the spaces defined in 2.g.3 we present the following factorization theorem.

Theorem 2.g.11 [24]. *Let V and X be Banach spaces. A bounded linear operator $T: V \to X$ is weakly compact if and only if T can be factored through a reflexive Banach space i.e. there exists a reflexive space W and bounded operators $T_1: V \to W$, $T_2: W \to X$ so that $T = T_2T_1$.*

Proof. The "if" assertion is trivial so we have to prove the "only if" assertion. Assume that T is weakly compact and let $B_1 = \overline{TB_V}$. Then B_1 is a w compact subset of X. Consider the subspace $X_1 = \bigcup_{n=1}^{\infty} nB_1$ of X. We norm X_1 by requiring that B_1 be its unit ball. Then X_1 becomes a Banach space which is continuously embedded in X. Let $\{a_n\}_{n=1}^\infty$ and $\{b_n\}_{n=1}^\infty$ be sequences of positive reals so that

$\sum_{n=1}^{\infty} a_n < \infty$ and $b_n \uparrow \infty$, and consider the space $W = K(X_1, X, l_2, \{a_n\}, \{b_n\})$. Clearly, $X_1 \subset W \subset X$, the identity mappings $j_1: X_1 \to W$ and $j_2: W \to X$ are continuous and $T = j_2(j_1 T)$. Thus in order to prove the theorem it suffices to show that W is reflexive.

Let $\{z_j\}_{j=1}^{\infty}$ be a sequence of vectors in W with $\|z_j\|_W < 1$ for all j. Then, in particular, for every j and n, $k(z_j, a_n, b_n) < 1$, i.e.

$$z_j = y_{j,n} + x_{j,n} \quad \text{with} \quad a_n \|y_{j,n}\|_{X_1} + b_n \|x_{j,n}\|_X < 1 .$$

By the weak compactness of $B_1 = B_{X_1}$, we may assume without loss of generality (by passing to a subsequence) that, for every n, the sequence $\{y_{j,n}\}_{j=1}^{\infty}$ converges weakly in X to some vector $y_n \in X$. Since $\|x_{j,n}\|_X \leq b_n^{-1}$ it follows that

$$\|y_n - y_m\|_X \leq b_n^{-1} + b_m^{-1}, \quad n, m = 1, 2, \ldots,$$

and thus $\{y_n\}_{n=1}^{\infty}$ converges in norm to some vector y in X. Clearly, the sequence $\{z_j\}_{j=1}^{\infty}$ converges weakly in X to y and $k(y, a_n, b_n) \leq \liminf_j k(z_j, a_n, b_n)$ for every n (reçall that $k(\cdot, a, b)$is a norm on X equivalent to the original one). Hence, $y \in W$ and $\|y\|_W \leq 1$. In order to show that z_j tends to y also weakly in W (and thus verify the reflexivity of W) we have to prove that

$$\lim_{j \to \infty} \sum_{n=1}^{\infty} x_n^*(z_j) = \sum_{n=1}^{\infty} x_n^*(y)$$

whenever $\{x_n^*\}_{n=1}^{\infty} \subset X^*$ and $\sum_{n=1}^{\infty} \|x_n^*\|_n^2 < \infty$, where

$$\|x^*\|_n = \sup \{|x^*(x)|;\ k(x, a_n, b_n) \leq 1\}, \ n = 1, 2, \ldots .$$

This however, is an immediate consequence of the facts that $\|z_j\|_W < 1$, for every j, and $\lim_{j \to \infty} x_n^*(z_j) = x_n^*(y)$ for every n. □

We pass now to the most important special class of the interpolation spaces defined in 2.g.3, namely to the Lions–Peetre interpolation spaces. Let us recall that if X is a Banach space we denote by $L_p(R, X)$ the space of all the measurable functions f from the real line R into X with $\|f\|_p = \left(\int_{-\infty}^{\infty} \|f(t)\|^p \, dt \right)^{1/p} < \infty$, respectively, $\|f\|_\infty = \text{ess sup}\, \|f(t) < \infty$ if $p = \infty$. (In case it is worthwhile to point out explicitly the underlying Banach space X we shall denote $\|f\|_p$ also by $\|f\|_{L_p(X)}$.)

Definition 2.g.12 [81]. Let (X_1, X_2) be an interpolation pair of Banach spaces. Let $0 < \theta < 1$ and $1 \leq p \leq \infty$. The space of all $z \in X_1 + X_2$ which admit a representation as $z = x_1(t) + x_2(t)$, $t \in R$, with

$$e^{\theta t} x_1(t) \in L_p(R, X_1) \quad \text{and} \quad e^{-(1-\theta)t} x_2(t) \in L_p(R, X_2),$$

will be denoted by $[X_1, X_2]_{\theta, p}$. The norm in this space is defined by

$$\|z\|_{\theta, p} = \inf \{\max (\|e^{\theta t} x_1(t)\|_p, \|e^{-(1-\theta)t} x_2(t)\|_p);\ z = x_1(t) + x_2(t),\ t \in \mathrm{R}\}.$$

It is easily verified that $[X_1, X_2]_{\theta, p}$ coincides with the space of all $z \in X_1 + X_2$ so that

$$\sum_{n=-\infty}^{\infty} k(z, e^{\theta n}, e^{-(1-\theta)n})^p < \infty,$$

respectively, $\sup_{-\infty < n < \infty} k(z, e^{\theta n}, e^{-(1-\theta)n}) < \infty$ if $p = \infty$, and is thus, in particular, one of the spaces defined in 2.g.3 (up to an equivalent norm). The "continuous" definition given in 2.g.10 is sometimes more convenient than the discrete version of 2.g.3 and it leads to better estimates of numerical constants which enter into the theory. One advantage of the continuous version is demonstrated by the following simple but useful observation.

Lemma 2.g.13. *With the notation of 2.g.12 we have*

$$\|z\|_{\theta, p} = \inf \{\|e^{\theta t} x_1(t)\|_p^{1-\theta} \|e^{-(1-\theta)t} x_2(t)\|_p^{\theta};\ z = x_1(t) + x_2(t), \quad t \in R\},$$

for every $z \in [X_1, X_2]_{\theta, p}$.

Proof. We have

$$\begin{aligned} \|z\|_{\theta, p} &= \inf \{\inf_{\tau \in R} \max (\|e^{\theta t} x_1(t-\tau)\|_p, \|e^{-(1-\theta)t} x_2(t-\tau)\|_p)\} \\ &= \inf \{\inf_{\tau \in R} \max (e^{\theta \tau} \|e^{\theta t} x_1(t)\|_p, e^{-(1-\theta)\tau} \|e^{-(1-\theta)t} x_2(t)\|_p)\} \\ &= \inf \{\|e^{\theta t} x_1(t)\|_p^{1-\theta} \|e^{-(1-\theta)t} x_2(t)\|_p^{\theta}\}, \end{aligned}$$

where the outer infima are taken over all possible representations of z as $x_1(t) + x_2(t)$, $t \in R$. $\square$

Corollary 2.g.14. *For every* $0 < \theta < 1$ *and* $1 \leq p \leq \infty$ *there is a constant* $C(\theta, p)$ *so that, for* $x \in X_1 \cap X_2$,

$$\|x\|_{\theta, p} \leq C(\theta, p) \|x\|_{X_1}^{1-\theta} \|x\|_{X_2}^{\theta}.$$

Proof. Pick a function $0 \leq \psi(t) \leq 1$ so that

$$e^{\theta t} \psi(t) \in L_p(-\infty, \infty) \quad \text{and} \quad e^{-(1-\theta)t}(1 - \psi(t)) \in L_p(-\infty, \infty)$$

and apply 2.g.13 to the representation $x = \psi(t)x + (1 - \psi(t))x$, $t \in R$. $\square$

Another consequence of 2.g.13 is the following sharper form of 2.g.4.

Proposition 2.g.15 [81]. *Let T be a bounded operator from the interpolation pair (X_1, X_2) to the interpolation pair (Y_1, Y_2). Then, for every $0<\theta<1$ and $1\leqslant p\leqslant\infty$, T maps $[X_1, X_2]_{\theta,p}$ into $[Y_1, Y_2]_{\theta,p}$ and*

$$\|T\|_{\theta,p}\leqslant\|T\|_1^{1-\theta}\|T\|_2^{\theta}.$$

Proof. Apply 2.g.13 and observe that

$$\|e^{\theta t}Tx_1(t)\|_p^{1-\theta}\|e^{-(1-\theta)t}Tx_2(t)\|_p^{\theta}$$
$$\leqslant\|T\|_1^{1-\theta}\|T\|_2^{\theta}\|e^{\theta t}x_1(t)\|_p^{1-\theta}\|e^{-(1-\theta)t}x_2(t)\|_p^{\theta}. \quad \square$$

We state now without proof some of the main results of Lions and Peetre on the properties of $[X_1, X_2]_{\theta,p}$. A detailed exposition of these results can be found (besides the original paper [81]) in [21] and especially in [10].

Theorem 2.g.16 [81] (cf. also [10] 3.7.1). *Let (X_1, X_2) be an interpolation pair so that $X_1\cap X_2$ is dense in both X_1 and X_2. Then, for every $0<\theta<1$ and $1\leqslant p<\infty$, the dual of $[X_1, X_2]_{\theta,p}$ is equal (up to an equivalent norm) to $[X_1^*, X_2^*]_{\theta,q}$, where $1/p+1/q=1$.*

Note that the pair (X_1^*, X_2^*) becomes in a natural way an interpolation pair if we consider these two spaces as embedded continuously in the Banach space $(X_1\cap X_2)^*$. An element in $[X_1^*, X_2^*]_{\theta,q}$ is thus, by definition, a continuous linear functional on $X_1\cap X_2$. One of the assertions of 2.g.16 is that this element extends to a continuous linear functional on $[X_1, X_2]_{\theta,p}$. The extension is unique in view of the density assumption in 2.g.16. A major step in the proof of 2.g.16 is the proof of the following proposition.

Proposition 2.g.17 [81] (cf. also [10] 3.3.1). *Let (X_1, X_2) be an interpolation pair and let $0<\theta<1$ and $1\leqslant p\leqslant\infty$. The space $[X_1, X_2]_{\theta,p}$ consists exactly of those elements $z\in X_1+X_2$ which can be represented as $z=\int_{-\infty}^{\infty} x(t)\,dt$ with x being a measurable function from R to $X_1\cap X_2$ so that $e^{\theta t}x(t)\in L_p(R, X_1)$ and $e^{-(1-\theta)t}x(t)\in L_p(R, X_2)$ (these conditions imply in particular that $\int_{-\infty}^{\infty}\|x(t)\|_{X_1+X_2}\,dt<\infty$). The norm on $[X_1, X_2]_{\theta,p}$ is equivalent to*

$$|||z|||_{\theta,p}=\inf\left\{\max\left(\|e^{\theta t}x(t)\|_{L_p(X_1)}, \|e^{-(1-\theta)t}x(t)\|_{L_p(X_2)}\right);\ z=\int_{-\infty}^{\infty} x(t)\,dt\right\}.$$

Remark. It is easily verified that analogues of 2.g.13 and 2.g.15 hold for $|||\cdot|||_{\theta,p}$.

In the applications of the Lions–Peetre interpolation to analysis it is of course important to identify explicitly the spaces $[X_1, X_2]_{\theta,p}$ if X_1 and X_2 are concrete function spaces. Much work has been done in this direction (cf. [10] and its references). We mention here the following result.

Theorem 2.g.18 [81] (cf. also [10] 5.2.1 and 5.3.1). *Let* (Ω, Σ, μ) *be a measure space.*

(i) *The space* $[L_{p_1}(\Omega, \Sigma, \mu), L_{p_2}(\Omega, \Sigma, \mu)]_{\theta,q}$ *with* $1 \leqslant p_1 < p_2 \leqslant \infty$, $0<\theta<1$, $1 \leqslant q \leqslant \infty$ *is equal, up to an equivalent norm, to the space* $L_{p,q}(\Omega, \Sigma, \mu)$ *(introduced in 2.b.8), where* $1/p=(1-\theta)/p_1+\theta/p_2$.

(ii) *The space* $[L_{p_1,q_1}(\Omega, \Sigma, \mu), L_{p_2,q_2}(\Omega, \Sigma, \mu)]_{\theta,q}$ *with* $1 \leqslant p_1 < p_2 \leqslant \infty$, $1 \leqslant q_1, q_2, q \leqslant \infty$, $0<\theta<1$, *is equal, up to an equivalent norm, to the space* $L_{p,q}(\Omega, \Sigma, \mu)$, *where* $1/p=(1-\theta)/p_1+\theta/p_2$.

The Marcinkiewicz interpolation theorem 2.b.15 is an immediate consequence of 2.g.18. If an operator T is of weak types (p_1, p_2) and (q_1, q_2) then it is, by definition, a bounded map from $L_{p_1,1}$ to $L_{p_2,\infty}$ and from $L_{q_1,1}$ to $L_{q_2,\infty}$. Hence, if $p_1 \neq p_2$ and $q_1 \neq q_2$ it follows from 2.g.18(ii) that, for every θ and q, T is a bounded map from $L_{r_1,q}$ into $L_{r_2,q}$, where $1/r_i=\theta/p_i+(1-\theta)/q_i$, $i=1, 2$. Hence, if $r_1 \leqslant r_2$ we get, by taking $q=r_1$, that T is bounded as an operator from L_{r_1} into L_{r_2,r_1} and thus in particular into L_{r_2} (use 2.b.9). This argument applies even if T is only quasilinear (observe that 2.g.4 is valid in general also for quasilinear T provided we replace $\|T\|_{1,2}$ by $C\|T\|_{1,2}$, where C is the constant appearing in the definition of quasilinearity).

An important tool in the proof of 2.g.18 is the following general reiteration theorem.

Theorem 2.g.19 [81] (cf. also [10] 3.5.3). *Let* (X_1, X_2) *be an interpolation pair, let* $0<\theta_i<1$, $1 \leqslant p_i \leqslant \infty$, $i=1, 2, 3$ *with* $\theta_1 \neq \theta_2$. *Then, up to an equivalent norm,*

$$[[X_1, X_2]_{\theta_1,p_1}, [X_1, X_2]_{\theta_2,p_2}]_{\theta_3,p_3}=[X_1, X_2]_{\theta,p_3},$$

where $\theta=(1-\theta_3)\theta_1+\theta_3\theta_2$.

This theorem enables the reduction of the proof of 2.g.18(i) to the case where $p_2=\infty$. Furthermore, 2.g.18(ii) is clearly a consequence of 2.g.19 and 2.g.18(i). The iteration theorem 2.g.19 is also used in the proof of the following proposition.

Proposition 2.g.20 [81] (cf. also [10] 5.6.2). *Let* (Ω, Σ, μ) *be a measure space and* (X_1, X_2) *an interpolation pair of Banach spaces. Let* $1 \leqslant p_1 < p_2 \leqslant \infty$, $0<\theta<1$, *and* $1/p=(1-\theta)/p_1+\theta/p_2$. *Then, up to equivalent norms,*

$$[L_{p_1}(\Omega, X_1), L_{p_2}(\Omega, X_2)]_{\theta,p}=L_p(\Omega, [X_1, X_2]_{\theta,p}).$$

We pass now to the study of some geometrical properties of the Lions–Peetre interpolation spaces. We have already mentioned above that, whenever X_1, X_2 and Y are uniformly convex, then $K(X_1, X_2, Y, \{a_n\}, \{b_n\})$ has an equivalent uniformly convex norm. It turns out that in the case of the Lions–Peetre interpolation spaces it is enough that one of the spaces X_1 and X_2 be uniformly convex for the interpolation space to be uniformly convex, too. This fact is of importance for several applications of the Lions–Peetre interpolation spaces in Banach space theory (see e.g. 2.g.23 below). Before stating this result precisely we mention that if X is uniformly convex the same is true for the space $L_p(R, X)$ which is isometric to $L_p([0, 1], X)$ (this space was denoted in 1.e also by $L_p(X)$), provided $1<p<\infty$. For $p=2$ we proved this in 1.e.9. Actually, we showed there that $\delta_{L_2(X)}(\varepsilon)$ is equivalent to $\delta_X(\varepsilon)$. A similar argument to that given in 1.e.9 shows that also for $1<p<2$, $\delta_{L_p(X)}(\varepsilon)$ is equivalent to $\delta_X(\varepsilon)$ and that, for $2<p<\infty$, $\delta_{L_p(X)}(\varepsilon)$ is equivalent to $\inf\{t^{-p}\delta_X(t\varepsilon);\ t\geqslant 1\}$ (cf. [40]). The fact that $L_p(X)$ is uniformly convex whenever X is and $1<p<\infty$ follows also from the computations done on pages 128–129 of Volume I (but this proof gives a rather poor estimate for $\delta_{L_p(X)}(\varepsilon)$). We state now the result of Beauzamy [5] concerning the uniform convexity of the Lions–Peetre spaces.

Theorem 2.g.21. *Let (X_1, X_2) be an interpolation pair and let $0<\theta<1$, $1<p<\infty$. Then $\delta_{[X_1, X_2]_{\theta,p}}(\varepsilon)$ dominates a function equivalent to $\max(\delta_{L_p(X_1)}(\varepsilon^{1/(1-\theta)}), \delta_{L_p(X_2)}(\varepsilon^{1/\theta}))$. In particular, $[X_1, X_2]_{\theta,p}$ is uniformly convex whenever X_1 or X_2 is uniformly convex.*

Proof. Let $u, v \in [X_1, X_2]_{\theta,p}$ satisfy $\|u\|_{\theta,p}<1$, $\|v\|_{\theta,p}<1$ and $\|u-v\|_{\theta,p}\geqslant\varepsilon$. Then there exist representations $u=u_1(t)+u_2(t)$, $v=v_1(t)+v_2(t)$, $t\in R$ so that

$$\max(\|e^{\theta t}u_1(t)\|_p, \|e^{\theta t}v_1(t)\|_p, \|e^{-(1-\theta)t}u_2(t)\|_p, \|e^{-(1-\theta)t}v_2(t)\|_p)\leqslant 1\,.$$

Since $u-v=(u_1(t)-v_1(t))+(u_2(t)-v_2(t))$, $t\in R$ we get, by 2.g.13, that

$$\varepsilon\leqslant\|u-v\|_{\theta,p}\leqslant\|e^{\theta t}(u_1(t)-v_1(t))\|_p^{1-\theta}\|e^{-(1-\theta)t}(u_2(t)-v_2(t))\|_p^{\theta}\,.$$

Consequently,

$$\|e^{\theta t}(u_1(t)-v_1(t))\|_p\geqslant(\varepsilon/2^{\theta})^{1/(1-\theta)},\ \|\varepsilon^{-(1-\theta)t}(u_2(t)-v_2(t))\|_p\geqslant(\varepsilon/2^{1-\theta})^{1/\theta}\,.$$

By using 2.g.13 once again, we deduce that

$$\begin{aligned}\|u+v\|_{\theta,p}&\leqslant\|e^{\theta t}(u_1(t)+v_1(t))\|_p^{1-\theta}\|e^{-(1-\theta)t}(u_2(t)+v_2(t))\|_p^{\theta}\\&\leqslant(2-2\delta_{L_p(X_1)}((\varepsilon/2^{\theta})^{1/(1-\theta)}))^{1-\theta}(2-2\delta_{L_p(X_2)}((\varepsilon/2^{1-\theta})^{1/\theta}))^{\theta}\,.\end{aligned}$$

Hence,

$$2-2\delta_{[X_1,X_2]_{\theta,p}}(\varepsilon)\leqslant(2-2\delta_{L_p(X_1)}((\varepsilon/2^{\theta})^{1/1-\theta)}))^{1-\theta}(2-2\delta_{L_p(X_2)}((\varepsilon/2^{1-\theta})^{1/\theta}))^{\theta}$$

and from this the desired result follows immediately. □

Remark. It follows from 2.g.21 by duality (in view of 1.e.2 and 2.g.16) that if either X_1 or X_2 is uniformly smooth then $[X_1, X_2]_{\theta,p}$ with $0<\theta<1$ and $1<p<\infty$ has an equivalent uniformly smooth norm whose modulus of smoothness can be estimated in terms of the moduli of smoothness of X_1 and X_2. The restriction in 2.g.16 that $X_1 \cap X_2$ be dense in X_1 and X_2 need not concern us in the present context. Indeed, for every interpolation pair (X_1, X_2), the space $X_1 \cap X_2$ is dense in $[X_1, X_2]_{\theta,p}$ (this is most easily verified by applying 2.g.17). Hence, $[X_1, X_2]_{\theta,p} = [X_1^0, X_2^0]_{\theta,p}$, where X_i^0 denotes the closure of $X_1 \cap X_2$ in X_i, $i=1, 2$.

We consider next the type of interpolation spaces.

Proposition 2.g.22 [5]. *Let (X_1, X_2) be an interpolation pair of Banach spaces with X_i of type p_i, $i=1, 2$. Then $[X_1, X_2]_{\theta,p}$ is of type p provided that $1/p=(1-\theta)/p_1+\theta/p_2$.*

Proof. Let $\{z_i\}_{i=1}^n \in [X_1, X_2]_{\theta,p}$ and let $z_i = x_{1,i}(t)+x_{2,i}(t)$ with $x_{1,i}(t) \in X_1$, $x_{2,i}(t) \in X_2$, $t \in R$ and $1 \leq i \leq n$. By 2.g.13 and Hölder's inequality

$$\begin{aligned}\int_0^1 \Big\|\sum_{i=1}^n r_i(s) z_i\Big\|_{\theta,p} ds &\leq \int_0^1 \Big\|\sum_{i=1}^n r_i(s) e^{\theta t} x_{1,i}(t)\Big\|_p^{1-\theta} \Big\|\sum_{i=1}^n r_i(s) e^{-(1-\theta)t} x_{2,i}(t)\Big\|_p^{\theta} ds \\ &\leq \Big(\int_0^1 \Big\|\sum_{i=1}^n r_i(s) e^{\theta t} x_{1,i}(t)\Big\|_p ds\Big)^{1-\theta} \\ &\qquad \times \Big(\int_0^1 \Big\|\sum_{i=1}^n r_i(s) e^{-(1-\theta)t} x_{2,i}(t)\Big\|_p ds\Big)^{\theta} \\ &\leq \Big(\int_0^1 \int_{-\infty}^{\infty} \Big\|\sum_{i=1}^n r_i(s) e^{\theta t} x_{1,i}(t)\Big\|_{X_1}^p dt\, ds\Big)^{(1-\theta)/p} \\ &\qquad \times \Big(\int_0^1 \int_{-\infty}^{\infty} \Big\|\sum_{i=1}^n r_i(s) e^{-(1-\theta)t} x_{2,i}(t)\Big\|_{X_2}^p dt\, ds\Big)^{\theta/p}.\end{aligned}$$

Hence, since X_1 is of type p_1 and X_2 of type p_2 it follows from 1.e.13 that, for some constant M (dependent only on X_1, X_2, p_1, p_2 and θ),

$$\begin{aligned}&\int_0^1 \Big\|\sum_{i=1}^n r_i(s) z_i\Big\|_{\theta,p} ds \\ &\leq M \Big(\int_{-\infty}^{\infty} \Big(\sum_{i=1}^n \|e^{\theta t} x_{1,i}(t)\|_{X_1}^{p_1}\Big)^{p/p_1} dt\Big)^{(1-\theta)/p} \\ &\qquad \times \Big(\int_{-\infty}^{\infty} \Big(\sum_{i=1}^n \|e^{-(1-\theta)t} x_{2,i}(t)\|_{X_2}^{p_2}\Big)^{p/p_2} dt\Big)^{\theta/p} \\ &= M \|e^{\theta t}(x_{1,1}(t), x_{1,2}(t), \ldots, x_{1,n}(t))\|_{L_p(l_{p_1}^n(X_1))}^{1-\theta} \\ &\qquad \times \|e^{-(1-\theta)t}(x_{2,1}(t), x_{2,2}(t), \ldots, x_{2,n}(t))\|_{L_p(l_{p_2}^n(X_2))}^{\theta}.\end{aligned}$$

Since $x_{1,i}(t)+x_{2,i}(t)$ are arbitrary representations of z_i, $i=1, \ldots, n$ it follows from 2.g.13 that

$$\int_0^1 \Big\|\sum_{i=1}^n r_i(s) z_i\Big\|_{\theta,p} ds \leq M \|(z_1, z_2, \ldots, z_n)\|_{[l_{p_1}^n(X_1),\, l_{p_2}^n(X_2)]_{\theta,p}}.$$

By 2.g.20, the space $[l_{p_1}^n(X_1), l_{p_2}^n(X_2)]_{\theta,p}$ coincides with $l_p^n([X_1, X_2]_{\theta,p})$, up to an equivalence of norm, and the constant C of this equivalence is independent of n. Consequently,

$$\int_0^1 \left\| \sum_{i=1}^n r_i(s) z_i \right\|_{\theta,p} ds \leqslant CM \left(\sum_{i=1}^n \|z_i\|_{\theta,p}^p \right)^{1/p}. \quad \square$$

Remarks. 1. Since, by 2.g.18, we have

$$[L_{p_1}(0,1), L_{p_2}(0,1)]_{\theta,p} = L_p(0,1),$$

up to equivalent norms, if $1/p=(1-\theta)/p_1+\theta/p_2$, it follows that 2.g.22 gives a sharp result in this case if $1 \leqslant p_1 < p_2 \leqslant 2$.

2. There seems to be no known result of a similar nature concerning the cotype. Note that in view of the lack of duality in the general case between type and cotype we cannot dualize 2.g.22.

3. By an argument very similar to the proof of 2.g.22 the following can be proved. If (X_1, X_2) is an interpolation pair of Banach lattices and if X_i satisfies an upper p_i-estimate, $i=1, 2$ then $[X_1, X_2]_{\theta,p}$ satisfies an upper p-estimate provided that $1/p=(1-\theta)/p_1+\theta/p_2$. By duality we infer that the same results are true for lower estimates. Also, from 1.f.7 we deduce that $[X_1, X_2]_{\theta,p}$ is $p-\varepsilon$ convex for every $\varepsilon>0$.

Further geometric properties of $[X_1, X_2]_{\theta,p}$ of an isomorphic as well as an isometric nature can be found in [6].

We conclude this section by presenting an application of the Lions–Peetre interpolation spaces. In I.d.6 we described an example of B. Maurey and H. P. Rosenthal of a normalized basic sequence $\{e_n\}_{n=1}^\infty$ in a Banach space E which tends weakly to 0 so that no subsequence of $\{e_n\}_{n=1}^\infty$ is unconditional. We shall now "uniformly convexify" this example.

Example 2.g.23 [97]. There is a uniformly convex Banach space $\tilde{E}$ with a normalized monotone basis $\{e_n\}_{n=1}^\infty$ so that no subsequence of $\{e_n\}_{n=1}^\infty$ is an unconditional basic sequence.

Observe that, since $\tilde{E}$ is uniformly convex and thus reflexive we have automatically in this case that $e_n \xrightarrow{w} 0$.

Proof. We recall first the pertinent facts concerning the space E defined in I.1.d.6. The definition of E depends on a collection Δ of sequences $\delta=\{\sigma_j\}_{j=1}^\infty$ of disjoint finite subsets of the integers. One of the properties of Δ was that, for every subsequence N_1 of the integers, there is a $\delta=\{\sigma_j\}_{j=1}^\infty$ in Δ so that $\sigma_j \subset N_1$ for every j. The space E was defined as the completion of the space of sequences of scalars $x=(a_1, a_2, \dots)$ which are eventually 0, with respect to the norm

$$\|x\|_E = \sup \left| \sum_{j=1}^\infty \left(\sum_{i \in \sigma_j} a_i \right) \bar{\sigma}_j^{-1/2} \right|,$$

where the supremum is taken over all sequences $\delta=(\sigma_j)_{j=1}^\infty$ in Δ. The unit vectors $\{e_n\}_{n=1}^\infty$ form a monotone basis of E and $\|e_n\|_E=\|e_n\|_{E^*}=1$ for every n. The family Δ was chosen so that, for every $\delta=\{\sigma_j\}_{j=1}^\infty$ in Δ, every n and every choice of scalars $\{c_j\}_{j=1}^n$,

$$\sup_{1\leq k\leq n}\left|\sum_{j=1}^k c_j\right|\leq\left\|\sum_{j=1}^n c_ju_j\right\|\leq 2\sup_{1\leq k\leq n}\left|\sum_{j=1}^k c_j\right|,$$

where $u_j=\bar{\bar{\sigma}}_j^{-1/2}\sum_{i\in\sigma_j} e_i$, $j=1,2,\ldots$.

Consider now the space $\tilde{E}=[E,l_2]_{1/2,2}$, By 2.g.21, $\tilde{E}$ is uniformly convex. By 2.g.14 and 2.g.16, there is a constant C so that for every sequence $x=(a_1,a_2,\ldots)$ of scalars which is eventually 0,

$$\|x\|_{\tilde{E}}\leq C\|x\|_E^{1/2}\|x\|_2^{1/2} \quad\text{and}\quad \|x\|_{\tilde{E}^*}\leq C\|x\|_{E^*}^{1/2}\|x\|_2^{1/2}.$$

The unit vectors $\{e_n\}_{n=1}^\infty$ form a monotone and normalized basis of $\tilde{E}$. (That $\|e_n\|_{\tilde{E}}=1$ for every n follows easily by a direct checking of the definition of the Lions–Peetre interpolation space. The inequalities above ensure only that $C^{-1}\leq\|e_n\|_{\tilde{E}}\leq C$ for every n which is, of course, also sufficient for our purposes.) For an arbitrary $\delta=\{\sigma_j\}_{j=1}^\infty$ in Δ and every integer n we have

$$\left\|\sum_{j=1}^n(-1)^ju_j\right\|_{\tilde{E}}\leq C\left\|\sum_{j=1}^n(-1)^ju_j\right\|_E^{1/2}\left\|\sum_{j=1}^n(-1)^ju_j\right\|_2^{1/2}\leq 2^{1/2}Cn^{1/4},$$

where $u_j=\bar{\bar{\sigma}}_j^{-1/2}\sum_{i\in\sigma_j} e_i$, $j=1,2,\ldots$. Since, by the definition of $\|\cdot\|_E$,

$$\left\|\sum_{j=1}^n u_j\right\|_{E^*}\leq 1,$$

for every n, we get that

$$\left\|\sum_{j=1}^n u_j\right\|_{\tilde{E}^*}\leq C\left\|\sum_{j=1}^n u_j\right\|_{E^*}^{1/2}\left\|\sum_{j=1}^n u_j\right\|_2^{1/2}\leq Cn^{1/4}$$

and hence,

$$\left\|\sum_{j=1}^n u_j\right\|_{\tilde{E}}\geq n\Big/\left\|\sum_{j=1}^n u_j\right\|_{\tilde{E}^*}\geq n^{3/4}/C.$$

Thus, $\{u_j\}_{j=1}^\infty$ is not unconditional. Since every subsequence $\{e_n\}_{n\in N_1}$ of $\{e_n\}_{n=1}^\infty$ has a block basis equal to such a $\{u_j\}_{j=1}^\infty$ it follows that $\{e_n\}_{n=1}^\infty$ has no unconditional subsequence. □

Remark. By an argument similar to the one used in the comments following 1.c.10 it follows that $\tilde{E}$ is not isomorphic to a subspace of an order continuous Banach lattice.

References

1. Altshuler, Z.: Uniform convexity in Lorentz sequence spaces. Israel J. Math. **20**, 260–275 (1975).
2. Alspach, D., Enflo, P., Odell, E.: On the structure of separable $\mathscr{L}_p$ spaces ($1<p<\infty$). Studia Math. **60**, 79–90 (1977).
3. Ando, T.: Banachverbände und positive Projektionen. Math. Z. **109**, 121–130 (1969).
4. Bade, W. G.: A multiplicity theory for Boolean algebras of projections in Banach spaces. Trans. Amer. Math. Soc. **92**, 508–530 (1958).
5. Beauzamy, B.: Propriétés geometriques des Espaces d'Interpolation, Séminaire Maurey Schwartz 1974/75 Exposé 14, École Polytechnique, Paris.
6. Beauzamy, B.: Espaces d'interpolation réels; topologie et géométrie. Lect. Notes in Math. **666**, Berlin-Heidelberg-New York, Springer Verlag 1978.
7. Beckner, W.: Inequalities in Fourier analysis. Ann. of Math. **102**, 159–182 (1975).
8. Benyamini, Y.: Separable G spaces are isomorphic to $C(K)$ spaces. Israel J. Math. **14**, 287–293 (1973).
9. Benyamini, Y.: An M space which is not isomorphic to a $C(K)$ space. Israel J. Math. **28**, 98–102 (1977).
10. Bergh, J., Löfström, J.: Interpolation spaces, an introduction. Berlin-Heidelberg-New York: Springer-Verlag 1976.
11. Birkhoff, G.: Three observations on linear algebra. Universidad Nacional de Tucuman Revista, Ser. A **5**, 147–151 (1946).
12. Boas, R. P.: Isomorphism between H_p and L_p. Amer. J. Math. **77**, 655–656 (1955).
13. Bohnenblust, H. F.: An axiomatic characterization of L_p-spaces. Duke Math. J. **6**, 627–640 (1940).
14. Borell, C.: A note on the integrability of Walsh polynomials (unpublished).
15. Boyd, D. W.: The spectral radius of averaging operators. Pacific J. Math. **24**, 19–28 (1968).
16. Boyd, D. W.: Indices of function spaces and their relationship to interpolation. Canadian J. Math. **21**, 1245–1254 (1969).
17. Boyd, D. W.: Indices for the Orlicz spaces. Pacific J. Math. **38**, 315–323 (1971).
18. Bretagnolle, J., Dacunha-Castelle, D., Krivine, J.L.: Lois stable et espaces L^p. Ann. Inst. H. Poincare **2**, 231–259 (1966).
19. Burkholder, D. L.: Distribution function inequalities for martingales. Ann. Probability **1**, 19–42 (1973).
20. Burkholder, D. L., Gundy, R. F.: Extrapolation and interpolation of quasi-linear operators on martingales. Acta. Math. **124**, 249–304 (1970).
21. Butzer, P. L., Behrens, H.: Semi-groups of operators and approximation. Berlin-Heidelberg-New York: Springer-Verlag 1967.
22. Calderon, A. P.: Spaces between L^1 and L^∞ and the theorem of Marcinkiewicz. Studia Math. **26**, 273–299 (1966).
23. Cotlar, M.: A unified theory of Hilbert transforms and ergodic theory. Rev. Mat. Cuyana I, 105–167 (1955).
24. Davis, W. J., Figiel, T., Johnson, W. B., Pelczynski, A.: Factoring weakly compact operators. J. Funct. Anal. **17**, 311–327 (1974).
25. Day, M. M.: Reflexive Banach spaces not isomorphic to uniformly convex spaces. Bull. Amer. Math. Soc. **47**, 313–317 (1941).

26. Day, M. M.: Uniform convexity in factor and conjugate spaces. Ann. of Math. **45**, 375–385 (1944).
27. Diestel, J.: Geometry of Banach spaces—Selected topics. Lect. Notes in Math. 485. Berlin-Heidelberg-New York: Springer Verlag 1975.
28. Diestel, J., Uhl, J. J.: Vector measures. Math. Surveys, No. 15. Amer. Math. Soc. 1977.
29. Doob, J. L.: Stochastic processes. New York: Wiley 1953.
30. Dor, L. E., Starbird, T.: Projections of L_p onto subspaces spanned by independent random variables. Compositio Math. (1979).
31. Dubinski, E., Pelczynski, A., Rosenthal, H. P.: On Banach spaces X for which $\prod_2(\mathscr{L}_\infty, X) = B(\mathscr{L}_\infty, X)$. Studia Math. **44**, 617–634 (1972).
32. Dunford, N., Schwartz, J.: Linear operators, Vol. I. New York: Interscience 1958.
33. Dunford, N., Schwartz, J.: Linear operators, Vol. II. New York: Interscience 1963.
34. Duren, P. L.: Theory of H^p-spaces. New York: Academic Press 1970.
35. Dvoretzky, A.: Some results on convex bodies and Banach spaces. Proc. Symp. on Linear Spaces, 123–160. Jerusalem 1961.
36. Ellis, H. W., Halperin, I.: Haar functions and the basis problem for Banach spaces. J. London Math. Soc. **31**, 28–39 (1956).
37. Enflo, P.: A counterexample to the approximation property in Banach spaces. Acta Math. **130**, 309–317 (1973).
38. Enflo, P., Starbird, T. W.: Subspaces of L^1 containing L^1. Studia Math. **65**, (1979).
39. Fehér, F., Gaspar, D., Johnen, H.: Normkonvergenz von Fourierreihen in rearrangement invarianten Banachräumen. J. Funct. Anal. **13**, 417–434 (1973).
40. Figiel, T.: On the moduli of convexity and smoothness. Studia Math. **56**, 121–155 (1976).
41. Figiel, T., Johnson, W. B.: A uniformly convex Banach space which contains no l_p. Compositio Math. **29**, 179–190 (1974).
42. Figiel, T., Johnson, W. B., Tzafriri, L.: On Banach lattices and spaces having local unconditional structure with applications to Lorentz function spaces. J. Approximation Theory **13**, 395–412 (1975).
43. Figiel, T., Pisier, G.: Séries aléatoires dans les espaces uniformément convexe ou uniformément lisses. C. R. Acad. Sci., Paris **279**, Serie A, 611–614 (1974).
44. Freudenthal, H.: Teilweise geordnete Moduln. Proc. Acad. Amsterdam **39**, 641–651 (1936).
45. Gamlen, J. L. B., Gaudet, R. J.: On subsequences of the Haar system in $L_p[0, 1]$ $(1 < p < \infty)$. Israel J. Math. **15**, 404–413 (1973).
46. Garsia, A. M.: Martingale inequalities. Seminar notes on recent progress. Reading, Mass.: W. A. Benjamin 1973.
47. Giesy, D. P.: On a convexity condition in normed linear spaces. Trans. Amer. Math. Soc. **125**, 114–146 (1966).
48. Gould, G.: On a class of integration spaces. J. London Math. Soc. **34**, 161–172 (1959).
49. Haagerup, U.: Les meilleures constantes de l'inégalité de Khintchine, C. R. Acad., Paris **286**, 259–262 (1978).
50. Halmos, P.: Measure theory. New York: D. Van Nostrand 1950.
51. Hardy, G. H., Littlewood, J. E., Polya, G.: Inequalities. Cambridge Univ. Press 1934.
52. Hoffman, K.: Banach spaces of analytic functions. Englewood Cliffs, New Jersey: Prentice Hall 1962.
53. Hoffmann-Jørgensen, J.: Sums of independent Banach space valued random variables. Studia Math. **52**, 159–186 (1974).
54. Hunt, R. A.: On $L(p, q)$ spaces. L'Enseignement mathématique **12**, 249–274 (1966).
55. James, R. C.: Nonreflexive spaces of type 2. Israel J. Math. **30**, 1–13 (1978).
56. Johnson, W. B.: On finite dimensional subspaces of Banach spaces with local unconditional structure. Studia Math. **51**, 223–238 (1974).
57. Johnson, W. B.: Banach spaces all of whose subspaces have the approximation property (to appear).
58. Johnson, W. B., Maurey, B., Schechtman, G., Tzafriri, L.: Symmetric structures in Banach spaces. Memoirs Amer. Math. Soc. 1979.
59. Johnson, W. B., Tzafriri, L.: Some more Banach spaces which do not have local unconditional structure. Houston J. Math. **3**, 55–60 (1977).

60. Kadec, M. I.: Unconditional convergence of series in uniformly convex spaces. Uspehi Mat. Nauk (N.S.) **11**, 185–190 (1956) (Russian).
61. Kadec, M. I., Pelczynski, A.: Bases, lacunary sequences and complemented subspaces in the spaces L_p. Studia Math. **21**, 161–176 (1962).
62. Kahane, J. P.: Series of random functions. Heath Math. Monographs. Lexington, Mass.: Heath and Co. 1968.
63. Kakutani, S.: Concrete representation of abstract *L*-spaces and the mean ergodic theorem. Ann. of Math. **42**, 523–537 (1941).
64. Kakutani, S.: Concrete representation of abstract *M*-spaces. Ann. of Math. **42**, 994–1024 (1941).
65. Köthe, G.: Topological vector spaces. Berlin-Heidelberg-New York: Springer-Verlag 1969.
66. Krivine, J. L.: Theoremes de factorisation dans les espaces reticules. Seminaire Maurey-Schwartz, 1973–74, Exposes 22–23, École Polytechnique, Paris.
67. Krivine, J. L.: Sous espaces de dimension finite des espaces de Banach reticulés. Ann. of Math. **104**, 1–29 (1976).
68. Kwapien, S.: On a theorem of L. Schwartz and its application to absolutely summing operators. Studia Math. **38**, 193–201 (1970).
69. Kwapien, S.: Isomorphic characterizations of inner product spaces by orthogonal series with vector valued coefficients. Studia Math. **44**, 583–595 (1972).
70. Kwapien, S.: A theorem on the Rademacher series with vector valued coefficients. Proc. of the Int. Conf. on Probability in Banach spaces. Lect. Notes in Math. **526**, pages 157–158. Springer-Verlag 1976.
71. Lacey, H. E.: The isometric theory of classical Banach spaces. Berlin-Heidelberg-New York: Springer-Verlag 1974.
72. Lewis, D. R.: Finite dimensional subspaces of L_p. Studia Math. (to appear).
73. Liapounoff, A. A.: On completely additive vector functions. Izv. Akad. Nauk SSSR **4**, 465–478 (1940).
74. Lindenstrauss, J.: On the modulus of smoothness and divergent series in Banach spaces. Mich. Math. J. **10**, 241–252 (1963).
75. Lindenstrauss, J.: A short proof of Liapounoff's convexity theorem. J. of Math. and Mech. **15**, 971–972 (1966).
76. Lindenstrauss, J., Pelczynski, A.: Contributions to the theory of the classical Banach spaces. J. Funct. Anal. **8**, 225–249 (1971).
77. Lindenstrauss, J., Tzafriri, L.: On the complemented subspaces problem. Israel J. Math. **9**, 263–269 (1971).
78. Lindenstrauss, J., Tzafriri, L.: On Orlicz sequence spaces III. Israel J. Math. **14**, 368–389 (1973).
79. Lindenstrauss, J., Tzafriri, L.: Classical Banach spaces I, Sequence spaces. Ergebnisse 92, Berlin-Heidelberg-New York: Springer-Verlag 1977.
80. Liokoumovich, V. I.: The existence of *B*-spaces with non-convex modules of convexity. Izv. Vysš. Učebn. Zaved. Matematika **12**, 43–50 (1973) (Russian).
81. Lions, J. L., Peetre, J.: Sur une classe d'espaces d'interpolation. Publ. Math. Inst. Hautes Etudes Sci. **19**, 5–68 (1964).
82. Lorentz, G. G.: Some new functional spaces. Ann. of Math. **51**, 37–55 (1950).
83. Lorentz, G. G., Shimogaki, T.: Interpolation theorems for the pairs of spaces (L^p, L^∞) and (L^1, L^q). Trans. Amer. Math. Soc. **159**, 207–221 (1971).
84. Lozanovskii, G. Ja.: Banach structures and bases. Funct. Anal. and its Appl. **1**, 294 (1967) (translated from Russian).
85. Lozanovskii, G. Ja., Mekler, A. A.: Completely linear functionals and reflexivity in normed linear lattices. Izv. Visš. Učebn. Zaved. **11**, 47–53 (1967) (Russian).
86. Luxemburg, W. A. J.: Banach function spaces. Thesis, Assen, Netherland 1955.
87. Luxenburg, W. A. J.: Rearrangement invariant Banach function spaces. Proc. Symp. Anal., Queen's University **10**, 83–144 (1967).
88. Luxemburg, W. A. J. , Zaanen, A. C.: Notes on Banach function spaces, VII, Indag. Math. **25**, 669–681 (1963).
89. Luxemburg, W. A. J., Zaanen, A. C.: Notes on Banach function spaces, XIII, Indag. Math. **26**, 530–543 (1964).

90. Luxemburg, W. A. J., Zaanen, A. C.: Riesz spaces I. Amsterdam: North-Holland 1971.
91. Maleev, R. P., Troyanski, S. L.: On the moduli of convexity and smoothness in Orlicz spaces. Studia Math. **54**, 131–141 (1975).
92. Maurey, B.: Sous-espaces complémentes de L_p, d'après P. Enflo. Seminaire Maurey-Schwartz, 1974–75, Exposé 3, École Polytechnique, Paris.
93. Maurey, B.: Un theorème de prolongement. C. R. Acad., Paris **279**, 329–332 (1974).
94. Maurey, B.: Type et cotype dans les espaces munis de structures locales inconditionnelles. Seminaire Maurey-Schwartz 1973–74, Exposes 24–25, École Polytechnique, Paris.
95. Maurey, B.: Théorèmes de factorisation pour les operateurs linéaires a valeurs dans un espace L^p, Asterisque, No. 11, Soc. Math. France 1974.
96. Maurey, B., Pisier, G.: Séries de variables aleatoires vectorielles independantes et propriétés geometriques des espaces de Banach. Studia Math. **58**, 45–90 (1976).
97. Maurey, B., Rosenthal, H. P.: Normalized weakly null sequences with no unconditional subsequence. Studia Math. **61**, 77–98 (1977).
98. Meyer-Nieberg, P.: Charakterisierung einiger topologischer und ordnungstheoretischer Eigenschaften von Banachverbänden mit Hilfe disjunkter Folgen. Arch. Math. **24**, 640–647 (1973).
99. Meyer-Nieberg, P.: Zur schwachen Kompaktheit in Banachverbänden. Math. Z. **134**, 303–315 (1973).
100. Milman, D. P.: On some criteria for the regularity of spaces of the type (B). Dokl. Akad. Nauk SSSR **20**, 243–246 (1938) (Russian).
101. Mitjagin, B. S.: An interpolation theorem for modular spaces. Mat. Sb. **66** (108), 473–482 (1965) (Russian).
102. Mitjagin, B. S.: The homotopy structure of the linear group of a Banach space. Russian Math. Surv. **25**, 59–103 (1970) (translated from Russian).
103. Nakano, H.: Über das System aller stetigen Funktionen auf einem topologischen Raum. Proc. Imp. Acad. Tokyo **17**, 308–310 (1941).
104. Nakano, H.: Modulared semi-ordered linear spaces. Tokyo: Maruzen 1950.
105. Nördlander, G.: The modulus of convexity in normed linear spaces. Arkiv. Mat. **4**, 15–17 (1960).
106. Ogasawara, T.: Vector Lattices. Tokyo 1948 (Japanese).
107. Olevskii, A. M.: Fourier series and Lebesgue functions. Uspehi Mat. Nauk **22**, 236–239 (1967) (Russian).
108. Paley, R. E.: A remarkable series of orthogonal functions. Proc. London Math. Soc. **34**, 241–264 (1932).
109. Peetre, J.: A theory of interpolation of normed spaces. Notas de matematica Brazil **39**, 1–86 (1968).
110. Pelczynski, A.: A connection between weakly unconditional convergence and weak completeness of Banach spaces. Bull. Acad. Polon. Sci. **6**, 251–253 (1958).
111. Pelczynski, A.: Banach spaces of analytic functions and absolutely summing operators. Regional conference series in mathematics, Vol. 30. Amer. Math. Soc. 1977.
112. Pelczynski, A., Singer, I.: On non-equivalent bases and conditional bases in Banach spaces. Studia Math. **25**, 5–25 (1964).
113. Pettis, B. J.: A proof that every uniformly convex space is reflexive. Duke Math. J. **5**, 249–253 (1939).
114. Rosenthal, H. P.: On the subspaces of $L^p(p>2)$ spanned by sequences of independent random variables. Israel J. Math. **8**, 273–303 (1970).
115. Rosenthal, H. P.: On subspaces of L^p. Ann of Math. **97**, 344–373 (1973).
116. Rosenthal, H. P.: On a theorem of J. L. Krivine concerning block finite-representability of l^p in general Banach spaces. J. Funct. Anal. **28**, 197–225 (1978).
117. Russu, G. I.: Intermediate symmetrically normed ideals. Math. Issled. Kišinef **4**, 74–89 (1969) (Russian).
118. Schaefer, H. H.: Banach lattices and positive operators. Berlin-Heidelberg-New York: Springer-Verlag 1974.
119. Schechtman, G.: A remark on unconditional basic sequences in $L_p(1<p<\infty)$. Israel J. Math. **19**, 220–224 (1974).
120. Shimogaki, T.: Exponents of norms in semi-ordered linear spaces. Bull. Acad. Polon. Sci. **13**,

121. Stone, M. H.: Applications of the theory of Boolean rings to general topology. Trans. Amer. Math. Soc. **41**, 376–481 (1937).
122. Stone, M. H.: Boundedness properties in function lattices. Canadian J. Math. **1**, 176–186 (1949).
123. Szankowski, A.: A Banach lattice without the approximation property. Israel J. Math. **24**, 329–337 (1976).
124. Szankowski, A.: Subspaces without approximation property. Israel J. Math. **30**, 123–129 (1978).
125. Szankowski, A.: To appear.
126. Szarek, S. J.: On the best constants in the Khintchine inequality. Studia Math. **58**, 197–208 (1976).
127. Tzafriri, L.: An isomorphic characterization of L_p and c_0 spaces. Studia Math. **31**, 195–304 (1969).
128. Tzafriri, L.: An isomorphic characterization of L_p and c_0 spaces, II. Mich. Math. J. **18**, 21–31 (1971).
129. Tzafriri, L.: Reflexivity of cyclic Banach spaces. Proc. Amer. Math. Soc. **22**, 61–68 (1969).
130. Tzafriri, L.: Reflexivity in Banach lattices and their subspaces. J. Funct. Anal. **10**, 1–18 (1972).
131. Yudin, A. J.: Solution of two problems on the theory of partially ordered spaces. Dokl. Akad. Nauk SSSR **23**, 418–422 (1939) (Russian).
132. Zygmund, A.: Trigonometric series I, II, second edition, Cambridge University Press 1959.
133. Davis, W. J., Ghoussoub, N., Lindenstrauss, J.: A lattice renorming theorem and applications to vector-valued processes (to appear).
134. Feder, M.: On subspaces of spaces with an unconditional basis and spaces of operators (to appear).
135. Feller, W.: An introduction to probability theory and its applications, Vol. II. New York: Wiley 1966.
136. Kalton, N. J.: Embedding L_1 in a Banach lattice. Israel J. Math. (to appear).
137. Rodin, V. A., Semyonov, E. M.: Rademacher series in symmetric spaces. Analysis Mathematika **1**, 207–222 (1975).

Subject Index

absolutely summing operators (see operators)
approximation property (A.P.) 33, 102, 113
— —, bounded (B.A.P.) 112
— —, compact (C.A.P.) 102, 103, 107, 111
— —, metric (M.A.P.) 33
averaging operators (see operators)

Banach lattice 1–4, 6, 7, 13, 18, 22, 29, 31, 45, 79, 93, 102
— —, atom of a 20
— —, band of a 3, 9, 10, 23, 24, 35, 85
— —, canonical embedding of a (in its second dual) 4, 17, 27, 28, 34, 35
— —, characterizations of an order continuous 7, 8, 10, 25, 27–29
— —, complexification of a real 43
— —, definition of 1
— —, disjoint elements in a 2, 6, 7, 9, 20–24, 38, 39, 52, 82, 84, 89, 91, 141, 157, 165, 181, 182, 188, 198–200
— —, dual of a 3, 4, 17, 24, 25, 29, 83, 84, 96, 97, 121
— —, functional representation of a 25, 27, 28, 35, 36, 39, 40, 52, 83
— —s, interpolation pair of (see interpolation)
— —, lower q-estimate (for disjoint elements) of a 79, 82–85, 88–92, 97–99, 130, 132, 139, 140, 231
— —, lower q-estimate constant of a 83
— —, monotone sequences in a 3
— —, norm-bounded sequences in a 34, 35
— —, order bounded sets in a 3–8, 29
— —, order complete 4, 6, 8, 10, 20, 35, 85
— —, σ-order complete 4–10, 12, 14, 15, 20, 25, 26, 29, 34, 38, 83, 89, 114, 118
— —, order continuous (norm in a) 7, 8, 10, 12, 14, 15, 19, 21, 25–29, 31, 32, 34–40, 83, 89, 118, 186, 232
— —, σ-order continuous (norm in a) 7, 9, 14, 22, 25, 26, 29, 34–36, 38, 83, 121
— —, p-convex 40, 45, 46, 49, 51–55, 57, 59, 73, 79–82, 85, 87–90, 95–99, 133, 181, 188, 191, 194–198, 209, 220–223, 231
Banach lattice, p-convexification of a 53, 55, 58, 81, 89, 98, 99, 201
— —, p-convexity constant of a 46, 51, 81, 99, 138, 189, 191, 194, 195, 197
— —, projection band of a 9–11, 25, 34, 35
— —, q-concave 40, 45, 46, 49–51, 53–56, 59, 73, 74, 79–82, 85, 88, 92, 93, 95–98, 126, 133, 162, 169, 170, 174–176, 181, 186–188, 195, 197, 198, 200, 201, 209, 220–223
— —, q-concavification of a 54, 55, 88, 201, 221, 222
— —, q-concavity constant of a 46, 51, 77, 81, 133, 189, 193
— —, second dual of a 4, 84, 85
— —, separable 7, 12, 25, 29, 38, 52, 83, 89, 118, 121
— —, sublattices of a 3, 19, 22, 51, 55, 59, 103
— —, subspaces of a 31, 34, 35, 37–39, 51, 98, 102, 189
— —, units of a (see strong unit and weak unit)
— —, upper p-estimate (for disjoint elements) of a 79, 82–85, 87, 88, 90, 96, 97, 130, 132, 139, 140, 195, 197, 200, 201, 231
— —, upper p-estimate constant of a 83, 202
band (see Banach lattices)
basis, block basis of a 158, 161, 198
—, conditional (see conditional basis)
—, monotone 150, 151, 231, 232
—, perfectly homogeneous 24
—, perturbation of a 38, 176, 179
—, precisely reproducible 158, 162, 189, 190, 195, 198, 200, 201
—, reproducible 150, 158, 162
—, subsymmetric (see subsymmetric basis)
—, symmetric (see symmetric basis)
—, unconditional (see unconditional basis)
Bohnenblust's characterization of abstract L_p and M spaces 18
Boolean algebras of projections (see projections)
— —, Stone's representation theorum for (see Stone)
— operations 114

Boyd indices 129–131, 134, 139, 142, 144, 147, 157, 162, 167, 168, 175, 179, 188, 199, 200, 202, 212, 219, 220
— interpolation theorem (see interpolation)

C^*-algebra, commutative 17
Caratheodory extension theorem 15
Cauchy–Schwarz inequality 155, 167
Central limit theorem 135, 136
concave Banach lattice (see Banach lattice, q-concave)
— operators (see operators, q-concave)
concavification of a Banach lattice (see Banach lattice, q-concavification of a)
condition Δ_2 64, 115, 120, 199
conditional basis 162
— expectation 122, 128, 151, 179, 215
cone, positive 2, 3, 7, 12, 17, 36, 44
convex Banach lattice (see Banach lattice, p-convex)
convex operators (see operators, p-convex)
convexification of a Banach lattice (see Banach lattice, p-convexification of a)
—, uniform 216, 231
convexity, modulus of 59, 63, 65–72, 78–80, 88, 90, 97, 99, 229
—, strict 60–62
—, uniform 40, 59–63, 69, 77, 79–81, 88, 90, 97, 102, 162, 199, 202, 228, 229, 231, 232
cotype 59, 72–74, 77–79, 82, 88, 90, 93, 96–99, 102, 111, 181, 189, 208, 231
— constant 73
cyclic space 12–16, 20, 26
— vector 12

decomposition method (see Pelczynski's decomposition method)
— property 2, 22, 26, 36, 88, 125, 221
differentiability, Fréchet 61
—, Gâteaux 61
distribution function 116, 117, 132, 141, 142, 157, 160, 170, 205–207, 209, 211
—, normal 135, 213
—, Poisson 204, 205, 214
—, uniform 205
—, vectors with the same 140–142, 157, 158, 160
Doob's maximal inequality 153, 155
Dvoretzky's theorem 63

Eberlein's theorem 35
equi-integrable sets 216, 223
estimate, lower (see Banach lattice, lower q-estimate of a)
—, upper (see Banach lattice, upper p-estimate of a)
extreme points 121, 124, 159

factorization theorems 57, 59, 95
— of weakly compact operators 216, 224
Fatou property 30, 118–121, 123, 128
Fourier transform 182
function, distribution (see distribution function)
— homogeneous of degree one 40, 42, 44
—, locally integrable 29, 142
—, spaces (see Köthe, Lorentz, Orlicz and rearrangement invariant function spaces)
—, square 126, 154, 156, 166
functional calculus 42, 43
—, Hahn–Banach 25
—, strictly positive 25–27

greatest lower bound (g.l.b) 1, 8, 21
Grothendieck's inequality 93, 94
— universal constant 93, 94, 173

Haar system 2, 150, 151, 154–158, 160–162, 165, 168, 175–179, 181, 182, 186, 187, 198–200, 223
Hölder inequality 44, 53, 56, 89, 106, 143, 154, 174, 209
Hölder type inequalities 43, 49, 76

ideal (see Banach lattices)
independent random variables (see random variables)
integrals (see also space X' of integrals) 29
interpolation method of Lions–Peetre 149, 216, 225, 227–229, 231, 232
— pair of spaces 216–218, 225, 227–230
— — — Banach lattices 218, 219, 221, 231, 232
—, real method in 216
—, spaces 215, 216, 218, 219, 227, 228, 230–232
— theorem, Boyd 144, 147, 156, 168, 209
— —, Marcinkiewicz 130, 149, 153, 228
— —, Riez–Thorin 149, 156, 166
— —s 114, 123, 124, 126, 127, 129, 130, 142, 144, 171, 215

Kakutani's representation theorem for abstract L_p spaces 15, 18, 24, 93, 95
— — — — — — M spaces 16, 18, 24, 93, 94
Khintchine's inequality 39, 49, 50, 70, 134, 169, 174, 186, 187, 204
— —, two dimensional 173, 174
Kolmogorov's consistency theorem 205
Köthe function spaces 14, 28–31, 36, 38, 83, 89, 104, 114–118. 121, 123, 127–130, 164, 188, 216, 218, 223
Krein–Milman theorem 121, 159
Krivine's theorem 72. 111, 141

lattices (see Banach and vector lattices)
least upper bound (l.u.b) 1, 3–6, 8, 29, 35, 42
Liapounoff's theorem 158
Lions–Peetre interpolation (see interpolation method of Lions–Peetre)
Lorentz spaces 64, 88, 97–99, 120, 121, 129, 132, 133, 142
— —, modulus of convexity of 67, 71, 72
lower q-estimate (see Banach lattice, lower q-estimate of a)

Marcinkiewicz interpolation theorem (see interpolation)
— weak type, operators of (see operators of Marcinkiewicz weak type)
martingale 150–156
matrices 123, 124, 138, 173
Maurey–Rosenthal example 39, 231
measure space, automorphism of a 114–116, 118, 127, 138, 149, 151, 186
measure space, complete 29
— —, separable 114
modulus of convexity (see convexity, modulus of)
— — smoothness (see smoothness, modulus of)

norming subspace 29, 30, 118

operators, automatic continuity of positive 2
—, averaging 122
—, compact 33
—, diagonal of 21
—, dilation 130, 131, 163, 170, 206, 219
—, finite rank 33
—, multiplicity theory for spectral 11
—, norm $\|T\|_{1,2}$ of 217–219
— of Marcinkiewicz weak type 144, 145
— of strong type 144, 149, 150, 156
— of weak type 130, 142, 144, 145, 147–149, 153, 156, 167, 228
—, order preserving 2
—, p-absolutely summing 56, 57, 93
—, p-convex 45, 47–49, 55, 57, 59, 95
—, positive 2, 4, 55–57, 59, 95, 115, 125
—, q-concave 46, 48, 49, 55–57, 59, 95
—, quasilinear 126, 127, 147, 149, 153, 228
—, simultaneous extension 21, 22
—, weakly compact 216, 224
order complete Banach lattice (see Banach lattice, order complete)
— continuous Banach lattice (see Banach lattice, order continuous)
— interval 28
— isometry 3, 4, 15–17, 19, 25, 29, 36, 41, 44, 58, 118, 190, 211
order isomorphism 2, 3, 22, 24, 26, 35, 121, 185
— like relation 123, 125
Orlicz functions 64, 65, 119, 120, 134, 164, 220
— spaces 67, 115, 120, 131, 134–136, 139, 140, 162, 163, 165, 181, 187, 188, 199, 200, 209, 223
— —, modulus of convexity of 67, 71, 72

p-absolutely summing operators (see operators, p-absolutely summing)
p-convex Banach lattice (see Banach lattice, p-convex)
— operators (see operators, p-convex)
p-convexification of a Banach lattice (see Banach lattice, p-convexification of a)
partition of unity 94
Pelczynski's decomposition method 172, 209, 220, 222
polar $Y^{\perp}$ of a band Y 9, 10, 20, 26, 27, 36
precisely reproducible basis (see basis, precisely reproducible)
primary space 168, 179
process, Poisson 202, 204–207, 212, 213
—, r-stable 213
—, stationary 204, 206, 213
—, symmetric 206, 213
— with independent increments 204, 206, 213
projections, band 10, 11, 23–25, 85
—, Boolean algebra (B.A.) of 11, 12, 14, 15, 20
—, σ-complete B.A. of 11–13, 26
—, contractive 8, 10, 19–21, 122, 123, 209
— of the form P_x 8–10, 12, 13, 15, 16
—, orthogonal 137, 138, 171, 172, 209
—, positive 8, 10, 19, 20
—, Riesz 166–168

q-concave Banach lattice (see Banach lattice, q-concave)
— operators (see operators, q-concave)
q-concavification of a Banach lattice (see Banach lattice, q-concavification of a)
quotient space 3

Rademacher elements 92, 93
— functions 29, 51, 72–74, 77, 85, 104, 106, 134, 136–138, 160, 162, 169, 198, 204, 213
Radon-Nikodym derivative 31
— — property 69
— — theorem 27, 29, 122
random variables, characteristic functions of 204, 205
— —, independent 85–87, 135, 160, 182, 185, 204–207, 213, 214
— —, p-stable 181, 182, 185, 213
— —, symmetric 206, 207, 214

rearrangement, decreasing 117, 123, 125
— invariant (r.i.) function space 40, 114, 115, 117–123, 125, 126, 130–134, 136, 138, 141, 145, 149–151, 156, 158, 161, 162, 165–170, 176, 178, 181–183, 190, 191, 198–200, 204, 207, 209, 211, 220
— — — —, analytic part of a 168
— — — —, complemented subspaces of a 168, 170, 172, 175, 176, 178, 195, 199, 200, 209
— — — —, minimal 118–121, 123, 126, 128, 133, 147, 204, 205, 211, 219, 220, 223
— — — —, maximal, 118–123, 126, 128, 136, 147, 204, 205, 211, 219
reflexivity 33, 35, 47, 61, 62, 69, 78, 224, 225, 231
— in Banach lattices 4, 22, 31, 35, 37
— in r.i. function spaces 119, 132, 181, 188, 199, 200
renorming theorems 54, 55, 80, 88, 90, 115
reproducible basis (see basis, reproducible)
Riemann–Stieltjes integral 12, 14
Riesz–Thorin interpolation theorem (see interpolation)

Schauder decomposition 111
— —, finite dimensional (F.D.D.) 33, 34
— —, unconditional 33
separation theorem 30, 58, 124
smoothness 60, 61
—, modulus of 59, 63, 64, 67, 68, 70, 71, 78–80, 88, 90, 229, 230
—, uniform 59–63, 69, 77, 80, 88, 90, 229
space, abstract L_1 17, 23, 24, 44, 95
—, — L_p 14, 15, 18, 21, 22, 24, 58
—, — M 9, 14, 16–22, 24, 41, 73
— c_0 6, 7, 16, 20, 22, 24, 27, 34–37, 46, 52, 57, 78, 89, 93, 96, 119, 127, 128, 158, 187
— $c_0(\Gamma)$ 7, 9, 20, 22, 24
— $c_0(X)$ 46, 47
— $C(0, 1)$ 2, 4, 10, 13, 36, 52, 62, 114, 162, 167
— $C(K)$ 2, 4–6, 14, 16, 17, 20, 41–43, 56, 57, 93, 94, 96
— —, sublattices of the 16, 20, 41
— $C(\beta N)$ 6
— H_p 168
—, Hilbert (l_2, $L_2(0, 1)$, etc) 52, 63, 73, 74, 96, 105, 106, 111, 134–137, 154, 162, 163, 165, 166, 181, 188, 190, 191, 195, 196, 198, 199, 202, 203, 209, 213
— J 36, 39
— $K(X_1, X_2, Y, \{a_n\}, \{b_n\})$ (also $\breve{K}(X_1, X_2, Y, \{a_n\}, \{b_n\})$) 218–221
— l_1 35, 36, 78, 82, 93, 118, 126, 132, 160, 161
— l_1^n 62, 74, 90, 92–94, 124, 141, 157, 197, 201
— l_p 15, 22–24, 46, 81, 87, 107, 108, 158, 178, 182, 195, 211, 213, 222
space l_p, modulus of convexity of the 63, 71, 72
— —, — — smoothness of the 63, 71
— l_p^n 20, 111, 112, 194
— $l_p(X)$ 46–48, 83, 193, 201
— l_2^n 63, 93, 112, 194, 200–202
— l_∞ 6, 7, 13, 30, 36, 115, 118
— l_∞^n 41, 74, 90, 92–94, 124, 141, 157, 182, 198
— $l_\infty(\Gamma)$ 6, 7, 26
— $l_\infty(X)$ 47
— $l_p \oplus l_2$ 213
— l_M (see Orlicz spaces)
— $L_1(0, 1)$ 2, 4, 78, 118–122, 125, 127, 128, 130, 156, 161, 167, 187, 191, 200, 215
— $L_1(-\infty, +\infty)$ 182
— $L_1(\mu)$ 17, 25–27, 29–31, 35, 38, 39, 44, 93, 94, 153, 208
— $L_1(X)$ 77
— $L_p(0, 1)$ 2, 52, 63, 75, 77, 102, 106, 129, 134, 140, 151, 156, 161, 162, 166, 168, 169, 171, 172, 175, 178, 181, 182, 188–191, 198–200, 202, 203, 209, 211–214, 223, 231
— $L_p(\mu)$ 2, 4, 9, 10, 12, 13, 15, 16, 19, 20, 22, 38, 43, 45, 53, 55, 57, 59, 73, 114, 120, 142–144, 153, 168, 182, 202, 207, 208
— —, modulus of convexity of the 63, 72, 81
— —, — — smoothness of the 63, 72
— —, sublattices of the 16
— $L_p(l_2)$ 171, 172
— $L_p(X)$ (also $L_p(R, X)$) 170, 225–227, 229
— $L_2(X)$ 67–69, 78, 229
— $L_\infty(0, 1)$ 7, 118–122, 126–127, 130, 134–136, 150, 160, 161, 183, 191, 215
— $L_\infty(\mu)$ 17, 25, 30, 44, 73, 96, 159, 214
— $L_1(0, \infty) \cap L_\infty(0, \infty)$ 118, 119, 123, 132
— $L_1(0, \infty) + L_\infty(0, \infty)$ 118–120, 123, 125, 127, 132, 217
— $L_p(0, \infty) \cap L_q(0, \infty)$ 132–134, 147, 189, 191, 202, 209, 213, 214, 218
— $L_p(0, \infty) + L_q(0, \infty)$ 132–134, 149, 191, 202, 209, 218
— $L_{p,q}$ 142–144, 148, 212, 228
— L_M (see Orlicz spaces)
—, Lorentz (see Lorentz spaces)
—, Orlicz (see Orlicz spaces)
—, partially ordered 1
— Rad X 169–172
— $U_{p,q}$ 222, 223
— $X(c_0)$ 46, 47, 85
— $X(l_p)$ (also $\widetilde{X(l_p)}$) 46–48
— $X(l_2)$ (aso $\widetilde{X(l_2)}$) 168–176, 179, 180
— $X(l_\infty)$ (also $\widetilde{X(l_\infty)}$) 46, 47
— X' of integrals 29–31, 117–121, 137
— $X_1 \cap X_2$ 217, 218, 226, 227
— $X_1 + X_2$ 217, 218
— $[X_1, X_2]_{\theta, p}$ 226–228, 230, 231

space $X_{p,2,w}$ (also $X_{p,2}$) 213, 214
— $X_{Y,\sigma}$ 214, 215
Stone's representation theorem for Boolean algebras 15
stopping time 152, 153
strong unit 9, 16, 17, 41
— type operators (see operators)
subsymmetric basis 158
symmetric basis 97, 99, 114, 115, 119, 122, 124, 127, 128, 160, 162, 188
— —, block bases generated by one vector of a 140
— —, — — with constant coefficients of a 121

topological space, basically disconnected 4, 5
— — βN 5
— —, extremally disconnected 4, 5
— —, totally disconnected 15, 26
tree of subsets 178, 179
trigonometric system 150, 165–167
Tsirelson's example 81
Tychonoff's theorem 20
type 59, 72–74, 77–79, 82, 88, 90, 92, 93, 96, 97, 102, 111, 112, 181, 188, 195, 197, 200, 201, 207, 209, 230, 231
— constant 73, 79, 112, 195
—, duality of the 79, 96
— power 63, 78–80, 88, 90

(u), property 31, 32, 34
ultrapower 213
unconditional basic sequences in Banach lattices 37–39, 50, 51, 174, 175, 188, 232
— — — — r.i. function spaces 126, 137, 168, 173, 175–177, 180, 188
— basis 2, 8, 12, 18, 31, 38, 45, 50, 51, 72, 78, 81, 150, 151, 155–158, 160–162, 168, 176, 191, 200, 207, 218–220, 222, 231, 232
— —, projections associated to a 8, 12, 156
— constant 2, 21, 50, 72, 78, 156, 161, 174, 177, 207, 222
— direct sum of disjoint ideals 9, 13, 15, 36
— F.D.D. (see Schauder decomposition)
unconditionally convergent series 69, 70
— — —, weakly (see weak unconditional convergence)
uniform convexity (see convexity, uniform)
— smoothness (see smoothness, uniform)
universal space U_1 for unconditional bases 219, 220, 222
upper p-estimate (see Banach lattices, upper p-estimate of a)

vector lattices 2, 14, 36

Walsh functions 104, 137, 138
weak compactness 35, 36, 216, 224, 225
— sequential completeness 31, 34–37, 47
— type operators (see operators)
— unconditional convergence (w.u.c.) 31, 33, 34
— unit 9, 12–16, 25–29, 35, 36, 38, 39

Zorn's lemma 9

∞ CIM

M. Aigner Combinatorial Theory ISBN 978-3-540-61787-7
A. L. Besse Einstein Manifolds ISBN 978-3-540-74120-6
N. P. Bhatia, G. P. Szegő Stability Theory of Dynamical Systems ISBN 978-3-540-42748-3
J. W. S. Cassels An Introduction to the Geometry of Numbers ISBN 978-3-540-61788-4
R. Courant, F. John Introduction to Calculus and Analysis I ISBN 978-3-540-65058-4
R. Courant, F. John Introduction to Calculus and Analysis II/1 ISBN 978-3-540-66569-4
R. Courant, F. John Introduction to Calculus and Analysis II/2 ISBN 978-3-540-66570-0
P. Dembowski Finite Geometries ISBN 978-3-540-61786-0
A. Dold Lectures on Algebraic Topology ISBN 978-3-540-58660-9
J. L. Doob Classical Potential Theory and Its Probabilistic Counterpart ISBN 978-3-540-41206-9
R. S. Ellis Entropy, Large Deviations, and Statistical Mechanics ISBN 978-3-540-29059-9
H. Federer Geometric Measure Theory ISBN 978-3-540-60656-7
S. Flügge Practical Quantum Mechanics ISBN 978-3-540-65035-5
L. D. Faddeev, L. A. Takhtajan Hamiltonian Methods in the Theory of Solitons ISBN 978-3-540-69843-2
I. I. Gikhman, A. V. Skorokhod The Theory of Stochastic Processes I ISBN 978-3-540-20284-4
I. I. Gikhman, A. V. Skorokhod The Theory of Stochastic Processes II ISBN 978-3-540-20285-1
I. I. Gikhman, A. V. Skorokhod The Theory of Stochastic Processes III ISBN 978-3-540-49940-4
D. Gilbarg, N. S. Trudinger Elliptic Partial Differential Equations of Second Order ISBN 978-3-540-41160-4
H. Grauert, R. Remmert Theory of Stein Spaces ISBN 978-3-540-00373-1
H. Hasse Number Theory ISBN 978-3-540-42749-0
F. Hirzebruch Topological Methods in Algebraic Geometry ISBN 978-3-540-58663-0
L. Hörmander The Analysis of Linear Partial Differential Operators I – Distribution Theory and Fourier Analysis ISBN 978-3-540-00662-6
L. Hörmander The Analysis of Linear Partial Differential Operators II – Differential Operators with Constant Coefficients ISBN 978-3-540-22516-4
L. Hörmander The Analysis of Linear Partial Differential Operators III – Pseudo-Differential Operators ISBN 978-3-540-49937-4
L. Hörmander The Analysis of Linear Partial Differential Operators IV – Fourier Integral Operators ISBN 978-3-642-00117-8
K. Itô, H. P. McKean, Jr. Diffusion Processes and Their Sample Paths ISBN 978-3-540-60629-1
T. Kato Perturbation Theory for Linear Operators ISBN 978-3-540-58661-6
S. Kobayashi Transformation Groups in Differential Geometry ISBN 978-3-540-58659-3
K. Kodaira Complex Manifolds and Deformation of Complex Structures ISBN 978-3-540-22614-7
Th. M. Liggett Interacting Particle Systems ISBN 978-3-540-22617-8
J. Lindenstrauss, L. Tzafriri Classical Banach Spaces I and II ISBN 978-3-540-60628-4
R. C. Lyndon, P. E Schupp Combinatorial Group Theory ISBN 978-3-540-41158-1
S. Mac Lane Homology ISBN 978-3-540-58662-3
C. B. Morrey Jr. Multiple Integrals in the Calculus of Variations ISBN 978-3-540-69915-6
D. Mumford Algebraic Geometry I – Complex Projective Varieties ISBN 978-3-540-58657-9
O. T. O'Meara Introduction to Quadratic Forms ISBN 978-3-540-66564-9
G. Pólya, G. Szegő Problems and Theorems in Analysis I – Series. Integral Calculus. Theory of Functions ISBN 978-3-540-63640-3
G. Pólya, G. Szegő Problems and Theorems in Analysis II – Theory of Functions. Zeros. Polynomials. Determinants. Number Theory. Geometry ISBN 978-3-540-63686-1
W. Rudin Function Theory in the Unit Ball of $\mathbb{C}^n$ ISBN 978-3-540-68272-1
S. Sakai C*-Algebras and W*-Algebras ISBN 978-3-540-63633-5
C. L. Siegel, J. K. Moser Lectures on Celestial Mechanics ISBN 978-3-540-58656-2
T. A. Springer Jordan Algebras and Algebraic Groups ISBN 978-3-540-63632-8
D. W. Stroock, S. R. S. Varadhan Multidimensional Diffusion Processes ISBN 978-3-540-28998-2
R. R. Switzer Algebraic Topology: Homology and Homotopy ISBN 978-3-540-42750-6
A. Weil Basic Number Theory ISBN 978-3-540-58655-5
A. Weil Elliptic Functions According to Eisenstein and Kronecker ISBN 978-3-540-65036-2
K. Yosida Functional Analysis ISBN 978-3-540-58654-8
O. Zariski Algebraic Surfaces ISBN 978-3-540-58658-6

Printed in the United States
by Baker & Taylor Publisher Services